Fundament
Digital Circuits

A. ANAND KUMAR

Academic Director
Sasi Institute of Technology & Engineering
Tadepalligudam, West Godavari Dist.
Andhra Pradesh

Prentice-Hall of India Private Limited
New Delhi - 110 001
2006

Rs. 295.00

FUNDAMENTALS OF DIGITAL CIRCUITS
by A. Anand Kumar

ISBN-81-203-1745-9

Eleventh Printing ... ... July, 2006

Published by Asoke K. Ghosh, Prentice-Hall of India Private Limited, M-97, Connaught Circus, New Delhi-110001 and Printed by V.K. Batra at Pearl Offset Press Private Limited, New Delhi-110015.

To

My Parents

Shri A. Nagabhushanam & Smt. A. Ushamani

CONTENTS

Preface *xv*

1. INTRODUCTION **1–16**

1.1 Digital and Analog Systems 1
1.2 Logic Levels and Pulse Waveforms 3
1.3 Elements of Digital Logic 4
1.4 Functions of Digital Logic 5
1.5 Digital Integrated Circuits 10
1.6 Microprocessors 11
1.7 Digital Computers 12
1.8 Types of Computers 13

Summary *14*
Questions *15*

2. NUMBER SYSTEMS **17–61**

2.1 The Decimal Number System 17
2.2 The Binary Number System 20
Counting in Binary 20
Binary to Decimal Conversion 21
Decimal to Binary Conversion 23
Binary Addition 26
Binary Subtraction 26
Binary Multiplication 28
Computer Method of Multiplication 28
Binary Division 29
Computer Method of Division 30
2.3 Representation of Signed Numbers and Binary Arithmetic in Computers 31
2.4 The Octal Number System 40
Usefulness of the Octal System 40
Octal to Binary Conversion 40
Binary to Octal Conversion 41

Octal to Decimal Conversion 41
Decimal to Octal Conversion 41
Octal Addition 43
Octal Subtraction 43
Octal Multiplication 45
Octal Division 45
2.5 The Hexadecimal Number System 45
Hexadecimal Counting Sequence 46
Binary to Hexadecimal Conversion 46
Hexadecimal to Binary 47
Hexadecimal to Decimal 47
Decimal to Hexadecimal 48
Octal to Hexadecimal Conversion 49
Hexadecimal to Octal 50
Hexadecimal Addition 50
Hexadecimal Subtraction 51
Hexadecimal Multiplication 54
Hexadecimal Division 54
2.6 Miscellaneous Examples 55
Summary 56
Questions 58
Problems 59

3. BINARY CODES 62–88

3.1 Weighted and Non-weighted Codes 62
Sequential Codes 63
Self-complementing Codes 63
Cyclic Codes 63
3.2 The 8421 BCD Code 64
3.3 The Excess Three (XS-3) Code 67
3.4 The Gray Code 71
Binary-to-Gray Conversion 73
Gray-to-Binary Conversion 74
The XS-3 Gray Code 74
3.5 Error-detecting Codes 75
3.6 Error-correcting Codes 79
The 7-bit Hamming Code 80
3.7 Alphanumeric Codes 82
The ASCII Code 82
The EBCDIC Code 84
Summary 84
Questions 86
Problems 87

4. LOGIC GATES **89–119**

4.1 Introduction 89
4.2 The AND Gate 90
4.3 The OR Gate 91
4.4 The NOT Gate 93
4.5 The Universal Gates 94
The NAND Gate 94
The NOR Gate 96
4.6 The Exclusive-OR (X-OR) Gate 98
4.7 The Exclusive-NOR (X-NOR) Gate 99
4.8 Inhibit Circuits 100
4.9 7400-series Integrated Circuits 104
4.10 ANSI/IEEE Standard Logic Symbols 104
4.11 Pulsed Operation of Logic Gates 109
Summary 116
Questions 117
Problems 119

5. BOOLEAN ALGEBRA **120–155**

5.1 Introduction 120
5.2 Logic Operations 121
AND Operation 121
OR Operation 121
NOT Operation 121
NAND Operation 121
NOR Operation 121
X-OR and X-NOR Operations 121
5.3 Axioms and Laws of Boolean Algebra 121
Complementation Laws 122
AND Laws 122
OR Laws 122
Commutative Laws 122
Associative Laws 123
Distributive Laws 124
Idempotence Laws 126
Complementation Laws or Negation Laws 127
Double Negation Law 127
Identity Laws 127
Null Laws 128
Absorption Laws 128
Consensus Theorem (Included Factor Theorem) 129
Transposition Theorem 130
De Morgan's Theorem 130
5.4 Duality 133
Duals 134

5.5 Reducing Boolean Expressions 135
5.6 Boolean Expressions and Logic Diagrams 137
5.7 Converting AND/OR/Invert Logic to NAND/NOR Logic 140
5.8 Miscellaneous Examples 148
5.9 Determination of Output Level from the Diagram 151
Summary *152*
Questions *153*
Problems *154*

6. THE KARNAUGH MAP AND QUINE-McCLUSKY METHODS 156–232

6.1 Introduction 156
6.2 Expansion of a Boolean Expression to SOP Form 157
6.3 Expansion of a Boolean Expression to POS Form 159
6.4 Computation of Total Gate Inputs 161
6.5 Two-variable K-map 161
Mapping of SOP Expressions 162
Minimization of SOP Expressions 163
Mapping of POS Expressions 165
Minimization of POS Expressions 166
6.6 Three-variable K-map 167
Minimization of SOP and POS Expressions 168
Reading the K-maps 168
6.7 Four-variable K-map 171
6.8 Five-variable K-map 174
6.9 Six-variable K-map 176
6.10 Don't Care Combinations 179
6.11 Hybrid Logic 181
6.12 Minimization of Multiple Output Circuits 183
6.13 Variable Mapping 189
Incompletely Specified Functions 193
6.14 Miscellaneous Examples 194
6.15 Quine-McClusky Method 206
6.16 Function Minimization of Multiple Output Circuits 220
Summary *227*
Questions *229*
Problems *230*

7. COMBINATIONAL CIRCUITS 233–308

7.1 The Half-adder 233
7.2 The Full-adder 234
7.3 The Half-subtractor 236
7.4 The Full-subtractor 238
7.5 Parallel Binary Adders 240
7.6 The Look-ahead Carry Adder 242

7.7 IC Parallel Adders 244
7.8 Two's Complement Addition and Subtraction Using Parallel Adders 246
7.9 Serial Adders 248
7.10 BCD Adder 249
7.11 Binary Multipliers 251
7.12 Code Converters 253
7.13 Parity Bit Generators/Checkers 259
7.14 Comparators 265
7.15 IC Comparator 268
7.16 Decoders 269
3-Line to 8-Line Decoder 270
Enable Inputs 271
8421 BCD-to-Decimal Decoder (4-Line to 10-Line Decoder) 273
7.17 BCD-to-Seven Segment Decoders 275
7.18 Display Devices 275
Classification of Displays 275
The Light Emitting Diode (LED) 278
The Liquid Crystal Display (LCD) 279
Incandescent Seven Segment Displays 281
7.19 Encoders 281
Octal-to-Binary Encoder 282
Decimal-to-BCD Encoder 283
7.20 Keyboard Encoders 284
7.21 Priority Encoders 285
Decimal-to-BCD Priority Encoder 285
Octal-to-Binary Priority Encoder 287
7.22 Multiplexers (Data Selectors) 290
Basic 2-input Multiplexer 290
The 74157 Quadruple 2-input Data Selector/Multiplexer 291
The 4-input Multiplexer 292
The 74151A 8-input Data Selector/Multiplexer 293
The 16-input Multiplexer from Two 8-input Multiplexers 293
7.23 Applications of Multiplexers 294
Logic Function Generator 294
Parallel-to-Serial Data Conversion 296
Multiplexing Seven Segment Displays 297
7.24 Demultiplexers (Data Distributors) 299
1-Line to 4-Line Demultiplexer 300
1-Line to 8-Line Demultiplexer 300
7.25 ANSI/IEEE Symbols 303
Summary 305
Questions 307
Problems 308

8. FLIP-FLOPS AND TIMING CIRCUITS 309–367

8.1 Introduction 309
8.2 The S-R Latch 310
The NOR Gate S-R Latch 311
The NAND Gate S-R Latch 313
8.3 Gated Latches 315
The Gated S-R Latch 315
The Gated D-latch 316
8.4 Edge-triggered Flip-flops 317
Generation of Narrow Spikes 318
The Edge-triggered S-R Flip-flop 319
The Edge-triggered D Flip-flop 321
The Edge-triggered J-K Flip-flop 323
The Edge-triggered T Flip-flop 326
8.5 Asynchronous Inputs 326
8.6 Flip-flop Operating Characteristics 329
8.7 Master-slave (Pulse-triggered) Flip-flops 333
The Master-slave (Pulse-triggered) S-R Flip-flop 334
The Master-slave (Pulse-triggered) D Flip-flop 336
The Master-slave (Pulse-triggered) J-K Flip-flop 337
The Data Lock-out Flip-flop 338
8.8 Conversion of Flip-flops 339
S-R Flip-flop to J-K Flip-flop 339
J-K Flip-flop to S-R Flip-flop 340
S-R Flip-flop to D Flip-flop 340
D Flip-flop to S-R Flip-flop 340
J-K Flip-flop to T Flip-flop 342
J-K Flip-flop to D Flip-flop 342
D Flip-flop to J-K Flip-flop 342
8.9 Applications of Flip-flops 343
Parallel Data Storage 343
Serial Data Storage 343
Transfer of Data 343
Serial-to-Parallel Conversion 344
Parallel-to-Serial Conversion 344
Counting 344
Frequency Division 344
8.10 ANSI/IEEE Symbols 344
8.11 Schmitt Trigger 347
8.12 Monostable Multivibrator (One-shot) 348
8.13 Astable Multivibrator 354
Astable Multivibrator Using Schmitt Trigger 354
Astable Multivibrator Using 555 Timer 355
Astable Multivibrator Using Inverters 355
Astable Multivibrator Using Op-amps 357

8.14 Crystal-controlled Clock Generators 358
Summary 362
Questions 363
Problems 365

9. SHIFT REGISTERS 368–388

9.1 Introduction 368
9.2 Buffer Register 369
9.3 Controlled Buffer Register 369
9.4 Data Transmission in Shift Registers 371
9.5 Serial-in, Serial-out, Shift Register 372
9.6 Serial-in, Parallel-out, Shift Register 376
9.7 Parallel-in, Serial-out, Shift Register 377
9.8 Parallel-in, Parallel-out, Shift Register 379
9.9 Bidirectional Shift Register 381
9.10 Universal Shift Registers 381
9.11 Dynamic Shift Registers 383
9.12 Applications of Shift Registers 384
9.13 ANSI/IEEE Standard Symbols 386
Summary 387
Questions 388

10. COUNTERS 389–465

10.1 Introduction 389
10.2 Asynchronous Counters 391
Two-bit Ripple Up-counter Using Negative Edge-triggered Flip-flops 391
Two-bit Ripple Down-counter Using Negative Edge-triggered Flip-flops 391
Two-bit Ripple Up/Down Counter Using Negative Edge-triggered Flip-flops 392
Two-bit Ripple Up-counter Using Positive Edge-triggered Flip-flops 393
Two-bit Ripple Down-counter Using Positive Edge-triggered Flip-flops 393
Two-bit Ripple Up/Down Counter Using Positive Edge-triggered Flip-flops 394
10.3 Design of Asynchronous Counters 394
10.4 Effects of Propagation Delay in Ripple Counters 397
10.5 Decoding of Ripple Counters 400
10.6 Integrated Circuit Ripple Counters 404
The 7493A 4-bit Binary Counter 404
The 7490 Decade Counter 407
10.7 Synchronous Counters 408
Four-bit Synchronous Up-counter—Propagation Delay 408
Four-bit Synchronous Down-counter 410

Four-bit Synchronous Up/Down (Bidirectional) Counter 411
Look-ahead Carry 411
10.8 Hybrid Counters 412
10.9 Design of Synchronous Counters 413
10.10 IC Synchronous Counters 428
The 74LS163A Synchronous 4-bit Binary Counter 428
The 74LS160A Synchronous Decade Counter 429
The 74190/74191 Series Up/Down Counters 430
The 74ALS560A/74ALS561A Counters 431
Programmable Counters 431
10.11 The 74193 (LS 193/HC 193) Counter 432
10.12 Shift Register Counters 437
Ring Counter 438
Twisted Ring Counter (Johnson Counter) 439
10.13 Pulse Train Generators 442
Direct Logic 442
Indirect Logic 445
10.14 Pulse Generators Using Shift Registers 447
Linear Sequence Generator 449
10.15 Cascading of Synchronous Counters 453
Cascaded IC Counters with Truncated Sequences 454
Counter Applications 455
Parallel-to-Serial Data Conversions 458
Summary 460
Questions 462
Problems 464

11. LOGIC FAMILIES **466–514**
11.1 Introduction 466
11.2 Digital IC Specification Terminology 466
11.3 Logic Families 471
11.4 Transistor Transistor Logic (TTL) 471
11.5 Open-collector Gates 479
11.6 TTL Subfamilies 483
11.7 Integrated Injection Logic (IIL or I^2L) 486
11.8 Emitter-coupled Logic (ECL) 488
11.9 Metal Oxide Semiconductor (MOS) Logic 494
11.10 Complementary Metal Oxide Semiconductor (CMOS) Logic 498
11.11 Dynamic MOS Logic 505
11.12 Interfacing 509
TTL to ECL 509
ECL to TTL 509
TTL to CMOS 509
CMOS to TTL 509
Summary 510
Questions 512

12. SEQUENTIAL MACHINES 515–545
12.1 Introduction 515
12.2 The Finite State Model 515
State Diagram 517
State Table 517
Transition and Output Table 517
Excitation Table 518
12.3 Memory Elements 518
D Flip-flop 518
T Flip-flop 519
S-R Flip-flop 520
J-K Flip-flop 521
12.4 Synthesis of Synchronous Sequential Circuits 522
12.5 Serial Binary Adder 522
12.6 The Sequence Detector 524
12.7 Parity-bit Generator 526
12.8 Counters 528
12.9 Iterative Networks 531
Summary 542
Questions 543
Problems 544

13. ANALOG-TO-DIGITAL AND DIGITAL-TO-ANALOG CONVERTERS 546–577
13.1 Introduction 546
13.2 Digital-to-Analog (D/A) Conversion 547
13.3 The *R*-2*R* Ladder Type DAC 552
13.4 The Weighted-resistor Type DAC 557
13.5 The Switched Current-source Type DAC 560
13.6 The Switched-capacitor Type DAC 562
13.7 Analog-to-Digital Conversion 563
13.8 The Counter-type A/D Converter 564
13.9 The Tracking-type A/D Converter 565
13.10 The Flash-type A/D Converter 567
13.11 The Dual-slope Type A/D Converter 569
13.12 The Successive-approximation Type ADC 571
Summary 575
Questions 576
Problems 577

14. MEMORIES 578–632
14.1 The Role of Memory in a Computer System 578
14.2 Memory Types and Terminology 579
Memory Organization and Operation 579
Reading and Writing 581
RAMs, ROMs, and PROMs 582
Constituents of Memories 583

14.3 Read Only Memory (ROM) 583
ROM Organization 583
ROM Timing 585
Types of ROMs 586
14.4 Semiconductor RAMs 592
Static RAMs (SRAMs) 593
ECL RAMs 596
Dynamic RAMs (DRAMs) 596
Address Multiplexing 598
DRAM Refreshing 598
14.5 Memory Expansion 598
14.6 Non-volatile RAMs 604
14.7 Sequential Memories 605
Recirculating Shift Registers 605
First In First Out (FIFO) Memories 606
14.8 Programmable Logic Devices 608
Programmable Array Logic (PAL) 608
Field Programmable Logic Array (FPLA) 611
Programmable ROM (PROM) 615
Other PLD Features 615
14.9 Magnetic Memories 617
Magnetic Core Memory 617
Magnetic Disk Memory 618
Magnetic Recording Formats 619
Floppy Disks 622
Hard Disk Systems 624
Magnetic Tape Memory 625
Magnetic Bubble Memory 625
14.10 Optical Disk Memory 627
14.11 Charge-coupled Devices 628
Summary 628
Questions 630

Appendix** Commonly Used TTL ICs* ***633–636
Glossary ***637–648***
Answers to Problems ***649–656***
Index ***657–664***

PREFACE

Reflecting over 27 years of experience in the classroom, this comprehensive textbook on digital circuits is developed to provide a solid grounding in the foundations of basic design techniques. Using the student-friendly writing style, the text introduces the reader to digital concepts in a simple and lucid manner with an emphasis on practical treatment and real-world applications. A large number of typical examples have been worked out, so that the reader can understand the related concepts clearly. Most of the problems in the book have been classroom tested. The book blends basic digital electronic theory with the latest in digital technology. The text is, therefore, suitable for use as course material by undergraduate students of electrical engineering, electronics, computer science, instrumentation, telecommunications, and information technology. As there is no specific prerequisite to understand the book except for an elementary knowledge of basic electronics, it can also be used by students of polytechnics and undergraduate science students pursuing courses in electronics and computer science.

The switching devices used in digital systems are generally two-state devices. So, it is natural to use binary numbers in digital systems. Human beings can interpret and understand data which are available in decimal form. Binary data can be represented concisely using the octal and hexadecimal notations. For this reason, decimal, binary, octal, and hexadecimal number systems, conversion of numbers from one system to another and arithmetic operations in those systems are discussed in Chapter 2.

To provide easy communication between man and machine, and also for ease of use in various devices and for transmission, decimal numbers, symbols, and alphabets are coded in various ways. Several codes and arithmetic operations involving some of those codes are presented in Chapter 3.

The basic building blocks used to construct combinational circuits are logic gates. Various logic gates and the functions performed by them are described in Chapter 4. Pin diagrams of various IC gates and the ANSI/IEEE logic symbols for the gates and the chips are also given.

The logic designer must determine how to interconnect the logic gates in order to convert the circuit input signals to the desired output signals. The relationship between the input and output signals can be described mathematically using Boolean algebra. Chapter 5 introduces the basic laws and theorems of Boolean algebra. It also deals with how to convert algebra to logic and logic to algebra.

Starting from a given problem statement, the first step in designing a combinational logic circuit is to derive a table or formulate algebraic logic equations which describe the circuit for the realization of the output function. The logic equations which describe the circuit output must generally be simplified. The simplification of logic equations using Boolean algebraic methods is presented in Chapter 5.

The simplification of complex functions cannot be performed by the algebraic methods. More systematic methods of simplification of logic expressions, such as the Karnaugh map method and the Quine-McClusky method are introduced in Chapter 6.

Various types of digital circuits used for processing and transmission of data such as arithmetic circuits, comparators, code converters, parity checkers/generators, encoders, decoders, multiplexers, and demultiplexers are discussed in detail in Chapter 7.

The basic memory elements used in the design of sequential circuits, called flip-flops, are introduced in Chapter 8. The flip-flops can be interconnected with gates to make registers for data storage and shifting. Shift registers are described in Chapter 9.

The flip-flops can be interconnected with gates to also form counters. Asynchronous, synchronous, and ring counters are discussed in Chapter 10.

Most of the gates, flip-flops, counters, shift registers, arithmetic circuits, encoders, decoders, etc. are available in several digital logic families. The TTL, ECL, IIL, MOS, and COMS class of logic families are introduced in Chapter 11.

Systematic design of sequential machines is very essential. The design procedures of synchronous sequential machines using state diagrams and state tables are outlined in Chapter 12.

Data processing requires conversion of signals from analog to digital form and from digital to analog form. Various types of analog to digital (A/D) and digital to analog (D/A) converters are explained in Chapter 13.

Modern data processing systems require the permanent or semi-permanent storage of large amounts of data. Both semiconductor and magnetic memories for this purpose are discussed in Chapter 14.

I express my profound gratitude to all those without whose assistance and cooperation, this book would not have been successfully completed. I thank my colleague Mr. N.S. Rane, who patiently drew all the figures in the book. I am appreciative of the help provided by Mr. L. Krishnananda, who typed most portions of the manuscript.

I am grateful to Mr. Ravi Yadahalli, Mr. Dharmaraj and Mr. A. Radhakrishna for their constant help and valuable suggestions. I acknowledge with gratitude the constant encouragement received from late Prof. B.N. Devaraj, in writing this book.

Finally, I am deeply indebted to my wife, A. Jhansi, for her moral support during the writing of this book. I affectionately appreciate my sons, Sunit and Anil, who inspired me to take up this venture.

The author will gratefully acknowledge suggestions from both students and teachers for further improvement of this book.

A. Anand Kumar

Chapter 1

INTRODUCTION

1.1 DIGITAL AND ANALOG SYSTEMS

Electronic circuits and systems are of two kinds—*analog* and *digital*. The distinction between them is not so much in the types of semiconductor devices used in these circuits as it is in voltage and current variations that occur when each type of circuit performs the function for which it is designed. Analog circuits are those in which voltages and currents vary continuously through the given range. They can take infinite values within the specified range. For example, the output voltage from an audio amplifier might be any one of the infinite values between – 10 V and + 10 V at any particular instant of time. Other examples of analog devices include signal generators, radio frequency transmitters and receivers, power supplies, electric motors and speed controllers, and many analog type instruments—those having *pointers* that move in a continuous arc across a calibrated scale. By contrast, a digital circuit is one in which the voltage levels assume a finite number of distinct values. In virtually all modern digital circuits, there are just two discrete voltage levels. However, each voltage level in a practical digital system can actually be a narrow *band* or *range* of voltages.

Digital circuits are often called switching circuits, because the voltage levels in a digital circuit are assumed to be switched from one value to another instantaneously, that is, the transition time is assumed to be zero.

Digital circuits are also called *logic* circuits, because each type of digital circuit obeys a certain set of logic rules. The manner in which a logic circuit responds to an input is referred to as the circuit's logic.

Digital systems are used extensively in computation and data processing, control systems, communications and measurement. Digital systems have a number of advantages over analog systems. Many tasks formally done by analog systems are now being performed digitally. The chief reasons for the shift to digital technology are summarized below:

Digital systems are easier to design. The switching circuits in which there are only two voltage levels, HIGH and LOW, are easier to design. The exact numerical values of voltages are not important because they have only logical significance; only the range in which they fall is important. In analog systems, signals have numerical significance; so, their design is more involved.

Information storage is easy. There are many types of semiconductor and magnetic memories of large capacity which can store data for periods as long as necessary.

Accuracy and precision are greater. Digital systems are much more accurate and precise than analog systems, because digital systems can be easily expanded to handle more digits by adding more switching circuits. Analog systems will be quite complex and costly for the same accuracy and precision.

Digital systems are more versatile. It is fairly easy to design digital systems whose operation is controlled by a set of stored instructions called the *program*. Any time the system operation is to be changed, it can easily be accomplished by modifying the program. Even though analog systems can also be programmed, the variety of the available operations is severely limited.

Digital circuits are less affected by noise. Unwanted electrical signals are called noise. Noise is unavoidable in any system. Since in analog systems the exact values of voltages are important and in digital systems only the range of values is important, the effect of noise is more severe in analog systems. In digital systems, noise is not critical as long as it is not large enough to prevent us from distinguishing a HIGH from a LOW.

More digital circuitry can be fabricated on IC chips. The fabrication of digital ICs is simpler and economical than that of analog ICs. Moreover, higher densities of integration can be achieved in digital ICs than in analog ICs, because digital design does not require high value capacitors, precision resistors, inductors and transformers (which cannot be integrated economically) like the analog design.

Limitations of digital techniques

Even though digital techniques have a number of advantages, they have only one major drawback.

THE REAL WORLD IS ANALOG

Most physical quantities are analog in nature, and it is these quantities that are often the inputs and outputs and continually monitored, operated and controlled by a system. When these quantities are processed and expressed digitally, we are really making a digital approximation to an inherently analog quantity. Instead of processing the analog information directly, it is first converted into digital form and then processed using digital techniques. The results of processing can be converted back to analog form for interpretation. Because of these conversions, the processing time increases and the system becomes more complex. In most cases, these disadvantages are outweighed by numerous advantages of digital techniques. However, there are situations where using only analog techniques is simpler and more economical. Both the analog and digital techniques can be employed in the same system to advantage. Such systems are called *hybrid systems*. But the tendency today is towards employing digital systems because the economic benefits of integration are of overriding importance.

The design of digital systems may be roughly divided into three stages—SYSTEM DESIGN, LOGIC DESIGN, and CIRCUIT DESIGN. System design involves breaking the overall system into subsystems and specifying the characteristics of each subsystem. For example, the system design of a digital computer involves specifying the number and type of memory units, arithmetic units and input-output devices, as well as specifying the interconnection and control of these subsystems. Logic design involves determining how to interconnect basic logic building blocks to perform a specific function. An example of logic design is determining the interconnection of logic gates and flip-flops required to perform binary addition. Circuit design involves specifying the interconnection of specific components such as resistors, diodes and transistors to form a gate, flip-flop or any other logic building block. This book is largely devoted to a study of logic design and the theory necessary for understanding the logic design process.

1.2 LOGIC LEVELS AND PULSE WAVEFORMS

Digital systems use the binary number system. Therefore, two-state devices are used to represent the two binary digits 1 and 0 by two different voltage levels, called HIGH and LOW. If the HIGH voltage level is used to represent 1 and the LOW voltage level to represent 0, the system is called the *positive logic system*. On the other hand, if the HIGH voltage level represents 0 and the LOW voltage level represents 1, the system is called the *negative logic system*. Normally, the binary 0 and 1 are represented by the logic voltage levels 0 V and + 5 V. So, in a positive logic system, 1 is represented by + 5 V (HIGH) and 0 is represented by 0 V (LOW); and in a negative logic system, 0 is represented by + 5 V (HIGH) and 1 is represented by 0 V (LOW). Both positive and negative logics are used in digital systems, but the positive logic is more common. For this reason, we will use only the positive logic system in this book.

In reality, because of circuit variations, the 0 and 1 would be represented by voltage ranges instead of particular voltage levels. Usually, any voltage between 0 V and 0.8 V represents the logic 0 and any voltage between 2 V and 5 V represents the logic 1. Normally, all input and output signals fall within one of these ranges except during transition from one level to another. The range between 0.8 V and 2 V is called the *indeterminate range*. If the signal falls between 0.8 V and 2 V, the response is not predictable.

Digital circuits are designed to respond predictably to input voltages that are within the specified range. That means, the exact values of voltages are not important and the circuit gives the same response for all input voltages in the allowed range, i.e. a voltage of 0 V gives the same response as a voltage of 0.4 V or 0.6 V or 0.8 V. Similarly, a voltage of 2 V gives the same response as a voltage of 2.8 V or 3.6 V or 4.7 V or 5 V.

In digital circuits and systems, the voltage levels are normally changing back and forth between the HIGH and LOW states. So, pulses are very important in their operation. A pulse may be a positive pulse or a negative pulse. A single positive pulse is generated when a normally LOW voltage goes to its HIGH level and then returns to its normal LOW level as shown in Fig. 1.1a. A single negative pulse is generated when a normally HIGH voltage goes to its LOW level and then returns to its normal HIGH level as shown in Fig. 1.1b.

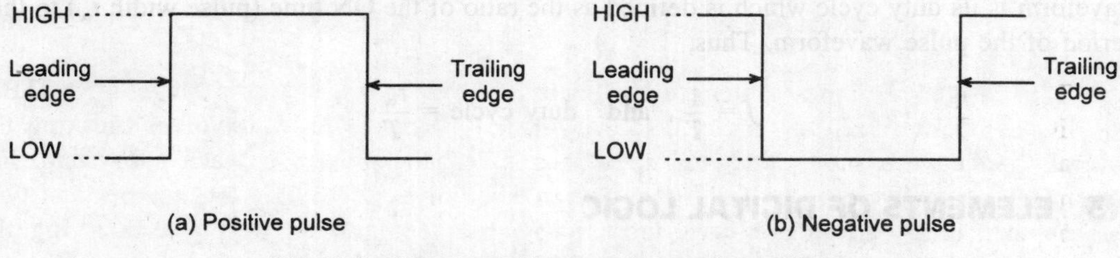

Fig. 1.1 Ideal positive and negative pulses.

As indicated in Fig. 1.1, a pulse has two edges: a leading edge and a trailing edge. For a positive pulse, the leading edge is a positive going transition (PGT or rising edge) and the trailing edge is a negative going transition (NGT or falling edge), whereas for a negative pulse, the leading edge is a negative going transition (NGT) and the trailing edge is a positive going transition (PGT). The pulses shown in Fig. 1.1 are ideal, because the rising and falling edges change instantaneously, i.e. in zero time. Practical pulses do not change instantaneously from LOW to HIGH or from HIGH to LOW.

A non-ideal pulse is shown in Fig. 1.2. It has finite rise and fall times. The time taken by the pulse to rise from LOW to HIGH is called the *rise time* and the time taken by the pulse to go from HIGH to LOW is called the *fall time*. Because of the nonlinearities that commonly occur at the bottom and top of the pulse, the rise time is defined as the time taken by the pulse to rise from 10% to 90% of the pulse amplitude, and the fall time is defined as the time taken by the pulse to fall from 90% to 10% of the pulse amplitude. The duration of the pulse is usually indicated by pulse width t_w which is defined as the time between the 50% points on the rising and falling edges.

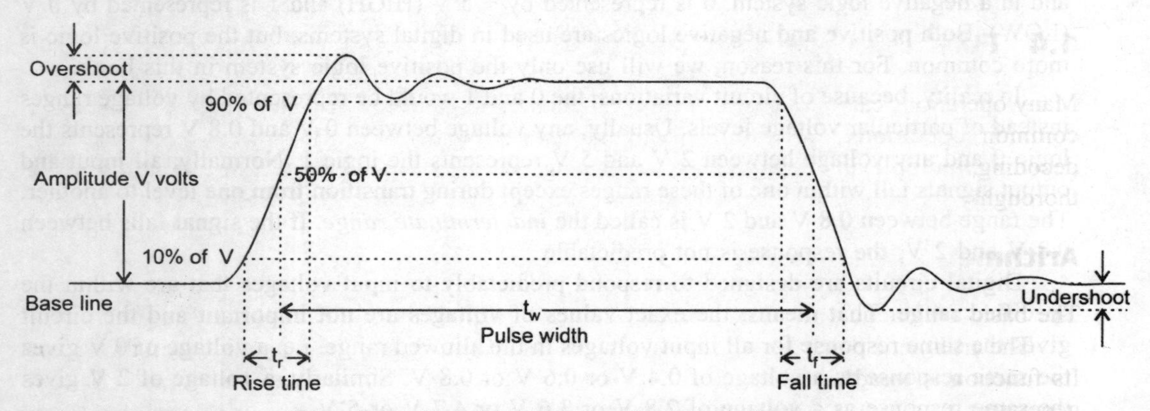

Fig. 1.2 Non-ideal pulse characteristics.

Most waveforms encountered in digital systems are composed of a series of pulses and can be classified as periodic waveforms and non-periodic waveforms. A *periodic waveform* is one which repeats itself at regular intervals of time called the period, *T*. A *non-periodic* waveform, of course, does not repeat itself at regular intervals and may be composed of pulses of different widths and/or differing time intervals between the pulses. The reciprocal of the period is called the *frequency* of the periodic waveform. Another important characteristic of the periodic pulse waveform is its duty cycle which is defined as the ratio of the ON time (pulse width t_w) to the period of the pulse waveform. Thus,

$$f = \frac{1}{T} \quad \text{and} \quad \text{duty cycle} = \frac{t_w}{T}$$

1.3 ELEMENTS OF DIGITAL LOGIC

In our daily life, we make many logic decisions. The term *logic* refers to something which can be reasoned out. In many situations, the problems and processes that we encounter, can be expressed in the form of propositional or logic functions. Since these functions are true/false, yes/no statements, digital circuits with their two-state characteristics are extremely useful. Several logic statements when combined form logic functions. These logic functions can be formulated mathematically using Boolean algebra (which is a system of mathematical logic) and the minimal expression for the function can be obtained using minimization techniques. There are four basic logic elements using which any digital system can be built. They are the three basic gates—NOT, AND and OR, and a flip-flop. In fact, a flip-flop can be constructed using gates. So, we can say that any digital circuit can be constructed using only gates. In addition to the three basic gates, there are two universal gates called NAND and NOR. They are called universal gates, because any circuit of any complexity can be constructed using only NAND gates or only NOR gates. In addition, there are two more gates called XOR and XNOR. We will learn about the characteristics of these gates and flip-flops in the later chapters.

Using logic gates and flip-flops, more complex logic circuits like counters, shift registers, arithmetic circuits, comparators, encoders, decoders, code converters, multiplexers, demultiplexers, memories, etc. can be constructed. These more complex logic functions can then be combined to form complete digital systems to perform specified tasks.

1.4 FUNCTIONS OF DIGITAL LOGIC

Many operations can be performed by combining logic gates and flip-flops. Some of the more common operations are arithmetic operations, comparison, code conversion, encoding, decoding, multiplexing, demultiplexing, shifting, counting and storing. These are all discussed thoroughly in the later chapters. The block diagram operation is given below.

Arithmetic Operations

The basic arithmetic operations are, addition, subtraction, multiplication and division.

The arithmetic operation of *addition* is performed by a digital logic circuit called the *adder*. Its function is to add two numbers *addend* (A) and *augend* (B) with a carry input (CI), and generate a sum term (S) and a carry output term (CO). Figure 1.3a is a block diagram of an adder. It illustrates the addition of the binary equivalents of 8 and 6 with a carry input of 1, which results in a binary sum term 5 and a carry output term 1.

The arithmetic operation of *subtraction* can be performed by a digital logic circuit called the *subtractor*. Its function is to subtract *subtrahend* (A) from *minuend* (B) considering the borrow input (BI) and to generate a difference term (D) and a borrow output term (BO). Since

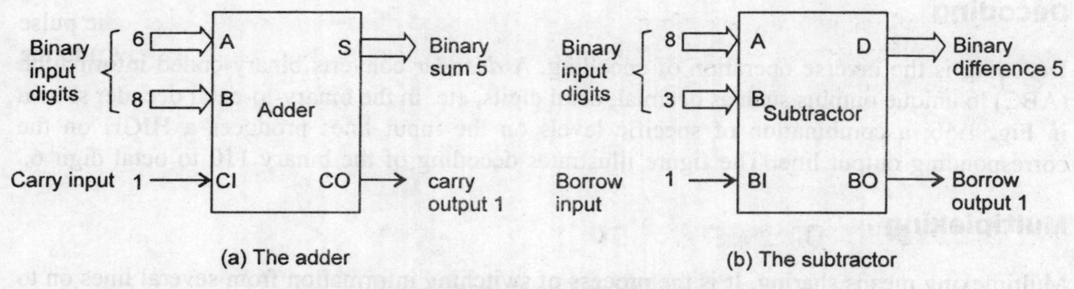

Fig. 1.3 The adder and the subtractor.

subtraction is equivalent to addition of a negative number, subtraction can be performed by using an adder. Figure 1.3b is a block diagram of a subtractor. It illustrates the subtraction of the binary equivalent of 3 from the binary equivalent of 8 with a borrow input of 1, which results in a binary difference term 5 and a borrow output term 1.

The arithmetic operation of *multiplication* can be performed by a digital logic circuit called the *multiplier*. Its function is to multiply *multiplicand* (A) by *multiplier* (B) and generate the product term (P). Since multiplication is simply a series of additions with shifts in the positions of the partial products, it can be performed using an adder. Figure 1.4a is a block diagram of a multiplier. It illustrates the multiplication of 6 by 4, which results in the product term 24.

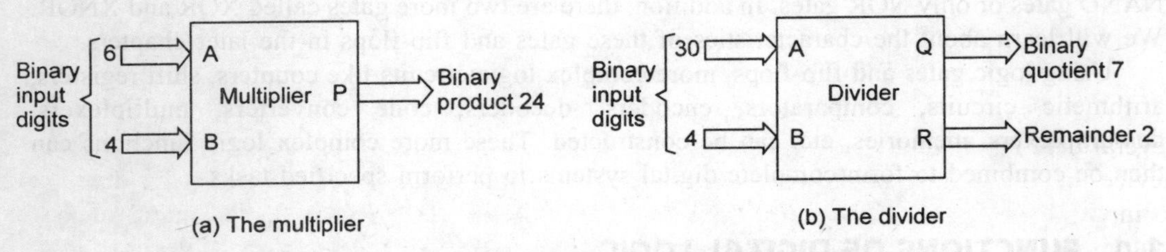

(a) The multiplier (b) The divider

Fig. 1.4 The multiplier and the divider.

The arithmetic operation of *division* can be performed by a digital logic circuit called the *divider*. Division can also be performed by an adder itself, since division involves a series of subtractions, comparisons and shifts. Its function is to divide *dividend* (A) by *divisor* (B) and generate a quotient term (Q) and a remainder term (R). Figure 1.4b is a block diagram of a divider. It illustrates the division of the binary equivalent of 30 by the binary equivalent of 4, which results in a binary quotient term 7 and a remainder term 2.

Encoding

Encoding is the process of converting a familiar number or symbol to some coded form. An *encoder* is a digital device that receives digits (decimal, octal, etc.), or alphabets, or special symbols and converts them to their respective binary codes. In the octal-to-binary encoder shown in Fig. 1.5a, a HIGH level on a given input corresponding to a specific octal digit produces the appropriate 3-bit code (ABC) on the output levels. The figure illustrates encoding of the octal digit 6 to binary 110.

Decoding

Decoding is the inverse operation of encoding. A *decoder* converts binary-coded information (ABC) to unique outputs such as decimal, octal digits, etc. In the binary-to-octal decoder shown in Fig. 1.5b, a combination of specific levels on the input lines produces a HIGH on the corresponding output line. The figure illustrates decoding of the binary 110 to octal digit 6.

Multiplexing

Multiplexing means sharing. It is the process of switching information from several lines on to a single line in a specified sequence. A multiplexer or data selector is a logic circuit that accepts

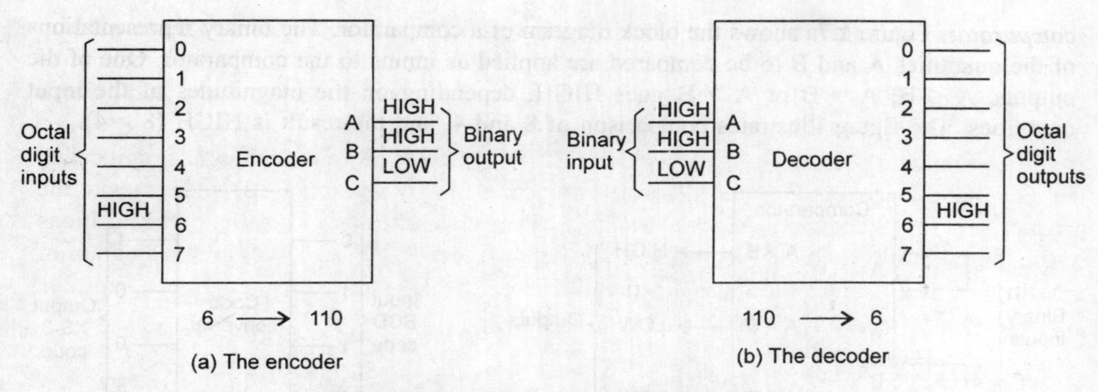

(a) The encoder (b) The decoder

Fig. 1.5 The encoder and the decoder.

several data inputs and allows only one of them to get through to the output. It is an *N*-to-1 device. In the multiplexer shown in Fig. 1.6a, if the switch is connected to input A for time t_1, to input B for time t_2, to input C for time t_3 and to input D for time t_4, the output of the multiplexer will be as shown in the figure. This figure illustrates a 4-to-1 multiplexer.

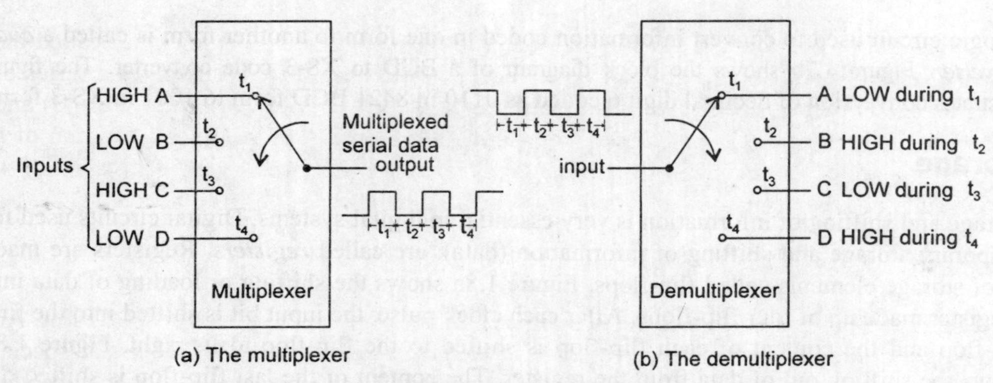

(a) The multiplexer (b) The demultiplexer

Fig. 1.6 The multiplexer and the demultiplexer.

Demultiplexing

Demultiplexing operation is the inverse of multiplexing. Demultiplexing is the process of switching information from one input line on to several output lines. A demultiplexer is a digital circuit that takes a single input and distributes it over several outputs. It is a 1-to-*N* device. In the demultiplexer shown in Fig. 1.6b, if the switch is connected to output A for time t_1, to output B for time t_2, to output C for time t_3 and to output D for time t_4, the output of the demultiplexer will be as shown in the figure. The figure illustrates a 1-to-4 demultiplexer.

Comparison

A logic circuit used to compare two quantities and give an output signal indicating whether the two input quantities are equal or not, and if not, which of them is greater, is called a *comparator*. Figure 1.7a shows the block diagram of a comparator. The binary representations of the quantities A and B to be compared are applied as inputs to the comparator. One of the outputs, A < B, A = B or A > B goes HIGH, depending on the magnitudes of the input quantities. The figure illustrates comparison of 8 and 4, and the result is HIGH (8 > 4).

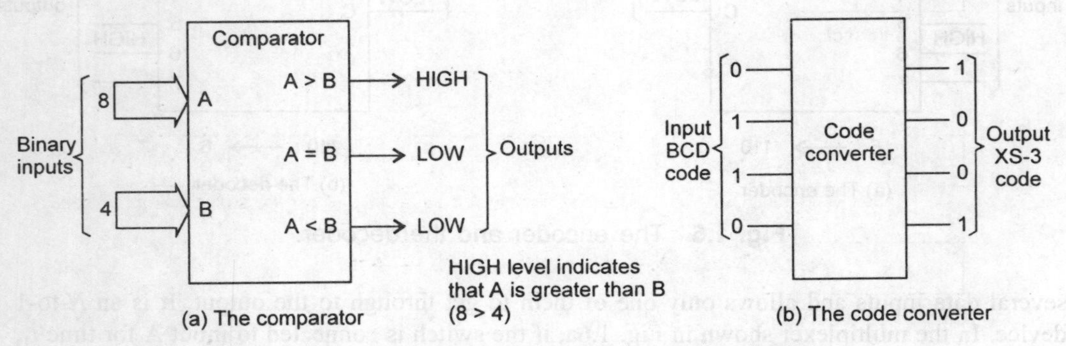

Fig. 1.7 The comparator and the code converter.

Code Conversion

A logic circuit used to convert information coded in one form to another form is called a *code converter*. Figure 1.7b shows the block diagram of a BCD to XS-3 code converter. The figure illustrates conversion of decimal digit 6 coded as 0110 in 8421 BCD form to 1001 in XS-3 form.

Storage

Storage and shifting of information is very essential in digital systems. Digital circuits used for temporary storage and shifting of information (data), are called *registers*. Registers are made up of storage elements called flip-flops. Figure 1.8a shows the shifting or loading of data into a register made up of four flip-flops. After each clock pulse, the input bit is shifted into the first flip-flop and the content of each flip-flop is shifted to the flip-flop to its right. Figure 1.8b shows the shifting out of data from the register. The content of the last flip-flop is shifted out and lost.

Counting

The counting operation is very important in digital systems. A logic circuit used to count the number of pulses inputted to it, is called a *counter*. The pulses may represent some events. In order to count, the counter must remember the present number, so that it can go to the next proper number in the sequence when the next pulse comes. So, storage elements, i.e. flip-flops are used to build counters too. Figure 1.9a shows the block diagram of a counter.

Frequency Division

A counter can also be used to perform *frequency division*. To divide a signal of frequency f by N, the signal is applied to a mod-N counter. The output of the counter will be of frequency f/N. Figure 1.9b shows the block diagram of a frequency divider.

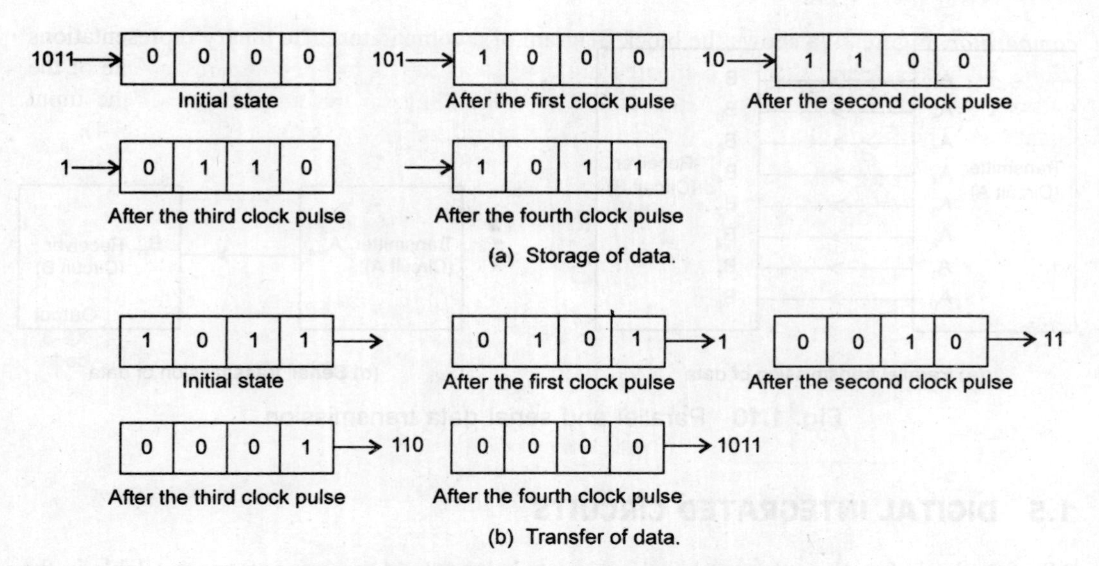

Fig. 1.8 Storage and transfer of data.

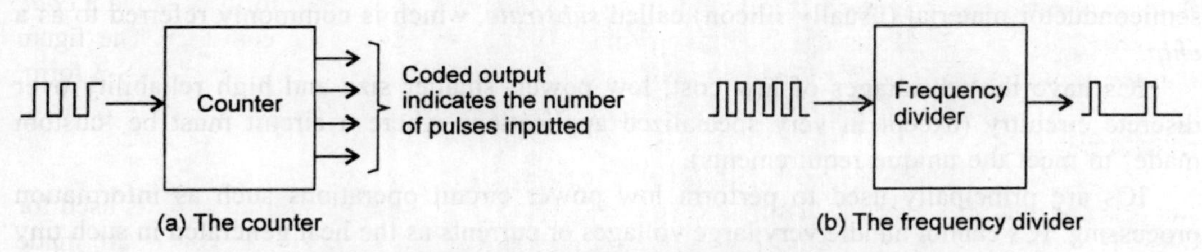

Fig. 1.9 The counter and the frequency divider.

Data Transmission

One of the most common operations that occurs in any digital system is the transmission of information (data) from one place to another. The distance over which information is transmitted may be very small or very large. The information transmitted is in binary form, representing voltages at the outputs of a sending circuit which are connected to the inputs of a receiving circuit. There are two basic methods for transmission of digital information: *parallel* and *serial*.

In parallel data transmission, all the bits are transmitted simultaneously. So, one connecting line is required for each bit. Though data transmission is faster, the number of lines used between the transmitter and the receiver is more. This system is therefore complex and costly. On the other hand, in serial transmission, the information is transmitted bit-by-bit. So, only one connecting line is sufficient between the transmitter and the receiver. Hence, a serial transmission is simpler and cheaper, but slower. The principal trade-off between parallel and serial transmissions is, therefore, one of speed versus circuit simplicity. Figure 1.10a shows parallel data transmission of 8 bits and Fig. 1.10b shows serial data transmission.

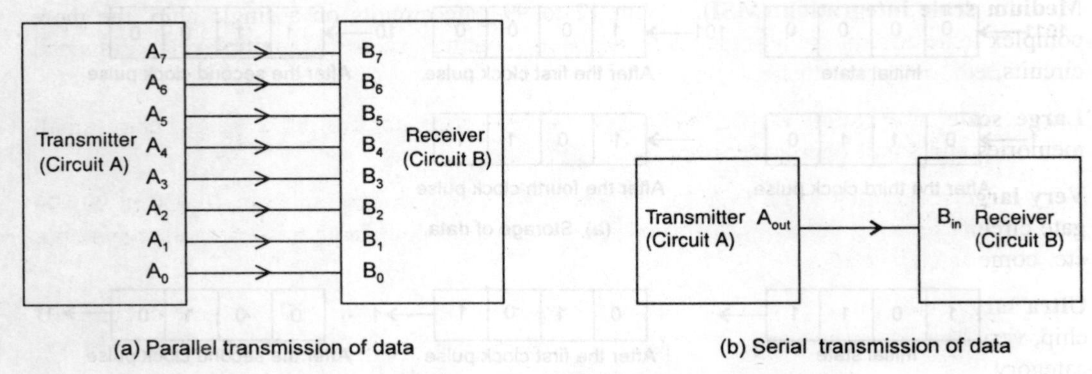

Fig. 1.10 Parallel and serial data transmission.

1.5 DIGITAL INTEGRATED CIRCUITS

All of the logic functions that we have enumerated above (and many more) are available in the integrated circuit (IC) form. Modern digital systems utilize integrated circuits in their design. A monolithic IC is an electronic circuit, that is constructed entirely on a single piece of semiconductor material (usually silicon) called *substrate*, which is commonly referred to as a *chip*.

ICs have the advantages of low cost, low power, smaller size and high reliability over discrete circuitry (except in very specialized applications where a circuit must be 'custom made' to meet the unique requirements).

ICs are principally used to perform low power circuit operations such as information processing. ICs cannot handle very large voltages or currents as the heat generated in such tiny devices would cause the temperature to rise beyond acceptable limits resulting in burning out of ICs. ICs cannot easily implement certain electrical devices such as precision resistors, inductors, transformers, and large capacitors.

ICs may be classified as analog (linear) and digital. Digital ICs are complete functioning blocks as no additional components are required for their operation. The output may be obtained by applying the input. The output is a logic level 0 or 1. For analog ICs, external components are required. Digital ICs are a collection of resistors, diodes, and transistors fabricated on a single chip. The chip is enclosed in a protective plastic or ceramic package from which pins extend for connecting ICs to other devices. There are two main types of packages: dual-in-line package (DIP) and the flat package.

Levels of Integration

Digital ICs are often categorized according to their circuit complexity as measured by the number of equivalent logic gates on the substrate. There are currently five standard levels of complexity.

Small scale integration (SSI). The least complex digital ICs with less than 12 gate circuits on a single chip. Logic gates and flip-flops belong to this category.

Medium scale integration (MSI). With 12 to 99 gate circuits on a single chip, the more complex logic circuits such as encoders, decoders, counters, registers, multiplexers, arithmetic circuits, etc. belong to this category.

Large scale integration (LSI). With 100 to 9999 gate circuits on a single chip, small memories and small microprocessors fall in this category.

Very large scale integration (VLSI). ICs with complexities ranging from 10,000 to 99,999 gate circuits per chip fall in this category. Large memories and large microprocessor systems, etc. come in this category.

Ultra large scale integration (ULSI). With complexities of over 100,000 gate circuits per chip, very large memories and microprocessor systems and single-chip computers come in this category.

Digital ICs can also be categorized according to the principal type of the electronic component used in their circuitry. They are:

(a) Bipolar ICs—which use BJTs

(b) Unipolar ICs—which use MOSFETs.

Several integrated-circuit fabrication technologies are used to produce digital ICs. Presently, digital ICs are fabricated using TTL, ECL, IIL, MOS and CMOS technologies. Each differs from the other in the type of circuitry used to provide the desired logic operation. While TTL, ECL and IIL use bipolar transistors as their main circuit elements, MOS and CMOS use MOSFETs as their main circuit elements. These technologies are also called *logic families*. Several sub-families of these main logic families are also available.

1.6 MICROPROCESSORS

A *microprocessor* is a LSI/VLSI device, that can be programmed to perform arithmetic and logic operations and other functions in a prescribed sequence for the movement and processing of data. Microprocessors are available in word lengths of 4, 8, 16, 32 and 64 bits. Presently, 128-bit microprocessors are being used in some prototype computers. The 4-bit processors are virtually obsolete. Because of their small size, low cost and low power consumption, microprocessors have revolutionized the digital computer system technology. The microprocessor is used as the central processing unit in microcomputer systems. The speed of the microprocessor determines the maximum speed of a microcomputer.

The arrangement of circuits within the microprocessor (called its architecture) permits the system to respond correctly to each of the many different instructions. In addition to arithmetic and logic operations, the microprocessor controls the flow of signals into and out of the computer, routing each to its proper destination in the required sequence to accomplish a specific task. The interconnections or paths along which signals flow are called *buses*. Figure 1.11 shows a block diagram of the microprocessor.

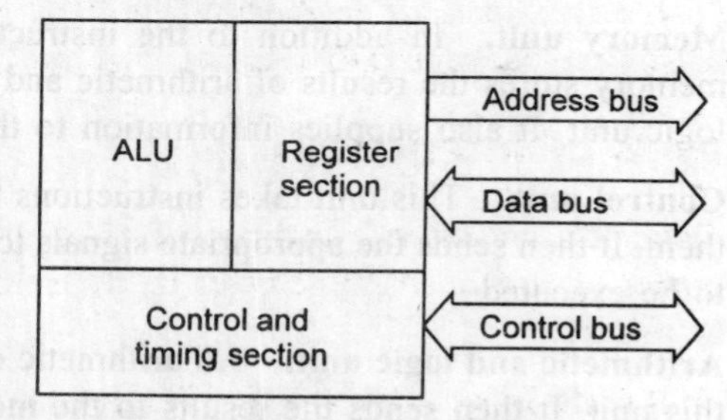

Fig. 1.11 Block diagram of the microprocessor.

1.7 DIGITAL COMPUTERS

The digital computer is a system of hardware that performs arithmetic operations, manipulates data (usually in binary form) and makes decisions. Even though human beings can do most of the things which a computer can do, the computer does the things with much greater speed and accuracy. The computer, however, has to be given a complete set of instructions that tell it exactly what to do at each step of its operation. This set of instructions is called a *program*. Programs are placed in the computer's memory unit in binary coded form with each instruction having a unique code. The computer takes these instruction codes from memory one at a time and performs the operation called for by the code.

Major Parts of a Computer

There are several types of computer systems, but each can be broken down into the same functional units. Each unit performs specific functions, and all the units function together to carry out the instructions given in the program. Figure 1.12 shows the five major functional units of the digital computer and their interconnections. The solid lines with arrows represent the flow of information. The dashed lines with arrows represent the flow of timing and control signals. The major functions of each unit are described below:

Input unit. Through this unit, a complete set of instructions and data is fed into the memory unit to be stored there until needed. The information typically enters the input unit by means of a magnetic tape, or a keyboard.

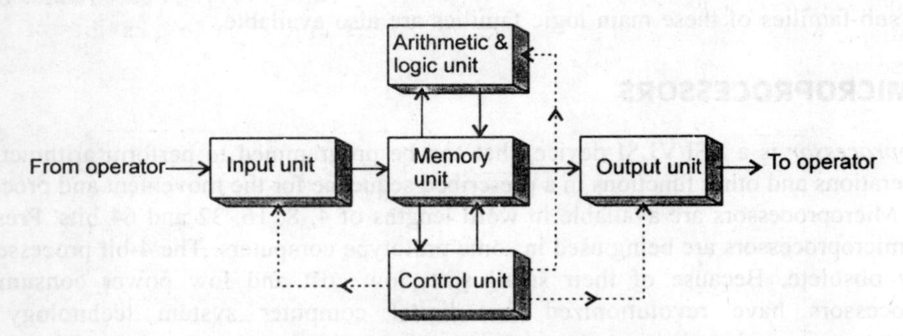

Fig. 1.12 Block diagram of the digital computer.

Memory unit. In addition to the instructions and data received from the input unit, the memory stores the results of arithmetic and logic operations received from the arithmetic and logic unit. It also supplies information to the output unit.

Control unit. This unit takes instructions from the memory unit one at a time and interprets them. It then sends the appropriate signals to all the other units to cause the specific instruction to be executed.

Arithmetic and logic unit. All arithmetic calculations and logical decisions are performed in this unit. It then sends the results to the memory unit to be stored there.

Output unit. This unit takes data from the memory unit and prints out, displays or otherwise presents information to the operator.

1.8 TYPES OF COMPUTERS

The number of computer types depends on the criteria used to classify them. Computers differ in their physical size, operating speed, memory capacity and processing capability as well as in respect of other characteristics. The most common way to classify computers is by their physical size, which usually but not always is indicative of their relative capabilities. The three basic classifications are: microcomputer, minicomputer, and mainframe.

The microcomputer is the smallest type of computer. It generally consists of several IC chips including a microprocessor chip, memory chips, and input-output interface chips along with input-output devices such as a keyboard and video display. Microcomputers resulted from the tremendous advances in IC fabrication technology that has made it possible to pack more and more circuitry into a small space. Figure 1.13 shows a block diagram of the microcomputer system.

Minicomputers are larger than microcomputers and are widely used in industrial control systems, research laboratories, etc. They are generally faster and possess more processing capabilities than microcomputers.

Mainframes are the largest computers. These maxicomputers include complete systems of peripheral equipment such as magnetic tape units, magnetic disk units, card punchers and readers (now obsolete), keyboards, printers and many more. Applications of mainframes range from computation-oriented science and engineering problem-solving to data-oriented business applications, where emphasis is on monitoring and updating of large quantities of data and information.

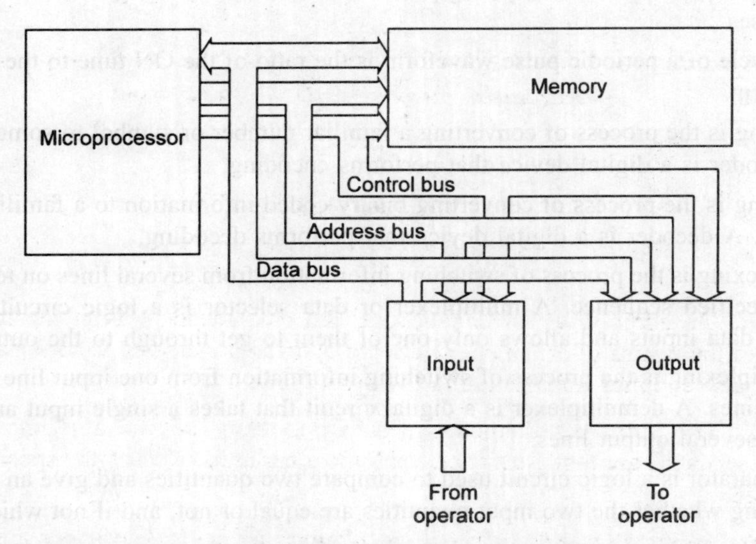

Fig. 1.13 Block diagram of thé microcomputer system.

SUMMARY

- Analog circuits are those in which voltages and currents vary continuously through the given range.
- Digital circuits are those in which the voltage levels can assume only a finite number of distinct values.
- Digital systems are more versatile, easier to design, less affected by noise, more accurate and precise than analog systems.
- The only one major drawback of digital techniques is that, "real world is not digital, it is analog".
- Hybrid systems employ both analog and digital techniques.
- In a positive logic system, the HIGH voltage level represents logic 1 and the LOW voltage level logic 0.
- In a negative logic system, the LOW voltage level represents logic 1 and the HIGH voltage level logic 0.
- The amplitude of a pulse is the height of the pulse, usually expressed in volts.
- Rise time is the time taken by the pulse to rise from 10% to 90% of its amplitude.
- Fall time is the time taken by the pulse to fall from 90% to 10% of its amplitude.
- Pulse width is the time interval between the 50% points on the leading and trailing edges.
- Pulse waveforms may be periodic or non-periodic.
- A periodic pulse waveform repeats itself at regular intervals of time called the *period* (T).
- A non-periodic pulse waveform does not repeat itself at regular intervals of time and may be composed of pulses of differing pulse widths and/or differing time intervals between the pulses.
- Duty cycle of a periodic pulse waveform is the ratio of the ON time to the period of the waveform.
- Encoding is the process of converting a familiar number or symbol to some coded form. An encoder is a digital device that performs encoding.
- Decoding is the process of converting binary coded information to a familiar number or symbol. A decoder is a digital device that performs decoding.
- Multiplexing is the process of switching information from several lines on to a single line in a specified sequence. A multiplexer or data selector is a logic circuit that accepts several data inputs and allows only one of them to get through to the output.
- Demultiplexing is the process of switching information from one input line on to several output lines. A demultiplexer is a digital circuit that takes a single input and distributes it over several output lines.
- A comparator is a logic circuit used to compare two quantities and give an output signal indicating whether the two input quantities are equal or not, and if not which of them is greater.
- A code converter is a logic circuit used to convert information coded in one form to another form.

- Digital circuits used for the temporary storage and shifting of information are called registers.
- ICs have the advantages of low cost, low power, small size and high reliability over discrete circuitry.
- ICs are principally used to perform low power circuit operations such as information processing.
- ICs cannot handle very large voltages or currents as the heat generated in such tiny devices will cause the temperature to rise beyond acceptable limits resulting in burning out of ICs.
- ICs cannot easily implement certain electrical devices such as precision resistors, inductors, transformers, and large capacitors.
- ICs may be analog or digital.
- Digital ICs are complete functional blocks. The output is obtained by applying the input. No other components are required.
- For analog ICs, external components are required to be connected to them for their operation.
- Dual-in-line package and the flat-package are the two main types of IC packages.
- Digital ICs may be classified as unipolar ICs or bipolar ICs depending on the principal component used.
- Digital ICs are also classified in terms of gate circuits per chip—SSI (< 12), MSI (12 to 99), LSI (100 to 9999), VLSI (10,000 to 99,999), and ULSI (> 100,000).
- An integrated circuit (IC) is an electronic circuit constructed entirely on a tiny silicon chip. All of the circuit components are an integral part of the chip.
- A microprocessor is a LSI/VLSI device that can be programmed to perform various digital functions in any specified sequence.
- Computers differ in their physical size, operating speed, memory capacity, and processing capability as well as in respect of other characteristics.
- Based on the physical size, computers may be classified into microcomputers, minicomputers, and mainframes.

QUESTIONS

1. What are the advantages of digital systems over the analog systems? What is the chief limitation to the use of digital techniques?
2. What are the two voltage levels normally used to represent binary digits 0 and 1?
3. Why are digital circuits also called (a) logic circuits, and (b) switching circuits?
4. Explain the following terms: (a) Pulse, (b) Leading edge of a pulse, (c) Trailing edge of a pulse, (d) Positive pulse, (e) Negative pulse, (f) Pulse width, (g) Amplitude of a pulse, (h) Rise time of a pulse, and (i) Fall time of a pulse.
5. What is the difference between periodic and non-periodic pulse waveforms?
6. What do you mean by the duty cycle of a periodic pulse waveform?

7. What do you mean by a positive logic system and a negative logic system?
8. What do you mean by positive going transition and negative going transition of a pulse?
9. Mention the various functions of digital logic.
10. Explain the following terms: (a) Integrated circuit, (b) Digital IC, (c) Analog IC, and (d) Monolithic IC.
11. How are ICs classified?
12. What are the advantages and drawbacks of ICs? Why ICs cannot handle very large voltages and currents?
13. Discuss the different levels of integration of ICs.
14. Name the two commonly used IC packages.
15. What is a microprocessor?
16. Name the five major functional units of a digital computer.
17. Name the three classes of computers based on physical size.

Chapter 2

NUMBER SYSTEMS

2.1 THE DECIMAL NUMBER SYSTEM

We begin our study of the number systems with the familiar decimal number system. The decimal system contains ten unique symbols, 0, 1, 2, 3, 4, 5, 6, 7, 8, and 9. Since counting in decimal involves ten symbols, we say that its *base* or *radix* is ten. There is no symbol for its base, i.e. for ten. It is a positional weighted system. It means that the value attached to a symbol depends on its location with respect to the decimal point. In this system, any number (integer, fraction, or mixed) of any magnitude can be represented by the use of these ten symbols only. Each symbol in the number is called a *digit*.

The left most digit in any number representation, which has the greatest positional weight out of all the digits present in that number, is called the most significant digit (MSD) and the right most digit which has the least positional weight out of all the digits present in that number, is called the least significant digit (LSD). The digits on the left side of the decimal point form the integer part of a decimal number and those on the right side form the fractional part. The digits to the right of the decimal point have weights which are negative powers of 10 and the digits to the left of the decimal point have weights which are positive powers of 10.

The value of a decimal number is the sum of the products of the digits of that number with their respective column weights. The weight of each column is 10 times greater than the weight of the column to its right.

The first digit to the left of the decimal point has a weight of unity or 10^0, the second digit to the left of the decimal point has a weight of 10 or 10^1, the third has a weight of 100 or 10^2, and so on. The first digit to the right of the decimal point has a weight of 1/10 or 10^{-1}, the second digit to the right has a weight of 1/100 or 10^{-2}, the third has a weight of 1/1000 or 10^{-3}, and so on.

In general, the value of any mixed decimal number

$$d_n \, d_{n-1} \, d_{n-2} \, . \, . \, . \, d_1 \, d_0 \, . \, d_{-1} \, d_{-2} \, d_{-3} \, . \, . \, . \, d_{-k}$$

is given by

$$(d_n \times 10^n) + (d_{n-1} \times 10^{n-1}) + \ldots + (d_1 \times 10^1) + (d_0 \times 10^0) + (d_{-1} \times 10^{-1}) + (d_{-2} \times 10^{-2}) + \ldots$$

Consider the decimal number 9256.26 using digits 2, 5, 6, 9. This is a mixed number.

Hence,

$$9256.26 = 9 \times 1000 + 2 \times 100 + 5 \times 10 + 6 \times 1 + 2 \times (1/10) + 6 \times (1/100)$$
$$= 9 \times 10^3 + 2 \times 10^2 + 5 \times 10^1 + 6 \times 10^0 + 2 \times 10^{-1} + 6 \times 10^{-2}$$

Consider another number 6592.69, using the same digits 2, 5, 6, 9. Here,

$$6592.69 = 6 \times 10^3 + 5 \times 10^2 + 9 \times 10^1 + 2 \times 10^0 + 6 \times 10^{-1} + 9 \times 10^{-2}$$

Note the difference in position values of the same digits, when placed in different positions.

Nine's and Ten's Complements

The 9's complement of a decimal number is obtained by subtracting each digit of that decimal number from 9. The 10's complement of a decimal number is obtained by adding a 1 to its 9's complement.

To perform decimal subtraction using the 9's complement method, obtain the 9's complement of the subtrahend and add it to the minuend. Call this number the intermediate result. If there is a carry, it indicates that the answer is positive. Add the carry to the LSD of this result to get the answer. The carry is called the *end around carry*. If there is no carry, it indicates that the answer is negative and the intermediate result is its 9's complement. Take the 9's complement of this result and place a negative sign in front to get the answer.

To perform decimal subtraction using the 10's complement method, obtain the 10's complement of the subtrahend and add it to the minuend. If there is a carry, ignore it. The presence of the carry indicates that the answer is positive; the result obtained is itself the answer. If there is no carry, it indicates that the answer is negative and the result obtained is its 10's complement. Obtain the 10's complement of the result and place a negative sign in front to get the answer.

EXAMPLE 2.1 Find the 9's complement of the following decimal numbers.

(a) 3465 (b) 782.54 (c) 4526.075

Solution

(a)
```
  9 9 9 9
– 3 4 6 5
---------
  6 5 3 4   (9's complement of 3465)
```

(b)
```
  9 9 9 . 9 9
– 7 8 2 . 5 4
-------------
  2 1 7 . 4 5   (9's complement of 782.54)
```

(c)
```
  9 9 9 9 . 9 9 9
– 4 5 2 6 . 0 7 5
-----------------
  5 4 7 3 . 9 2 4   (9's complement of 4526.075)
```

EXAMPLE 2.2 Find the 10's complement of the following decimal numbers.

(a) 4069 (b) 1056.074

Solution

(a)
```
  9 9 9 9
− 4 0 6 9
  5 9 3 0     (9's complement of 4069)
      + 1       Add 1
  5 9 3 1     (10's complement of 4069)
```

(b)
```
  9 9 9 9 . 9 9 9
− 1 0 5 6 . 0 7 4
  8 9 4 3 . 9 2 5     (9's complement of 1056.074)
              + 1       Add 1
  8 9 4 3 . 9 2 6     (10's complement of 1056.074)
```

EXAMPLE 2.3 Subtract the following numbers using the 9's complement method.

(a) 745.81 − 436.62 (b) 436.62 − 745.81

Solution

(a)
```
  7 4 5 . 8 1          7 4 5 . 8 1
− 4 3 6 . 6 2   ⇒    + 5 6 3 . 3 7     (9's complement of 436.62)
  3 0 9 . 1 9        ❶ 3 0 9 . 1 8     (Intermediate result)
                     ↳         + 1     (End around carry)
                       3 0 9 . 1 9     (Answer)
```

(b)
```
  4 3 6 . 6 2          4 3 6 . 6 2
− 7 4 5 . 8 1   ⇒    + 2 5 4 . 1 8     (9's complement of 745.81)
− 3 0 9 . 1 9          6 9 0 . 8 0     (Intermediate result with no carry)
```

The 9's complement of 690.80 is 309.19.
Therefore, the answer is − 309.19.

EXAMPLE 2.4 Subtract the following numbers using the 10's complement method.

(a) 2928.54 − 416.73 (b) 416.73 − 2928.54

Solution

(a)
```
  2 9 2 8 . 5 4          2 9 2 8 . 5 4
− 0 4 1 6 . 7 3   ⇒    + 9 5 8 3 . 2 7     (10's complement of 416.73)
  2 5 1 1 . 8 1        ❶ 2 5 1 1 . 8 1     (Ignore the carry)
```

The answer is 2511.81.

(b)
```
  0 4 1 6 . 7 3          0 4 1 6 . 7 3
− 2 9 2 8 . 5 4   ⇒    + 7 0 7 1 . 4 6     (10's complement of 2928.54)
− 2 5 1 1 . 8 1          7 4 8 8 . 1 9     (No carry)
```

The 10's complement of 7488.19 is 2511.81.
Therefore, the answer is − 2511.81.

2.2 THE BINARY NUMBER SYSTEM

The binary number system is a positional weighted system. The base or radix of this number system is 2. Hence, it has two independent symbols. The base itself cannot be a symbol. The symbols used are 0 and 1. A binary digit is called a bit. A binary number consists of a sequence of bits, each of which is either a 0 or a 1. The binary point separates the integer and fraction parts. Each digit (bit) carries a weight based on its position relative to the binary point. The weight of each bit position is one power of 2 greater than the weight of the position to its immediate right. The first bit to the left of the binary point has a weight of 2^0 and that column is called the units column. The second bit to the left has a weight of 2^1 and it is in the 2's column. The third bit to the left has a weight of 2^2 and it is in the 4's column, and so on. The first bit to the right of the binary point has a weight of 2^{-1} and it is said to be in the 1/2's column, the next right bit with a weight of 2^{-2} is in the 1/4's column, and so on.

The decimal value of the binary number is the sum of the products of all its bits multiplied by the weights of their respective positions. In general, a binary number with an integer part of $(n + 1)$ bits and a fraction part of k bits can be written as

$$d_n \, d_{n-1} \, d_{n-2} \ldots d_1 \, d_0 \, . \, d_{-1} \, d_{-2} \, d_{-3} \ldots d_{-k}$$

Its decimal equivalent is

$$(d_n \times 2^n) + (d_{n-1} \times 2^{n-1}) + \ldots + (d_1 \times 2^1) + (d_0 \times 2^0) + (d_{-1} \times 2^{-1}) + (d_{-2} \times 2^{-2}) + \ldots$$

In general, the decimal equivalent of the number $d_n d_{n-1} \ldots d_1 d_0 \, . \, d_{-1} d_{-2} \ldots$ in any number system with base b is given by

$$(d_n \times b^n) + (d_{n-1} \times b^{n-1}) + \ldots + (d_1 \times b^1) + (d_0 \times b^0) + (d_{-1} \times b^{-1}) + (d_{-2} \times b^{-2}) + \ldots$$

The binary number system is used in digital computers because the switching circuits used in these computers use two-state devices such as transistors, diodes, etc. A transistor can be OFF or ON, a switch can be OPEN or CLOSED, a diode can be OFF or ON, etc. These devices have to exist in one of the two possible states. So, these two states can be represented by the symbols 0 and 1, respectively.

Counting in Binary

Counting in binary is very much similar to decimal counting as shown in Table 2.1. Start counting with 0, the next count is 1. We have now exhausted all symbols; therefore we put a 1 in the column to the left and continue to get 10, 11. Thus, 11 is the maximum we can count using two bits. So, put a 1 in the next column to the left and continue counting; we can count 100, 101, 110, 111. The largest number we can count using three bits is 111. Put a 1 to the left and continue; we get, 1000, 1001, 1010, 1011, 1100, 1101, 1110, 1111. The maximum count we can get using four bits is 1111. Continue counting with 5, 6, . . . bits as shown in Table 2.1.

Table 2.1 Counting in binary

Decimal number	Binary number	Decimal number	Binary number
0	0	20	1 0 1 0 0
1	1	21	1 0 1 0 1
2	1 0	22	1 0 1 1 0
3	1 1	23	1 0 1 1 1
4	1 0 0	24	1 1 0 0 0
5	1 0 1	25	1 1 0 0 1
6	1 1 0	26	1 1 0 1 0
7	1 1 1	27	1 1 0 1 1
8	1 0 0 0	28	1 1 1 0 0
9	1 0 0 1	29	1 1 1 0 1
10	1 0 1 0	30	1 1 1 1 0
11	1 0 1 1	31	1 1 1 1 1
12	1 1 0 0	32	1 0 0 0 0 0
13	1 1 0 1	33	1 0 0 0 0 1
14	1 1 1 0	34	1 0 0 0 1 0
15	1 1 1 1	35	1 0 0 0 1 1
16	1 0 0 0 0	36	1 0 0 1 0 0
17	1 0 0 0 1	37	1 0 0 1 0 1
18	1 0 0 1 0	38	1 0 0 1 1 0
19	1 0 0 1 1	39	1 0 0 1 1 1

An easy way to remember to write a binary sequence of n bits is:

The right most column in the binary number begins with a 0 and alternates between 0 and 1.

The second column begins with $2(= 2^1)$ zeros and alternates between the groups of 0 0 and 1 1.

The third column begins with $4(= 2^2)$ zeros and alternates between the groups of 0 0 0 0 and 1 1 1 1.

The nth column begins with 2^{n-1} zeros and alternates between the groups of 2^{n-1} zeros and 2^{n-1} ones.

Binary To Decimal Conversion

Binary numbers may be converted to their decimal equivalents by the positional weights method. In this method, each binary digit of the number is multiplied by its position weight and the product terms are added to obtain the decimal number.

EXAMPLE 2.5 Convert 10101_2 to decimal.

Solution

(Positional weights) $2^4\ 2^3\ 2^2\ 2^1\ 2^0$

(Binary number) $1\ 0\ 1\ 0\ 1 = (1 \times 2^4) + (0 \times 2^3) + (1 \times 2^2) + (0 \times 2^1) + (1 \times 2^0)$

$$= 16 + 0 + 4 + 0 + 1$$
$$= 21_{10}$$

EXAMPLE 2.6 Convert 11011.101_2 to decimal.

Solution

$$\begin{array}{cccccccc} 2^4 & 2^3 & 2^2 & 2^1 & 2^0 & 2^{-1} & 2^{-2} & 2^{-3} \\ 1 & 1 & 0 & 1 & 1 \cdot & 1 & 0 & 1 \end{array} = (1 \times 2^4) + (1 \times 2^3) + (0 \times 2^2) + (1 \times 2^1) + (1 \times 2^0)$$
$$+ (1 \times 2^{-1}) + (0 \times 2^{-2}) + (1 \times 2^{-3})$$
$$= 16 + 8 + 0 + 2 + 1 + 0.5 + 0 + 0.125$$
$$= 27.625_{10}$$

EXAMPLE 2.7 Convert 1001011_2 to decimal.

Solution

An integer binary number can also be converted to an integer decimal number as follows. Starting with the extreme left bit, i.e. MSB, multiply this bit by 2 and add the product to the next bit to the right.
Multiply the result obtained in the previous step by 2 and add the product to the next bit to the right.
Continue this process as shown below till all the bits in the number are exhausted.

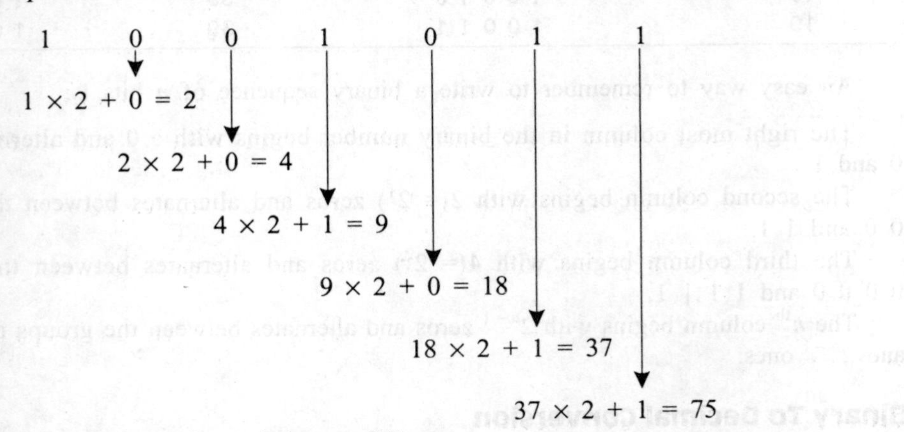

The above steps can be written down as follows:

Copy down the extreme left bit, i.e. MSB = 1
Multiply by 2, add the next bit $(1 \times 2) + 0 = 2$
Multiply by 2, add the next bit $(2 \times 2) + 0 = 4$
Multiply by 2, add the next bit $(4 \times 2) + 1 = 9$
Multiply by 2, add the next bit $(9 \times 2) + 0 = 18$
Multiply by 2, add the next bit $(18 \times 2) + 1 = 37$
Multiply by 2, add the next bit $(37 \times 2) + 1 = 75$
The result is, $1\ 0\ 0\ 1\ 0\ 1\ 1_2 = 75_{10}$

Decimal to Binary Conversion

There are two methods to convert a decimal number to a binary number. These are the reverse processes of the two methods used to convert a binary number to a decimal number.

First method. In this method, which is normally used only for small numbers, the values of various powers of 2 need to be remembered. For conversion of large numbers, we should have a table of powers of 2. In this method, known as the *sum-of-weights* method, the set of binary weight values whose sum is equal to the decimal number is determined. To convert a given decimal integer number to binary, first obtain the largest decimal number which is a power of 2 not exceeding the given decimal number and record it. Subtract this number from the given number and obtain the remainder. Once again obtain the largest decimal number which is a power of 2 not exceeding this remainder and record it. Subtract this number from the remainder to obtain the next remainder. Repeat the above process on the successive remainders till you get a 0 remainder. The sum of these powers of 2 expressed in binary is the binary equivalent of the original decimal number. In a similar manner, this sum-of-weights method can also be applied to convert decimal fractions to binary. To convert a decimal mixed number to binary, convert the integer and fraction parts separately into binary. In general, the sum-of-weights method can be used to convert a decimal number to a number in any other base system.

Second method. In this method, the decimal integer number is converted to binary integer number by successive division by 2, and the decimal fraction is converted to binary fraction by successive multiplication by 2. This is also known as *double-dabble* method. In the successive division-by-2 method, the given decimal integer number is successively divided by 2 till the quotient is zero. The last remainder is the MSB. The remainders read from bottom to top give the equivalent binary integer number. In the successive multiplication-by-2 method, the given decimal fraction and the subsequent decimal fractions are successively multiplied by 2, till the fraction part of the product is 0 or till the desired accuracy is obtained. The first integer obtained is the MSB. Thus, the integers read from top to bottom give the equivalent binary fraction. To convert a mixed number to binary, convert the integer and fraction parts separately to binary and then combine them.

In general, these methods can be used for converting a decimal number to an equivalent number in any other base system by just replacing 2 by the base b of that number system. That is, any decimal number can be converted to an equivalent number in any other base system by the sum-of-weights method, or by the double-dabble method—repeated division-by-b for integers and repeated multiplication-by-b for fractions.

EXAMPLE 2.8 Convert 163.875_{10} to binary.

Solution

The given decimal number is a mixed number. We, therefore, convert its integer and fraction parts separately.

The integer part is 163_{10}.

The largest number, which is a power of 2, not exceeding 163 is 128.

$$128 = 2^7 = 10000000_2$$

The remainder is $163 - 128 = 35$

The largest number, which a power of 2, not exceeding 35 is 32.

$$32 = 2^5 = 100000_2$$

The remainder is $35 - 32 = 3$

The largest number, which a power of 2, not exceeding 3 is 2.

$$2 = 2^1 = 10_2$$

The remainder is $3 - 2 = 1$

$$1 = 2^0 = 1_2$$

Therefore, $163_{10} = 10000000_2 + 100000_2 + 10_2 + 1_2 = 10100011_2$

The fraction part is 0.875_{10}

The largest fraction, which is a power of 2, not exceeding 0.875 is 0.5.

$$0.5 = 2^{-1} = 0.100_2$$

The remainder is $0.875 \;\; 0.5 = 0.375$

The largest fraction, which is a power of 2, not exceeding 0.375 is 0.25.

$$0.25 = 2^{-2} = 0.01_2$$

The remainder is $0.375 - 0.25 = 0.125$

The largest fraction, which a power of 2, not exceeding 0.125 is 0.125 itself.

$$0.125 = 2^{-3} = 0.001_2$$

Therefore, $0.875_{10} = 0.100_2 + 0.01_2 + 0.001_2 = 0.111_2$

The final result is, $163.875_{10} = 10100011.111_2$.

EXAMPLE 2.9 Convert 52_{10} to binary using the double-dabble method.

Solution

We divide the given decimal number successively by 2 and read the remainders upwards to get the equivalent binary number.

Successive division		*Remainder*
2	52	
2	26	0
2	13	0
2	6	1
2	3	↑ 0
2	1	1
	0	1

Reading the remainders from bottom to top, the result is $52_{10} = 110100_2$.

EXAMPLE 2.10 Convert 0.75_{10} to binary using the double-dabble method.

Solution

Multiply the given fraction by 2. Keep the integer in the product as it is and multiply the new fraction in the product by 2. Continue this process and read the integers in the products from top to bottom.

Given fraction	0.75
Multiply 0.75 by 2	↓ 1.50
Multiply 0.50 by 2	↓ 1.00

Reading the integers from top to bottom, $0.75_{10} = 0.11_2$.

EXAMPLE 2.11 Convert 105.15_{10} to binary.

Solution

Conversion of integer 105.

Successive division		*Remainder*
2	105	
2	52	1
2	26	0
2	13	0
2	6	1
2	3	↑ 0
2	1	1
	0	1

Reading the remainders from bottom to top, $105_{10} = 1101001_2$.

Conversion of fraction 0.15_{10}

Given fraction		0.15
Multiply 0.15	by 2	0.30
Multiply 0.30	by 2	0.60
Multiply 0.60	by 2	1.20
Multiply 0.20	by 2	↓ 0.40
Multiply 0.40	by 2	0.80
Multiply 0.80	by 2	1.60

This particular fraction can never be expressed exactly in binary. This process may be terminated after a few steps.

Reading the integers from top to bottom, $0.15_{10} = 0.001001_2$.

Therefore, the final result is, $105.15_{10} = 1101001.001001_2$.

Binary Addition

The rules for binary addition are the following:

0 + 0 = 0; 0 + 1 = 1; 1 + 0 = 1; 1 + 1 = 10, i.e. 0 with a carry of 1.

EXAMPLE 2.12 Add the binary numbers 1010 and 111.

Solution

```
      16 8 4 2 1   (Column numbers)
         1 0 1 0
+          1 1 1
      ----------
       1 0 0 0 1
```

In the 1's column 0 + 1 = 1
In the 2's column 1 + 1 = 0, with a carry of 1 to the 4's column
In the 4's column 0 + 1 + 1 = 0, with a carry of 1 to the 8's column
In the 8's column 1 + 1 = 0, with a carry of 1 to the 16's column.

EXAMPLE 2.13 Add the binary numbers 1101.101 and 111.011.

Solution

	8	4	2	1		2^{-1}	2^{-2}	2^{-3}	(Column numbers)
	1	1	0	1	·	1	0	1	
+		1	1	1	·	0	1	1	
1	0	1	0	1	·	0	0	0	

In the 2^{-3}'s column 1 + 1 = 0, with a carry of 1 to the 2^{-2} column
In the 2^{-2}'s column 0 + 1 + 1 = 0, with a carry of 1 to the 2^{-1} column
In the 2^{-1}'s column 1 + 0 + 1 = 0, with a carry of 1 to the 1's column
In the 1's column 1 + 1 + 1 = 1, with a carry of 1 to the 2's column
In the 2's column 0 + 1 + 1 = 0, with a carry of 1 to the 4's column
In the 4's column 1 + 1 + 1 = 1, with a carry of 1 to the 8's column
In the 8's column 1 + 1 = 0, with a carry of 1 to the 16's column.

Binary Subtraction

The binary subtraction is performed in a manner similar to that in decimal subtraction. The rules for binary subtraction are:

0 − 0 = 0; 1 − 1 = 0; 1 − 0 = 1; 0 − 1 = 1, with a borrow of 1.

EXAMPLE 2.14 Subtract 10_2 from 1000_2.

Solution

In the 1's column 0 − 0 = 0.

In the 2's column, a 1 cannot be subtracted from a 0; so, a 1 must be borrowed from the 4's column. But the 4's column also has a 0; so, a 1 must be borrowed from the 8's column, making the 8's column 0. The 1 borrowed from the 8's column becomes 10 in the 4's column. Keep one 1 in the 4's column and bring the remaining 1 to the 2's column. This becomes 10 in the 2's column, so, 10 − 1 = 1 in the 2's column. Therefore,

		8	4	2	1	(Column numbers)
In the 4's column	1 − 0 = 1	1	0	0	0	
In the 8's column	0 − 0 = 0	−		1	0	
		0	1	1	0	

Hence, the result is 0110_2.

EXAMPLE 2.15 Subtract 111.111_2 from 1010.01_2.

Solution

8	4	2	1		2^{-1}	2^{-2}	2^{-3}	(Column numbers)
1	0	1	0	.	0	1	0	
	1	1	1	.	1	1	1	
0	0	1	0	.	0	1	1	

In the 2^{-3} column, a 1 cannot be subtracted from a 0. So, borrow a 1 from the 2^{-2} column making the 2^{-2} column 0. The 1 borrowed from the 2^{-2} column becomes 10 in the 2^{-3} column. Therefore, in the 2^{-3} column, 10 − 1 = 1.

In the 2^{-2} column, a 1 cannot be subtracted from a 0. So, borrow a 1 from the 2^{-1} column, but it is also a 0. So, borrow a 1 from the 1's column. That is also a 0, so borrow a 1 from the 2's column making the 2's column 0. This 1 borrowed from the 2's column becomes 10 in the 1's column. Keep one 1 in the 1's column, bring the other 1 to the 2^{-1} column, which becomes 10 in this column. Keep one 1 in the 2^{-1} column and bring the other 1 to the 2^{-2} column, which becomes 10 in this column. Therefore,

In the 2^{-2}'s column 10 − 1 = 1
In the 2^{-1}'s column 1 − 1 = 0
In the 1's column 1 − 1 = 0

Now, in the 2's column, a 1 cannot be subtracted from a 0; so, borrow a 1 from the 4's column. But the 4's column has a 0. So, borrow a 1 from the 8's column, making the 8's column 0, and bring it to the 4's column. It becomes 10 in the 4's column. Keep one 1 in the 4's column and bring the second 1 to the 2's column making it 10 in the 2's column. Therefore,

In the 2's column 10 − 1 = 1
In the 4's column 1 − 1 = 0
In the 8's column 0 − 0 = 0

Hence, the result is 0010.011_2.

Binary Multiplication

There are two methods of binary multiplication—the *paper* method and the *computer* method. Both the methods obey the following multiplication rules:

$$0 \times 0 = 0; \quad 1 \times 1 = 1; \quad 1 \times 0 = 0; \quad 0 \times 1 = 0$$

The paper method is similar to the multiplication of decimal numbers on paper.

Multiply the multiplicand with each bit of the multiplier, and add the partial products. The partial product is the same as the multiplicand if the multiplier bit is a 1 and is zero if the multiplier bit is a 0.

EXAMPLE 2.16 Multiply 1101_2 by 110_2.

Solution

The LSB of the multiplier is a 0. So, the first partial product is a 0. The next two bits of the multiplier are 1s. So, the next two partial products are equal to the multiplicand itself. The sum of the partial products gives the answer.

```
          1 1 0 1
        ×   1 1 0
        ---------
          0 0 0 0
        1 1 0 1
      1 1 0 1
      -----------
      1 0 0 1 1 1 0
```

EXAMPLE 2.17 Multiply 1011.101_2 by 101.01_2.

Solution

```
              1 0 1 1 . 1 0 1
            ×   1 0 1 . 0 1 0
            -----------------
                1 0 1 1 1 0 1
              0 0 0 0 0 0 0
            1 0 1 1 1 0 1
          0 0 0 0 0 0 0
        1 0 1 1 1 0 1
        -----------------------
        1 1 1 1 0 1 . 0 0 0 0 1
```

Computer Method of Multiplication

A computer can add only two numbers at a time with a carry. So, the paper method cannot be used by the digital computer. Besides, the computer has only a fixed number of bit positions available. A p-bit number multiplied by another p-bit number yields a $2p$-bit number. To multiply a p-bit number by another p-bit number, a p-bit multiplicand register and a $2p$-bit multiplier/result register are required. The multiplier is placed in the left p-bits of the multiplier/quotient (MQ) register and the right p-bits are set to zero. This permits the MQ to be used as a dual register, which holds the multiplier and the partial (then final) results. The multiplicand is placed in the M register. The MQ register is then shifted one bit to the left. If a 1 is shifted out, M is added to the MQ register. If a 0 is shifted out, a 0 is added to the MQ register. The entire process is repeated p (the number of bits in the multiplier) times. The result appears in the MQ register.

EXAMPLE 2.18 Multiply 1100_2 by 1001_2 using the computer method.

Solution

The computer method of multiplication is performed as shown below.

MQ register	1 0 0 1 0 0 0 0	
Shift MQ left	1 0 0 1 0 0 0 0 0	(A 1 is shifted out. So, add M to MQ)
Add M	1 1 0 0	
Partial sum in MQ	0 0 1 0 1 1 0 0	
Shift MQ left	0 0 1 0 1 1 0 0 0	(A 0 is shifted out. So, add 0 to MQ)
Add 0	0 0 0 0	
Partial sum in MQ	0 1 0 1 1 0 0 0	
Shift MQ left	0 1 0 1 1 0 0 0 0	(A 0 is shifted out. So, add 0 to MQ)
Add 0	0 0 0 0	
Partial sum in MQ	1 0 1 1 0 0 0 0	
Shift MQ left	1 0 1 1 0 0 0 0 0	(A 1 is shifted out. So, add M to MQ)
Add M	1 1 0 0	
Final sum in MQ	0 1 1 0 1 1 0 0	

Binary Division

Like multiplication, division too can be performed by two methods—the paper method and the computer method. In the paper method, long-division procedures similar to those in decimal are used.

EXAMPLE 2.19 Divide 101101_2 by 110_2.

Solution

Divisor 110 cannot go in 101. So, consider the first 4 bits 1011 of the dividend. 110 can go in 1011, one time with a remainder of 101. Next, 110 can go in 1010, one time with a remainder of 100. Next, 110 can go in 1001, one time with a remainder of 11. Finally, 110 can go in 110 with a remainder of 0.

```
1 1 0 ) 1 0 1 1 0 1  ( 1 1 1 . 1
        1 1 0
        -------
        1 0 1 0
          1 1 0
          -------
          1 0 0 1
            1 1 0
            -------
              1 1 0
              1 1 0
              -------
              0 0 0
```

Therefore, 101101 ÷ 110 = 111.1

EXAMPLE 2.20 Divide 110101.11_2 by 101_2.

Solution

```
101 )1 1 0 1 0 1 . 1 1 ( 1 0 1 0 . 1 1
     1 0 1
     -----
     0 1 1 0
       1 0 1
       -----
         1 1 1
         1 0 1
         -----
           1 0 1
           1 0 1
           -----
           0 0 0
```

Therefore, 110101.11 ÷ 101 = 1010.11

Computer Method of Division

The computer method of division involves successive subtraction. Suppose we divide an eight-bit dividend by a four-bit divisor; the quotient will be formed in the right-half of the MQ register and the remainder in the left-half. The dividend is first placed in the MQ register, and the divisor in the D register. The divisor is then subtracted from the dividend. The result is considered negative if a borrow is required, and positive if no borrow is required. If this result is positive, the quotient will be greater than four bits; so, an error has occurred. We cannot complete the division with those registers. If the result is negative, then the quotient will be four or less bits, sufficiently small to be contained in the right-half of the MQ register.

The MQ register is next shifted left one bit. The bit shifted out, a 0 or a 1 is not lost. It is stored in a carry flag, a single flip-flop, and is used for the subtraction following the left shift. If the result of subtraction is positive, a 1 is added to the LSB of the MQ register where the quotient is accumulated; if it is negative, the divisor is added to the MQ, and MQ is shifted left one bit. This effectively puts a 0 in this bit of the quotient. The process is continued until the MQ register has been shifted left four bits (the number of bits in the D register). The remainder is then in the left-half of the MQ register and the quotient in the right-half.

EXAMPLE 2.21 Divide 32 by 5 in binary using the computer method.

Solution

Dividend = 32 = 100000 = 00100000
Divisor = 5 = 101 = 0101

Computer division

MQ	0 0 1 0	0 0 0 0	
Subtract D	0 1 0 1		
MQ	1 1 0 1	0 0 0 0	(Borrow is required; the result is negative; division is valid)

Contd.

Contd.

Add D		0 1 0 1		
MQ		0 0 1 0	0 0 0 0	(Original number)
Shift MQ left	0	0 1 0 0	0 0 0 0	
Subtract D		0 1 0 1		
MQ		1 1 1 1	0 0 0 0	(Result is negative)
Add D		0 1 0 1		
MQ		0 1 0 0	0 0 0 0	
Shift MQ left	0	1 0 0 0	0 0 0 0	
Subtract D		0 1 0 1		
MQ		0 0 1 1	0 0 0 0	(Result is positive)
Add 1			1	
MQ		0 0 1 1	0 0 0 1	
Shift MQ left	0	0 1 1 0	0 0 1 0	
Subtract D		0 1 0 1		
MQ		0 0 0 1	0 0 1 0	(Result is positive)
Add 1			1	
MQ		0 0 0 1	0 0 1 1	
Shift MQ left	0	0 0 1 0	0 1 1 0	
Subtract D		0 1 0 1		
MQ		1 1 0 1	0 1 1 0	(Result is negative)
Add D		0 1 0 1		
MQ		0 0 1 0	0 1 1 0	(Final answer)

Result: Remainder = 0 0 1 0 Quotient = 0 1 1 0

2.3 REPRESENTATION OF SIGNED NUMBERS AND BINARY ARITHMETIC IN COMPUTERS

So far, we have considered only positive numbers. The representation of negative numbers is also equally important. There are two ways of representing signed numbers—sign-magnitude form and complement form. There are two complement forms: 1's complement form and 2's complement form.

Most digital computers do subtraction by the 2's complement method, but some do it by the 1's complement method. The advantage of performing subtraction by the complement method is reduction in the hardware. Instead of having separate digital circuits for addition and subtraction, only adding circuits are needed. That is, subtraction is also performed by adders only. Instead of subtracting one number from the other, the complement of the subtrahend is added to the minuend.

In sign-magnitude form, an additional bit called the *sign bit* is placed in front of the number. If the sign bit is a 0, the number is positive. If it is a 1, the number is negative. For example,

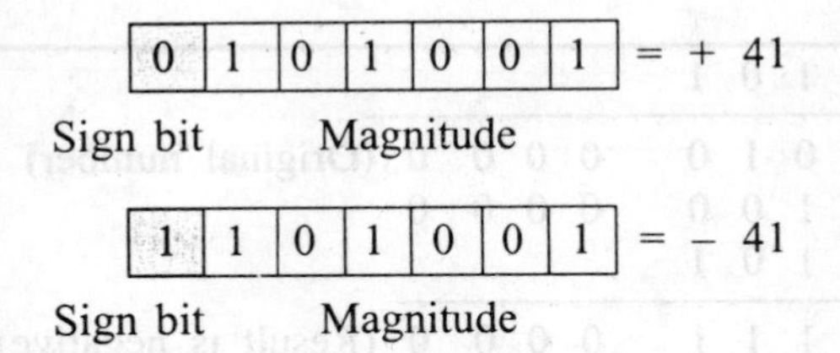

Under the signed-magnitude system, a great deal of manipulation is necessary to add a positive number to a negative number. Thus, though the signed-magnitude number system is possible, it is impractical.

Representation of Signed Numbers Using the 2's (or 1's) Complement Method

The 2's (or 1's) complement system for representing signed numbers works like this:

1. If the number is positive, the magnitude is represented in its true binary form and a sign bit 0 is placed in front of the MSB.
2. If the number is negative, the magnitude is represented in its 2's (or 1's) complement form and a sign bit 1 is placed in front of the MSB.

That is, to represent the numbers in sign 2's (or 1's) complement form, determine the 2's (or 1's) complement of the magnitude of the number and then attach the sign bit.

The 2's (or 1's) complement operation on a signed number will change a positive number to a negative number and vice versa. The conversion of complement to true binary is the same as the process used to convert true binary to complement. The representation of + 51 and − 51 in both 2's and 1's complement forms is shown below:

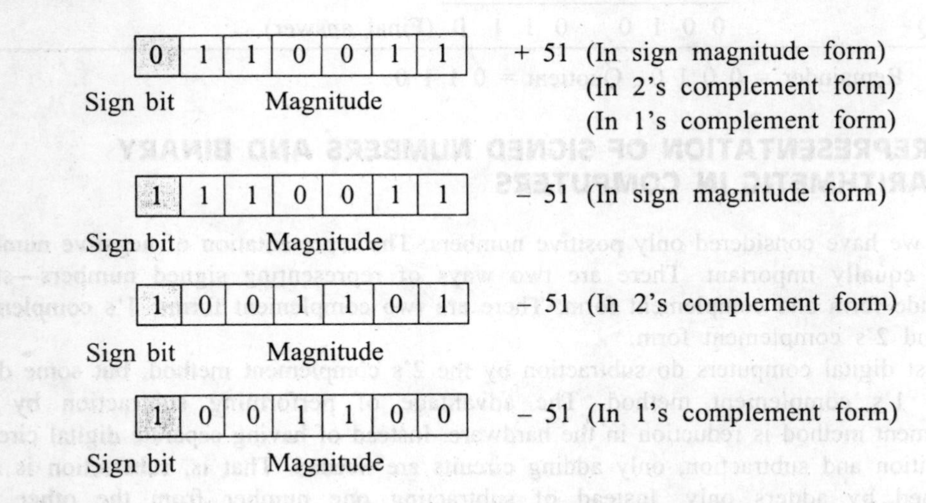

EXAMPLE 2.22 Each of the following numbers is a signed binary number. Determine the decimal value in each case, if they are in (i) sign magnitude form, (ii) 2's complement form, and (iii) 1's complement form.

(a) 01101 (b) 010111 (c) 10111 (d) 1101010

Solution

Given number	*Sign magnitude form*	*2's complement form*	*1's complement form*
0 1 1 0 1	+ 13	+ 13	+ 13
0 1 0 1 1 1	+ 23	+ 23	+ 23
1 0 1 1 1	− 7	− 9	− 8
1 1 0 1 0 1 0	− 42	− 22	− 21

To subtract using the 2's (or 1's) complement method, represent both the subtrahend and the minuend by the same number of bits. Take the 2's (or 1's) complement of the subtrahend including the sign bit. Keep the minuend in its original form and add the 2's (or 1's) complement of the subtrahend to it.

The choice of 0 for positive sign, and 1 for negative sign is not arbitrary. In fact, this choice makes it possible to add the sign bits in binary addition just as other bits are added. When the sign bit is a 0, the remaining bits represent magnitude, and when the sign bit is a 1, the remaining bits represent 2's or 1's complement of the number. The polarity of the signed number can be changed simply by performing the complement on the complete number.

Special case in 2's complement representation. Whenever a signed number has a 1 in the sign bit and all 0s for the magnitude bits, the decimal equivalent is $-\ 2^n$, where n is the number of bits in the magnitude. For example, 1000 = − 8 and 10000 = − 16.

Characteristics of the 2's complement numbers. The 2's complement numbers have the following properties:

1. There is one unique zero.
2. The 2's complement of 0 is 0.
3. The left most bit cannot be used to express a quantity. It is a sign bit. If it is a 1, the number is negative and if it is a 0, the number is positive.
4. For an n-bit word which includes the sign bit, there are $2^{n-1} - 1$ positive integers, 2^{n-1} negative integers and one 0, for a total of 2^n unique states.
5. Significant information is contained in the 1s of the positive numbers and 0s of the negative numbers.
6. A negative number may be converted into a positive number by finding its 2's complement.

Methods of obtaining the 2's complement of a number. The 2's complement of a number can be obtained in three ways as given below.

1. By obtaining the 1's complement of the given number (by changing all 0s to 1s and 1s to 0s) and then adding 1.
2. By subtracting the given n-bit number N from 2^n.

3. Starting at the LSB, copying down each bit up to and including the first 1 bit encountered, and complementing the remaining bits.

EXAMPLE 2.23 Express – 45 in 8-bit 2's complement form.

Solution

+ 45 in 8-bit form is 00101101.

First method

Obtain the 1's complement of 00101101 and then add 1.

Positive expression of the given number	0 0 1 0 1 1 0 1
1's complement of it	1 1 0 1 0 0 1 0
Add 1	+ 1
Thus, the 2's complement form of – 45 is	1 1 0 1 0 0 1 1

Second method

Subtract the given number N from 2^n

2^n =	1 0 0 0 0 0 0 0 0
Subtract 45 =	– 0 0 1 0 1 1 0 1
Thus, the 2's complement form of – 45 is	1 1 0 1 0 0 1 1

Third method

Copy down the bits starting from LSB up to and including the first 1, and then complement the remaining bits.

Original number	0 0 1 0 1 1 0 1
Copy up to the first 1 bit	1
Complement the remaining bits	1 1 0 1 0 0 1
Thus, the 2's complement form of – 45 is	1 1 0 1 0 0 1 1

EXAMPLE 2.24 Express – 73.75 in 12-bit 2's complement form.

Solution

+ 73.75 = N = 01001001.1100

First method

Positive expression of the given number	0 1 0 0 1 0 0 1 1 1 0 0
1's complement of it	1 0 1 1 0 1 1 0 . 0 0 1 1
Add 1	+ 1
Thus, the 2's complement of – 73.75 is	1 0 1 1 0 1 1 0 0 1 0 0

Second method

2^8 =	1 0 0 0 0 0 0 0 0 . 0 0 0 0
Subtract 73.75 = –	0 1 0 0 1 0 0 1 . 1 1 0 0
Thus, the 2's complement of – 73.75 is	1 0 1 1 0 1 1 0 . 0 1 0 0

Third method

Original number	0 1 0 0 1 0 0 1 . 1 1 0 0
Copy up to the first 1 bit	1 0 0
Complement the remaining bits	1 0 1 1 0 1 1 0 . 0
Thus, – 73.75 in 2's complement form is	1 0 1 1 0 1 1 0 . 0 1 0 0

Two's Complement Arithmetic

The 2's complement system is used to represent negative numbers using modulus arithmetic. The word length of a computer is fixed. That means, if a 4-bit number is added to another 4-bit number, the result will be only of 4 bits. Carry, if any, from the fourth bit will overflow. This is called the *modulus arithmetic*. For example: 1100 + 1111 = 1011.

In the 2's complement subtraction, add the 2's complement of the subtrahend to the minuend. If there is a carry out, ignore it. Look at the sign bit, i.e. MSB of the sum term. If the MSB is a 0, the result is positive and is in true binary form. If the MSB is a 1 (whether there is a carry or no carry at all) the result is negative and is in its 2's complement form. Take its 2's complement to find its magnitude in binary.

EXAMPLE 2.25 Subtract 14 from 46 using the 8-bit 2's complement arithmetic.

Solution

+ 1 4	=	0 0 0 0 1 1 1 0	
– 1 4	=	1 1 1 1 0 0 1 0	(In 2's complement form)
+ 4 6		0 0 1 0 1 1 1 0	
– 1 4	⇒	+ 1 1 1 1 0 0 1 0	(2's complement form of – 14)
+ 3 2		❶ 0 0 1 0 0 0 0 0	(Ignore the carry)

There is a carry, ignore it. The MSB is 0; so, the result is positive and is in normal binary form. Therefore, the result is + 00100000 = + 32

EXAMPLE 2.26 Add – 75 to + 26 using the 8-bit 2's complement arithmetic.

Solution

+ 7 5	=	0 1 0 0 1 0 1 1	
– 7 5	=	1 0 1 1 0 1 0 1	(In 2's complement form)
+ 2 6		0 0 0 1 1 0 1 0	
– 7 5	⇒	+ 1 0 1 1 0 1 0 1	(2's complement form of – 75)
– 4 9		1 1 0 0 1 1 1 1	(No carry)

There is no carry, the MSB is a 1. So, the result is negative and is in 2's complement form. The magnitude is 2's complement of 11001111, that is, 00110001 = 49. Therefore, the result is – 49.

EXAMPLE 2.27 Add – 45.75 to + 87.5 using the 12-bit 2's complement arithmetic.

Solution

```
+ 8 7 . 5          0 1 0 1 0 1 1 1 . 1 0 0 0
− 4 5 . 7 5  ⇒   + 1 1 0 1 0 0 1 0 . 0 1 0 0  (– 45.75 in 2's complement form)
+ 4 1 . 7 5      ❶ 0 0 1 0 1 0 0 1 . 1 1 0 0  (Ignore the carry)
```

There is a carry, ignore it. The MSB is 0; so, the result is positive and is in normal binary form. Therefore, the result is + 41.75

EXAMPLE 2.28 Add 27.125 to – 79.625 using the 12-bit 2's complement arithmetic.

Solution

```
+ 2 7 . 1 2 5        0 0 0 1 1 0 1 1 . 0 0 1 0
− 7 9 . 6 2 5  ⇒   + 1 0 1 1 0 0 0 0 . 0 1 1 0  (– 79.625 in 2's
                                                 complement form)
− 5 2 . 5 0 0        1 1 0 0 1 0 1 1 . 1 0 0 0  (No carry)
```

There is no carry, indicating that the result is negative and is in its 2's complement form. The 2's complement of 11001011.1000 is 00110100.1000. Therefore, the result is – 52.5.

EXAMPLE 2.29 Add – 31.5 to – 93.125 using the 12-bit 2's complement arithmetic.

Solution

```
 − 9 3 . 1 2 5        1 0 1 0 0 0 1 0 . 1 1 1 0  (– 93.125 in 2's complement form)
 − 3 1 . 5 0 0  ⇒   + 1 1 1 0 0 0 0 0 . 1 0 0 0  (– 31.5 in 2's complement form)
− 1 2 4 . 6 2 5     ❶ 1 0 0 0 0 0 1 1 . 0 1 1 0  (Ignore the carry)
```

There is a carry, ignore it. The MSB is a 1; so, the result is negative and is in its 2's complement form. The 2's complement of 1000 0011.0110 is 0111 1100.1010. Therefore, the result is – 124.625.

EXAMPLE 2.30 Add 47.25 to 55.75 using the 2's complement method.

Solution

```
  4 7 . 2 5          0 0 1 0 1 1 1 1 . 0 1 0 0
  5 5 . 7 5    ⇒   + 0 0 1 1 0 1 1 1 . 1 1 0 0
1 0 3 . 0 0          0 1 1 0 0 1 1 1 . 0 0 0 0   (No carry)
```

There is no carry, and the MSB is 0. Therefore, the result is positive and is in its true binary form. Hence, it is equal to + 103.0.

EXAMPLE 2.31 Add + 40.75 to – 40.75 using the 12-bit 2's complement arithmetic.

Solution

```
+ 4 0 . 7 5        0 0 1 0 1 0 0 0 . 1 1 0 0
− 4 0 . 7 5  ⇒   + 1 1 0 1 0 1 1 1 . 0 1 0 0  (– 40.75 in 2's complement form)
  0 0 . 0 0      ❶ 0 0 0 0 0 0 0 0 . 0 0 0 0  (Ignore the carry)
```

There is a carry, ignore it. The result is 0.

One's Complement Arithmetic

The 1's complement of a number is obtained by simply complementing each bit of the number, that is, by changing all the 0s to 1s and all the 1s to 0s. We can also say that the 1's complement of a number is obtained by subtracting each bit of the number from 1. This complemented value represents the negative of the original number. This system is very easy to implement in hardware by simply feeding all bits through inverters. One of the difficulties of using 1's complement is its representation of zero. Both 00000000 and its 1's complement 11111111 represent zero. The 00000000 is called *positive zero* and the 11111111 is called *negative zero.*

EXAMPLE 2.32 Represent – 99 and – 77.25 in 8-bit 1's complement form.

Solution

We first write the positive representation of the given number in binary form and then complement each of its bits to represent the negative of the number.

(a) + 9 9 = 0 1 1 0 0 0 1 1

– 9 9 = 1 0 0 1 1 1 0 0 (In 1's complement form)

(b) + 7 7 . 2 5 = 0 1 0 0 1 1 0 1 . 0 1 0 0

– 7 7 . 2 5 = 1 0 1 1 0 0 1 0 . 1 0 1 1 (In 1's complement form)

In 1's complement subtraction, add the 1's complement of the subtrahend to the minuend. If there is a carry out, bring the carry around and add it to the LSB. This is called the *end around carry.* Look at the sign bit (MSB); if this is a 0, the result is positive and is in true binary. If the MSB is a 1 (whether there is a carry or no carry at all), the result is negative and is in its 1's complement form. Take its 1's complement to get the magnitude in binary.

EXAMPLE 2.33 Subtract 14 from 25 using the 8-bit 1's complement arithmetic.

Solution

```
  2 5          0 0 0 1 1 0 0 1
– 1 4   ⇒    + 1 1 1 1 0 0 0 1    (In 1's complement form)
-----        -----------------
+ 1 1      ➊ 0 0 0 0 1 0 1 0
           ↳                + 1    (Add the end around carry)
             -----------------
               0 0 0 0 1 0 1 1    = + 11₁₀
```

EXAMPLE 2.34 Add – 25 to + 14 using the 8-bit 1's complement method.

Solution

```
+ 1 4          0 0 0 0 1 1 1 0
– 2 5   ⇒    + 1 1 1 0 0 1 1 0    (In 1's complement form)
-----        -----------------
– 1 1          1 1 1 1 0 1 0 0    (No carry)
```

There is no carry and the MSB is a 1. So, the result is negative and is in its 1's complement form. The 1's complement of 11110100 is 00001011. The result is, therefore, – 11_{10}.

EXAMPLE 2.35 Add – 25 to – 14 using the 8-bit 1's complement method.

Solution

– 2 5		1 1 1 0 0 1 1 0	(In 1's complement form)
– 1 4	⇒	+ 1 1 1 1 0 0 0 1	(In 1's complement form)
– 3 9		➊ 1 1 0 1 0 1 1 1	
		↳ + 1	(Add the end around carry)
		1 1 0 1 1 0 0 0	

The MSB is a 1. So, the result is negative and is in its 1's complement form. The 1's complement of 11011000 is 00100111. Therefore, the result is – 39.

EXAMPLE 2.36 Add + 25 to + 14 using the 8-bit 1's complement arithmetic.

Solution

+ 2 5		0 0 0 1 1 0 0 1
+ 1 4	⇒	+ 0 0 0 0 1 1 1 0
+ 3 9		0 0 1 0 0 1 1 1

There is no carry. The MSB is a 0. So, the result is positive and is in pure binary. Therefore, the result is, 00100111 = + 39.

EXAMPLE 2.37 Add + 25 to – 25 using the 8-bit 1's complement method.

Solution

+ 2 5		0 0 0 1 1 0 0 1	
– 2 5	⇒	+ 1 1 1 0 0 1 1 0	(In 1's complement form)
0 0		1 1 1 1 1 1 1 1	

There is no carry. The MSB is a 1. So, the result is negative and is in its 1's complement form. The 1's complement of 11111111 is 00000000. Therefore, the result is – 0.

EXAMPLE 2.38 Subtract 27.50 from 68.75 using the 12 bit 1's complement arithmetic.

Solution

+ 6 8 . 7 5		0 1 0 0 0 1 0 0 . 1 1 0 0	
– 2 7 . 5 0	⇒	+ 1 1 1 0 0 1 0 0 . 0 1 1 1	(In 1's complement form)
+ 4 1 . 2 5		➊ 0 0 1 0 1 0 0 1 . 0 0 1 1	
		↳ + 1	(Add the end around carry)
		0 0 1 0 1 0 0 1 . 0 1 0 0	

The MSB is a 0. So, the result is positive and is in its normal binary form. Therefore, the result is + 41.25.

EXAMPLE 2.39 Add – 89.75 to + 43.25 using the 12-bit 1's complement method.

Solution

+ 4 3 . 2 5		0 0 1 0 1 0 1 1 . 0 1 0 0	
– 8 9 . 7 5	⇒	+ 1 0 1 0 0 1 1 0 . 0 0 1 1	(In 1's complement form)
– 4 6 . 5 0		1 1 0 1 0 0 0 1 . 0 1 1 1	

There is no carry. The MSB is a 1. So, the result is negative and is in its 1's complement form. The 1's complement of 11010001.0111 is 00101110.1000. Therefore, the result is – 46.50.

Double Precision Numbers

For any computer the word length is fixed. In a 16-bit computer, that is, in a computer with 16-bit word length, only numbers from $+ 2^{16-1} - 1$ (+ 32,767) to $- 2^{16-1}$ (– 32,768) can be expressed in each register. If numbers greater than this are to be expressed, two storage locations need to be used, that is, each such number has to be stored in two registers. This is called *double precision*. Leaving the MSB which is the sign bit, this allows a 31-bit number length with two 16-bit registers. If still larger numbers are to be expressed, three registers are used to store each number. This is called *triple precision*.

Floating Point Numbers

In the decimal system, very large and very small numbers are expressed in scientific notation, by stating a number (mantissa) and an exponent of 10. Examples are 6.53×10^{-27} and 1.58×10^{21}. Binary numbers can also be expressed in the same notation by stating a number and an exponent of 2. However, the format for a computer may be as shown below.

Mantissa	Exponent
0110000000	100101

In this machine, the 16-bit word consists of two parts, a 10-bit mantissa and a 6-bit exponent. The mantissa is in 2's complement form. So, the MSB can be thought of as the sign bit. The binary point is assumed to be to the right of this sign bit. The 6 bits of the exponent could represent 0 through 63. However, to express negative exponents, the number 32_{10} (100000_2) has been added to the desired exponent.

The actual exponent of the number, therefore, is equal to the exponent given by the 6 bits minus 32. This is a common system used in floating point formats. It is called the excess-32 notation. According to these definitions, the floating-point number shown above is

The mantissa portion = + 0 . 1 1 0 0 0 0 0 0 0
The exponent portion = 1 0 0 1 0 1
The actual exponent = 1 0 0 1 0 1 – 1 0 0 0 0 0 = 0 0 0 1 0 1
The entire number = $N = + 0.1100_2 \times 2^5 = 11000_2 = 24_{10}$

There are many formats of floating point numbers, each computer having its own. Some use two words for the mantissa, and one for the exponent; others use one and half words for the mantissa, and a half word for the exponent. On some machines the programmer can select from several formats, depending on the accuracy desired. Some use excess-*n* notation for the exponent, some use 2's complement, some even use sign-magnitude for both the mantissa and the exponent.

2.4 THE OCTAL NUMBER SYSTEM

The octal number system was extensively used by early minicomputers. It is also a positional weighted system. Its base or radix is 8. It has 8 independent symbols 0, 1, 2, 3, 4, 5, 6, and 7. Since its base $8 = 2^3$, every 3-bit group of binary can be represented by an octal digit. An octal number is, thus, 1/3rd the length of the corresponding binary number.

Usefulness of the Octal System

The ease with which conversions can be made between octal and binary makes the octal system more attractive as a *shorthand* means of expressing large binary numbers. In computer work, binary numbers with up to 64-bits are not uncommon. These binary numbers do not always represent a numerical quantity; they often represent some type of code, which conveys non-numerical information. In computers, binary numbers might represent (a) the actual numerical data, (b) the numbers corresponding to a location (address) in memory, (c) an instruction code, (d) a code expressing alphabetic and other non-numerical characters, or (e) a group of bits representing the status of devices internal or external to the computer.

When dealing with large binary numbers of many bits, it is convenient and more efficient for us to write the numbers in octal rather than binary. However, the digital circuits and systems work strictly in binary; we use octal only for the convenience of the operators of the system. Table 2.2 shows octal counting.

Table 2.2 Octal counting

Decimal number	Octal number	Decimal number	Octal number	Decimal number	Octal number	Decimal number	Octal number
0	0	10	12	20	24	30	36
1	1	11	13	21	25	31	37
2	2	12	14	22	26	32	40
3	3	13	15	23	27	33	41
4	4	14	16	24	30	34	42
5	5	15	17	25	31	35	43
6	6	16	20	26	32	36	44
7	7	17	21	27	33	37	45
8	10	18	22	28	34	38	46
9	11	19	23	29	35	39	47

Octal to Binary Conversion

To convert a given octal number to a binary, just replace each octal digit by its 3-bit binary equivalent.

EXAMPLE 2.40 Convert 367.52_8 to binary.

Solution

Given octal number is	3	6	7	.	5	2
Convert each octal digit to binary	011	110	111	.	101	010
The result is		011110111		.	101010_2	

Binary to Octal Conversion

To convert a binary number to an octal number, starting from the binary point make groups of 3 bits each, on either side of the binary point, and replace each 3-bit binary group by the equivalent octal digit.

EXAMPLE 2.41 Convert 110101.101010_2 to octal.

Solution

Groups of three bits are	110	101	.	101	010
Convert each group to octal	6	5	.	5	2
The result is		65	.	52_8	

EXAMPLE 2.42 Convert 10101111001.0111_2 to octal.

Solution

Groups of three bits are	10	101	111	001	.	011	1
=	010	101	111	001	.	011	100
Convert each group into octal	2	5	7	1	.	3	4
The result is				2571	.	34_8	

Octal to Decimal Conversion

To convert an octal number to a decimal number, multiply each digit in the octal number by the weight of its position and add all the product terms.

The decimal value of the octal number $d_n d_{n-1} d_{n-2} \cdots d_1 d_0 \cdot d_{-1} d_{-2} \cdots d_{-k}$ is

$$(d_n \times 8^n) + (d_{n-1} \times 8^{n-1}) + \cdots + (d_1 \times 8^1) + (d_0 \times 8^0) + (d_{-1} \times 8^{-1}) + (d_{-2} \times 8^{-2}) + \cdots$$

EXAMPLE 2.43 Convert 4057.06_8 to decimal.

Solution

$$\begin{aligned} 4057.06_8 &= 4 \times 8^3 + 0 \times 8^2 + 5 \times 8^1 + 7 \times 8^0 + 0 \times 8^{-1} + 6 \times 8^{-2} \\ &= 2048 + 0 + 40 + 7 + 0 + 0.0937 \\ &= 2095.0937_{10} \end{aligned}$$

Decimal to Octal Conversion

To convert a mixed decimal number to a mixed octal number, convert the integer and fraction parts separately.

To convert the given decimal integer number to octal, successively divide the given number by 8 till the quotient is 0. The last remainder is the MSD. The remainders read upwards give the equivalent octal integer number.

To convert the given decimal fraction to octal, successively multiply the decimal fraction by 8 till the product is 0 or till the required accuracy is obtained. The first integer from the top is the MSD. The integers to the left of the octal point read downwards give the octal fraction.

EXAMPLE 2.44 Convert 378.93_{10} to octal.

Solution

Conversion of 378_{10} *to octal*

Successive division		*Remainders*
8	378	
8	47	↑ 2
8	5	7
	0	5

Read the remainders from bottom to top. Therefore, $378_{10} = 572_8$

Conversion of 0.93_{10} *to octal*

0.93×8	7.44
0.44×8	3.52
0.52×8	↓ 4.16
0.16×8	1.28

Read the integers to the left of the octal point downwards.
Therefore, $0.93_{10} = 0.7341_8$. Hence $378.93_{10} = 572.7341_8$.

Conversion of large decimal numbers to binary and large binary numbers to decimal can be conveniently and quickly performed via octal as shown below.

EXAMPLE 2.45 Convert 5497_{10} to binary.

Solution

Since the given decimal number is large, we first convert this number to octal and then convert the octal number to binary.

Successive division		*Remainders*
8	5497	
8	687	↑ 1
8	85	7
8	10	5
8	1	2
	0	1

Therefore, $5497_{10} = 12571_8 = 001010101111001_2$

EXAMPLE 2.46 Convert 101111010001_2 to decimal.

Solution

Since the given binary number is large, we first convert this number to octal and then convert the octal number to decimal.

$$101111010001_2 = 5721_8 = 5 \times 8^3 + 7 \times 8^2 + 2 \times 8^1 + 1 \times 8^0 = 2560 + 448 + 16 + 1 = 3025_{10}.$$

Octal Addition

The octal addition is performed by the decimal method. Add the digits in each column in decimal and convert this sum into octal. Record the octal sum term in that column and carry the carry term to the next column.

EXAMPLE 2.47 Add 327.54_8 to 665.37_8.

Solution

	8^3	8^2	8^1	8^0		8^{-1}	8^{-2}
		3	2	7	.	5	4
+		6	6	5	.	3	7
	1	2	1	5	.	1	3

In the 8^{-2}'s column, $4 + 7 = 11_{10} = 13_8$ (sum 3, carry 1)
In the 8^{-1}'s column, $5 + 3 + 1 = 9_{10} = 11_8$ (sum 1, carry 1)
In the 8^0's column, $7 + 5 + 1 = 13_{10} = 15_8$ (sum 5, carry 1)
In the 8^1's column, $2 + 6 + 1 = 9_{10} = 11_8$ (sum 1, carry 1)
In the 8^2's column, $3 + 6 + 1 = 10_{10} = 12_8$ (sum 2, carry 1)

Octal Subtraction

The octal subtraction is similar to decimal subtraction.

EXAMPLE 2.48 Subtract 16.47_8 from 20.14_8.

Solution

	8^1	8^0		8^{-1}	8^{-2}
	2	0	.	1	4
−	1	6	.	4	7
	0	1	.	4	5

In the 8^{-2}'s column, 7 cannot be subtracted from 4. So, borrow a 1 from the 8^{-1}'s column leaving a 0 there. A 1 borrowed from the 8^{-1}'s column becomes 8 in the 8^{-2}'s column. So, in the 8^{-2}'s column $8 + 4 - 7 = 5$.

In the 8^{-1}'s column, 4 cannot be subtracted from 0. So, borrow a 1 from the 8^0's column, but the 8^0's column has a 0. So, borrow a 1 from the 8^1's column leaving a 1 there. The 1 borrowed from the 8^1's column becomes 8 in 8^0's column. Leave 7 there and bring 1 to the 8^{-1}'s column. This becomes 8 in this column.

So, in the column 8^{-1} $\quad 8 - 4 = 4$
in the column 8^0 $\quad 7 - 6 = 1$
in the column 8^1 $\quad 1 - 1 = 0$

The result is 01.45_8.

Subtraction of octal numbers can also be accomplished by the 7's and 8's complement methods similar to 1's and 2's complement methods. The 7's complement of an octal number is obtained by subtracting each digit of the number from 7 and its 8's complement is obtained by adding a 1 to its 7's complement.

EXAMPLE 2.49 Find the 7's and 8's complement of the following octal numbers.

(a) 4015 (b) 2057.106

Solution

To find the 7's complement of an octal number, subtract each digit of that number from 7.

```
(a)     7 7 7 7
      − 4 0 1 5
      ---------
        3 7 6 2   (7's complement of 4015)
              1   (Add 1)
      ---------
        3 7 6 3   (8's complement of 4015)
```

```
(b)     7 7 7 7 . 7 7 7
      − 2 0 5 7 . 1 0 6
      -----------------
        5 7 2 0 . 6 7 1   (7's complement of 2057.106)
                      1   (Add 1)
      -----------------
        5 7 2 0 . 6 7 2   (8's complement of 2057.106)
```

EXAMPLE 2.50 Subtract 236.43_8 from 5427.65_8 using the 7's complement arithmetic.

Solution

```
   5 4 2 7 . 6 5              5 4 2 7 . 6 5
 −   2 3 6 . 4 3     ⇒    +   7 5 4 1 . 3 4   (7's complement of 0236.43)
 ---------------          -----------------
 + 5 1 7 1 . 2 2          ❶  5 1 7 1 . 2 1
                           ↳            + 1   (End around carry)
                          -----------------
                              5 1 7 1 . 2 2
```

The carry indicates that the answer is positive and is in octal itself. Hence, the result is $+\ 5171.22_8$

EXAMPLE 2.51 Subtract 5427.65_8 from 236.43_8 using the 8's complement method.

Solution

```
     2 3 6 . 4 3            0 2 3 6 . 4 3
 − 5 4 2 7 . 6 5     ⇒    + 2 3 5 0 . 1 3   (8's complement of 5427.65)
 ---------------          ---------------
 − 5 1 7 1 . 2 2            2 6 0 6 . 5 6   (No carry)
```

No carry indicates that the result is negative and is in 8's complement form. The 8's complement of 2606.56_8 is 5171.22_8. Hence, the result is $-\ 5171.22_8$.

Octal Multiplication

To perform multiplication in octal, multiply the digits in decimal, convert the product to octal, record the sum term, and carry the carry term to the next column. Add all partial products.

EXAMPLE 2.52 Multiply 2763.5_8 by 6_8.

Solution

2 7 6 3 . 5	$5 \times 6 = 30_{10} = 36_8$ (sum 6, carry 3)
$\times 6$	$3 \times 6 = 18_{10} = 22_8$; $22_8 + 3_8 = 25_8$ (sum 5, carry 2)
2 1 6 6 5 . 6	$6 \times 6 = 36_{10} = 44_8$; $44_8 + 2_8 = 46_8$ (sum 6, carry 4)
	$7 \times 6 = 42_{10} = 52_8$; $52_8 + 4_8 = 56_8$ (sum 6, carry 5)
	$2 \times 6 = 12_{10} = 14_8$; $14_8 + 5_8 = 21_8$ (sum 1, carry 2)

Octal Division

The octal division is accomplished by the decimal method.

EXAMPLE 2.53 Divide 4570.32_8 by 6_8.

Solution

6)4570.32(624.042	
44	
17	6 cannot go in 4, So, consider 45_8
14	6 can go in 45_8 (37_{10}) 6 times with a remainder of 1
30	6 can go in 17_8 (15_{10}) 2 times with a remainder of 3
30	6 can go in 30_8 (24_{10}) 4 times with a remainder of 0
032	6 can go in 03_8 (03_{10}) 0 times with a remainder of 3 so, consider 032
30	6 can go in 32_8 (26_{10}) 4 times with a remainder of 2
20	6 can go in 20_8 (16_{10}) 2 times with a remainder of 4
14	The quotient is 624.042_8 and the remainder is 4_8.
4	

2.5 THE HEXADECIMAL NUMBER SYSTEM

Binary numbers are long. These numbers are fine for machines but are too lengthy to be handled by human beings. So, there is a need to represent the binary numbers concisely. One number system developed with this objective is the hexadecimal number system (or Hex). Although it is somewhat more difficult to interpret than the octal number system, it has become the most popular means of direct data entry and retrieval in digital systems. The hexadecimal number system is a positional-weighted system. The base or radix of this number system is 16, that means, it has 16 independent symbols. The symbols used are 0, 1, 2, 3, 4, 5, 6, 7, 8, 9, A, B, C, D, E, and F. Since its base is $16 = 2^4$, every 4 binary digit combination can be represented by one hexadecimal digit. So, a hexadecimal number is 1/4th the length of the corresponding binary number, yet it provides the same information as the binary number. A 4-bit group is called a *nibble*. Since computer words come in 8 bits, 16 bits, 32 bits and so on, that is, multiples of 4 bits, they can be easily represented in hexadecimal. The hexadecimal system is particularly useful for human communications with computers. By far, this is the most commonly used number system in computer literature. It is used both in large and small computers.

Hexadecimal Counting Sequence

0	1	2	3	4	5	6	7	8	9	A	B	C	D	E	F
10	11	12	13	14	15	16	17	18	19	1A	1B	1C	1D	1E	1F
.	.	.	.	.	.	.	.	.	.	.	.	.	.	.	.
.	.	.	.	.	.	.	.	.	.	.	.	.	.	.	.
F0	F1	F2	F3	F4	F5	F6	F7	F8	F9	FA	FB	FC	FD	FE	FF
100	101	102	103	104	105	106	107	108	109	10A	10B	10C	10D	10E	10F
.	.	.	.	.	.	.	.	.	.	.	.	.	.	.	.
.	.	.	.	.	.	.	.	.	.	.	.	.	.	.	.
1F0	1F1	1F2	1F3	1F4	1F5	1F6	1F7	1F8	1F9	1FA	1FB	1FC	1FD	1FE	1FF

Binary to Hexadecimal Conversion

To convert a binary number to a hexadecimal number, starting from the binary point, make groups of 4 bits each, on either side of the binary point and replace each 4-bit group by the equivalent hexadecimal digit as shown in Table 2.3.

Table 2.3 Binary to hexadecimal conversion

Hexadecimal	Binary	Hexadecimal	Binary
0	0 0 0 0	8	1 0 0 0
1	0 0 0 1	9	1 0 0 1
2	0 0 1 0	A	1 0 1 0
3	0 0 1 1	B	1 0 1 1
4	0 1 0 0	C	1 1 0 0
5	0 1 0 1	D	1 1 0 1
6	0 1 1 0	E	1 1 1 0
7	0 1 1 1	F	1 1 1 1

EXAMPLE 2.54 Convert 1011011011_2 to hexadecimal.

Solution

Make groups of 4 bits, and replace each 4-bit group by a hex digit.

Given binary number is 1011011011

Groups of four bits are 0010 1101 1011

Convert each group to hex	2 D B
The result is	$2DB_{16}$

EXAMPLE 2.55 Convert 01011111011.011111_2 to hexadecimal.

Solution

Given binary number is	01011111011.011111
Groups of four bits are	0010 1111 1011 . 0111 1100
Convert each group to hex	2 F B . 7 C
The result is	$2FB.7C_{16}$

Hexadecimal to Binary

To convert a hexadecimal number to binary, replace each hex digit by its 4-bit binary group.

EXAMPLE 2.56 Convert $4BAC_{16}$ to binary.

Solution

Given hex number is	4 B A C
Convert each hex digit to 4-bit binary	0100 1011 1010 1100
The result is	0100101110101100_2

EXAMPLE 2.57 Convert $3A9E.B0D_{16}$ to binary.

Solution

Given hex number is	3 A 9 E . B 0 D
Convert each hex digit to 4-bit binary	0011 1010 1001 1110 . 1011 0000 1101
The result is	$0011101010011110.101100001101_2$

Hexadecimal to Decimal

To convert a hexadecimal number to decimal, multiply each digit in the hex number by its position weight and add all those product terms.

If the hex number is $d_n d_{n-1} \ldots d_1 d_0 \,.\, d_{-1} d_{-2} \ldots d_{-k}$, its decimal equivalent is $(d_n \times 16^n) + (d_{n-1} \times 16^{n-1}) + \ldots + (d_1 \times 16^1) + (d_0 \times 16^0) + (d_{-1} \times 16^{-1}) + (d_{-2} \times 16^{-2}) + \ldots$

EXAMPLE 2.58 Convert $5C7_{16}$ to decimal.

Solution

Multiply each digit of 5C7 by its position weight and add the product terms.

$$
\begin{aligned}
5C7_{16} &= (5 \times 16^2) + (12 \times 16^1) + (7 \times 16^0) \\
&= 1280 + 192 + 7 \\
&= 1479_{10}
\end{aligned}
$$

EXAMPLE 2.59 Convert $A0F9.0EB_{16}$ to decimal.

Solution

$$\begin{aligned} A0F9.0EB_{16} &= (10 \times 16^3) + (0 \times 16^2) + (15 \times 16^1) + (9 \times 16^0) + (0 \times 16^{-1}) + (14 \times 16^{-2}) \\ &\quad + (11 \times 16^{-3}) \\ &= 40960 + 0 + 240 + 9 + 0 + 0.0546 + 0.0026 \\ &= 41209.0572_{10} \end{aligned}$$

Decimal To Hexadecimal

To convert a decimal integer number to hexadecimal, successively divide the given decimal number by 16 till the quotient is zero. The last remainder is the MSB. The remainders read from bottom to top give the equivalent hexadecimal integer.

To convert a decimal fraction to hexadecimal, successively multiply the given decimal fraction by 16, till the product is zero or till the required accuracy is obtained, and collect all the integers to the left of decimal point. The first integer is the MSB and the integers read from top to bottom give the hexadecimal fraction. This is known as the *hex dabble method.*

EXAMPLE 2.60 Convert 2598.675_{10} to hex.

Solution

The given decimal number is a mixed number. Convert the integer and the fraction parts separately to hex.

Conversion of 2598_{10}

Successive division		Remainder: Decimal	Remainder: Hex
16	2598		
16	162	6	↑ 6
16	10	2	2
	0	10	A

Reading the remainders upwards, $2598_{10} = A26_{16}$.

Conversion of 0.675_{10}

Given fraction is 0.675

0.675 × 16	10.8
0.800 × 16	12.8
0.800 × 16	12.8
0.800 × 16	↓ 12.8

Reading the integers to the left of hexadecimal point downwards, $0.675_{10} = 0.ACCC_{16}$.
Therefore, $2598.675_{10} = A26.ACCC_{16}$.

Conversion of very large decimal numbers to binary and very large binary numbers to decimal is very much simplified if it is done via the hex route.

EXAMPLE 2.61 Convert 49056_{10} to binary.

Solution

The given decimal number is very large. It is tedious to convert this number to binary directly. So, convert this to hex first, and then convert the hex to binary.

Successive division	*Remainder*		
	Decimal	*Hex*	*Binary group*
16 | 49056			
16 | 3066	0	↑ 0	0 0 0 0
16 | 191	10	A	1 0 1 0
16 | 11	15	F	1 1 1 1
0	11	B	1 0 1 1

Therefore, $49056_{10} = \text{BFA0}_{16} = 1011111110100000_2$.

EXAMPLE 2.62 Convert 1011011101101110_2 to decimal.

Solution

The given binary number is very large. So, perform the binary to decimal conversion via the hex route.

$$1011\ 0111\ 0110\ 1110_2 = \text{B 7 6 E}_{16}$$

$$\text{B76E}_{16} = (11 \times 16^3) + (7 \times 16^2) + (6 \times 16^1) + (14 \times 16^0)$$

$$= 45056 + 1792 + 96 + 14$$

$$= 46958_{10}$$

Octal To Hexadecimal Conversion

To convert an octal number to hexadecimal, the simplest way is to first convert to binary and then the binary to hexadecimal.

EXAMPLE 2.63 Convert 756.603_8 to hex.

Solution

Given octal number is	7	5	6	.	6	0	3
Convert each octal digit to binary	111	101	110	.	110	000	011
Groups of four bits are	0001	1110	1110	.	1100	0001	1000
Convert each four-bit group to hex	1	E	E	.	C	1	8
The result is			1EE	.	C18$_{16}$		

Hexadecimal to Octal

To convert a hexadecimal number to octal, the simplest way is to first convert the given hexadecimal number to binary and then the binary number to octal.

EXAMPLE 2.64 Convert $B9F.AE_{16}$ to octal.

Solution

Given hex number is	B 9 F . A E
Convert each hex digit to binary	1011 1001 1111 . 1010 1110
Groups of three bits are	101 110 011 111 . 101 011 100
Convert each three-bit group to octal	5 6 3 7 . 5 3 4
The result is	5637.534_8.

Hexadecimal Addition

Addition in hexadecimal is performed in a manner similar to that in decimal. To add in hex, all the digits in each column along with the carry if any, are added in decimal and its hex equivalent obtained. The sum term in hex is recorded in that column, and the carry term is carried to the next column.

EXAMPLE 2.65 Add $2A7C.30D_{16}$ and $8D9.E8B_{16}$.

Solution

16^3	16^2	16^1	16^0		16^{-1}	16^{-2}	16^{-3}
2	A	7	C	.	3	0	D
+	8	D	9	.	E	8	B
3	3	5	6	.	1	9	8

Starting from the LSD

$D + B = 13_{10} + 11_{10} = 24_{10} = 18_{16}$ (sum 8, carry 1)
$0 + 8 + 1 = 9_{10} = 9_{16}$ (sum 9, carry 0)
$3 + E = 3_{10} + 14_{10} = 17_{10} = 11_{16}$ (sum 1, carry 1)
$C + 9 + 1 = 12_{10} + 9_{10} + 1_{10} = 22_{10} = 16_{16}$ (sum 6, carry 1)
$7 + D + 1 = 7_{10} + 13_{10} + 1_{10} = 21_{10} = 15_{16}$ (sum 5, carry 1)
$A + 8 + 1 = 10_{10} + 8_{10} + 1_{10} = 19_{10} = 13_{16}$ (sum 3, carry 1)
$2 + 1 = 3_{10} = 3_{16}$ (sum 3, carry 0)

The result is 3356.198_{16}.

EXAMPLE 2.66 Add $3BCA.5078_{16} + 9EBD.97F3_{16} + 5FB.E2C_{16}$.

Solution

	3	B	C	A	.	5	0	7	8
+	9	E	B	D	.	9	7	F	3
		5	F	B	.	E	2	C	
	E	0	8	3	.	C	B	2	B

Starting from the LSD

$8 + 3 = 11_{10} = B_{16}$	(sum B, carry 0)
$7 + F + C = 7_{10} + 15_{10} + 12_{10} = 34_{10} = 22_{16}$	(sum 2, carry 2)
$0 + 7 + 2 + 2 = 11_{10} = B_{16}$	(sum B, carry 0)
$5 + 9 + E = 5_{10} + 9_{10} + 14_{10} = 28_{10} = 1C_{16}$	(sum C, carry 1)
$A + D + B + 1 = 10_{10} + 13_{10} + 11_{10} + 1_{10} = 35_{10} = 23_{16}$	(sum 3, carry 2)
$C + B + F + 2 = 12_{10} + 11_{10} + 15_{10} + 2_{10} = 40_{10} = 28_{16}$	(sum 8, carry 2)
$B + E + 5 + 2 = 11_{10} + 14_{10} + 5_{10} + 2_{10} = 32_{10} = 20_{16}$	(sum 0, carry 2)
$3 + 9 + 2 = 14_{10} = E_{16}$	

Hexadecimal Subtraction

Subtraction in hexadecimal is similar to that in decimal.

EXAMPLE 2.67 Subtract $78D6.3B_{16}$ from $B08E.A1_{16}$.

Solution

The hex subtraction is performed as explained below.

	16^3	16^2	16^1	16^0		16^{-1}	16^{-2}
	B	0	8	E	.	A	1
−	7	8	D	6	.	3	B
	3	7	B	8	.	6	6

In the 16^{-2}'s column, B cannot be subtracted from 1. So, borrow a 1 from the 16^{-1}'s column leaving 9 there; the 1 borrowed from the 16^{-1}'s column becomes 16 in the 16^{-2}'s column. So,

In the 16^{-2}'s column, $16 + 1 - B = 16_{10} + 1_{10} - 11_{10} = 6$

In the 16^{-1}'s column, $9 - 3 = 6$

In the 16^0's column, $E - 6 = 14_{10} - 6_{10} = 8$

In the 16^1's column, D cannot be subtracted from 8. So, borrow a 1 from the 16^2's column. But there is a 0 in that column, so, borrow a 1 from the 16^3's column leaving A there. The 1 borrowed from the 16^3's column becomes 16 in the 16^2's column, leave 15 there and bring 1 to the 16^1's column. It becomes 16 in that column.

In the 16^1's column, $16 + 8 - D = 24_{10} - 13_{10} = 11_{10} = B$

In the 16^2's column, $15 - 8 = 7$

In the 16^3's column, $A - 7 = 10_{10} - 7_{10} = 3$

The result is $37B8.66_{16}$.

Subtraction in hexadecimal can also be accomplished by the 15's and 16's complement methods, similar to 1's and 2's complement methods of binary.

The 15's complement of a hex number is obtained by subtracting each digit of the number from 15 (F) and its 16's complement is obtained by adding a 1 to its 15's complement. The 15's complement of a hex number is the same as the 1's complement of the number expressed in binary and its 16's complement is the same as the 2's complement of the number expressed in binary.

EXAMPLE 2.68 Find the 15's and 16's complements of the following hexadecimal numbers.

(a) 76 (b) 4A9 (c) 7B.A (d) F20.3AE

Solution

(a)

```
   F F
 - 7 6
 -----
   8 9        (15's complement of 76)
     1        (Add 1)
 -----
   8 A        (16's complement of 76)
```

(b)

```
   F F F
 - 4 A 9
 -------
   B 5 6      (15's complement of 4A9)
       1      (Add 1)
 -------
   B 5 7      (16's complement of 4A9)
```

(c)

```
   F F . F
 - 7 B . A
 ---------
   8 4 . 5    (15's complement of 7B.A)
         1    (Add 1)
 ---------
   8 4 . 6    (16's complement of 7B.A)
```

(d)

```
   F F F . F F F
 - F 2 0 . 3 A E
 ---------------
   0 D F . C 5 1    (15's complement of F20.3AE)
               1    (Add 1)
 ---------------
   0 D F . C 5 2    (16's complement of F20.3AE)
```

EXAMPLE 2.69 Subtract $4AB.6B_{16}$ from 5074.56_{16} using the 15's complement method.

Solution

```
   5 0 7 4 . 5 6             5 0 7 4 . 5 6
 - 0 4 A B . 6 B   ⇒       + F B 5 4 . 9 4    (15's complement of 04AB.6B)
 ---------------          ----------------
 + 4 B C 8 . E B         (1) 4 B C 8 . E A
                           ↳           + 1    (End around carry)
                          ----------------
                             4 B C 8 . E B
```

The carry indicates that the result obtained is positive and is in hex itself.

EXAMPLE 2.70 Subtract $507D.56_{16}$ from $4AB.68_{16}$ using the 16's complement method.

Solution

0 4 A B . 6 8		0 4 A B . 6 8	
− 5 0 7 D . 5 6	⇒	+ A F 8 2 . A A	(16's complement of 507D.56)
− 4 B D 1 . E E		B 4 2 E . 1 2	(No carry)

There is no carry, indicating that the result obtained is negative and therefore in its 16's complement form. The 16's complement of $B42E.12_{16}$ is $4BD1.EE_{16}$. Therefore, the final result is $-\ 4BD1.EE_{16}$.

Hexadecimal numbers can also be subtracted using the 1's or 2's complement method, because the 15's complement of a hex number is the same as the 1's complement of the equivalent binary number and the 16's complement of a hex number is the same as the 2's complement of the equivalent binary number.

EXAMPLE 2.71 Subtract the following hexadecimal numbers using the 1's complement arithmetic.

(a) $48_{16} - 26_{16}$ (b) $45_{16} - 74_{16}$

Solution

Find the 1's complement of the subtrahend and express it in hex and add it to the minuend.

(a) 26_{16} in binary = 0010 0110
26_{16} in 1's complement form = 1101 1001 = $D9_{16}$

48_{16}		4 8	
$-\ 26_{16}$	⇒	+ D 9	(1's complement of 26_{16})
22_{16}		❶ 2 1	
		↳ + 1	(End around carry)
		2 2	

The carry indicates that the result obtained is positive and is in hex itself.

(b) 74_{16} in binary = 0111 0100
74_{16} in 1's complement form = 1000 1011 = $8B_{16}$

45_{16}		4 5	
$-\ 74_{16}$	⇒	+ 8 B	(1's complement of 74_{16})
$-\ 2F_{16}$		D 0	(No carry)

There is no carry. So, the result obtained is negative and is in the 1's complement form. The 1's complement of $D0_{16}$ = $2F_{16}$. So, the final result is $-\ 2F_{16}$.

EXAMPLE 2.72 Subtract (a) $69_{16} - 43_{16}$ and (b) $27_{16} - 73_{16}$ using the 2's complement arithmetic.

Solution

Find the 2's complement of the subtrahend and add it to the minuend.

(a) 43_{16} in binary = 0100 0011
43_{16} in 2's complement form = 1011 1101 = BD_{16}

$$\begin{array}{r} 69_{16} \\ -\,43_{16} \\ \hline 26_{16} \end{array} \quad \Rightarrow \quad \begin{array}{r} 6\ 9 \\ +\ B\ D \\ \hline \textcircled{1}\ 2\ 6 \end{array} \quad \begin{array}{l} \\ \text{(2's complement of } 43_{16}\text{)} \\ \text{(Ignore carry)} \end{array}$$

The carry indicates that the result obtained is positive and is in hex itself.

(b) 73_{16} in binary = 0111 0011
73_{16} in 2's complement form = 1000 1101 = $8D_{16}$

$$\begin{array}{r} 27_{16} \\ -\,73_{16} \\ \hline -\,4C_{16} \end{array} \quad \Rightarrow \quad \begin{array}{r} 2\ 7 \\ +\ 8\ D \\ \hline B\ 4 \end{array} \quad \begin{array}{l} \\ \text{(2's complement of } 73_{16}\text{)} \\ \text{(No carry)} \end{array}$$

No carry indicates that the answer is negative and is in its 2's complement form. The 2's complement of B4 is 4C. Therefore, the final result is $-\ 4C_{16}$.

Hexadecimal Multiplication

The hexadecimal multiplication is similar to decimal multiplication. To perform multiplication in hexadecimal, multiply the digits in decimal and convert each product term into hex. Record the sum term in hex, and carry the carry term to the next column.

EXAMPLE 2.73 Multiply $2A8_{16}$ by $B6_{16}$.

Solution

$$\begin{array}{r} 2\ A\ 8 \\ \times\quad B\ 6 \\ \hline F\ F\ 0 \\ 1\ D\ 3\ 8\ \ \\ \hline 1\ E\ 3\ 7\ 0 \end{array}$$

$8 \times 6 = 48_{10} = 30_{16}$
$A \times 6 = 10 \times 6 = 60_{10};\ 60_{10} + 3_{10} = 63_{10} = 3F_{16}$
$2 \times 6 = 12_{10};\ 12_{10} + 3_{10} = 15_{10} = F_{16}$
$8 \times B = 8 \times 11 = 88_{10} = 58_{16}$
$A \times B = 10 \times 11 = 110_{10};\ 110_{10} + 5_{10} = 115_{10} = 73_{16}$
$2 \times B = 2 \times 11 = 22_{10};\ 22_{10} + 7_{10} = 29_{10} = 1D_{16}$

Hexadecimal Division

Hexadecimal division is similar to decimal division.

EXAMPLE 2.74 Divide $2C0BE_{16}$ by $2A_{16}$.

Solution

2A)2C0BE(10C7.924	
2A	$2A(42_{10})$ can go in $2C(44_{10})$ 1 time, with a remainder of 2.
20B	2A cannot go in 20. $2A(42_{10})$ can go in $20B(523_{10})$ $12_{10}(C)$ times, with a remainder of $19_{10}(13_{16})$.
1F8	
13E	$2A(42_{10})$ can go 7 times in $13E(318_{10})$, with a remainder of $24_{10}(18_{16})$.
126	
180	$2A(42_{10})$ can go 9 times in $180_{16}(384_{10})$, with a remainder of $6_{10}(6_{16})$.
17A	
60	$2A(42_{10})$ can go 2 times in $60_{16}(96_{10})$, with a remainder of $12_{10}(C_{16})$.
54	
C0	$2A(42_{10})$ can go 4 times in $C0_{16}(192_{10})$, with a remainder of $24_{10}(18_{16})$.
A8	
18	

2.6 MISCELLANEOUS EXAMPLES

EXAMPLE 2.75 Given that $16_{10} = 100_b$, find the value of b.

Solution

Given $16_{10} = 100_b$

Convert 100_b to decimal.

Therefore, $16 = 1 \times b^2 + 0 \times b^1 + 0 \times b^0 = b^2$, or $b = 4$.

EXAMPLE 2.76 Given that $292_{10} = 1204$ in some number system, find the base of that system.

Solution

Let the base be b. Then

$$292_{10} = 1204_b = 1 \times b^3 + 2 \times b^2 + 0 \times b^1 + 4 \times b^0$$

or $$292_{10} = b^3 + 2b^2 + 4$$

Since 4 is the largest digit in the given number, $b \geq 5$, by trial and error, $b = 6$.

EXAMPLE 2.77 In the following series, the same integer is expressed in different number systems. Determine the missing number of the series: 10000, 121, 100, ?, 24, 22, 20.

Solution

If we take the first number to be in binary, its equivalent is 16_{10}. The next number is 121; it is equal to 16_{10} in base 3. Similarly 100 is equal to 16_{10} in base 4, and 24 is equal to 16_{10} in base 6. So, the missing number is 31 in base 5.

EXAMPLE 2.78 Each of the following arithmetic operations is correct in at least one number system. Determine the possible bases in each operation.

(a) $1234 + 5432 = 6666$ (b) $\frac{41}{3} = 13$

(c) $\frac{33}{3} = 11$ (d) $23 + 44 + 14 + 32 = 223$

(e) $\frac{302}{20} = 12.1$ (f) $\sqrt{41} = 5$

Solution

(a)
$$\begin{array}{r} 1\,2\,3\,4 \\ +\ 5\,4\,3\,2 \\ \hline 6\,6\,6\,6 \end{array}$$
It is valid in any number system with base ≥ 7 (since the largest digit used is 6).

(b) Let the base be b. Express both sides in decimal.

$$\frac{4b + 1}{3} = b + 3 \quad \text{or} \quad 4b + 1 = 3b + 9 \quad \text{or} \quad b = 8$$

Therefore, the above equation is valid in base 8 system.

(c) $\frac{33}{3} = 11$

Let the base be b. Converting to decimal,

$$3b + 3 = 3(b + 1) = 3b + 3.$$

It is valid for any value of b. Since the largest digit in the number is 3. The base can be anything greater than 3.

(d) Let the base be b. Expressing in decimal,

$$2b + 3 + 4b + 4 + b + 4 + 3b + 2 = 2b^2 + 2b + 3 \quad \text{or} \quad b^2 - 4b - 5 = 0 \quad \text{or} \quad b = 5$$

The relation is therefore valid in base 5 system.

(e) Let the base be b. Expressing in decimal,

$$\frac{3b^2 + 2}{2b} = b + 2 + \frac{1}{b}$$

Solving for b, we get $b = 4$.

(f) Let the base be b. Expressing in decimal, $\sqrt{4b + 1} = 5$.

Squaring both sides, $4b + 1 = 25$ or $b = 6$

Therefore, the base is 6.

SUMMARY

- Binary, octal, hexadecimal and decimal number systems are positional weighted, which means that the values attached to the symbols depend on their location with respect to the radix point.
- The base or radix of a number system indicates the number of unique symbols used in that system.

- The base or radix point separates the integer and fraction parts.
- The extreme right digit in any number is the LSD and the extreme left digit is the MSD.
- A decimal number can be converted to a number in any other system with base b by using the sum-of-weights method or by repeated division by b. In repeated division by b, the remainders are read from bottom to top.
- A decimal fraction can be converted to a fraction in any other number system with base b by using the sum-of-weights method or by repeated multiplication by b. In repeated multiplication by b, the integers to the left of the radix point are read from top to bottom.
- The binary system has two symbols 0 and 1. A binary digit is also called a bit.
- Each 4-bit binary group is called a nibble. Binary word lengths are multiples of four bits.
- Conversion of large decimal numbers into binary and vice versa is simpler via the hexadecimal system.
- Conversion of octal to hexadecimal and vice versa is simpler via the binary system.
- The computer method of multiplication requires repeated addition.
- The computer method of division requires successive subtraction.
- Subtraction in any number system with base b can be done by the b's complement method or (b – 1)'s complement method.
- In b's complement subtraction, the carry is ignored and in (b – 1)'s complement method the carry is added to the LSD.
- Negative numbers can be represented in sign magnitude form, or in 1's complement form, or in 2's complement form.
- In signed magnitude form, the MSB represents the sign bit (0 for positive and 1 for negative) and the remaining bits represent the magnitude of the number.
- If b is the base of a number system, (b – 1)'s complement of a number can be obtained by subtracting each digit of the number from (b – 1) and b's complement of any number can be obtained by adding 1 to its (b – 1)'s complement.
- The complement of the complement gives the original number.
- The 2's complement system has a unique zero. But the 1's complement system has two zeros, a positive 0 (all 0s) and a negative 0 (all 1s).
- The 1's complement of a binary number is obtained by complementing each bit of the number.
- The 2's complement of a binary number can be obtained (a) by finding its 1's complement and adding 1 to it, or (b) by subtracting the given N-bit binary number from 2^N, or (c) by copying down, starting from the LSB, all bits up to and including the first 1 and then complementing the remaining bits.
- Large numbers can be represented using double precision or triple precision.
- Double precision notation requires two storage locations (registers) to represent each binary number.

- In the floating point notation, both very large and very small binary numbers can be represented very conveniently.
- The main advantage of octal and hexadecimal systems is their ease of conversion to and from binary.
- An octal number is 1/3rd the length of the corresponding binary number and a hex number is 1/4th the length of the corresponding binary number.
- To convert an octal number into binary, replace each octal digit by its 3-bit binary equivalent.
- To convert a binary number into octal, starting from the binary point make groups of 3-bits on either side of the binary point and replace each 3-bit group by an octal digit.
- To convert a binary number into hex, starting from the binary point, make groups of 4-bits on either side of binary point and replace each 4-bit group by a hex digit.
- To convert a hex number into binary, replace each hex digit by its 4-bit binary equivalent.
- The hex system is particularly useful for human communication with computers.

QUESTIONS

1. What is meant by a positional weighted system?
2. What is meant by 'base' or 'radix' of a number system?
3. How do you convert a decimal number to an equivalent number in any other base system?
4. How do you convert a number in any base system to an equivalent decimal number?
5. Explain the following terms: (a) bit, (b) byte, (c) nibble, and (d) word length.
6. What are the advantages of (a) the binary number system, and (b) the hexadecimal number system?
7. What do you mean by the sum-of-weights method and successive division-by-b method?
8. What do you mean by successive multiplication-by-b method?
9. Discuss the computer method of multiplication and division.
10. What do you mean by double precision numbers and floating point numbers?
11. How can negative numbers be represented?
12. What do you mean by the 'signed magnitude' form of representation?
13. How do you obtain the 1's complement and 2's complement of a binary number?
14. Compare the 1's complement and 2's complement methods of representation of negative numbers.
15. What are the characteristics of the 2's complement method?
16. How do you perform binary subtraction using the 1's and 2's complement methods?
17. What do you mean by end around carry? When does it come into picture?

18. How are binary and octal numbers related?

19. What are the advantages of octal and hexadecimal systems? Where are they used?

20. How do you convert a hexadecimal number to octal and vice versa?

21. Which of the following systems have two 0s?
(a) Sign-magnitude (b) 1's complement (c) 2's complement.

22. What weight does the digit 6 have in each of the following numbers?
(a) 216 (b) 460 (c) 658 (d) 115.67

23. Express each of the following decimal numbers as a sum of the products of each digit and its appropriate weight.
(a) 63 (b) 157 (c) 125.64 (d) 6015.75

24. What are the weights of the MSB of (a) a 10-bit binary number and (b) an 8-bit hex number?

PROBLEMS

2.1 Subtract the following decimal numbers by the 9's and 10's complement methods.
(a) 274 – 86 (b) 93 – 615 (c) 574.6 – 279.7 (d) 376.3 – 765.6

2.2 Convert the following binary numbers to decimal.
(a) 1011 (b) 1101101 (c) 1101.11 (d) 1101110.011

2.3 Convert the following decimal numbers to binary.
(a) 37 (b) 128 (c) 197.56 (d) 205.05

2.4 Add the following binary numbers.
(a) 11011 + 1101 (b) 1011 + 1101 + 1001 + 1111
(c) 10111.101 + 110111.01 (d) 1010.11 + 1101.10 + 1001.11 + 1111.11

2.5 Subtract the following binary numbers.
(a) 1011 – 101 (b) 10110 – 1011
(c) 1100.10 – 111.01 (d) 10001.01 – 1111.11

2.6 Multiply the following binary numbers.
(a) 1101 × 101 (b) 11001 × 10
(c) 1101.11 × 101.1 (d) 10110 × 10.1

2.7 Divide the following binary numbers.
(a) 1010 by 11 (b) 11110 by 101
(c) 11011 by 10.1 (d) 110111.1 by 101

2.8 Multiply the following binary numbers by the computer method.
(a) 1011 × 11 (b) 1001 × 101 (c) 1100 × 110 (d) 1111 × 100

2.9 Divide the following binary numbers by the computer method.
(a) 1011 by 10 (b) 110010 by 101
(c) 1100100 by 1101 (d) 100010 by 1010.

2.10 Find the 12 bit 2's complement form of the following decimal numbers.
(a) – 37 (b) – 173 (c) – 65.5 (d) – 197.5

2.11 Find the 12-bit 1's complement form of the following decimal numbers.
(a) – 97 (b) – 224 (c) – 205.75 (d) – 29.375

2.12 Subtract the following decimal numbers using the 12-bit 2's complement arithmetic.
(a) 46 – 19 (b) 27 – 75 (c) 125.3 – 46.7 (d) 36.75 – 89.5

2.13 Subtract the following decimal numbers using the 12-bit 1's complement arithmetic.
(a) 52 – 17 (b) 46 – 84 (c) 63.75 – 17.5 (d) 73.5 – 112.75

2.14 Represent the following decimal numbers in 8-bit, (i) sign magnitude form, (ii) sign 1's complement form, and (iii) sign 2's complement form.
(a) + 14 (b) + 27 (c) + 45 (d) – 17
(e) – 37 (f) – 76

2.15 Convert the following octal numbers to hexadecimal.
(a) 256 (b) 2035 (c) 1762.46 (d) 6054.263

2.16 Convert the following hexadecimal numbers to octal.
(a) 2AB (b) 42FD (c) 4F7.A8 (d) BC70.0E

2.17 Convert the following octal numbers to decimal.
(a) 463 (b) 2056 (c) 2057.64 (d) 6534.04

2.18 Convert the following decimal numbers to octal.
(a) 287 (b) 3956 (c) 420.6 (d) 8476.47

2.19 Add the following octal numbers.
(a) 173 + 265 (b) 1247 + 2053 (c) 25.76 + 16.57 (d) 273.56 + 425.07

2.20 Subtract the following octal numbers.
(a) 64 – 37 (b) 462 – 175
(c) 175.6 – 47.7 (d) 3006.05 – 2657.16

2.21 Multiply the following octal numbers.
(a) 46 × 4 (b) 267 × 5 (c) 26.5 × 2.5 (d) 647.2 × 5.4

2.22 Divide the following octal numbers.
(a) 420 by 5 (b) 2567 by 6 (c) 153.6 by 7 (d) 4634.62 by 12

2.23 Subtract the following octal numbers by the 7's and 8's complement methods.

(a) 76 − 25 (b) 256 − 643 (c) 173.5 − 66.6 (d) 243.6 − 705.64

2.24 Convert the following hexadecimal numbers to binary.

(a) C20 (b) F297 (c) AF9.B0D (d) E79A.6A4

2.25 Convert the following binary numbers to hexadecimal.

(a) 10110 (b) 1011011011
(c) 110110111.01111 (d) 1101101101101.101101

2.26 Convert the following hexadecimal numbers to decimal.

(a) AB6 (b) 2EB7 (c) A08F.EA (d) 8E47.AB

2.27 Convert the following decimal numbers to hexadecimal.

(a) 452 (b) 4796 (c) 1248.56 (d) 8957.75

2.28 Add the following hexadecimal numbers.

(a) AC6 + B59 (b) A0FC + B75F
(c) E0F3.5D + 49E6.F7 (d) ABC7.54 + 26F3.AB + DAC9.6F

2.29 Subtract the following hexadecimal numbers.

(a) BC5 − A2B (b) F27 − B9E
(c) CDF7.52 − AB5.8 (d) 67F2.6E − 4A0E.A9

2.30 Multiply the following hexadecimal numbers.

(a) 28A × B (b) 5A9B × 7 (c) 5032.6E × 6 (d) 92.5 × B.3

2.31 Divide the following hexadecimal numbers.

(a) AB9 by C (b) 4056 by 9 (c) 6ABF.6D by 1A (d) 3781.4 by 5.8

2.32 Subtract the following hexadecimal numbers by the 15's and 16's complement methods.

(a) BE − 5E (b) AB6 − 745
(c) BC7.5 − A53E.4B (d) 4230.5 − 73A.3.

Chapter 3

BINARY CODES

3.1 WEIGHTED AND NON-WEIGHTED CODES

Binary codes can be classified as numeric codes and alphanumeric codes. Alphanumeric codes represent alphanumeric information, i.e. letters of the alphabet and decimal numbers as a sequence of 0s and 1s. Numeric codes represent numeric information, i.e. only numbers as a series of 0s and 1s. Numeric codes used to represent decimal digits are called Binary Coded Decimal (BCD) codes.

We are very comfortable with the decimal number system, but digital systems force us to use the binary system. Although the binary number system has many practical advantages and is widely used in digital computers, in many cases it is very convenient to work with decimal numbers, especially when communication between man and machine is extensive. Since most of the numerical data generated by man are in decimal numbers, to simplify the communication process between man and machine, several systems of numeric codes have been devised to represent decimal numbers as a series of BCD codes.

A BCD code is one, in which the digits of a decimal number are encoded—one at a time—into groups of four binary digits. These codes combine the features of decimal and binary numbers. There are a large number of BCD codes. In order to represent decimal digits 0, 1, 2,..., 9, it is necessary to use a sequence of at least four binary digits. Such a sequence of binary digits which represents a decimal digit is called a *code word*.

The BCD codes may be weighted or non-weighted. The weighted codes are those which obey the position-weighting principle. Each position of the number represents a specific weight. For each group of four bits, the sum of the weights of those positions where the binary digit is 1 is equal to the decimal digit which the group represents. There are several weighted codes. The weighted codes may be either *positively-weighted* or *negatively-weighted*.

Positively-weighted codes are those in which all the weights assigned to the binary digits are positive. There are only 17 positively-weighted codes. In every positively-weighted code, the first weight must be 1, the second weight must be either 1 or 2, and the sum of all the weights must be equal to or greater than 9. In negatively-weighted codes, some of the weights assigned to the binary digits must be negative. The codes 8421, 2421, 5211, 3321 and 4311 are some of the positively-weighted codes available. The codes 642–3, 631–1, 84–2–1 and 74–2–1 are some of the negatively-weighted codes. Non-weighted codes are codes, which do not obey the position-weighting principle. Excess-3 (XS-3) code and Gray code are non-weighted codes. Table 3.1 shows some of the positively-weighted, negatively-weighted, and non-weighted codes.

Table 3.1 Binary coded decimal codes

Decimal digit	8 4 2 1	2 4 2 1	5 2 1 1	5 4 2 1	6 4 2 –3	8 4 –2 –1	XS-3
0	0 0 0 0	0 0 0 0	0 0 0 0	0 0 0 0	0 0 0 0	0 0 0 0	0 0 1 1
1	0 0 0 1	0 0 0 1	0 0 0 1	0 0 0 1	0 1 0 1	0 1 1 1	0 1 0 0
2	0 0 1 0	0 0 1 0	0 0 1 1	0 0 1 0	0 0 1 0	0 1 1 0	0 1 0 1
3	0 0 1 1	0 0 1 1	0 1 0 1	0 0 1 1	1 0 0 1	0 1 0 1	0 1 1 0
4	0 1 0 0	0 1 0 0	0 1 1 1	0 1 0 0	0 1 0 0	0 1 0 0	0 1 1 1
5	0 1 0 1	1 0 1 1	1 0 0 0	1 0 0 0	1 0 1 1	1 0 1 1	1 0 0 0
6	0 1 1 0	1 1 0 0	1 0 1 0	1 0 0 1	0 1 1 0	1 0 1 0	1 0 0 1
7	0 1 1 1	1 1 0 1	1 1 0 0	1 0 1 0	1 1 0 1	1 0 0 1	1 0 1 0
8	1 0 0 0	1 1 1 0	1 1 1 0	1 0 1 1	1 0 1 0	1 0 0 0	1 0 1 1
9	1 0 0 1	1 1 1 1	1 1 1 1	1 1 0 0	1 1 1 1	1 1 1 1	1 1 0 0

Sequential Codes

A sequential code is one, in which each succeeding code word is one binary number greater than its preceding code word. Such a code facilitates mathematical manipulation of data. The 8421 and XS-3 codes are sequential. The codes 5211, 2421 and 642–3 are not sequential.

Self-complementing Codes

A code is said to be self-complementing, if the code word of the 9's complement of *N*, i.e. of 9 – *N* can be obtained from the code word of *N* by interchanging all the 0s and 1s. Therefore, in a self-complementing code, the code for 9 is the complement of the code for 0, 8 for 1, 7 for 2, 6 for 3, and 5 for 4. The 2421, 5211, 642–3, 84–2–1 and XS-3 are self-complementing codes. The 8421 and 5421 codes are not self-complementing. The self-complementing property is desirable in a code when the 9's complement must be found such as in 9's complement subtraction. For a code to be self-complementing, the sum of all its weights must be 9. This is because whatever may be the weights, 0 is to be represented by 0000 and since in a self-complementing code, the code for 9 is the complement of the code for 0, 9 has to be represented by 1111. There are only four (2421, 5211, 3321, 4311) positively-weighted self-complementing codes. There are 13 negatively-weighted self-complementing codes.

Cyclic Codes

Cyclic codes are those in which each successive code word differs from the preceding one in only one bit position. They are also called *unit distance codes*. The Gray code is a cyclic code. It is often used for translating an analog quantity such as shaft position into a digital form.

3.2 THE 8421 BCD CODE

The 8421 BCD code is so widely used that it is a common practice to refer to it simply as BCD code. In this code, each decimal digit, 0 through 9, is coded by a 4-bit binary number. It is also called the *natural binary code* because of the 8, 4, 2 and 1 weights attached to it. It is a weighted code and is also sequential. Therefore, it is useful for mathematical operations. The main advantage of this code is its ease of conversion to and from decimal. It is less efficient than the pure binary, in the sense that it requires more bits. For example, the decimal number 14 can be represented as 1110 in pure binary but as 0001 0100 in 8421 code. Another disadvantage of the BCD code is that, arithmetic operations are more complex than they are in pure binary. There are six illegal combinations 1010, 1011, 1100, 1101, 1110 and 1111 in this code, i.e. they are not part of the 8421 BCD code system.

BCD Arithmetic

A disadvantage of the 8421 code is that, the rules of binary addition and subtraction do not apply to the entire 8421 number but only to the individual 4-bit groups. The BCD addition is, therefore, performed by individually adding the corresponding digits of the decimal numbers expressed in 4-bit binary groups starting from the LSD. If there is a carry out of one group to the next group, or if the result is an illegal code, then 6_{10} (0110) is added to the sum term of that group and the resulting carry is added to the next group. (This is done to skip the six illegal states.) The BCD subtraction is performed by subtracting the digits of each 4-bit group of the subtrahend from the corresponding 4-bit group of the minuend in binary starting from the LSD. If there is a borrow from the next group, then 6_{10} (0110) is subtracted from the difference term of this group. (This is done to skip the six illegal states.) In practice, subtraction is performed by the complement method. Since we are subtracting decimal digits, we must form the 9's or 10's complement of the decimal subtrahend and encode that number in the 8421 code. The resulting BCD numbers are then added.

EXAMPLE 3.1 Perform the following decimal additions in the 8421 code.

(a) 25 + 13 (b) 679.6 + 536.8

Solution

```
(a)   25         0010   0101   (25 in BCD)
    + 13   ⇨   + 0001   0011   (13 in BCD)
    ----       -------------
      38         0011   1000   (No carry, no illegal code. So, this is the correct sum)

(b)   679.6          0110        0111        1001       .0110     (679.6 in BCD)
    + 536.8   ⇨    + 0101        0011        0110       .1000     (536.8 in BCD)
    -------        ---------------------------------------------
     1216.4          1011        1010        1111       .1110     (All are illegal codes)
                   + 0110      + 0110      + 0110     + .0110     (Add 0110 to each)
                   ---------------------------------------------
                   ❶ 0001    ❶ 0000    ❶ 0101    ❶ .0100     (Propagate carry)
               +1 ↲       +1 ↲       +1 ↲       +1 ↲
               ------------------------------------------------
               0001      0010        0001        0110       .0100     (Corrected sum)
                  1         2           1           6          .4
```

EXAMPLE 3.2 Perform the following decimal subtractions in the 8421 BCD code.

(a) 38 – 15 (b) 206.7 – 147.8

Solution

```
(a)   38     ⇨     0011    1000     (38 in BCD)
    – 15         – 0001    0101     (15 in BCD)
    ----         ----------------
      23           0010    0011     (No borrow. So this is the correct difference.)

(b)  206.7   ⇨      0010    0000    0110   .0111    (206.7 in BCD)
   – 147.8        – 0001    0100    0111   .1000    (147.8 in BCD)
   -------        ---------------------------------
      58.9          0000    1011    1110   .1111    (Borrows are present, subtract 0110)
                          – 0110  – 0110  –.0110
                  ---------------------------------
                            0101    1000   .1001    (Corrected difference)
```

EXAMPLE 3.3 Perform the following decimal subtractions in BCD by the 9's complement method.

(a) 85 – 24 (b) 305.5 – 168.8 (c) 679.6 – 885.9

Solution

```
(a)   85     ⇨     85
    – 24         + 75     (9's complement of 24)
    ----         ❶ 60
      61         ↳ +1     (End around carry)
                 ----
                   61

     85₁₀     ⇨     1000         0101     (85 in BCD)
   + 75₁₀         + 0111         0101     (9's complement of 24 in BCD)
                  -----------------------
                    1111         1010     (Both are illegal codes, add 0110 to each)
                  + 0110       + 0110
                  -----------------------
                  ❶ 0101       ❶ 0000     (Propagate carry)
          + 1 ↲     + 1 ↲
          ---------------
            1       0110         0000
            ↳                     + 1     (End around carry)
          -------------------------------
                    0110         0001     (Corrected difference)

(b)  305.5   ⇨     305.5
   – 168.8       + 831.1    (9's complement of 168.8)
   -------       ❶ 136.6
     136.7       ↳    +1    (End around carry)
                 -------
                   136.7

     305.5₁₀   ⇨      0011    0000    0101    .0101    (305.5 in BCD)
   + 831.1₁₀        + 1000    0011    0001    .0001    (9's complement of 168.8 in BCD)
                    + 1011    0011    0110    .0110    (1011 is an illegal code, add 0110)
                    + 0110
                    -----------------------------------
```

```
        ❶ 0001   0011   0110   .0110
          ↳                       +1   (End around carry)
          0001   0011   0110   .0111   (Corrected difference)
```

```
(c)    679.6            679.6
     – 885.9     ⇨    + 114.0    (9's complement of 885.9)
     – 206.3            793.6    (No carry)
```

No carry. So the result is negative and is in 9's complement form. So the result is – 206.3.

```
   679.6₁₀          0110   0111     1001   .0110   (679.6 in BCD)
 + 114.0₁₀    ⇨   + 0001   0001     0100   .0000   (9's complement of 885.9 in BCD)
                    0111   1000     1101   .0110   (1101 is an illegal code, add 0110)
                                  + 0110
                    0111   1000 ❶  0011   .0110   (Propagate carry)
                           + 1 ↲
                    0111   1001     0011   .0110   (No carry; 793.6 in BCD)
```

There is no carry and, therefore, the result is negative and is in its 9's complement form. The 9's complement of 793.6 is 206.3. So the result is – 206.3.

EXAMPLE 3.4 Perform the following subtractions in 8421 code using the 10's complement method.

(a) 342.7 – 108.9 (b) 206.4 – 507.6

Solution

```
(a)    342.7            342.7
     – 108.9     ⇨    + 891.1    (10's complement of 108.9)
       233.8          ❶ 233.8    (Ignore carry)
```

```
   342.7          0011     0100   0010   .0111   (342.7 in BCD)
 + 891.1    ⇨   + 1000     1001   0001   .0001   (10's complement of 108.9 in BCD)
                  1011     1101   0011   .1000   (1011 and 1101 are illegal codes, add 0110
                                                  to each.)
                + 0110   + 0110
                ❶ 0001  ❶ 0011   0011   .1000   (Propagate carry)
         + 1 ↲      + 1 ↲
              ❶   0010     0011   0011   .1000   (Ignore carry)
                  0010     0011   0011   .1000   (Corrected difference)
```

```
(b)    206.4             206.4
     – 507.6     ⇨     + 492.4   (10's complement of 507.6)
     – 301.2             698.8   (No carry)
```

No carry. So the answer is negative and is in 10's complement form. The 10's complement of 698.8 is – 301.2.

```
  206.4   ⇨    0010   0000   0110  .0100  (206.4 in BCD)
+ 492.4      + 0100   1001   0010  .0100  (10's complement of 507.6 in BCD)
               ---------------------------
               0110   1001   1000  .1000  (No illegal codes, no carry)
```

As there is no carry, the result is negative and is in its 10's complement form. The 10's complement of 698.8 is 301.2. So the corrected difference is – 301.2.

3.3 THE EXCESS THREE (XS-3) CODE

The Excess-3 code, also called XS-3, is a non-weighted BCD code. This code derives its name from the fact that each binary code word is the corresponding 8421 code word plus 0011(3). It is a sequential code and, therefore, can be used for arithmetic operations. It is a self-complementing code. Therefore, subtraction by the method of complement addition is more direct in XS-3 code than that in 8421 code. The XS-3 code has six invalid states 0000, 0001, 0010, 1101, 1110 and 1111.

XS-3 Arithmetic

The XS-3 code has some very interesting properties when used in addition and subtraction. To add in XS-3, add the XS-3 numbers by adding the 4-bit groups in each column starting from the LSD. If there is no carry out from the addition of any of the 4-bit groups, subtract 0011 from the sum term of those groups (because when two decimal digits are added in XS-3 and there is no carry, the result is in XS-6). If there is a carry out, add 0011 to the sum term of those groups (because when there is a carry, the invalid states are skipped and the result is in normal binary).

To subtract in XS-3, subtract the XS-3 numbers by subtracting each 4-bit group of the subtrahend from the corresponding 4-bit group of the minuend starting from the LSD. If there is no borrow from the next 4-bit group, add 0011 to the difference term of such groups (because when decimal digits are subtracted in XS-3 and there is no borrow, the result is in normal binary). If there is a borrow, subtract 0011 from the difference term (because taking a borrow is equivalent to adding six invalid states, so the result is in XS-6). In practice, subtraction is performed by the 9's complement or the 10's complement method.

EXAMPLE 3.5 Perform the following additions in XS-3 code.

(a) 5 + 3 (b) 37 + 28 (c) 247.6 + 359.4

Solution

```
(a)    5    ⇨      1000  (5 in XS-3)
     + 3         + 0110  (3 in XS-3)
     ---         ------
       8           1110  (No carry)
                 – 0011  (Subtract 0011)
                 ------
                   1011  (Corrected sum in XS-3)
```

```
(b)    37           0110      1010     (37 in XS-3)
     + 28    ⇨    + 0101      1011     (28 in XS-3)
     ----         ----------------
       65           1011   ❶ 0101     (Carry generated)
                    + 1    ↵           (Propagate carry)
                  ----------------
                    1100      0101     (Add 0011 to correct 0101)
                  – 0011    + 0011     (Subtract 0011 to correct 1100)
                  ----------------
                    1001      1000     (Corrected sum in XS-3)

(c)   247.6         0101     0111      1010       .1001   (247.6 in XS-3)
    + 359.4  ⇨    + 0110     1000      1100       .0111   (359.4 in XS-3)
    -------       -------------------------------------
      607.0         1011     1111   ❶ 0110    ❶ .0000   (Carry generated)
                             + 1    ↵ + 1     ↵          (Propagate carry)
                  -------------------------------------
                    1011  ❶ 0000      0111       .0000
                    + 1  ↵
                  -------------------------------------
                    1100     0000      0111       .0000   (Add 0011 to 0000, 0111 and
                                                           .0000)
                  – 0011   + 0011    + 0011     + .0011   (Subtract 0011 from 1100)
                  -------------------------------------
                    1001     0011      1010       .0011   (Corrected sum in XS-3)
```

EXAMPLE 3.6 Perform the following subtractions in XS-3 code.

(a) 267 – 175 (b) 57.6 – 27.8

Solution

```
(a)   267          0101      1001      1010    (267 in XS-3)
    – 175   ⇨    – 0100      1010      1000    (175 in XS-3)
    -----        ----------------------------
      092          0000      1111      0010    (Correct 0010 and 0000 by adding 0011)
                 + 0011    – 0011    + 0011    (Correct 1111 by subtracting 0011)
                 ----------------------------
                   0011      1100      0101    (Corrected difference in XS-3)

(b)   57.6         1000      1010     .1001    (57.6 in XS-3)
    – 27.8  ⇨    – 0101      1010     .1011    (27.8 in XS-3)
    ------       ----------------------------
      29.8         0010      1111     .1110    (Correct 0010 by adding 0011 and correct
                                                .1110 and 1111 by subtracting 0011)
                 + 0011    – 0011   – .0011
                 ----------------------------
                   0101      1110     .1011    (Corrected difference in XS-3)
```

EXAMPLE 3.7 Perform the following subtractions in XS-3 code using the 9's complement method.

(a) 687 – 348 (b) 246 – 592

Solution

(a) The subtrahend (348) in XS-3 code and its complements are:
9's complement of 348 = 651

XS-3 code of 348 = 0110 0111 1011
I's complement of 348 in XS-3 = 1001 1000 0100
XS-3 code of 687 = 1001 1011 1010

```
  687             687
– 348    ⇨      + 651     (9's complement of 348)
 ----            ----
  339          ➊  338
               ↳   +1     (End around carry)
                 ----
                  339     (Corrected difference in decimal)
```

```
              1001        1011        1010     (687 in XS-3)
            + 1001        1000        0100     (1's complement of 348 in XS-3)
          ------------------------------------
            ➊ 0010      ➊ 0011        1110     (Carry generated)
     + 1  ↲   + 1    ↲                          (Propagate carry)
          ------------------------------------
              0011        0011        1110
        ↳                               + 1     (End around carry)
          ------------------------------------
              0011        0011        1111     (Correct 1111 by subtracting 0011 and cor-
            + 0011      + 0011      – 0011      rect both groups of 0011 by adding 0011)
          ------------------------------------
              0110        0110        1100     (Corrected difference in XS-3)
```

(b) The subtrahend (592) in XS-3 and its complements are:
9's complement of 592 = 407
XS-3 code of 592 = 1000 1100 0101
1's complement of 592 in XS-3 = 0111 0011 1010
XS-3 code of 246 = 0101 0111 1001

```
  246             246
– 592    ⇨      + 407   (9's complement of 592)
 ----            ----
– 346             653   (No carry)
```

As there is no carry, the result is negative and is in 9's complement form. The 9's complement of 653 is 346. So, the result is – 346.

```
     0101        0111          1001      (246 in XS-3)
   + 0111        0011          1010      (1's complement of 592 in XS-3)
  -------------------------------------
     1100        1010        ➊ 0011      (Propagate carry)
                  + 1    ↲
  -------------------------------------
     1100        1011          0011      (Correct 0011 by adding 0011 and correct
   – 0011      – 0011        + 0011      1011 and 1100 by subtracting 0011)
  -------------------------------------
     1001        1000          0110      (No carry)
```

The sum is the XS-3 form of 653. There is no carry. So the result is negative and is in 1's complement form. The 1's complement of the sum is 0110 0111 1001 in XS-3 code (346_{10}). So, the answer is – 346.

EXAMPLE 3.8 Perform the following subtractions in XS-3 code using the 10's complement method.

(a) 597 − 239 (b) 354 − 672

Solution

(a) 10's complement of 239 = 761
XS-3 code of 239 = 0101 0110 1100
2's complement of 239 in XS-3 code = 1010 1001 0100
XS-3 code of 597 = 1000 1100 1010

597		597	
− 239	⇨	+ 761	(10's complement of 239)
358		❶ 358	(Ignore carry)
		358	(Corrected difference in decimal)

	1000	1100	1010	(597 in XS-3)
	+ 1010	1001	0100	(2's complement of 239 in XS-3)
	❶ 0010	❶ 0101	1110	(Propagate carry)
+ 1	↲ + 1	↲		
1	0011	0101	1110	(Ignore carry)
	+ 0011	+ 0011	− 0011	(Correct 1110 by subtracting 0011 and correct 0101 and 0011 by adding 0011)
	0110	1000	1011	(Corrected difference in XS-3 code)

(b) 10's complement of 672 = 328
XS-3 code of 672 = 1001 1010 0101
2's complement of 672 in XS-3 code = 0110 0101 1011
XS-3 code of 354 = 0110 1000 0111

354		354	
− 672	⇨	+ 328	(10's complement of 672)
− 318		682	(No carry)

No carry. So the result is negative and is in its 10's complement form. The 10's complement of 682 is 318. So, the answer is − 318.

0110	1000	0111	(354 in XS-3)
+ 0110	0101	1011	(2's complement of 672 in XS-3)
1100	1101	❶ 0010	(Propagate carry)
	+ 1 ↲		
1100	1110	0010	(Correct 0010 by adding 0011 and correct 1110 and 1100 by subtracting 0011)
− 0011	− 0011	+ 0011	
1001	1011	0101	(No carry)

The sum is the XS-3 form of 682. There is no carry. So the result is negative and is in the 2's complement form. The 2's complement of the sum is 0110 0100 1011 in XS-3 code (318_{10}). So, the answer is − 318.

EXAMPLE 3.9 Encode the decimal digits 0, 1, 2, . . ., 9 by means of weighted codes 3321, 4221, 731–2, 631–1, 5311, 74–2–1 and 7421.

Solution

The encoding is shown in Table 3.2.

Table 3.2 Encoded decimal digits

Decimal	3 3 2 1	4 2 2 1	7 3 1 –2	6 3 1 –1	5 3 1 1	7 4 –2 –1	7 4 2 1
0	0 0 0 0	0 0 0 0	0 0 0 0	0 0 0 0	0 0 0 0	0 0 0 0	0 0 0 0
1	0 0 0 1	0 0 0 1	0 0 1 0	0 0 1 0	0 0 0 1	0 1 1 1	0 0 0 1
2	0 0 1 0	0 0 1 0	0 1 1 1	0 1 0 1	0 0 1 1	0 1 1 0	0 0 1 0
3	0 0 1 1	0 0 1 1	0 1 0 0	0 1 0 0	0 1 0 0	0 1 0 1	0 0 1 1
4	0 1 0 1	1 0 0 0	0 1 1 0	0 1 1 0	0 1 0 1	0 1 0 0	0 1 0 0
5	1 0 1 0	0 1 1 1	1 0 0 1	1 0 0 1	1 0 0 0	1 0 1 0	0 1 0 1
6	1 1 0 0	1 1 0 0	1 0 1 1	1 0 1 1	1 0 0 1	1 0 0 1	0 1 1 0
7	1 1 0 1	1 1 0 1	1 0 0 0	1 0 1 0	1 0 1 1	1 0 0 0	1 0 0 0
8	1 1 1 0	1 1 1 0	1 1 0 1	1 1 0 1	1 1 0 0	1 1 1 1	1 0 0 1
9	1 1 1 1	1 1 1 1	1 1 1 1	1 1 1 1	1 1 0 1	1 1 1 0	1 0 1 0

3.4 THE GRAY CODE

The Gray code is a non-weighted code, and is not suitable for arithmetic operations. It is not a BCD code. It is a cyclic code because successive code words in this code differ in one bit position only, i.e. it is a unit distance code. It is the most popular of the unit distance codes. It is also a reflective code. The n least significant bits for 2^n through $2^{n+1} - 1$ are the mirror images of those for 0 through $2^n - 1$. An N-bit Gray code can be obtained by reflecting an $N - 1$ bit code about an axis at the end of the code, and putting the MSB of 0 above the axis and the MSB of 1 below the axis. Reflection of Gray codes is shown in Table 3.3. Another property of the Gray code is that the Gray-coded number corresponding to the decimal number 2^{n-1} for any n differs from Gray-coded 0 in one bit position only. This property places the Gray code for the largest N-bit binary number at unit distance from 0. It is easier to determine the pattern assignment corresponding to an arbitrary number, or decode an arbitrary number Gray code pattern using the conversions to and from binary. In fact, one reason for the popularity of the Gray code is its ease of conversion to and from binary. Gray codes are used in instrumentation and data acquisition systems where linear or angular displacement is measured. They are also used in shaft encoders, I/O devices, A/D converters and other peripheral equipment.

Consider a rotating disk that provides an output of its position in 3-bit binary (Fig. 3.1). When the brushes are on the black part, they output a 1, and when they are on a white sector they output a 0. Suppose the disk is coded in binary; consider now what happens when the brushes are on the 111 sector and almost ready to enter the 000 sector. If one brush were slightly ahead of the other, say the 4's brush, the position would be indicated by a 011 instead of a 111 or 000. Therefore, a 180° error in disk position would result. Since it is physically impossible to have all the brushes precisely aligned, some error would always

Table 3.3 Reflection of Gray codes

Gray code				Decimal	4-bit binary
1-bit	2-bit	3-bit	4-bit		
0	0 0	0 0 0	0 0 0 0	0	0 0 0 0
1	0 1	0 0 1	0 0 0 1	1	0 0 0 1
	1 1	0 1 1	0 0 1 1	2	0 0 1 0
	1 0	0 1 0	0 0 1 0	3	0 0 1 1
		1 1 0	0 1 1 0	4	0 1 0 0
		1 1 1	0 1 1 1	5	0 1 0 1
		1 0 1	0 1 0 1	6	0 1 1 0
		1 0 0	0 1 0 0	7	0 1 1 1
			1 1 0 0	8	1 0 0 0
			1 1 0 1	9	1 0 0 1
			1 1 1 1	10	1 0 1 0
			1 1 1 0	11	1 0 1 1
			1 0 1 0	12	1 1 0 0
			1 0 1 1	13	1 1 0 1
			1 0 0 1	14	1 1 1 0
			1 0 0 0	15	1 1 1 1

be present at the edges of the sectors. If the disks were coded in gray, a similar error would make the disk to be read as 110 instead of 010 causing a very small error, i.e. if the brushes are on the sector 010 and almost ready to enter the 110 sector and if the 4's brush is slightly ahead, the position would be indicated by 110 instead of 010 resulting in a very small error. Figure 3.1 illustrates this operation.

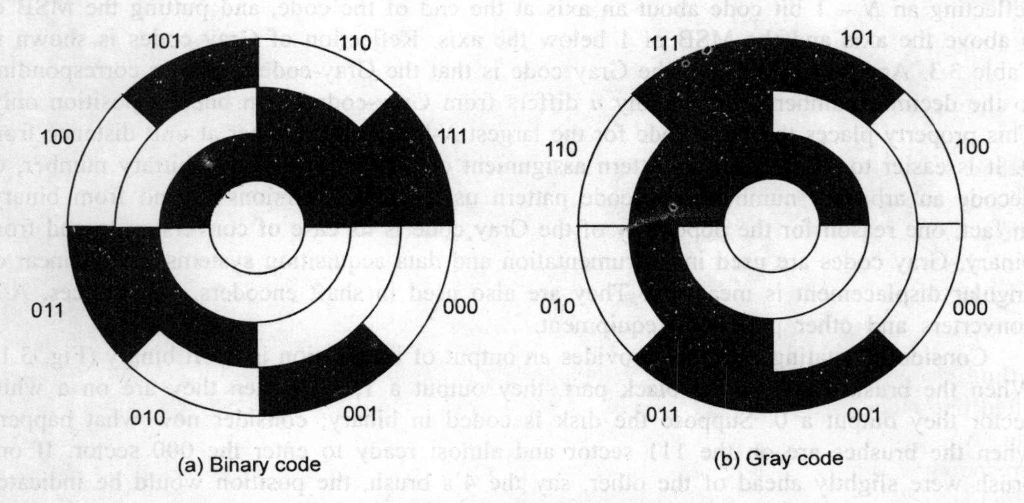

Fig. 3.1 Position indicating system.

Binary-to-Gray Conversion

If an n-bit binary number is represented by $B_n B_{n-1} \ldots B_1$ and its Gray code equivalent by $G_n G_{n-1} \ldots G_1$, where B_n and G_n are the MSBs, then the Gray code bits are obtained from the binary code as follows.

$$
\begin{aligned}
G_n &= B_n \\
G_{n-1} &= B_n \oplus B_{n-1} \\
G_{n-2} &= B_{n-1} \oplus B_{n-2} \\
&\;\vdots \\
G_1 &= B_2 \oplus B_1
\end{aligned}
$$

where the symbol $\oplus$ stands for the Exclusive OR (XOR) operation explained below.

The conversion procedure is as follows:

1. Record the MSB of the binary as the MSB of the Gray code.
2. Add the MSB of the binary to the next bit in binary, recording the sum and ignoring the carry, if any, i.e. XOR the bits. This sum is the next bit of the Gray code.
3. Add the 2nd bit of the binary to the 3rd bit of the binary, the 3rd bit to the 4th bit, and so on.
4. Record the successive sums as the successive bits of the Gray code until all the bits of the binary number are exhausted.

Another way to convert a binary number to its Gray code is to Exclusive OR (i.e. to take the modulo sum of) the bits of the binary number with those of the binary number shifted one position to the right. The LSB of the shifted number is discarded and the MSB of the Gray code number is the same as the MSB of the original binary number.

EXAMPLE 3.10 Convert the binary 1001 to Gray code.

Solution

(a)

Binary	1	—⊕→	0	—⊕→	0	—⊕→	1
	‖		‖		‖		‖
Gray	1		1		0		1

(b)

Binary	1 0 0 1
Shifted binary	1 0 0 ❶
Gray	1 1 0 1

Method I. As shown in (a), record the 8's bit '1' (MSB) of the binary as the 8's bit of the Gray code. Then add the 8's bit of the binary to the 4's bit of the binary (1 + 0 = 1). Record the sum as the 4's bit of the Gray code. Add the 4's bit of the binary to the 2's bit of the binary (0 + 0 = 0). Record the sum as the 2's bit of the Gray code. Add the 2's bit of the binary to the 1's bit of the binary (0 + 1 = 1). Record the sum as the 1's bit of the Gray code. The resultant Gray code is 1101.

Method II. As shown in (b), write the given binary number and add it to the same number shifted one place to the right. Record the MSB of the binary, i.e. 1 as the MSB of the Gray code. Add the subsequent columns (0 + 1 = 1; 0 + 0 = 0; 1 + 0 = 1) and record the corresponding sums as the subsequent significant bits of the Gray code. Ignore the bit shifted out. The resultant Gray code is 1101.

Gray-to-Binary Conversion

If an n-bit Gray number is represented by $G_n G_{n-1} \ldots G_1$ and its binary equivalent by $B_n B_{n-1} \ldots B_1$, then binary bits are obtained from Gray bits as follows:

$$\begin{aligned} B_n &= G_n \\ B_{n-1} &= B_n \oplus G_{n-1} \\ B_{n-2} &= B_{n-1} \oplus G_{n-2} \\ &\vdots \\ B_1 &= B_2 \oplus G_1 \end{aligned}$$

The conversion procedure is:

1. The MSB of the binary number is the same as the MSB of the Gray code number; record it.
2. Add the MSB of the binary to the next significant bit of the Gray code, i.e. XOR them; record the sum and ignore the carry.
3. Add the 2nd bit of the binary to the 3rd bit of the Gray; the 3rd bit of the binary to the 4th bit of the Gray code, and so on.
4. Continue this till all the Gray bits are exhausted. The sequence of bits that has been written down is the binary equivalent of the Gray code number.

EXAMPLE 3.11 Convert the Gray code 1101 to binary.

Solution

The conversion is done as shown below:

Gray	1	1	0	1
	‖ ⊕↗	‖ ⊕↗	‖ ⊕↗	‖
Binary	1	0	0	1

Record the 8's bit (MSB) of the Gray code as the 8's bit of the binary. Add the 8's bit of the binary to the 4's bit of the Gray code $(1 + 1 = 10)$ and record the sum (0) as the 4's bit of the binary (ignore the carry bit). Add the 4's bit of the binary to the 2's bit of the Gray code $(0 + 0 = 0)$ and record the sum as the 2's bit of the binary. Add the 2's bit of the binary to the 1's bit of Gray code $(0 + 1 = 1)$ and record the sum as the 1's bit of the binary. The resultant binary number is 1001.

The XS-3 Gray Code

In a normal Gray code, the bit patterns for 0 (0000) and 9 (1101) do not have a unit distance between them. That is, they differ in more than one position. In XS-3 Gray code, each decimal digit is encoded with the Gray code pattern of the decimal digit that is greater by 3. It has a unit distance between the patterns for 0 and 9. Table 3.4 shows the XS-3 Gray code for decimal digits 0 through 9.

Table 3.4 The XS-3 Gray code for decimal digits 0–9

Decimal digit	XS-3 Gray code				Decimal digit	XS-3 Gray code			
0	0	0	1	0	5	1	1	0	0
1	0	1	1	0	6	1	1	0	1
2	0	1	1	1	7	1	1	1	1
3	0	1	0	1	8	1	1	1	0
4	0	1	0	0	9	1	0	1	0

3.5 ERROR-DETECTING CODES

When a binary data is transmitted and processed, it is susceptible to noise that can alter or distort its contents. The 1s may get changed to 0s and 0s to 1s. Because digital systems must be accurate to the digit, errors can pose a serious problem. Several schemes have been devised to detect the occurrence of a single-bit error in a binary word, so that whenever such an error occurs the concerned binary word can be corrected and retransmitted.

Parity

The simplest technique for detecting errors is that of adding an extra bit, known as the *parity* bit, to each word being transmitted. There are two types of parity—odd parity and even parity. For odd parity, the parity bit is set to a 0 or a 1 at the transmitter such that the total number of 1 bits in the word including the parity bit is an odd number. For even parity, the parity bit is set to a 0 or a 1 at the transmitter such that the total number of 1 bits in the word including the parity bit is an even number. Table 3.5 shows the parity bits to be added to transmit decimal digits 0 through 9 in the 8421 code.

Table 3.5 Odd and even parity in the 8421 BCD code

Decimal	8421 BCD				Odd parity	Even parity
0	0	0	0	0	1	0
1	0	0	0	1	0	1
2	0	0	1	0	0	1
3	0	0	1	1	1	0
4	0	1	0	0	0	1
5	0	1	0	1	1	0
6	0	1	1	0	1	0
7	0	1	1	1	0	1
8	1	0	0	0	0	1
9	1	0	0	1	1	0

When the digital data is received, a parity checking circuit generates an error signal if the total number of 1s is even in an odd-parity system or odd in an even-parity system. This parity check can always detect a single-bit error but cannot detect two or more errors within the same word. In any practical system, there is always a finite probability of the occurrence of single error. The probability that two or more errors will occur simultaneously, although non-zero, is substantially smaller. Odd parity is used more often than even parity because even parity does not detect the situation where all 0s are created by a short-circuit or some other fault condition.

If the code possesses the property by which the occurrence of any single-bit error transforms a valid code word into an invalid one, it is said to be an error-detecting (single bit) code. In general, to obtain an N bit error-detecting code, no more than half of the possible 2^N combinations of digits can be used. Thus to obtain an error-detecting code for 10 decimal digits, at least 5 binary digits are to be used. The code words are chosen in such a manner that in order to change one valid code word into another valid code word, at least two digits must be complemented. A code is an error-detecting code, if and only if its minimum distance is two or more. The *distance* between two words is defined as the number of digits that must change in a word so that the other word results. For example, the distance between 0011 and 1010 is 2, and the distance between 0111 and 1000 is 4.

EXAMPLE 3.12 In an even-parity scheme, which of the following words contain an error?

(a) 10101010 (b) 11110110 (c) 10111001

Solution

(a) The number of 1s in the word is even (4). Therefore, there is no error.
(b) The number of 1s in the word is even (6). Therefore, there is no error.
(c) The number of 1s in the word is odd (5). So, this word has an error.

EXAMPLE 3.13 In an odd-parity scheme, which of the following words contain an error?

(a) 10110111 (b) 10011010 (c) 11101010

Solution

(a) The number of 1s in the word is even (6). So, this word has an error.
(b) The number of 1s in the word is even (4). So, this word has an error.
(c) The number of 1s in the word is odd (5). Therefore, there is no error.

Check Sums

Simple parity cannot detect two errors within the same word. One way of overcoming this difficulty is to use a sort of two-dimensional parity. As each word is transmitted, it is added to the sum of the previously transmitted words, and the sum retained at the transmitter end. At the end of transmission, the sum (called the *check sum*) up to that time is sent to the receiver. The receiver can check its sum with the transmitted sum. If the two sums are the same, then no errors were detected at the receiver end. If there is an error, the receiving location can ask for retransmission of the entire data. This is the type of transmission used in teleprocessing systems.

Block Parity

When several binary words are transmitted or stored in succession, the resulting collection of bits can be regarded as a block of data, having rows and columns. Parity bits can then be assigned to both rows and columns. This scheme makes it possible to *correct* any single error occurring in a data word and to *detect* any two errors in a word. The parity row is often called a parity word. Such a block parity technique, also called word parity, is widely used for data stored on magnetic tapes.

For example, six 8-bit words in succession can be formed into a 6 × 8 block for transmission. Parity bits are added so that odd parity is maintained both row-wise and column-wise and the block is transmitted as a 7 × 9 block as shown in Table A. At the receiving end, parity is checked both row-wise and column-wise and suppose errors are detected as shown in Table B. These single-bit errors detected can be corrected by complementing the error bit. In Table B, parity errors in the 3rd row and 5th column mean that the 5th bit in the 3rd row is in error. It can be corrected by complementing it.

Two errors as shown in Table C can only be detected but not corrected. In Table C, parity errors are observed in both columns 2 and 4. It indicates that in one row there are two errors.

Table A

	0	1	0	1	1	0	1	1	0
	1	0	0	1	0	1	0	1	1
	0	1	1	0	1	1	1	0	0
	1	1	0	1	0	0	1	1	0
	1	0	0	0	1	1	0	1	1
	0	1	1	1	0	1	1	1	1
Parity row →	0	1	1	1	0	1	1	0	0
									↑ Parity column

Table B

0	1	0	1	1	0	1	1	0	
1	0	0	1	0	1	0	1	1	
0	1	1	0	0	1	1	0	0	← Parity error in 3rd row
1	1	0	1	0	0	1	1	0	
1	0	0	0	1	1	0	1	1	
0	1	1	1	0	1	1	1	1	
0	1	1	1	0	1	1	0	0	
				↑ Parity error in 5th column					

Table C

0	1	0	1	1	0	1	1	0
1	0	0	1	0	1	0	1	1
0	1	1	0	1	1	1	0	0
1	0	0	0	0	0	1	1	0
1	0	0	0	1	1	0	1	1
0	1	1	1	0	1	1	1	1
0	1	1	1	0	1	1	0	0
	↑		↑					

Parity errors in 2nd and 4th columns

Five-bit Codes

Some 5-bit BCD codes that have parity contained within each code for ease of error detection are shown in Table 3.6. The 63210 is a weighted code (except for the decimal digit 0). It has the useful error-detecting property that there are exactly two 1s in each code group. This code is used for storing data on magnetic tapes. The 2-out-of-5 code is a non-weighted code. It also has exactly two 1s in each code group. This code is used in the telephone and communication industries. At the receiving end, the receiver can check the number of 1s in each character received. The shift-counter code, also called the Johnson code, has the bit patterns produced by a 5-bit Johnson counter. The 51111 code is similar to Johnson code but is weighted.

Table 3.6 Five-bit BCD codes

Decimal	6 3 2 1 0	2-out-of-5	Shift-counter	5 1 1 1 1
0	0 0 1 1 0	0 0 0 1 1	0 0 0 0 0	0 0 0 0 0
1	0 0 0 1 1	0 0 1 0 1	0 0 0 0 1	0 0 0 0 1
2	0 0 1 0 1	0 0 1 1 0	0 0 0 1 1	0 0 0 1 1
3	0 1 0 0 1	0 1 0 0 1	0 0 1 1 1	0 0 1 1 1
4	0 1 0 1 0	0 1 0 1 0	0 1 1 1 1	0 1 1 1 1
5	0 1 1 0 0	0 1 1 0 0	1 1 1 1 1	1 0 0 0 0
6	1 0 0 0 1	1 0 0 0 1	1 1 1 1 0	1 1 0 0 0
7	1 0 0 1 0	1 0 0 1 0	1 1 1 0 0	1 1 1 0 0
8	1 0 1 0 0	1 0 1 0 0	1 1 0 0 0	1 1 1 1 0
9	1 1 0 0 0	1 1 0 0 0	1 0 0 0 0	1 1 1 1 1

The Biquinary Code

The biquinary code shown in Table 3.7 is a weighted 7-bit BCD code. It is a parity data code. Note that each code group can be regarded as consisting of a 2-bit subgroup and a

Table 3.7 The biquinary code

Decimal digit	Biquinary code						
	5	0	4	3	2	1	0
0	0	1	0	0	0	0	1
1	0	1	0	0	0	1	0
2	0	1	0	0	1	0	0
3	0	1	0	1	0	0	0
4	0	1	1	0	0	0	0
5	1	0	0	0	0	0	1
6	1	0	0	0	0	1	0
7	1	0	0	0	1	0	0
8	1	0	0	1	0	0	0
9	1	0	1	0	0	0	0

5-bit subgroup, and each of these subgroups contains a single 1. Thus it has the error-checking feature, for each code group has exactly two 1s and each subgroup has exactly one 1. The weights of the bit positions are 50 43210. Since there are two positions with weight 0, it is possible to encode decimal 0 with a group containing 1s, unlike other weighted codes. The biquinary code is used in the Abacus.

The Ring-counter Code

A 10-bit ring counter will produce a sequence of 10-bit groups having the property that each group has a single 1. The ring-counter code shown in Table 3.8 is the code obtained by assigning a decimal digit to each of those ten patterns. It is a weighted code (9 8 7 6 5 4 3 2 1 0) because each bit position has a weight equal to one of the 10 decimal digits. Although this code is inefficient (for 1024 numbers can be encoded in pure binary with 10 bits as against 10 numbers only in ring-counter code), it has excellent error-detecting properties and is easier to implement.

Table 3.8 The ring-counter code

Decimal digit	Ring-counter code
	9 8 7 6 5 4 3 2 1 0
0	0 0 0 0 0 0 0 0 0 1
1	0 0 0 0 0 0 0 0 1 0
2	0 0 0 0 0 0 0 1 0 0
3	0 0 0 0 0 0 1 0 0 0
4	0 0 0 0 0 1 0 0 0 0
5	0 0 0 0 1 0 0 0 0 0
6	0 0 0 1 0 0 0 0 0 0
7	0 0 1 0 0 0 0 0 0 0
8	0 1 0 0 0 0 0 0 0 0
9	1 0 0 0 0 0 0 0 0 0

3.6 ERROR-CORRECTING CODES

A code is said to be an error-correcting code, if the correct code word can always be deduced from an erroneous word. For a code to be a single-bit error-correcting code, the minimum distance of that code must be three. The minimum distance of a code is the smallest number of bits by which any two code words must differ. A code with minimum distance of three can not only correct single-bit errors, but also detect (but cannot correct) two-bit errors. The key to error correction is that it must be possible to detect and locate erroneous digits. If the location of an error has been determined, then by complementing the erroneous digit, the message can be corrected. One type of error-correcting code is the Hamming code. In this code, to each group of *m* information or message or data bits, *k* parity-checking bits denoted by $P_1, P_2, \ldots, P_k$ located at positions 2^{k-1} are added to form an $(m + k)$-bit code word. To correct the error, *k* parity checks are performed on selected digits of each code word, and the position of the error bit is located by forming an error word, and the error bit is then complemented. The k-bit error word is generated by putting a 0 or a 1 in the 2^{k-1}th position depending upon whether the check for parity involving the parity bit P_k is satisfied or not.

The 7-bit Hamming Code

To transmit four data bits, three parity bits located at positions 2^0, 2^1, and 2^2 are added to make a 7-bit code word which is then transmitted. The word format would be as shown below:

$$D_7 \quad D_6 \quad D_5 \quad P_4 \quad D_3 \quad P_2 \quad P_1$$

where the D bits are the data bits and the P bits are the parity bits. P_1 is set to a 0 or 1 so that it establishes even parity over bits 1, 3, 5, and 7 (P_1 D_3 D_5 D_7). P_2 is set to a 0 or a 1 to establish even parity over bits 2, 3, 6 and 7 (P_2 D_3 D_6 D_7). P_4 is set to a 0 or a 1 to establish even parity over bits 4, 5, 6, and 7 (P_4 D_5 D_6 D_7). The 7-bit Hamming code for the decimal digits coded in BCD is shown in Table 3.9.

Table 3.9 The 7-bit Hamming code

Decimal digit	Hamming code bits						
	D_7	D_6	D_5	P_4	D_3	P_2	P_1
0	0	0	0	0	0	0	0
1	0	0	0	0	1	1	1
2	0	0	1	1	0	0	1
3	0	0	1	1	1	1	0
4	0	1	0	1	0	1	0
5	0	1	0	1	1	0	1
6	0	1	1	0	0	1	1
7	0	1	1	0	1	0	0
8	1	0	0	1	0	1	1
9	1	0	0	1	1	0	0

EXAMPLE 3.14 Encode data bits 0011 into the 7-bit even-parity Hamming code.

Solution

The bit pattern is

D_7	D_6	D_5	P_4	D_3	P_2	P_1
0	0	1		1		

Bits 1, 3, 5, 7 (i.e. P_1 1 1 0) must have even parity. So, P_1 must be a 0.
Bits 2, 3, 6, 7 (i.e. P_2 1 0 0) must have even parity. So, P_2 must be a 1.
Bits 4, 5, 6, 7 (i.e. P_4 1 0 0) must have even parity. So, P_4 must be a 1.

Therefore, the final code is 0011110.

At the receiving end, the message received in Hamming code is decoded to see if any errors have occurred. Bits 1, 3, 5, 7, bits 2, 3, 6, 7, and bits 4, 5, 6, 7 are all checked for even parity. If they all check out, there is no error. If there is an error, the error bit can be located by forming a 3-bit binary number out of the three parity checks and the error bit is then corrected by complementing it.

Assume that a 7-bit Hamming code is received as 1111001. Analyzing the bits 1, 3, 5, 7 (1 0 1 1) of which P_1 is a parity bit, an error is detected. So, put a 1 in the 1's position of the error word. Analyzing the bits 2, 3, 6, 7 (0 0 1 1) of which P_2 is a parity bit, no error is detected. So, put a 0 in the 2's position of the error word. Analyzing the bits 4, 5, 6, 7 (1 1 1 1) of which P_4 is a parity bit, no error is detected. So, put a 0 in the 4's position of the error word. So the error word is 001, a decimal 1. Therefore, the bit in position 1 is in error; complement it. The corrected code should be read as 1111000.

The concept of 7-bit Hamming code discussed above can be extended to any number of bits by inserting parity bits at 2^n bit positions. The format for a 15-bit Hamming code would be as shown below.

$$D_{15} \quad D_{14} \quad D_{13} \quad D_{12} \quad D_{11} \quad D_{10} \quad D_9 \quad P_8 \quad D_7 \quad D_6 \quad D_5 \quad P_4 \quad D_3 \quad P_2 \quad P_1.$$

EXAMPLE 3.15 The message below coded in the 7-bit Hamming code is transmitted through a noisy channel. Decode the message assuming that at most a single error occurred in each code word.

1001001011100111101100011011

Solution

The given data is of 28 bits; split it into four groups of 7 bits each and correct for the error, if any, in each group and write the corrected data.

First group is 1 0 0 1 0 0 1

1, 3, 5, 7 (1001)	→	no error	→	put a 0 in the 1's position
2, 3, 6, 7 (0001)	→	error	→	put a 1 in the 2's position
4, 5, 6, 7 (1001)	→	no error	→	put a 0 in the 4's position

The error word is 010. So, complement the 2nd bit. Therefore, the correct code is 1 0 0 1 0 1 1.

Second group is 0 1 1 1 0 0 1

1, 3, 5, 7 (1010)	→	no error	→	put a 0 in the 1's position
2, 3, 6, 7 (0010)	→	error	→	put a 1 in the 2's position
4, 5, 6, 7 (1110)	→	error	→	put a 1 in the 4's position

The error word is 110. So, complement the 6th bit. Therefore, the correct code is 0 0 1 1 0 0 1.

Third group is 1 1 1 0 1 1 0

1, 3, 5, 7 (0111) → error → put a 1 in the 1's position

2, 3, 6, 7 (1111) → no error → put a 0 in the 2's position

4, 5, 6, 7 (0111) → error → put a 1 in the 4's position

The error word is 101. So, complement the 5th bit. Therefore, the correct code is 1 1 0 0 1 1 0

Fourth group is 0 0 1 1 0 1 1

1, 3, 5, 7 (1010) → no error → put a 0 in the 1's position

2, 3, 6, 7 (1000) → error → put a 1 in the 2's position

4, 5, 6, 7 (1100) → no error → put a 0 in the 4's position

The error word is 010. So, complement the 2nd bit. Therefore, the correct code is 0 0 1 1 0 0 1.
Therefore, the corrected message is 1001011001100111001100011001.

3.7 ALPHANUMERIC CODES

Alphanumeric codes are codes used to encode the characters of alphabet in addition to the decimal digits. They are used primarily for transmitting data between computers and its I/O devices such as printers, keyboards and video display terminals. Because the number of bits used in most alphanumeric codes is much more than those required to encode 10 decimal digits and 26 alphabetic characters, these codes include bit patterns for a wide range of other symbols and functions as well. The most popular modern alphanumeric codes are the ASCII code and the EBCDIC code.

The ASCII Code

The American Standard Code for Information Interchange (ASCII) pronounced as 'ASKEE' is a widely used alphanumeric code. This is basically a 7-bit code. Since the number of different bit patterns that can be created with 7 bits is $2^7 = 128$, the ASCII can be used to encode both the lowercase and uppercase characters of the alphabet (52 symbols) and some special symbols as well, in addition to the 10 decimal digits. It is used extensively for printers and terminals that interface with small computer systems. Many large systems also make provisions for its accommodation. Because characters are assigned in ascending binary numbers, ASCII is very easy for a computer to alphabetize and sort.

Table 3.10 shows the ASCII code groups. For ease of presentation, they are listed with the three most significant bits of each group along the top row and the four least significant bits along the left column.

Table 3.10 The ASCII code

		MSBs							
		0 0 0	0 0 1	0 1 0	0 1 1	1 0 0	1 0 1	1 1 0	1 1 1
LSBs	0 0 0 0	NUL	DEL	Space	0	@	P		p
	0 0 0 1	SOH	DC1	!	1	A	Q	a	q
	0 0 1 0	STX	DC2	“	2	B	R	b	r
	0 0 1 1	ETX	DC3	#	3	C	S	c	s
	0 1 0 0	EOT	DC4	$	4	D	T	d	t
	0 1 0 1	END	NAK	%	5	E	U	e	u
	0 1 1 0	ACK	SYN	&	6	F	V	f	v
	0 1 1 1	BEL	ETB	‘	7	G	W	g	w
	1 0 0 0	BS	CAN	(	8	H	X	h	x
	1 0 0 1	HT	EM	)	9	I	Y	i	y
	1 0 1 0	LF	SUB	*	:	J	Z	j	z
	1 0 1 1	VT	ESC	+	;	K	[	k	{
	1 1 0 0	FF	FS	,	<	L	\	l	|
	1 1 0 1	CR	GS	–	=	M	]	m	}
	1 1 1 0	SO	RS	.	>	N	^	n	~
	1 1 1 1	SI	US	/	?	O	_	o	DLE

Abbreviations

ACK	Acknowledge	EM	End of medium	NAK	Negative acknowledge
BEL	Bell	ENQ	Enquiry	NUL	Null
BS	Backspace	EOT	End of transmission	RS	Record separator
CAN	Cancel	ESC	Escape	SI	Shift in
CR	Carriage return	ETB	End of transmission block	SO	Shift out
DC1	Direct control 1	EXT	End of text	SOH	Start of heading
DC2	Direct control 2	FF	Form feed	STX	Start text
DC3	Direct control 3	FS	Form separator	SUB	Substitute
DC4	Direct control 4	GS	Group separator	SYN	Synchronous idle
DEL	Delete idle	HT	Horizontal tab	US	Unit separator
DLE	Data link escape	LF	Line feed	VT	Vertical tab

The EBCDIC Code

The Extended Binary Coded Decimal Interchange Code (EBCDIC) pronounced as 'eb-si-dik' is an 8-bit alphanumeric code. Since 2^8 (= 256) bit patterns can be formed with 8-bits, the EBCDIC code can be used to encode all the symbols and control characters found in ASCII. It encodes many other symbols too. In fact, many of the bit patterns in the EBCDIC code are unassigned. It is used by most large computers to communicate in alphanumeric data. Unlike ASCII, which uses a straight binary sequence for representing characters, this code uses BCD as the basis of binary assignment. Table 3.11 shows the EBCDIC code.

Table 3.11 The EBCDIC code

MSD (Hex)

LSD (Hex)	0	1	2	3	4	5	6	7	8	9	A	B	C	D	E	F
0	NUL	DLE	DS		SP	&							[	]	\	0
1	SOH	DC1	SOS				/		a	j	~		A	J		1
2	STX	DC2	FS	SYN					b	k	s		B	K	S	2
3	ETX	DC3							c	l	t		C	L	T	3
4	PF	RES	BYP	PN					d	m	u		D	M	U	4
5	HT	NL	LF	RS					e	n	v		E	N	V	5
6	LC	BS	EOB	YC					f	o	w		F	O	W	6
7	DEL	IL	PRE	EOT					g	p	x		G	P	X	7
8		CAN							h	q	y		H	Q	Y	8
9		EM							i	r	z		I	R	Z	9
A	SMM	CC	SM		Ø	!	|	:								
B	VT				.	$	,	#								
C	FF	IFS		DC4	<	*	%	@								
D	CR	IGS	ENQ	NAK	(	)	–	'								
E	SO	IRS	ACK		+	;	>	=								
F	SI	IUS	BEL	SUB	|	⌈	?	'								

SUMMARY

- Binary codes may be numeric or alphanumeric.
- Alphanumeric codes represent the letters of the alphabet and decimal numbers as a sequence of 0s and 1s.
- A sequence of binary digits such as that representing a decimal digit, is called a code word.
- Numeric codes used to represent decimal numbers are called BCD codes.

- The BCD codes may be weighted or non-weighted. The weighted codes obey the position-weighting principle.
- In every positively-weighted code, one of the weights must be 1, the second must be either 1 or 2 and the sum of all the weights must be equal to or greater than 9.
- Positively-weighted codes are those codes in which all the weights assigned to the binary digits are positive only.
- Negatively-weighted codes are codes in which at least one of the weights assigned to the binary digits is negative.
- There are 17 positively-weighted codes, out of which only four are self-complementing.
- There are 13 negatively-weighted self-complementing codes.
- In a sequential code, each succeeding code word is one binary number greater than its preceding code word.
- In a self-complementing code, the code word for the 9's complement of N is the 1's complement of the code word for N.
- For a code to be self-complementing, the sum of positional weights must be 9.
- In a cyclic code each successive code word differrs in only one digit position from its preceding code word.
- The 8421 BCD is less efficient than pure binary, but its main advantage is its ease of conversion to and from decimal.
- Each one of the 4-bit BCD codes has 6 illegal states.
- The XS-3 BCD code is a non-weighted code, but is used for mathematical operations because it is a sequential code.
- In the XS-3 code, each binary code word is the corresponding 8421 code word plus 0011.
- Where special additive requirements exist, the XS-3 code can be used; and where mechanical systems must be used, binary coded Gray codes are helpful.
- The Gray code is a unit distance code and its main advantage is its ease of conversion to and from binary.
- The Gray code is not a BCD code; it is not sequential, but cyclic and reflective.
- The Gray code cannot be used for mathematical operations. It is used in instrumentation and data acquisition systems where linear or angular displacement is measured. It is used in shaft encoders, I/O devices, A/D converters and other peripheral devices.
- The simplest technique for detecting a single error is that of adding an extra bit, known as parity bit, to each word being transmitted.
- For even parity, the parity bit should be set to a 0 or a 1, so that the total number of 1s in the word to be transmitted including the parity bit is even.
- For odd parity the parity bit should be set to a 0 or a 1, so that the total number of 1s in the word to be transmitted including the parity bit is odd.
- Odd parity is used more often than even parity, because even parity does not detect a situation where all 0s are created by a short circuit or any other fault condition.

- A code is said to be error detecting, if it possesses the property such that the occurrence of a single error transforms a valid code word into an invalid one.
- Check sums provide a sort of two-dimensional parity.
- Block parity can *correct* any single error in a data word and *detect* any two errors in a data word.
- A code is said to be error correcting, if the correct code word can always be deduced from an erroneous word.
- The minimum distance of a code is the smallest number of bits by which any two code words differ.
- The minimum distance of both BCD and XS-3 codes is 1.
- The number of digits that must change in a word to get the other word is the distance between those two words.
- For single-bit error detection, the minimum distance between code words must be two or more, and for single-bit error correction it must be three or more.
- Hamming codes can, not only detect errors but also correct them.
- In the Hamming code, to each group of m data bits, k parity bits located at positions 2^{k-1} are added to form an $(m + k)$-bit code word.
- The ASCII code and the EBCDIC code are the most popular, modern alphanumeric codes.
- The EBCDIC code does not use a straight binary sequence for representing characters, but uses BCD as the basis of binary assignment.

QUESTIONS

1. What is the difference between (a) a code and a code word, (b) a weighted code and a non-weighted code, and (c) a positively-weighted code and a negatively-weighted code?
2. What do you mean by straight binary coding?
3. What do you mean by a sequential code? Write two sequential codes.
4. What do you mean by a self-complementing code? Write two self-complementing codes.
5. What is the condition for a code to be a self-complementing code?
6. What do you mean by a cyclic code? Give an example of a cyclic code.
7. What is a BCD code? What are its advantages and disadvantages?
8. Compare BCD and binary systems of coding.
9. What do you mean by an invalid state? Give examples.
10. Define XS-3 code and XS-3 Gray code. What are their salient features? Write the XS-3 code.
11. What are the rules of (a) BCD addition and subtraction, and (b) XS-3 addition and subtraction?

12. What do you mean by the Gray code? What are its applications?

13. How do you convert a binary number to a gray number and vice versa?

14. What do you mean by a reflected code? Write an example.

15. What do you mean by a unit distance code? Write one such 4-bit code.

16. What is a parity bit? What do you mean by odd and even parity systems? What is block parity? Why is odd parity used more often than even parity?

17. What do you mean by an error-detecting code and an error-correcting code? Enumerate their requirements.

18. What do you mean by the minimum distance of a code?

19. Define check-sums?

20. Name some 5-bit codes.

21. What is the error checking property of the biquinary code?

22. What are the advantages and disadvantages of the ring-counter code?

23. What is Hamming code? How is the Hamming code word tested and corrected?

24. What do you mean by an alphanumeric code? Name some of them. What are alphanumeric codes used for?

PROBLEMS

3.1 Express the following decimal numbers in the 8421 BCD code.

(a) 296 (b) 157.5 (c) 4228.5
(d) 809 (e) 37.52 (f) 2040.08

3.2 Express the following decimal numbers in 2421, 5211, and 4311 codes.

(a) 205 (b) 2630 (c) 697 (d) 4568

3.3 Express the following 8421 BCD numbers as decimals.

(a) 1000 0011 1001 (b) 0101 0111 1000 . 1001
(c) 0110 1001 0111 . 0100 (d) 1000 0110 0000 0011 . 1000

3.4 Express the following decimal numbers in the XS-3 code.

(a) 97 (b) 653 (c) 812.9
(d) 19 (e) 2515 (f) 2053.76

3.5 Express the following XS-3 numbers as decimals.

(a) 1011 1000 1100 (b) 0110 1010 0111 . 1000
(c) 1001 1100 1000 . 0111 (d) 0111 1000 0101 1001 . 1000

3.6 Add the following in (i) BCD and (ii) XS-3 codes.

(a) 275 + 496 (b) 108 + 789
(c) 89.6 + 273.7 (d) 205.7 + 193.65

3.7 Subtract the following in (i) BCD and (ii) XS-3 codes.

(a) 920 – 265 (b) 86 – 24
(c) 467.6 – 258.9 (d) 635.7 – 419.8

3.8 Convert the following decimal numbers to Gray code.

(a) 6 (b) 12 (c) 20
(d) 96 (e) 286

3.9 Convert the following binary numbers to Gray code.

(a) 1010 (b) 1101010 (c) 110110110
(d) 10001 (e) 11001100

3.10 Convert the following Gray codes to binary.

(a) 1111 (b) 101101 (c) 100011101
(d) 1001 (e) 1110111

3.11 Which of the following words contain an error for odd parity?

(a) 10010101 (b) 11010101 (c) 10110110
(d) 1010 (e) 110101

3.12 Convert the following decimal numbers to (i) 2-out-of-5, (ii) biquinary (iii) shift counter, and (iv) ring-counter codes.

(a) 29 (b) 153 (c) 697

3.13 Encode all the BCD words into a 7-bit even-parity Hamming code.

3.14 Detect and correct errors, if any, in the following even-parity Hamming code words.

(a) 1101010 (b) 0101101 (c) 0111101
(d) 1001011 (e) 1110111

3.15 The message below was coded in the Hamming code and transmitted through a noisy channel. Decode the message assuming that a single error has occurred in each code word: 1011011011100111000111010111.

3.16 Code the following in ASCII and EBCDIC.

(a) BIRTH (b) AK47

Chapter 4

LOGIC GATES

4.1 INTRODUCTION

Logic gates are the fundamental building blocks of digital systems. The name logic gate is derived from the ability of such a device to make decisions, in the sense that it produces one output level when some combinations of input levels are present, and a different output level when other combinations of input levels are present. There are just three basic types of gates—AND, OR and NOT. The fact that computers are able to perform very complex logic operations, stems from the way these elementary gates are interconnected. The interconnection of gates to perform a variety of logical operations is called *logic design*.

Logic gates are electronic circuits because they are made up of a number of electronic devices and components. They are constructed in a wide variety of forms. They are usually embedded in large-scale integrated circuits (LSI) and very large-scale integrated circuits (VLSI) along with a large number of other devices, and are not easily accessible or identifiable. Each gate is dedicated to a specific logic operation. Logic gates are also constructed in small-scale integrated circuits (SSI), where they appear with few others of the same type. In these integrated devices, the inputs and outputs of all the gates are accessible, that is, external connections can be made to them just like discrete logic gates.

Inputs and outputs of logic gates can occur only in two levels. These two levels are termed HIGH and LOW, or TRUE and FALSE, or ON and OFF, or simply 1 and 0.

A table which lists all the possible combinations of input variables and the corresponding outputs is called a *truth table*. It shows how the logic circuit's output responds to various combinations of logic levels at the inputs.

In this book, we use *level logic*, a logic in which the voltage levels represent logic 1 and logic 0. Level logic may be *positive logic* or *negative logic*. A positive logic system is the one in which the higher of the two voltage levels represents the logic 1 and the lower of the two voltage levels represents the logic 0. A negative logic system is the one in which the lower of the two voltage levels represents the logic 1 and the higher of the two voltage levels represents the logic 0. In transistor-transistor logic (TTL, the most widely used logic family), the voltage levels are + 5 V and 0 V. In the following discussion in this chapter, logic 1 corresponds to + 5 V and logic 0 to 0 V.

4.2 THE AND GATE

An AND gate has two or more inputs but only one output. The output assumes the logic 1 state, only when each one of its inputs is at logic 1 state. The output assumes the logic 0 state even if one of its inputs is at logic 0 state. The AND gate may, therefore, be defined as a device whose output is 1, if and only if all its inputs are 1. Hence the AND gate is also called an *all* or *nothing* gate.

The logic symbols and truth tables of two-input and three-input AND gates are shown in Figs. 4.1a and 4.1b, respectively. Note that the output is 1 only when all the inputs are 1. The symbol for the AND operation is '·', or we use no symbol at all.

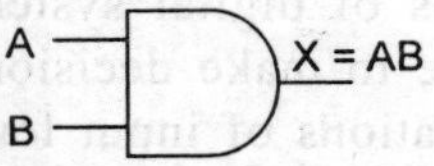

A, B → X = AB

Logic symbol

Truth table

Inputs		Output
A	B	X
0	0	0
0	1	0
1	0	0
1	1	1

(a) Two-input AND gate

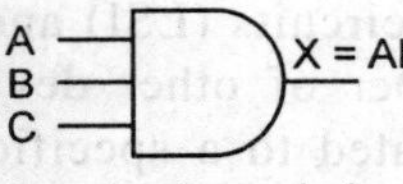

A, B, C → X = ABC

Logic symbol

Truth table

Inputs			Output
A	B	C	X
0	0	0	0
0	0	1	0
0	1	0	0
0	1	1	0
1	0	0	0
1	0	1	0
1	1	0	0
1	1	1	1

(b) Three-input AND gate

Fig. 4.1 The AND gate.

With the input variables to the AND gate are represented by A, B, C, . . ., the Boolean expression for the output can be written as $X = A \cdot B \cdot C \ldots$, which is read as "X is equal to A and B and C . . . " or "X is equal to ABC . . . ", or "X is equal to A dot B dot C . . . "

Discrete AND gates may be realized by using diodes or transistors as shown in Figs. 4.2a and 4.2b, respectively. The inputs A and B to the gates may be either 0 V or + 5 V.

In the diode AND gate, when A = + 5 V and B = + 5 V, both the diodes D1 and D2 are OFF. So, no current flows through R and, therefore, no voltage drop occurs across R. Hence, the output $X \approx 5$ V. When A = 0 V or B = 0 V or when both A and B are equal to 0 V, the corresponding diode D1 or D2 is ON or both diodes are ON and act as short-circuits (ideal case), and, therefore, the output $X \approx 0$ V. In practical circuits, X = 0.6 V or 0.7 V which is treated as logic 0.

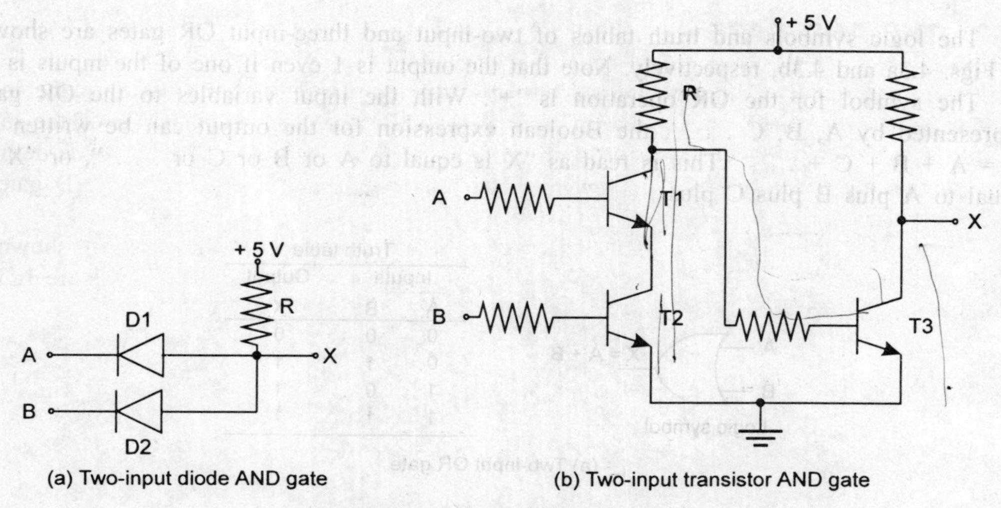

(a) Two-input diode AND gate (b) Two-input transistor AND gate

Fig. 4.2 Discrete AND gates.

In the transistor AND gate, when A = 0 V and B = 0 V or when A = 0 V and B = + 5 V or when A = + 5 V and B = 0 V, both the transistors T1 and T2 are OFF. Transistor T3 gets enough base drive from the supply through R and so, T3 will be ON. Hence, the output voltage $X = V_{ce\,(sat)} \approx 0$ V. When both A and B are equal to + 5 V, both the transistors T1 and T2 will be ON and, therefore, the voltage at the collector of transistor T1 will drop. So, T3 does not get enough base drive and, therefore, remains OFF. Hence no current flows through the collector resistor of T3 and, therefore, no voltage drop occurs across it. Hence output voltage, $X \approx 5$ V. The truth table for the above gate circuits is as shown below.

Truth table

Inputs		Output
A	B	X
0 V	0 V	0 V
0 V	5 V	0 V
5 V	0 V	0 V
5 V	5 V	5 V

The IC 7408 contains four two-input AND gates, the IC 7411 contains three three-input AND gates, and the IC 7421 contains two four-input AND gates.

4.3 THE OR GATE

Like an AND gate, an OR gate may have two or more inputs but only one output. The output assumes the logic 1 state, even if one of its inputs is in logic 1 state. Its output assumes the logic 0 state, only when each one of its inputs is in logic 0 state. An OR gate may, therefore, be defined as a device whose output is 1, even if one of its inputs is 1. Hence an OR gate is also called an *any* or *all* gate. It can also be called an inclusive OR gate because it includes the condition 'both the inputs can be present'.

The logic symbols and truth tables of two-input and three-input OR gates are shown in Figs. 4.3a and 4.3b, respectively. Note that the output is 1 even if one of the inputs is 1.

The symbol for the OR operation is '+'. With the input variables to the OR gate represented by A, B, C . . . , the Boolean expression for the output can be written as X = A + B + C + ... This is read as "X is equal to A or B or C or . . . ", or "X is equal to A plus B plus C plus . . . ."

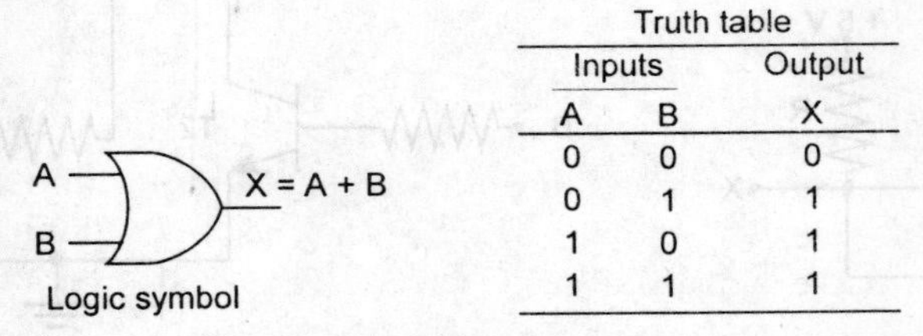

Truth table

Inputs		Output
A	B	X
0	0	0
0	1	1
1	0	1
1	1	1

(a) Two-input OR gate

A
B
C
X = A + B + C
Logic symbol

Truth table

Inputs			Output
A	B	C	X
0	0	0	0
0	0	1	1
0	1	0	1
0	1	1	1
1	0	0	1
1	0	1	1
1	1	0	1
1	1	1	1

(b) Three-input OR gate

Fig. 4.3 The OR gate.

Discrete OR gates may be realized by using diodes or transistors as shown in Figs. 4.4a and 4.4b, respectively. The inputs A and B to the gates may be either 0 V or + 5 V.

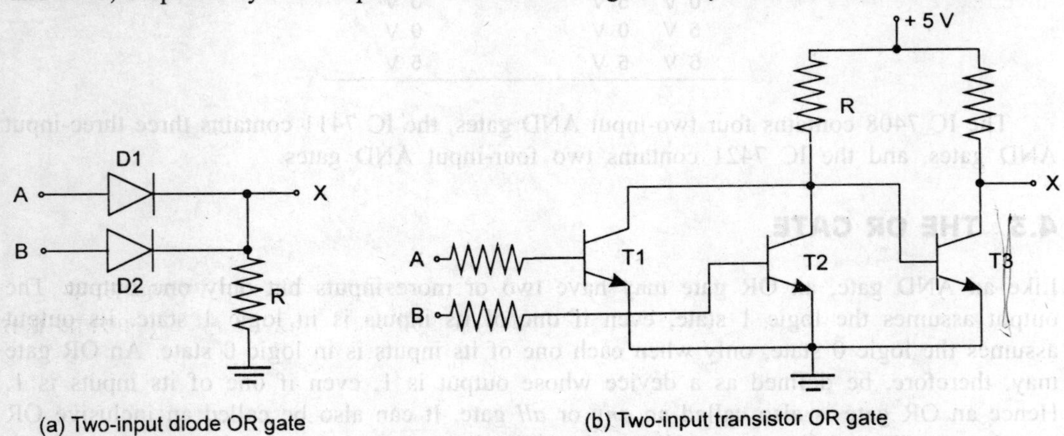

(a) Two-input diode OR gate

(b) Two-input transistor OR gate

Fig. 4.4 Discrete OR gates.

In the diode OR gate, when A = 0 V and B = 0 V, both the diodes D1 and D2 are OFF. No current flows through R, and so, no voltage drop occurs across R. Hence, the output voltage X = 0 V. When either A = + 5 V or B = + 5 V or when both A and B are equal to + 5 V, the corresponding diode D1 or D2 is ON or both D1 and D2 are ON and act as short-circuits (ideal case) and, therefore, output X ≈ 5 V. In practice, X = + 5 V – diode drop = + 5 V – 0.7 V = 4.3 V, which is regarded as logic 1.

In the transistor OR gate, when A = 0 V and B = 0 V, both the transistors T1 and T2 are OFF. Transistor T3 gets enough base drive from + 5 V through R and, therefore, it will be ON. The output voltage, $X = V_{ce(sat)} \approx 0$ V. When either A = + 5 V or B = + 5 V or when both A and B are equal to + 5 V, the corresponding transistor T1 or T2 is ON or both T1 and T2 will be ON and, therefore, the voltage at the collector of T1 is = $V_{ce(sat)} \approx 0$ V. This cannot forward bias the base-emitter junction of T3 and, therefore, it will remain OFF. Hence, the output voltage will be X = 5 V (logic 1 level).

The truth table for the above OR gate circuits is as shown below.

Truth table

Inputs		Output
A	B	X
0 V	0 V	0 V
0 V	5 V	5 V
5 V	0 V	5 V
5 V	5 V	5 V

The IC 7432 contains four two-input OR gates.

4.4 THE NOT GATE

A NOT gate, also called an *inverter*, has only one input and, of course, only one output. It is a device whose output is always the complement of its input. That is, the output of a NOT gate assumes the logic 1 state when its input is in logic 0 state and assumes the logic 0 state when its input is in logic 1 state. The logic symbol and the truth table of an inverter are shown in Figs. 4.5a and 4.5b, respectively.

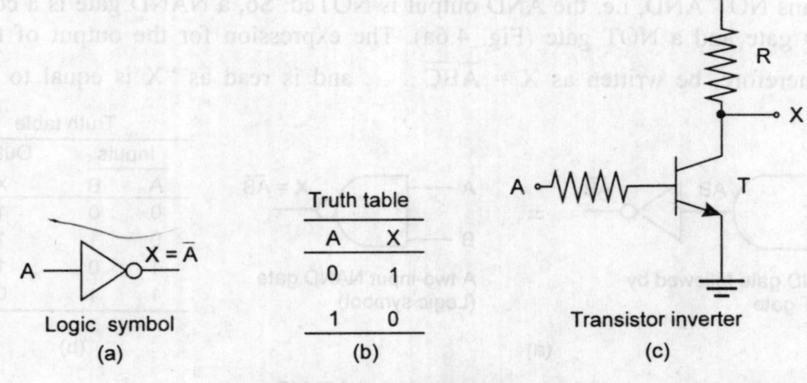

Fig. 4.5 The inverter.

The symbol for NOT operation is '–' (bar). When the input variable to the NOT gate is represented by A and the output variable by X, the expression for the output is $X = \overline{A}$. This is read as "X is equal to A bar".

A discrete NOT gate may be realized using a transistor as shown in Fig. 4.5c. The input to the gate may be 0 V or + 5 V. When A = 0 V, transistor T is reverse biased and, therefore, remains OFF. As no current flows through R, no voltage drop occurs across R. Hence, the output voltage X = + 5 V. When the input A = + 5 V, T is ON and the output voltage $X = V_{ce\,(sat)} \approx 0$ V. The truth table for the NOT gate circuit is as shown below. The IC 7404 contains six inverters.

Truth table

Input A	Output X
0 V	5 V
5 V	0 V

Logic circuits of any complexity can be realized using only AND, OR and NOT gates. Logic circuits which use these three gates only are called AND/OR/INVERT, i.e. AOI logic circuits. Logic circuits which use AND gates and OR gates only are called AND/OR, i.e. AO logic circuits.

4.5 THE UNIVERSAL GATES

Though logic circuits of any complexity can be realized by using only the three basic gates (AND, OR and NOT), there are two universal gates (NAND and NOR), each of which can also realize logic circuits single-handedly. The NAND and NOR gates are also, therefore, called universal building blocks. Both NAND and NOR gates can perform all the three basic logic functions (AND, OR and NOT). Therefore, AOI logic can be converted to NAND logic or NOR logic.

The NAND Gate

NAND means NOT AND, i.e. the AND output is NOTed. So, a NAND gate is a combination of an AND gate and a NOT gate (Fig. 4.6a). The expression for the output of the NAND gate can, therefore, be written as $X = \overline{ABC}$ . . . and is read as "X is equal to A · B · C

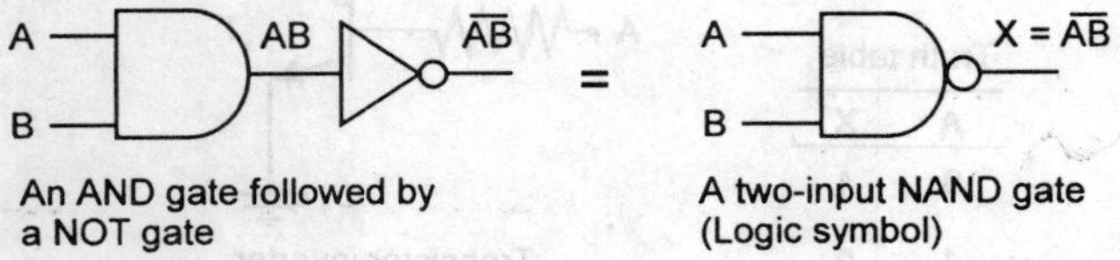

(a)

Truth table

Inputs A	Inputs B	Output X
0	0	1
0	1	1
1	0	1
1	1	0

(b)

Fig. 4.6 A two-input NAND gate.

... whole bar." The output is logic 0 level, only when each of the inputs assumes a logic 1 level. For any other combination of inputs, the output is a logic 1 level. The truth table of a two-input NAND gate is shown in Fig. 4.6b.

The logic symbol and truth table for a three-input NAND gate are shown in Figs. 4.7a and 4.7b, respectively.

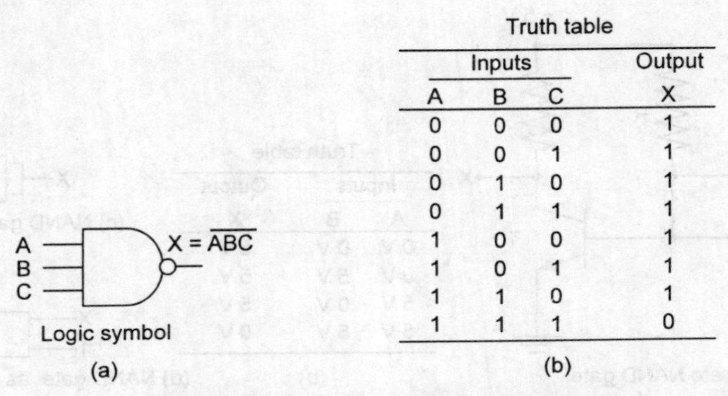

Truth table

Inputs			Output
A	B	C	X
0	0	0	1
0	0	1	1
0	1	0	1
0	1	1	1
1	0	0	1
1	0	1	1
1	1	0	1
1	1	1	0

(b)

Fig. 4.7 A three-input NAND gate.

Looking at the truth table of a two-input NAND gate, we see that the output X is 1 when either A = 0 or B = 0 or when both A and B are equal to 0, i.e. if either $\overline{A}$ = 1 or $\overline{B}$ = 1 or both $\overline{A}$ and $\overline{B}$ are equal to 1. Therefore, the NAND gate can perform the OR function. The corresponding output expression is, X = $\overline{A} + \overline{B}$. So, a NAND function can also be realized by first inverting the inputs and then ORing the inverted inputs. Thus, a NAND gate is a combination of two NOT gates and an OR gate (see Fig. 4.8a). Hence, from Figs. 4.6a and 4.8a, we can express the output of a two-input NAND gate as

$$X = \overline{AB} = \overline{A} + \overline{B}$$

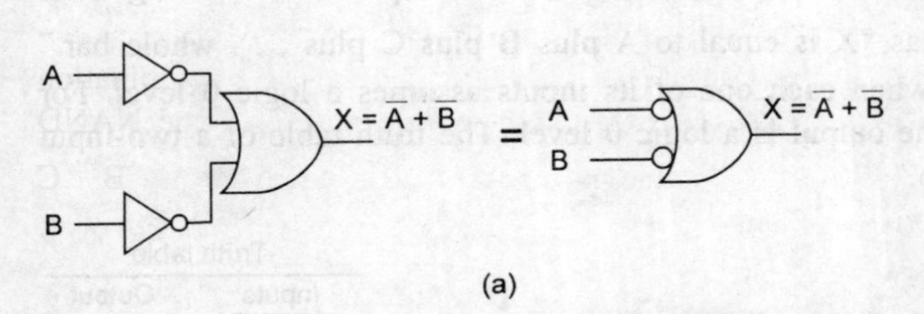

(a)

Truth table

Inputs		Inverted Inputs		Output
A	B	$\overline{A}$	$\overline{B}$	X
0	0	1	1	1
0	1	1	0	1
1	0	0	1	1
1	1	0	0	0

(b)

Fig. 4.8 Bubbled OR gate.

The OR gate with inverted inputs (Fig. 4.8a) is called a *bubbled* OR gate. So, a NAND gate is equivalent to a bubbled OR gate whose truth table is shown in Fig. 4.8b. A bubbled OR gate is also called a *negative* OR gate. Since its output assumes the HIGH state even if any one of its inputs is 0, the NAND gate is also called an active-LOW OR gate.

NAND = -ve OR
NAND = 'NOT' AND

A discrete two-input NAND gate is shown in Fig. 4.9a. When A = + 5 V and B = + 5 V, both the diodes D1 and D2 are OFF. The transistor T gets enough base drive from the supply through R and, therefore, T is ON and the output $X = V_{ce(sat)} \approx 0$ V. When A = 0 V or B = 0 V or when both A and B are equal to 0 V, the transistor T is OFF and, therefore, output ≈ + 5 V. The truth table is as shown in Fig. 4.9b.

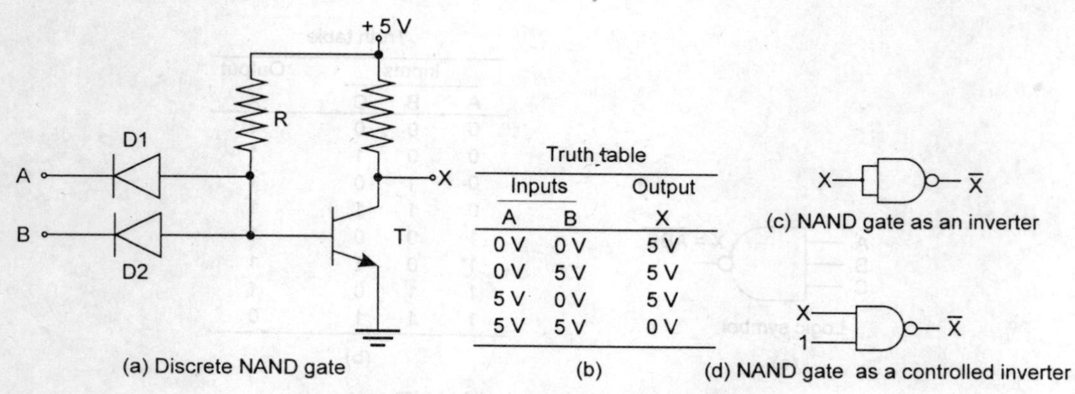

Truth table

Inputs		Output
A	B	X
0 V	0 V	5 V
0 V	5 V	5 V
5 V	0 V	5 V
5 V	5 V	0 V

(a) Discrete NAND gate (b) (d) NAND gate as a controlled inverter

Fig. 4.9 Discrete two-input NAND gate.

The IC 7400 contains four two-input NAND gates; the IC 7410 contains three three-input NAND gates; the IC 7420 contains two four-input NAND gates; and the IC 7430 contains one eight-input NAND gate.

A NAND gate can also be used as an inverter by tying all its input terminals together and applying the signal to be inverted to the common terminal (Fig. 4.9c), or by connecting all its input terminals except one, to logic 1 and applying the signal to be inverted to the remaining terminal as shown in Fig. 4.9d. In the latter form, it is said to act as a controlled inverter.

The NOR gate

NOR means NOT OR, i.e. the OR output is NOTed. So, a NOR gate is a combination of an OR gate and a NOT gate (Fig. 4.10a). The expression for the output of the NOR gate is, $X = \overline{A + B + C + \ldots}$ and is read as "X is equal to A plus B plus C plus . . . whole bar." The output is logic 1 level, only when each one of its inputs assumes a logic 0 level. For any other combination of inputs, the output is a logic 0 level. The truth table of a two-input NOR gate is shown in Fig. 4.10b.

A, B → A+B → $\overline{A+B}$ = A, B → $\overline{A+B}$

An OR gate followed by a NOT gate A two-input NOR gate (logic symbol)

(a)

Truth table

Inputs		Output
A	B	X
0	0	1
0	1	0
1	0	0
1	1	0

(b)

Fig. 4.10 A two-input NOR gate.

The logic symbol and truth table for a three-input NOR gates are shown in Figs. 4.11a and 4.11b, respectively.

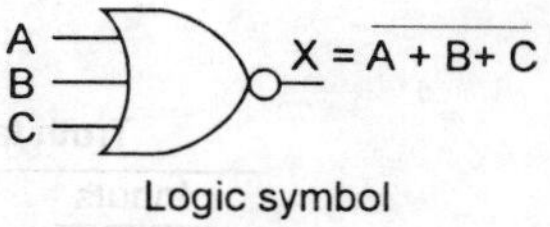

Logic symbol
(a)

Truth table			
Inputs			Output
A	B	C	X
0	0	0	1
0	0	1	0
0	1	0	0
0	1	1	0
1	0	0	0
1	0	1	0
1	1	0	0
1	1	1	0

(b)

Fig. 4.11 A three-input NOR gate.

Looking at the truth table of a two-input NOR gate, we see that the output X is 1 only when both A and B are equal to 0, i.e. only when both $\overline{A}$ and $\overline{B}$ are equal to 1. That means, a NOR gate is equivalent to an AND gate with inverted inputs and the corresponding output expression is, $X = \overline{A}\,\overline{B}$. So, a NOR function can also be realized by first inverting the inputs and then ANDing those inverted inputs. Thus, a NOR gate is a combination of two NOT gates and an AND gate (see Fig. 4.12a). Hence, from Figs. 4.10a and 4.12a, we can see that the output of a two-input NOR gate is, $X = \overline{A + B} = \overline{A}\,\overline{B}$.

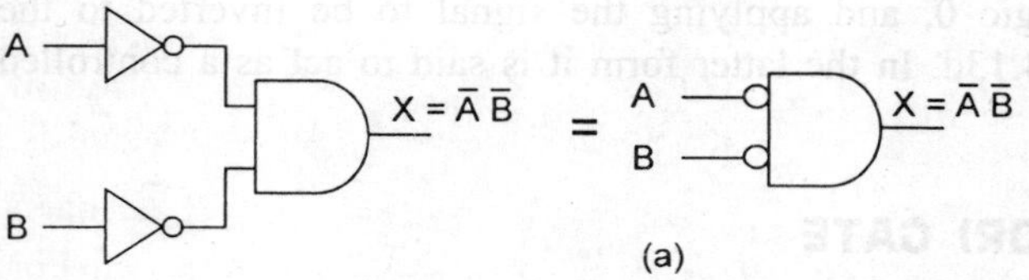

(a)

Truth table				
Inputs		Inverted inputs		Output
A	B	$\overline{A}$	$\overline{B}$	X
0	0	1	1	1
0	1	1	0	0
1	0	0	1	0
1	1	0	0	0

(b)

Fig. 4.12 Bubbled AND gate.

The AND gate with inverted inputs (Fig. 4.12a) is called a *bubbled* AND gate. So, a NOR gate is equivalent to a bubbled AND gate whose truth table is shown in Fig. 4.12b. A bubbled AND gate is also called a *negative* AND gate. Since its output assumes the HIGH state only when all its inputs are in LOW states, a NOR gate is also called an active-LOW AND gate.

A discrete two-input NOR gate is shown in Fig. 4.13a. When A = 0 V and B = 0 V, both transistors T1 and T2 are OFF; so, no current flows through R and, therefore, no voltage drop occurs across R. Hence, the output voltage X ≈ + 5 V (logic 1). When either A = + 5 V or B = + 5 V or when both A and B are equal to + 5 V, the corresponding

NOR = -ve AND = Bubbled AND = Active-Low AND

transistor T1 or T2 or both T1 and T2 are ON. Therefore, X is at $V_{ce(sat)}$ with respect to ground and equal to 0 V (logic 0). The truth table shown in Fig. 4.13b.

The IC 7402 contains four two-input NOR gates; the IC 7427 contains three three-input NOR gates; and the IC 7425 contains two four-input NOR gates.

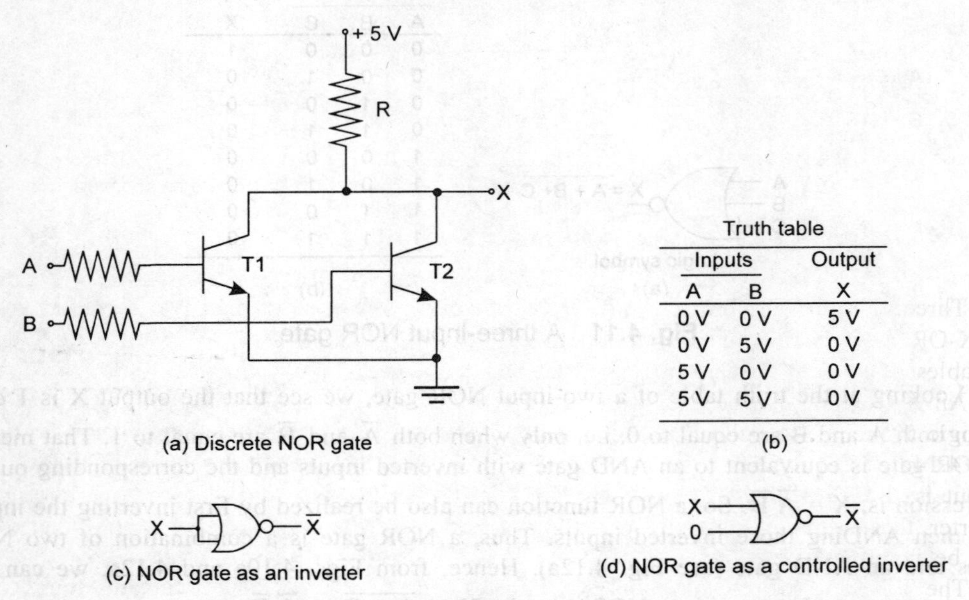

Truth table

Inputs		Output
A	B	X
0 V	0 V	5 V
0 V	5 V	0 V
5 V	0 V	0 V
5 V	5 V	0 V

(a) Discrete NOR gate

(b)

(c) NOR gate as an inverter

(d) NOR gate as a controlled inverter

Fig. 4.13 Discrete two-input NOR gate.

A NOR gate can also be used as an inverter, by tying all its input terminals together and applying the signal to be inverted to the common terminal (Fig. 4.13c) or by connecting all its input terminals except one to logic 0, and applying the signal to be inverted to the remaining terminal as shown in Fig. 4.13d. In the latter form it is said to act as a controlled inverter.

4.6 THE EXCLUSIVE-OR (X-OR) GATE

An X-OR gate is a two input, one output logic circuit, whose output assumes a logic 1 state when one and only one of its two inputs assumes a logic 1 state. Under the conditions when both the inputs assume the logic 0 state, or when both the inputs assume the logic 1 state, the output assumes a logic 0 state.

Since an X-OR gate produces an output 1 only when the inputs are not equal, it is called an *anti-coincidence* gate or *inequality detector*. The output of an X-OR gate is the modulo sum of its two inputs. The name Exclusive-OR is derived from the fact that its output is a 1, only when exclusively one of its inputs is a 1 (it excludes the condition when both the inputs are 1).

The logic symbol and truth table of a two-input X-OR gate are shown in Figs. 4.14a and 4.14b, respectively. If the input variables are represented by A and B and the output variable by X, the expression for the output of this gate is written as $X = A \oplus B = A\overline{B} + \overline{A}B$ and read as "X is equal to A ex-or B".

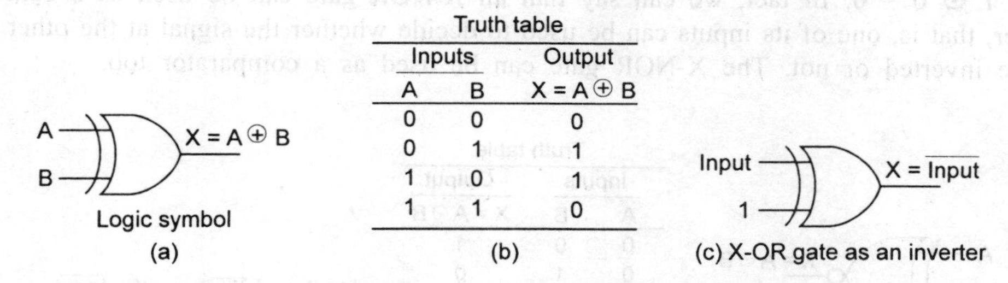

Fig. 4.14 Exclusive-OR gate.

Three or more variable X-OR gates do not exist. When more than two variables are to be X-ORed, a number of two-input X-OR gates will be used. The X-OR of a number of variables assumes a 1 state only when an odd number of input variables assume a 1 state.

An X-OR gate can be used as an inverter by connecting one of the two input terminals to logic 1 and feeding the input sequence to be inverted to the other terminal as shown in Fig. 4.14c. If the input bit is a 0, the output is, $0 \oplus 1 = 1$, and if the input bit is a 1, the output is, $1 \oplus 1 = 0$. In fact, we can say that an X-OR gate can be used as a controlled inverter, that is, one of its inputs can be used to decide whether the signal at the other input will be inverted or not.

The TTL IC 7486 contains four X-OR gates. The CMOS IC 74C86 contains four X-OR gates. High speed CMOS IC 74HC86 contains four X-OR gates.

4.7 THE EXCLUSIVE-NOR (X-NOR) GATE

An X-NOR gate is a combination of an X-OR gate and a NOT gate. The X-NOR gate is a two input, one output logic circuit, whose output assumes a 1 state only when both the inputs assume a 0 state or when both the inputs assume a 1 state. The output assumes a 0 state, when one of the inputs assumes a 0 state and the other a 1 state. It is also called a *coincidence* gate, because its output is 1 only when its inputs coincide. It can be used as an *equality detector* because it outputs a 1 only when its inputs are equal.

The logic symbol and truth table of a two-input X-NOR gate are shown in Figs. 4.15a and 4.15b, respectively. If the input variables are represented by A and B and the output variable by X, the expression for the output of this gate is written as

$$X = A \odot B = AB + \overline{A}\,\overline{B} = \overline{A \oplus B} = \overline{A\overline{B} + \overline{A}B}$$

and read as "X is equal to A ex-nor B".

Three or more variable X-NOR gates do not exist. When a number of variables are to be X-NORed, a number of two-input X-NOR gates can be used. The X-NOR of a number of variables assumes a 1 state, only when an even number (including zero) of input variables assume a 0 state.

An X-NOR gate can be used as an inverter by connecting one of the two input terminals to logic 0 and feeding the input sequence to be inverted to the other terminal as shown in Fig. 4.15c. If the input bit is a 0, the output is, $0 \odot 0 = 1$, and if the input bit is a 1, the output $1 \odot 0 = 0$. In fact, we can say that an X-NOR gate can be used as a controlled inverter, that is, one of its inputs can be used to decide whether the signal at the other input will be inverted or not. The X-NOR gate can be used as a comparator too.

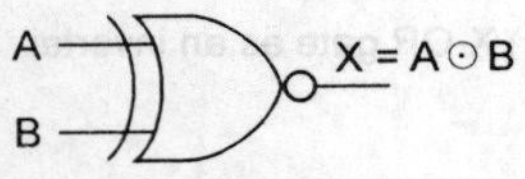

A, B, $X = A \odot B$

Logic symbol

(a)

Truth table

Inputs		Output
A	B	$X = A \odot B$
0	0	1
0	1	0
1	0	0
1	1	1

(b)

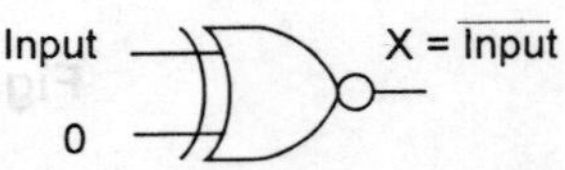

Input, 0, $X = \overline{\text{Input}}$

(c) X-NOR gate as an inverter

Fig. 4.15 Exclusive-NOR gate.

The X-NOR of two variables A and B is the complement of the X-OR of those two variables. That is,

$$A \odot B = \overline{A \oplus B}$$

But the X-NOR of three variables A, B and C is not equal to the complement of the X-OR of A, B, and C. That is,

$$A \odot B \odot C \neq \overline{A \oplus B \oplus C}$$

However, the X-NOR of a number of variables is equal to the complement of the X-OR of those variables only when the number of variables involved is even.

The TTL IC 74LS266, the CMOS IC 74C266 and the high speed CMOS IC 74HC266 contain four each X-NOR gates.

4.8 INHIBIT CIRCUITS

AND, OR, NAND, and NOR gates can be used to control the passage of an input logic signal through to the output. This is shown in Fig. 4.16 where a logic signal A is applied to one input of each of the above mentioned logic gates. The other input of each gate is the control input B. The logic level at this control input will determine whether the input signal is enabled to reach the output or inhibited from reaching the output.

We notice that when AND and OR gates are enabled, the output follows the A input exactly. Conversely, when NAND and NOR gates are enabled, the output is exactly the inverse of the A input.

We also see that when AND and NOR gates are inhibited, they produce a constant LOW output. Conversely, the NAND and OR gates produce a constant HIGH output in the inhibited condition.

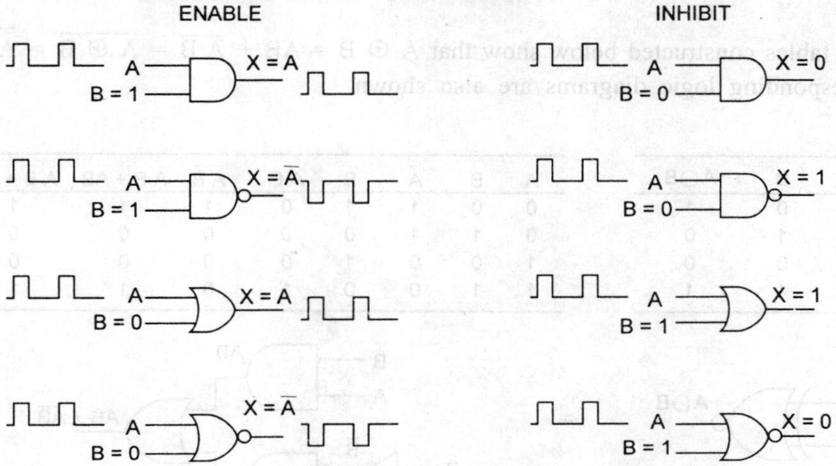

Fig. 4.16 Enable and inhibit circuits.

There will be many situations in digital circuit design, where the passage of a logic signal is either enabled or inhibited depending on the conditions present at one or more control inputs.

EXAMPLE 4.1 Show that $A \oplus B = A\bar{B} + \bar{A}B$ and construct the corresponding logic diagrams.

Solution

The truth tables constructed below show that $A \oplus B = A\bar{B} + \bar{A}B$. The corresponding logic diagrams are also shown.

A	B	$A \oplus B$
0	0	0
0	1	1
1	0	1
1	1	0

A	B	$\bar{A}$	$\bar{B}$	$\bar{A}B$	$A\bar{B}$	$A\bar{B} + \bar{A}B$
0	0	1	1	0	0	0
0	1	1	0	1	0	1
1	0	0	1	0	1	1
1	1	0	0	0	0	0

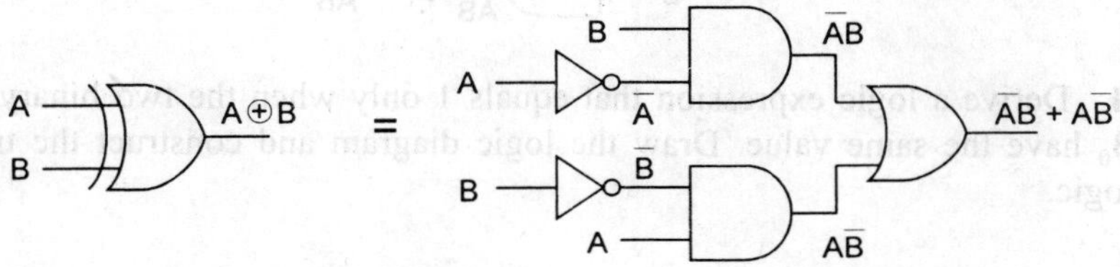

EXAMPLE 4.2 Show that $A \odot B = AB + \bar{A}\,\bar{B} = \overline{A \oplus B} = \overline{A\bar{B} + \bar{A}B}$. Also, construct the corresponding logic diagrams.

Solution

The truth tables constructed below show that $A \odot B = AB + \bar{A}\bar{B} = \overline{A \oplus B} = \overline{A\bar{B} + \bar{A}B}$. The corresponding logic diagrams are also shown.

A	B	$A \odot B$
0	0	1
0	1	0
1	0	0
1	1	1

A	B	$\bar{A}$	$\bar{B}$	AB	$\bar{A}\bar{B}$	$AB + \bar{A}\bar{B}$	$\overline{A\bar{B} + \bar{A}B}$
0	0	1	1	0	1	1	1
0	1	1	0	0	0	0	0
1	0	0	1	0	0	0	0
1	1	0	0	1	0	1	1

EXAMPLE 4.3 Show that $(A + B)\,\overline{AB}$ is equivalent to $A \oplus B$.

Solution

The truth tables constructed below show that $(A + B)\,\overline{AB} = A \oplus B$. The logic diagrams are also shown.

A	B	$A \oplus B$
0	0	0
0	1	1
1	0	1
1	1	0

A	B	A + B	AB	$\overline{AB}$	$(A + B)(\overline{AB})$
0	0	0	0	1	0
0	1	1	0	1	1
1	0	1	0	1	1
1	1	1	1	0	0

EXAMPLE 4.4 Derive a logic expression that equals 1 only when the two binary numbers A_1A_0 and B_1B_0 have the same value. Draw the logic diagram and construct the truth table to verify the logic.

Solution

The numbers A_1A_0 and B_1B_0 have the same value when their MSBs (A_1 and B_1) coincide and their LSBs (A_0 and B_0) coincide. So, we must AND the coincidence ($A_1 \odot B_1$) with the coincidence ($A_0 \odot B_0$). Therefore, the desired logic expression is $(A_1 \odot B_1)\,(A_0 \odot B_0)$. The logic diagram and truth table are shown below. Four variables can combine in 16 ways.

B_1	B_0	A_1	A_0	$A_1 \odot B_1$	$A_0 \odot B_0$	$(A_1 \odot B_1)(A_0 \odot B_0)$
0	0	0	0	1	1	1
0	0	0	1	1	0	0
0	0	1	0	0	1	0
0	0	1	1	0	0	0
0	1	0	0	1	0	0
0	1	0	1	1	1	1
0	1	1	0	0	0	0
0	1	1	1	0	1	0
1	0	0	0	0	1	0
1	0	0	1	0	0	0
1	0	1	0	1	1	1
1	0	1	1	1	0	0
1	1	0	0	0	0	0
1	1	0	1	0	1	0
1	1	1	0	1	0	0
1	1	1	1	1	1	1

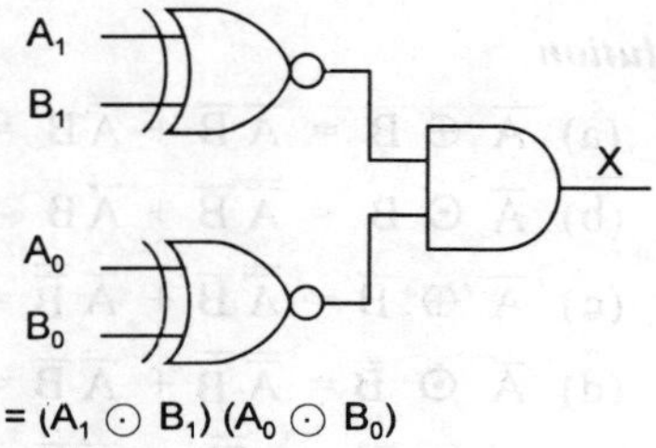

$X = (A_1 \odot B_1)(A_0 \odot B_0)$

EXAMPLE 4.5 Show that $AB + \overline{(A + B)}$ is equivalent to $A \odot B$.

Solution

The truth tables constructed below show that $AB + \overline{(A + B)} = A \odot B$.

A	B	$A \odot B$
0	0	1
0	1	0
1	0	0
1	1	1

A	B	A + B	AB	$\overline{A+B}$	$AB + \overline{A+B}$
0	0	0	0	1	1
0	1	1	0	0	0
1	0	1	0	0	0
1	1	1	1	0	1

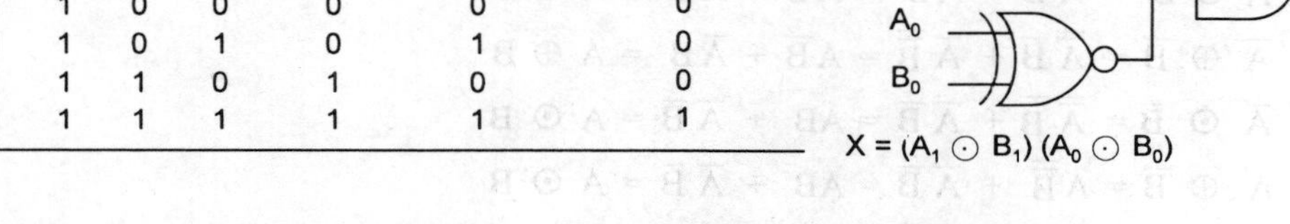

EXAMPLE 4.6 Find the logical equivalent of the following expressions.

(a) $A \oplus 0$ (b) $A \oplus 1$ (c) $A \odot 0$

(d) $A \odot 1$ (e) $1 \oplus \overline{A}$ (f) $0 \oplus \overline{A}$

Solution

(a) $A \oplus 0 = A \,.\, \overline{0} + \overline{A} \,.\, 0 = A \,.\, 1 + \overline{A} \,.\, 0 = A + 0 = A$

(b) $A \oplus 1 = A \,.\, \overline{1} + \overline{A} \,.\, 1 = A \,.\, 0 + \overline{A} \,.\, 1 = 0 + \overline{A} = \overline{A}$

(c) $A \odot 0 = \overline{A} \,.\, \overline{0} + A \,.\, 0 = \overline{A} \,.\, 1 + A \,.\, 0 = \overline{A} + 0 = \overline{A}$

(d) $A \odot 1 = \overline{A} \,.\, \overline{1} + A \,.\, 1 = \overline{A} \,.\, 0 + A \,.\, 1 = 0 + A = A$

(e) $1 \oplus \overline{A} = 1 . \overline{\overline{A}} + \overline{1} . \overline{A} = 1 . A + 0 . \overline{A} = A + 0 = A$

(f) $0 \oplus \overline{A} = 0 . \overline{\overline{A}} + \overline{0} . \overline{A} = 0 . A + 1 . \overline{A} = 0 + \overline{A} = \overline{A}$

EXAMPLE 4.7 Determine which of the following expressions are equivalent to $A \oplus B$ and which to $A \odot B$?

(a) $\overline{A} \oplus B$ (b) $\overline{A} \odot B$ (c) $\overline{A} \oplus \overline{B}$

(d) $\overline{A} \odot \overline{B}$ (e) $A \oplus \overline{B}$ (f) $A \odot \overline{B}$

Solution

(a) $\overline{A} \oplus B = \overline{A}\,\overline{B} + \overline{\overline{A}}B = \overline{A}\,\overline{B} + AB = A \odot B$

(b) $\overline{A} \odot B = \overline{\overline{A}}\,\overline{B} + \overline{A}B = A\overline{B} + \overline{A}B = A \oplus B$

(c) $\overline{A} \oplus \overline{B} = \overline{\overline{A}}\,\overline{B} + \overline{A}\,\overline{\overline{B}} = A\overline{B} + \overline{A}B = A \oplus B$

(d) $\overline{A} \odot \overline{B} = \overline{\overline{A}}\,\overline{\overline{B}} + \overline{A}\,\overline{B} = AB + \overline{A}\,\overline{B} = A \odot B$

(e) $A \oplus \overline{B} = A\overline{\overline{B}} + \overline{A}\,\overline{B} = AB + \overline{A}\,\overline{B} = A \odot B$

(f) $A \odot \overline{B} = A\overline{B} + \overline{A}\,\overline{\overline{B}} = A\overline{B} + \overline{A}B = A \oplus B$

4.9 7400-SERIES INTEGRATED CIRCUITS

There are a number of logic families. The logic gates discussed so far and many other logic circuits, are available in each one of those logic families. Each logic family has its own merits and demerits. Logic families are discussed in detail in Chapter 11.

TTL (Transistor-Transistor-Logic) is the most popular of the logic families. This popular series of chips contains a variety of gates whose inputs and outputs are accessible through connections to external pins (IC terminals). The chips are designated by 7400-series numbers, each of which refers to a particular set or combination of logic gates implemented by a chip. For example, IC–7400 contains four two-input NAND gates, IC–7410 contains three three–input NAND gates.

The voltage levels used to represent 0 and 1 logic levels in 7400 series logic are 0 V and + 5 V, respectively. The 0 V level means ground. It does not mean open circuit or no connection. In fact, if the input to a TTL gate is left open, the gate behaves as if a logic 1 were connected to the input. All 7400 series chips require + 5 volts DC power connection designated V_{CC}, and ground connection. Figure 4.17 shows some of the most commonly used 7400 series chips and their pin connections. The pins or terminals are numbered in a standard sequence. The groove or dot on the chip identifies the top of a chip and the numbers begin from the upper left-hand pin, run down and across (anticlockwise) and end up in the upper right-hand pin. Each pin is the terminal for an input, or output of a gate, or power supply connection. ICs are available in different package styles, and the pin assignments vary with the style selected.

4.10 ANSI/IEEE STANDARD LOGIC SYMBOLS

The symbols used to represent logic gates in this chapter have been in use for many years and will, no doubt, continue to be used. However, a new standard issued jointly by the

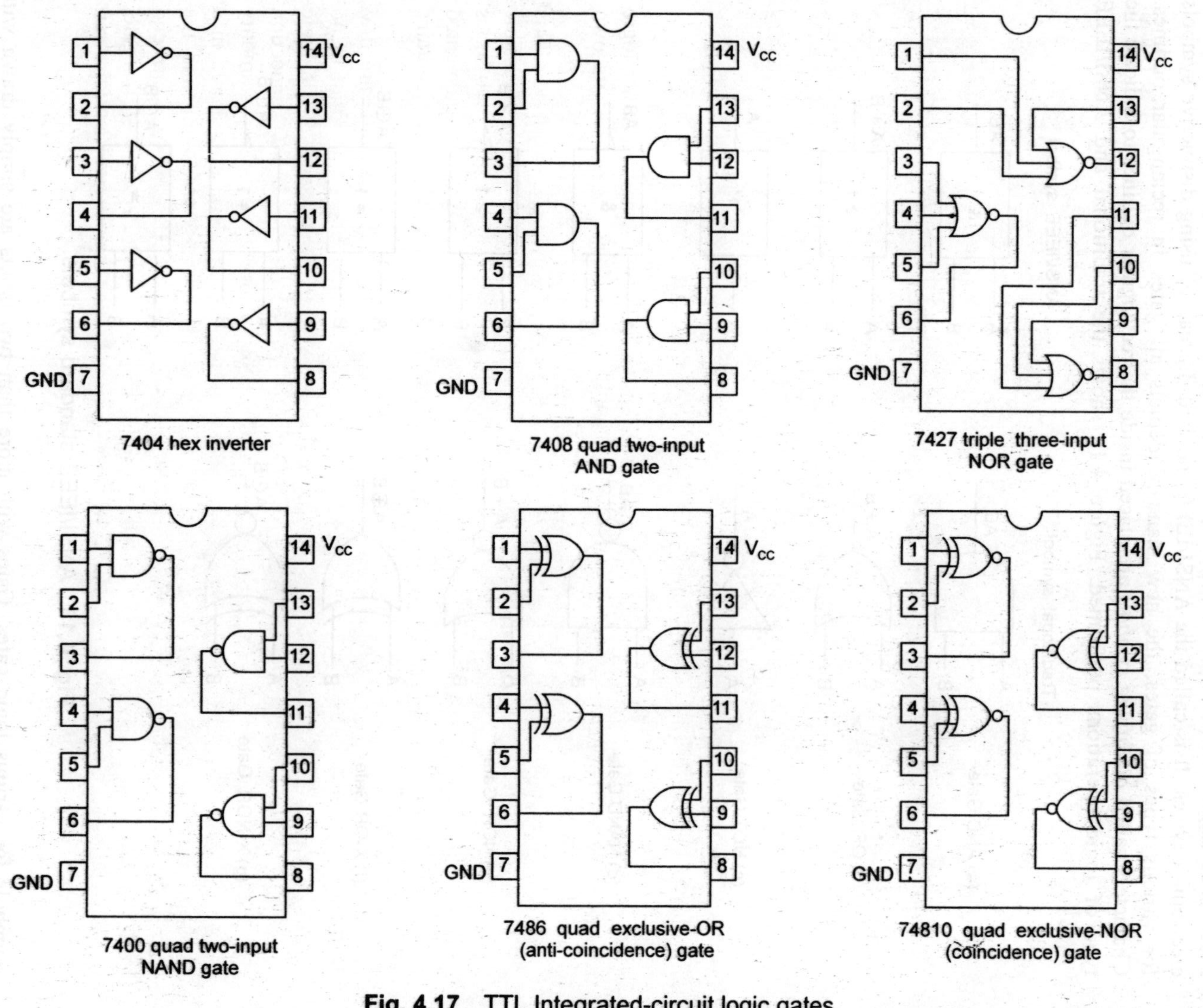

Fig. 4.17 TTL Integrated-circuit logic gates.

American National Standards Institute (ANSI) and The Institute for Electrical and Electronics Engineers (IEEE) is now being used by many industries and manufacturers of electronic devices. It is called the ANSI/IEEE standard. Instead of using distinctive symbols for various types of gates, the new standard depicts all gates in rectangular outlines. Characters called *qualifying symbols* are placed inside the rectangular outlines to indicate the type of logic operations performed. Figure 4.18 shows the traditional and ANSI/IEEE

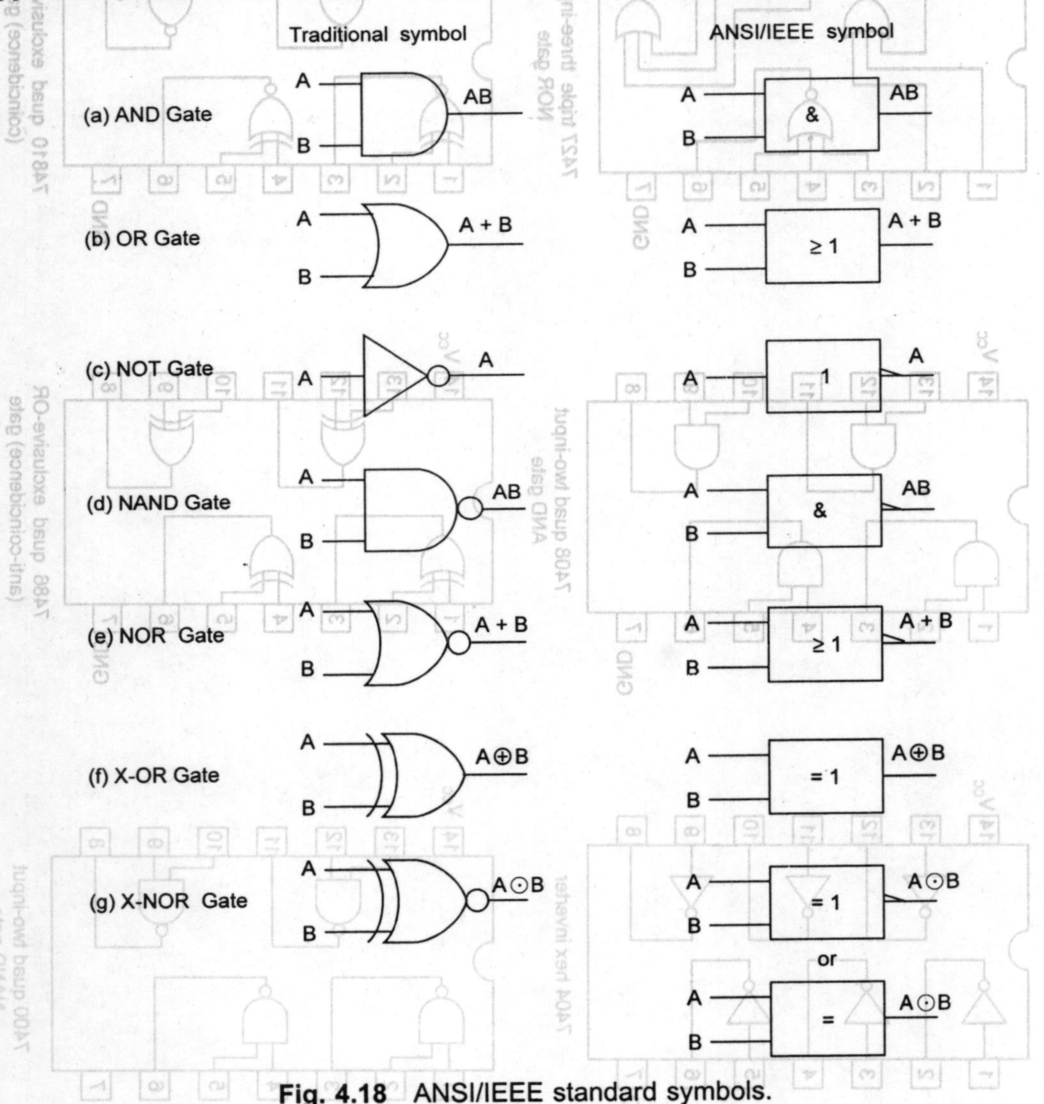

Fig. 4.18 ANSI/IEEE standard symbols.

symbols for various logic gates. Gates with more than two inputs are simply drawn with additional input lines, as in the traditional notation. The direction of signal flow is assumed to be from left to right, i.e. inputs on the left and outputs on the right. So, no arrowheads are required.

The symbol '&' represents the AND operation. The symbol '≥1' represents the OR operation. The symbol '1' with a triangle drawn at the output represents inversion. The NAND and NOR gates are represented by combinations of AND, OR, and INVERT symbols. The X-OR gate has the symbol '= 1' drawn inside, since the output is a 1, if and only if exactly one input is a 1. Since the coincidence or X-NOR function is the complement of the X-OR, it can be depicted as a '= 1' gate with inverted output. It is also represented by the symbol '=' meaning thereby that the output is a 1, if both inputs are equal (i.e. both 0 or 1).

The ANSI/IEEE logic symbols for integrated circuits containing multiple gates are drawn by stacking rectangles on top of each other, one rectangle for each gate. Figure 4.19 shows the ANSI/IEEE symbols for each of the 7400-series integrated circuit chips whose pin diagrams are shown in Fig. 4.17.

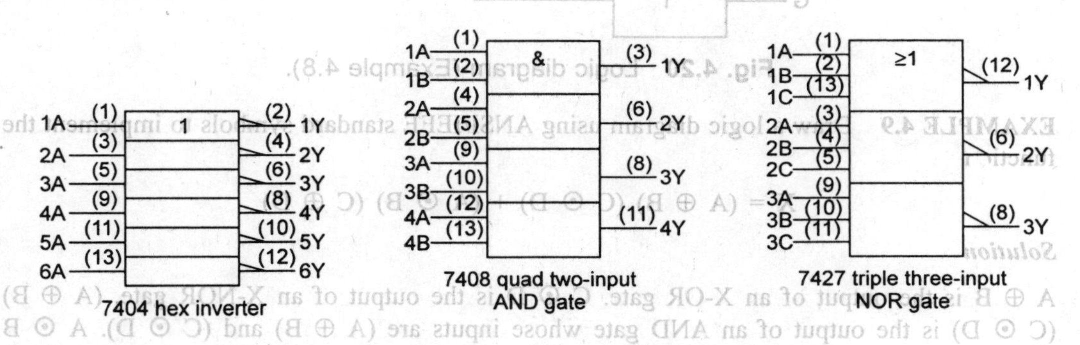

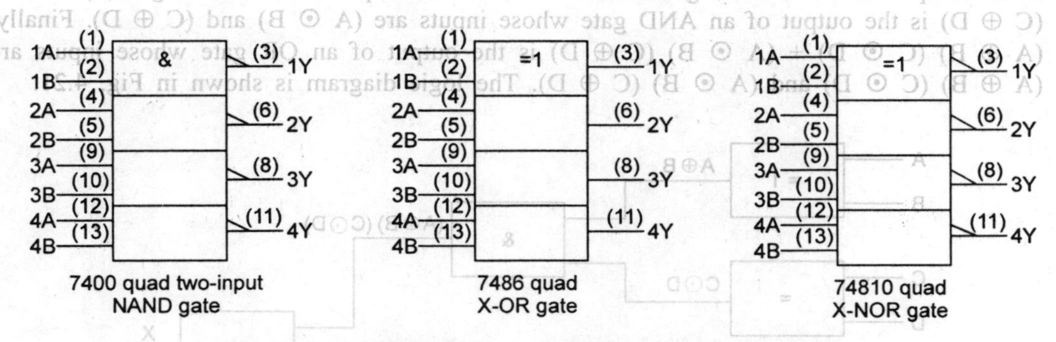

Fig. 4.19 ANSI/IEEE symbols for 7400 series ICs.

EXAMPLE 4.8 Draw a logic diagram using ANSI/IEEE standard symbols to implement the function

$$X = (A + B)\,(\overline{CD})\,(\overline{E + F})\overline{G}$$

Solution

The given expression is ANDing of four functions: A + B, $\overline{CD}$, $\overline{E + F}$ and $\overline{G}$. Now, A + B is ORing of A and B; $\overline{CD}$ is NANDing of C and D; $\overline{E + F}$ is NORing of E and F; $\overline{G}$ is inversion of G. The logic diagram using ANSI/IEEE symbols is shown in Fig. 4.20.

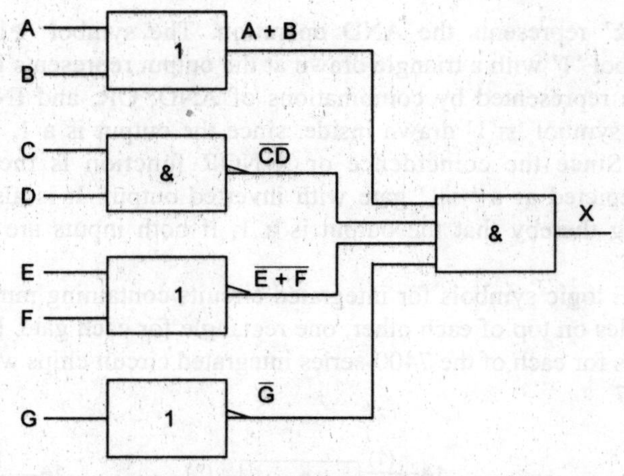

Fig. 4.20 Logic diagram (Example 4.8).

EXAMPLE 4.9 Draw a logic diagram using ANSI/IEEE standard symbols to implement the function

$$X = (A \oplus B)(C \odot D) + (A \odot B)(C \oplus D)$$

Solution

$A \oplus B$ is the output of an X-OR gate. $C \odot D$ is the output of an X-NOR gate. $(A \oplus B)(C \odot D)$ is the output of an AND gate whose inputs are $(A \oplus B)$ and $(C \odot D)$. $A \odot B$ is the output of an X-NOR gate and $C \oplus D$ is the output of an X-OR gate. $(A \odot B)(C \oplus D)$ is the output of an AND gate whose inputs are $(A \odot B)$ and $(C \oplus D)$. Finally, $(A \oplus B)(C \odot D) + (A \odot B)(C \oplus D)$ is the output of an OR gate whose inputs are $(A \oplus B)(C \odot D)$ and $(A \odot B)(C \oplus D)$. The logic diagram is shown in Fig. 4.21.

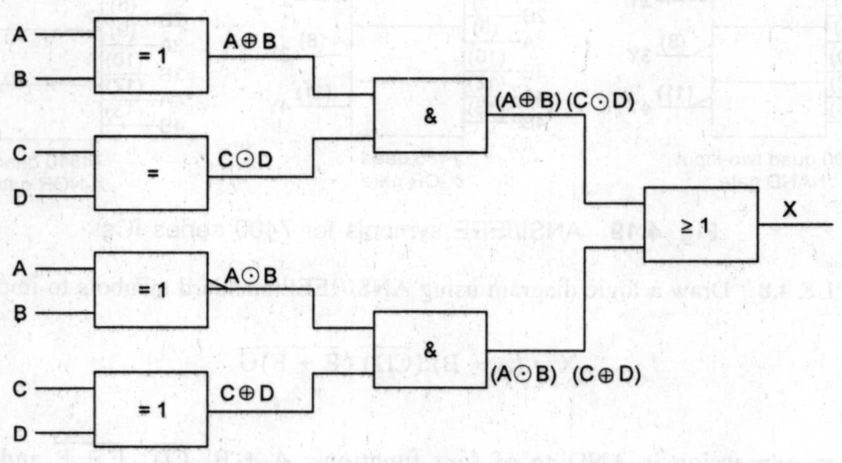

Fig. 4.21 Logic diagram (Example 4.9).

4.11 PULSED OPERATION OF LOGIC GATES

In a majority of applications, the inputs to a gate are not stationary levels, but are voltages that change frequently between two logic levels and can be classified as pulse waveforms. We will now look at the operation of all logic gates discussed till now with pulsed waveforms. All the logic gates obey the truth table operation, regardless of whether the inputs are constant levels or pulsed levels. The timing diagrams of various gates for different inputs are shown in the following examples.

EXAMPLE 4.10 For a two-input AND gate, determine its output waveform in relation to input waveforms of Fig. 4.22a.

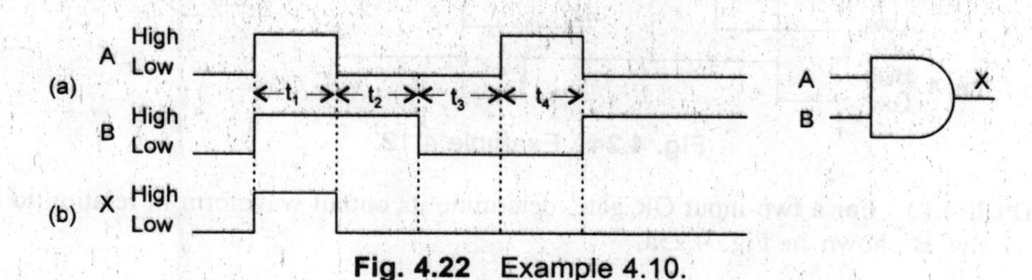

Fig. 4.22 Example 4.10.

Solution

In the interval t_1, A is HIGH and B is HIGH, so X is HIGH.
In the interval t_2, A is LOW and B is HIGH, so X is LOW.
In the interval t_3, A is LOW and B is LOW, so X is LOW.
In the interval t_4, A is HIGH and B is LOW, so X is LOW.
The output waveform X is shown in Fig. 4.22b.

EXAMPLE 4.11 If the three waveforms A, B, and C shown in Fig. 4.23a are applied to a three-input AND gate, determine the resulting output waveform.

Solution

The output waveform X is shown in Fig. 4.23b.

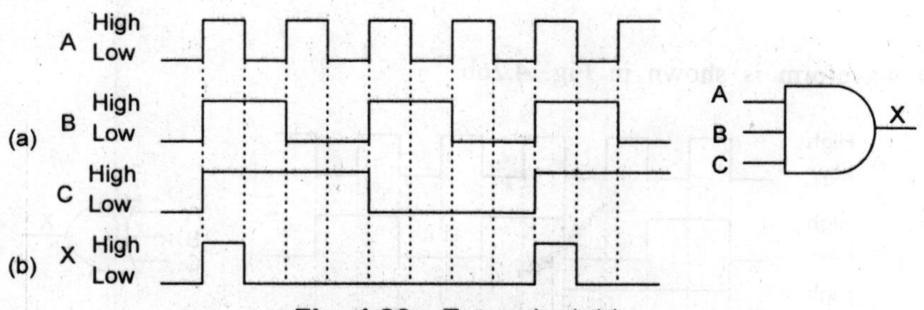

Fig. 4.23 Example 4.11.

EXAMPLE 4.12 Obtain the output waveform of a two-input OR gate in relation to its inputs A and B shown in Fig. 4.24a.

Solution

In the interval t_1, A is HIGH and B is LOW, so X is HIGH.
In the interval t_2, A is LOW and B is HIGH, so X is HIGH.
In the interval t_3, A is HIGH and B is HIGH, so X is HIGH.
In the interval t_4, A is HIGH and B is LOW, so X is HIGH.
In the interval t_5, A is LOW and B is LOW, so X is LOW.
The output waveform is shown in Fig. 4.24b.

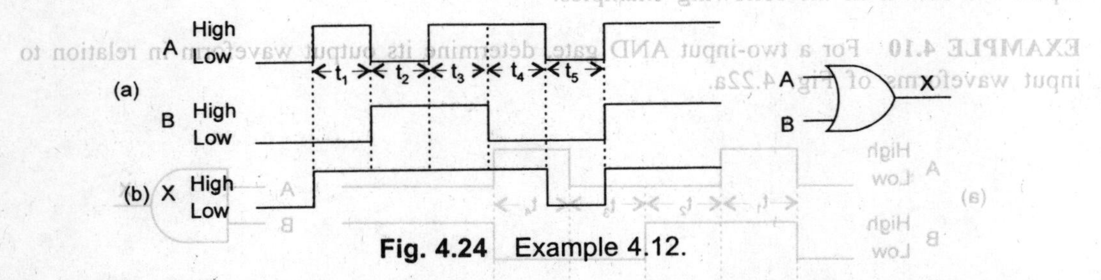

Fig. 4.24 Example 4.12.

EXAMPLE 4.13 For a two-input OR gate, determine its output waveform in relation to the inputs A and B shown in Fig. 4.25a.

Solution

The output waveform is shown in Fig. 4.25b.

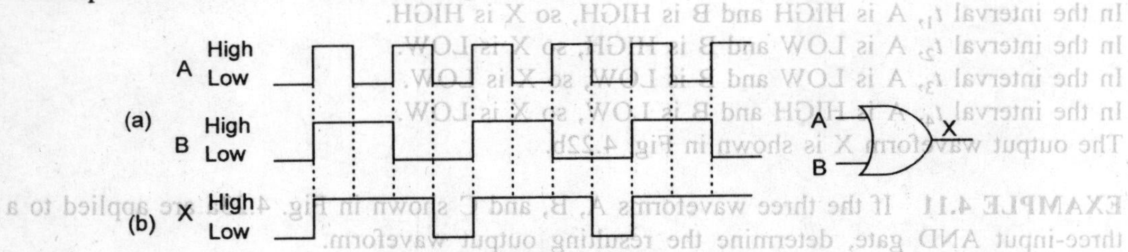

Fig. 4.25 Example 4.13.

EXAMPLE 4.14 For a three-input OR gate, determine its output waveform in proper relation to the inputs A, B and C shown in Fig. 4.26a.

Solution

The output waveform is shown in Fig. 4.26b.

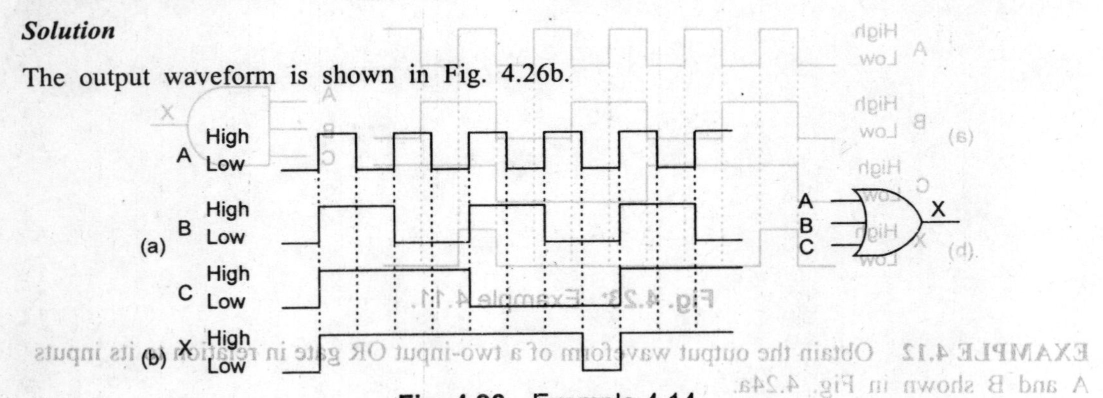

Fig. 4.26 Example 4.14.

EXAMPLE 4.15 If the waveform shown in Fig. 4.27a is applied to an inverter, what is the resulting output waveform?

Solution

The output of an inverter is the complement of its input.
In the interval t_1, A is HIGH, so X is LOW.
In the interval t_2, A is LOW, so X is HIGH.
In the interval t_3, A is HIGH, so X is LOW.
In the interval t_4, A is LOW, so X is HIGH.
The output waveform is shown in Fig. 4.27b.

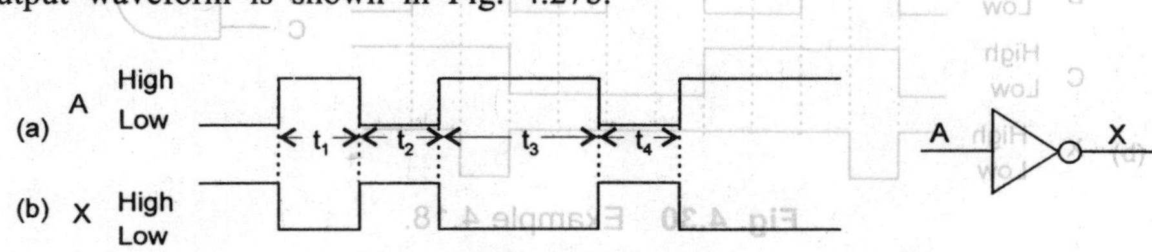

Fig. 4.27 Example 4.15.

EXAMPLE 4.16 If the waveform shown in Fig. 4.28a is applied to an inverter, determine the resulting output waveform.

Solution

The output waveform is shown in Fig. 4.28b.

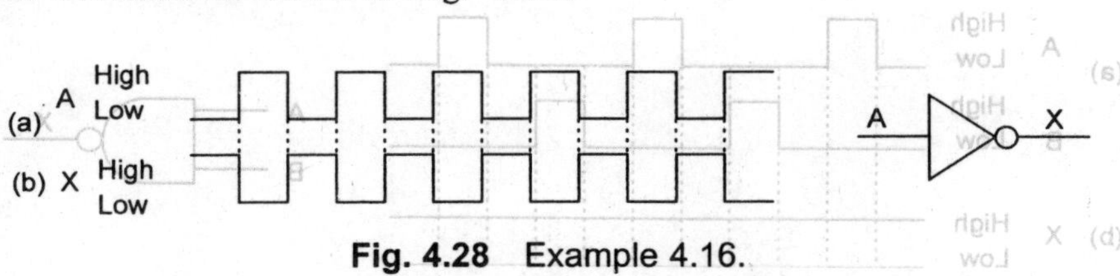

Fig. 4.28 Example 4.16.

EXAMPLE 4.17 The waveforms A and B shown in Fig. 4.29a are applied to a two-input NAND gate. Determine the output waveform.

Solution

The output of a NAND gate is LOW only when all its inputs are HIGH. The output waveform is shown in Fig. 4.29b.

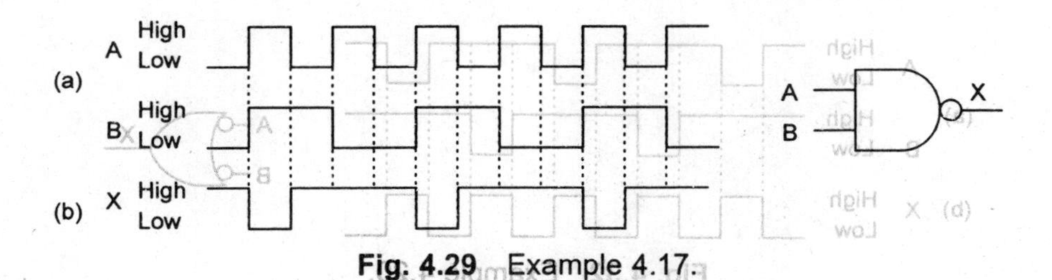

Fig. 4.29 Example 4.17.

EXAMPLE 4.18 The waveforms A, B and C shown in Fig. 4.30a are applied to a three-input NAND gate. Determine the output waveform.

Solution

The output waveform is shown in Fig. 4.30b.

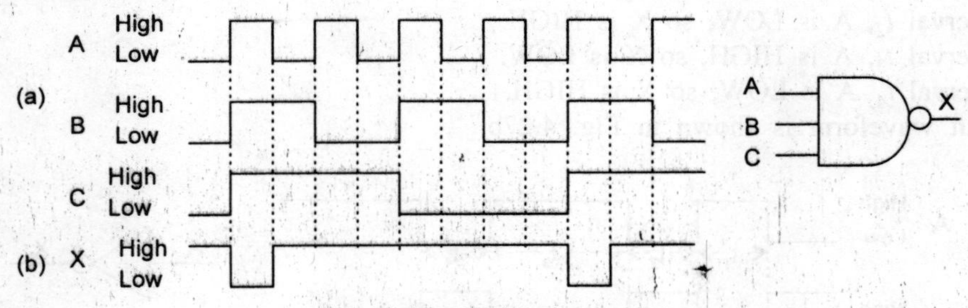

Fig. 4.30 Example 4.18.

EXAMPLE 4.19 The waveforms A and B shown in Fig. 4.31a are applied to a two-input NAND gate. Determine the output waveform.

Solution

The output waveform is shown in Fig. 4.31b. In this case, the two inputs of the NAND gate are never HIGH simultaneously. So, the output is never LOW. It is always HIGH.

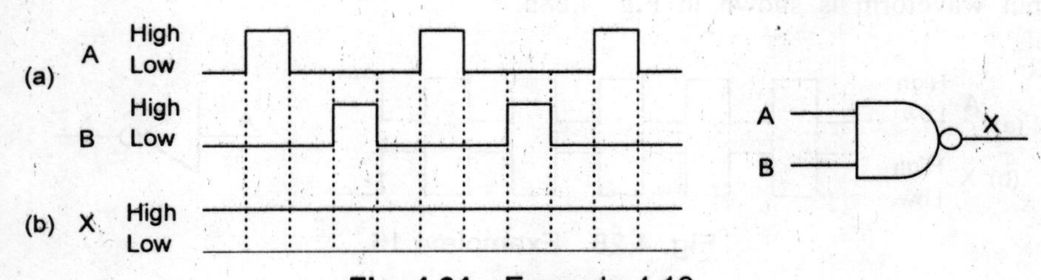

Fig. 4.31 Example 4.19.

EXAMPLE 4.20 For the two-input NAND gate operating as a negative OR gate, determine the output waveform when the input waveforms A and B are as shown in Fig. 4.32a.

Solution

The output of an active-LOW OR gate is HIGH, if either A is LOW or B is LOW or both A and B are LOW. The output waveform is shown in Fig. 4.32b.

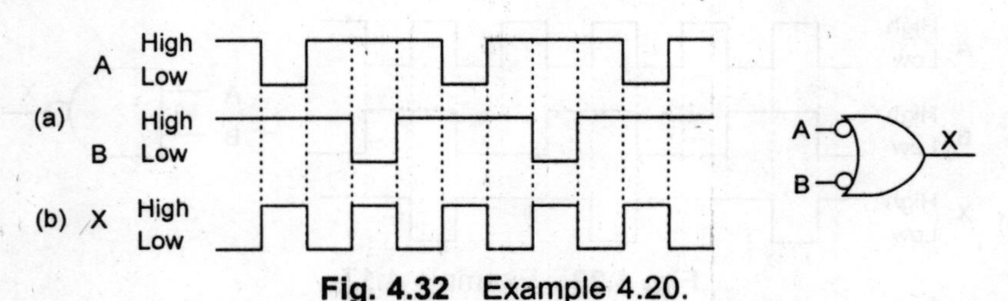

Fig. 4.32 Example 4.20.

EXAMPLE 4.21 For the three-input NAND gate operating as a negative OR gate, determine the output waveform when the input waveforms A, B and C are as shown in Fig. 4.33a.

Solution

The output of an active-low OR gate is HIGH, even if one of its inputs is LOW. The output waveform is shown in Fig. 4.33b. In this case, at no time, all the 3 inputs are HIGH simultaneously. So, the output is never LOW, it is always HIGH.

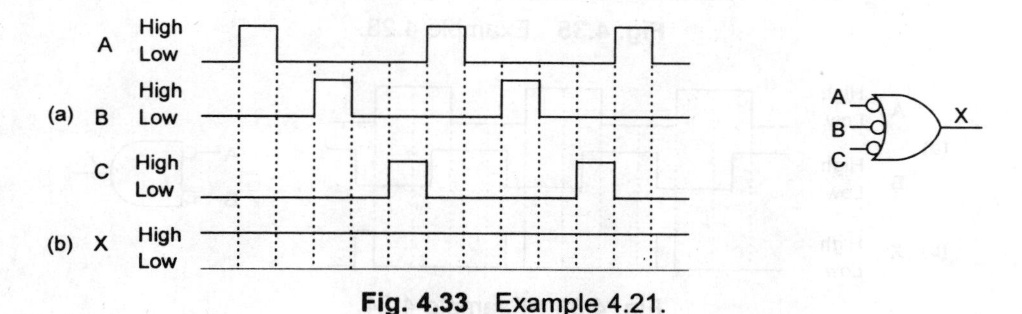

Fig. 4.33 Example 4.21.

EXAMPLE 4.22 If the waveforms A and B shown in Fig. 4.34a are applied to a two-input NOR gate, determine the resulting output waveform.

Solution

The output of a NOR gate is HIGH only when all its inputs are LOW. The output waveform is shown in Fig. 4.34b.

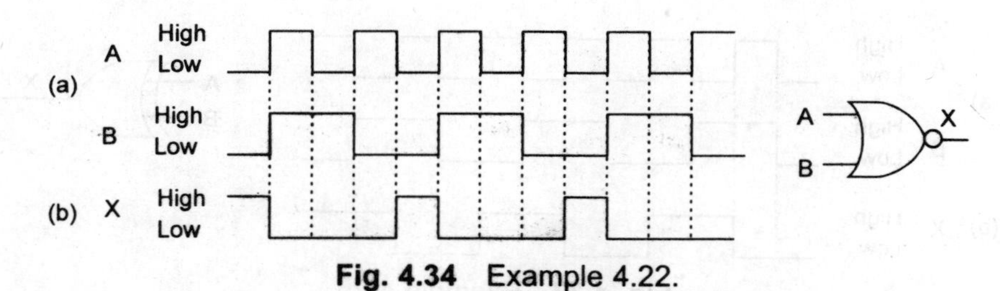

Fig. 4.34 Example 4.22.

EXAMPLE 4.23 If the waveforms A, B and C shown in Fig. 4.35a are applied to a three-input NOR gate, determine the resulting output waveform.

Solution

The output waveform is shown in Fig. 4.35b.

EXAMPLE 4.24 The waveforms A and B shown in Fig. 4.36a are applied to an active-LOW AND gate. What is the output waveform?

Solution

The NOR gate is an active-LOW AND gate. The output of an active-LOW AND gate is HIGH, only when all its inputs are LOW. The output waveform is shown in Fig. 4.36b.

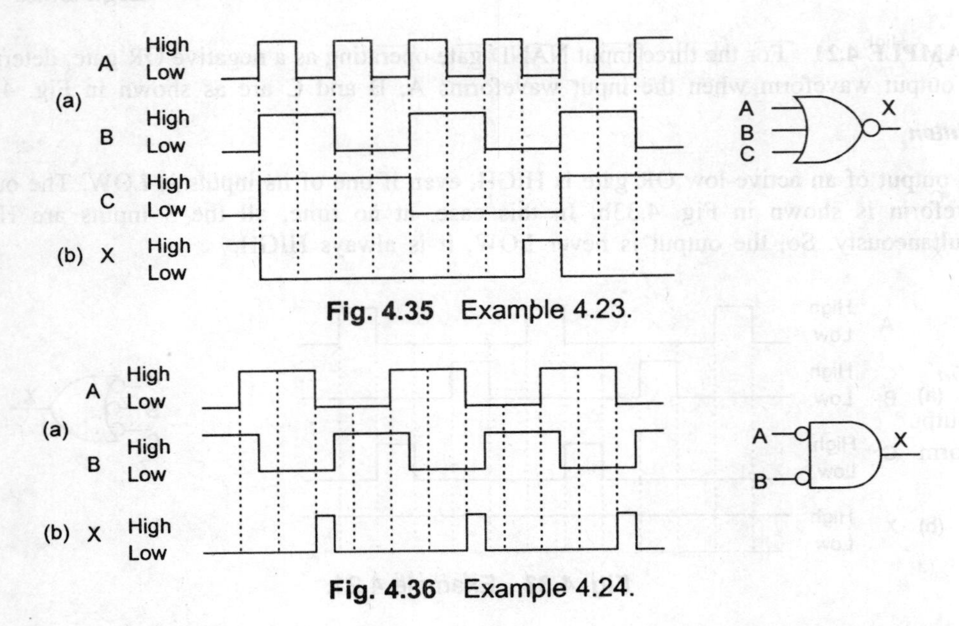

Fig. 4.35 Example 4.23.

Fig. 4.36 Example 4.24.

EXAMPLE 4.25 If the waveforms A and B shown in Fig. 4.37a are applied to a two-input X-OR gate, determine the output waveform.

Solution

The output of an X-OR gate is HIGH, only when the inputs are not equal. The output waveform is shown in Fig. 4.37b.

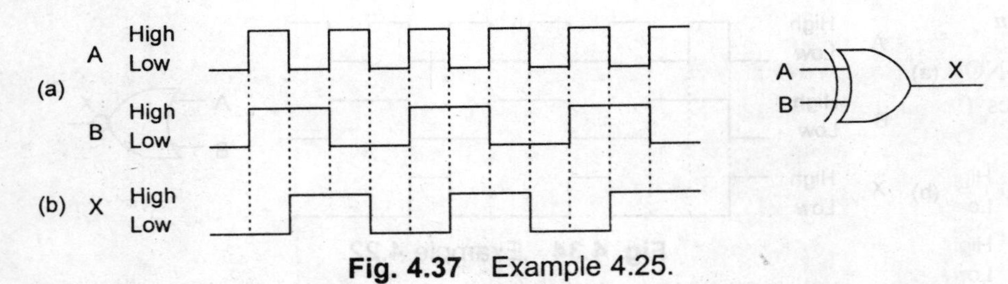

Fig. 4.37 Example 4.25.

EXAMPLE 4.26 If the three waveforms A, B and C, shown in Fig. 4.38a, are to be X-ORed, determine the output waveform.

Solution

The X-OR output of a number of variables is HIGH, only when an odd number of input variables are HIGH. The output waveform is shown in Fig. 4.38b.

EXAMPLE 4.27 If the waveforms A and B shown in Fig. 4.39a are applied to a two-input X-NOR gate, determine the output waveform.

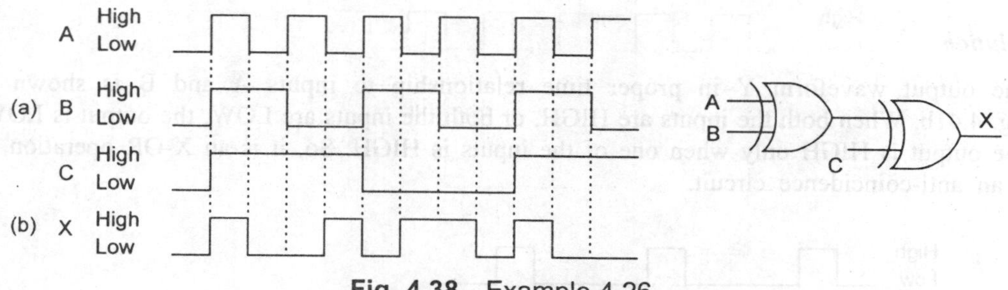

Fig. 4.38 Example 4.26.

Solution

The output of an X-NOR gate is HIGH, only when the inputs are equal. The output waveform is shown in Fig. 4.39b.

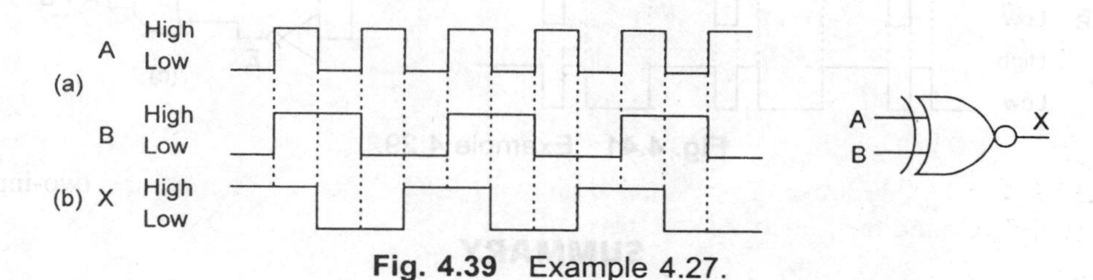

Fig. 4.39 Example 4.27.

EXAMPLE 4.28 If the waveforms A, B, and C shown in Fig. 4.40a are X-NORed, determine the output waveform.

Solution

The X-NOR output of a number of variables is HIGH, only when an even number of input variables (including 0) are LOW. The output waveform is shown in Fig. 4.40b.

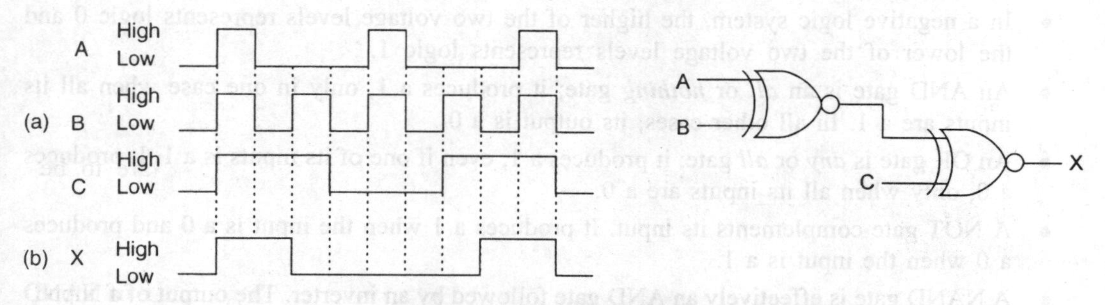

Fig. 4.40 Example 4.28.

EXAMPLE 4.29 Determine the output waveform for the circuit shown in Fig. 4.41c, when the inputs A and B shown in Fig. 4.41a are applied to it.

Solution

The output waveform Y in proper time relationship to inputs A and B is shown in Fig. 4.41b. When both the inputs are HIGH, or both the inputs are LOW, the output is LOW. The output is HIGH only when one of the inputs is HIGH. So, it is an X-OR operation. It is an anti-coincidence circuit.

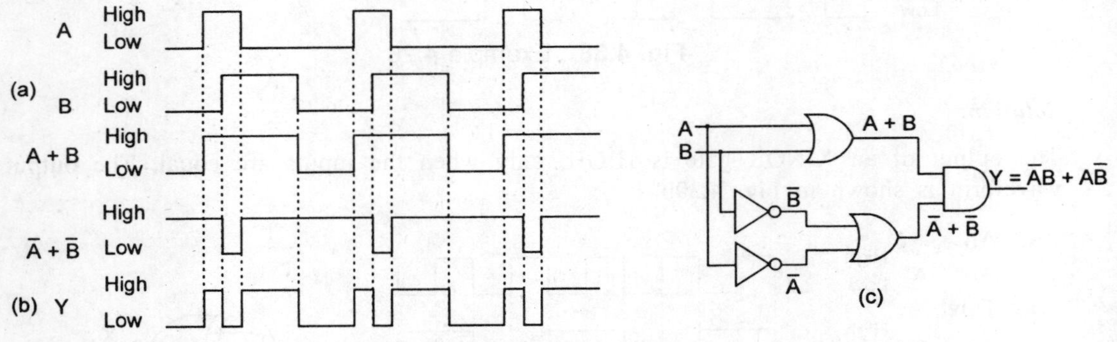

Fig. 4.41 Example 4.29.

SUMMARY

- Logic gates are the fundamental building blocks of digital systems.
- AND, OR and NOT gates are the basic types of gates. The interconnection of gates to perform a variety of logical operations is called logic design.
- A truth table lists all possible combinations of inputs and the corresponding outputs.
- In a positive logic system, the higher of the two voltage levels represents logic 1 and the lower of the two voltage levels represents logic 0.
- In a negative logic system, the higher of the two voltage levels represents logic 0 and the lower of the two voltage levels represents logic 1.
- An AND gate is an *all* or *nothing* gate; it produces a 1, only in one case when all its inputs are a 1. In all other cases, its output is a 0.
- An OR gate is *any* or *all* gate; it produces a 1, even if one of its inputs is a 1. It produces a 0, only when all its inputs are a 0.
- A NOT gate complements its input. It produces a 1 when the input is a 0 and produces a 0 when the input is a 1.
- A NAND gate is effectively an AND gate followed by an inverter. The output of a NAND gate is a 0, only when each one of its inputs is a 1. It produces a 1 when any of the inputs is a 0. It therefore also acts as a negative OR gate.
- A NAND gate can be used as an inverter by tying all its input terminals together and feeding the signal to be inverted to the common terminal, or by tying all but one input terminal to logic 1 and feeding the signal to be inverted to the remaining terminal.

- A NOR gate is effectively an OR gate followed by an inverter. It produces a 1 when all of its inputs are a 0. It produces a 0 when any of its inputs is a 1. It therefore also acts as a negative AND gate.
- A NOR gate can be used as an inverter by tying all its input terminals together and feeding the signal to be inverted to the common terminal, or by tying all but one input terminal to logic 0 and feeding the signal to be inverted to the remaining terminal.
- NAND and NOR gates are called universal gates. Any circuit of any complexity can be realized by using only NAND gates or only NOR gates.
- AND, OR, NAND, and NOR gates can have any number of inputs.
- An X-OR gate is an anti-coincidence gate. It is an inequality detector. It produces a 1, only when its two inputs are not equal, i.e. when one is a 0 and the other is a 1.
- An X-OR gate can be used as an inverter by connecting one input terminal to logic 1 and feeding the signal to be inverted to the other terminal. It is a controlled inverter.
- An X-NOR gate is a coincidence gate. It is an equality detector. It produces a 1, only when its two inputs are equal, i.e. when both inputs are a 0 or a 1.
- Three or more input X-OR or X-NOR gates do not exist.
- An X-NOR gate can be used as an inverter by connecting one input terminal to logic 0 and feeding the signal to be inverted to the other terminal. It is a controlled inverter.
- When three or more variables are to be X-ORed or X-NORed, a number of two-input X-OR or X-NOR gates may be used.
- All logic gates have only one output.
- AND, OR, NAND, and NOR gates can be used either to enable or inhibit the passage of an input signal.
- When AND and OR gates are enabled, the output follows the input exactly.
- When NAND and NOR gates are enabled, the output is the exact inverse of the input signal.
- AND and NOR gates produce a constant LOW output when they are in the inhibited condition.
- NAND and OR gates produce a constant HIGH output when they are in the inhibited condition.
- All logic gates obey their truth table operations regardless of whether their inputs are constant levels or pulsed levels.

QUESTIONS

1. What do you mean by a logic gate? Name the basic gates.
2. What do you mean by logic design?
3. What do you mean by discrete logic gates?
4. What is a truth table?

5. What do you mean by (a) level logic, (b) positive logic system, and (c) negative logic system?
6. What are the logic levels used in TTL logic system?
7. Define the function performed by the following gates. (a) AND (b) OR (c) NOT.
8. What is the only input combination that will produce a HIGH at the output of a six-input AND gate?
9. If one of the inputs to an AND gate is permanently kept LOW, what would be the shape of the output waveform when the remaining inputs are applied?
10. What is the only input combination that will produce a LOW at the output of a five-input OR gate?
11. If one of the inputs to an OR gate is permanently kept HIGH, what would be the shape of the output waveform when the remaining inputs are applied?
12. An AND gate output always differs from an OR gate output. True or false?
13. Which logic gate is called (a) any or all gate (b) all or nothing gate (c) inverter?
14. How is logical addition different from ordinary addition?
15. How is logical multiplication different from ordinary multiplication?
16. What is the maximum number of inputs a NOT gate can have?
17. Name the universal gates. Define the functions performed by them.
18. What do you mean by (a) bubbled AND gate (b) bubbled OR gate?
19. What is the only set of input conditions that will produce a HIGH output from a three-input NOR gate?
20. What is the only set of input conditions that will produce a LOW output from a three-input NAND gate?
21. What type of gate is equivalent to a NAND gate followed by an inverter?
22. What type of gate is equivalent to a NOR gate followed by an inverter?
23. Which logic gate is called (a) coincidence gate (b) anti-coincidence gate?
24. Which gates can be used as inverters in addition to the NOT gate and how?
25. Give the IC numbers of the following TTL gates: AND, OR, NOT, X-OR, X-NOR, NOR and NAND.
26. What is the maximum number of outputs of any logic gate?
27. Draw logic diagrams to realize the following expressions: (a) $A \oplus B \oplus C \oplus D$ (b) $A \odot B \odot C$.
28. What is the equivalent of an active-LOW input AND gate?
29. What is the equivalent of an active-LOW input OR gate?
30. What do you mean by an active-LOW input gate?
31. What do you mean by pulsed operation of a logic gate?

PROBLEMS

4.1 Draw the logic symbols and construct the truth tables of the following gates:
(a) AND (b) OR (c) NAND (d) NOR

4.2 Show an arrangement to X-OR and X-NOR the inputs A, B, C and D.

4.3 Draw the logic diagram and construct the truth table for each of the following expressions:

(a) $X = A + B + \overline{CD}$ (b) $Y = (AB)\,\overline{(A + B)} + \overline{EF}$ (c) $Z = \overline{\overline{A}B + C\overline{D} + ABC}$

4.4 Draw the logic diagram using the ANSI/IEEE symbols for each of the following expressions:

(a) $X = \overline{A} + BC + \overline{CD}$ (b) $Y = \overline{AB}\,(C + D) + EF$ (c) $Z = \overline{A\overline{B} + \overline{C}D + BCD}$

4.5 Draw a logic diagram that implements:
(a) $A = (Y_1 \oplus Y_2)\,(Y_3 \odot Y_4) + (Y_5 \oplus Y_6 \oplus Y_7)$
(b) $A = (X_1 \odot X_2) \oplus (X_3 \odot X_4) + (X_4 \oplus X_5) \odot (X_6 \oplus X_7)$

4.6. Two square waves, A of 1 kHz and B of 2 kHz frequency, are applied as inputs to the following logic gates. Draw the output waveform in each case.
(a) AND (b) OR (c) NAND
(d) NOR (e) X-OR (f) X-NOR

4.7 Three square waves A, B and C of frequency 1, 2 and 4 kHz, respectively, are to be
(a) ANDed (b) ORed (c) NANDed
(d) NORed (e) X-ORed (f) X-NORed
Draw the resultant waveform in each case.

Chapter 5

BOOLEAN ALGEBRA

5.1 INTRODUCTION

Boolean algebra is a system of mathematical logic. Any complex logic statement can be expressed by a Boolean function. The Boolean algebra is governed by certain well-developed rules and laws. In the applications of Boolean algebra in this book, we use capital letters to represent the variables. Any single variable, or a function of the variables can have a value of either a 0 or a 1. The binary digits 0 and 1 are used to represent the two voltage levels that occur within the digital logic circuit. In this book, we follow positive logic. Hence binary 1 represents the higher of the two voltage levels (+ 5 V), and binary 0 represents the lower of the two voltage levels (0 V). Ideally, no other voltages ever occur at the inputs or outputs. In actual practice, however, any voltage above some level (2 V, for example) is treated as logic 1 (TRUE, ON, HIGH) and any voltage below some level (0.8 V, for example) is treated as logic 0 (FALSE, OFF, LOW).

Boolean algebra differs from both the ordinary algebra and the binary number system. In Boolean algebra, $A + A = A$ and $A \cdot A = A$, because the variable A has only a logical value. It doesn't have any numerical significance. In ordinary algebra, $A + A = 2A$ and $A \cdot A = A^2$, because the variable A has a numerical value here. In Boolean algebra, $1 + 1 = 1$, whereas in the binary number system, $1 + 1 = 10$, and in ordinary algebra, $1 + 1 = 2$. There is nothing like subtraction or division in Boolean algebra. Also, there are no negative or fractional numbers in Boolean algebra. In Boolean algebra, the multiplication and addition of the variables and functions are also only logical. They actually represent logic operations. Logical multiplication is the same as the AND operation, and logical addition is the same as the OR operation. There are only two constants 0 and 1 within the Boolean system, whereas in ordinary algebra, you can have any number of constants. A variable or function of variables in Boolean algebra can assume only two values, either a 0 or a 1, whereas the variables or functions in ordinary algebra can assume an infinite number of values.

Thus, in Boolean algebra

$$\text{If } A = 1, \quad \text{then } A \neq 0$$
$$\text{If } A = 0, \quad \text{then } A \neq 1.$$

Any functional relation in Boolean algebra can be proved by the method of *perfect induction*. Perfect induction is a method of proof, whereby a functional relation is verified for every possible combination of values that the variables may assume. This can be done by forming a truth table. A truth table shows how a logic circuit responds to various combinations of logic levels at its inputs.

5.2 LOGIC OPERATIONS

The AND, OR and NOT are the three basic operations or functions that are performed in Boolean algebra. In addition, there are some derived operations such as NAND, NOR, X-OR and X-NOR that are also performed in Boolean algebra. These operations have been described in detail in Chapter 4 in the context of logic gates.

AND Operation

The AND operation in Boolean algebra is similar to multiplication in ordinary algebra. In fact, it is logical multiplication as performed by the AND gate.

OR Operation

The OR operation in Boolean algebra is similar to addition in ordinary algebra. In fact, it is logical addition as performed by the OR gate.

NOT Operation

The NOT operation in Boolean algebra is nothing but complementation or inversion, that is, negation as performed by the NOT gate. The NOT operation is indicated by a bar '–' over the variable.

NAND Operation

The NAND operation in Boolean algebra is equivalent to AND operation plus NOT operation, i.e. it is the negation of the AND operation as performed by the NAND gate.

NOR Operation

The NOR operation in Boolean algebra is equivalent to OR operation plus NOT operation, i.e. it is the negation of the OR operation as performed by the NOR gate.

X-OR and X-NOR Operations

The X-OR and X-NOR operations on variables A and B in Boolean algebra are denoted by $A \oplus B$ (= $A\overline{B} + \overline{A}B$) and $A \odot B$ (= $AB + \overline{A}\,\overline{B}$), respectively. These operations have been described in detail in Chapter 3.

The X-OR operation is also called the modulo-2 addition since it assigns to each pair of elements its modulo-2 sum.

5.3 AXIOMS AND LAWS OF BOOLEAN ALGEBRA

Axioms or *postulates* of Boolean algebra are a set of logical expressions that we accept without proof and upon which we can build a set of useful theorems. Actually, axioms are nothing more than the definitions of the three basic logic operations that we have already discussed: AND, OR, and INVERT. Each axiom can be interpreted as the outcome of an operation performed by a logic gate.

Axiom 1 :	$0 \cdot 0 = 0$	Axiom 6 :	$0 + 1 = 1$
Axiom 2 :	$0 \cdot 1 = 0$	Axiom 7 :	$1 + 0 = 1$
Axiom 3 :	$1 \cdot 0 = 0$	Axiom 8 :	$1 + 1 = 1$
Axiom 4 :	$1 \cdot 1 = 1$	Axiom 9 :	$\overline{1} = 0$
Axiom 5 :	$0 + 0 = 0$	Axiom 10 :	$\overline{0} = 1$

Complementation Laws

The term *complement* simply means to invert, i.e. to change 0s to 1s and 1s to 0s. The five laws of complementation are as follows:

Law 1 : $\overline{0} = 1$

Law 2 : $\overline{1} = 0$

Law 3 : If $A = 0$, then $\overline{A} = 1$

Law 4 : If $A = 1$, then $\overline{A} = 0$

Law 5 : $\overline{\overline{A}} = A$

AND Laws

The four AND laws are as follows:

Law 1 : $A \cdot 0 = 0$

Law 2 : $A \cdot 1 = A$

Law 3 : $A \cdot A = A$

Law 4 : $A \cdot \overline{A} = 0$

OR Laws

The four OR laws are as follows:

Law 1 : $A + 0 = A$

Law 2 : $A + 1 = 1$

Law 3 : $A + A = A$

Law 4 : $A + \overline{A} = 1$

Commutative Laws

Commutative laws allow change in position of AND or OR variables. There are two commutative laws.

Law 1 : $A + B = B + A$

This law states that, A OR B is thé same as B OR A, i.e. the order in which the variables are ORed is immaterial. This means that it makes no difference which input of an OR gate is connected to A and which to B. We give below the truth tables illustrating this law.

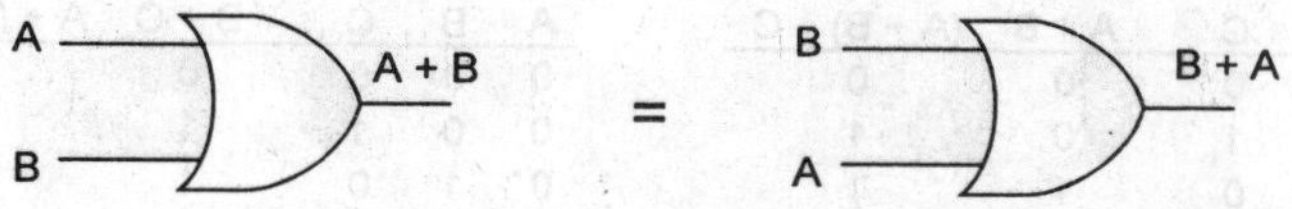

A, B → A + B = B, A → B + A

A	B	A + B
0	0	0
0	1	1
1	0	1
1	1	1

=

B	A	B + A
0	0	0
0	1	1
1	0	1
1	1	1

This law can be extended to any number of variables. For example,

$$A + B + C = B + C + A = C + A + B = B + A + C$$

$$\text{Law 2 :} \qquad A \cdot B = B \cdot A$$

This law states that, A AND B is the same as B AND A, i.e. the order in which the variables are ANDed is immaterial. This means that it makes no difference which input of an AND gate is connected to A and which to B. The truth tables given below illustrate this law.

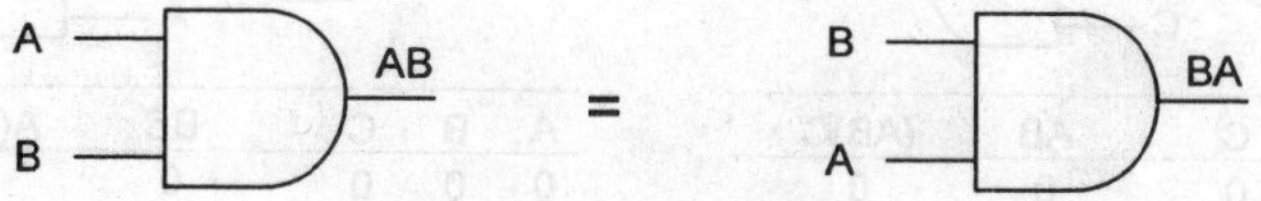

A, B → AB = B, A → BA

A	B	A · B
0	0	0
0	1	0
1	0	0
1	1	1

=

B	A	B · A
0	0	0
0	1	0
1	0	0
1	1	1

This law can be extended to any number of variables. For example,

$$A \cdot B \cdot C = B \cdot C \cdot A = C \cdot A \cdot B = B \cdot A \cdot C$$

Associative Laws

The associative laws allow grouping of variables. There are two associative laws.

$$\text{Law 1 :} \qquad (A + B) + C = A + (B + C)$$

A OR B ORed with C is the same as A ORed with B OR C. This law states that the way the variables are grouped and ORed is immaterial. The truth tables given next illustrate this law.

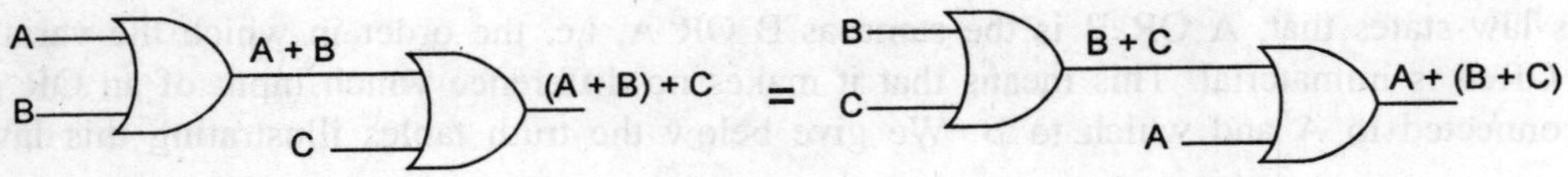

A	B	C	A + B	(A + B) + C
0	0	0	0	0
0	0	1	0	1
0	1	0	1	1
0	1	1	1	1
1	0	0	1	1
1	0	1	1	1
1	1	0	1	1
1	1	1	1	1

=

A	B	C	B + C	A + (B + C)
0	0	0	0	0
0	0	1	1	1
0	1	0	1	1
0	1	1	1	1
1	0	0	0	1
1	0	1	1	1
1	1	0	1	1
1	1	1	1	1

This law can be extended to any number of variables. For example,

$$A + (B + C + D) = (A + B + C) + D = (A + B) + (C + D).$$

Law 2 : $(A \cdot B)C = A(B \cdot C)$

A AND B ANDed with C is thé same as A ANDed with B AND C. This law states that the way the variables are grouped and ANDed is immaterial. See the truth tables below:

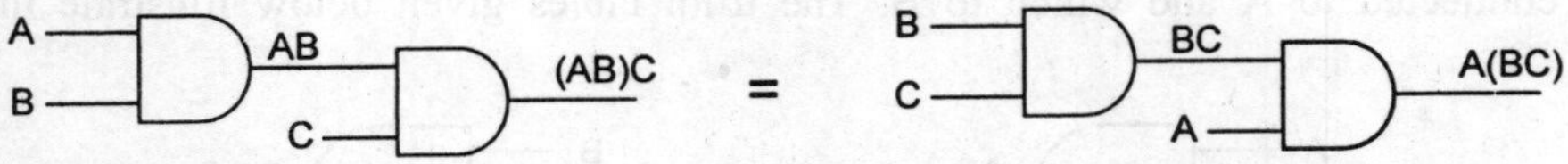

A	B	C	AB	(AB)C
0	0	0	0	0
0	0	1	0	0
0	1	0	0	0
0	1	1	0	0
1	0	0	0	0
1	0	1	0	0
1	1	0	1	0
1	1	1	1	1

=

A	B	C	BC	A(BC)
0	0	0	0	0
0	0	1	0	0
0	1	0	0	0
0	1	1	1	0
1	0	0	0	0
1	0	1	0	0
1	1	0	0	0
1	1	1	1	1

This law can be extended to any number of variables. For example,

$$A(BCD) = (ABC)D = (AB)\ (CD)$$

Distributive Laws

The distributive laws allow factoring or multiplying out of expressions. There are three distributive laws.

Law 1 : $A(B + C) = AB + AC$

This law states that ORing of several variables and ANDing the result with a single variable is equivalent to ANDing that single variable with each of the several variables and then ORing the products. The truth tables given below illustrate this law.

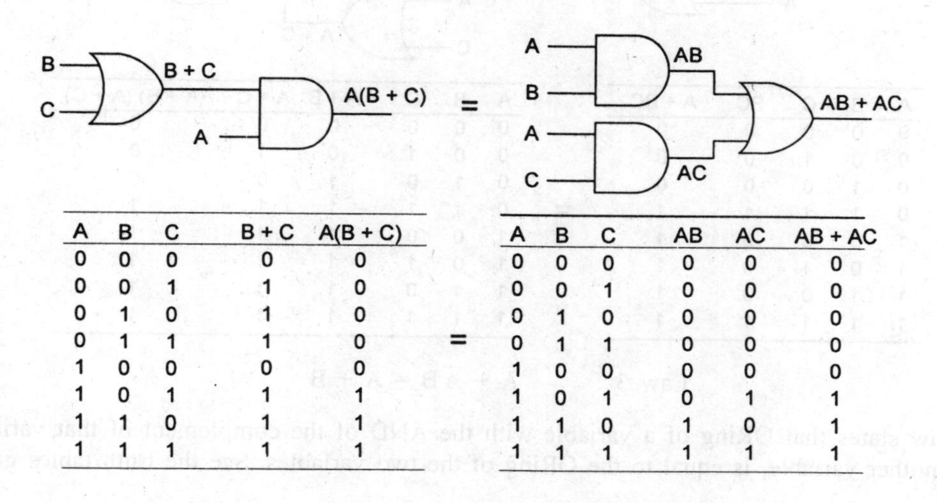

A	B	C	B + C	A(B + C)
0	0	0	0	0
0	0	1	1	0
0	1	0	1	0
0	1	1	1	0
1	0	0	0	0
1	0	1	1	1
1	1	0	1	1
1	1	1	1	1

=

A	B	C	AB	AC	AB + AC
0	0	0	0	0	0
0	0	1	0	0	0
0	1	0	0	0	0
0	1	1	0	0	0
1	0	0	0	0	0
1	0	1	0	1	1
1	1	0	1	0	1
1	1	1	1	1	1

This law applies to single variables as well as combinations of variables. For example,

$$ABC(D + E) = ABCD + ABCE$$

$$AB(CD + EF) = ABCD + ABEF$$

The distributive property is often used in the reverse. That is, given $AB + AC$, we replace it by $A(B + C)$; and $ABC + ABD$ by $AB(C + D)$.

$$\text{Law 2 :} \quad A + BC = (A + B)\,(A + C)$$

This law states that ANDing of several variables and ORing the result with a single variable is equivalent to ORing that single variable with each of the several variables and then ANDing the sums. This can be proved algebraically as shown below. Also, the truth tables given next illustrate this law.

$$\begin{aligned}
\text{RHS} &= (A + B)\,(A + C) \\
&= AA + AC + BA + BC \\
&= A + AC + AB + BC \\
&= A(1 + C + B) + BC \\
&= A \cdot 1 + BC \qquad (\because 1 + C + B = 1 + B = 1) \\
&= A + BC \\
&= \text{LHS}
\end{aligned}$$

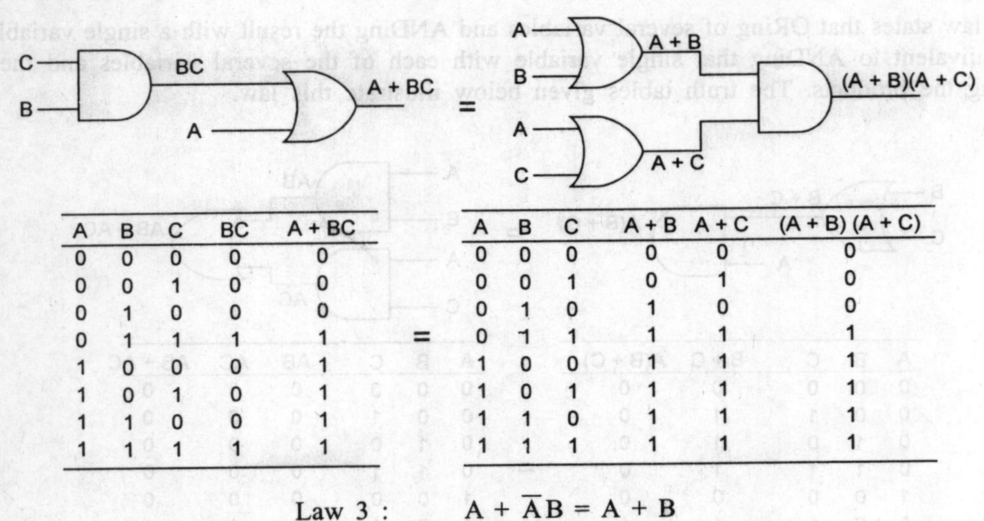

A	B	C	BC	A + BC
0	0	0	0	0
0	0	1	0	0
0	1	0	0	0
0	1	1	1	1
1	0	0	0	1
1	0	1	0	1
1	1	0	0	1
1	1	1	1	1

=

A	B	C	A + B	A + C	(A + B) (A + C)
0	0	0	0	0	0
0	0	1	0	1	0
0	1	0	1	0	0
0	1	1	1	1	1
1	0	0	1	1	1
1	0	1	1	1	1
1	1	0	1	1	1
1	1	1	1	1	1

Law 3 : $A + \bar{A}B = A + B$

This law states that ORing of a variable with the AND of the complement of that variable with another variable, is equal to the ORing of the two variables. See the truth tables given below.

A	B	$\bar{A}B$	$A + \bar{A}B$
0	0	0	0
0	1	1	1
1	0	0	1
1	1	0	1

=

A	B	A + B
0	0	0
0	1	1
1	0	1
1	1	1

This law can also be proved algebraically as shown below.

$$A + \bar{A}B = (A + \bar{A})(A + B)$$
$$= 1 \cdot (A + B)$$
$$= A + B$$

Idempotence Laws

Law 1 : $A \cdot A = A$

Idempotence means the same value. We are already familiar with the following laws:

If $A = 0$, then $A \cdot A = 0 \cdot 0 = 0 = A$

If $A = 1$, then $A \cdot A = 1 \cdot 1 = 1 = A$

This law states that ANDing of a variable with itself is equal to that variable only.

$$\text{Law 2 :} \qquad A + A = A$$

$$\text{If } A = 0, \quad \text{then} \quad A + A = 0 + 0 = 0 = A$$

$$\text{If } A = 1, \quad \text{then} \quad A + A = 1 + 1 = 1 = A$$

This law states that ORing of a variable with itself is equal to that variable only.

Complementation Laws or Negation Laws

There are two complementation or negation laws.

$$\text{Law 1 :} \qquad A \,.\, \overline{A} = 0$$

This law states that a variable ANDed with its complement is always equal to zero. That is,

$$\text{If } A = 0, \quad \text{then} \quad \overline{A} = 1 \quad \text{and} \quad A \cdot \overline{A} = 0 \cdot 1 = 0$$

$$\text{If } A = 1, \quad \text{then} \quad \overline{A} = 0 \quad \text{and} \quad A \cdot \overline{A} = 1 \cdot 0 = 0$$

$$\text{Law 2 :} \qquad A + \overline{A} = 1$$

This law states that a variable ORed with its complement is always equal to 1. That is,

$$\text{If } A = 0, \quad \text{then} \quad \overline{A} = 1 \quad \text{and} \quad A + \overline{A} = 0 + 1 = 1$$

$$\text{If } A = 1, \quad \text{then} \quad \overline{A} = 0 \quad \text{and} \quad A + \overline{A} = 1 + 0 = 1$$

Double Negation Law

This law states that double negation of a variable is equal to the variable itself. That is,

$$\overline{\overline{A}} = A$$

$$\text{If } A = 0, \quad \text{then} \quad \overline{\overline{A}} = \overline{\overline{0}} = \overline{1} = 0 = A$$

$$\text{If } A = 1, \quad \text{then} \quad \overline{\overline{A}} = \overline{\overline{1}} = \overline{0} = 1 = A$$

Any odd number of inversions is equivalent to a single inversion and any even number of inversions is equivalent to no inversion at all.

Identity Laws

There are two identity laws.

$$\text{Law 1 :} \qquad A \cdot 1 = A$$

This law states that a variable ANDed with 1 is equal to the variable itself. That is,

$$\text{if } A = 0, \quad \text{then} \quad A \cdot 1 = 0 \cdot 1 = 0 = A$$

$$\text{if } A = 1, \quad \text{then} \quad A \cdot 1 = 1 \cdot 1 = 1 = A$$

$$\text{Law 2 :} \qquad A + 1 = 1$$

This law states that a variable ORed with 1 is always equal to 1. That is,

$$\text{if } A = 0, \quad \text{then} \quad A + 1 = 0 + 1 = 1$$
$$\text{if } A = 1, \quad \text{then} \quad A + 1 = 1 + 1 = 1$$

Null Laws

There are two null laws.

$$\text{Law 1 :} \qquad A \cdot 0 = 0$$

This law states that a variable ANDed with 0 is always equal to 0. That is,

$$\text{if } A = 0, \quad \text{then} \quad A \cdot 0 = 0 \cdot 0 = 0$$
$$\text{if } A = 1, \quad \text{then} \quad A \cdot 0 = 1 \cdot 0 = 0$$

$$\text{Law 2 :} \qquad A + 0 = A$$

This law states that a variable ORed with 0 is equal to the variable itself. That is,

$$\text{if } A = 0, \quad \text{then} \quad A + 0 = 0 + 0 = 0 = A$$
$$\text{if } A = 1, \quad \text{then} \quad A + 0 = 1 + 0 = 1 = A$$

Absorption Laws

There are two laws:

$$\text{Law 1 :} \qquad A + A \cdot B = A$$

This law states that ORing of a variable (A) with the AND of that variable (A) and another variable (B) is equal to that variable itself (A).

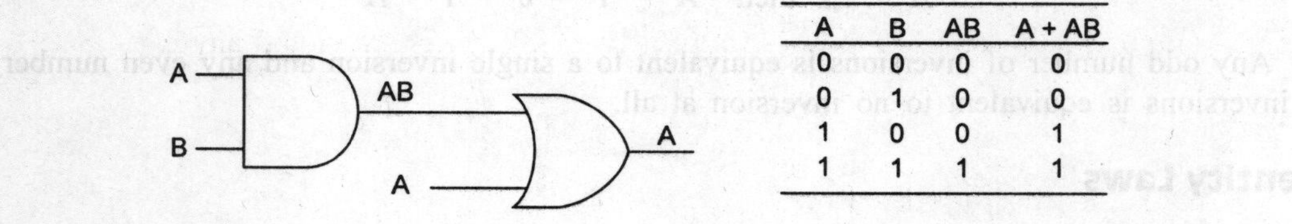

A	B	AB	A + AB
0	0	0	0
0	1	0	0
1	0	0	1
1	1	1	1

Algebraically, we have

$$A + A \cdot B = A(1 + B) = A \cdot 1 = A$$

Therefore,

$$A + A \cdot \text{Any term} = A$$

Law 2 : $\quad A(A + B) = A$

This law states that ANDing of a variable (A) with the OR of that variable (A) and another variable (B) is equal to that variable itself (A).

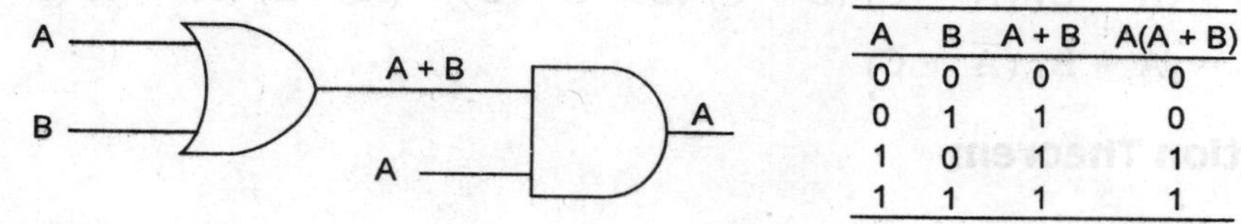

A	B	A + B	A(A + B)
0	0	0	0
0	1	1	0
1	0	1	1
1	1	1	1

Algebraically, we have

$$A(A + B) = A \cdot A + A \cdot B = A + AB = A(1 + B) = A \cdot 1 = A$$

Therefore,

$$A(A + \text{Any term}) = A$$

Consensus Theorem (Included Factor Theorem)

Theorem 1: $\quad AB + \overline{A}C + BC = AB + \overline{A}C$

Proof:

$$\begin{aligned} \text{LHS} &= AB + \overline{A}C + BC \\ &= AB + \overline{A}C + BC(A + \overline{A}) \\ &= AB + \overline{A}C + BCA + BC\overline{A} \\ &= AB(1 + C) + \overline{A}C(1 + B) \\ &= AB(1) + \overline{A}C(1) \\ &= AB + \overline{A}C \\ &= \text{RHS} \end{aligned}$$

This theorem can be extended to any number of variables. For example,

$$AB + \overline{A}C + BCD = AB + \overline{A}C$$

$$\text{LHS} = AB + \overline{A}C + BCD = AB + \overline{A}C + BC + BCD = AB + \overline{A}C + BC = AB + \overline{A}C = \text{RHS}$$

Theorem 2: $\quad (A + B)(\overline{A} + C)(B + C) = (A + B)(\overline{A} + C)$

Proof:

$$\begin{aligned} \text{LHS} &= (A + B)(\overline{A} + C)(B + C) = (A\overline{A} + AC + B\overline{A} + BC)(B + C) \\ &= (AC + BC + \overline{A}B)(B + C) \\ &= ABC + BC + \overline{A}B + BC + AC + \overline{A}BC = AC + BC + \overline{A}B \end{aligned}$$

$$\begin{aligned} \text{RHS} &= (A + B)(\overline{A} + C) \\ &= A\overline{A} + AC + BC + \overline{A}B \\ &= AC + BC + \overline{A}B = \text{LHS} \end{aligned}$$

This theorem can be extended to any number of variables. For example,

$$(A + B)(\overline{A} + C)(B + C + D) = (A + B)(\overline{A} + C)$$

$$\text{LHS} = (A + B)(\overline{A} + C)(B + C)(B + C + D) = (A + B)(\overline{A} + C)(B + C)$$
$$= (A + B)(\overline{A} + C)$$

Transposition Theorem

Theorem: $AB + \overline{A}C = (A + C)(\overline{A} + B)$

Proof:

$$\begin{aligned} \text{RHS} &= (A + C)(\overline{A} + B) \\ &= A\overline{A} + C\overline{A} + AB + CB \\ &= 0 + \overline{A}C + AB + BC \\ &= \overline{A}C + AB + BC(A + \overline{A}) \\ &= AB + ABC + \overline{A}C + \overline{A}BC \\ &= AB + \overline{A}C \\ &= \text{LHS} \end{aligned}$$

De Morgan's Theorem

De Morgan's theorem represents two of the most powerful laws in Boolean algebra.

Law 1: $\overline{A + B} = \overline{A}\,\overline{B}$

This law states that the complement of a sum of variables is equal to the product of their individual complements. What it means is that the complement of two or more variables ORed together, is the same as the AND of the complements of each of the individual variables. Schematically, each side of this law can be represented as:

LHS

NOR gate

RHS

Bubbled AND gate

A	B	A + B	$\overline{A + B}$
0	0	0	1
0	1	1	0
1	0	1	0
1	1	1	0

=

A	B	$\overline{A}$	$\overline{B}$	$\overline{A} \cdot \overline{B}$
0	0	1	1	1
0	1	1	0	0
1	0	0	1	0
1	1	0	0	0

It shows that the NOR gate is equivalent to a bubbled AND gate. This has also been shown quite simply by truth tables.

This law can be extended to any number of variables or combinations of variables. For example,

$$\overline{A + B + C + D + \cdots} = \overline{A}\,\overline{B}\,\overline{C}\,\overline{D} \cdots$$

$$\overline{AB + CD + EFG + \cdots} = \overline{AB}\ \overline{CD}\ \overline{EFG} \cdots$$

It may thus be seen that this law permits removal of individual variables from under a NOT sign and transformation from a sum-of-products form to a product-of-sums form.

Law 2: $\overline{AB} = \overline{A} + \overline{B}$

This law states that the complement of the product of variables is equal to the sum of their individual complements. That is, the complement of two or more variables ANDed together, is equal to the sum of the complements of each of the individual variables. Schematically, we have

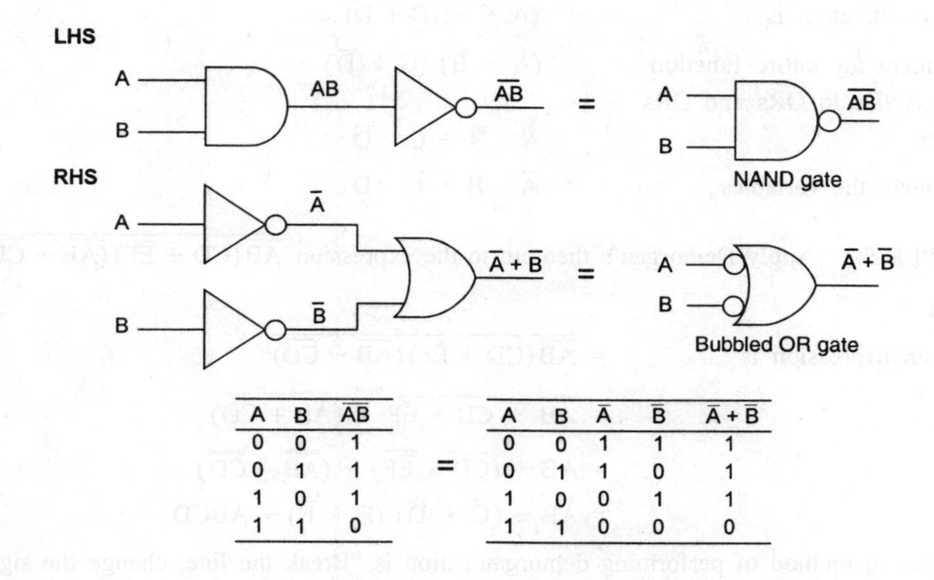

A	B	$\overline{AB}$
0	0	1
0	1	1
1	0	1
1	1	0

=

A	B	$\overline{A}$	$\overline{B}$	$\overline{A} + \overline{B}$
0	0	1	1	1
0	1	1	0	1
1	0	0	1	1
1	1	0	0	0

It shows that the NAND gate is equivalent to a bubbled OR gate. This has also been shown quite simply by truth tables.

This law can be extended to any number of variables or combinations of variables. For example,

$$\overline{ABCD\ldots} = \overline{A} + \overline{B} + \overline{C} + \overline{D} + \ldots$$

$$\overline{(AB)(CD)(EFG)\ldots} = \overline{AB} + \overline{CD} + \overline{(EFG)} + \ldots$$

It may also be seen that like law 1, the law 2 also permits removal of individual variables from under a NOT sign, and transformation from a product-of-sums form to a sum-of-products form.

It may be seen that the transformations

$$\overline{A + B} = \overline{A}\,\overline{B}$$

$$\overline{AB} = \overline{A} + \overline{B}$$

can be extended to complicated expressions by the following three steps:

1. Complement the entire given function.
2. Change all the ANDs to ORs and all the ORs to ANDs.
3. Complement each of the individual variables.

This procedure is called *demorganization.*

EXAMPLE 5.1 Demorganize $\overline{(A + \overline{B})(C + \overline{D})}$

Solution

The given function is $\overline{(A + \overline{B})(C + \overline{D})}$

Complement the entire function $(A + \overline{B})(C + \overline{D})$

Change ANDs to ORs and ORs to ANDs $A \cdot \overline{B} + C \cdot \overline{D}$

Complement the variables $\overline{A} \cdot B + \overline{C} \cdot D$

EXAMPLE 5.2 Apply Demorgan's theorem to the expression $\overline{\overline{\overline{AB}}(CD + \overline{\overline{E}}F)(\overline{\overline{AB}} + \overline{\overline{CD}})}$

Solution

The given expression is

$$= \overline{\overline{\overline{AB}}(CD + \overline{\overline{E}}F)(\overline{\overline{AB}} + \overline{\overline{CD}})}$$

$$= \overline{\overline{\overline{AB}}} + \overline{CD + \overline{\overline{E}}F} + (\overline{\overline{\overline{AB}} + \overline{\overline{CD}}})$$

$$= AB + (\overline{CD} \cdot \overline{\overline{\overline{E}}F}) + (\overline{\overline{\overline{AB}}} \cdot \overline{\overline{\overline{CD}}})$$

$$= AB + (\overline{C} + \overline{D})(E + \overline{F}) + ABCD$$

A second method of performing demorganization is "Break the line, change the sign". For example, if we wish to demorganize the expression $\overline{AB} + \overline{CDE}$, we can break the line between A and B, and the line between C and D, and that between D and E and change the sign from ANDing to ORing. This yields $\overline{A} + \overline{B} + \overline{C} + \overline{D} + \overline{E}$.

EXAMPLE 5.3 Reduce the expression $\overline{\overline{AB} + \overline{A} + AB}$

Solution

The given expression is $\overline{\overline{AB} + \overline{A} + AB}$

Break the upper line between $\overline{AB}$ and $\overline{A}$, and that between $\overline{A}$ and AB, and change the signs between $\overline{AB}$, $\overline{A}$, AB.

$$= \overline{\overline{AB}} \cdot \overline{\overline{A}} \cdot \overline{AB}$$

$$= AB \cdot A \cdot \overline{AB}$$

Simplify

$$= AB \cdot \overline{AB}$$

$$= 0$$

Alternatively:
Break the lower line between A and B and change the sign between them.

$$= \overline{\overline{A} + \overline{B} + \overline{A} + AB}$$

Simplify

$$= \overline{\overline{A} + \overline{B} + AB}$$

Break the line between $\overline{A}$ and $\overline{B}$, and that between $\overline{B}$ and AB, and change the sign between them.

$$= \overline{\overline{A}} \cdot \overline{\overline{B}} \cdot \overline{AB} = AB \cdot \overline{AB} = 0$$

Additional Theorems

Theorem 1: $X \cdot f(X, \overline{X}, Y, \ldots, Z) = X \cdot f(1, 0, Y, \ldots, Z)$

This theorem states that if a function containing expressions/terms with X and $\overline{X}$ is multiplied by X, then all the Xs and $\overline{X}$s in the function can be replaced by 1s and 0s, respectively. This is permissible because,

$$X \cdot X = X = X \cdot 1 \quad \text{and} \quad X \cdot \overline{X} = 0 = X \cdot 0$$

Theorem 2: $X + f(X, \overline{X}, Y, \ldots, Z) = X + f(0, 1, Y, \ldots, Z)$

This theorem states that if a function containing expressions/terms with X and $\overline{X}$ is added to X, then all the Xs and $\overline{X}$s in the function can be replaced by 0s and 1s, respectively. This is permissible because,

$$X + X = X = X + 0 \quad \text{and} \quad X + \overline{X} = 1 = X + 1.$$

Theorem 3: $f(X, \overline{X}, Y, \ldots, Z) = X \cdot f(1, 0, Y, \ldots, Z) + \overline{X} \cdot f(0, 1, Y, \ldots Z),$

Theorem 4: $f(X, \overline{X}, Y, \ldots, Z) = [X + f(0, 1, Y, \ldots, Z)] \cdot [\overline{X} + f(1, 0, Y, \ldots, Z)]$

5.4 DUALITY

We know that in a positive logic system the more positive of the two voltage levels is represented by a 1 and the more negative by a 0. In a negative logic system the more positive of the two voltage levels is represented by a 0 and the more negative by a 1. This distinction between positive and negative logic systems is important because an OR gate in the positive logic system becomes an AND gate in the negative logic system, and vice versa. Positive and negative logics thus give rise to a basic duality in all Boolean identities. When changing from one logic system to another, 0 becomes 1 and 1 becomes 0. Furthermore, an AND gate becomes an OR gate and an OR gate becomes an AND gate. Given a Boolean identity, we can produce a dual identity by changing all '+' signs to '·' signs, all '.' signs to '+' signs, and complementing all 0s and 1s. The variables are not complemented in this process. Some dual identities are given below:

Duals

	Given Expression	*Dual*
1.	$\overline{0} = 1$	$\overline{1} = 0$
2.	$0 \cdot 1 = 0$	$1 + 0 = 1$
3.	$0 \cdot 0 = 0$	$1 + 1 = 1$
4.	$1 \cdot 1 = 1$	$0 + 0 = 0$
5.	$A \cdot 0 = 0$	$A + 1 = 1$
6.	$A \cdot 1 = A$	$A + 0 = A$
7.	$A \cdot A = A$	$A + A = A$
8.	$A \cdot \overline{A} = 0$	$A + \overline{A} = 1$
9.	$A \cdot B = B \cdot A$	$A + B = B + A$
10.	$A \cdot (B \cdot C) = (A \cdot B) \cdot C$	$A + (B + C) = (A + B) + C$
11.	$A \cdot (B + C) = AB + AC$	$A + BC = (A + B)(A + C)$
12.	$A(A + B) = A$	$A + AB = A$
13.	$A \cdot (A \cdot B) = A \cdot B$	$A + A + B = A + B$
14.	$\overline{AB} = \overline{A} + \overline{B}$	$\overline{A + B} = \overline{A}\ \overline{B}$
15.	$(A + B)(\overline{A} + C)(B + C) = A + B(\overline{A} + C)$	$AB + \overline{A}C + BC = AB + \overline{A}C$
16.	$(A + C)(\overline{A} + B) = AB + \overline{A}C$	$AC + \overline{A}B = (A + B)(\overline{A} + C)$
17.	$A + \overline{B}C = (A + \overline{B})(A + C)$	$A(\overline{B} + C) = (A\overline{B} + AC)$
18.	$(A + B)(C + D) = AC + AD + BC + BD$	$(AB + CD) = (A + C)(A + D)(B + C)(B + D)$
19.	$A + B = AB + \overline{A}B + A\overline{B}$	$AB = (A + B)(\overline{A} + B)(A + \overline{B})$
20.	$A + B\overline{(C + \overline{\overline{DE}})} = A + B\overline{C}DE$	$A[B + \overline{(C \cdot \overline{\overline{D + E}})}] = A \cdot (B + \overline{C} + D + E)$
21.	$\overline{\overline{AB} + \overline{A} + AB} = 0$	$\overline{\overline{A + B} \cdot \overline{A} \cdot (A + B)} = 1$
22.	$AB + \overline{AC} + A\overline{B}C(AB + C) = 1$	$(A + B)(\overline{A + C}) \cdot [(A + \overline{B} + C) + (A + B)C] = 0$

23. $ABD + ABCD = ABD$	$(A + B + D)\,(A + B + C + D) = (A + B + D)$
24. $\overline{A\overline{B} + ABC} + A(B + A\overline{B}) = 0$	$\overline{(A + \overline{B}) \cdot (A + B + C)} \,.\, (A + [B(A + \overline{B})]) = 1$
25. $A + \overline{B}C(A + \overline{B}C) = A + \overline{B}C$	$A \cdot [(\overline{B} + C) + A \cdot (\overline{B} + C)] = A \cdot (\overline{B} + C)$

5.5 REDUCING BOOLEAN EXPRESSIONS

Every Boolean expression must be reduced to as simple a form as possible before realization, because every logic operation in the expression represents a corresponding element of hardware. Realization of a digital circuit with the minimal expression, therefore, results in reduction of cost and complexity and the corresponding increase in reliability. To reduce Boolean expressions, all the laws of Boolean algebra may be used. The techniques used for these reductions are similar to those used in ordinary algebra. The procedure is:

(a) Multiply all variables necessary to remove parentheses.

(b) Look for identical terms. Only one of those terms be retained and all others dropped. For example,

$$AB + AB + AB + AB = AB$$

(c) Look for a variable and its negation in the same term. This term can be dropped. For example,

$$A \cdot B\overline{B} = A \cdot 0 = 0; \qquad ABC\overline{C} = AB \cdot 0 = 0$$

(d) Look for pairs of terms that are identical except for one variable which may be missing in one of the terms. The larger term can be dropped. For example,

$$AB\overline{C}\,\overline{D} + AB\overline{C} = AB\overline{C}(\overline{D} + 1) = AB\overline{C} \cdot 1 = AB\overline{C}$$

(e) Look for the pairs of terms which have the same variables, with one or more variables complemented. If a variable in one term of such a pair is complemented while in the second term it is not, then such terms can be combined into a single term with that variable dropped. For example,

$$AB\overline{C}\,\overline{D} + AB\overline{C}D = AB\overline{C}(\overline{D} + D) = AB\overline{C} \cdot 1 = AB\overline{C}$$

$$AB(C + D) + AB(\overline{C + D}) = AB[(C + D) + (\overline{C + D})] = AB \cdot 1 = AB$$

EXAMPLE 5.4 Reduce the expression $A[B + \overline{C}(\overline{AB + A\overline{C}})]$.

Solution

The given expression is $A[B + \overline{C}(\overline{AB + A\overline{C}})]$

Demorganize $\overline{AB + A\overline{C}}$ $= A[B + \overline{C}(\overline{AB}\ \overline{A\overline{C}})]$

Demorganize $\overline{AB}$ and $\overline{A\overline{C}}$ $= A[B + \overline{C}(\overline{A} + \overline{B})(\overline{A} + C)]$

Multiply $(\overline{A} + \overline{B})(\overline{A} + C)$ $= A[B + \overline{C}(\overline{A}\,\overline{A} + \overline{A}C + \overline{B}\,\overline{A} + \overline{B}C)]$

Simplify $= A(B + \overline{C}\,\overline{A} + \overline{C}\,\overline{A}C + \overline{C}\,\overline{B}\,\overline{A} + \overline{C}\,\overline{B}C)$

Simplify $= A(B + \overline{C}\,\overline{A} + 0 + \overline{C}\,\overline{B}\,\overline{A} + 0)$

Simplify $= AB + A\overline{C}\,\overline{A} + A\overline{C}\,\overline{B}\,\overline{A}$

Simplify $= AB + 0 + 0$

Simplify $= AB$

EXAMPLE 5.5 Reduce the expression $A + B[AC + (B + \overline{C})D]$

Solution

The given expression is $A + B[AC + (B + \overline{C})D]$

Expand $(B + \overline{C})D$ $= A + B(AC + BD + \overline{C}D)$

Expand $B(AC + BD + \overline{C}D)$ $= A + BAC + BBD + B\overline{C}D$

Write in order $= A + ABC + BD + BD\overline{C}$

Factor $= A(1 + BC) + BD(1 + \overline{C})$

Reduce $= A \cdot 1 + BD \cdot 1$

Simplify $= A + BD$

EXAMPLE 5.6 Reduce the expression $(\overline{A + \overline{\overline{BC}}})(A\overline{B} + ABC)$

Solution

The given expression is $(\overline{A + \overline{\overline{BC}}})(A\overline{B} + ABC)$

Demorganize $(\overline{A + \overline{\overline{BC}}})$ $= (\overline{A}\,\overline{\overline{BC}})(A\overline{B} + ABC)$

Simplify $= (\overline{A}BC)(A\overline{B} + ABC)$

Multiply $= \overline{A}BC\,A\overline{B} + \overline{A}BC\,ABC$

Rearrange $= A\overline{A}B\overline{B}C + A\overline{A}BBCC$

$= 0 + 0 = 0$

EXAMPLE 5.7 Reduce the expression $(B + BC)(B + \overline{\overline{B}}C)(B + D)$

Solution

The given expression is $(B + BC)(B + \overline{\overline{B}}C)(B + D)$

Multiply the first two terms $= (BB + BCB + B\overline{\overline{B}}C + BC\overline{\overline{B}}C)(B + D)$

Reduce $= (B + BC + 0 + 0)(B + D)$

Factor	$= B(1 + C)(B + D)$
Reduce	$= B(B + D)$
Expand	$= BB + BD$
Simplify	$= B(1 + D) = B$

EXAMPLE 5.8 Show that $AB + A\overline{B}C + B\overline{C} = AC + B\overline{C}$

Solution

$$\begin{aligned} AB + A\overline{B}C + B\overline{C} &= A(B + \overline{B}C) + B\overline{C} \\ &= A(B + \overline{B})(B + C) + B\overline{C} \\ &= AB + AC + B\overline{C} \\ &= AB(C + \overline{C}) + AC + B\overline{C} \\ &= ABC + AB\overline{C} + AC + B\overline{C} \\ &= AC(1 + B) + B\overline{C}(1 + A) \\ &= AC + B\overline{C} \end{aligned}$$

EXAMPLE 5.9 Show that $A\overline{B}C + B + B\overline{D} + AB\overline{D} + \overline{A}C = B + C$

Solution

$$\begin{aligned} A\overline{B}C + B + B\overline{D} + AB\overline{D} + \overline{A}C &= A\overline{B}C + \overline{A}C + B(1 + \overline{D} + A\overline{D}) \\ &= C(\overline{A} + A\overline{B}) + B \\ &= C(\overline{A} + A)(\overline{A} + \overline{B}) + B \\ &= C\overline{A} + C\overline{B} + B \\ &= (B + C)(B + \overline{B}) + C\overline{A} \\ &= B + C + C\overline{A} \\ &= B + C(1 + \overline{A}) \\ &= B + C \end{aligned}$$

5.6 BOOLEAN EXPRESSIONS AND LOGIC DIAGRAMS

Boolean expressions can be realized as hardware using logic gates. Conversely, hardware can be translated into Boolean expressions for the analysis of existing circuits.

Converting Boolean Expressions to Logic

The easiest way to convert a Boolean expression to a logic circuit is to start with the output and work towards the input. Assume that the expression $\overline{AB} + A + \overline{B + C}$ is to be realized using AOI logic. Start with the final expression $\overline{AB} + A + \overline{B + C}$. Since it is a summation of three terms, it must be the output of a three-input OR gate. So, draw an OR gate with three inputs as shown below.

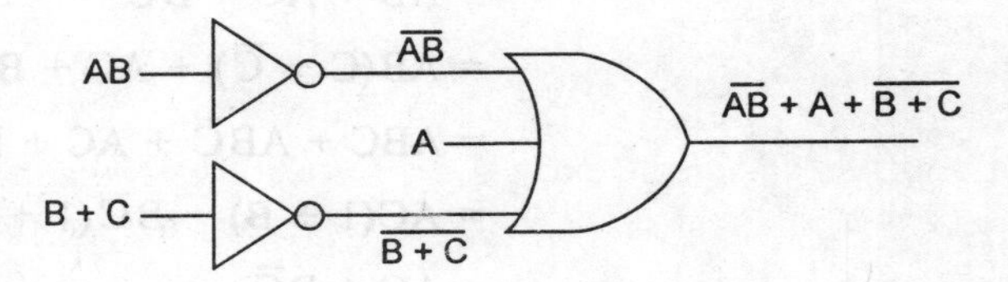

$\overline{AB}$ must be the output of an inverter whose input is AB. And $\overline{B + C}$ must be the output of an inverter whose input is B + C. So, we introduce two inverters as shown below.

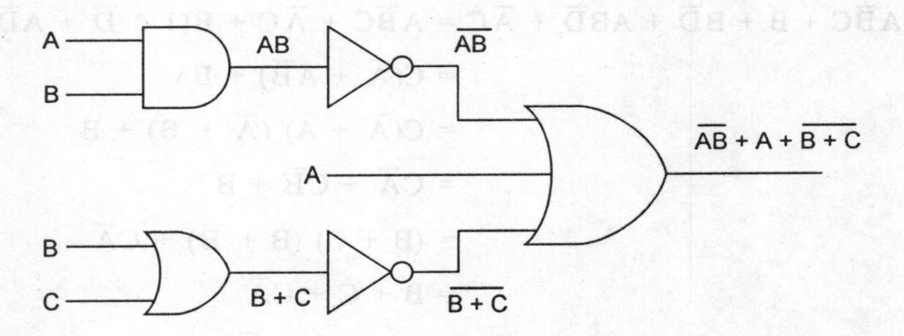

Now AB must be the output of a two-input AND gate whose inputs are A and B. And B + C must be the output of a two-input OR gate whose inputs are B and C. So, we introduce an AND gate and an OR gate as shown below.

Converting Logic to Boolean Expressions

To convert logic to algebra, start with the input signals and develop the terms of the Boolean expression until the output is reached. Consider the logic diagram shown below.

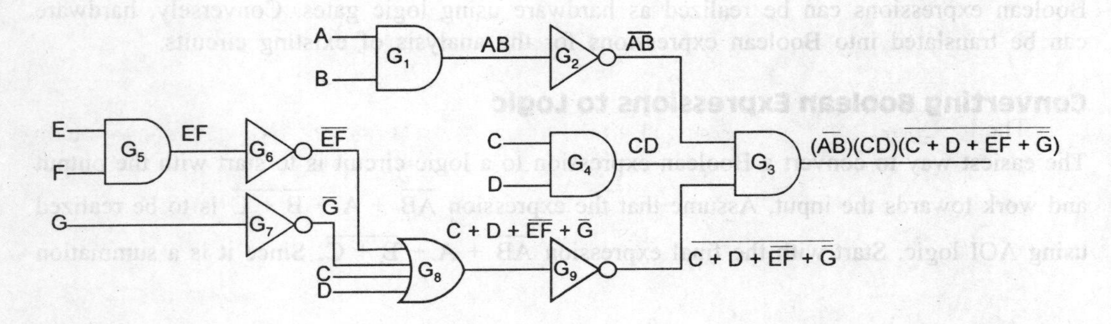

First label all the inputs. The signals A and B feed the AND gate G_1. The output of this gate is, therefore, AB. This is the input to NOT gate G_2. The output of the NOT gate G_2 will, therefore, be $\overline{AB}$. The C and D are the inputs to AND gate G_4. The output of this gate is, therefore, CD. The E and F are the inputs to AND gate G_5. The output of this gate is, therefore, EF. This signal EF is the input to inverter G_6. Its output is, therefore, $\overline{EF}$. The input to inverter G_7 is G. So, its output is $\overline{G}$. Now C, D, $\overline{EF}$, and $\overline{G}$ are the inputs to OR gate G_8. The output of the OR gate G_8 will, therefore, be $(C + D + \overline{EF} + \overline{G})$. This is the input to inverter G_9. The output of G_9 is, therefore, $\overline{(C + D + \overline{EF} + \overline{G})}$. The inputs to the AND gate G_3 are, $\overline{AB}$, CD and $\overline{(C + D + \overline{EF} + \overline{G})}$. The output of G_3, which is also the output of the logic circuit is, therefore, equal to $(\overline{AB}) \cdot (CD) \cdot \overline{(C + D + \overline{EF} + \overline{G})}$.

EXAMPLE 5.10 Write the Boolean expression for the logic diagram given below and simplify it as much as possible and draw the logic diagram that implements the simplified expression.

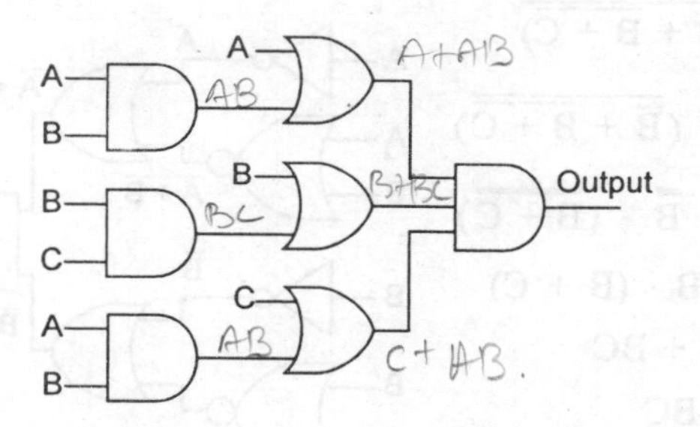

Solution

Starting from the input side and writing the expressions for the outputs of the individual gates, we can easily show that the

$$\begin{aligned}
\text{Output} &= (A + AB)\,(B + BC)\,(C + AB)\\
&= A(1 + B)\,B(1 + C)\,(C + AB)\\
&= AB(C + AB)\\
&= ABC + AB\\
&= AB(1 + C)\\
&= AB
\end{aligned}$$

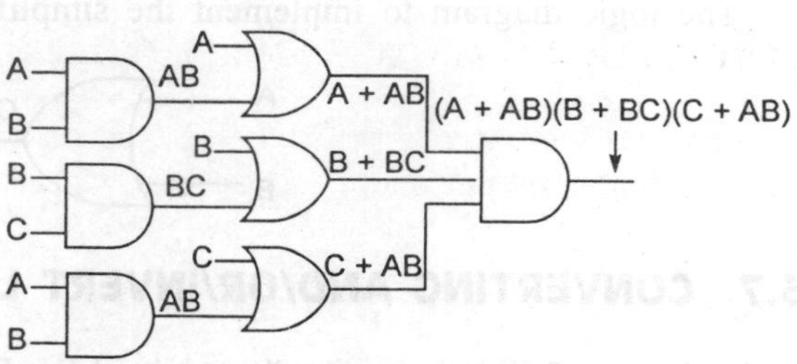

The logic diagram to realize the simplified expression is just an AND gate shown below.

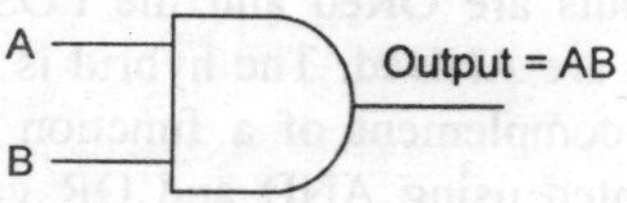

EXAMPLE 5.11 Draw the simplest possible logic diagram that implements the output of the logic diagram shown below.

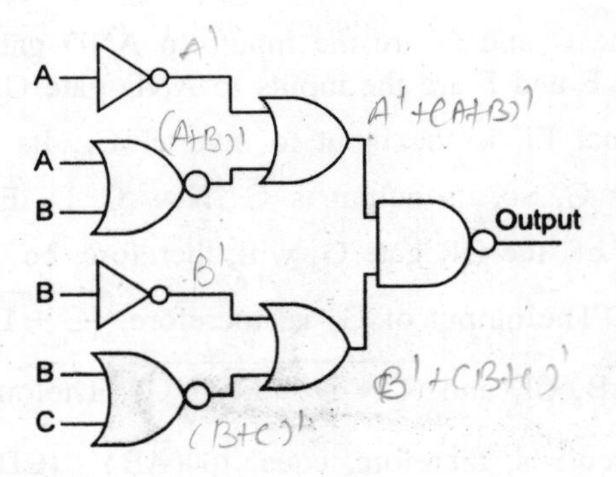

Solution

Starting from the input side and writing the expressions for the outputs of the individual gates as shown in the diagram below, we have

$$\text{Output} = \overline{(\bar{A} + \overline{A + B})(\bar{B} + \overline{B + C})}$$

$$= \overline{(\bar{A} + \overline{A + B})} + \overline{(\bar{B} + \overline{B + C})}$$

$$= \bar{\bar{A}} \cdot \overline{\overline{(A + B)}} + \bar{\bar{B}} \cdot \overline{\overline{(B + C)}}$$

$$= A \cdot (A + B) + B \cdot (B + C)$$

$$= AA + AB + BB + BC$$

$$= A + AB + B + BC$$

$$= A(1 + B) + B(1 + C)$$

$$= A + B$$

A, B, B, C; $\bar{A}$; $\overline{A + B}$; $\bar{A} + (\overline{A + B})$; $\bar{B}$; $\overline{B + C}$; $\bar{B} + (\overline{B + C})$; $\overline{(\bar{A} + \overline{A + B})(\bar{B} + \overline{B + C})}$

The logic diagram to implement the simplified expression is shown below.

A, B; Output = A + B

5.7 CONVERTING AND/OR/INVERT LOGIC TO NAND/NOR LOGIC

In the design of digital circuits, the minimal Boolean expressions are usually obtained in SOP (sum-of-products) form or POS (product-of-sums) form. Sometimes the minimal expressions may also be expressed in hybrid form. The SOP form is implemented using a group of AND gates whose outputs are ORed and the POS form is implemented using a group of OR gates whose outputs are ANDed. The hybrid is a combination of both the SOP and POS forms. Sometimes the complement of a function is to be implemented. So, the designed circuit can be implemented using AND and OR gates only called the A/O logic, or using AND/OR/NOT gates called the AOI logic. In the realization of Boolean expressions, the variable and its complement are assumed to be available. For example, the SOP expression $ABC + A\overline{B} + \overline{A}BC$ can be implemented in A/O logic as shown below.

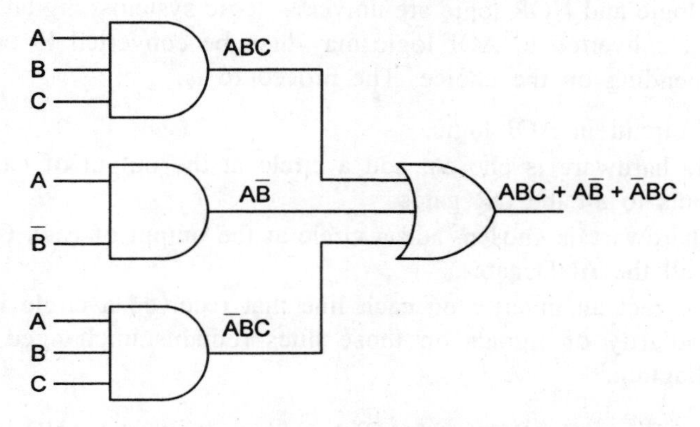

The POS expression $(A + B + \overline{C})(B + C)(\overline{A} + \overline{B} + C)$ can be implemented using OR and AND gates as shown below.

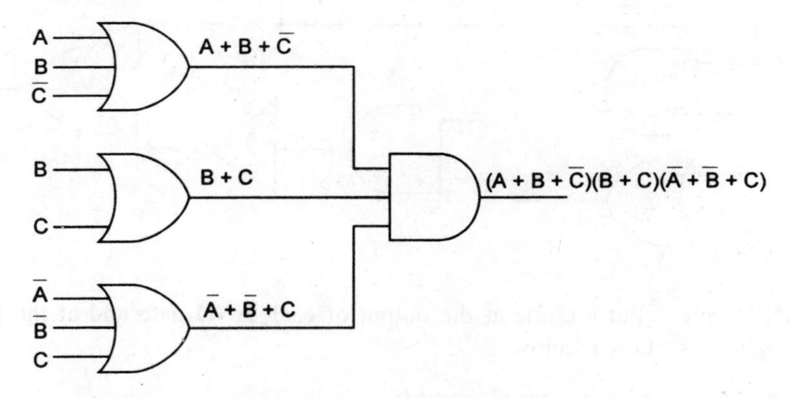

The expression $AB\overline{C} + \overline{A}B$ [= $B(\overline{A} + A\overline{C})$] can be implemented in hybrid form as shown below.

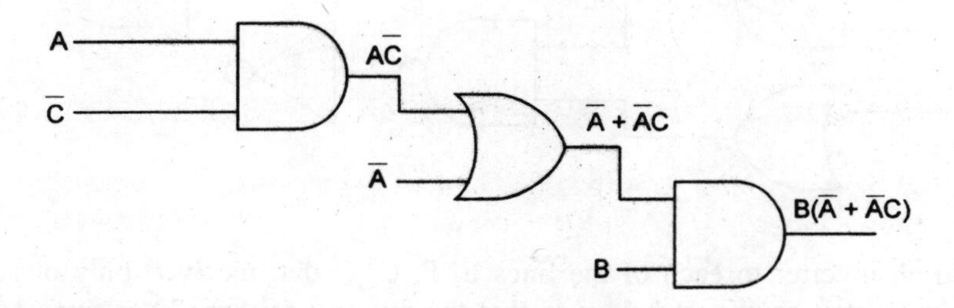

Hybrid logic reduces the number of gate inputs required for realization (from 7 to 6 in this case), but results in multilevel logic. Different inputs pass through different number of gates to reach the output. It leads to non-uniform propagation delay between input and output and may give rise to logic race. The SOP and POS realizations give rise to two-level logic. The two-level logic provides uniform time delay between input and output, because each input signal has to pass through two gates to reach the output. So, it does not suffer from the problem of logic race.

Since NAND logic and NOR logic are universal logic systems, digital circuits which are first computed and converted to AOI logic may then be converted to either NAND logic or NOR logic depending on the choice. The procedure is:

1. Draw the circuit in AOI logic.
2. If NAND hardware is chosen, add a circle at the output of each AND gate and at the inputs to all the OR gates.
3. If NOR hardware is chosen, add a circle at the output of each OR gate and at the inputs to all the AND gates.
4. Add or subtract an inverter on each line that received a circle in steps 2 or 3 so that the polarity of signals on those lines remains unchanged from that of the original diagram.

EXAMPLE 5.12 Convert the following AOI logic circuit to (a) NAND logic, and (b) NOR logic.

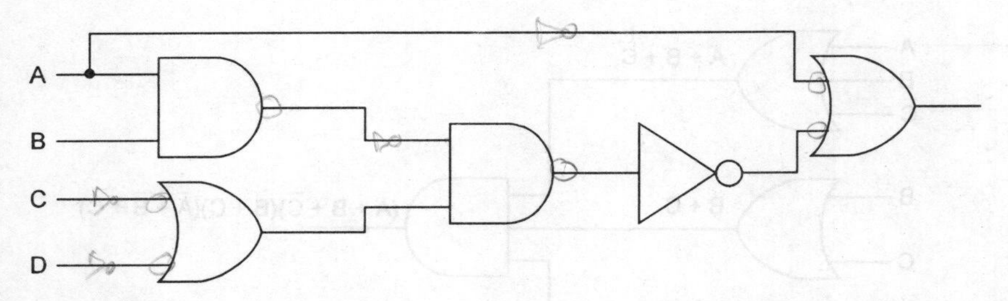

Solution

(a) NAND logic. Put a circle at the output of each AND gate and at the inputs to all OR gates as shown below.

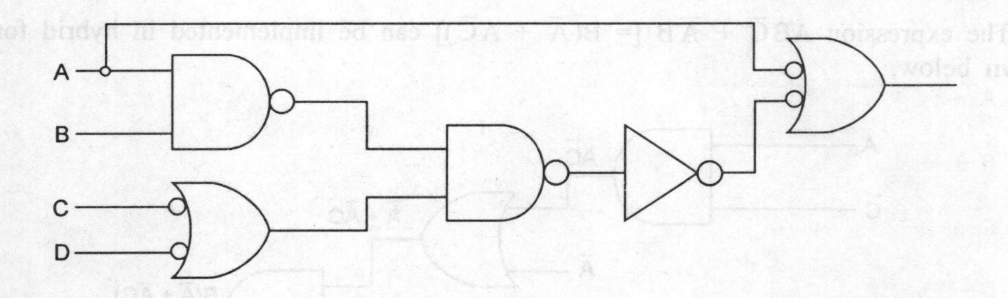

Add an inverter to each of the lines E, F, C, D that received only one circle in the previous step as shown below so that the polarity of these lines remains unchanged.

Inverters in lines C and D can be removed, if C and D are replaced by $\overline{C}$ and $\overline{D}$. Line H received two circles. So, no change is required.

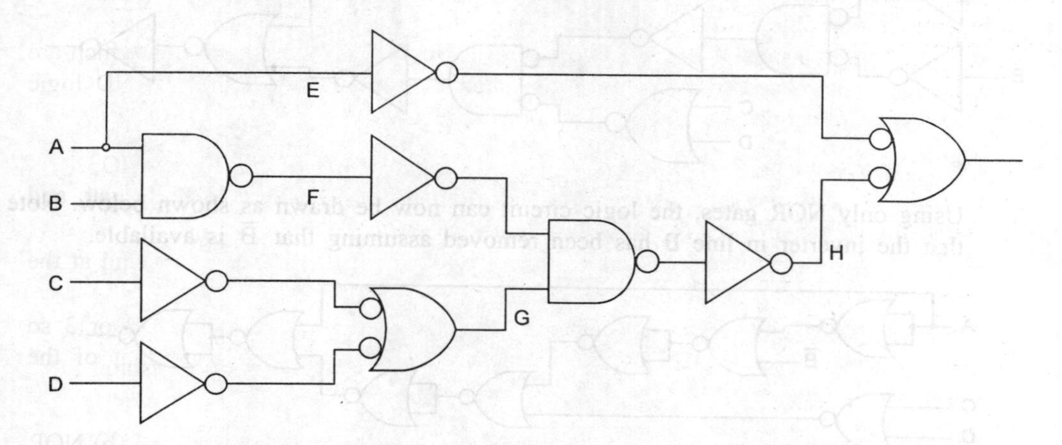

Using only NAND gates, the logic circuit can now be drawn as shown below.

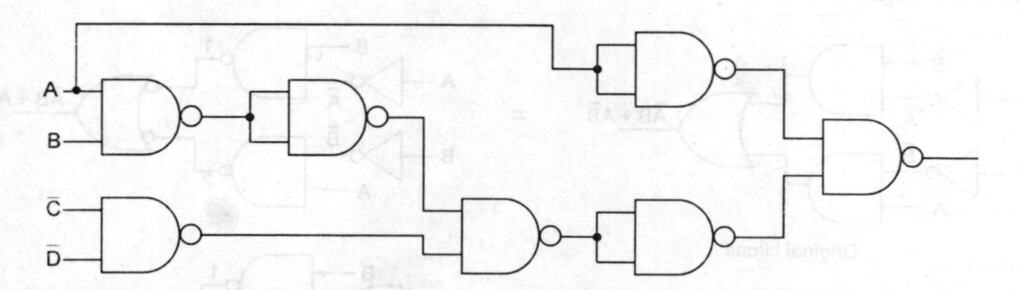

(b) NOR logic. Put a circle at the output of each OR gate and at the inputs to all AND gates as shown below.

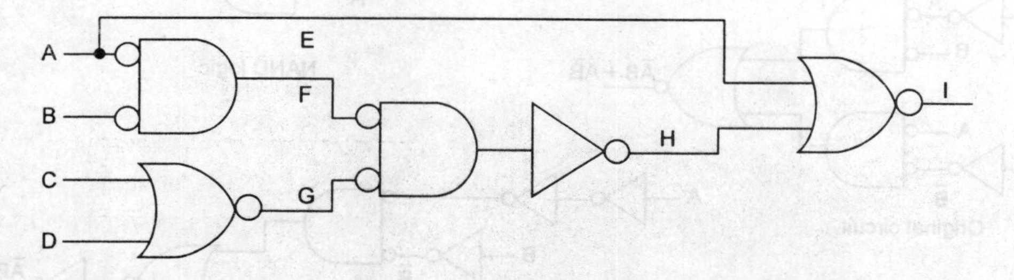

Add an inverter in each of the lines A, B, F, and I that received only one circle in the previous step, so that the polarity of these lines remains unchanged. Line G received two circles. So, no change is required.

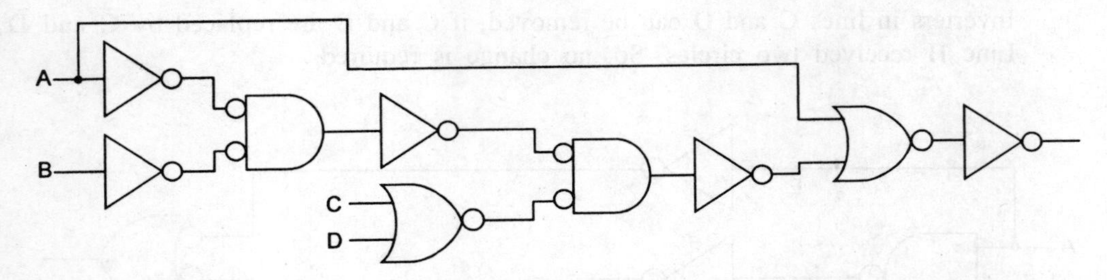

Using only NOR gates, the logic circuit can now be drawn as shown below. Note that the inverter in line B has been removed assuming that $\overline{B}$ is available.

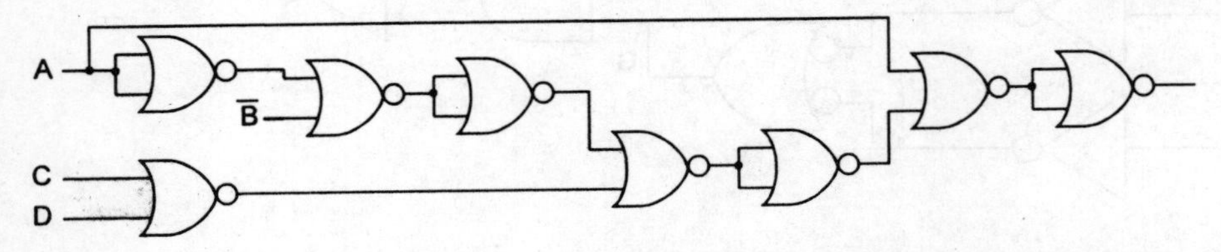

Now study the conversion of the following AOI circuit to (a) NAND logic, and (b) NOR logic.

B 1 A Ā B̄ B A 2 $\overline{A}B + A\overline{B}$ = B 1 A Ā B̄ B A 2 $\overline{A}B + A\overline{B}$

Original circuit

= B 1 A Ā B̄ B A 2 $\overline{A}B + A\overline{B}$

A Ā B A B B̄ $\overline{A}B + A\overline{B}$

NAND logic

Original circuit

= A B B̄ A Ā B $\overline{A}B + A\overline{B}$

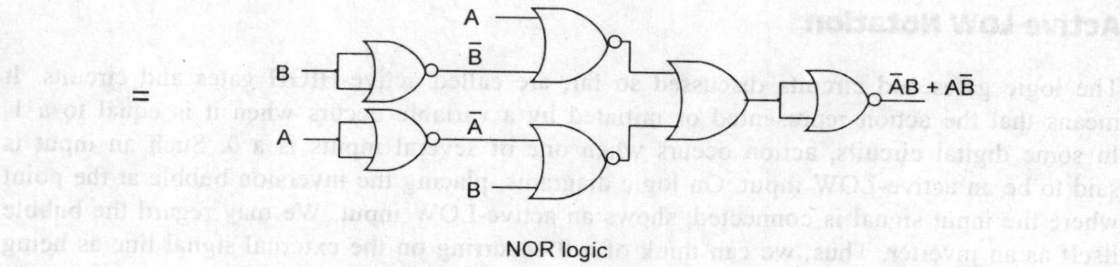

NOR logic

The A/O gates and AOI gates are available in IC form. A circuit having *n* AND gates is said to be *n*-wide. The A/O gates in which an additional variable or a combination of variables can be included in the logic operation are called *expandable gates*. Some AOI gates and expandable A/O gates available in IC form are shown below.

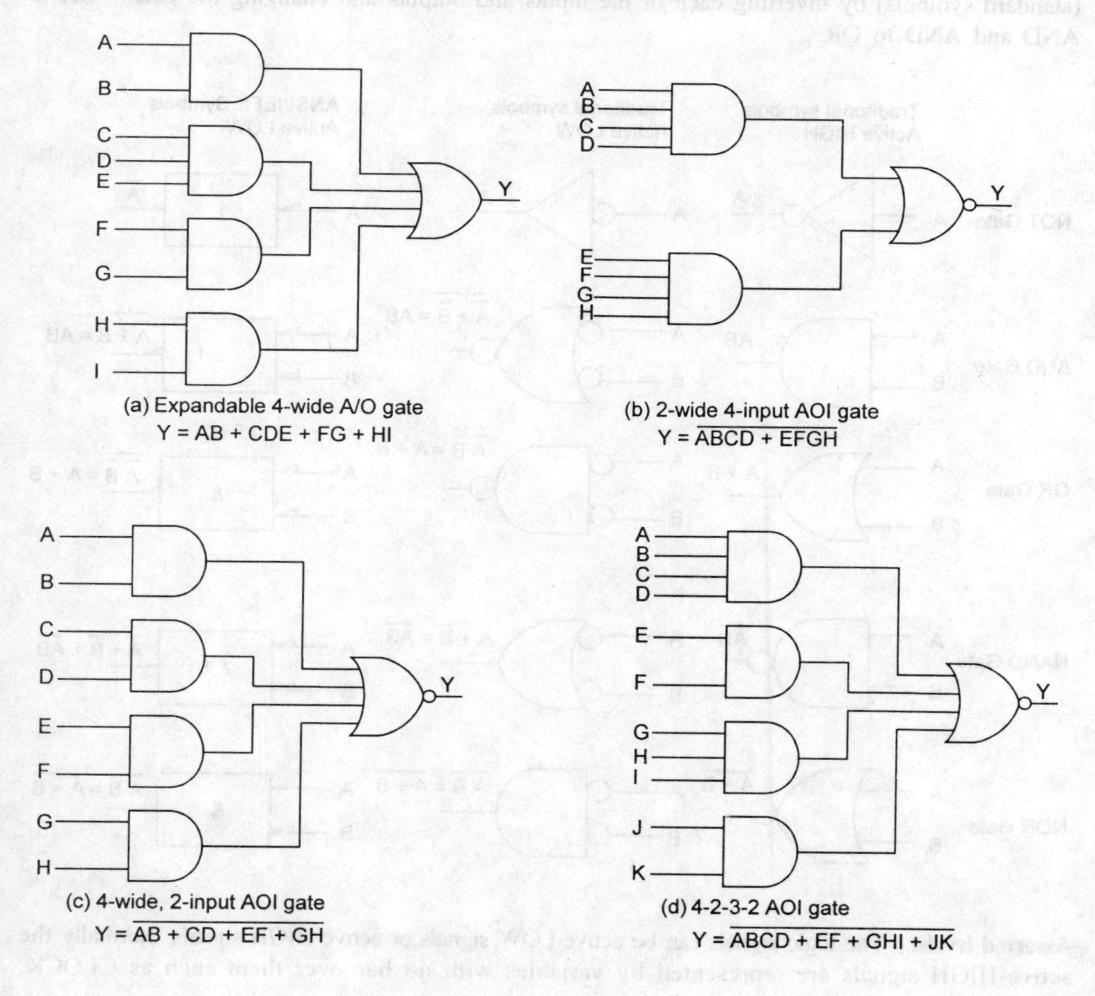

(a) Expandable 4-wide A/O gate

$Y = AB + CDE + FG + HI$

(b) 2-wide 4-input AOI gate

$Y = \overline{ABCD + EFGH}$

(c) 4-wide, 2-input AOI gate

$Y = \overline{AB + CD + EF + GH}$

(d) 4-2-3-2 AOI gate

$Y = \overline{ABCD + EF + GHI + JK}$

Active-LOW Notation

The logic gates and circuits discussed so far, are called active-HIGH gates and circuits. It means that the action represented or initiated by a variable occurs when it is equal to a 1. In some digital circuits, action occurs when one of several inputs is a 0. Such an input is said to be an active-LOW input. On logic diagrams, placing the inversion bubble at the point where the input signal is connected, shows an active-LOW input. We may regard the bubble itself as an inverter. Thus, we can think of a 0 occurring on the external signal line as being inverted and producing a 1 internal to the device and vice versa.

The logic symbols of active-HIGH logic gates, their active-LOW versions, and ANSI/IEEE symbols for active-lOW notation are shown below. Note that the NAND gate is equivalent to an active-LOW input OR gate. The NOR gate is equivalent to an active-LOW input AND gate. The active-LOW symbols (alternate symbols) can be obtained from active-HIGH symbols (standard symbols) by inverting each of the inputs and outputs and changing the gates—OR to AND and AND to OR.

	Traditonal symbols Active HIGH	Traditonal symbols Active LOW	ANSI/IEEE Symbols Active LOW
NOT Gate	A, $\overline{A}$	A, $\overline{A}$	A, 1, $\overline{A}$
AND Gate	A, B, AB	A, B, $\overline{\overline{A}+\overline{B}} = AB$	A, B, ≥ 1, $\overline{\overline{A}+\overline{B}} = AB$
OR Gate	A, B, A + B	A, B, $\overline{\overline{A}\,\overline{B}} = A + B$	A, B, &, $\overline{\overline{A}\,\overline{B}} = A + B$
NAND Gate	A, B, $\overline{AB}$	A, B, $\overline{A} + \overline{B} = \overline{AB}$	A, B, ≥ 1, $\overline{A} + \overline{B} = \overline{AB}$
NOR Gate	A, B, $\overline{A + B}$	A, B, $\overline{A}\,\overline{B} = \overline{A + B}$	A, B, &, $\overline{A}\,\overline{B} = \overline{A + B}$

Asserted levels. The logic signals can be active-LOW signals or active-HIGH signals. Normally the active-HIGH signals are represented by variables with no bar over them such as CLOCK,

A, MEM, etc. whereas the active-LOW signals are represented by variables with bar over them such as $\overline{X}$, $\overline{CLR}$, $\overline{MEM}$, etc. The bar simply emphasizes that a particular signal is active-LOW. When a signal is in its active state, it is said to be *asserted*. When it is not in its active state, i.e. when it is inactive, it is said to be *unasserted*. The terms 'asserted' and 'unasserted' are synonymous with 'active' and 'inactive', respectively.

Negative logic. The assertion level refers to the signal level necessary to cause an event to occur. Till now, we have assumed that logic 1 is + 5 V and logic 0 is 0 V. In this notation, events occur when inputs are + 5 V, i.e. the assertion level is a 1. This is called *positive logic*, because a 1 is more positive than a 0. In some systems, it is convenient to define the ground level as a 1 and + 5 V as a 0. That means that the assertion level is a 0. That is, the level required to do something is a 0. This is called *negative logic*, because a 0 is more negative than a 1. In a negative logic system, the more positive of the two voltage levels is represented by a logic 0. An AND gate in the positive logic system becomes an OR gate in the negative logic system and vice versa.

Active-LOW bubbles can be very useful when analyzing logic diagrams, because they can be placed in such a way that they effectively cancel out one another. This eliminates the necessity for writing numerous inversion bars over compound logic expressions.

The circuit below implements the expression $\overline{\overline{D(\overline{\overline{C}+\overline{AB}})}}$. We can use Boolean algebra to show that this expression is equivalent to DC(A + B).

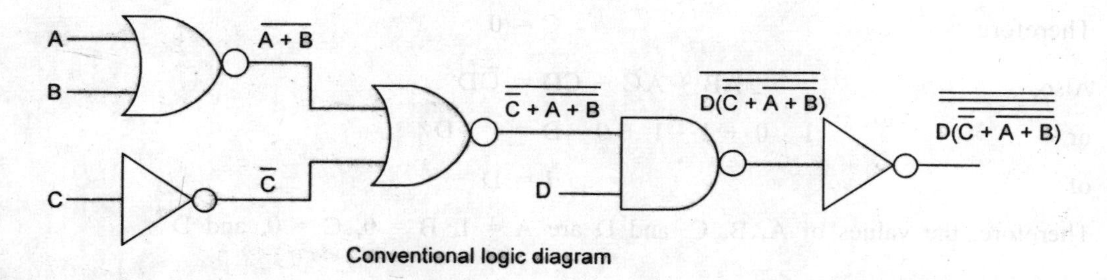

Conventional logic diagram

The same implementation with active-LOW equivalents by replacing one NOR gate and one inverter is shown below. Since consequent inversion bubbles cancel out as shown, it is readily apparent in this diagram that the output is DC(A + B).

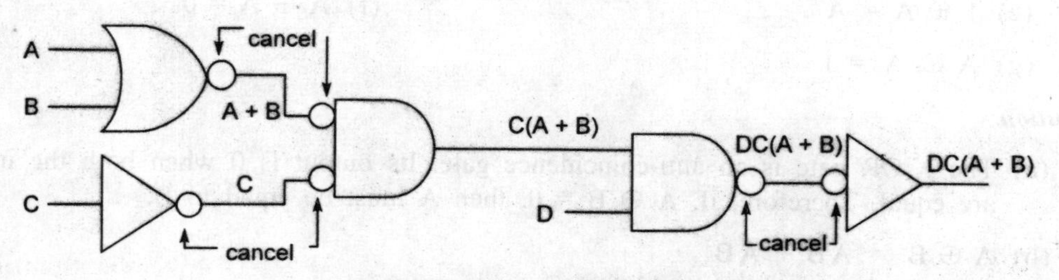

Logic diagram using active-LOW notation

5.8 MISCELLANEOUS EXAMPLES

EXAMPLE 5.13 Find the values of the two-valued variables A, B, C, and D by solving the set of simultaneous equations.

$$\bar{A} + AB = 0$$

$$AB = AC$$

$$AB + A\bar{C} + CD = \bar{C}D$$

Solution

Given $\bar{A} + AB = 0$

or $(\bar{A} + A)(\bar{A} + B) = 0$

or $(1)(\bar{A} + B) = 0$

or $\bar{A} + B = 0$

i.e. $\bar{A}$ must be 0 and B must be 0. Therefore, A = 1 and B = 0.

Also $AB = AC$

or $1 \cdot 0 = 1 \cdot C$

Therefore $C = 0$

Also, $AB + A\bar{C} + CD = \bar{C}D$

or $1 \cdot 0 + 1 \cdot 1 + 0 \cdot D = 1 \cdot D$

or $1 = D$

Therefore, the values of A, B, C, and D are A = 1, B = 0, C = 0, and D = 1.

EXAMPLE 5.14 Prove that

(a) If $A \oplus B = 0$, then $A = B$ (b) $A \oplus B = \bar{A} \oplus \bar{B}$

(c) $\overline{A \oplus B} = \bar{A} \oplus B = A \oplus \bar{B}$ (d) $0 \oplus A = A$

(e) $1 \oplus A = \bar{A}$ (f) $A \oplus A = 0$

(g) $A \oplus \bar{A} = 1$

Solution

(a) The X-OR gate is an anti-coincidence gate. Its output is 0 when both the inputs are equal. Therefore, if, $A \oplus B = 0$, then A must be equal to B.

(b) $A \oplus B = A\bar{B} + \bar{A}B$

$\bar{A} \oplus \bar{B} = \bar{\bar{A}} \cdot \bar{B} + \bar{A} \cdot \bar{\bar{B}} = A\bar{B} + \bar{A}B$

$\therefore A \oplus B = \bar{A} \oplus \bar{B}$

(c) $\overline{A} \oplus B = \overline{\overline{A}} \cdot B + \overline{A} \cdot \overline{B} = AB + \overline{A}\,\overline{B} = \overline{A \oplus B}$

$A \oplus \overline{B} = \overline{A} \cdot \overline{B} + A \cdot \overline{\overline{B}} = \overline{A}\,\overline{B} + AB = \overline{A \oplus B}$

$\therefore \overline{A \oplus B} = \overline{A} \oplus B = A \oplus \overline{B}$

(d) $0 \oplus A = 0 \cdot \overline{A} + \overline{0} \cdot A = 0 + 1 \cdot A = 0 + A = A$

(e) $1 \oplus A = 1 \cdot \overline{A} + \overline{1} \cdot A = \overline{A} + 0 \cdot A = \overline{A} + 0 = \overline{A}$

(f) $A \oplus A = A \cdot \overline{A} + \overline{A} \cdot A = 0 + 0 = 0$

(g) $A \oplus \overline{A} = A \cdot \overline{\overline{A}} + \overline{A} \cdot \overline{A} = A \cdot A + \overline{A} \cdot \overline{A} = A + \overline{A} = 1$

EXAMPLE 5.15 Show that NAND and NOR gates are universal gates.

Solution

We know that AND, OR, and NOT gates are the basic building blocks of a digital computer. They are called the basic gates. Any digital circuit of any complexity can be built using only these three gates. A universal gate is a gate which alone can be used to build any logic circuit. So, to show that the NAND gate and the NOR gate are universal gates, we have to show that all the three basic logic gates can be realized using only NAND gates or using only NOR gates. The diagrams given below show the realization of AND, OR, and NOT functions using either only NAND gates or only NOR gates.

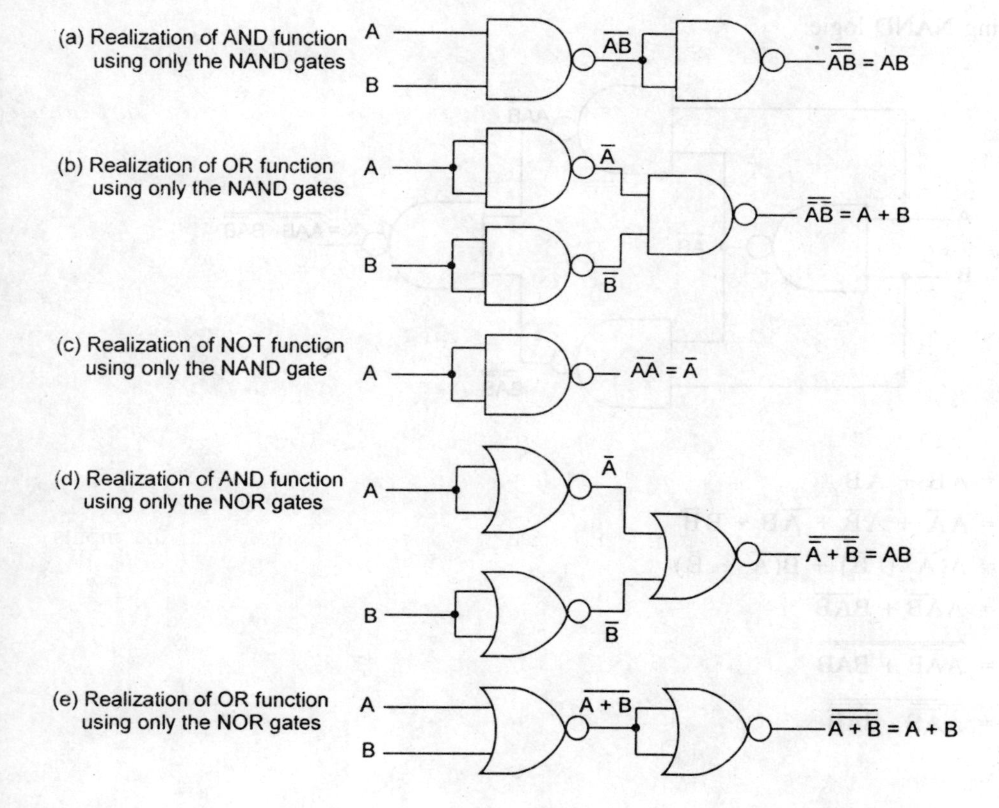

(f) Realization of NOT function using only the NOR gate A —— $\overline{A + A} = \overline{A}$

EXAMPLE 5.16 Realize the X-OR function using (a) AOI logic, (b) NAND logic, and (c) NOR logic.

Solution

(a) Using AOI logic:

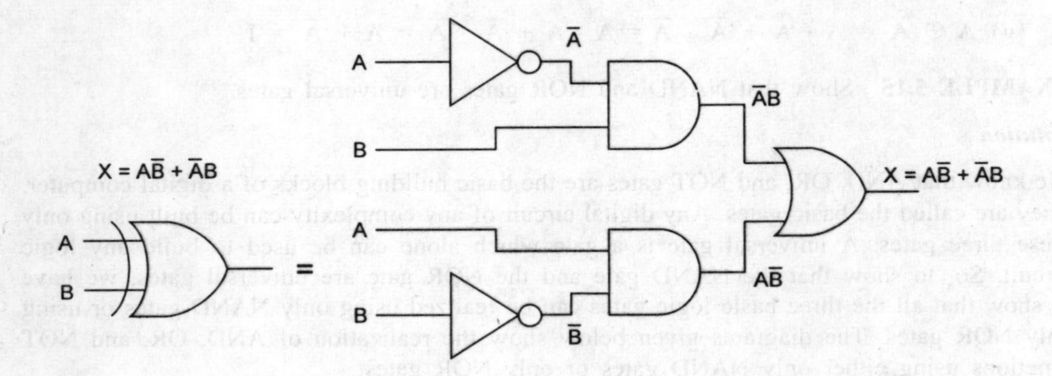

(b) Using NAND logic:

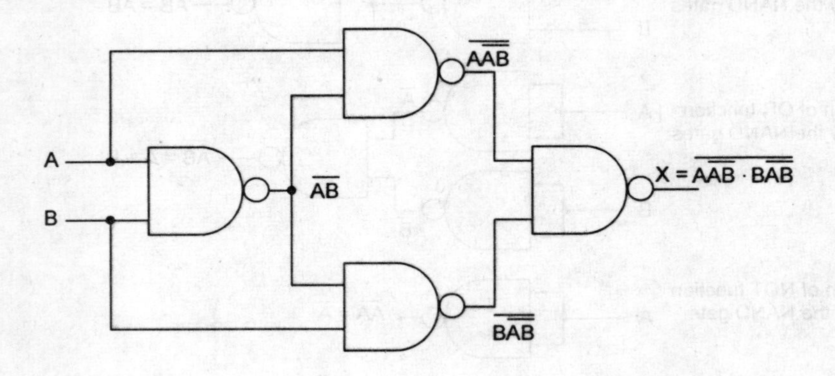

$$X = A\overline{B} + \overline{A}B$$
$$= A\overline{A} + A\overline{B} + \overline{A}B + B\overline{B}$$
$$= A(\overline{A} + \overline{B}) + B(\overline{A} + \overline{B})$$
$$= A\overline{AB} + B\overline{AB}$$
$$= \overline{\overline{A\overline{AB} + B\overline{AB}}}$$
$$= \overline{\overline{A\overline{AB}} \cdot \overline{B\overline{AB}}}$$

(c) Using NOR logic:

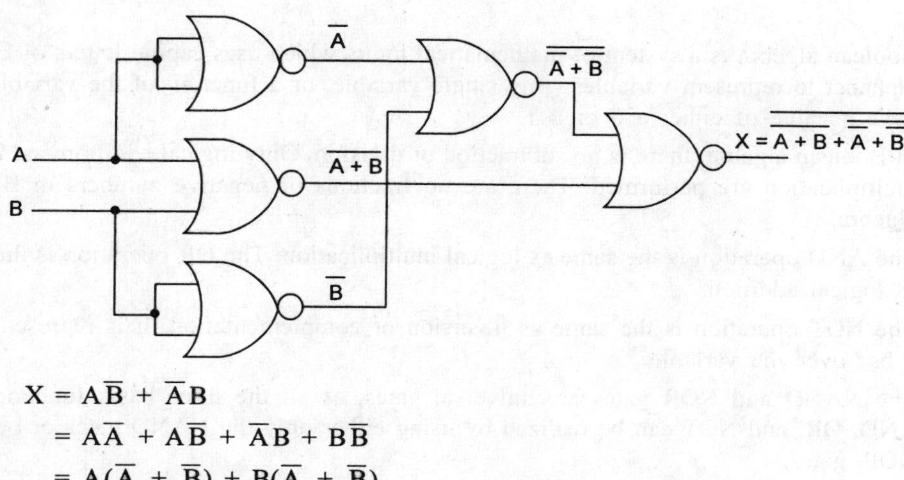

$$
\begin{aligned}
X &= A\overline{B} + \overline{A}B \\
&= A\overline{A} + A\overline{B} + \overline{A}B + B\overline{B} \\
&= A(\overline{A} + \overline{B}) + B(\overline{A} + \overline{B}) \\
&= (A + B)(\overline{A} + \overline{B}) \\
&= \overline{\overline{(A + B)(\overline{A} + \overline{B})}} \\
&= \overline{\overline{(A + B)} + \overline{(\overline{A} + \overline{B})}}
\end{aligned}
$$

5.9 DETERMINATION OF OUTPUT LEVEL FROM THE DIAGRAM

The output logic level for the given input levels can be determined by obtaining the Boolean expression for the output of the circuit and then substituting the values of the inputs in that expression. This can also be determined directly from the circuit diagram by writing down the output level of each gate starting from the input side till the final output is reached. This technique is often used for troubleshooting or testing of a logic system. This procedure is self-evident from the illustration given below.

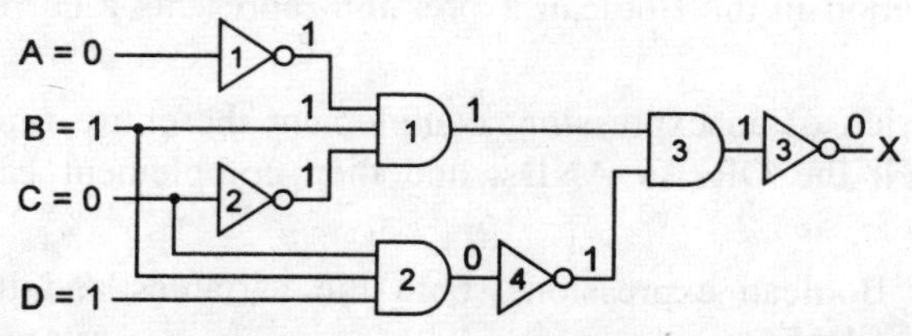

The same result can be obtained by writing the output expression and substituting the values of the inputs in it. Thus,

$$
\begin{aligned}
X &= \overline{(\overline{A}\ B\ \overline{C})\overline{(BCD)}} \\
&= \overline{(\overline{0}\cdot 1\cdot \overline{0})\overline{(1\cdot 0\, 1)}} \\
&= \overline{1\cdot \overline{0}} \\
&= 0
\end{aligned}
$$

SUMMARY

- Boolean algebra is a system of mathematical logic, which uses capital letters of English alphabet to represent variables. Any single variable, or a function of the variables can have a value of either a 0 or a 1.
- In Boolean algebra, there is no subtraction or division. Only logical addition and logical multiplication are performed. There are no fractions or negative numbers in Boolean algebra.
- The AND operation is the same as logical multiplication. The OR operation is the same as logical addition.
- The NOT operation is the same as inversion or complementation. It is represented by a bar over the variable.
- The NAND and NOR gates are universal gates, as all the three basic functions, i.e. AND, OR, and NOT can be realized by using either only the NAND gates or only the NOR gates.
- In universal logic, logic circuits are realized using either only the NAND gates or only the NOR gates.
- The X-OR operation is called the modulo-2 addition operation because it assigns to each pair of elements its modulo-2 sum.
- Axioms or postulates of Boolean algebra are a set of logical expressions that we accept without proof and upon which we can build a set of useful theorems.
- De Morgan's theorem states that the complement of a sum of variables is equal to the product of their individual complements, and the complement of a product of variables is equal to the sum of their individual complements.
- De Morgan's theorem allows removal of variables from under a NOT sign. It allows transformation from SOP form to POS form.
- The dual of an expression is obtained by changing ANDs to ORs, ORs to ANDs, 0s to 1s, and 1s to 0s. The variables are not complemented.
- Every logic operation in the Boolean expression represents a corresponding element of hardware.
- For demorganization of an expression, complement the entire function, change all the ANDs to ORs, all the ORs to ANDs, and then complement each of the individual variables.
- In realization of Boolean expressions, both the variables and their complement are assumed to be available.
- In AOI logic, the circuits are realized using AND, OR, and NOT gates only.
- There are two basic forms of Boolean expressions—SOP form and POS form.
- The hybrid form of realization is a combination of both SOP and POS forms.
- A/O gates in which an additional variable or a combination of variables can be included in the logic operation, are called expandable gates.
- A circuit having n AND gates feeding an OR gate is called an n-wide A/O gate.

- Active-LOW gates and circuits are those in which the action represented or initiated by a variable occurs when it is equal to a 0.
- Active-HIGH gates and circuits are those in which the action represented or initiated by a variable occurs when it is equal to a 1.
- On logic diagrams, an inversion bubble at the point where the input is connected to a gate, indicates an active-LOW input.
- An OR gate in the positive logic system becomes an AND gate in the negative logic system and vice versa.
- *Laws of Boolean Algebra at a glance*

Laws of complementation	$\overline{0} = 1$; $\overline{1} = 0$; $\overline{\overline{A}} = A$
	If A = 0, then $\overline{A} = 1$; if A = 1, then $\overline{A} = 0$
AND laws	$A \cdot 0 = 0$; $A \cdot 1 = A$; $A \cdot A = A$; $A \cdot \overline{A} = 0$
OR laws	$A + 0 = A$; $A + 1 = 1$; $A + A = A$; $A + \overline{A} = 1$
Commutative laws	$A + B = B + A$; $AB = BA$
Associative laws	$(A + B) + C = A + (B + C)$; $(AB) \cdot C = A \cdot (BC)$
Distributive laws	$A \cdot (B + C) = AB + AC$; $A + BC = (A + B)(A + C)$
Idempotence laws	$A \cdot A = A$; $A + A = A$
Negation laws	$A \cdot \overline{A} = 0$; $A + \overline{A} = 1$
Double negation law	$\overline{\overline{A}} = A$
Identity laws	$A \cdot 1 = A$; $A + 1 = 1$
Null laws	$A \cdot 0 = 0$; $A + 0 = A$
Absorption laws	$A + AB = A$; $A(A + B) = A$
Consensus theorem (Included factor theorem)	$AB + \overline{A}C + BC = AB + \overline{A}C$
Transposition theorem	$AB + \overline{A}C = (A + C)(\overline{A} + B)$
De Morgan's theorem	$\overline{A + B} = \overline{A} \cdot \overline{B}$; $\overline{AB} = \overline{A} + \overline{B}$

QUESTIONS

1. What do you mean by the following logic operations?

   (a) AND (b) OR (c) NOT (d) NAND
   (e) NOR (f) X-OR (g) X-NOR

2. What do you mean by axioms or postulates of Boolean algebra?
3. State and prove commutative, associative, distributive, idempotence, negation, identity, null and absorption laws of Boolean algebra.

4. State and prove consensus theorem and transposition theorem.

5. State and prove De Morgan's theorem.

6. How do you demorganize a Boolean expression?

7. How do you obtain the dual of a Boolean function?

8. What are the steps followed in the reduction of Boolean expressions?

9. What do you mean by (a) active-HIGH gates or circuits and (b) active-LOW gates or circuits?

10. What is (a) AOI logic and (b) universal logic?

11. How do you convert AOI logic to (a) NAND logic and (b) NOR logic?

12. What are the two basic forms of Boolean expressions?

13. From positive to negative logic system, what does an OR gate become and what does an AND gate become?

14. Show that both NAND gate and NOR gate are universal gates.

15. Realize X-OR operation using (a) only NAND gates, (b) only NOR gates, and (c) AOI logic.

PROBLEMS

5.1 Verify by the truth table method.

(a) $A + \overline{A}B + AB = A + B$ (b) $(A + \overline{B})(\overline{A} + B) = AB + \overline{A}\,\overline{B}$

5.2 Reduce the following Boolean expressions.

(a) $AABBC$ (b) $ABCBAC$ (c) $A \cdot I \cdot A \cdot A$

(d) $ABC\overline{A}\,\overline{B}$ (e) $AB\overline{B}AC$ (f) $BB\overline{B}B$

5.3 Reduce the following Boolean expressions.

(a) $P + Q + P$ (b) $P + Q + R + \overline{P}$ (c) $P + Q + R + R + \overline{R}$

(d) $P + Q + R + 1$ (e) $0 + P + Q + 1$ (f) $P + P + P + P$

5.4 Reduce the following Boolean expressions.

(a) $XY + XY + Y + Y$ (b) $X\overline{X} + YY\overline{Y}$ (c) $X\overline{X}Y + XY\overline{Y}$

(d) $XY + XY\overline{Y}$ (e) $XY + XY + \overline{XY}$ (f) $XYZ + YYZ + \overline{Y}Z + X\overline{Y}$

5.5 Reduce the following Boolean expressions.

(a) $X(\overline{X} + YZ)$ (b) $X(YZ + \overline{Y}Z)$ (c) $X(\overline{X}Y + \overline{X}Z)$ (d) $AAB(\overline{A}BC + BBC)$

5.6 Reduce the following Boolean expressions.

(a) $AB + A(B + C) + \overline{B}(B + D)$ (b) $(X + Y + Z)(\overline{X} + \overline{Y} + \overline{Z})X$

(c) $ABEF + AB\overline{EF} + \overline{AB}EF$ (d) $ABC[AB + \overline{C}(BC + AC)]$

(e) $\overline{A}B + \overline{A}B\overline{C} + \overline{A}BCD + \overline{A}B\overline{C}\,\overline{D}E$ (f) $A + B + \overline{A}BC$

(g) $\overline{B}\,\overline{C}D + \overline{(B + C + D)} + \overline{B}\,\overline{C}\,\overline{D}E$ (h) $\overline{(A\overline{B}C)(\overline{\overline{AB}}) + BC}$

(i) $(WX + W\overline{Y})(X + W) + WX(\overline{X} + \overline{Y})$ (j) $AB + \overline{AC} + A\overline{B}C(AB + C)$

(k) $\overline{A}\,\overline{B}\,\overline{C} + \overline{A}B\overline{C} + A\overline{B}\,\overline{C} + AB\overline{C}$ (l) $AB + \overline{AC} + A\overline{B}C(AB + C)$

5.7 Prove that

(a) $\overline{\overline{AB} + \overline{A} + AB} = 0$ (b) $AB + \overline{AC} + A\overline{B}C(AB + C) = 1$

(c) $\overline{\overline{A\overline{B} + ABC} + A(B + A\overline{B})} = 0$ (d) $AB + A(B + C) + B(B + C) = B + AC$

(e) $A\overline{B}(C + BD) + \overline{A}\,\overline{B} = \overline{B}C$

(f) $\overline{A}\,\overline{B}C + \overline{(A + B + \overline{C})} + \overline{A}\,\overline{B}\,\overline{C}D = A\overline{B}(C + D)$

(g) $ABCD + AB(\overline{CD}) + (\overline{AB})CD = AB + CD$

(h) $(A + \overline{A})(AB + AB\overline{C}) = AB$ (i) $(A\overline{B} + A\overline{C})(BC + B\overline{C})(ABC) = 0$

(j) $A\overline{B}C + \overline{A}BC + ABC = AC + AB$ (k) $A[B + C(\overline{AB + AC}) = AB$

(l) $(\overline{A + \overline{BC}})(A\overline{B} + \overline{ABC}) = \overline{A}BC$ (m) $A + \overline{B}C(A + \overline{\overline{BC}}) = A$

(n) $\overline{ABC}\,\overline{(A + B + C)} = ABC$ (o) $\overline{\overline{ABC + \overline{A}\,\overline{B}} + BC} = \overline{A}\,\overline{B}$

5.8 Apply De Morgan's theorem to each of the following expressions.

(a) $\overline{P(Q + R)}$ (b) $\overline{(P + \overline{Q})(\overline{R} + S)}$

(c) $\overline{\overline{\overline{(A + B)}\,\overline{(C + D)}}\;\overline{\overline{(E + F)}\,\overline{(G + H)}}}$ (d) $(\overline{A} = B + \overline{C} + D)(0 + \overline{A}BC\overline{D})$

5.9 Without reducing, convert the following expressions to NAND logic.

(a) $(A + B)(C + D)$ (b) $(A + C)(ABC + ACD)$

(c) $(A + \overline{B}C)D$ (d) $AB + CD(A\overline{B} + CD)$

5.10 Without reducing, convert the following expressions to NAND logic.

(a) $A + BC + ABC$ (b) $(XY + Z)(XY + P)$ (c) $(1 + A)(ABC)$

5.11 Without reducing, convert the following expressions to NOR logic.

(a) $X + Y + XY$ (b) $(X\overline{Y} + X + \overline{X + Y})$ (c) $(1 + A)(AC)$

5.12 Without reducing, implement the following expressions in AOI logic and then convert them into (1) NAND logic and (2) NOR logic.

(a) $A + BC = (A + \overline{B}C) + D$ (b) $A + B\overline{C} + \overline{B + C} + \overline{B}\,\overline{C}$

5.13 Prove that

(a) If $A + B = A + C$ and $\overline{A} + B = \overline{A} + C$, then $B = C$.

(b) If $A + B = A + C$ and $AB = AC$, then $B = C$.

5.14 Given $A\overline{B} + \overline{A}B = C$, show that $A\overline{C} + \overline{A}C = B$.

Chapter 6

THE KARNAUGH MAP AND QUINE-McCLUSKY METHODS

6.1 INTRODUCTION

We have seen how Boolean expressions can be simplified algebraically, but being not a systematic method we can never be sure whether the minimal expression obtained is the real minimal. The effectiveness of algebraic simplification depends on our familiarity with, and ability to apply Boolean algebraic rules, laws and theorems. The Karnaugh map (K-map), on the other hand, is a systematic method of simplifying Boolean expressions. The K-map is a chart or a graph, composed of an arrangement of adjacent cells, each representing a particular combination of variables in sum or product form. Like a truth table, it is a means of showing the relationship between the logic inputs and the desired output. Although a K-map can be used for problems involving any number of variables, it becomes tedious for problems involving five or more variables. Usually it is limited to six variables. An n variable function can have 2^n possible combinations of product terms in SOP form, or 2^n possible combinations of sum terms in POS form. Since the K-map is a graphical representation of Boolean expressions, a two-variable K-map will have $2^2 = 4$ cells or squares, a three variable map will have $2^3 = 8$ cells or squares, and a four variable map will have $2^4 = 16$ cells, and so on.

Any Boolean expression can be expressed in a *standard* or *canonical* or *expanded sum* (OR) *of products* (AND)—*SOP form*—or in a *standard* or *canonical* or *expanded product* (AND) *of sums* (OR)—*POS form*. A standard SOP form is one in which a number of product terms, each one of which contains all the variables of the function either in complemented or non-complemented form, are summed together. A standard POS form is one in which a number of sum terms, each one of which contains all the variables of the function either in complemented or non-complemented form, are multiplied together. Each of the product terms in the standard SOP form is called a *minterm* and each of the sum terms in the standard POS form is called a *maxterm*. For simplicity, the minterms and maxterms are usually represented as binary words in terms of 0s and 1s, instead of actual variables. For minterms, the binary words are formed by representing each non-complemented variable by a 1 and each complemented variable by a 0, and the decimal equivalent of this binary word is expressed as a subscript of lower case m, i.e. m_0, m_2, m_5, m_7, etc. For maxterms, the binary words are formed by representing each non-complemented variable by a 0 and each complemented variable by a 1, and the decimal equivalent of this binary word is expressed as a subscript of the upper case letter M, i.e. M_0, M_1, etc. Any given function which is not in the standard form, can always be converted to standard form by *unreducing*, that is, *expanding* the function.

A standard SOP form can always be converted to a standard POS form, by treating the missing minterms of the SOP form as the maxterms of the POS form. Similarly, a standard POS form can always be converted to a standard SOP form, by treating the missing maxterms of the POS form as the minterms of the corresponding SOP form.

6.2 EXPANSION OF A BOOLEAN EXPRESSION TO SOP FORM

The following steps are followed for the expansion of a Boolean expression in SOP form to the standard SOP form:

1. Write down all the terms.
2. If one or more variables are missing in any term, expand that term by multiplying it with the sum of each one of the missing variable and its complement.
3. Drop out the redundant terms.

Also, the given expression can be directly written in terms of its minterms by using the following procedure:

1. Write down all the terms.
2. Put Xs in terms where variables must be inserted to form a minterm.
3. Replace the non-complemented variables by 1s and the complemented variables by 0s, and use all combinations of Xs in terms of 0s and 1s to generate minterms.
4. Drop out all the redundant terms.

EXAMPLE 6.1 Expand $\overline{A} + \overline{B}$ to minterms and maxterms.

Solution

The given expression is a two-variable function. In the first term $\overline{A}$, the variable B is missing; so, multiply it by $(B + \overline{B})$. In the second term $\overline{B}$, the variable A is missing; so, multiply it by $(A + \overline{A})$. Therefore,

$$
\begin{aligned}
\overline{A} + \overline{B} &= \overline{A}(B + \overline{B}) + \overline{B}(A + \overline{A}) \\
&= \overline{A}B + \overline{A}\,\overline{B} + \overline{B}A + \overline{B}\,\overline{A} \\
&= \overline{A}B + \overline{A}\,\overline{B} + A\overline{B} + \overline{A}\,\overline{B} \\
&= \overline{A}B + \overline{A}\,\overline{B} + A\overline{B} \\
&= 01 + 00 + 10 \\
&= m_1 + m_0 + m_2 \\
&= \Sigma\, m(0, 1, 2)
\end{aligned}
$$

The minterm m_3 is missing in the SOP form. Therefore, the maxterm M_3 will be present in the POS form. Hence the POS form is ΠM_3, i.e. $\overline{A} + \overline{B}$. Also,

$$\begin{aligned}\overline{A} + \overline{B} &= \overline{A}\cdot X + X\cdot\overline{B}\\ &= 0X + X0\\ &= 00 + 01 + 00 + 10\\ &= 00 + 01 + 10\\ &= \Sigma\ m(0,\ 1,\ 2)\end{aligned}$$

EXAMPLE 6.2 Expand $A + B\overline{C} + AB\overline{D} + ABCD$ to minterms and maxterms.

Solution

The given expression is a four-variable function. In the first term A, the variables B, C, and D are missing. So, multiply it by $(B + \overline{B})\ (C + \overline{C})\ (D + \overline{D})$. In the second term $B\overline{C}$, the variables A and D are missing. So, multiply it by $(A + \overline{A})\ (D + \overline{D})$. In the third term, $AB\overline{D}$, the variable C is missing. So, multiply it by $(C + \overline{C})$. In the fourth term ABCD, all the variables are present. So, leave it as it is. Therefore,

$$A = A(B + \overline{B})\ (C + \overline{C})\ (D + \overline{D})$$

$$= ABCD + ABC\overline{D} + AB\overline{C}D + AB\overline{C}\,\overline{D} + A\overline{B}CD + A\overline{B}C\overline{D} + A\overline{B}\,\overline{C}D + A\overline{B}\,\overline{C}\,\overline{D}$$

$$B\overline{C} = B\overline{C}(A + \overline{A})\ (D + \overline{D}) = AB\overline{C}D + AB\overline{C}\,\overline{D} + \overline{A}B\overline{C}D + \overline{A}B\overline{C}\,\overline{D}$$

$$AB\overline{D} = AB\overline{D}(C + \overline{C}) = ABC\overline{D} + AB\overline{C}\,\overline{D}$$

or

$$\begin{aligned}A + B\overline{C} + AB\overline{D} + ABCD = {} & ABCD + ABC\overline{D} + AB\overline{C}D + AB\overline{C}\,\overline{D} + A\overline{B}CD + A\overline{B}C\overline{D}\\ & + A\overline{B}\,\overline{C}D + A\overline{B}\,\overline{C}\,\overline{D} + \overline{A}B\overline{C}D + \overline{A}B\overline{C}\,\overline{D}\\ = {} & m_{15} + m_{14} + m_{13} + m_{12} + m_{11} + m_{10} + m_9 + m_8 + m_5 + m_4\\ = {} & \Sigma\ m(4,\ 5,\ 8,\ 9,\ 10,\ 11,\ 12,\ 13,\ 14,\ 15)\end{aligned}$$

In the SOP form, the minterms 0, 1, 2, 3, 6, and 7 are missing. So in the POS form, the maxterms 0, 1, 2, 3, 6, and 7 will be present. Therefore, the POS form is

$$\Pi\ M(0,\ 1,\ 2,\ 3,\ 6,\ 7)$$

Also,

$$\begin{aligned}A &= AXXX = 1XXX\\ &= 1000 + 1001 + 1010 + 1011 + 1100 + 1101 + 1110 + 1111\\ &= m_8 + m_9 + m_{10} + m_{11} + m_{12} + m_{13} + m_{14} + m_{15}\end{aligned}$$

$$\begin{aligned}B\overline{C} &= XB\overline{C}X = X10X\\ &= 0100 + 0101 + 1100 + 1101\\ &= m_4 + m_5 + m_{12} + m_{13}\end{aligned}$$

$$AB\overline{D} = ABX\overline{D} = 11X0 = 1100 + 1110 = m_{12} + m_{14}$$

Dropping out the redundant terms, we get

$$A + B\overline{C} + AB\overline{D} + ABCD = \Sigma\, m(4, 5, 8, 9, 10, 11, 12, 13, 14, 15)$$

6.3 EXPANSION OF A BOOLEAN EXPRESSION TO POS FORM

The expansion of a Boolean expression to the standard POS form is conducted as follows:

1. If one or more variables are missing in any sum term, expand that term by adding the products of each of the missing term and its complement.
2. Drop out the redundant terms.

The given expression can also be written in terms of maxterms by using the following procedure:

(a) Put Xs in terms wherever variables must be inserted to form a maxterm.

(b) Replace thé complemented variables by 1s and thé non-complemented variables by 0s and use all combinations of Xs in terms of 0s and 1s to generate maxterms.

(c) Drop out the redundant terms.

EXAMPLE 6.3 Expand $A(\overline{B} + A)B$ to maxterms and minterms.

Solution

The given expression is a two-variable function in the POS form. The variable B is missing in the first term A. So, add $B\overline{B}$ to it. The second term contains all the variables. So, leave it as it is. The variable A is missing in the third term B. So, add $A\overline{A}$ to it. Therefore,

$$A = A + B\overline{B} = (A + B)\,(A + \overline{B})$$

$$B = B + A\overline{A} = (B + A)\,(B + \overline{A})$$

or

$$\begin{aligned} A(\overline{B} + A)B &= (A + B)\,(A + \overline{B})\,(A + \overline{B})\,(A + B)\,(\overline{A} + B) \\ &= (A + B)\,(A + \overline{B})\,(\overline{A} + B) \\ &= (00)\;(01)\;(10) \\ &= M_0 \cdot M_1 \cdot M_2 \\ &= \Pi\, M(0, 1, 2) \end{aligned}$$

The maxterm M_3 is missing in the POS form. So, the SOP form will contain only the minterm m_3.

Also,

$$A \rightarrow 0X = (00)\;(01) = M_0 \cdot M_1$$

$$(A + \overline{B}) \rightarrow (01) = M_1$$

$$B \rightarrow X0 = (00)\;(10) = M_0 \cdot M_2$$

Therefore,

$$A(A + \overline{B})B = \prod M(0, 1, 2)$$

EXAMPLE 6.4 Expand $A(\overline{A} + B)(\overline{A} + B + \overline{C})$ to maxterms and minterms.

Solution

The given expression is a three-variable function in the POS form. The variables B and C are missing in the first term A. So, add $B\overline{B}$ and $C\overline{C}$ to it. The variable C is missing in the second term $(\overline{A} + B)$. So, add $C\overline{C}$ to it. The third term $(\overline{A} + B + \overline{C})$ contains all the three variables. So, leave it as it is. Therefore,

$$A = A + B\overline{B} + C\overline{C} = (A + B)(A + \overline{B}) + C\overline{C} = (A + B + C\overline{C})(A + \overline{B} + C\overline{C})$$

$$= (A + B + C)(A + B + \overline{C})(A + \overline{B} + C)(A + \overline{B} + \overline{C})$$

$$\overline{A} + B = \overline{A} + B + C\overline{C} = (\overline{A} + B + C)(\overline{A} + B + \overline{C})$$

Therefore,

$$A(\overline{A} + B)(\overline{A} + B + \overline{C}) = (A + B + C)(A + B + \overline{C})(A + \overline{B} + C)(A + \overline{B} + \overline{C})$$

$$(\overline{A} + B + C)(\overline{A} + B + \overline{C})$$

$$= (000)\ (001)\ (010)\ (011)\ (100)\ (101)$$

$$= M_0 \cdot M_1 \cdot M_2 \cdot M_3 \cdot M_4 \cdot M_5$$

$$= \prod M(0, 1, 2, 3, 4, 5)$$

The maxterms M_6 and M_7 are missing in the POS form. So, the SOP form will contain the minterms 6 and 7. Therefore, the given expression in the SOP form is $\Sigma\ m(6, 7)$.

Also,

$$A \rightarrow 0XX = (000)\ (001)\ (010)\ (011)$$

$$= M_0 \cdot M_1 \cdot M_2 \cdot M_3$$

$$\overline{A} + B \rightarrow 10X = (100)\ (101) = M_4 \cdot M_5$$

$$\overline{A} + B + \overline{C} = 101 = M_5$$

Therefore,

$$A(\overline{A} + B)(\overline{A} + B + \overline{C}) = \prod M(0, 1, 2, 3, 4, 5)$$

EXAMPLE 6.5 Write the algebraic terms of a four-variable expression having the following minterms.

(a) m_0 (b) m_5 (c) m_9 (d) m_{14}

Solution

Given minterm	m_0	m_5	m_9	m_{14}
Binary form	0000	0101	1001	1110
Product term	$\overline{A}\,\overline{B}\,\overline{C}\,\overline{D}$	$\overline{A}B\overline{C}D$	$A\overline{B}\,\overline{C}D$	$ABC\overline{D}$

EXAMPLE 6.6 Write the algebraic terms of a four-variable expression having the following maxterms.

(a) M_3 (b) M_9 (c) M_{11} (d) M_{14}

Solution

Given maxterm	M_3	M_9	M_{11}	M_{14}
Binary form	0011	1001	1011	1110
Sum term	$A + B + \overline{C} + \overline{D}$	$\overline{A} + B + C + \overline{D}$	$\overline{A} + B + \overline{C} + \overline{D}$	$\overline{A} + \overline{B} + \overline{C} + D$

6.4 COMPUTATION OF TOTAL GATE INPUTS

The total number of gate inputs required to realize a Boolean expression is computed as follows:

If the expression is in the SOP form, count the number of AND inputs and the number of AND gates feeding the OR gate. If the expression is in the POS form, count the number of OR inputs and the number of OR gates feeding the AND gate. If it is in hybrid form, count the gate inputs and the gates feeding other gates.

The cost of implementing a circuit is roughly proportional to the number of gate inputs required.

EXAMPLE 6.7 How many gate inputs are required to realize the following expressions?

(a) $f_1 = ABC + A\overline{B}CD + E\overline{F} + AD$

(b) $f_2 = A(B + C + \overline{D})\,(\overline{B} + C + \overline{E})\,(A + \overline{B} + C + E)$

Solution

(a)

Write the expression	ABC	+	$A\overline{B}CD$	+	$E\overline{F}$	+	AD		
Count the AND inputs	3	+	4	+	2	+	2	=	11
Count the AND gates feeding the OR gate	1	+	1	+	1	+	1	=	4
Total gate inputs								=	15

(b)

Write the expression	A	$\cdot$	$(B + C + \overline{D})$	$\cdot$	$(\overline{B} + C + \overline{E})$	$\cdot$	$(A + \overline{B} + C + E)$		
Count the OR inputs	0	+	3	+	3	+	4	=	10
Count the OR gates feeding the AND gate	1	+	1	+	1	+	1	=	4
Total gate inputs								=	14

6.5 TWO-VARIABLE K-MAP

A two-variable expression can have $2^2 = 4$ possible combinations of the input variables A and B. Each of these combinations, $\overline{A}\,\overline{B}$, $\overline{A}B$, $A\overline{B}$, and AB (in the SOP form) is called a minterm. Instead of representing the minterms in terms of the input variables, using the shorthand notation the minterms may be represented in terms of their decimal designations—m_0 for $\overline{A}\,\overline{B}$, m_1 for $\overline{A}B$, m_2 for $A\overline{B}$, and m_3 for AB, assuming that A represents the MSB. The letter m stands for minterm and the subscript represents the decimal designation of the minterm.

The presence (absence) of a minterm in the expression indicates that the output of the logic circuit assumes a logic 1 (logic 0) level for that combination of input variables.

Consider the expression, $\overline{A}\,\overline{B} + A\overline{B} + AB$. It can be expressed using minterms as

$$m_0 + m_2 + m_3 = \Sigma\, m(0, 2, 3)$$

and should be read as the sum of minterms 0, 2, and 3. It can also be represented in terms of a truth table as shown below.

Minterm	Inputs		Output
	A	B	f
0	0	0	1
1	0	1	0
2	1	0	1
3	1	1	1

The first column indicates the minterm designation, the second column indicates the input combinations, and the last column indicates the presence or absence of that minterm in the output expression. A 1 in the output column indicates that the output contains that particular minterm in its sum and a 0 in that column indicates that the particular minterm does not appear in the expression for output. Such information about the two-variable expression can also be indicated by a two-variable K-map.

Mapping of SOP Expressions

A two-variable K-map has $2^2 = 4$ squares. These squares are called cells. Each square on the K-map represents a unique minterm. The minterm designations of the squares are shown in Fig. 6.1. A 1 placed in any square indicates that the corresponding minterm is included in the output expression, and a 0 or no entry in any square indicates that the corresponding minterm does not appear in the expression for output.

A \ B	0	1
0	0: $\overline{A}\,\overline{B}$	1: $\overline{A}B$
1	2: $A\overline{B}$	3: AB

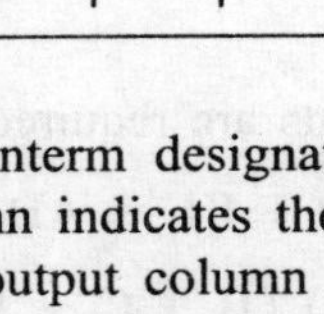

Fig. 6.1 Two-variable K-map.

The mapping of the expression $\Sigma\, m(0, 2, 3)$ is shown in Fig. 6.2.

A \ B	0	1
0	1 (0)	0 (1)
1	1 (2)	1 (3)

Fig. 6.2 K-map of Σ m(0, 2, 3).

EXAMPLE 6.8 Map the expression $\overline{A}B + A\overline{B}$.

Solution

The given expression in minterms is

$$m_1 + m_2 = \Sigma\, m(1, 2).$$

The K-map is shown below.

A \ B	0	1
0	0 (0)	1 (1)
1	1 (2)	0 (3)

Minimization of SOP Expressions

To minimize a Boolean expression given in the SOP form by using the K-map, we have to look for adjacent squares having 1s, that is, minterms adjacent to each other, and combine them to form larger squares to eliminate some variables. Two squares are said to be adjacent to each other, if their minterms differ in only one variable. For example, in a two-variable K-map, m_0 and m_1, i.e. $\overline{A}\,\overline{B}$ and $\overline{A}B$ differ only in variable B ($\overline{A}$ is common to both of them). So, they may be combined to form a 2-square to eliminate the variable B.

Similarly, minterms m_0 ($\overline{A}\,\overline{B}$) and m_2 ($A\overline{B}$); m_1 ($\overline{A}B$) and m_3 (AB); and m_2 ($A\overline{B}$) and m_3 (AB) are adjacent to each other. However, minterms m_0 ($\overline{A}\,\overline{B}$) and m_3 (AB), and m_1 ($\overline{A}B$) and m_2 ($A\overline{B}$) are not adjacent to each other, because they differ in more than one variable.

The necessary (but not sufficient) condition for adjacency of minterms is that their decimal designations must differ by a power of 2. A minterm can be combined with any number of minterms adjacent to it to form larger squares.

Two minterms, which are adjacent to each other, can be combined to form a bigger square called a 2-square or a pair. This eliminates one variable—the variable that is not common to both the minterms. For example in Fig. 6.3,

m_0 and m_1 can be combined to yield, $f_1 = m_0 + m_1 = \overline{A}\,\overline{B} + \overline{A}B = \overline{A}$

m_0 and m_2 can be combined to yield, $f_2 = m_0 + m_2 = \overline{A}\,\overline{B} + A\overline{B} = \overline{B}$

m_1 and m_3 can be combined to yield, $f_3 = m_1 + m_3 = \overline{A}B + AB = B$

m_2 and m_3 can be combined to yield, $f_4 = m_2 + m_3 = A\overline{B} + AB = A$

m_1, m_2, m_3, and m_4 can be combined to yield,

$$f_5 = m_0 + m_1 + m_2 + m_3$$
$$= \overline{A}\,\overline{B} + \overline{A}B + A\overline{B} + AB$$
$$= \overline{A} + A$$
$$= 1$$

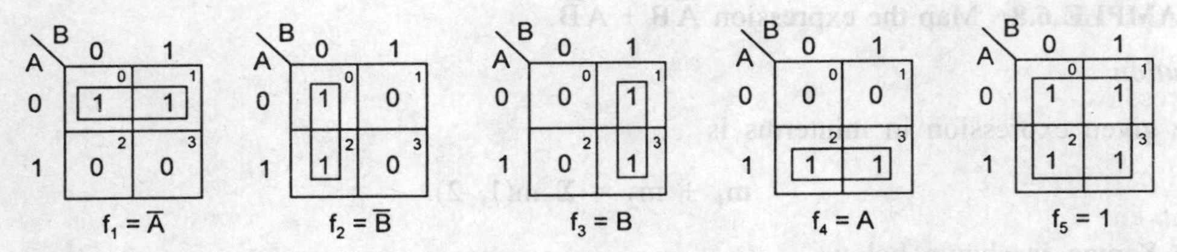

Fig. 6.3 The possible minterm groupings in a two-variable K-map.

Two 2-squares adjacent to each other can be combined to form a 4-square. A 4-square eliminates 2 variables. A 4-square is called a *quad*.

To read the squares on the map after minimization, consider only those variables which remain constant throughout the square, and ignore the variables which are varying. Write the non-complemented variable if the variable remaining constant is a 1, and the complemented variable if the variable remaining constant is a 0, and write the variables as a product term. In Fig. 6.3, f_1 is read as $\overline{A}$, because, along the square, A remains constant as a 0, that is, as $\overline{A}$, whereas B is changing from 0 to 1. f_3 is read as B, because, along the square, B remains constant as a 1, whereas A is changing from 0 to 1. f_5 is read as a 1, because, no variable remains constant throughout the square, which means that the output is a 1 for any combination of inputs.

EXAMPLE 6.9 Reduce the expression $\overline{A}\,\overline{B} + \overline{A}B + AB$ using mapping.

Solution

Expressed in terms of minterms, the given expression is

$$m_0 + m_1 + m_3 = \Sigma\, m(0, 1, 3)$$

The diagram below shows the K-map and its reduction. In one 2-square, A is constant as a 0 but B varys from a 0 to a 1, and in the other 2-square, B is constant as a 1 but A varys from a 0 to a 1. So, the reduced expression is $\overline{A} + B$. It requires two gate inputs for realization as shown below.

$f = \overline{A} + B$

The main criteria in the design of a digital circuit is that its cost should be as low as possible. To design a circuit with the least cost, the expression used to realize that circuit must be minimal. Since the cost is roughly proportional to the number of gate inputs in the circuit, an expression is considered minimal only if it corresponds to the least possible number of gate inputs. There is no guarantee that the minimal expression obtained from the K-map in the SOP form is the real minimal. To obtain the real minimal expression, we obtain the minimal expressions for any problem in both the SOP and POS forms by using the K-maps and then take the minimal of these two minimals.

We know that the 1s on the K-map indicate the presence of minterms in the output expression, whereas the 0s indicate the absence of minterms. Since the absence of a minterm in the SOP expression means the presence of the corresponding maxterm in the POS expression of the same problem, when a SOP expression is plotted on the K-map, 0s or no entries on the K-map represent the maxterms. To obtain the minimal expression in the POS form, consider the 0s on the K-map and follow the procedure used for combining 1s. Also, since the absence of a maxterm in the POS expression means the presence of the corresponding minterm in the SOP expression of the same problem, when a POS expression is plotted on the K-map, 1s or no entries on the K-map represent the minterms.

Mapping of POS Expressions

Each sum term in the standard POS expression is called a maxterm. A function in two variables (A, B) has four possible maxterms, $A + B$, $A + \overline{B}$, $\overline{A} + B$, and $\overline{A} + \overline{B}$. They are represented as M_0, M_1, M_2, and M_3 respectively. The upper case letter M stands for maxterm and its subscript denotes the decimal designation of that maxterm obtained by treating the non-complemented variable as a 0 and the complemented variable as a 1 and putting them side by side for reading the decimal equivalent of the binary number so formed.

For mapping a POS expression on to the K-map, 0s are placed in the squares corresponding to the maxterms which are present in the expression and 1s are placed (or no entries are made) in the squares corresponding to the maxterms which are not present in the expression. The decimal designation of the squares for maxterms is the same as that for the minterms. A two-variable K-map and the associated maxterms are shown in Fig. 6.4.

A \ B	0	1
0	0: $A + B$	1: $A + \overline{B}$
1	2: $\overline{A} + B$	3: $\overline{A} + \overline{B}$

Fig. 6.4 The maxterms of a two-variable K-map.

EXAMPLE 6.10 Plot the expression $(A + B)(\overline{A} + B)(\overline{A} + \overline{B})$ on the K-map.

Solution

The given expression in terms of maxterms is $\prod M(0, 2, 3)$. The corresponding K-map is shown on the following page:

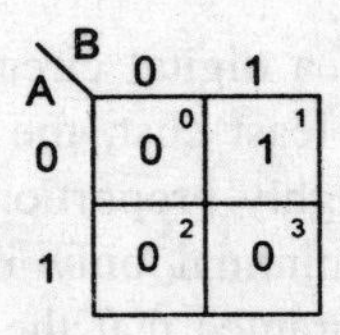

Minimization of POS Expressions

To obtain the minimal expression in the POS form, map the given POS expression on to the K-map and combine the adjacent 0s into as large squares as possible. Read the squares putting the complemented variable if its value remains constant as a 1 and the non-complemented variable if its value remains constant as a 0 along the entire square (ignoring the variables which do not remain constant throughout the square) and then write them as a sum term.

Various maxterm combinations and the corresponding reduced expressions are shown in Fig. 6.5. f_1 is read as A because A remains constant as a 0 throughout the square, and B changes from a 0 to a 1. f_2 is read as $\overline{B}$ because B remains constant along the square as a 1 and A changes from a 0 to a 1. f_5 is read as a 0 because both the variables are changing along the square.

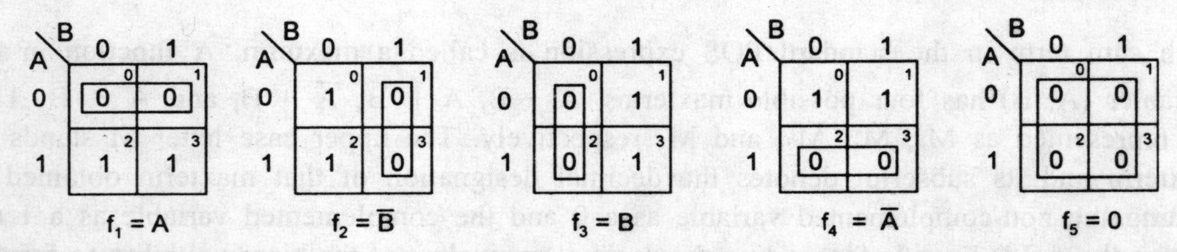

Fig. 6.5 The possible maxterm groupings in a two-variable K-map.

EXAMPLE 6.11 Reduce the expression $(A + B)(A + \overline{B})(\overline{A} + \overline{B})$ using mapping.

Solution

The given expression in terms of maxterms is $\prod M(0, 1, 3)$. The diagram below shows the K-map and its reduction. It requires two gate inputs for realization of the reduced expression.

$f = A\overline{B}$

In the given expression, the maxterm M_2 is absent. This is indicated by a 1 on the K-map. The corresponding SOP expression is $\Sigma\, m_2$ or $A\overline{B}$. This realization is the same as that for the POS form.

6.6 THREE-VARIABLE K-MAP

A function in three variables (A, B, C) expressed in the SOP form can have eight possible combinations: $\overline{A}\,\overline{B}\,\overline{C}$, $\overline{A}\,\overline{B}C$, $\overline{A}B\overline{C}$, $\overline{A}BC$, $A\overline{B}\,\overline{C}$, $A\overline{B}C$, $AB\overline{C}$, and ABC. Each one of these combinations designated by m_0, m_1, m_2, m_3, m_4, m_5, m_6, and m_7, respectively, is called a minterm. A is the MSB of the minterm designator and C is the LSB.

In the POS form, the eight possible combinations are: $A + B + C$, $A + B + \overline{C}$, $A + \overline{B} + C$, $A + \overline{B} + \overline{C}$, $\overline{A} + B + C$, $\overline{A} + B + \overline{C}$, $\overline{A} + \overline{B} + C$, and $\overline{A} + \overline{B} + \overline{C}$. Each one of these combinations designated by M_0, M_1, M_2, M_3, M_4, M_5, M_6, and M_7, respectively, is called a maxterm. A is the MSB of the maxterm designator and C is the LSB.

A three-variable K-map has, therefore, $8(= 2^3)$ squares or cells, and each square on the map represents a minterm or maxterm as shown in Figs. 6.6a and b.

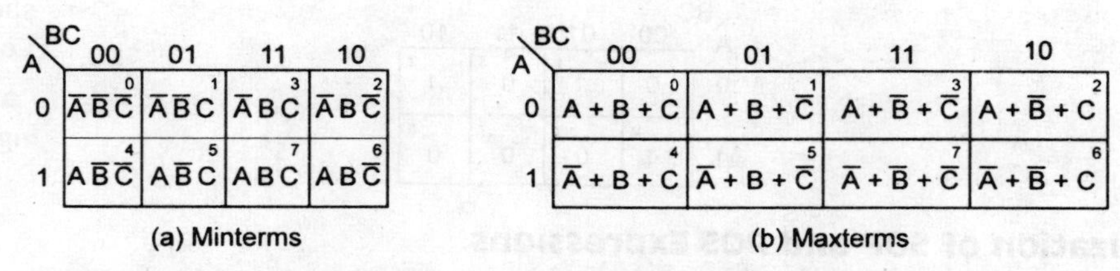

Fig. 6.6 The minterms and maxterms of a three-variable K-map.

The binary numbers along the top of the map indicate the condition of B and C for each column. The binary number along the left side of the map against each row indicates the condition of A for that row. For example, the binary number 01 on top of the second column in Fig. 6.6a indicates that the variable B appears in complemented form and the variable C in non-complemented form in all the minterms in that column. The binary number 0 on the left of the first row indicates that the variable A appears in complemented form in all the minterms in that row. Similarly, the binary number 01 on top of the second column in Fig. 6.6b indicates that the variable B appears in non-complemented form and the variable C in complemented form in all the maxterms in that column. The binary number 0 on the left of the first row indicates that the variable A appears in non-complemented form in all the maxterms in that row. Observe that the binary numbers along the top of the K-map are not in normal binary order. They are, in fact, in Gray code. This is to ensure that two physically adjacent squares are really adjacent, i.e. their minterms or maxterms differ by only one variable.

EXAMPLE 6.12 Map the expression $\overline{A}\,\overline{B}C + A\overline{B}C + \overline{A}B\overline{C} + AB\overline{C} + ABC$.

Solution

The minterms are: $\overline{A}\,\overline{B}C = 001 = m_1$; $A\overline{B}C = 101 = m_5$; $\overline{A}B\overline{C} = 010 = m_2$; $AB\overline{C} = 110 = m_6$; $ABC = 111 = m_7$. The K-map is shown on the following page.

A \ BC	00	01	11	10
0	0 (0)	1 (1)	0 (3)	1 (2)
1	0 (4)	1 (5)	1 (7)	1 (6)

EXAMPLE 6.13 Map the expression $(A + B + C)(\overline{A} + B + \overline{C})(\overline{A} + \overline{B} + \overline{C})(A + \overline{B} + \overline{C})(\overline{A} + \overline{B} + C)$.

Solution

The maxterms are: $A + B + C = 000 = M_0$; $\overline{A} + B + \overline{C} = 101 = M_5$; $\overline{A} + \overline{B} + \overline{C} = 111 = M_7$; $A + \overline{B} + \overline{C} = 011 = M_3$; $\overline{A} + \overline{B} + C = 110 = M_6$

A \ BC	00	01	11	10
0	0 (0)	1 (1)	0 (3)	1 (2)
1	1 (4)	0 (5)	0 (7)	0 (6)

Minimization of SOP and POS Expressions

For reducing the Boolean expressions in SOP form plotted on the K-map, look at the 1s present on the map. These represent the minterms. Look for the minterms adjacent to each other, in order to combine them into larger squares. Combining of adjacent squares in a K-map containing 1s (or 0s) for the purpose of simplification of a SOP (or POS) expression is called *looping*. Some of the minterms may have many adjacencies. Always start with the minterm with the least number of adjacencies and try to form as large a square as possible. The larger squares must form a geometric square or rectangle. They can be formed even by wrapping around, but cannot be formed by using diagonal configurations. Next consider the minterm with next to the least number of adjacencies and form as large a square as possible. Continue this till all the minterms are taken care of. Read the minimal expression from the K-map, corresponding to the squares formed. There can be more than one minimal expression.

Two squares are said to be adjacent to each other (since the binary designations along the top of the map and those along the left side of the map are in Gray code), if they are physically adjacent to each other, or can be made adjacent to each other by wrapping around.

Reading the K-maps

Some possible combinations of minterms and the corresponding minimal expressions read from the K-maps are shown in Fig. 6.7. Here f_6 is read as 1, because along the 8-square no variable remains constant. f_5 is read as $\overline{A}$, because, along the 4-square formed by m_0, m_1, m_2, and m_3, the variables B and C are changing, and A remains constant as a 0. Algebraically, we have

$$
\begin{aligned}
f_5 = m_0 + m_1 + m_2 + m_3 &= \overline{A}\,\overline{B}\,\overline{C} + \overline{A}\,\overline{B}C + \overline{A}B\overline{C} + \overline{A}BC \\
&= \overline{A}\,\overline{B}(\overline{C} + C) + \overline{A}B(C + \overline{C}) \\
&= \overline{A}\,\overline{B} + \overline{A}B = \overline{A}(\overline{B} + B) = \overline{A}
\end{aligned}
$$

$f_1 = \overline{B}\,\overline{C} + \overline{A}\,\overline{B} + \overline{A}\,\overline{C}$ $\quad$ $f_2 = \overline{A}\,\overline{B} + B\overline{C} + \overline{A}\,\overline{C}$ $\quad$ $f_3 = \overline{C} + \overline{B}$

$f_4 = \overline{B} + C$ $\quad$ $f_5 = \overline{A}$ $\quad$ $f_6 = 1$

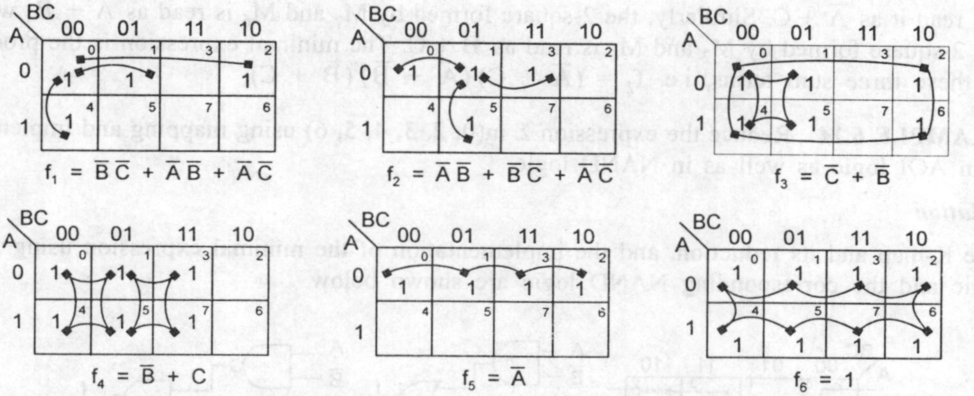

Fig. 6.7 Some possible combinations of minterms in a three-variable K-map (in the SOP form).

f_3 is read as $\overline{C} + \overline{B}$, because in the 4-square formed by m_0, m_2, m_6, and m_4, the variables A and B are changing, whereas the variable C remains constant as a 0. So it is read as $\overline{C}$. In the 4-square formed by m_0, m_1, m_4 and m_5, A and C are changing but B remains constant as a 0. So it is read as $\overline{B}$. So, the resultant expression for f_3 is the sum of these two, i.e. $\overline{C} + \overline{B}$.

f_1 is read as $\overline{B}\,\overline{C} + \overline{A}\,\overline{B} + \overline{A}\,\overline{C}$, because in the 2-square formed by m_0 and m_4, A is changing from a 0 to a 1, whereas B and C remain constant as a 0. So, it is read as $\overline{B}\,\overline{C}$. In the 2-square formed by m_0 and m_1, C is changing from a 0 to a 1, whereas A and B remain constant as a 0. So, it is read as $\overline{A}\,\overline{B}$. In the 2-square formed by m_0 and m_2, B is changing from a 0 to a 1 whereas A and C remain constant as a 0. So, it is read as $\overline{A}\,\overline{C}$. Therefore, the resultant SOP expression is, $\overline{B}\,\overline{C} + \overline{A}\,\overline{B} + \overline{A}\,\overline{C}$.

Some possible maxterm groupings and the corresponding minimal POS expressions read from the K-map are shown in Fig. 6.8.

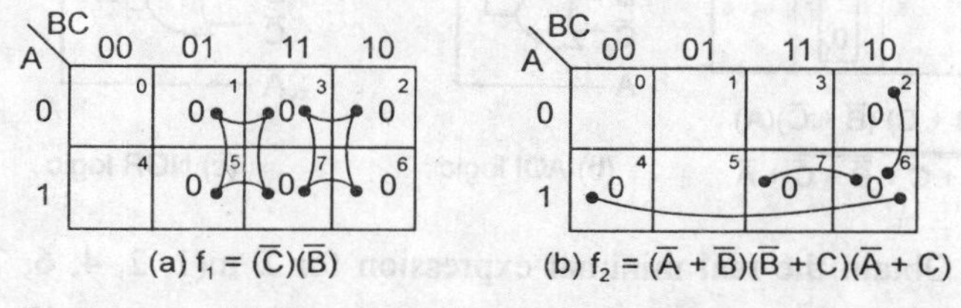

(a) $f_1 = (\overline{C})(\overline{B})$ $\quad$ (b) $f_2 = (\overline{A} + \overline{B})(\overline{B} + C)(\overline{A} + C)$

Fig. 6.8 Some possible combinations of maxterms in the three-variable K-map (in the POS form).

In Fig. 6.8a, along the 4-square formed by M_1, M_3, M_7 and M_5, A and B are changing from a 0 to a 1, whereas C remains constant as a 1. So it is read as $\overline{C}$. Along the 4-square formed by M_3, M_2, M_7 and M_6, variables A and C are changing from a 0 to a 1, but B remains constant as a 1. So it is read as $\overline{B}$. The minimal expression is the product of these two terms, i.e. $f_1 = (\overline{C})(\overline{B})$.

In Fig. 6.8b, along the 2-square formed by M_4 and M_6, variable B is changing from a 0 to a 1, while variable A remains constant as a 1 and variable C remains constant as a 0. So, read it as $\overline{A} + C$. Similarly, the 2-square formed by M_7 and M_6 is read as $\overline{A} + \overline{B}$, while the 2-square formed by M_2 and M_6 is read as $\overline{B} + C$. The minimal expression is the product of these three sum terms, i.e. $f_2 = (\overline{A} + C)(\overline{A} + \overline{B})(\overline{B} + C)$.

EXAMPLE 6.14 Reduce the expression Σ m(0, 2, 3, 4, 5, 6) using mapping and implement it in AOI logic as well as in NAND logic.

Solution

The K-map and its reduction, and the implementation of the minimal expression using AOI logic and the corresponding NAND logic are shown below.

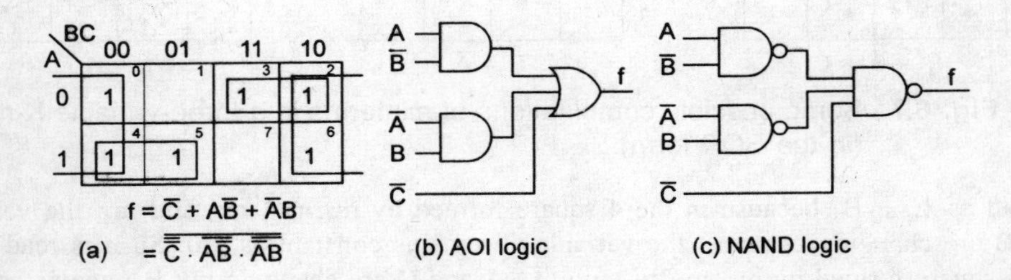

$f = \overline{C} + A\overline{B} + \overline{A}B$

(a) $= \overline{\overline{\overline{C}} \cdot \overline{A\overline{B}} \cdot \overline{\overline{A}B}}$ (b) AOI logic (c) NAND logic

EXAMPLE 6.15 Reduce the expression Π M(0, 1, 2, 3, 4, 7) using mapping and implement it in AOI logic as well as in NOR logic.

Solution

The K-map and its reduction and the implementation of the minimal expression using AOI logic and the corresponding NOR logic are shown below.

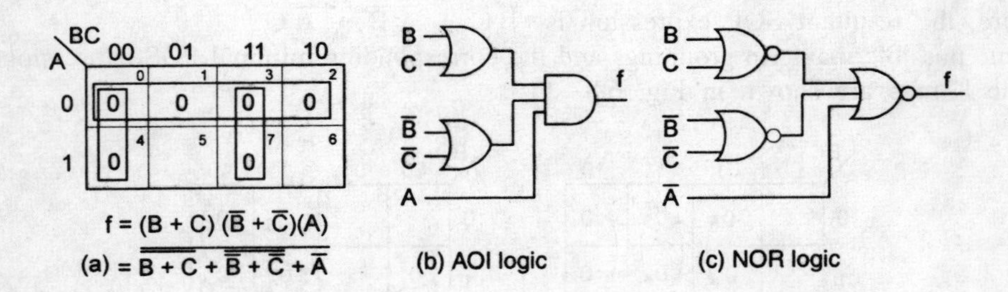

$f = (B + C)(\overline{B} + \overline{C})(A)$

(a) $= \overline{\overline{B + C} + \overline{\overline{B} + \overline{C}} + \overline{A}}$ (b) AOI logic (c) NOR logic

EXAMPLE 6.16 Obtain the real minimal expression for Σ m(1, 2, 4, 6, 7) and implement it using universal gates.

Solution

In the given expression, minterms m_0, m_3, and m_5 are missing. They, therefore, become the maxterms for the POS expression, Π M(0, 3, 5).

To obtain the real minimal expression, obtain the minimal expressions in both the SOP and POS forms and then take the minimal of those two. The respective K-maps with their minimal expressions are shown below.

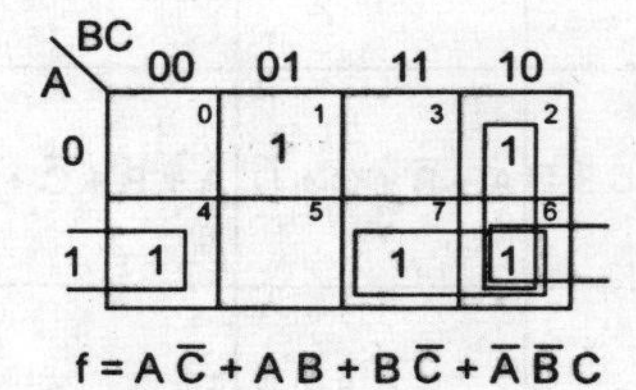

$f = A\bar{C} + AB + B\bar{C} + \bar{A}\bar{B}C$

A \ BC	00	01	11	10
0	0 (0)	(1)	0 (3)	(2)
1	(4)	0 (5)	(7)	(6)

$f = (A + B + C)(A + \bar{B} + \bar{C})(\bar{A} + B + \bar{C})$

In the POS form, no minimization is possible. The SOP form requires 13 gate inputs whereas the POS form requires 12. So, the POS form is preferred. Thus, the real minimal expression is

$$(A + B + C)\,(A + \bar{B} + \bar{C})\,(\bar{A} + B + \bar{C})$$
$$= \overline{\overline{(A + B + C)} + \overline{(A + \bar{B} + \bar{C})} + \overline{(\bar{A} + B + \bar{C})}}$$

The logic diagram using NOR gates is shown below.

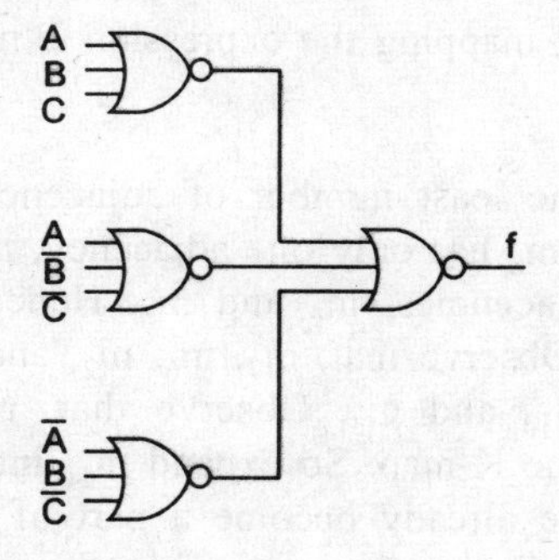

6.7 FOUR-VARIABLE K-MAP

A four-variable (A, B, C, D) expression can have $2^4 = 16$ possible combinations of input variables such as $\bar{A}\bar{B}\bar{C}\bar{D}$, $\bar{A}\bar{B}\bar{C}D$, . . ., ABCD with minterm designations m_0, m_1, . . ., m_{15}, respectively, in the SOP form and as A + B + C + D, . . ., $\bar{A} + \bar{B} + \bar{C} + \bar{D}$ with maxterm designations M_0, M_1, . . ., M_{15}, respectively, in the POS form.

A four-variable K-map has $2^4 = 16$ squares or cells and each square on the map represents either a minterm or a maxterm as shown in Fig. 6.9.

The binary number designations of the rows and columns are in Gray code. The binary numbers along the top of the map indicate the conditions of C and D along any column and binary numbers along the left side indicate the conditions of A and B along any row. The numbers in the top right corners of the squares indicate the minterm or maxterm designations as usual.

SOP form

AB \ CD	00	01	11	10
00	0: $\bar{A}\bar{B}\bar{C}\bar{D}$	1: $\bar{A}\bar{B}\bar{C}D$	3: $\bar{A}\bar{B}CD$	2: $\bar{A}\bar{B}C\bar{D}$
01	4: $\bar{A}B\bar{C}\bar{D}$	5: $\bar{A}B\bar{C}D$	7: $\bar{A}BCD$	6: $\bar{A}BC\bar{D}$
11	12: $AB\bar{C}\bar{D}$	13: $AB\bar{C}D$	15: $ABCD$	14: $ABC\bar{D}$
10	8: $A\bar{B}\bar{C}\bar{D}$	9: $A\bar{B}\bar{C}D$	11: $A\bar{B}CD$	10: $A\bar{B}C\bar{D}$

POS form

AB \ CD	00	01	11	10
00	0: $A+B+C+D$	1: $A+B+C+\bar{D}$	3: $A+B+\bar{C}+\bar{D}$	2: $A+B+\bar{C}+D$
01	4: $A+\bar{B}+C+D$	5: $A+\bar{B}+C+\bar{D}$	7: $A+\bar{B}+\bar{C}+\bar{D}$	6: $A+\bar{B}+\bar{C}+D$
11	12: $\bar{A}+\bar{B}+C+D$	13: $\bar{A}+\bar{B}+C+\bar{D}$	15: $\bar{A}+\bar{B}+\bar{C}+\bar{D}$	14: $\bar{A}+\bar{B}+\bar{C}+D$
10	8: $\bar{A}+B+C+D$	9: $\bar{A}+B+C+\bar{D}$	11: $\bar{A}+B+\bar{C}+\bar{D}$	10: $\bar{A}+B+\bar{C}+D$

Fig. 6.9 The minterms and maxterms of a four-variable K-map.

EXAMPLE 6.17 Reduce using mapping the expression Σ m(2, 3, 6, 7, 8, 10, 11, 13, 14).

Solution

Start with the minterm with the least number of adjacencies. The minterm m_{13} has no adjacency. Keep it as it is. The m_8 has only one adjacency, m_{10}. Expand m_8 into a 2-square with m_{10}. The m_7 has two adjacencies, m_6 and m_3. Hence m_7 can be expanded into a 4-square with m_6, m_3 and m_2. Observe that, m_7, m_6, m_2, and m_3 form a geometric square. The m_{11} has 2 adjacencies, m_{10} and m_3. Observe that, m_{11}, m_{10}, m_3, and m_2 form a geometric square on wrapping the K-map. So expand m_{11} into a 4-square with m_{10}, m_3 and m_2. Note that, m_2 and m_3, have already become a part of the 4-square m_7, m_6, m_2, and m_3. But if m_{11} is expanded only into a 2-square with m_{10}, only one variable is eliminated. So m_2 and m_3 are used again to make another 4-square with m_{11} and m_{10} to eliminate two variables. Now only m_6 and m_{14} are left uncovered. They can form a 2-square that eliminates only one variable. Don't do that. See whether they can be expanded into a larger square. Observe that, m_2, m_6, m_{14}, and m_{10} form a rectangle. So m_6 and m_{14} can be expanded into a 4-square with m_2 and m_{10}. This eliminates two variables.

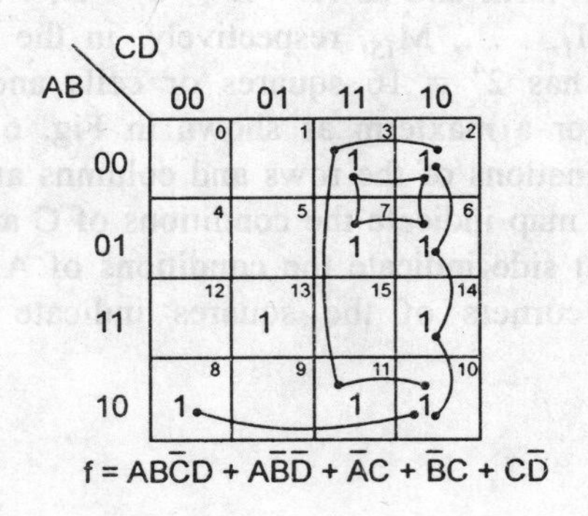

The K-map above produces the following terms.

(a) m_{13} is to be read as $AB\overline{C}D$

(b) 2-square of m_8 and m_{10} yields $A\overline{B}\,\overline{D}$

(c) 4-square of m_7, m_6, m_2, and m_3 yields $\overline{A}C$

(d) 4-square of m_2, m_3, m_{11}, and m_{10} yields $\overline{B}C$

(e) 4-square of m_2, m_6, m_{14}, and m_{10} yields $C\overline{D}$

Therefore, the reduced expression is $AB\overline{C}D + A\overline{B}\,\overline{D} + \overline{A}C + \overline{B}C + C\overline{D}$ (18 inputs).

EXAMPLE 6.18 Reduce using mapping the expression Σ m(0, 1, 2, 3, 5, 7, 8, 9, 10, 12, 13) and implement it in universal logic.

Solution

The given problem in the POS form is $\prod$ M(4, 6, 11, 14, 15). The K-maps for the SOP and POS forms and their reductions are shown below. The SOP form of realization is more economical. Now,

$$\overline{B}\,\overline{D} + A\overline{C} + \overline{A}D = \overline{\overline{\overline{B}\,\overline{D}} \cdot \overline{A\overline{C}} \cdot \overline{\overline{A}D}}$$

The implementation of the minimal expression using NAND logic is shown below.

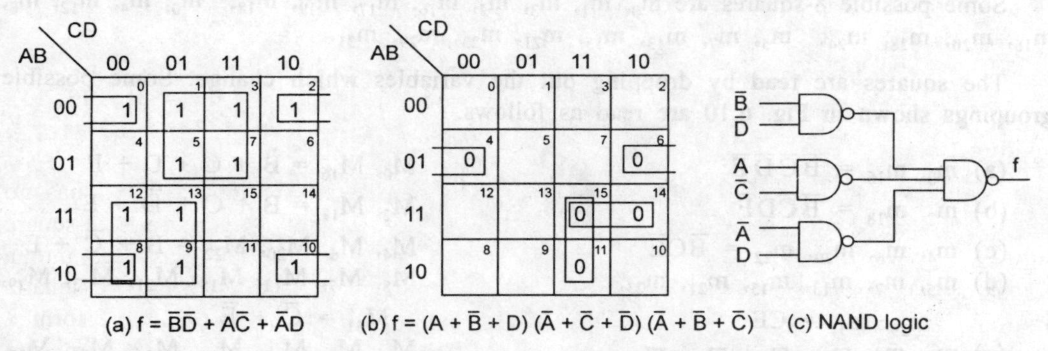

(a) $f = \overline{B}\,\overline{D} + A\overline{C} + \overline{A}D$ (b) $f = (A + \overline{B} + D)(\overline{A} + \overline{C} + \overline{D})(\overline{A} + \overline{B} + \overline{C})$ (c) NAND logic

EXAMPLE 6.19 Reduce using mapping the expression $\prod$ M(2, 8, 9, 10, 11, 12, 14) and implement it in universal logic.

Solution

The given expression in the SOP form is Σ m (0, 1, 3, 4, 5, 6, 7, 13, 15). The K-maps for the SOP and POS forms and their reductions are shown below. The POS form is more economical. Its implementation is also given below.

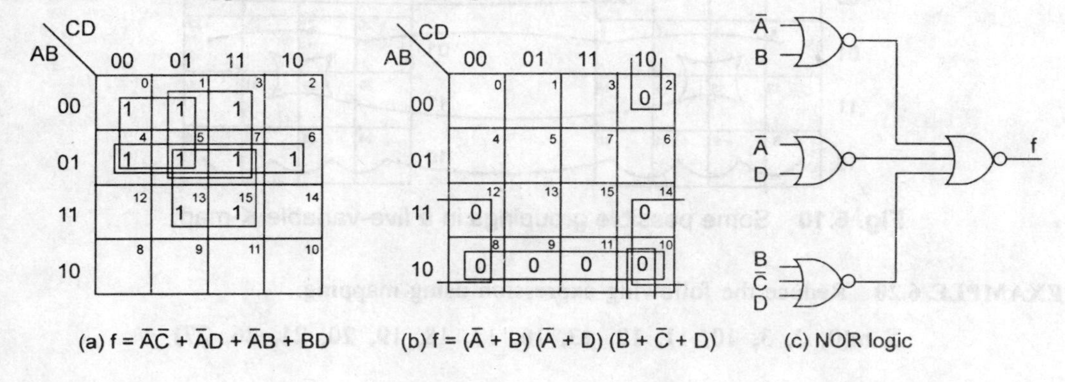

(a) $f = \overline{A}\,\overline{C} + \overline{A}D + \overline{A}B + BD$ (b) $f = (\overline{A} + B)(\overline{A} + D)(B + \overline{C} + D)$ (c) NOR logic

6.8 FIVE-VARIABLE K-MAP

A five-variable (A, B, C, D, E) expression can have $2^5 = 32$ possible combinations of input variables such as $\overline{A}\,\overline{B}\,\overline{C}\,\overline{D}\,\overline{E}$, E, . . ., ABCDE, with minterm designations m_0, m_1, . . . , m_{31}, respectively, in SOP form and as A + B + C + D + E, A + B + C + D + $\overline{E}$, . . ., $\overline{A}$ + $\overline{B}$ + $\overline{C}$ + $\overline{D}$ + $\overline{E}$, with maxterm designations M_0, M_1, . . ., M_{31}, respectively, in POS form. The 32 squares of the K-map are divided into 2 blocks of 16 squares each. The left block represents minterms from m_0 to m_{15} in which A is a 0, and the right block represents minterms from m_{16} to m_{31} in which A is 1. The five-variable K-map may contain 2-squares, 4-squares, 8-squares, or other combinations involving these two blocks. Squares are also considered adjacent in these two blocks, if when superimposing one block on top of another, the squares coincide with one another.

Some possible 2-squares in a five-variable map are m_0, m_{16}; m_2, m_{18}; m_5, m_{21}; m_{15}, m_{31}; m_{11}, m_{27}.

Some possible 4-squares are m_0, m_2, m_{16}, m_{18}; m_0, m_1, m_{16}, m_{17}; m_0, m_4, m_{16}, m_{20}; m_{13}, m_{15}, m_{29}, m_{31}; m_5, m_{13}, m_{21}, m_{29}.

Some possible 8-squares are m_0, m_1, m_3, m_2, m_{16}, m_{17}, m_{19}, m_{18}; m_0, m_4, m_{12}, m_8, m_{16}, m_{20}, m_{28}, m_{24}; m_5, m_7, m_{13}, m_{15}, m_{21}, m_{23}, m_{29}, m_{31}.

The squares are read by dropping out the variables which change. Some possible groupings shown in Fig. 6.10 are read as follows.

(a) m_0, m_{16} = $\overline{B}\,\overline{C}\,\overline{D}\,\overline{E}$ $\qquad$ M_0, M_{16} = B + C + D + E

(b) m_2, m_{18} = $\overline{B}\,\overline{C}D\overline{E}$ $\qquad$ M_2, M_{18} = B + C + $\overline{D}$ + E

(c) m_4, m_6, m_{20}, m_{22} = $\overline{B}C\overline{E}$ $\qquad$ M_4, M_6, M_{20}, M_{22} = B + $\overline{C}$ + E

(d) m_5, m_7, m_{13}, m_{15}, m_{21}, m_{23}, m_{29}, m_{31} = CE $\qquad$ M_5, M_7, M_{13}, M_{15}, M_{21}, M_{23}, M_{29}, M_{31} = $\overline{C}$ + $\overline{E}$

(e) m_8, m_9, m_{10}, m_{11}, m_{24}, m_{25}, m_{26}, m_{27} = $B\overline{C}$ $\qquad$ M_8, M_9, M_{10}, M_{11}, M_{24}, M_{25}, M_{26}, M_{27} = $\overline{B}$ + C

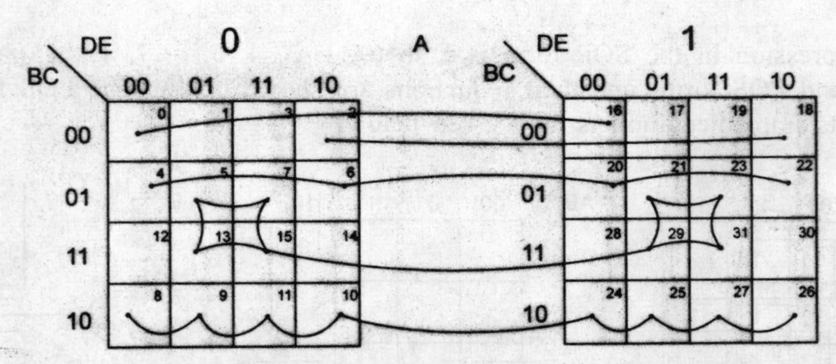

Fig. 6.10 Some possible groupings in a five-variable K-map.

EXAMPLE 6.20 Reduce the following expression using mapping:

$$\Sigma\, m(0, 2, 3, 10, 11, 12, 13, 16, 17, 18, 19, 20, 21, 26, 27)$$

Solution

The given expression in the POS form is

$$\prod M(1, 4, 5, 6, 7, 8, 9, 14, 15, 22, 23, 24, 25, 28, 29, 30, 31)$$

The real minimal expression is the minimal of the SOP and POS forms.

SOP form

In the K-map shown below:

m_{12} can go only with m_{13}. Form a 2-square which is read as $\overline{A}BC\overline{D}$.

m_0 can go with m_2, m_{16} and m_{18}. So, form a 4-square which is read as $\overline{B}\,\overline{C}\,\overline{E}$.

m_{20}, m_{21}, m_{17} and m_{16} form a 4-square which is read as $A\overline{B}\,\overline{D}$.

m_2, m_3, m_{18}, m_{19}, m_{10}, m_{11}, m_{26}, and m_{27} form an 8-square which is read as $\overline{C}D$.

The minimal expression is $\overline{A}BC\overline{D} + \overline{B}\,\overline{C}\,\overline{E} + A\overline{B}\,\overline{D} + \overline{C}D$ (16 inputs)

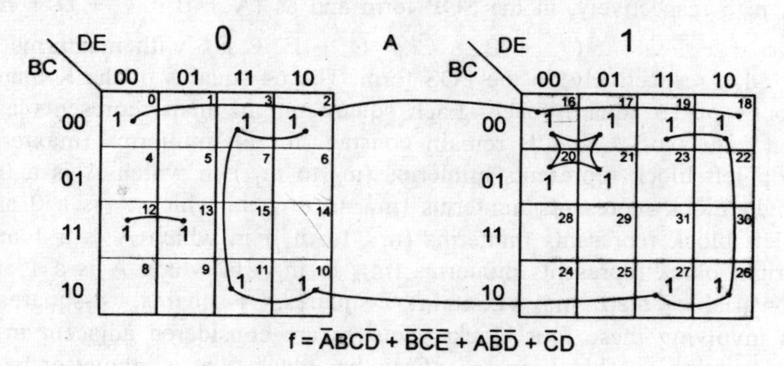

$f = \overline{A}BC\overline{D} + \overline{B}\overline{C}\overline{E} + A\overline{B}\overline{D} + \overline{C}D$

POS form

In the K-map shown below:

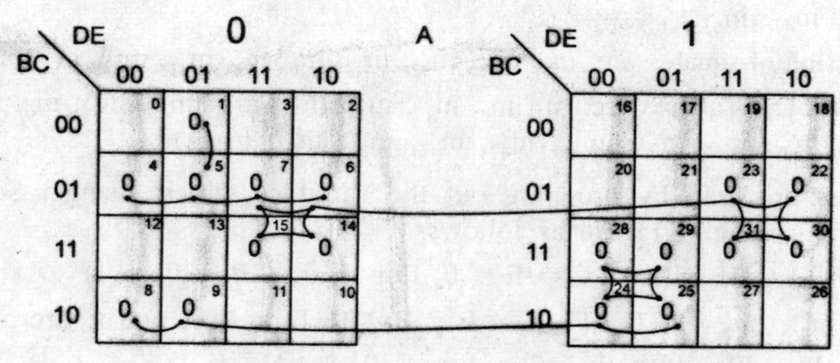

$f = (A + B + D + \overline{E})(A + B + \overline{C})(\overline{B} + C + D)(\overline{A} + \overline{B} + D)(\overline{C} + \overline{D})$

M_1 can go only with M_5. So, make a 2-square, which is read as $(A + B + D + \overline{E})$.

M_4 can go with M_5, M_7, and M_6 to form a 4-square, which is read as $(A + B + \overline{C})$.

M_8 can go with M_9, M_{24}, and M_{25} to form a 4-square, which is read as $(\overline{B} + C + D)$.
M_{28} can go with M_{29}, M_{24}, and M_{25} to form a 4-square, which is read as $(\overline{A} + \overline{B} + D)$.
M_{30} can make a 4-square with M_{31}, M_{29}, and M_{28} or with M_{31}, M_{14}, and M_{15}. Don't do that. Note that it can make an 8-square with M_{31}, M_{23}, M_{22}, M_6, M_7, M_{14} and M_{15}, which is read as $(\overline{C} + \overline{D})$.

The minimal expression is

$$(A + B + D + \overline{E})\,(A + B + \overline{C})\,(\overline{B} + C + D)\,(\overline{A} + \overline{B} + D)\,(\overline{C} + \overline{D}) \quad \text{(20 inputs)}$$

The SOP form requires less number of gate inputs. The real minimal expression is, therefore,

$$\overline{A}BC\overline{D} + \overline{B}\,\overline{C}\,\overline{E} + A\overline{B}\,\overline{D} + \overline{C}D$$

6.9 SIX-VARIABLE K-MAP

A six-variable (A, B, C, D, E, F) expression can have $2^6 = 64$ possible combinations of input variables such as $\overline{A}\,\overline{B}\,\overline{C}\,\overline{D}\,\overline{E}\,\overline{F}$, $\overline{A}\,\overline{B}\,\overline{C}\,\overline{D}\,\overline{E}F$, . . ., ABCDEF, with minterm designations m_0, m_1, . . ., m_{63}, respectively, in the SOP form and as $(A + B + C + D + E + F)$, $(A + B + C + D + E + \overline{F})$,. . ., $(\overline{A} + \overline{B} + \overline{C} + \overline{D} + \overline{E} + \overline{F})$, with maxterms designations M_0, M_1, . . ., M_{63}, respectively, in the POS form. The 64 squares of the K-map are divided into four blocks of 16 squares each. Each square on the map represents a minterm or maxterm. The values of A and B remain constant for all minterms (maxterms) in each block. The top left block represents minterms (m_0 to m_{15}) in which A is a 0 and B is a 0. The top right block represents minterms (m_{16} to m_{31}) in which A is a 0 and B is a 1. The bottom left block represents minterms (m_{32} to m_{47}) in which A is a 1 and B is a 0. The bottom right block represents minterms (m_{48} to m_{63}) in which A is a 1 and B is a 1.

The six-variable map may contain 2-squares, 4-squares, 8-squares, or other combinations involving these four blocks. Squares are considered adjacent in two blocks, if upon superimposing one block on top of another block, that is, above or beside the first block, the squares coincide with each other. Diagonal elements like m_{10}, m_{58}; m_{15}, m_{63}; m_{18}, m_{34}; m_{29}, m_{45} are not adjacent to each other.

Some possible 2-squares are: m_0, m_{16}; m_{10}, m_{42}; m_{16}, m_{48}; m_7, m_{23}; m_7, m_{39}; m_{23}, m_{55}; m_{47}, m_{63};

Some possible 4-squares are: m_0, m_{16}, m_{32}, m_{48}; m_0, m_1, m_{32}, m_{33}; m_{32}, m_{33}, m_{48}, m_{49};

Some possible 8-squares are: m_1, m_3, m_{17}, m_{19}, m_{33}, m_{35}, m_{49}, m_{51}; m_0, m_2, m_{16}, m_{18}, m_{32}, m_{34}, m_{48}, m_{50}; m_{39}, m_{38}, m_{47}, m_{46}, m_{55}, m_{54}, m_{63}, m_{62}.

The squares are read by dropping out the variables which change. Some possible groupings shown in Fig. 6.11 are as follows.

m_5, m_{21} $= \overline{A}\,\overline{C}D\overline{E}F$ (A = C = E = 0, D = F = 1, B = 0 or 1)

m_4, m_{12}, m_{36}, m_{44} $= \overline{B}D\overline{E}\,\overline{F}$ (B = E = F = 0, D = 1, A and C are a 0 or a 1)

m_{45}, m_{47}, m_{41}, m_{43}, m_{61}, m_{63}, m_{57}, m_{59} = ACF (A = C = F = 1, B, C, and E are a 0 or a 1)

m_0, m_1, m_2, m_3, m_{16}, m_{17}, m_{18}, m_{19}, m_{32}, m_{33}, m_{34}, m_{35}, m_{48}, m_{49}, m_{50}, m_{51} $= \overline{C}\,\overline{D}$ (C = D = 0, A, B, E, and F are a 0 or a 1).

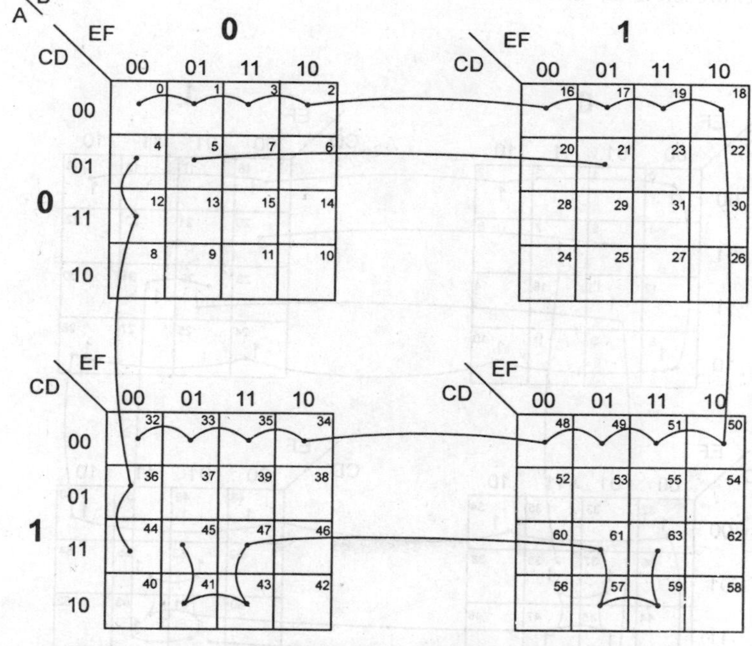

Fig. 6.11 Some possible groupings in a six-variable K-map.

EXAMPLE 6.21 Reduce the expression

Σ m(0, 2, 7, 8, 10, 13, 16, 18, 24, 26, 29, 31, 32, 34, 37, 39, 40, 42, 45, 47, 48, 50, 53, 55, 56, 58, 61, 63) using mapping.

Solution

The given expression in the POS form is

Π M(1, 3, 4, 5, 6, 9, 11, 12, 14, 15, 17, 19, 20, 21, 22, 23, 25, 27, 28, 30, 33, 35, 36, 38, 41, 43, 44, 46, 49, 51, 52, 54, 57, 59, 60, 62).

The real minimal expression is the minimal of the SOP and POS forms.

SOP form

In the K-map shown on the following page:

m_7 has only one adjacency m_{39}. It can form a 2-square with m_{39}. Read it as $\overline{B}\,\overline{C}DEF$.

m_{13} can make a 4-square with m_{29}, m_{45}, m_{61}. Read it as $CD\overline{E}F$.

m_{31} can make a 4-square with m_{29}, m_{63}, m_{61}. Read it as BCDF.

m_{55} can make an 8-square with m_{53}, m_{61}, m_{63}, m_{37}, m_{39}, m_{45}, m_{47}. Read it as ADF.

m_0 can make a 16-square with m_2, m_{16}, m_{18}, m_8, m_{10}, m_{24}, m_{26}, m_{32}, m_{34}, m_{40}, m_{42}, m_{56}, m_{58}, m_{50}, m_{48}. Read it as $\overline{D}\,\overline{F}$.

The minimal expression is, therefore: $\overline{B}\,\overline{C}DEF + CD\overline{E}F + BCDF + ADF + \overline{D}\,\overline{F}$ (23 inputs).

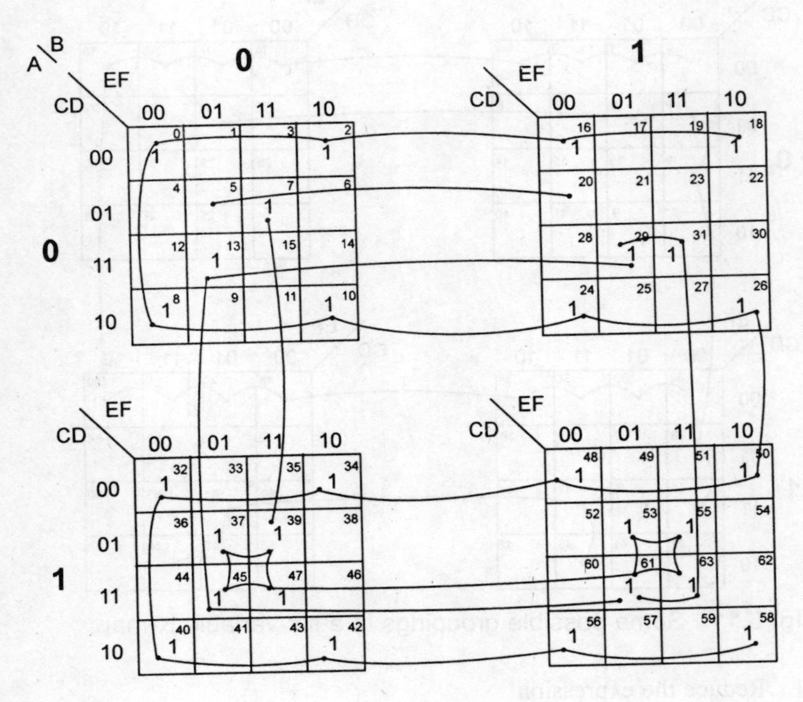

POS form

In the K-map shown on the following page:

M_{15} has only two adjacencies M_{14} and M_{11}. It can make a 2-square with any one of them. Make a 2-square of M_{15}, M_{14}. Read it as $(A + B + \overline{C} + \overline{D} + \overline{E})$.

M_5 can make a 4-square with M_4, M_{20}, M_{21} or M_1, M_{17}, M_{21}. Don't take a decision yet.

M_4 can be expanded into a 16-square with M_6, M_{12}, M_{14}, M_{20}, M_{22}, M_{28}, M_{30}, M_{36}, M_{44}, M_{38}, M_{46}, M_{52}, M_{54}, M_{60}, and M_{62}. Read it as $(\overline{D} + F)$.

M_1 can be expanded into a 16-square with M_3, M_9, M_{11}, M_{17}, M_{19}, M_{25}, M_{27}, M_{33}, M_{35}, M_{41}, M_{43}, M_{49}, M_{51}, M_{57}, and M_{59}. Read it as $(D + \overline{F})$.

Only M_5, M_{21}, and M_{23} are left uncovered. M_{21}, and M_{23} can form a 4-square with M_{20}, M_{22} or with M_{17}, M_{19} which are already taken care of. Form a 4-square of M_{21}, M_{23}, M_{17} and M_{19}. Read it as $(A + \overline{B} + C + \overline{F})$.

Only M_5 is left. Make a 4-square, say with M_4, M_{20}, and M_{21}. Read it as $(A + C + \overline{D} + E)$.

The minimal expression is, therefore: $(A + B + \overline{C} + \overline{D} + \overline{E})\,(\overline{D} + F)\,(D + \overline{F})\,(A + \overline{B} + C + \overline{F})\,(A + C + \overline{D} + E)$ (22 inputs)

The POS form is thus less expensive. So the real minimal expression is the POS form.

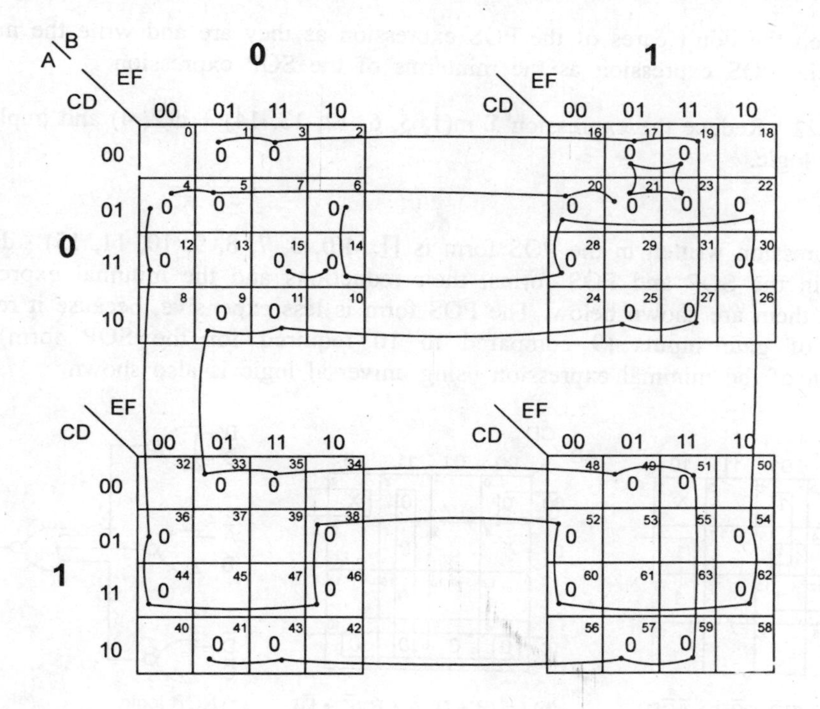

6.10 DON'T CARE COMBINATIONS

So far, the expressions considered have been completely specified for every combination of the input variables, that is, each minterm (maxterm) has been specified as a 1 or a 0. It often occurs that for certain input combinations, the value of the output is unspecified either because the input combinations are invalid or because the precise value of the output is of no consequence. The combinations for which the values of the expression are not specified are called ***don't care*** combinations and such expressions, therefore, stand incompletely specified. The output is a don't care for these invalid combinations. For example, in Excess-3 code system, the binary states 0000, 0001, 0010, 1101, 1110, and 1111 are unspecified and never occur. These are called don't cares. Similarly in 8421 code, the binary states 1001, 1010, 1011, 1100, 1101, 1110, and 1111 are invalid and the corresponding outputs are don't cares. The don't care terms are denoted by d, X or ϕ. During the process of design using an SOP map, each don't care is treated as a 1 if it is helpful in map reduction, otherwise it is treated as a 0 and left alone. During the process of design using a POS map, each don't care is treated as a 0 if it is useful in map reduction, otherwise it is treated as a 1 and left alone.

An SOP expression with don't cares can be converted into a POS form by keeping the don't cares as they are, and writing the missing minterms of the SOP form as the maxterms of the POS form. Similarly, to convert a POS expression with don't cares into an SOP expression, keep the don't cares of the POS expression as they are and write the missing maxterms of the POS expression as the minterms of the SOP expression.

EXAMPLE 6.22 Reduce the expression Σ m(1, 5, 6, 12, 13, 14) + d(2, 4) and implement it in universal logic.

Solution

The given expression written in the POS form is $\prod$ M(0, 3, 7, 8, 9, 10, 11, 15) · d(2, 4). The K-maps in the SOP and POS forms, their reductions and the minimal expressions obtained from them are shown below. The POS form is less expensive, because it requires less number of gate inputs (9 compared to 10 required for the SOP form). The implementation of the minimal expression using universal logic is also shown.

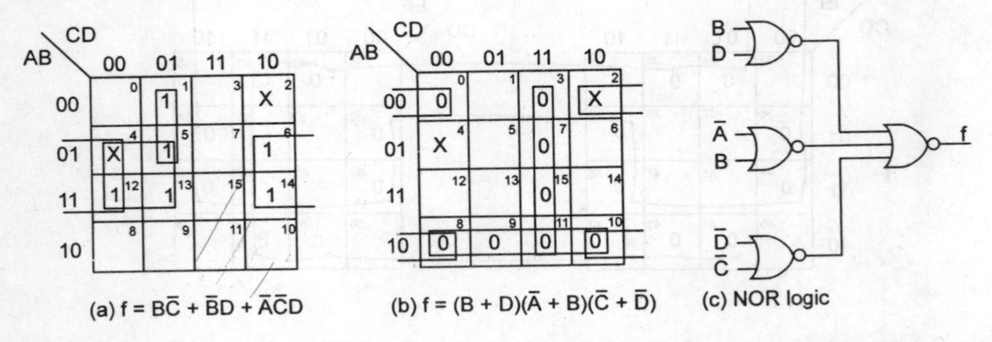

(a) $f = B\overline{C} + \overline{B}D + \overline{A}\,\overline{C}D$ (b) $f = (B + D)(\overline{A} + B)(\overline{C} + \overline{D})$ (c) NOR logic

EXAMPLE 6.23 Reduce the following expression to the simplest possible POS and SOP forms.

$$\Sigma \text{ m}(6, 9, 13, 18, 19, 25, 27, 29, 31) + \text{d}(2, 3, 11, 15, 17, 24, 28)$$

Solution

The given expression written in the POS form is

$\prod$ M(0, 1, 4, 5, 7, 8, 10, 12, 14, 16, 20, 21, 22, 23, 26, 30) · d (2, 3, 11, 15, 17, 24, 28)

The K-maps, their reductions and the minimal expressions in the SOP and POS forms obtained from them are shown below:

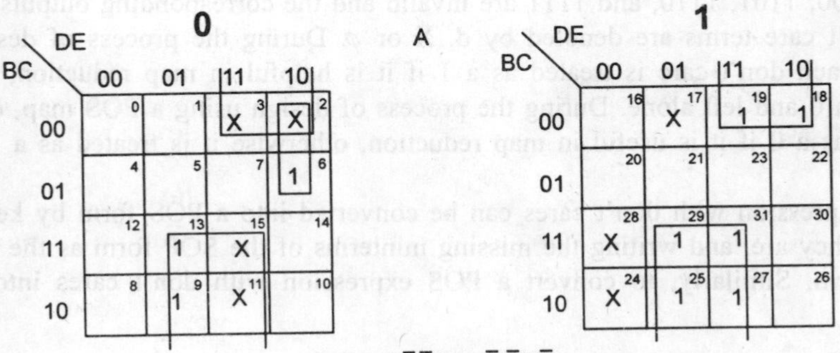

$f = BE + \overline{B}\,\overline{C}D + \overline{A}\,\overline{B}D\overline{E}$

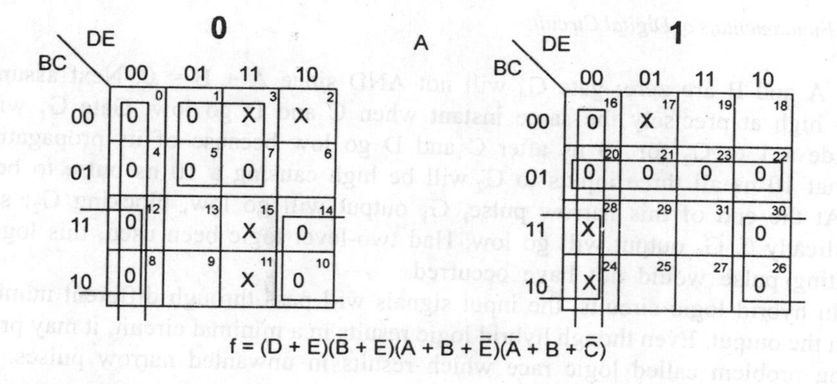

6.11 HYBRID LOGIC

Both SOP and POS reductions result in a logic circuit in which each input signal has to pass through two gates to reach the output. It is, therefore, called a two-level logic and has the advantage of providing uniform time delay between input signals and the output. But the disadvantage is that the minimal expression obtained by either SOP reduction or POS reduction may not be the actual minimal. In fact, the actual minimal may be obtained by manipulating the minimals of SOP and POS forms into a hybrid form. For example, the minimal of the expression Σ m(0, 1, 2, 3, 5, 7, 8, 9, 10, 12, 13) in the SOP form is given by $f = A\overline{C} + \overline{A}D + \overline{A}B + BC$ (12 inputs). But this can be written as $f = \overline{A}(\overline{C} + D + B) + BC$ (9 inputs).

Also, the expression ABC + ABD + ACD + BCD is in minimum SOP form and requires 16 inputs. It can, however, be reduced by factoring to: AB(C + D) + CD(A + B) and implemented as shown in Fig. 6.12 with 12 inputs.

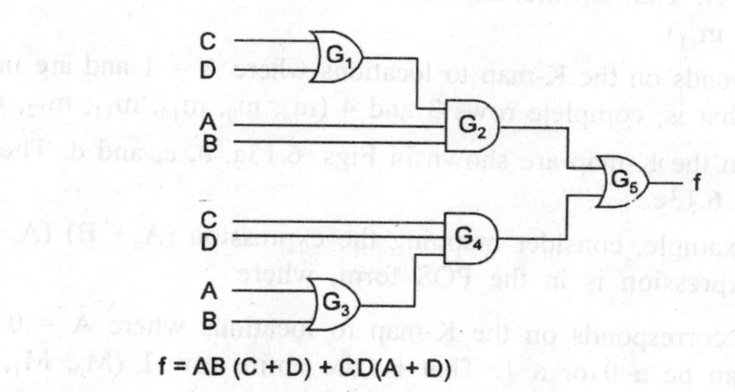

Fig. 6.12 Hybrid logic.

Figure 6.12 shows that we have reduced the number of inputs from 16 to 12. Note, however, that the C input to the OR gate must go through three levels of logic before reaching the output, whereas the C input to the AND gate must only go through two levels. This can result in a critical timing problem called *logic race*. Assume, for example, that each gate has a 10 ns delay and that A = 0, B = 0, C = 1, and D = 1. Gate G_2 will not AND since A and B are zero; gate G_4 will not AND since A + B = 0. Next assume that A and B go high at precisely the same instant when C and D go low. Gate G_1 will continue to provide a 1 to G_2 for 10 ns after C and D go low because of its propagation delay, and for that 10 ns all three inputs to G_2 will be high causing a 10 ns pulse to be outputted by G_5. At the end of this narrow pulse, G_1 output will go low, blocking G_2; since C and D are already 0, G_5 output will go low. Had two-level logic been used, this logic race and its resulting pulse would not have occurred.

In hybrid logic circuits, the input signals will pass through different number of gates to reach the output. Even though hybrid logic results in a minimal circuit, it may provide a critical timing problem called logic race which results in unwanted narrow pulses. The two-level logic is free from logic race.

Mapping when function is not expressed in minterms (maxterms)

Our discussion of mapping suggests that if an expression is to be entered on a K-map, it must be available as a sum (product) of minterms (maxterms). However, if not so expressed, it is not necessary to expand the expression algebraically into its minterms (maxterms). Instead, the expansion into minterms (maxterms) can be accomplished in the process of entering the terms of the expression on the K-map.

For example, let us enter on K-map the expression $\overline{A}\,\overline{B}CD + \overline{A}C\overline{D} + \overline{B}C + A$.

(a) $\overline{A}\,\overline{B}CD$ is minterm m_3. Enter it as it is.

(b) $\overline{A}\,\overline{C}\,\overline{D}$ corresponds on the K-map to locations where A = C = D = 0 and are independent of B. That is, intersections of rows 1 and 2 with column 1 (m_0 and m_4).

(c) $\overline{B}C$ corresponds on the K-map to locations where B = 0, C = 1 and are independent of A and D. That is, intersections of rows 1 and 4 with columns 3 and 4 (m_2, m_3, m_{10}, and m_{11}).

(d) A corresponds on the K-map to locations where A = 1 and are independent of B, C, and D. That is, complete rows 3 and 4 (m_8, m_9, m_{10}, m_{11}, m_{12}, m_{13}, m_{14}, and m_{15}).

The entries on the K-map are shown in Figs. 6.13a, b, c, and d. The complete mapping is shown in Fig. 6.13e.

As another example, consider mapping the expression $(A + B)(A + \overline{B} + C)(A + \overline{C})$. The given expression is in the POS form, where

(a) (A + B) corresponds on the K-map to locations where A = 0 and B = 0, and C and D can be a 0 or a 1. That is, the entire row 1 (M_0, M_1, M_2, and M_3).

(b) $(A + \overline{B} + C)$ corresponds on the K-map to locations where A = 0, B = 1, and C = 0, and are independent of D. That is, the intersection of row 2 with columns, 1 and 2 (M_4 and M_5).

(c) $(A + \overline{C})$ corresponds on the K-map to locations where A = 0 and C = 1, and are independent of B and D. That is, the intersection of rows 1 and 2 with columns 3 and 4 (M_2, M_3, M_6, and M_7).

Hence the given expression can be mapped as $\prod M(0, 1, 2, 3, 4, 5, 2, 3, 6, 7)$, i.e. $\prod M(0, 1, 2, 3, 4, 5, 6, 7)$. The mapping is shown in Fig. 6.14.

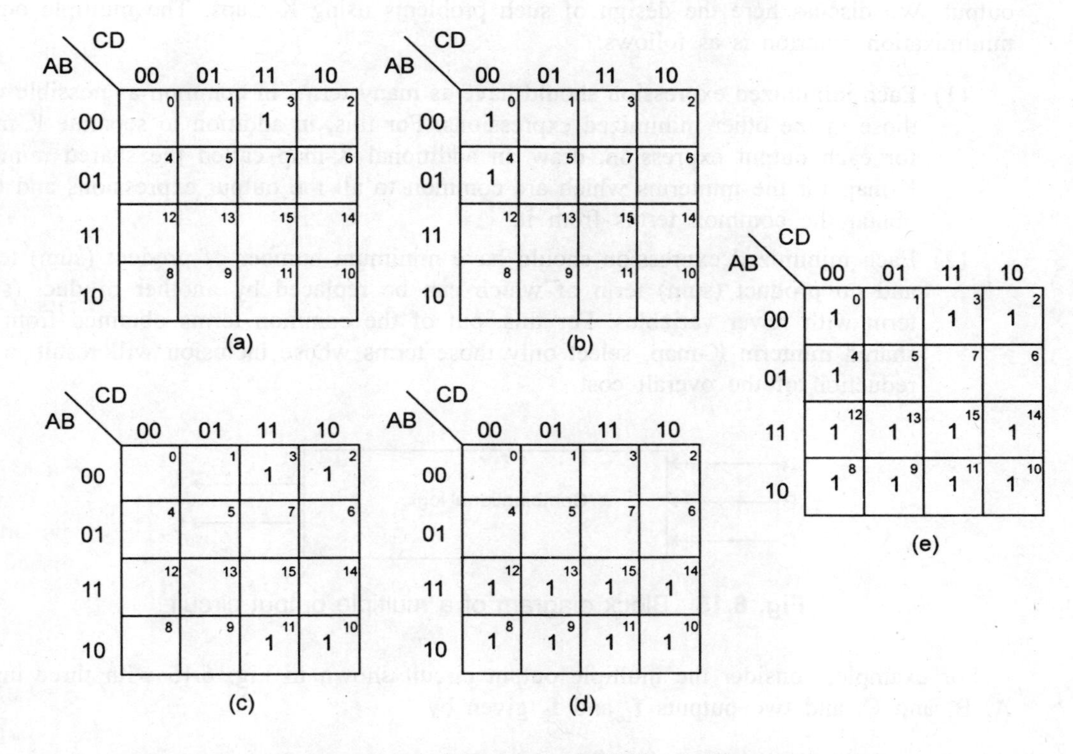

Fig. 6.13 Mapping of (a) $\overline{A}\,\overline{B}CD$, (b) $\overline{A}\,\overline{C}\,\overline{D}$, (c) $\overline{B}C$, (d) A, and (e) $\overline{A}\,\overline{B}CD + \overline{A}\,\overline{C}\,\overline{D} + \overline{B}C + A$.

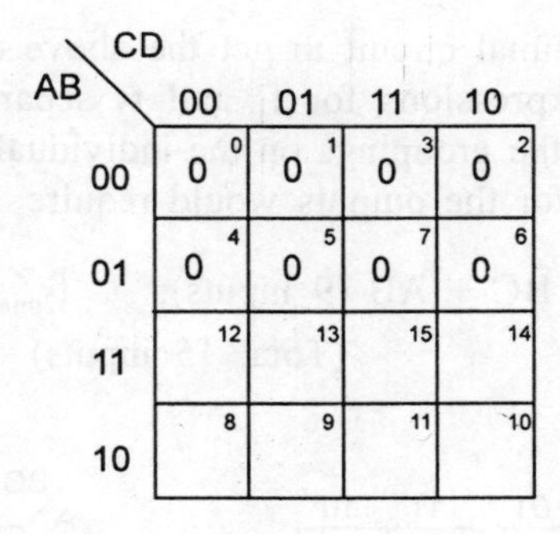

Fig. 6.14 Mapping of $(A + B)(A + \overline{B} + C)(A + \overline{C})$.

6.12 MINIMIZATION OF MULTIPLE OUTPUT CIRCUITS

So far, we have discussed the minimization of single expressions by the K-map method. In practice, many logic design problems involve designing of circuits with more than one output. We discuss here the design of such problems using K-maps. The multiple output minimization criterion is as follows:

(1) Each minimized expression should have as many terms in common as possible with those in the other minimized expressions. For this, in addition to separate K-maps for each output expression, draw an additional K-map called the shared minterm K-map for the minterms which are common to all the output expressions and then obtain the common terms from it.

(2) Each minimized expression should have minimum number of product (sum) terms and no product (sum) term of which can be replaced by another product (sum) term with fewer variables. For this, out of the common terms obtained from the shared minterm K-map, select only those terms whose inclusion will result in the reduction of the overall cost.

Fig. 6.15 Block diagram of a multiple output circuit.

For example, consider the multiple output circuit shown in Fig. 6.15 with three inputs A, B, and C, and two outputs f_1 and f_2 given by

$$f_1(A, B, C) = \Sigma\, m(0, 1, 2, 5, 6, 7)$$
$$f_2(A, B, C) = \Sigma\, m(2, 4, 5, 6)$$

We want to design a minimal circuit to get the above outputs. If we design the circuit by obtaining the minimal expressions for f_1 and f_2 separately, we may not get the overall minimal circuit. Suppose the groupings on the individual K-maps are as shown in Fig. 6.16; the minimal expressions for the outputs would require 15 gate inputs for realization. Here,

$$f_{1\min} = \overline{A}\,\overline{C} + \overline{B}C + AB \text{ (9 inputs)}; \qquad f_{2\min} = A\overline{B} + B\overline{C} \text{ (6 inputs)}$$
$$\text{(Total 15 inputs)}$$

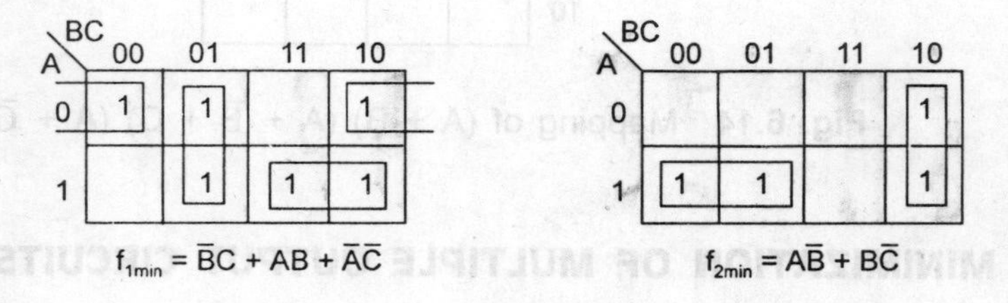

Fig. 6.16 One way of grouping f_1 and f_2.

An alternative way of grouping the outputs f_1 and f_2 is shown in Fig. 6.17.

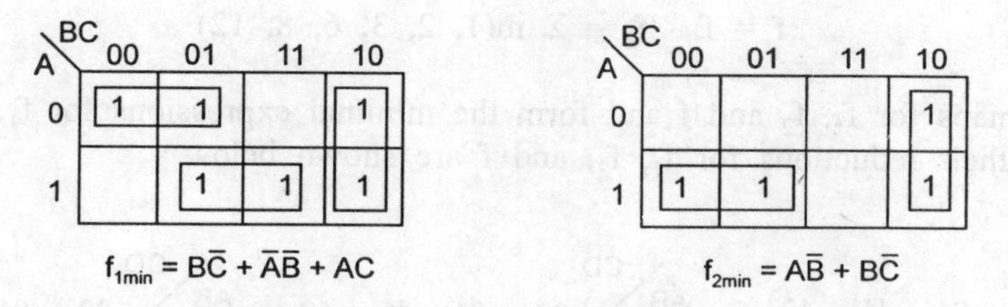

Fig. 6.17 Alternative way of grouping f_1 and f_2.

Now,

$$f_{1min} = B\overline{C} + \overline{A}\,\overline{B} + AC \text{ (9 inputs)}; \qquad f_{2min} = A\overline{B} + B\overline{C} \text{ (6 inputs)}$$

(Total 13 inputs because the term $B\overline{C}$ is common to f_{1min} and f_{2min})

As the term $B\overline{C}$ is present in both the expressions, it can be generated only once and utilized for generating both the expressions, thus reducing the cost and complexity. The realization is shown in Fig. 6.18. So the first step in multiple output minimization is to find the common terms and try to utilize them so that the overall cost is reduced.

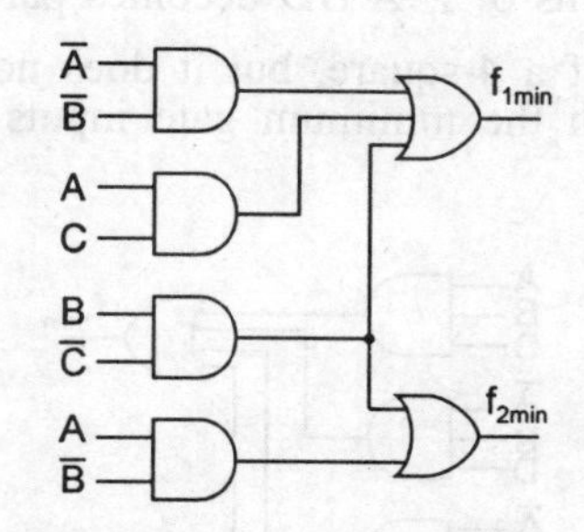

Fig. 6.18 Minimal circuit.

EXAMPLE 6.24 Minimize and implement the following multiple output functions.

$$f_1 = \Sigma\, m(1, 2, 3, 6, 8, 12, 14, 15)$$
$$f_2 = \Pi\, M(0, 4, 9, 10, 11, 14, 15)$$

Solution

Here f_2 is in the POS form and f_1 in the SOP form. Express both f_1 and f_2 either in the SOP form or in the POS form and obtain the minimal expressions. Therefore, in the SOP form, we have

$$f_1 = \Sigma\, m(1, 2, 3, 6, 8, 12, 14, 15); \qquad f_2 = \Sigma\, m(1, 2, 3, 5, 6, 7, 8, 12, 13)$$

First form a function f with the minterms common to both the functions, i.e. $f = f_1 \cdot f_2$. Therefore,

$$f = f_1 \cdot f_2 = \Sigma\, m(1, 2, 3, 6, 8, 12)$$

Draw the K-maps for f_1, f_2 and f and form the minimal expressions for f_1, f_2, and f. The K-maps and their reductions for f_1, f_2, and f are shown below.

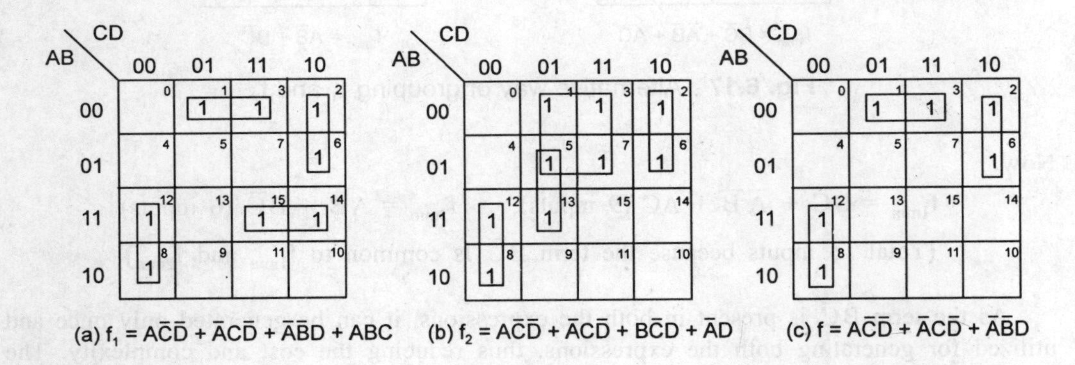

(a) $f_1 = A\bar{C}\bar{D} + \bar{A}C\bar{D} + \bar{A}\bar{B}D + ABC$ (b) $f_2 = A\bar{C}\bar{D} + \bar{A}C\bar{D} + B\bar{C}D + \bar{A}D$ (c) $f = A\bar{C}\bar{D} + \bar{A}C\bar{D} + \bar{A}\bar{B}D$

In f_1, all the terms of f are present. We cannot make any larger square using any of these terms. In f_2, out of the three terms of f, $\bar{A}\bar{B}D$ becomes part of a 4-square, so, $\bar{A}D$ is read. $\bar{A}C\bar{D}$ can also be made part of a 4-square, but it does not reduce the hardware. So it is not considered. The circuit with the minimum gate inputs is shown below.

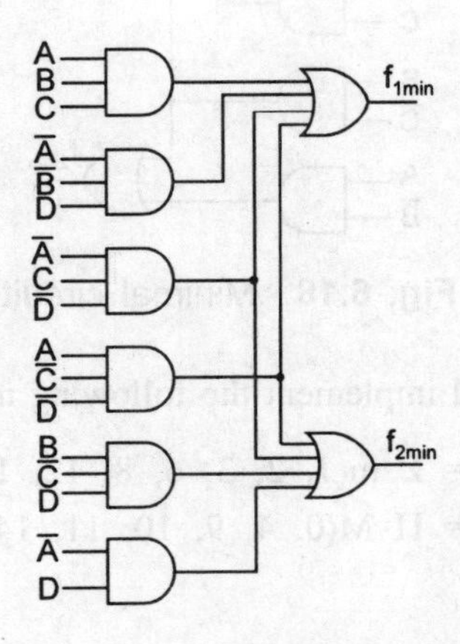

Don't Care Conditions

When there are incompletely specified functions, the minterms for the shared map are generated by the rules listed in Table 6.1.

Table 6.1 Generating minterms for a shared map

f_1	f_2	$f = f_1 \; f_2$
0	0	0
0	1	0
1	0	0
0	X	0
X	0	0
1	X	1
X	1	1
1	1	1
X	X	X

EXAMPLE 6.25 Minimize the following multiple output functions.

$$f_1 = \Sigma\, m(0, 2, 6, 10, 11, 12, 13) + d(3, 4, 5, 14, 15)$$
$$f_2 = \Sigma\, m(1, 2, 6, 7, 8, 13, 14, 15) + d(3, 5, 12)$$

Solution

The shared minterm function generated according to the rules listed in Table 6.1 is

$$f = \Sigma\, m(2, 6, 12, 13, 14, 15) + d(3, 5)$$

The K-maps for f_1, f_2, and f and their minimization are shown as:

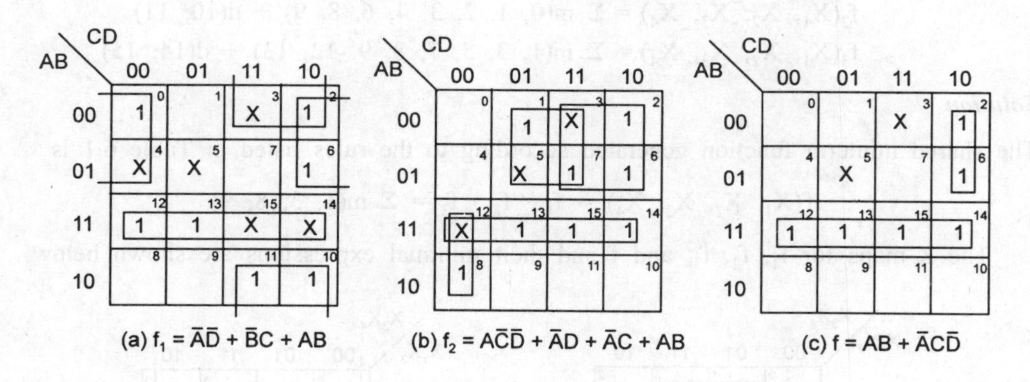

(a) $f_1 = \bar{A}\bar{D} + \bar{B}C + AB$ (b) $f_2 = AC\bar{D} + \bar{A}D + \bar{A}C + AB$ (c) $f = AB + \bar{A}C\bar{D}$

The shared mintern function f has two terms AB and $\bar{A}C\bar{D}$. Out of these two, only AB is utilized, because $\bar{A}C\bar{D}$ can be merged into a bigger square.

EXAMPLE 6.26 Find the minimal expressions for the multiple output functions

$$f_1(X_1, X_2, X_3, X_4) = \Pi\, M(3, 4, 5, 7, 11, 13, 15) \cdot d(6, 8, 10, 12)$$
$$f_2(X_1, X_2, X_3, X_4) = \Pi\, M(2, 7, 9, 10, 11, 12, 14, 15) \cdot d(0, 4, 6, 8)$$

Solution

The given expressions are in the POS form. Generate the shared maxterm function using the same rules as for the minterms. Therefore,

$$f = f_1 \cdot f_2 = \prod M(4, 7, 10, 11, 12, 15) \cdot d(6, 8)$$

The K-maps for f_1, f_2 and f and their reductions are shown below.

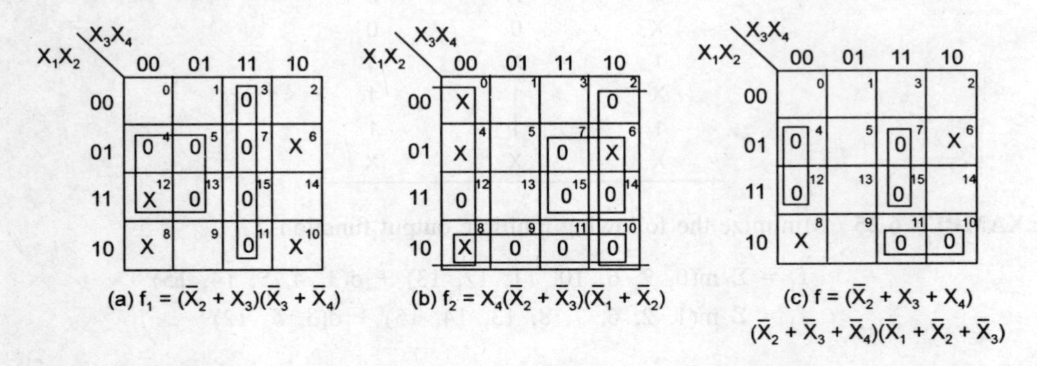

(a) $f_1 = (\bar{X}_2 + X_3)(\bar{X}_3 + \bar{X}_4)$ (b) $f_2 = X_4(\bar{X}_2 + \bar{X}_3)(\bar{X}_1 + \bar{X}_2)$ (c) $f = (\bar{X}_2 + X_3 + X_4)(\bar{X}_2 + \bar{X}_3 + \bar{X}_4)(\bar{X}_1 + \bar{X}_2 + \bar{X}_3)$

None of the terms of f can be used in the minimal expressions for f_1 and f_2 because these terms are combined into bigger ones in f_1 and f_2.

The problem may be converted to and solved in SOP form too.

EXAMPLE 6.27 Minimize the following multiple output functions using K-map.

$$f_1(X_1, X_2, X_3, X_4) = \Sigma\, m(1, 2, 3, 5, 7, 8, 9) + d(12, 14)$$
$$f_2(X_1, X_2, X_3, X_4) = \Sigma\, m(0, 1, 2, 3, 4, 6, 8, 9) + d(10, 11)$$
$$f_3(X_1, X_2, X_3, X_4) = \Sigma\, m(1, 3, 5, 7, 8, 9, 12, 13) + d(14, 15)$$

Solution

The shared minterm function generated according to the rules listed in Table 6.1 is

$$f(X_1, X_2, X_3, X_4) = f_1 \cdot f_2 \cdot f_3 = \Sigma\, m(1, 3, 8, 9)$$

The K-maps for f_1, f_2, f_3, and f and their minimal expressions are shown below.

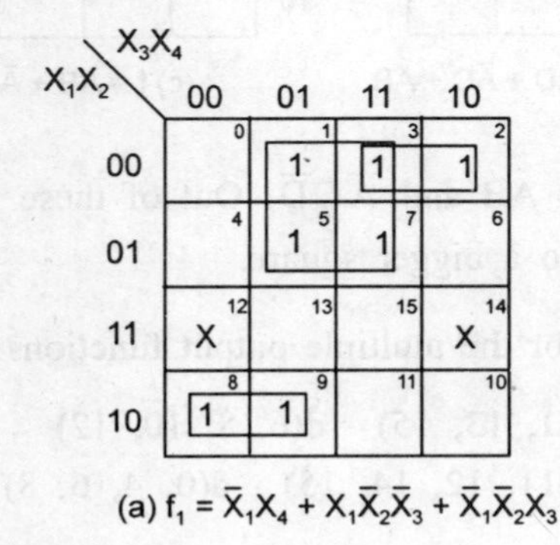

(a) $f_1 = \bar{X}_1X_4 + X_1\bar{X}_2\bar{X}_3 + \bar{X}_1\bar{X}_2X_3$

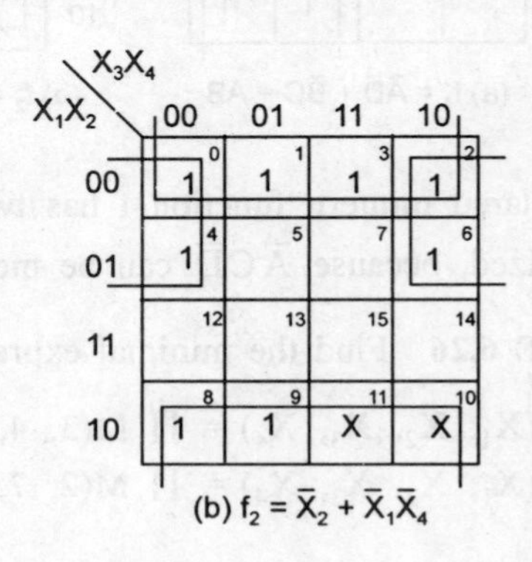

(b) $f_2 = \bar{X}_2 + \bar{X}_1\bar{X}_4$

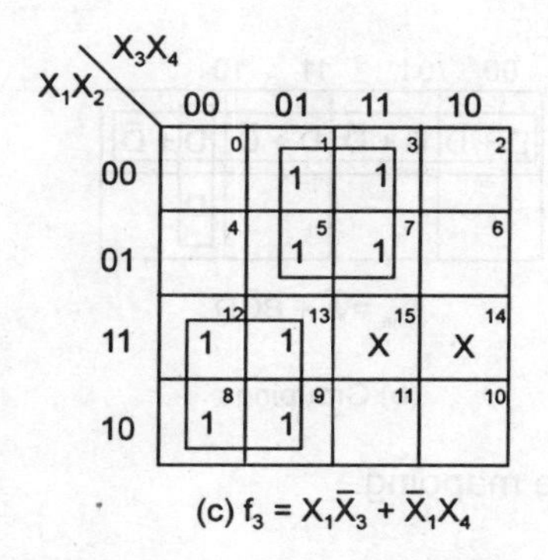

(c) $f_3 = X_1\overline{X}_3 + \overline{X}_1X_4$

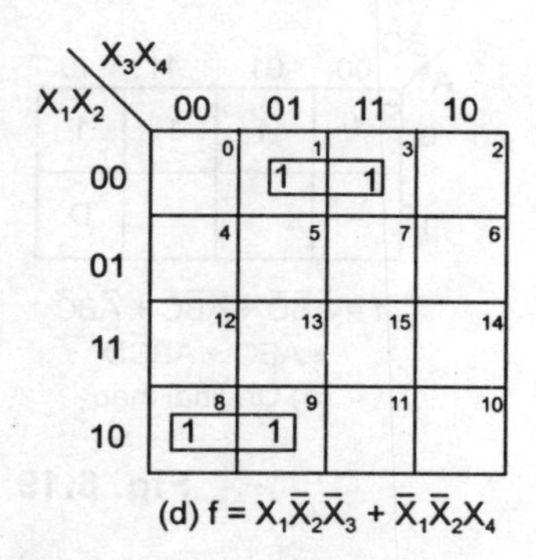

(d) $f = X_1\overline{X}_2\overline{X}_3 + \overline{X}_1\overline{X}_2X_4$

None of the terms in the minimal expression of the shared map is useful for overall reduction. However, we can see that $\overline{X}_1X_4$ is common to f_1 and f_3.

6.13 VARIABLE MAPPING

Variable mapping is a powerful and useful tool that can be used in a wide variety of problems. The design of sequential circuits is greatly simplified by this technique. Variable mapping can also be used to minimize Boolean expressions, which involve infrequently used variables. If properly used, it can reduce the work required for plotting and reading maps. It allows us to reduce a large mapping problem to one that uses just a small map. This technique can reduce the map size for 3, 4, 5, 6, 7, and 8 variable maps. It is especially useful in those problems which have a few isolated variables among more frequently used variables. Consider the equation

$$f = \overline{A}\,\overline{B}\,\overline{C} + \overline{A}\,\overline{B}C + \overline{A}B\overline{C} + \overline{A}BC + AB\overline{C}D$$

Normally, this would be a four-variable problem. Notice that the variable D is used only once whereas variables A, B, and C are used many times. Using variable mapping, we can make it a three-variable problem. Assuming this to be a three-variable problem in A, B, and C, we have

$$f = m_0 + m_1 + m_2 + m_3 + m_6D$$

For mapping such a function, put a 1 on the map where a minterm appears and a 0 (or no entry) where it does not. Since each 1 represents a minterm, the entry $1 \cdot D$ represents the minterm multiplied by D. In fact, each 1 entered onto the map represents $D + \overline{D}$, because if it were to be treated as a four-variable problem instead of a three-variable one, each term in the function is to be multiplied by $(D + \overline{D}) = 1$ to convert it to the standard SOP form. For minimization, we must cover each of the individual variables. We can make a 4-square of minterms m_0, m_1, m_2, and m_3 [covering $(D + \overline{D})$ of all the terms] and can read it as usual as $\overline{A}(D + \overline{D}) = \overline{A}$. We can also make a 2-square of the D in m_6 and D in $(D + \overline{D})$ of m_2 and read it as $(B\overline{C})D = B\overline{C}D$. This yields

$$f_{min} = \overline{A} + B\overline{C}D$$

as shown in Fig. 6.19.

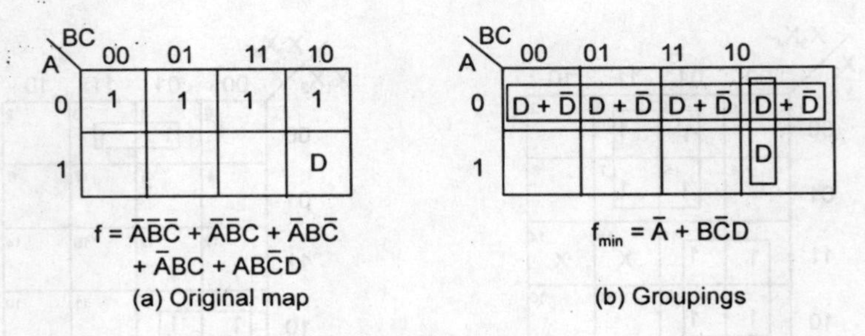

Fig. 6.19 Variable mapping.

Note that the D in m_2 is covered twice. However, our objective is to cover all the variables at least once.

EXAMPLE 6.28 Reduce by mapping:

$$f = \bar{A}\bar{B}CD + \bar{A}\bar{B}C\bar{D} + AB\bar{C}D + AB\bar{C}\bar{D} + ABC\bar{D} + ABCD + A\bar{B}C\bar{D} + \bar{A}BCD$$
$$+ \bar{A}B\bar{C}\bar{D} + A\bar{B}\bar{C}D$$

Solution

Although this is a four-variable problem, it can be treated as a three-variable one and plotted on a three-variable K-map and reduced as shown below. Thus, the three-variable problem in A, B, C would be:

$$f = m_1(D + \bar{D}) + m_6(D + \bar{D}) + m_7(D + \bar{D}) + m_5\bar{D} + m_3D + m_2\bar{D} + m_4D$$

Note that all parts of $1(D + \bar{D})$ must be covered.

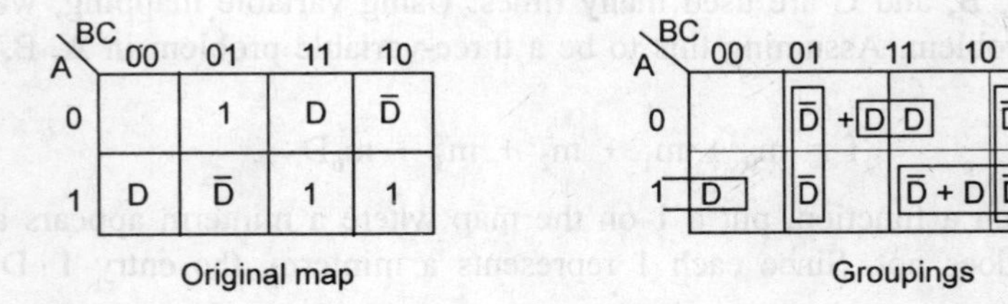

As seen from the K-maps, the $\bar{D}$'s of m_1 and m_5 form a 2-square which is read as $\bar{B}C\bar{D}$. The D's of m_1 and m_3 form a 2-square which is read as $\bar{A}CD$. The D's of m_4 and m_6 form a 2-square which is read as $A\bar{C}D$. The $\bar{D}$'s of m_2 and m_6 form a 2-square which is read as $B\bar{C}\bar{D}$. The m_1 and m_6 are fully covered. Also, m_7 can be combined with m_6 to form a 2-square which is read as AB. So, the reduced expression is

$$f = \bar{A}CD + \bar{B}C\bar{D} + A\bar{C}D + B\bar{C}\bar{D} + AB$$

EXAMPLE 6.29 Reduce by mapping:

$$f = \bar{A}\,\bar{B}\,\bar{C} + \bar{A}\,\bar{B}CD + \bar{A}BC\bar{E} + \bar{A}B\bar{C}E + A\bar{B}C + ABC + AB\bar{C}\,\bar{D}$$

Solution

Although this is a five-variable problem, it can be treated as a three-variable one and plotted on a three-variable K-map and then reduced as shown below.

The given function in three-variable form may be expressed as

$$f = m_0 + m_1D + m_3\bar{E} + m_2E + m_5 + m_7 + m_6\bar{D}$$

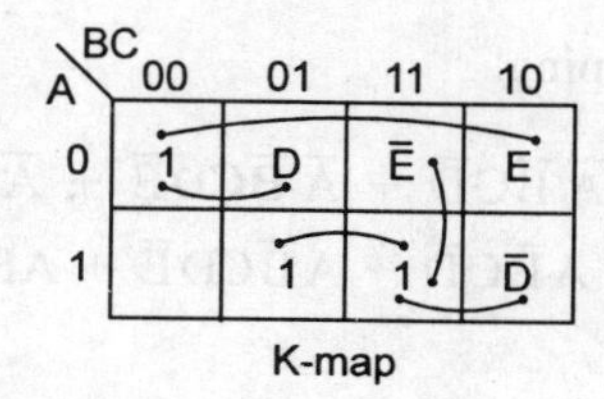

K-map

As seen from the K-map,

D of m_1 forms a 2-square with D present in m_0 which is read as $\bar{A}\,\bar{B}D$.

E of m_2 forms a 2-square with E present in m_0 which is read as $\bar{A}\,\bar{C}E$.

$\bar{E}$ of m_3 forms a 2-square with $\bar{E}$ present in m_7 which is read as $BC\bar{E}$.

$\bar{D}$ of m_6 forms a 2-square with $\bar{D}$ present in m_7 which is read as $AB\bar{D}$.

A minterm is said to be fully covered, only when at least any one of the variables as well as its complement are covered. Only $\bar{D}$ and $\bar{E}$ of m_7 are covered. So, m_7 is not fully covered. Also, m_5 is not covered at all. So, make a 2-square with m_5 and m_7 which is read as AC.

All parts of m_0 are not covered. Only D and E of m_0 are covered and m_0 cannot also be combined with any other minterms. So, it appears in the reduced expression.

From the K-map, the reduced expression is

$$f = AC + AB\bar{D} + BC\bar{E} + \bar{A}\,\bar{B}D + \bar{A}\,\bar{C}E + \bar{A}\,\bar{B}\,\bar{C}$$

EXAMPLE 6.30 Reduce by mapping:

$$f = \bar{A}\,\bar{B}\,\bar{C}E + \bar{A}\,\bar{B}C + \bar{A}BCF + \bar{A}B\bar{C} + A\bar{B}CD + AB\bar{C}\,\bar{F}$$

Solution

Although this is a six-variable problem, it can be treated as a three-variable one and plotted on a three-variable K-map and then reduced as shown below.

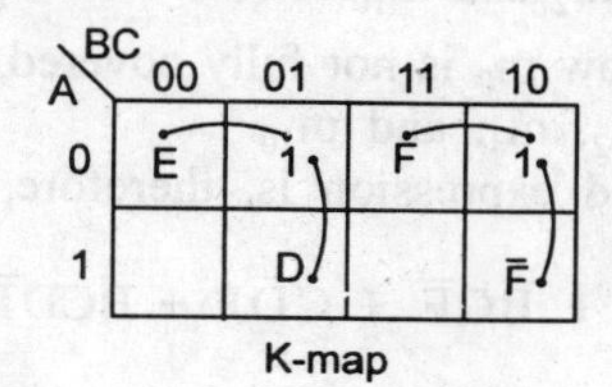

K-map

The given function may be expressed in three variable form as

$$f = m_0E + m_1 + m_2 + m_3F + m_5D + m_6\overline{F}$$

Here m_1 is not fully covered, because only D and E of m_1 are covered. So it will appear in the final expression. The m_2 is fully covered because both F and $\overline{F}$ of it are covered. So, it will not appear in the final expression.

From the K-map the reduced expression is, therefore, given by

$$f = \overline{A}\,\overline{B}E + \overline{B}CD + B\overline{C}\,\overline{F} + \overline{A}BF + \overline{A}\,\overline{B}C$$

EXAMPLE 6.31 Reduce by mapping:

$$f = \overline{A}\,\overline{B}\,\overline{C}\,\overline{D} + \overline{A}\,\overline{B}\,\overline{C}DE + \overline{A}\,\overline{B}C\overline{D} + \overline{A}\,\overline{B}CD\overline{E} + \overline{A}B\overline{C}DF + \overline{A}BC\overline{D}E + \overline{A}BCD$$
$$+ A\overline{B}\,\overline{C}\,\overline{D} + A\overline{B}\,\overline{C}DE + A\overline{B}C\overline{D} + A\overline{B}CD\overline{E} + AB\overline{C}\,\overline{D} + ABC\overline{D}E + ABCD\overline{F}$$

Solution

Although this is a six-variable problem, it can be treated as a four-variable one and plotted on a four-variable K-map and then reduced as shown below.

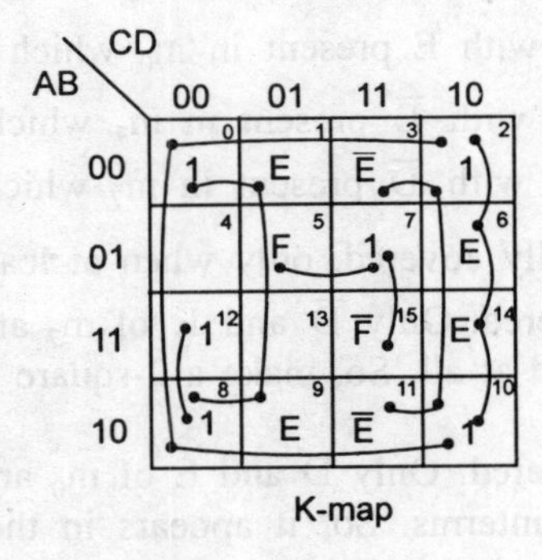

K-map

Expressed as a four-variable problem, we have

$$f = m_0 + m_1E + m_2 + m_3\overline{E} + m_5F + m_6E + m_7 + m_8 + m_9E + m_{10} + m_{11}\overline{E}$$
$$+ m_{12} + m_{14}E + m_{15}\overline{F}$$

The m_7 is fully covered because it combines with F and $\overline{F}$; likewise, m_2 and m_{10} are also fully covered because each one of them combines with E and $\overline{E}$. The E's of m_1 and m_9 form a 4-square with the E's present in m_0 and m_8. The $\overline{E}$'s of m_3 and m_{11} form a 4-square with the $\overline{E}$'s present in m_2 and m_{10}. The E's of m_6 and m_{14} form a 4-square with the E's present in m_2 and m_{10}. Now m_0 is not fully covered, because $\overline{E}$ of this is not used. So, form a 4-square with m_0, m_2, m_8, and m_{10}.

From the K-map, the reduced expression is, therefore, obtained as

$$f = \overline{A}BDF + \overline{B}\,\overline{C}E + \overline{B}C\overline{E} + C\overline{D}E + BCD\overline{F} + A\overline{C}\,\overline{D} + \overline{B}\,\overline{D}$$

Incompletely Specified Functions

The procedure used for previous don't care problems has to be modified only slightly to accommodate variable mapping. Suppose a four-variable problem is treated as a three-variable (A, B, C) one and entered on a three-variable K-map. Each 1 in the three-variable map represents that the corresponding minterm is multiplied by $(D + \overline{D})$. In fact, it may be entered as a 1 or $(D + \overline{D})$. If this minterm is a don't care, it is entered as $XD + X\overline{D}$ or simply as X. Suppose the term is $ABC\overline{D}$. It is entered as a $\overline{D}$ in the cell for m_7 ($ABC\overline{D} = m_7\overline{D}$). If $ABC\overline{D}$ is a don't care, it is entered as an $X\overline{D}$ in the cell for m_7. Similarly, when a five-variable problem is treated as a three-variable one, each 1 on the map represents the corresponding minterm multiplied by $(D + \overline{D})(E + \overline{E})$. If it is a don't care, it contains the terms $X\overline{D}$, XD, XE, and $X\overline{E}$. These don't cares may or may not be covered. They should be used to make 2^n squares when covering other *do care* terms. Any minterm is considered to be completely covered only if any of its variable plus its complement present in that minterm are both covered. Other variables can be covered partially or fully, if required or not covered at all.

EXAMPLE 6.32 Reduce by mapping:

$$f = \overline{A}\,\overline{B}\,\overline{C} + \overline{A}\,\overline{B}CD + \overline{A}B\overline{C}\,\overline{D} + A\overline{B}\,\overline{C}D + ABCE + ABC\overline{E} + d(A\overline{B}CD + A\overline{B}CE)$$

Solution

The given problem is actually a five-variable one. Treating it as a three-variable one, we get

$$f = m_0 + m_1D + m_2\overline{D} + m_4D + 1m_7\overline{E} + m_7\overline{E} + d(m_5D + m_5E)$$

It is mapped onto a K-map and reduced as shown below.

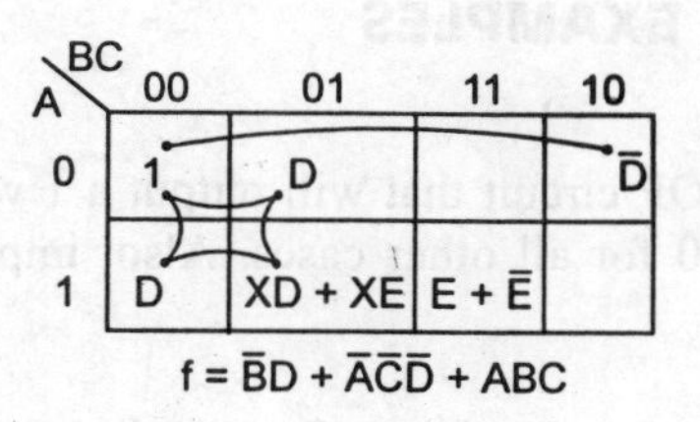

$f = \overline{B}D + \overline{A}C\overline{D} + ABC$

The D's of m_1 and m_4 form a 4-square with the D present in m_0 and the XD of m_5. The $\overline{D}$ of m_2 combines with the $\overline{D}$ present in m_0 to form a 2-square. So, m_0 is fully covered. The E of m_7 can combine with the XE of m_5. This is not done, because there is no way to combine the $\overline{E}$ of m_7 with anything else, and so, m_7 has to appear in the final expression. From the K-map, the reduced expression is, therefore, obtained as

$$f = \overline{B}D + \overline{A}\,\overline{C}\,\overline{D} + ABC$$

EXAMPLE 6.33 Reduce by mapping:

$$f = m_0 + m_1F + m_2 + m_4E + m_6(E + \overline{E}) + m_7F + m_{10}E + m_{12} + m_{15}F + d(m_5F + m_9\overline{F} + m_{11}\overline{E} + m_8E)$$

Solution

The given problem is actually a six-variable one. The four-variable K-map and its minimization are shown below.

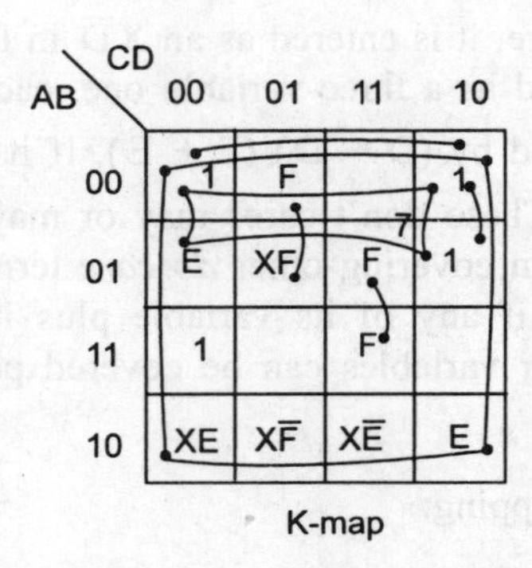

K-map

The minterm m_6 contains $E + \overline{E}$. Its E is used to make a 4-square with the E's of m_0, m_2, and m_4. Since m_6 is not fully covered, it is combined with m_2 to make a 2-square. The E of m_{12} can be combined with the E's of m_0, m_4, and m_8 but this is not done because m_{12} cannot be fully covered by this operation. So, m_{12} has to be read anyway. The minterm m_0 is not fully covered. Only E of it is used. So, combine it with m_2 to form a 2-square.

From the K-map, the reduced expression is, therefore, obtained as

$$f = \overline{B}\,\overline{D}E + \overline{A}\,\overline{D}E + \overline{A}\,\overline{B}\,\overline{D} + \overline{A}\,\overline{C}DF + BCDF + AB\overline{C}\,\overline{D} + \overline{A}C\overline{D}$$

6.14 MISCELLANEOUS EXAMPLES

EXAMPLE 6.34 Design an SOP circuit that will output a 1 when the Gray codes 5 through 12 appear at the inputs, and a 0 for all other cases. Also, implement it in the NAND logic.

Solution

The input is a 4-bit Gray code. Let the input Gray code be ABCD. There are 16 possible combinations of 4-bit Gray code. The truth table of the SOP circuit is shown below.

Decimal number	4-bit Gray code A B C D	Output	Decimal number	4-bit Gray code A B C D	Output
0	0 0 0 0	0	8	1 1 0 0	1
1	0 0 0 1	0	9	1 1 0 1	1
2	0 0 1 1	0	10	1 1 1 1	1

Contd.

Contd.

3	0 0 1 0	0	11	1 1 1 0	1
4	0 1 1 0	0	12	1 0 1 0	1
5	0 1 1 1	1	13	1 0 1 1	0
6	0 1 0 1	1	14	1 0 0 1	0
7	0 1 0 0	1	15	1 0 0 0	0

The K-map, its minimization, the minimal expression and its logic diagram in NAND logic are shown below.

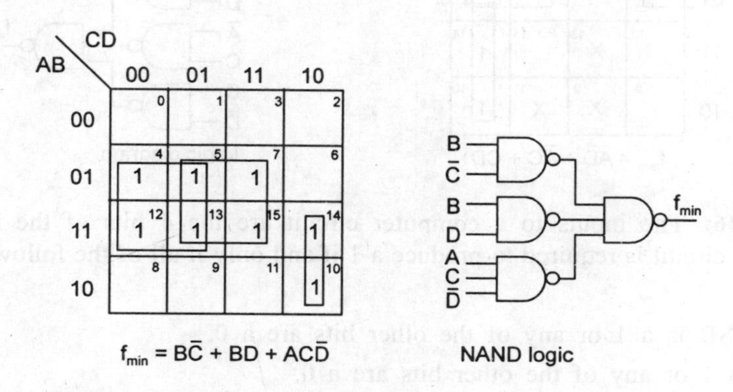

$f_{min} = B\bar{C} + BD + AC\bar{D}$

EXAMPLE 6.35 A circuit receives a 4-bit 5211 BCD code. Design the minimum SOP circuit to detect the decimal numbers 0, 2, 4, 6, and 8.

Solution

The input is a 5211 BCD code. Let it be ABCD. It is a 4-bit input. Therefore, there are 16 possible combinations of inputs, out of which only 10 combinations are used to code the decimal digits. The remaining 6 combinations 0010, 0100, 0110, 1001, 1011, and 1101 are invalid because they do not belong to 5211 code. So, the corresponding outputs are don't cares. The truth table of the SOP circuit is shown below.

Decimal number	5211 Code A B C D	Output	Decimal number	5211 Code A B C D	Output
0	0 0 0 0	1	5	1 0 0 0	0
1	0 0 0 1	0	6	1 0 1 0	1
2	0 0 1 1	1	7	1 1 0 0	0
3	0 1 0 1	0	8	1 1 1 0	1
4	0 1 1 1	1	9	1 1 1 1	0

The problem may be stated as

$$f = \Sigma\ m(0, 3, 7, 10, 14) + d(2, 4, 6, 9, 11, 13)$$

The K-map, its minimization, and the realization of the minimal expression in NAND logic are shown below.

$$f_{min} = \bar{A}\bar{D} + \bar{A}C + C\bar{D} = \overline{\overline{\bar{A}\bar{D}}\cdot\overline{\bar{A}C}\cdot\overline{C\bar{D}}}$$

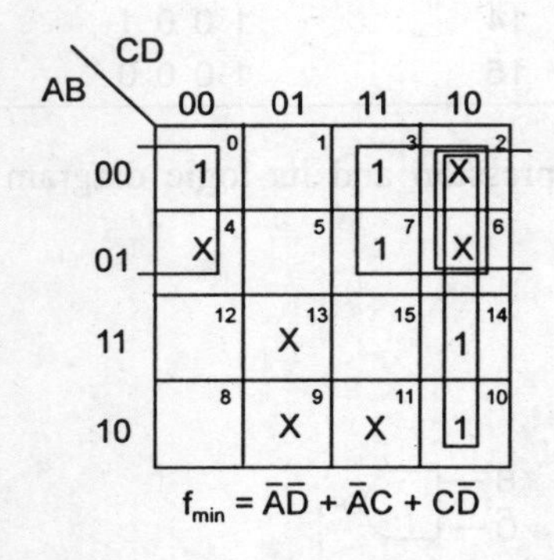

$f_{min} = \bar{A}\bar{D} + \bar{A}C + C\bar{D}$

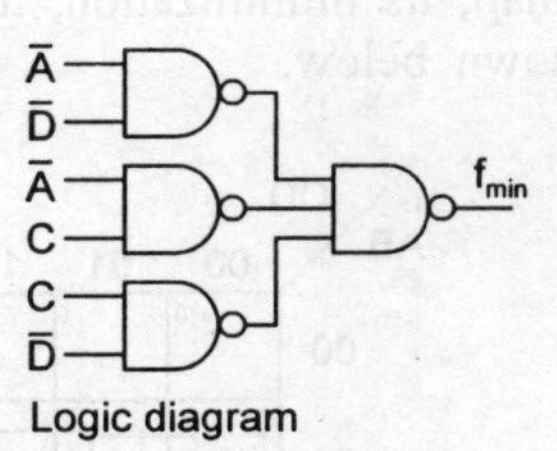

Logic diagram

EXAMPLE 6.36 The inputs to a computer circuit are the 4 bits of the binary number $A_3A_2A_1A_0$. The circuit is required to produce a 1 if and only if all of the following conditions hold.

1. The MSB is a 1 or any of the other bits are a 0.
2. A_2 is a 1 or any of the other bits are a 0.
3. Any of the 4 bits are a 0.

Obtain a minimal expression.

Solution

From the statement, the Boolean expression must be in the POS form given by

$$f = (A_3 + \bar{A}_2 + \bar{A}_1 + \bar{A}_0)\,(\bar{A}_3 + A_2 + \bar{A}_1 + \bar{A}_0)\,(\bar{A}_3 + \bar{A}_2 + \bar{A}_1 + \bar{A}_0)$$

where each non-complemented variable represents a 1 and the complemented variable a 0. Since X · X = X, the minimal expression is given by

$$f_{min} = (A_3 + \bar{A}_2 + \bar{A}_1 + \bar{A}_0)\,(\bar{A}_3 + A_2 + \bar{A}_1 + \bar{A}_0)\,(\bar{A}_3 + \bar{A}_2 + \bar{A}_1 + \bar{A}_0)$$

$$= (\bar{A}_2 + \bar{A}_1 + \bar{A}_0)\,(\bar{A}_3 + \bar{A}_1 + \bar{A}_0)$$

$$= (\bar{A}_2\bar{A}_3 + \bar{A}_1 + \bar{A}_0)$$

EXAMPLE 6.37 Design each of the following circuits that can be built using AOI logic and outputs a 1 when:

(a) A 4-bit hexadecimal input is an odd number from 0 to 9.

(b) A 4-bit BCD code translated to a number that uses the upper right segment of a seven-segment display.

Solution

(a) The output is a 1 when the input is a 4-bit hexadecimal odd number from 0 to 9. There are 16 possible combinations of inputs, and all are valid. The output is a 1 only for the input combinations 0001, 0011, 0101, 0111, and 1001. For all other combinations of inputs, the output is a 0. The problem may thus be stated as $f = \Sigma\, m(1, 3, 5, 7, 9)$. The K-map, its minimization, and the realization of the minimal expression in AOI logic are shown below.

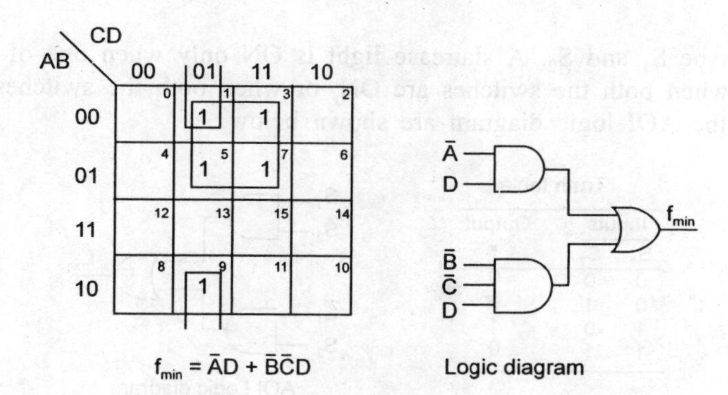

(b) We see from the figure below that display of digits 0, 1, 2, 3, 4, 7, 8, and 9 requires the upper-right segment of the seven-segment display. Since the input is a 4-bit BCD, inputs 1010 though 1111 are invalid, and therefore, the corresponding outputs are don't cares. The problem may be stated as

$$f = \Sigma\, m(0, 1, 2, 3, 4, 7, 8, 9) + d(10, 11, 12, 13, 14, 15)$$

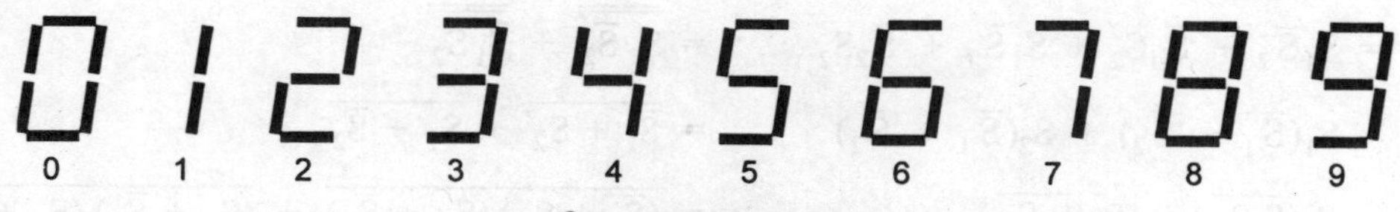

Seven-segment display

The K-map, its minimization, and the realization of the minimal expression using AOI logic are shown below.

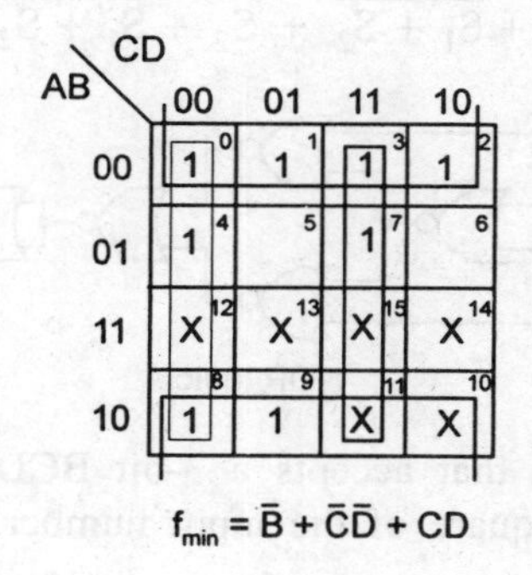

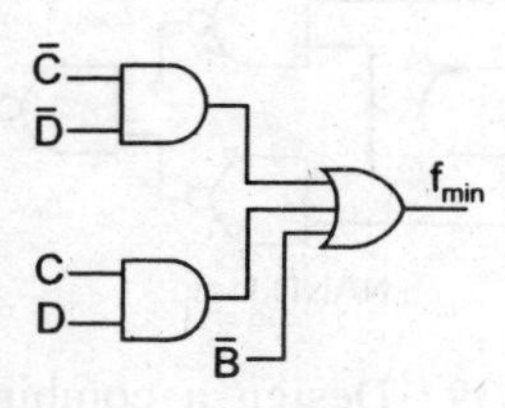

EXAMPLE 6.38 A staircase light is controlled by two switches, one is at the top of the stairs and the other at the bottom of the stairs.

(a) Make a truth table for this system.
(b) Write the logic equation in the SOP form.
(c) Realize the circuit using AOI logic.
(d) Realize the circuit using minimum number of (i) NAND gates, (ii) NOR gates.

Solution

Let the switches be S_1 and S_2. A staircase light is ON only when one of the switches is ON. It is OFF when both the switches are ON, or when both the switches are OFF. The truth table and the AOI logic diagram are shown below.

Truth table

Inputs		Output
S_1	S_2	f
0	0	0
0	1	1
1	0	1
1	1	0

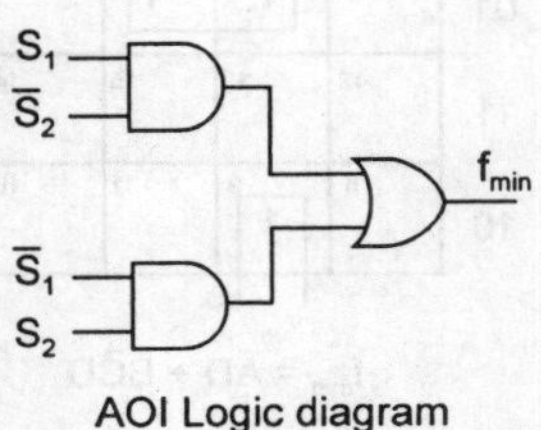

AOI Logic diagram

The logic diagrams using NAND gates and NOR gates, respectively, are also shown below.

The given operation is XOR of S_1 and S_2. Therefore, the logic equation is $f = S_1\bar{S}_2 + \bar{S}_1 S_2$.

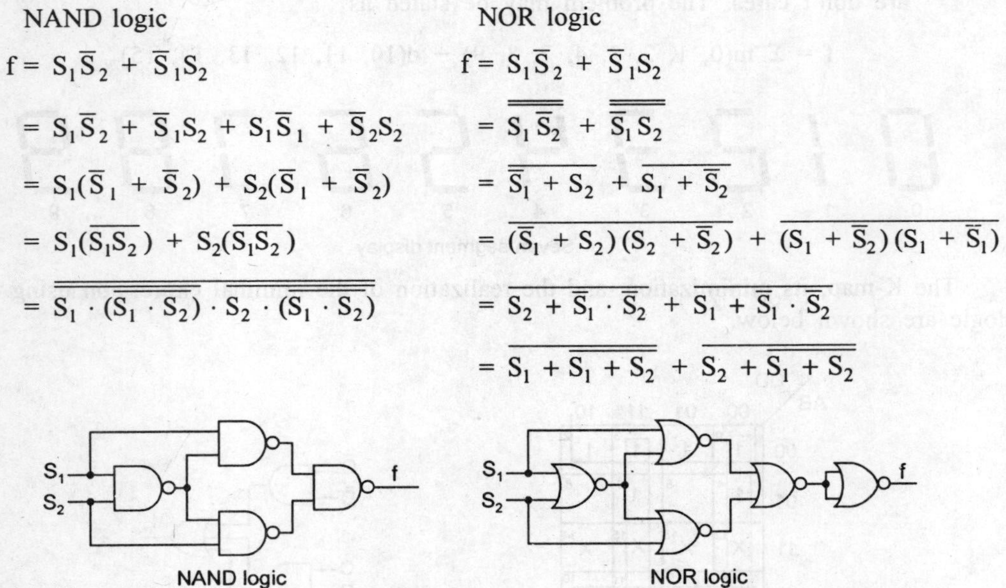

NAND logic

$$f = S_1\bar{S}_2 + \bar{S}_1 S_2$$

$$= S_1\bar{S}_2 + \bar{S}_1 S_2 + S_1\bar{S}_1 + \bar{S}_2 S_2$$

$$= S_1(\bar{S}_1 + \bar{S}_2) + S_2(\bar{S}_1 + \bar{S}_2)$$

$$= S_1(\overline{S_1 S_2}) + S_2(\overline{S_1 S_2})$$

$$= \overline{\overline{S_1 \cdot \overline{(S_1 \cdot S_2)}} \cdot \overline{S_2 \cdot \overline{(S_1 \cdot S_2)}}}$$

NOR logic

$$f = S_1\bar{S}_2 + \bar{S}_1 S_2$$

$$= \overline{\overline{S_1\bar{S}_2}} + \overline{\overline{\bar{S}_1 S_2}}$$

$$= \overline{\bar{S}_1 + S_2} + \overline{S_1 + \bar{S}_2}$$

$$= \overline{(\bar{S}_1 + S_2)(S_2 + \bar{S}_2)} + \overline{(S_1 + \bar{S}_2)(S_1 + \bar{S}_1)}$$

$$= \overline{S_2 + \bar{S}_1 \cdot \bar{S}_2} + \overline{S_1 + \bar{S}_1 \cdot \bar{S}_2}$$

$$= \overline{S_1 + \overline{S_1 + S_2}} + \overline{S_2 + \overline{S_1 + S_2}}$$

NAND logic

NOR logic

EXAMPLE 6.39 Design a combinational circuit that accepts a 3-bit BCD number and generates an output binary number equal to the square of the input number.

Solution

The square of a 3-bit number is a 6-bit number. The truth table of the combinational circuit is shown below.

Truth table

Inputs			Ouputs					
A	B	C	X_6	X_5	X_4	X_3	X_2	X_1
0	0	0	0	0	0	0	0	0
0	0	1	0	0	0	0	0	1
0	1	0	0	0	0	1	0	0
0	1	1	0	0	1	0	0	1
1	0	0	0	1	0	0	0	0
1	0	1	0	1	1	0	0	1
1	1	0	1	0	0	1	0	0
1	1	1	1	1	0	0	0	1

The minimal expressions for the outputs as seen from the table are:

$$X_1 = \overline{A}\,\overline{B}C + \overline{A}BC + A\overline{B}C + ABC = C$$

$$X_2 = 0$$

$$X_3 = \overline{A}B\overline{C} + AB\overline{C} = B\overline{C}$$

$$X_4 = \overline{A}BC + A\overline{B}C = C(A \oplus B)$$

$$X_5 = A\overline{B}\,\overline{C} + A\overline{B}C + ABC = A(\overline{B} + C)$$

$$X_6 = AB\overline{C} + ABC = AB$$

A logic diagram can be drawn based on the above equations.

EXAMPLE 6.40 A safe has 5 locks v, w, x, y, and z; all of which must be unlocked for the safe to open. The keys to the locks are distributed among five executives in the following manner.

Mr. A has keys for locks v and x.
Mr. B has keys for locks v and y.
Mr. C has keys for locks w and y.
Mr. D has keys for locks x and z.
Mr. E has keys for locks v and z.

(a) Determine the minimal number of executives required to open the safe.

(b) Find all the combinations of executives that can open the safe; write an expression f(A, B, C, D, E) which specifies when the safe can be opened as a function of what executives are present.

(c) Who is the essential executive?

Solution

The following table indicates the executives and the locks they can open.

Executive	Keys for locks				
	v	w	x	y	z
Mr. A	•		•		
Mr. B	•			•	
Mr. C		•		•	
Mr. D			•		•
Mr. E	•				•

We see that the key for lock w is only with Mr. C. So, Mr. C is the essential executive, without whom the safe cannot be opened. Once C is present, he can open lock y too. As seen from the table, the remaining locks v, x, and z can be opened by A and D or A and E or B and D or D and E. So the combinations of executives who can open the locks are CAD or CAE or CBD or CDE.

The Boolean expression corresponding to the above statement is

$$f(A, B, C, D, E) = CAD + CAE + CBD + CDE$$

The minimal number of executives required is 3.

EXAMPLE 6.41 You are presented with a set of requirements under which an insurance policy can be issued. The applicant must be:

1. a married female 25 years old or over, or
2. a female under 25, or
3. a married male under 25 who has not been involved in a car accident, or
4. a married male who has been involved in a car accident, or
5. a married male 25 years or over who has not been involved in a car accident.

Find an algebraic expression which assumes a value 1 whenever the policy is issued. Simplify the expression obtained.

Solution

Let the variables w, x, y, and z assume the truth value in the following cases.

w = 1, if the applicant has been involved in a car accident.
x = 1, if the applicant is married.
y = 1, if the applicant is a male.
z = 1, if the applicant is under 25.

The policy can be issued when any one of the conditions 1, 2, 3, 4, or 5 is met. The conditions 1, 2, 3, 4, and 5 are represented algebraically by $x\bar{y}\bar{z}$, $\bar{y}z$, $xyz\bar{w}$, xyw, $xy\bar{z}\bar{w}$. Therefore,

$$f(w, x, y, z) = x\bar{y}\bar{z} + \bar{y}z + xyz\bar{w} + xyw + xy\bar{z}\bar{w}$$

$$= xy\bar{w}(z + \bar{z}) + xyw + \bar{y}(z + x\bar{z})$$

$$= xy\overline{w} + xyw + \overline{y}(z + \overline{z})(z + x)$$
$$= xy(w + \overline{w}) + \overline{y}(z + x)$$
$$= xy + x\overline{y} + \overline{y}z$$
$$= x(y + \overline{y}) + \overline{y}z$$
$$= x + \overline{y}z$$

So the policy can be issued if the applicant is either married or is a female under 25.

EXAMPLE 6.42 An air-conditioning unit is controlled by four variables: temperature T, humidity H, the time of the day D, and the day of the week W. The unit is turned on under any of the following circumstances.

1. The temperature exceeds 78° F, and the time of the day is between 8 a.m. and 5 p.m.
2. The humidity exceeds 85%, the temperature exceeds 78° F, and it is a weekend.
3. The humidity exceeds 85%, the temperature exceeds 78° F, and the time of day is between 8 a.m. and 5 p.m.
4. It is Saturday or Sunday and humidity exceeds 85%.

Write a logic expression for controlling the air-conditioning unit. Simplify the expression obtained as far as possible.

Solution

Define the variables:

(a) T = 1, if the temperature exceeds 78° F.

(b) H = 1, if the humidity exceeds 85%.

(c) D = 1, if the time of the day is between 8 a.m. and 5 p.m.

(d) W = 1, if it is weekend, i.e. Saturday or Sunday.

The circumstances 1, 2, 3, and 4, respectively, are then given algebraically as, TD, HTD, HTW, and WH. Therefore, the Boolean expression for turning on the machine is

$$f = TD + HTD + HTW + WH$$
$$= TD(1 + H) + HW(1 + T) = TD + HW$$

So the air-conditioning unit is turned on, if the temperature exceeds 78° F and the time of the day is between 8 a.m. and 5 p.m, or if it is a weekend and humidity exceeds 85%.

EXAMPLE 6.43 Five soldiers A, B, C, D, and E volunteer to perform an important military task if their following conditions are satisfied.

1. Either A or B or both must go.
2. Either C or E but not both must go.
3. Either both A and C go or neither goes.
4. If D goes, then E must also go.
5. If B goes, then A and C must also go.

Define the variables A, B, C, D, and E, so that an unprimed variable will mean that the corresponding soldier has been selected to go. Determine the expression which specifies the combinations of volunteers who can get the assignment.

Solution

Analyzing the problem to perform the task, the first condition is, either A or B or both must go.

Case 1. Suppose A goes, then according to condition 3, C must also go. If C goes, then according to condition 2, E cannot go. Then according to condition 4 when E is not going, D also cannot go. So D does not go. So A and C can go to perform the task.

Case 2. When B goes, according to condition 5, A and C must go. When C goes, E cannot go, and when E cannot go, D also cannot go. So, the second combination of soldiers who can perform the task is ABC.

Case 3. When both A and B go, C has to go. When C goes, E and therefore, D cannot go. This is the same as the second combination ABC.

So, the conclusion is either A and C, or A, B and C can go and perform the military task. Therefore,

$$f = AC + ABC$$
$$= AC\,(1 + B) = AC$$

So, the minimal combination of soldiers who can get the assignment is A and C.

EXAMPLE 6.44 A circuit receives a 4-bit excess-3 code. Design a minimal circuit to detect the decimal numbers 0, 1, 4, 6, 7, and 8.

Solution

The input is a 4-bit excess-3 code. There are 16 possible combinations of 4-bit inputs. Out of these, 10 combinations represent valid excess-3 code. The remaining 6 are invalid. Hence, the corresponding outputs are don't cares. For a minimal design, we find the minimum SOP form, the minimum POS form and then we take the minimal of these two minimums. The Boolean expression in terms of minterms is

$$f = \Sigma\, m(3, 4, 7, 9, 10, 11) + d(0, 1, 2, 13, 14, 15)$$

The truth table is shown below.

Truth table

4-bit excess-3				Decimal number	Output	4-bit excess-3				Decimal number	Output
A	B	C	D			A	B	C	D		
0	0	1	1	0	1	1	0	0	0	5	0
0	1	0	0	1	1	1	0	0	1	6	1
0	1	0	1	2	0	1	0	1	0	7	1
0	1	1	0	3	0	1	0	1	1	8	1
0	1	1	1	4	1	1	1	0	0	9	0

The K-maps in the SOP and POS forms, their minimization, the minimal expressions obtained from each and the actual minimal circuit in the NOR logic are all shown below.

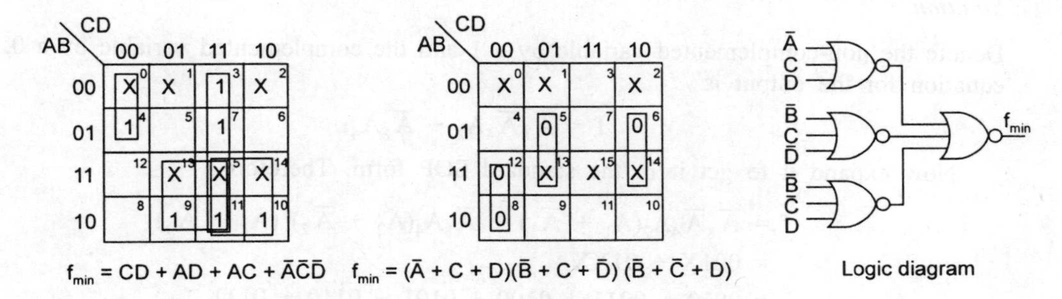

EXAMPLE 6.45 Design a minimal circuit to produce an output of 1, when its input is a 2421 code representing an even decimal number less than 10.

Solution

For a 2421 code, the input combinations 0101, 0110, 0111, 1000, 1001 and 1010 are invalid and the corresponding outputs are don't cares. The Boolean expression for this problem is

$$f = \Sigma\ m(0, 2, 4, 12, 14) + d(5, 6, 7, 8, 9, 10)$$

The truth table is shown below.

Truth Table

Decimal number	2421 code A	B	C	D	Output	Decimal number	2421 code A	B	C	D	Output
0	0	0	0	0	1	5	1	0	1	1	0
1	0	0	0	1	0	6	1	1	0	0	1
2	0	0	1	0	1	7	1	1	0	1	0
3	0	0	1	1	0	8	1	1	1	0	1
4	0	1	0	0	1	9	1	1	1	1	0

The K-maps in SOP and POS forms, their minimization, the minimal expressions obtained from each, and the logic diagram based on the real minimal expression are all shown below. In this case, both the SOP and POS minimal forms are the same.

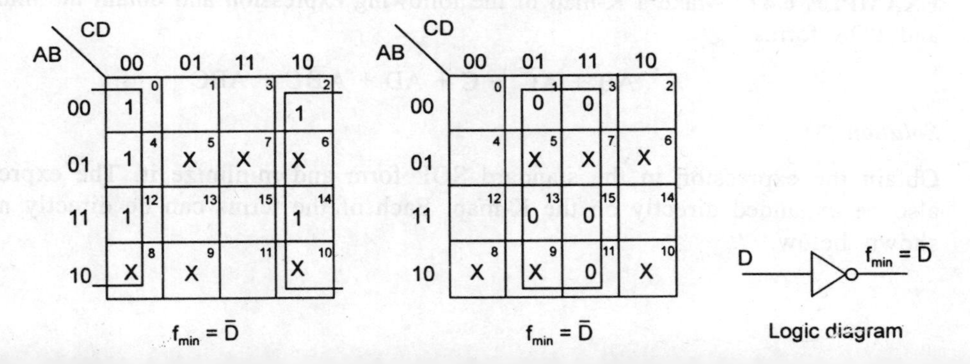

EXAMPLE 6.46 $A_8A_4A_2A_1$ is an 8421 BCD input to a logic circuit whose output is a 1 when $A_8 = 0$, $A_4 = 0$ and $A_2 = 1$, or when $A_8 = 0$ and $A_4 = 1$. Design the simplest possible logic circuit.

Solution

Denote the non-complemented variable by a 1 and the complemented variable by a 0. The equation for the output is

$$f = \overline{A}_8\overline{A}_4A_2 + \overline{A}_8A_4$$

Now expand it to get it in the standard SOP form. Therefore,

$$\begin{aligned} f &= \overline{A}_8\overline{A}_4A_2(A_1 + \overline{A}_1) + \overline{A}_8A_4(A_2 + \overline{A}_2)\,(A_1 + \overline{A}_1) \\ &= 001X + 01XX \\ &= 0010 + 0011 + 0100 + 0101 + 0110 + 0111 \\ &= \Sigma\, m(2, 3, 4, 5, 6, 7) \end{aligned}$$

The input is a 4-bit BCD. So there are 6 invalid combinations (10, 11, 12, 13, 14, 15) and the corresponding outputs are don't cares. So, the Boolean equation is

$$f = \Sigma\, m(2, 3, 4, 5, 6, 7) + d(10, 11, 12, 13, 14, 15).$$

Obtain the minimal expressions in SOP and POS forms and implement the minimal of these minimals. The K-maps in SOP and POS forms, their minimization, the minimal expressions obtained from each, and the logic diagram based on the minimal expression are all shown below.

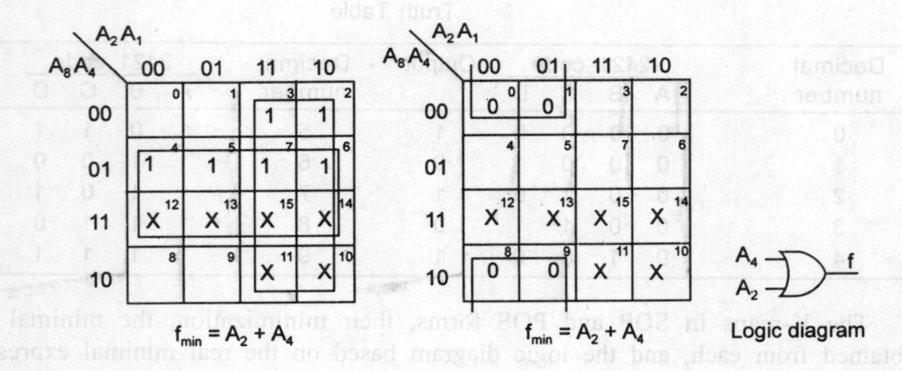

EXAMPLE 6.47 Make a K-map of the following expression and obtain the minimal SOP and POS forms.

$$AB + A\overline{C} + C + AD + A\overline{B}C + ABC$$

Solution

Obtain the expression in the standard SOP form and minimize it. The expression can also be expanded directly on the K-map. Each of the terms can be directly mapped as shown below.

AB = ABXX (1 1 X X) → A = 1, B = 1; C and D can be a 0 or a 1, i.e. the entire third row (m_{12}, m_{13}, m_{14}, and m_{15}).

$A\bar{C} = AX\bar{C}X$(1 X 0 X) → A = 1, C = 0; B and D can be a 0 or a 1, i.e. the intersections of the third and fourth rows with the first and second columns (m_8, m_9 and m_{12}, m_{13}).

C = XXCX (X X 1 X) → C = 1, A, B, and D can be a 0 or a 1, i.e. the entire third and fourth columns (m_2, m_3, m_6, m_7 and m_{10}, m_{11}, m_{14}, m_{15}).

AD = AXXD (1 X X 1) → A = 1, D = 1; B and C can be a 0 or a 1, i.e. the intersection of the third and fourth rows with the second and third columns (m_9, m_{11} and m_{13}, m_{15}).

$A\bar{B}C = A\bar{B}CX$ (1 0 1 X) → A = 1, B = 0, C = 1; D can be a 0 or a 1, i.e. the intersection of the fourth row with the third and fourth columns (m_{10} and m_{11}).

ABC = ABCX (1 1 1 X) → A = 1, B = 1, C = 1; D can be a 0 or a 1, i.e. the intersection of the third row with the third and fourth columns (m_{14} and m_{15}).

The K-maps, the minimal SOP and POS expressions, and the logic diagram are shown below. Both the SOP and POS forms give the same minimal expression.

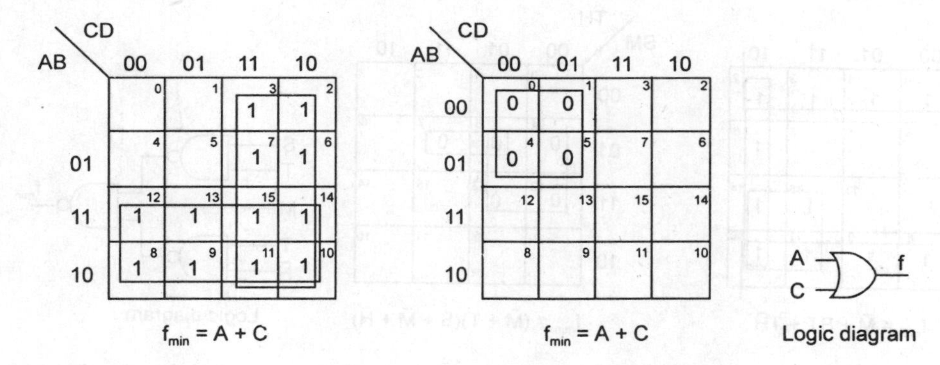

f_{min} = A + C f_{min} = A + C Logic diagram

EXAMPLE 6.48 A lawn sprinkling system is controlled automatically by certain combinations of the following variables.

Season (S = 1, if summer; 0, otherwise)
Moisture content of soil (M = 1, if high; 0, if low)
Outside temperature (T = 1, if high; 0, if low)
Outside humidity (H = 1, if high; 0, if low)

The sprinkler is turned on under any of the following circumstances.

1. The moisture content is low in winter.
2. The temperature is high and the moisture content is low in summer.
3. The temperature is high and the humidity is high in summer.
4. The temperature is low and the moisture content is low in summer.
5. The temperature is high and the humidity is low.

Use a K-map to find the simplest possible logic expression involving the variables S, M, T, and H for turning on the sprinkler system.

Solution

The given circumstances 1, 2, 3, 4, and 5 are expressed in terms of the defined variables S, M, T, and H as $\overline{M}\,\overline{S}$, $T\overline{M}S$, THS, $\overline{T}\,\overline{M}S$, and $T\overline{H}$, respectively.

The Boolean expression is

$$\overline{S}\,\overline{M} + S\overline{M}T + STH + S\overline{M}\,\overline{T} + T\overline{H}$$

$$= 0\,0\,X\,X + 1\,0\,1\,X + 1\,X\,1\,1 + 1\,0\,0\,X + X\,X\,1\,0$$

The expressions in terms of minterms and maxterms are

$$\Sigma\, m(0, 1, 2, 3, 6, 8, 9, 10, 11, 14, 15)$$

$$\Pi\, M(4, 5, 7, 12, 13)$$

The K-maps in the SOP and POS forms, their minimization, the minimal expressions obtained from each, and the logic diagram in the SOP form are all shown below. Both the SOP and POS forms give the same minimum.

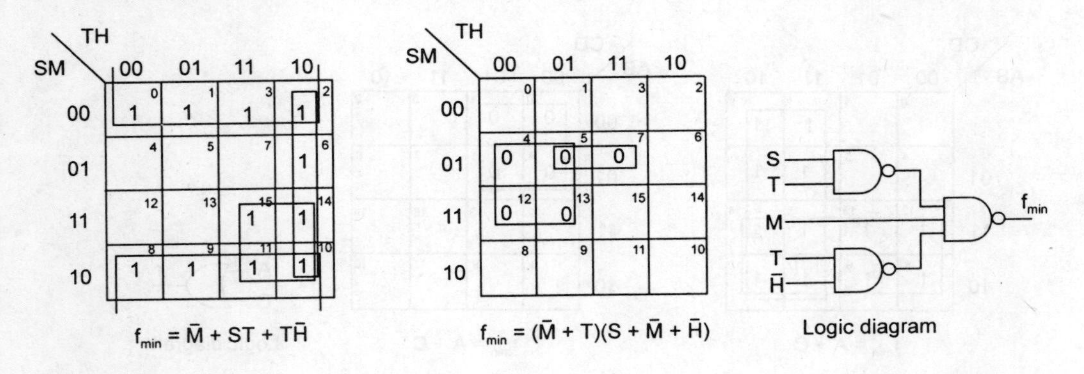

6.15 QUINE-McCLUSKY METHOD

The minimization of Boolean expressions using K-maps is usually limited to a maximum of six variables. The Quine-McClusky method, also known as the tabular method, is a more systematic method of minimizing expressions of even larger number of variables. This is suitable for hand computation as well as computation by machines.

The fundamental idea on which this tabulation procedure is based is that, repeated application of the combining theorem $PA + P\overline{A} = P$ (where P is a set of literals) on all adjacent pairs of terms, yields the set of all prime implicants, from which a minimal sum may be selected.

Consider the minimization of the expression

$$\Sigma\, m(0, 1, 4, 5) = \overline{A}\,\overline{B}\,\overline{C} + \overline{A}\,\overline{B}C + A\overline{B}\,\overline{C} + A\overline{B}C$$

The first two terms, and the third and fourth terms can be combined to yield

$$\overline{A}\,\overline{B}(C + \overline{C}) + A\overline{B}(C + \overline{C}) = \overline{A}\,\overline{B} + A\overline{B}$$

This expression can further be reduced to

$$\overline{B}(\overline{A} + A) = \overline{B}$$

In the first step, we combined two pairs of adjacent terms, each of 3 literals per term, into two terms each of two literals. In the second step, these two terms are combined again and reduced to one term of a single variable.

The same result can be obtained by combining m_0 and m_4 and m_1 and m_5 in the first step and the resulting terms in the second step. Minterms $m_0(\overline{A}\,\overline{B}\,\overline{C})$ and $m_1(\overline{A}\,\overline{B}C)$ are adjacent to each other because they differ in only literal C. Similarly, minterms $m_4(A\overline{B}\,\overline{C})$ and $m_5(A\overline{B}C)$ are adjacent to each other because they differ in only one literal C. Minterms $m_0(\overline{A}\,\overline{B}\,\overline{C})$ and $m_5(A\overline{B}C)$ or $m_1(\overline{A}\,\overline{B}C)$ and $m_4(A\overline{B}\,\overline{C})$ cannot be combined, being not adjacent to each other since they differ in more than one variable. If we consider the binary representation of minterms, $m_0(0\ 0\ 0)$ and $m_1(0\ 0\ 1)$, i.e. 0 0 0 and 0 0 1, they differ in only one position. When combined they result in 0 0 –, i.e. variable C is absorbed. Similarly, $m_4(1\ 0\ 0)$ and $m_5(1\ 0\ 1)$, i.e. 1 0 0 and 1 0 1 differ in only one position. So, when combined, they result in 10 –. Now 0 0 – and 1 0 – can be combined because they differ in only one position. The result is a – 0 –.

For the binary representation of two minterms to be different in just one position, it is necessary (but not sufficient) that the number of 1s in those two minterms differ exactly by one. Consequently, to facilitate the combination process, the minterms are arranged in groups according to the number of 1s in their binary representation.

The procedure for the minimization of a Boolean expression by the tabular method may, therefore, be described as below.

1. Arrange all minterms in groups of the same number of 1s in their binary representation. Start with the least number of 1s group and continue with groups of increasing number of 1s. The number of 1s in a term is called the ***index*** of the term.
2. Compare each term of the lowest index group with every term in the succeeding group. Whenever possible, combine the two terms being compared by means of the combining theorem. Repeat this by comparing each term in a group of index *i* with every term in the group of index *i* + 1, until all possible applications of the combining theorem have been exhausted.

   Two terms from adjacent groups are combinable, if their binary representations differ by just a single digit in the same position; the combined terms consist of the original fixed representation with the differing one replaced by a dash (–). Place a check mark (✓) next to every term, which has been combined with at least one term (each term may be combined with several terms, but only a single check is required).
3. Compare the terms generated in step 2 in the same fashion; combine two terms which differ by only a single 1 and whose dashes are in the same position to

generate a new term. Continue the process until no further combinations are possible. The remaining unchecked terms constitute the set of prime implicants of the expression. They are called ***prime implicants*** because they are not covered by any other term with fewer literals.

The Decimal Representation

The tabulation procedure can be further simplified by adopting the decimal code for the minterms, rather than the binary representation. Two minterms can be combined, only if they differ by a power of 2; that is, the difference between their decimal codes is 2^i. The combined term consists of the same literals as the minterms with the exception of the variable whose weight is 2^i, being deleted. For example, minterms m_0 (0000) and m_8 (1000) differ by $8 = 2^3$. So, they can be combined and the combined term can be written as 0, 8 (8), instead of – 0 0 0. The 8 in the parentheses indicates that the variable A whose weight is 8 can be deleted.

The condition that the decimal codes of two combinable terms must differ by a power of 2 is necessary but not sufficient. The terms whose codes differ by a power of 2, but which have the same index cannot be combined, since they differ by more than one variable. For example 2 and 4, 2 and 8, 4 and 8, 10 and 12, etc. cannot be combined. Similarly, if a term with a smaller index has a higher decimal value than another term whose index is higher, then the two terms cannot be combined although they may differ by a power of 2; for example, minterms 9 and 7, 17 and 13, 20 and 19, 25 and 23, etc. cannot be combined although they differ by a power of 2. Except for the above phenomenon, the tabulation procedure using decimal representation is completely analogous to that using binary representation. In practice, only the decimal representations are used.

EXAMPLE 6.49 Obtain the set of prime implicants for Σ m(0, 1, 6, 7, 8, 9, 13, 14, 15) using the binary designations of minterms.

Solution

Group the minterms in terms of the number of 1s present in them and write their binary designations. The procedure to obtain the prime implicants is shown in the table below.

	Column 1		Column 2		Column 3		
	Minterm	Binary designation		A B C D		A B C D	
Index 0	0	0 0 0 0 ✓	0, 1 (1)	0 0 0 – ✓	0, 1, 8, 9 (1, 8)	– 0 0 –	Q
Index 1	1	0 0 0 1 ✓	0, 8 (8)	– 0 0 0 ✓		... ...	
	8	1 0 0 0 ✓	1, 9 (8)	– 0 0 1 ✓			
Index 2	6	0 1 1 0 ✓	8, 9 (1)	1 0 0 – ✓	6, 7, 14, 15 (1, 8)	– 1 1 –	P
	9	1 0 0 1 ✓	6, 7 (1)	0 1 1 – ✓			
Index 3	7	0 1 1 1 ✓	6, 14 (8)	– 1 1 0 ✓			
	13	1 1 0 1 ✓	9, 13 (4)	1 – 0 1 S			
	14	1 1 1 0 ✓	7, 15 (8)	– 1 1 1 ✓			
Index 4	15	1 1 1 1 ✓	13, 15 (2)	1 1 – 1 R			
			14, 15 (1)	1 1 1 – ✓			

Starting with the index 0 group, compare each term in this group with every term in the index 1 group. Repeat this procedure with the index groups 1 and 2, 2 and 3, and 3 and 4. Combine the terms if they differ in only one position and generate a new term and record it in column 2. Check off the terms of column 1 which are combined. Repeat the entire procedure for the groups formed in columns 2 and 3 too. The detailed procedure is given below:

Comparing the terms of index 0 with the terms of index 1 of column 1, $m_0(0000)$ is combined with $m_1(0001)$ to yield 0, 1 (1), i.e. 000 –. This is recorded in column 2 and 0000 and 0001 are checked off in column 1. $m_0(0000)$ is combined with $m_8(1000)$ to yield 0, 8 (8), i.e. – 000. This is recorded in column 2 and 1000 is checked off in column 1. Note that 0000 of column 1 has already been checked off. No more combinations of terms of index 0 and index 1 are possible. So, draw a line below the last combination of these groups, i.e. below 0, 8 (8), – 000 in column 2. Now 0, 1 (1), i.e. 000 – and 0, 8 (8), i.e. – 000 are the terms in the first group of column 2.

Comparing the terms of index 1 with the terms of index 2 in column 1, $m_1(0001)$ is combined with $m_9(1001)$ to yield 1, 9 (8), i.e. – 001. This is recorded in column 2 and 1001 is checked off in column 1 because 0001 has already been checked off. $m_8(1000)$ is combined with $m_9(1001)$ to yield 8, 9 (1), i.e. 100 –. This is recorded in column 2. 1000 and 1001 of column 1 have already been checked off. So, no need to check them off again. No more combinations of terms of index 1 and index 2 are possible. So, draw a line below the last combination of these groups, i.e. 8, 9 (1), – 001 in column 2. Now 1, 9 (8), i.e. – 001 and 8, 9 (1), i.e. 100– are the terms in the second group of column 2.

Similarly, comparing the terms of index 2 with the terms of index 3 in column 1,

$m_6(0110)$ and $m_7(0111)$ yield 6, 7 (1), i.e. 011–. Record it in column 2 and check off 6(0110) and 7(0111).

$m_6(0110)$ and $m_{14}(1110)$ yield 6, 14 (8), i.e. –110. Record it in column 2 and check off 6(0110) and 14(1110).

$m_9(1001)$ and $m_{13}(1101)$ yield 9, 13 (4), i.e. 1–01. Record it in column 2 and check off 9(1001) and 13(1101).

So, 6, 7 (1), i.e. 011–, and 6, 14 (8), i.e. –110 and 9, 13 (4), i.e. 1–01 are the terms in group 3 of column 2. Draw a line at the end of 9, 13 (4), i.e. 1–01.

Also, comparing the terms of index 3 with the terms of index 4 in column 1,

$m_7(0111)$ and $m_{15}(1111)$ yield 7, 15 (8), i.e. –111. Record it in column 2 and check off 7(0111) and 15(1111).

$m_{13}(1101)$ and $m_{15}(1111)$ yield 13, 15 (2), i.e. 11–1. Record it in column 2 and check off 13 and 15.

$m_{14}(1110)$ and $m_{15}(1111)$ yield 14, 15 (1), i.e. 111–. Record it in column 2 and check off 14 and 15.

So, 7, 15 (8), i.e. –111, and 13, 15 (2), i.e. 11–1 and 14, 15 (1), i.e. 111– are the terms in group 4 of column 2. Column 2 is completed now.

Comparing the terms of group 1 with the terms of group 2 in column 2, the terms 0, 1 (1), i.e. 000– and 8, 9 (1), i.e. 100– are combined to form 0, 1, 8, 9 (1, 8), i.e. –00–. Record it in group 1 of column 3 and check off 0, 1 (1), i.e. 000–, and 8, 9 (1), i.e. 100–

of column 2. The terms 0, 8 (8), i.e. –000 and 1, 9 (8), i.e. –001 are combined to form 0, 1, 8, 9 (1, 8), i.e. –00–. This has already been recorded in column 3. So, no need to record again. Check off 0, 8 (8), i.e. –000 and 1, 9 (8), i.e. –001 of column 2. Draw a line below 0, 1, 8, 9 (1, 8), i.e. –00–. This is the only term in group 1 of column 3. No term of group 2 of column 2 can be combined with any term of group 3 of column 2. So, no entries are made in group 3 of column 2.

Comparing the terms of group 3 of column 2 with the terms of group 4 of column 2, the terms 6, 7 (1), i.e. 011–, and 14, 15 (1), i.e. 111– are combined to form 6, 7, 14, 15 (1, 8), i.e. –11–. Record it in group 3 of column 3 and check off 6, 7 (1), i.e. 011– and 14, 15 (1), i.e. 111– of column 2. The terms 6, 14 (8), i.e. –110 and 7, 15 (8), i.e. –111 are combined to form 6, 7, 14, 15 (1, 8), i.e. –11–. This has already been recorded in column 3; so, check off 6, 14 (8), i.e. –110 and 7, 15 (8), i.e. –111 of column 2.

Observe that the terms 9, 13 (4), i.e. 1–01 and 13, 15 (2), i.e. 11–1 cannot be combined with any other terms. Similarly in column 3, the terms 0, 1, 8, 9 (1, 8), i.e. –00– and 6, 7, 14, 15 (1, 8), i.e. –11– cannot also be combined with any other terms. So, these 4 terms are the prime implicants.

The terms, which cannot be combined further, are labelled as P, Q, R, and S. These form the set of prime implicants.

EXAMPLE 6.50 Obtain the set of prime implicants for Σ m(1, 2, 3, 5, 6, 7, 8, 9, 12, 13, 15) using the decimal designations of minterms.

Solution

The procedure to obtain the set of prime implicants is illustrated in the table below.

Step 1		Step 2	Step 3	
Index 1	1 ✓	1, 3 (2) ✓	1, 3, 5, 7 (2, 4)	T
	2 ✓	1, 5 (4) ✓	1, 5, 9, 13 (4, 8)	S
	8 ✓	1, 9 (8) ✓	2, 3, 6, 7 (1, 4)	R
Index 2	3 ✓	2, 3 (1) ✓	8, 9, 12, 13 (1, 4)	Q
	5 ✓	2, 6 (4) ✓	5, 7, 13, 15 (2, 8)	P
	6 ✓	8, 9 (1) ✓		
	9 ✓	8, 12 (4) ✓		
	12 ✓	3, 7 (4) ✓		
Index 3	7 ✓	5, 7 (2) ✓		
	13 ✓	5, 13 (8) ✓		
Index 4	15 ✓	6, 7 (1) ✓		
		9, 13 (4) ✓		
		12, 13 (1) ✓		
		7, 15 (8) ✓		
		13, 15 (2) ✓		

The non-combinable terms P, Q, R, S and T are recorded as prime implicants.

$$P \rightarrow 5, 7, 13, 15\ (2, 8) = X\ 1\ X\ 1 = BD$$

(Literals with weights 2 and 8, i.e. C and A are deleted. The lowest minterm is 5 = (4 + 1). So, literals with weights 4 and 1, i.e. B and D are present in non-complemented form. So, read it as BD.)

$$Q \rightarrow 8, 9, 12, 13\ (1, 4) = 1\ X\ 0\ X = A\overline{C}$$

(Literals with weights 1 and 4, i.e. D and B are deleted. The lowest minterm is 8 = (8 + 0). So, literal with weight 8 is present in non-complemented form and literal with weight 2 is present in complemented form. So, read it as $A\overline{C}$.)

$$R \rightarrow 2, 3, 6, 7\ (1, 4) = 0\ X\ 1\ X = \overline{A}C$$

(Literals with weights 1 and 4, i.e. D and B are deleted. The lowest minterm is 2. So, literal with weight 2 is present in non-complemented form and literal with weight 8 is present in complemented form. So, read it as $\overline{A}C$.)

$$S \rightarrow 1, 5, 9, 13\ (4, 8) = X\ X\ 0\ 1 = \overline{C}D$$

(Literals with weights 4 and 8, i.e. A and B are deleted. The lowest minterm is 1. So, literal with weight 1 is present in non-complemented form and literal with weight 2 is present in complemented form. So, read it as $\overline{C}D$.)

$$T \rightarrow 1, 3, 5, 7\ (2, 4) = 0\ X\ X\ 1 = \overline{A}D$$

(Literals with weights 2 and 4, i.e. C and B are deleted. The lowest minterm is 1. So, literal with weight 1 is present in non-complemented form and literal with weight 8 is present in complemented form. So, read it as $\overline{A}D$.)

Don't Cares

When don't care terms are present in an expression, during the process of generating the set of prime implicants, don't care combinations are regarded as true combinations, i.e. combinations for which the expression assumes a 1. This, in effect, increases to the maximum, the number of possible prime implicants. The don't care terms are, however, not considered in the next step of selecting a minimum set of prime implicants.

EXAMPLE 6.51 Obtain the set of prime implicants for the following expression:

$$\Sigma\ m(6, 7, 8, 9) + d(10, 11, 12, 13, 14, 15)$$

Solution

Treat the don't cares as minterms and apply the usual procedure as shown in the table on the next page. From this above table, we see that the prime implicants are P → 8, 9, 10, 11, 12, 13, 14, 15 (1, 2, 4) and Q → 6, 7, 14, 15 (1, 8). The term 6, 7, 14, 15 (1, 8) means that literals with weights 1 and 8, i.e. D and A are deleted and the lowest designated minterm is $m_6(4 + 2)$, i.e. literals with weights 4 and 2 are present in non-complemented form. So it is read as BC. The term 8, 9, 10, 11, 12, 13, 14, 15 (1, 2, 4) means that literals

Step 1	Step 2	Step 3	Step 4
8 ✓	8, 9 (1) ✓	8, 9, 10, 11 (1, 2) ✓	8, 9, 10, 11, 12, 13, 14, 15 (1, 2, 4) P
6 ✓	8, 10 (2) ✓	8, 9, 12, 13 (1, 4) ✓	
9 ✓	8, 12 (4) ✓	8, 10, 12, 14 (2, 4) ✓	
10 ✓	6, 7 (1) ✓	6, 7, 14, 15 (1, 8) Q	
12 ✓	6, 14 (8) ✓	9, 11, 13, 15 (2, 4) ✓	
7 ✓	9, 11 (2) ✓	10, 11, 14, 15 (1, 4) ✓	
11 ✓	9, 13 (4) ✓	12, 13, 14, 15 (1, 2) ✓	
13 ✓	10, 11 (1) ✓		
14 ✓	10, 14 (4) ✓		
15 ✓	12, 13 (1) ✓		
	12, 14 (2) ✓		
	7, 15 (8) ✓		
	11, 15 (4) ✓		
	13, 15 (2) ✓		
	14, 15 (1) ✓		

with weights 1, 2, and 4, i.e. D, C, and B are deleted. The lowest designated minterm is therefore, m_8. So, literal with weight 8 is present in non-complemented form. So, it is read as A.

The Prime Implicant Chart

The prime implicant chart is a pictorial representation of the relationships between the prime implicants and thé minterms of the expression. It consists of an array of *u* columns and *v* rows where *u* and *v* designate the number of minterms for which the expression takes on the value of a 1 and the number of prime implicants, respectively. The entries of the *i*th row consist of xs placed at the intersections with the columns corresponding to minterms covered by the *i*th prime implicant. For example, the prime implicant chart of the expression

$$\Sigma\ m(1, 2, 3, 5, 6, 7, 8, 9, 12, 13, 15)$$

is as shown in the table below. It consists of 11 columns corresponding to the number of minterms and 5 rows corresponding to the prime implicants P, Q, R, S, and T generated (see Example 6.50). Row R contains four xs at the intersections with columns 2, 3, 6, and 7, because these minterms are covered by the prime implicant R. A row is said to cover the columns in which it has xs. The problem now is to select a minimal subset of prime implicants, such that each column contains at least one x in the rows corresponding to the selected subset and the total number of literals in the prime implicants selected is as small as possible. These requirements guarantee that the number of unions of the selected prime implicants is equal to the original number of minterms and that, no other expression containing fewer literals can be found.

		✓	✓	✓	✓	✓	✓	✓	✓	✓	✓
	1	2	3	5	6	7	8	9	12	13	15
*P → 5, 7, 13, 15 (2, 8)				×		×				×	×
*Q → 8, 9, 12, 13 (1, 4)							×	×	×	×	
*R → 2, 3, 6, 7, (1, 4)		×	×		×	×					
S → 1, 5, 9, 13 (4, 8)	×			×				×		×	
T → 1, 3, 5, 7 (2, 4)	×		×	×		×					

Essential Prime Implicants

Essential prime implicants are the implicants which will definitely occur in the final expression. Any row in a prime implicant chart which has at least one minterm that is not present in any other row is called the ***essential row*** and the corresponding prime implicant is called the ***essential prime implicant***. In other words, if any column contains a single ×, the prime implicant corresponding to the row in which this × appears is the essential prime implicant. Once an essential prime implicant has been selected, all the minterms it covers are checked off. After all the essential prime implicants and their corresponding columns have been checked, the union of all the essential prime implicants yields the minimal expression. If this is not the case, additional prime implicants are necessary.

In the preceding prime implicant chart, m_2 and m_6 are covered by R only. So, R is an essential prime implicant. So, check off all the minterms covered by it, i.e. m_2, m_3, m_6, and m_7. Q is also an essential prime implicant because only Q covers m_8 and m_{12}. Check off all the minterms covered by it, i.e. m_8, m_9, m_{12}, and m_{13}. P is also an essential prime implicant, because m_{15} is covered only by P. So check off m_{15}, m_5, m_7, and m_{13} covered by it. Thus, only minterm 1 is not covered. Either row S or row T can cover it and both have the same number of literals. Thus, two minimal expressions are possible.

$$P + Q + R + S = BD + A\overline{C} + \overline{A}C + \overline{C}D$$

or

$$P + Q + R + T = BD + A\overline{C} + \overline{A}C + \overline{A}D$$

Don't Care Combinations

Don't care minterms need not be listed as column headings in the prime implicant chart since they do not have to be covered by the minimal expression. By not listing them, we actually leave the specification of don't care terms open. The prime implicant chart, thus, yields a minimal of an expression, which covers all the specified minterms. For example, in the prime implicant chart of Σ m(6, 7, 8, 9) + d(10, 11, 12, 13, 14, 15) shown in the table below, all the don't care minterms are omitted.

	✓	✓	✓	✓
	6	7	8	9
*P → 8, 9, 10, 11, 12, 13, 14, 15 (1, 2, 4)			×	×
*Q → 6, 7, 14, 15 (1, 8)	×	×		

As seen from the above table, P and Q are both essential prime implicants. So, the minimal expression is A + BC.

Dominating Rows and Columns

Two rows (or columns) I and J of a prime implicant table, which have ×s in exactly the same columns (or rows) are said to be equal (written I = J).

A column I in a prime implicant chart is said to dominate another column J of that chart, if column I has an × in every row in which column J has an ×. Any minimal expression derived from a chart which contains both columns I and J can be obtained from a chart containing the dominated column. Hence, if column I dominates column J, then column I can be deleted from the chart without affecting the search for a minimal expression.

A row I in a prime implicant chart is said to dominate another row J, if row I has an × in every column in which row J has an ×. Any minimal expression derived from a chart which contains both rows I and J can be derived from a chart which contains only the dominating row. Hence, if row I dominates row J, then row J can be deleted from the chart without affecting the search for a minimal expression.

Determination of Minimal Expressions in Complex Cases

In simple cases, we can determine the minimal expression by simply inspecting the prime implicant chart. In more complex cases, however, the inspection method becomes prohibitive. Here, the procedure is: First determine the essential prime implicants from the prime implicant chart. Second, form a reduced chart by removing all essential prime implicants and the columns covered by them. Although, none of the rows in the reduced chart is essential, only some of them may be removed. Third, remove all the dominating columns and the dominated rows of this reduced chart and form a new reduced chart. Fourth, look for the secondary essential prime implicants in the new reduced chart, and form another chart by removing the secondary essential prime implicants and the columns covered by them and write the minimal expression in SOP form. Continue the process, if required.

EXAMPLE 6.52 Minimize the following expression:

$$\Sigma\, m(0, 1, 2, 8, 9, 15, 17, 21, 24, 25, 27, 31)$$

Solution

The following table shows the procedure for obtaining all the prime implicants.

Step 1			Step 2			Step 3		
Index 0	0	✓	0, 1	(1)	✓	0, 1, 8, 9	(1, 8)	R
	1	✓	0, 2	(2)	X	1, 9, 17, 25	(8, 16)	Q
Index 1	2	✓	0, 8	(8)	✓	8, 9, 24, 25	(1, 16)	P
	8	✓	1, 9	(8)	✓			
	9	✓	1, 17	(16)	✓			

Contd.

Contd.

Step 1		Step 2		Step 3
Index 2	17 ✓	8, 9 (1)	✓	
	24 ✓	8, 24 (16)	✓	
Index 3	21 ✓	9, 25 (16)	✓	
	25 ✓	17, 21 (4)	W	
Index 4	15 ✓	17, 25 (8)	✓	
	27 ✓	24, 25 (1)	V	
Index 5	31 ✓	25, 27 (2)	U	
		15, 31 (16)	T	
		27, 31 (4)	S	

From the above table, we see that the prime implicants are P, Q, R, S, T, U, V, W, and X. The prime implicant chart is shown below.

	✓		✓			✓	✓	✓				✓
	0	1	2	8	9	15	17	21	24	25	27	31
P → 8, 9, 24, 25				×	×				×	×		
Q → 1, 9, 17, 25		×			×		×			×		
R → 0, 1, 8, 9	×	×		×	×							
S → 27, 31											×	×
*T → 15, 31						×						×
U → 25, 27										×	×	
V → 24, 25									×	×		
*W → 17, 21							×	×				
*X → 0, 2	×		×									

T, W, and X are the essential prime implicants because m_{15} is covered by T only, m_{21} is covered by W only and m_2 is covered by X only. Delete the essential prime implicants and the columns covered by them and form the reduced prime implicant chart as shown below.

	1	8	9	24	25	27
P		×	×	×	×	
Q	×		×		×	
R	×	×	×			
S						×
U					×	×
V				×	×	

Here, 9 ⊃ 8, i.e. column 9 dominates column 8. So, remove column 9. Also 25 ⊃ 24, i.e. column 25 dominates column 24. So remove column 25. Further, U ⊃ S, i.e. row U dominates row S. So, remove row S. And P ⊃ V, i.e. row P dominates row V. So, remove row V.

Draw the new reduced prime implicant chart after removing the dominating columns 9 and 25 and the dominated rows S and V as shown below.

	✓	✓	✓	
	1	8	24	27
*P		×	×	
Q	×			
R	×	×		
*U				×

From the above new reduced prime implicant chart, we observe that P and U are the secondary essential prime implicants. Write another reduced table removing the secondary essential prime implicants and the columns covered by them. The further reduced chart is as shown below.

	1
Q	×
R	×

We can select Q or R because both have the same number of literals. Thus, we get two minimal expressions as follows:

$$T + W + X + P + U + Q = BCDE + A\bar{B}\bar{D}E + \bar{A}\bar{B}\bar{C}\bar{E} + B\bar{C}\bar{D} + AB\bar{C}E + \bar{C}\bar{D}E$$

or

$$T + W + X + P + U + R = BCDE + A\bar{B}\bar{D}E + \bar{A}\bar{B}\bar{C}\bar{E} + B\bar{C}\bar{D} + AB\bar{C}E + \bar{A}\bar{C}\bar{D}$$

Determination of All Possible Minimal Expressions

To get all the possible minimal expressions after getting the essential prime implicants, derive an appropriate expression from the reduced prime implicant chart by multiplying the sums of the rows that take care of each minterm. For the above Example 6.52, from the reduced prime implicant chart, we get

$$(Q + R)\,(P + R)\,(P + Q + R)\,(P + V)\,(P + Q + U + V)(S + U)$$
$$= PQS + PRS + RSV + PRU + PQU + RUV$$

Out of these terms, PQS, PRS, PRU, and PQU yield the same minimum literals. So, there are four minimal expressions given as

1. $T + W + X + P + Q + S = BCDE + A\bar{B}\bar{D}E + \bar{A}\bar{B}\bar{C}\bar{E} + B\bar{C}\bar{D} + \bar{C}\bar{D}E + ABDE$
2. $T + W + X + P + R + S = BCDE + A\bar{B}\bar{D}E + \bar{A}\bar{B}\bar{C}\bar{E} + B\bar{C}\bar{D} + \bar{A}\bar{C}\bar{D} + ABDE$
3. $T + W + X + P + R + U = BCDE + A\bar{B}\bar{D}E + \bar{A}\bar{B}\bar{C}\bar{E} + B\bar{C}\bar{D} + \bar{A}\bar{C}\bar{D} + AB\bar{C}E$
4. $T + W + X + P + Q + U = BCDE + A\bar{B}\bar{D}E + \bar{A}\bar{B}\bar{C}\bar{E} + B\bar{C}\bar{D} + \bar{C}\bar{D}E + AB\bar{C}E$

The Branching Method

If the prime implicant chart has no essential prime implicants, dominated rows and dominated columns, the minimal expression can be obtained by a different approach called the *branching method*. Here, we consider any column and note the rows which cover that column. Make an arbitrary selection of one of those rows and apply the normal reduction procedure for the prime implicant chart without this row and the selected column. The entire procedure is repeated for each row. Take the minimal of all such expressions obtained.

EXAMPLE 6.53 Find the minimal expression for

$$\prod M(2, 3, 8, 12, 13) \cdot d(10, 14)$$

Solution

The tabulation method for the POS form is exactly the same as that for the SOP form. Treat the maxterms as if they are minterms and complete the process. While writing the minimal expression in the POS form, treat the non-complemented variable as a 0 and the complemented variable as a 1. The prime implicants are obtained as shown below.

Step 1	Step 2	Step 3
2 ✓	2, 3 (1) S	8, 10, 12, 14 (2, 4) P
8 ✓	2, 10 (8) R	
3 ✓	8, 10 (2) ✓	
10 ✓	8, 12 (4) ✓	
12 ✓	10, 14 (4) ✓	
13 ✓	12, 13 (1) Q	
14 ✓	12, 14 (2) ✓	

There are four prime implicants P, Q, R, and S. Draw the prime implicant chart as shown below.

	✓	✓	✓	✓	✓
	2	3	8	12	13
*P → 8, 10, 12, 14 (2, 4)			×	×	
*Q → 12, 13 (1)				×	×
R → 2, 10 (8)	×				
*S → 2, 3 (1)	×	×			

Here m_8 is covered by P only. Further, m_{13} is covered by Q only and m_3 is covered by S only. Therefore, P, Q, and S are the essential prime implicants. Check them out and the minterms covered by them. They cover all the minterms. So, the final expression is given by

$$PQS = (\overline{A} + D)(\overline{A} + \overline{B} + C)(A + B + \overline{C}) \qquad \text{(11 gate inputs)}$$

Therefore, the minimal of the POS and SOP forms is the SOP with 10 gate inputs. So, the actual minimal expression is $\overline{A}\,\overline{C} + \overline{B}\,\overline{C} + A\overline{B}\,\overline{D}$.

EXAMPLE 6.54 Obtain the minimal POS expression for

$$\prod M(0, 1, 4, 5, 9, 11, 13, 15, 16, 17, 25, 27, 28, 29, 31) \cdot \prod d(20, 21, 22, 30)$$

Solution

The prime implicants are obtained as shown below.

Step 1	Step 2	Step 3	Step 4
0 ✓	0, 1 (1) ✓	0, 1, 4, 5 (1, 4) ✓	0, 1, 4, 5, 16, 17, 20, 21 (1, 4, 16) R
1 ✓	0, 4 (4) ✓	0, 1, 16, 17 (1, 16) ✓	1, 5, 9, 13, 17, 21, 25, 29 (4, 8, 16) Q
4 ✓	0, 16 (16) ✓	0, 4, 16, 20 (4, 16) ✓	9, 11, 13, 15, 25, 27, 29, 31 (2,4,16) P
16 ✓	1, 5 (4) ✓	1, 5, 9, 13 (4, 8) ✓	
5 ✓	1, 9 (8) ✓	1, 5, 17, 21 (4, 16) ✓	
9 ✓	1, 17 (16) ✓	1, 9, 17, 25 (8, 16) ✓	
17 ✓	4, 5 (1) ✓	4, 5, 20, 21 (1, 16) ✓	
20 ✓	4, 20 (16) ✓	16, 17, 20, 21 (1, 4) ✓	
11 ✓	16, 17 (1) ✓	5, 13, 21, 29 (8, 16) ✓	
13 ✓	16, 20 (4) ✓	9, 11, 13, 15 (2, 4) ✓	
21 ✓	5, 13 (8) ✓	9, 11, 25, 27 (2, 16) ✓	
22 ✓	5, 21 (16) ✓	9, 13, 25, 29 (4, 16) ✓	
25 ✓	9, 11 (2) ✓	17, 21, 25, 29 (4, 16) ✓	
28 ✓	9, 13 (4) ✓	20, 21, 28, 29 (1, 8) V	
15 ✓	9, 25 (16) ✓	20, 22, 28, 30 (2, 8) U	
27 ✓	17, 21 (4) ✓	11, 15, 27, 31 (4, 16) T	
29 ✓	17, 25 (8) ✓	13, 15, 29, 31 (2, 16) ✓	
30 ✓	20, 21 (1) ✓	25, 27, 29, 31 (2, 4) ✓	
31 ✓	20, 22 (2) ✓	28, 29, 30, 31 (1, 2) S	
	20, 28 (8) ✓		
	11, 15 (4) ✓		
	11, 27 (16) ✓		
	13, 15 (2) ✓		
	13, 29 (16) ✓		
	21, 29 (8) ✓		
	22, 30 (8) ✓		
	25, 27 (2) ✓		
	25, 29 (4) ✓		
	28, 29 (1) ✓		
	28, 30 (2) ✓		
	15, 31 (16) ✓		
	27, 31 (4) ✓		
	29, 31 (2) ✓		
	30, 31 (1) ✓		

From the preceding table we see that there are seven prime implicants P, Q, R, S, T, U, and V. The prime implicant chart is shown below.

	0	1	4	5	9	11	13	15	16	17	25	27	28	29	31
	✓	✓	✓	✓					✓	✓					
P					×	×	×	×			×	×		×	×
Q		×		×	×		×			×	×			×	
*R	×	×	×	×					×	×					
S													×	×	×
T						×		×				×			×
U													×		
V													×	×	

Here, m_0, m_4 and m_{16} are covered by R only. So, R is an essential prime implicant. Remove row R and the minterms covered by it and form the reduced prime implicant chart as shown below.

	9	11	13	15	25	27	28	29	31
P	×	×	×	×	×	×		×	×
Q	×		×		×			×	
S							×	×	×
T		×		×		×			×
U							×		
V							×	×	

Remove the dominating columns and the dominated rows.
Column 31 ⊃ 11 = 15 = 27; 29 ⊃ 25 = 13 = 9. Columns 31, 15, and 27 dominate or equal column 11. Columns 29, 25, and 13 dominate or equal column 9. So, delete columns 31, 15, 27, 29, 25, and 23. Row P ⊃ Q, P ⊃ T, S ⊃ U, S ⊃ V. That is, rows Q and T are dominated by row P and rows U and V are dominated by row S. So, delete rows Q, T, U and V. Draw the secondary reduced prime implicant chart as shown below.

	9	11	28
	✓	✓	✓
*P	×	×	
*S			×

P and S are the secondary essential prime implicants. The minimal POS expression is, therefore, given by

$$\text{RPS} = (\text{B} + \text{D})\ (\overline{\text{B}} + \overline{\text{E}})\ (\overline{\text{A}} + \overline{\text{B}} + \overline{\text{C}})$$

and which requires 10 gate inputs.

6.16 FUNCTION MINIMIZATION OF MULTIPLE OUTPUT CIRCUITS

So far, we have discussed the minimization of the single switching functions by the Quine-McClusky method. In practice, many logic design problems have more than one output. For these problems, which are too large or complicated to be solved by K-maps, it is necessary to turn to the tabular method. The multiple switching-function minimization criterion is as follows:

1. Each minimized function should have as many terms in common as possible with other minimized functions.
2. Each minimized function should have a minimum number of product (sum) terms and no product (sum) term of it should be replaceable by a product (sum) term with fewer variables.

The procedure for minimizing a set of *m* switching functions according to the above minimization criterion using the Quine-McClusky method is described below.

Step 1: Find the prime implicants of the *m* switching functions. We use a single table to find the prime implicants of the *m* functions simultaneously by using the tag. Suppose we want to minimize the set of three output functions:

$$f_1(x_1, x_2, x_3, x_4) = \Sigma\, m(1, 2, 3, 5, 7, 8, 9) + \Sigma\, d(12, 14)$$
$$f_2(x_1, x_2, x_3, x_4) = \Sigma\, m(0, 1, 2, 3, 4, 6, 8, 9) + \Sigma\, d(10, 11)$$
$$f_3(x_1, x_2, x_3, x_4) = \Sigma\, m(1, 3, 5, 7, 8, 9, 12, 13) + \Sigma\, d(14, 15)$$

The single tabulation that can be used to find the prime implicants of the three functions simultaneously, is shown in Table 6.2.

Note that the first column of step 1 of this table includes the minterms and the don't cares (represented in decimal numbers) of all three functions. They are grouped into groups as usual, according to the number of 1s in their binary representation. The second column of step 1 consists of a tag for each minterm (row) to indicate which of the output functions include the minterm. Each symbol of the tag corresponds to one of the output functions and is either a 0 or a 1 depending on whether the corresponding minterm is absent in the output function or present in the output function. For example, the tag for the first row is

f_1	f_2	f_3
0	1	0

which indicates that the minterm 0 is included in the output function f_2, but is not included in the output functions f_1 and f_3.

The 1st, 2nd and 3rd reductions of the implicants to form prime implicants are the same as those in the single output technique. The tags of each row of steps 2, 3, and 4 in Table 6.2 are obtained by multiplying the tags of the corresponding implicants by combining which, that row is obtained. Note that, there is a 1 in the new tag only when there are 1s in both the original implicants. The tag of the implicant (0, 1) in the first row

Table 6.2 To find the prime implicants of three functions simultaneously

Step 1

Minterm	f_1 f_2 f_3 (tag)
0	0 1 0 ✓
1	1 1 1 ✓
2	1 1 0 ✓
4	0 1 0 ✓
8	1 1 1 ✓
3	1 1 1 ✓
5	1 0 1 ✓
6	0 1 0 ✓
9	1 1 1 ✓
10	0 1 0 ✓
12	1 0 1 ✓
7	1 0 1 ✓
11	0 1 0 ✓
13	0 0 1 ✓
14	1 0 1 ✓
15	0 0 1 ✓

Step 2

Implicant	f_1 f_2 f_3 (tag)
0, 1 (1)	0 1 0 ✓
0, 2 (2)	0 1 0 ✓
0, 4 (4)	0 1 0 ✓
0, 8 (8)	0 1 0 ✓
1, 3 (2)	1 1 1 A
1, 5 (4)	1 0 1 ✓
1, 9 (8)	1 1 1 B
2, 3 (1)	1 1 0 C
2, 6 (4)	0 1 0 ✓
2, 10 (8)	0 1 0 ✓
4, 6 (2)	0 1 0 ✓
8, 9 (1)	1 1 1 D
8, 10 (2)	0 1 0 ✓
8, 12 (4)	1 0 1 E
3, 7 (4)	1 0 1 ✓
3, 11 (8)	0 1 0 ✓
5, 7 (2)	1 0 1 ✓
5, 13 (8)	0 0 1 ✓
9, 11 2)	0 1 0 ✓
9, 13 (4)	0 0 1 ✓
10, 11 (1)	0 1 0 ✓
12, 13 (1)	0 0 1 ✓
12, 14 (2)	1 0 1 F
7, 15 (8)	0 0 ✓
13, 15 (2)	0 0 1 ✓
14, 15 (1)	0 0 1 ✓

Step 3

Implicant	f_1 f_2 f_3 (tag)
0, 1, 2, 3 (1, 2)	0 1 0 ✓
0, 1, 8, 9 (1, 8)	0 1 0 ✓
0, 2, 8, 10 (2, 8)	0 1 0 ✓
0, 2, 4, 6 (2, 4)	0 1 0 G
1, 3, 5, 7 (2, 4)	1 0 1 H
1, 3, 9, 11 (2, 8)	0 1 0 ✓
1, 5, 9, 13 (4, 8)	0 0 1 I
2, 3, 10, 11 (1, 8)	0 1 0 ✓
8, 9, 10, 11 (1, 2)	0 1 0 ✓
8, 9, 12, 13 (1, 4)	0 0 1 J
5, 7, 13, 15 (2, 8)	0 0 1 K
12, 13, 14, 15 (1, 2)	0 0 1 L

Step 4

Implicant	f_1 f_2 f_3 (tag)
0,1,2,3,8,9,10,11 (1,2,8)	0 0 M

of step 2 in Table 6.2 is

f_1	f_2	f_3
0	1	0

This is because, only the symbols of the tags of the minterms 0 and 1 under f_2 are both a 1. It should be noted that when the tag of the new cell implicant is (0, 0, 0), it means that none of the output functions includes this new implicant. This new implicant should not therefore be included. For example, in Table 6.2, minterms 4 and 5 are combinable. Since their tags are (0 1 0) and (1 0 1), the resulting tag is (0 0 0). Thus, the implicant (4, 5) does not appear in step 2 of Table 6.2. For the same reason, the implicant (11, 15) is also not included in the table.

It is important to point out a difference between the single function minimization and the multiple function minimization in determination of prime implicants in the function reduction table. For example, in the first reduction column, i.e. step 2 of Table 6.2, the implicants (0, 2) and (1, 3) are combined into the implicant (0, 1, 2, 3) whose tag is (0 1 0) which is same as that of the implicant (0, 2), but not that of the implicant (1, 3); hence the implicant (0, 2) is 'checked out', but implicant (1, 3) is not.

The rule for checking out an implicant in a multiple function reduction table is: an implicant is checked out (✓) if and only if its tag is identical to that of the newly formed implicant.

From Table 6.2 we see that the prime implicants are A, B, C, D, E, F, G, H, I, J, K, L, and M.

Step 2: Construct the multiple output prime implicant table as shown in Table 6.3. The row dominance, row I ⊃ row J, in a multiple output prime implicant table is defined as row I ⊃ row J for every output function. The definition of column dominance for the multiple output case is the same as that for the single output case.

Step 3: Remove all the dominating columns from the table. Delete columns 1, 3, and 5 of f_1, because they are dominating column 7. Delete columns 0, 2, and 4 of f_2, because they are dominating column 6. Delete column 9 of f_2, because it is dominating column 8 of f_2. Delete columns 1 and 5 of f_3, because they are dominating columns 3 and 7, respectively, of f_3.

Step 4: Remove all the dominated rows from the table. Delete row F because it is dominated by row E. Remove row L, because it is dominated by row J.

Step 5: Remove the essential prime implicants from the table and include them in their respective functions. Form the reduced prime implicant chart after removing the essential prime implicants and the minterms covered by them.

The prime implicant chart is as shown in Table 6.3.

Table 6.3 Prime implicant chart

	f_1							f_2								f_3							
	✓	✓	✓	✓	✓			✓		✓	✓	✓	✓			✓	✓	✓	✓				
	1	2	3	5	7	8	9	0	1	2	3	4	6	8	9	1	3	5	7	8	9	12	13
A	×		×						×		×					×	×						
B	×						×		×						×	×					×		
*C		×	×							×	×												
D						×	×							×	×					×	×		
E						×														×		×	
F																						×	
*G								×		×		×	×										
*H	×		×	×	×											×	×	×	×				
I																×		×			×		×
J																				×	×	×	×
K																		×	×				×
L																						×	×
M								×	×	×	×			×	×								

From the multiple output prime implicant table (Table 6.3), we observe that,

In f_1, $5 \supset 7,\ 3 \supset 7,\ 1 \supset 7,\ 3 \supset 2$

In f_2, $0 \supset 6,\ 4 \supset 6,\ 2 \supset 6,\ 9 \supset 8$

In f_3, $1 \supset 3,\ 5 \supset 7$ and $E \supset F,\ J \supset L$

C is an essential prime implicant for f_1, because m_2 is covered by C only. G is an essential prime implicant for f_2, because m_6 and m_4 are covered by G only. H is an essential prime implicant for f_1, because m_5 and m_7 of f_1 are covered by H only. H is also an essential prime implicant for f_3, because its m_3 is covered by H only.

The reduced prime implicant chart is shown in Table 6.4.

Table 6.4 Reduced prime implicant chart

	f_1		f_2		f_3			
	8	9	1	8	8	9	12	13
A			x					
B		x	x			x		
D	x	x		x	x	x		
E	x				x		x	
I						x		x
J					x	x	x	x
K								x
M			x	x				
	f_1 = C + D + H		f_2 = G + M		f_3 = H + J			

Step 6: Repeat steps 3 to 5 as many times as needed until every minterm of each function is covered by an essential prime implicant. The set of minimized functions is

$$f_1(x_1, x_2, x_3, x_4) = C + D + H = (\bar{x}_1 \cdot \bar{x}_2 \cdot x_3) + (x_1 \cdot \bar{x}_2 \cdot \bar{x}_3) + (\bar{x}_1 \cdot x_4)$$

$$f_2(x_1, x_2, x_3, x_4) = G + M = (\bar{x}_1 \cdot \bar{x}_4) + \bar{x}_2$$

$$f_3(x_1, x_2, x_3, x_4) = H + J = (\bar{x}_1 \cdot x_4) + (x_1 \cdot \bar{x}_3)$$

EXAMPLE 6.55 Find the minimal expressions for the following multiple output functions.

$$f_1(x_1, x_2, x_3, x_4) = \prod(3, 4, 5, 7, 11, 13, 15) \cdot d(6, 8, 10, 12)$$

$$f_2(x_1, x_2, x_3, x_4) = \prod(2, 7, 9, 10, 11, 12, 14, 15) \cdot d(0, 4, 6, 8)$$

Solution

The prime implicants are obtained as shown in the table on the next page.

Step 1

Maxterm	f_1	f_2	
0	0	1	✓
2	0	1	✓
4	1	1	✓
8	1	1	✓
3	1	0	✓
5	1	0	✓
6	1	1	✓
9	0	1	✓
10	1	1	✓
12	1	1	✓
7	1	1	✓
11	1	1	✓
13	1	0	✓
14	0	1	✓
15	1	1	✓

Step 2

	f_1	f_2	
0, 2 (2)	0	1	✓
0, 4 (4)	0	1	✓
0, 8 (8)	0	1	✓
2, 6 (4)	0	1	✓
2, 10 (8)	0	1	✓
4, 5 (1)	1	0	✓
4, 6 (2)	1	1	P
4, 12 (8)	1	1	O
8, 9 (1)	0	1	✓
8, 10 (2)	1	1	N
8, 12 (4)	1	1	M
3, 7 (4)	1	0	✓
3, 11 (8)	1	0	✓
5, 7 (2)	1	0	✓
5, 13 (8)	1	0	✓
6, 7 (1)	1	1	L
6, 14 (8)	0	1	✓
9, 11 (2)	0	1	✓
10, 11 (1)	1	1	K
10, 14 (4)	0	1	✓
12, 13 (1)	1	0	✓
12,14 (2)	0	1	✓
7, 15 (8)	1	1	J
11, 15 (4)	1	1	I
13, 15 (2)	1	0	✓
14, 15 (1)	0	1	✓

Step 3

	f_1	f_2	
0, 2, 4, 6 (2, 4)	0	1	✓
0, 2, 8, 10 (2, 8)	0	1	✓
0, 4, 8, 12 (4, 8)	0	1	✓
2, 6, 10, 14 (4, 8)	0	1	✓
4, 5, 6, 7 (1, 2)	1	0	H
4, 5, 12, 13 (1, 8)	1	0	G
4, 6, 12, 14 (2, 8)	0	1	✓
8, 9, 10, 11 (1, 2)	0	1	F
8, 10, 12, 14 (2, 4)	0	1	✓
3, 7, 11, 15 (4, 8)	1	0	E
5, 7, 13, 15 (2, 8)	1	0	D
6, 7, 14, 15 (1, 8)	0	1	C
10, 11, 14, 15 (1,4)	0	1	B

Step 4

	f_1	f_2	
0, 2, 4, 6, 8, 10, 12, 14 (2, 4, 8)	0,	1	A

From the preceding table, the prime implicants are, A, B, C, D, E, F, G, H, I, J, K, L, M, N, O, and P. The prime implicant chart is as shown below.

	f_1							f_2							
	✓			✓	✓		✓	✓		✓	✓	✓	✓	✓	
	3	4	5	7	11	13	15	2	7	9	10	11	12	14	15
*A								×			×		×	×	
B											×	×		×	×
C									×					×	×
D			×	×		×	×								
*E	×			×	×		×								
*F										×	×	×			

Contd.

Contd.

	f_1							f_2							
	✓			✓	✓		✓	✓		✓	✓	✓	✓	✓	
	3	4	5	7	11	13	15	2	7	9	10	11	12	14	15
G		×	×			×									
H		×	×	×											
I					×		×					×			×
J				×			×		×						×
K					×						×	×			
L				×					×						
M													×		
N											×				
O		×											×		
P		×													

For f_1, E is an essential prime implicant because m_3 is covered by E only. Columns 7 $\supset$ 3, 11 $\supset$ 3, 15 $\supset$ 3. So, remove columns 7, 11, and 15. After removal of the dominating columns 7, 11, and 15, the essential prime implicants and the minterms covered by them, we observe that only minterms 4, 5, and 13 are left uncovered. Columns 4, 5, and 13 are covered by G. So, E and G cover all minterms of f_1. Therefore,

$$f_1 = EG = (\bar{x}_3 + \bar{x}_4)(\bar{x}_2 + x_3)$$

For f_2, A is an essential prime implicant because m_2 is covered by A only. F is also an essential prime implicant because m_9 is covered by F only. Columns 10 $\supset$ 2, 12 $\supset$ 2, 14 $\supset$ 2, 11 $\supset$ 9. So, remove columns 10, 12, 14, and 11. After removal of the dominating columns 10, 12, 14, and 11 and the essential prime implicants and the minterms covered by them, we observe that only minterms m_7 and m_{15} are left uncovered. By inspection, 7 and 15 are covered by both C and J. Out of that C is minimal. Therefore,

$$f_2 = AFC = x_4(\bar{x}_1 + x_2)(\bar{x}_2 + \bar{x}_3)$$

No term is common to f_1 and f_2. So, there is no advantage in simplifying them together as the same result would be obtained when they are minimized separately.

EXAMPLE 6.56 Design a multiple output logic circuit for the following functions using the tabular method.

$$f_1 = \Sigma\, m(2, 3, 7, 10, 11, 14) + d(1, 5, 15)$$
$$f_2 = \Sigma\, m(0, 1, 4, 7, 13, 14) + d(5, 8, 15)$$

Solution

The prime implicants are obtained as shown in the table on the next page.

Minterm	Step 1 f_1	f_2		Step 2	f_1	f_2		Step 3	f_1	f_2	
0	0	1	✓	0, 1 (1)	0	1	✓	0, 1, 4, 5 (1, 4)	0	1	F
1	1	1	✓	0, 4 (4)	0	1	✓	1, 3, 5, 7 (2, 4)	1	0	E
2	1	0	✓	0, 8 (8)	0	1	K	2, 3, 10, 11 (1, 8)	1	0	D
4	0	1	✓	1, 3 (2)	1	0	✓	3, 7, 11, 15 (4, 8)	1	0	C
8	0	1	✓	1, 5 (4)	1	1	J	5, 7, 13, 15 (2, 8)	0	1	B
3	1	0	✓	2, 3 (1)	1	0	✓	10, 11, 14, 15 (1, 4)	1	0	A
5	1	1	✓	2, 10 (8)	1	0	✓				
10	1	0	✓	4, 5 (1)	0	1	✓				
7	1	1	✓	3, 7 (4)	1	0	✓				
11	1	0	✓	3, 11 (8)	1	0	✓				
13	0	1	✓	5, 7 (2)	1	1	I				
14	1	1	✓	5, 13 (8)	0	1	✓				
15	1	1	✓	10, 11 (1)	1	0	✓				
				10, 14 (4)	1	0	✓				
				7, 15 (8)	1	1	H				
				11, 15 (4)	1	0	✓				
				13, 15 (2)	0	1	✓				
				14, 15 (1)	1	1	G				

From the above table, the prime implicants are, A, B, C, D, E, F, G, H, I, J, and K. The prime implicant chart is as shown in the table below.

	f_1						f_2					
	✓ 2	✓ 3	7	✓ 10	✓ 11	✓ 14	✓ 0	✓ 1	✓ 4	✓ 7	✓ 13	✓ 14
A				×	×	×						
*B										×	×	
C		×	×		×							
*D	×	×		×	×							
E		×	×									
*F							×	×	×			
*G						×						×
H			×							×		
I			×							×		
J								×				
K							×					

For f_1, D is an essential prime implicant because m_2 is covered by D only. Check out all the minterms covered by it, i.e. minterms 2, 3, 10, and 11. G is already there in f_2 as an essential prime implicant to cover 14. So, only 7 of f_1 is left uncovered. This 7 can be covered by C, E, H, or I. Out of these, C or E can be chosen. Both give the same minimum. Therefore,

$$f_1 = D + G + C = \bar{x}_2 x_3 + x_1 x_2 x_3 + x_3 x_4$$

For f_2, $0 \supset 4$, $1 \supset 4$, $7 \supset 13$. So, remove columns 0, 1, and 7. G, F, and B are the essential prime implicants because minterms 14, 4, and 13, respectively, are covered by them only. Also, they cover all minterms. Therefore,

$$f_2 = G + F + B = x_1 x_2 x_3 + \bar{x}_1 \bar{x}_3 + x_2 x_4$$

Looking at the expressions for f_1 and f_2, we observe that the term $x_1 x_2 x_3$ is common to both f_1 and f_2. The logic diagram based on this minimization is shown below.

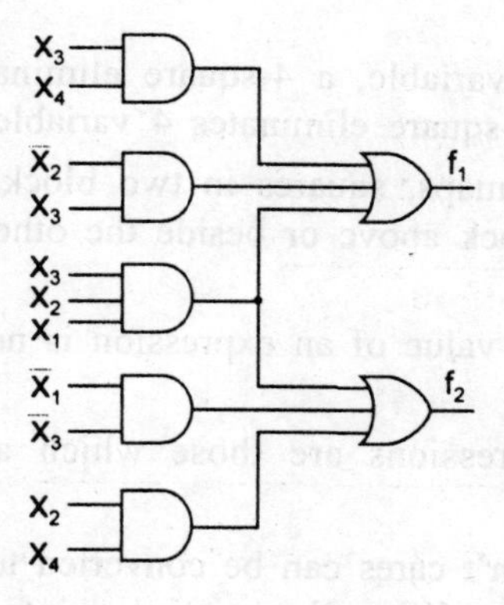

SUMMARY

- The K-map is a chart or a graph composed of an arrangement of adjacent cells, each representing a particular combination of variables in a sum or product term.
- An *n*-variable K-map will have 2^n cells or squares.
- The use of K-maps is usually limited to six variables.
- A standard or canonical or expanded SOP form is one in which each one of the product terms contains all the variables of the expression either in complemented or non-complemented form.
- A standard or canonical or expanded POS form is one in which each one of the sum terms contain all the variables of the expression either in complemented or non-complemented form.
- Each one of the product terms in the canonical SOP form is called a minterm.
- Each one of the sum terms in the canonical POS form is called a maxterm.
- A standard SOP form can always be converted into a standard POS form by treating the missing minterms of the SOP form as the maxterms of the POS form.
- A standard POS form can always be converted to a standard SOP form by treating the missing maxterms of the POS form as the minterms of the SOP form.

- Two squares are said to be adjacent to each other, if their min (max) terms differ in only one variable.
- Two minterms or maxterms can be combined, only if they are adjacent to each other.
- A real minimal expression is the minimal of the SOP and POS minimals.
- The binary number designations of the rows and columns of the K-map are in Gray code.
- The main criteria in the design of digital circuits is the cost (indicated by the number of gate inputs) which must be kept to a minimum.
- The squares may also be formed by wrapping around the K-maps.
- 2-squares, 4-squares, or 8-squares, etc. form either geometric squares or rectangles.
- A 2-square is called a pair, a 4-square is called a quad, and an 8-square is called an octet.
- A 2-square eliminates one variable, a 4-square eliminates 2 variables, an 8-square eliminates 3 variables, a 16-square eliminates 4 variables, and so on.
- In five- and six-variable K-maps, squares in two blocks are considered adjacent, if, when superimposing one block above or beside the other block, the squares coincide with one another.
- Combinations for which the value of an expression is not specified, are called "don't care" combinations.
- Incompletely specified expressions are those which are not specified for certain combinations.
- An SOP expression with don't cares can be converted to a POS form by keeping the don't cares as they are and writing the missing minterms of the SOP form as the maxterms of the POS form, and vice versa.
- In K-maps, don't care terms are used only if they help in reducing the expression. Otherwise, they need not be considered.
- If each one of the input signals has to pass through two gates to reach the output, it is called the two-level logic. The SOP and POS forms of realization give two-level logic.
- Two-level logic provides uniform propagation delay between the input and the output, but may not yield the real minimal.
- In hybrid logic, different input signals pass through different number of gates to reach the output.
- Hybrid logic does not produce uniform time delay and may suffer from the problem of 'logic race', but results in a circuit with the least number of gate inputs.
- Variable mapping (variable entry mapping) technique can be used to minimize the given Boolean expressions which involve infrequently used variables.
- Variable mapping technique allows us to reduce a large mapping problem to one that uses just a small map.
- The Quine-McClusky method, also known as the tabular method, is a more systematic method than the K-map for minimizing functions of a large number of variables.

- In the tabular method, the condition that the binary designations of two combinable terms must differ by a power of 2, is necessary but not sufficient.
- Two terms whose codes differ by a power of 2 but which have the same index cannot be combined because these terms differ in more than one variable.
- The 'index' of a term indicates the total number of 1s present in that term.
- A term with a smaller index but having a higher decimal value cannot be combined with a term whose index is higher even though they may differ by a power of 2, because they would differ in more than one variable.
- The terms which cannot be combined further in the tabular method are called prime implicants. These terms may occur in the final expression.
- Essential prime implicants are the implicants which will definitely occur in the final expression.
- Each essential prime implicant will cover at least one minterm which is not covered by any other prime implicant.
- The prime implicant chart is a pictorial representation of the relationships between the prime implicants and the minterms of the expression.
- Don't care minterms are used in the table, only to obtain the set of prime implicants. They are not used in the prime implicant chart to obtain the essential prime implicants.
- Any row in a prime implicant chart is said to dominate any other row, if the first row has a x in every column in which the second row has a x.
- Any column in a prime implicant chart is said to dominate any other column, if the first column has a x in every row in which the second column has an x.
- All dominating columns and dominated rows can be removed while drawing the reduced prime implicant chart.
- If the prime implicant chart has no essential prime implicants and dominated rows and dominating columns, the minimal expressions can be obtained by the branching method
- A minterm fills with 1s the minimum possible area of the K-map, short of filling no area at all. Hence, the name minterm.
- A maxterm fills with 1s the maximum possible area of the K-map, short of filling the entire area. Hence, the name maxterm.

QUESTIONS

1. What do you mean by a K-map? Name its advantages and disadvantages.
2. How many cells are there on an *n*-variable K-map?
3. What are SOP and POS forms of Boolean expressions?
4. How do you determine the total number of gate inputs to realize a Boolean expression?
5. How do you convert an SOP form to a POS form and vice versa?
6. What do you mean by minterms and maxterms of Boolean expressions?
7. What is cell of a K-map? What is meant by a pair, a quad, and an octet of a K-map?
8. What do 1s and 0s on the SOP K-map represent? What do 0s and 1s on the POS K-map represent?

9. What do you mean by adjacent squares of a K-map?
10. Write a procedure to reduce K-maps.
11. What do you mean by a real minimal expression?
12. What is the main criteria for the design of digital circuits?
13. How do you compare the cost of realizing a circuit from the Boolean expressions?
14. Write the codes of binary designations of rows and columns of a four-variable K-map.
15. What do you mean by don't care combinations?
16. What are incompletely specified functions?
17. What do you mean by two-level logic? What is its main advantage?
18. What is hybrid logic? What are its main advantages and disadvantages?
19. What is the criteria for the minimization of multiple output switching functions?
20. What are the advantages and disadvantages of the tabular method vis-a-vis the K-map?
21. What do you mean by variable mapping? What are its advantages?
22. What is meant by 'index' of a term in the tabular method?
23. When can two minterms or maxterms be combined?
24. In the tabular method, why two terms whose codes differ by a power of 2 but which have the same index cannot be combined?
25. Why a term with a smaller index but having a higher decimal value cannot be combined with a term whose index is higher, even though they may differ by a power of 2?
26. What do you mean by (a) a prime implicant chart and (b) a reduced prime implicant chart?
27. What is meant by a prime implicant, an essential prime implicant, and a secondary essential prime implicant?
28. What are dominating rows and dominating columns?
29. Does elimination of dominating columns and dominated rows end the search for a minimal expression? If not, why not?
30. When do you use the branching method?
31. How do you get all the possible minimal expressions by the tabular method?
32. What does 'tag' indicate in multiple output minimization using tabular method?
33. How many variables are eliminated by (i) a 2-square, (ii) a 4-square, (iii) an 8-square, and (iv) a 16-square?

PROBLEMS

6.1 Convert to minterms:

(a) $A + \overline{B}\,\overline{C}$ (b) $\overline{A} + B + CA$

(c) $ABC + AB + DC + \overline{D}$ (d) $ABCDE + AB\overline{E} + ACD$

6.2 Convert to maxterms:

(a) $A(B + \overline{C})$ (b) $(A + \overline{B})(\overline{A} + D)$

(c) $(A + B + \overline{D})(\overline{A} + C + D)$ (d) $A(\overline{A} + B)(\overline{C})$

6.3 How many gate inputs are required to realize the following expressions?

(a) $AB\overline{C} + \overline{A}BC + ABC\overline{D} + ABD$

(b) $WX\overline{Y} + WXZ + VUX + XY\overline{Z}W$

(c) $A + BC + \overline{D}EF$

(d) $(A + C)(A + \overline{B} + C)(A + C + \overline{D})(A + B + C + \overline{D})$

(e) $A(B + \overline{D})(A + C + E)(B + \overline{C} + \overline{D} + E)$

(f) $(A + B)(\overline{C} + D)(E + F + \overline{G})\overline{D}$

6.4 Reduce the following expressions using K-map.

(a) $AB + A\overline{B}C + \overline{A}B\overline{C} + B\overline{C}$

(b) $AB\overline{C} + AB + C + B\overline{C} + D\overline{B}$

(c) $AB + A\overline{C} + C + AD + A\overline{B}C + ABC$

(d) $A\overline{B}C + B + B\overline{D} + AB\overline{D} + \overline{A}C$

6.5 Reduce the following expressions using K-map.

(a) $(A + B)(A + \overline{B} + C)(A + \overline{C})$

(b) $A(B + \overline{C})(A + \overline{B})(B + C + \overline{D})$

(c) $\overline{A} + B(A + B + \overline{D})(B + \overline{C})(B + C + D)$

6.6 Obtain the minimal SOP expression for Σ m(2, 3, 5, 7, 9, 11, 12, 13, 14, 15) and implement it in NAND logic.

6.7 Obtain the minimal POS expression for Π M(0, 1, 2, 4, 5, 6, 9, 11, 12, 13, 14, 15) and implement it in NOR logic.

6.8 Reduce Π M(1, 2, 3, 5, 6, 7, 8, 9, 12, 13) and implement it in universal logic.

6.9 Reduce the following expressions using K-map and implement them in universal logic.

(a) Σ m(5, 6, 7, 9, 10, 11, 13, 14, 15)

(b) Σ m(0, 1, 2, 3, 4, 6, 8, 9, 10, 11)

(c) Π M(1, 4, 5, 11, 12, 14) · d(6, 7, 15)

(d) Π M(3, 6, 8, 11, 13, 14) · d(1, 5, 7, 10)

(e) Σ m(0, 1, 4, 5, 6, 7, 9, 11, 15) + d(10, 14)

(f) Σ m(9, 10, 12) + d(3, 5, 6, 7, 11, 13, 14, 15)

6.10 Simplify the following logic expressions and realize them using NOR gates.

(a) Σ m(6, 9, 13, 18, 19, 25, 27, 29, 31)

(b) Σ m(0, 2, 3, 10, 12, 16, 17, 18, 21, 26, 27) + d(11, 13, 19, 20)

(c) Σ m(0, 1, 2, 4, 5, 7, 8, 9, 10, 14, 15, 17, 19, 20, 28, 29, 34, 36, 40, 41, 42, 43)

(d) Σ m(4, 6, 8, 9, 10, 12, 13, 18, 19, 25, 26, 29, 33, 35, 36, 41, 42, 48, 49, 50, 56, 57) + d(0, 1, 11, 15, 30, 38, 40)

6.11 Obtain the minimal expression using thé tabular method.

(a) Σ m(0, 1, 3, 4, 5, 7, 10, 13, 14, 15)
(b) Σ m(0, 2, 3, 6, 7, 8, 10, 11, 12, 15)
(c) Σ m(0, 1, 3, 4, 5, 6, 7, 13, 15)
(d) Π M(6, 7, 8, 9) $\cdot$ d(10, 11, 12, 13, 14, 15)
(e) Π M(1, 5, 6, 7, 11, 12, 13, 15)
(f) Σ m(1, 5, 6, 12, 13, 14) + d(2, 4)

6.12 Minimize and implement the following multiple output functions in SOP form using (1) K-maps and (2) the tabular method.

(a) $f_1 = \Sigma$ m(1, 2, 5, 6, 8, 9, 10)
$f_2 = \Sigma$ m(2, 4, 6, 8, 10, 12, 15)

(b) $f_1 = \Sigma$ m(0, 1, 4, 6, 8, 9, 11) + d(2, 7, 13)
$f_2 = \Sigma$ m(2, 4, 5, 7, 9, 12) + d(0, 1, 6)

(c) $f_1 = \Sigma$ m(0, 1, 2, 4, 6, 7, 10, 14, 15)
$f_2 = \Sigma$ m(3, 4, 5, 9, 10, 11, 14)

(d) $f_1 = \Sigma$ m(2, 3, 7, 10, 11, 14) + d(1, 5, 15)
$f_2 = \Sigma$ m(0, 1, 4, 7, 13, 14) + d(5, 8, 15)

6.13 Reduce the following expressions using a three-variable map.

(a) $A\overline{B}C + \overline{A}BC\overline{D} + AB\overline{C}D + ABC$
(b) $AB\overline{E}\,\overline{C} + ABC\overline{D} + ABC\overline{E} + A\overline{B}C\overline{D} + \overline{A}BC$
(c) $\overline{A}\,\overline{B}C\overline{D} + \overline{A}\,\overline{B}CD + A\overline{B}C\overline{D} + A\overline{B}CD + \overline{A}BCD + \overline{A}B\overline{C}\,\overline{D} + AB\overline{C}D + AB\overline{C}\,\overline{D}$
(d) $m_1 + Dm_2 + m_3 + \overline{D}m_5 + m_7 + d(m_0 + Dm_6)$
(e) $Em_2 + m_3 + m_4 + Dm_5 + \overline{D}m_7 + d(m_0 + Em_1 + \overline{E}m_6)$
(f) $Dm_2 + Dm_5 + m_6 + d(m_1 + \overline{D}m_7)$

6.14 Solve the following equations using a four-variable map.

(a) $Q = m_0 + m_2 + Fm_6 + m_7 + Em_8 + Em_{10} + m_{12} + m_{14} + Fm_{15}$
(b) $R = m_0 + m_2 + Fm_4 + Gm_7 + \overline{F}m_9 + m_{10} + \overline{E}m_{11} + m_{12} + Em_{14} + m_{15}$
(c) $S = m_1 + Fm_2 + Gm_4 + m_6 + \overline{F}m_7 + Hm_9 + (F + G)m_{10} + d[(F + \overline{G})m_0 + Gm_5 + m_7 + m_{14} + \overline{F}m_{15}]$
(d) $T = Em_0 + Hm_1 + Fm_2 + (\overline{E} + G)m_6 + m_8 + d(m_{10} + \overline{E}m_{14} + Gm_7)$
(e) $U = Em_3 + m_5 + m_7 + Fm_{10} + m_{12} + m_{14}$
(f) $V = m_0 + (E + G)m_2 + \overline{E}m_5 + Gm_{10} + Fm_{13} + m_{14} + m_{15} + d(m_1 + Em_4 + \overline{F}m_8 + \overline{F}m_9)$.

Chapter 7

COMBINATIONAL CIRCUITS

7.1 THE HALF-ADDER

Arithmetic circuits are the circuits that perform arithmetic operations. A half-adder arithmetic circuit adds two binary digits, giving a sum bit and a carry bit.

The sum (S) bit and the carry (C) bit, according to the rules of binary addition, are given by:

Inputs		Outputs	
A	B	S	C
0	0	0	0
0	1	1	0
1	0	1	0
1	1	0	1

The sum (S) is the X-OR of A and B. Therefore,

$$S = A\overline{B} + B\overline{A} = A \oplus B$$

The carry (C) is the AND of A and B. Therefore,

$$C = AB$$

A half-adder can, therefore, be realized by using one X-OR gate and one AND gate as shown in Fig. 7.1.

A half-adder can also be realized in universal logic by using either only NAND gates or only NOR gates as explained below.

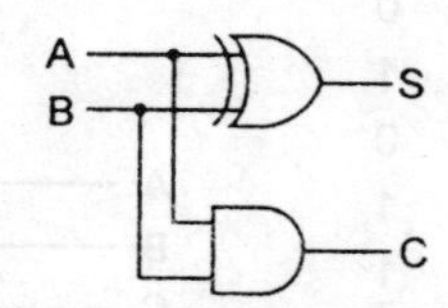

Fig. 7.1 Logic diagram of a half-adder.

NAND logic

$$S = A\bar{B} + \bar{A}B = A\bar{B} + A\bar{A} + \bar{A}B + B\bar{B}$$
$$= A(\bar{A} + \bar{B}) + B(\bar{A} + \bar{B})$$
$$= A \cdot \overline{AB} + B \cdot \overline{AB}$$
$$= \overline{\overline{A \cdot \overline{AB}} \cdot \overline{B \cdot \overline{AB}}}$$
$$C = AB = \overline{\overline{AB}}$$

Fig. 7.2 Half-adder using NAND logic.

NOR logic

$$S = A\bar{B} + \bar{A}B = A\bar{B} + A\bar{A} + \bar{A}B + B\bar{B}$$
$$= A(\bar{A} + \bar{B}) + B(\bar{A} + \bar{B})$$
$$= (A + B)(\bar{A} + \bar{B})$$
$$= \overline{\overline{A + B} + \overline{\bar{A} \cdot B}}$$
$$C = AB = \overline{\overline{AB}} = \overline{\bar{A} + \bar{B}}$$

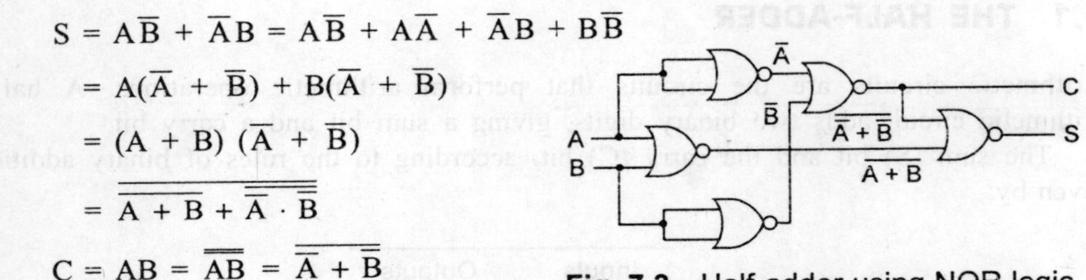

Fig. 7.3 Half-adder using NOR logic.

7.2 THE FULL-ADDER

A full-adder is an arithmetic circuit that adds two bits and a carry and outputs a sum bit and a carry bit. When we want to add two binary numbers, each having two or more bits, the LSBs can be added by using a half-adder. The carry resulted from the addition of the LSBs is carried over to the next significant column and added to the two bits in that column. So, in the second and higher columns, the two data bits of that column and the carry bit generated from the addition in the previous column need to be added.

The full-adder adds the bits A and B and the carry from the previous column called the carry-in C_{in} and outputs the sum bit S and the carry bit called the carry-out C_{out}. The block diagram and the truth table of a full-adder are shown below.

Inputs			Sum	Carry
A	B	C_{in}	S	C_{out}
0	0	0	0	0
0	0	1	1	0
0	1	0	1	0
0	1	1	0	1
1	0	0	1	0
1	0	1	0	1
1	1	0	0	1
1	1	1	1	1

(a) Turth table

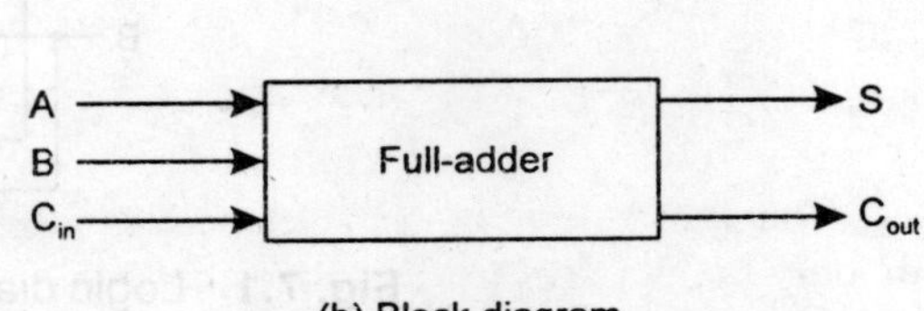

(b) Block diagram

From the truth table, a circuit that will produce the correct sum and carry bits in response to every possible combination of A, B, and C_{in} is described by

$$S = \overline{A}\,\overline{B}C_{in} + \overline{A}B\overline{C}_{in} + A\overline{B}\,\overline{C}_{in} + ABC_{in} = A \oplus B \oplus C_{in}$$

and

$$C_{out} = \overline{A}BC_{in} + A\overline{B}C_{in} + AB\overline{C}_{in} + ABC_{in} = AB + (A \oplus B)C_{in} = AB + AC_{in} + BC_{in}$$

The sum term of the full-adder is the X-OR of A, B, and C_{in}, i.e. the sum bit is the modulo sum of the data bits in that column and the carry from the previous column. The logic diagram of the full-adder using two X-OR gates and two AND gates (i.e. two half-adders) and one OR gate is shown in Fig. 7.4.

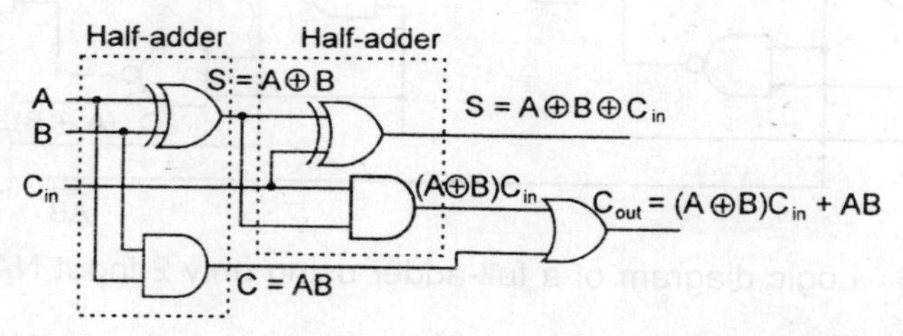

Fig. 7.4 Logic diagram of a full-adder using two half-adders.

Even though a full-adder can be constructed using two half-adders as shown in Fig. 7.4, the disadvantage is that the bits must propagate through several gates in succession, which makes the total propagation delay greater than that of the full-adder circuit using AOI logic as shown in Fig. 7.5.

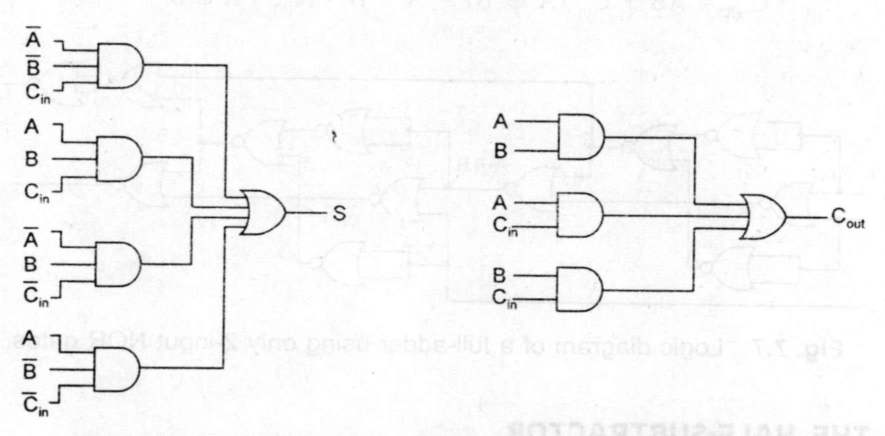

Fig. 7.5 Sum and carry bits of a full-adder using AOI logic.

The full-adder can also be realized using universal logic, i.e. either only NAND gates or only NOR gates as explained below.

NAND logic

Let $\quad A \oplus B = \overline{\overline{A \cdot \overline{AB}} \cdot \overline{B \cdot \overline{AB}}} = X$. Then

$$S = A \oplus B \oplus C_{in} = \overline{\overline{X \cdot \overline{XC_{in}}} \cdot \overline{C_{in} \cdot \overline{XC_{in}}}} = X \oplus C_{in}$$

$$C_{out} = C_{in}\,(A \oplus B) + AB = \overline{\overline{C_{in}\,(A \oplus B)} \cdot \overline{AB}}$$

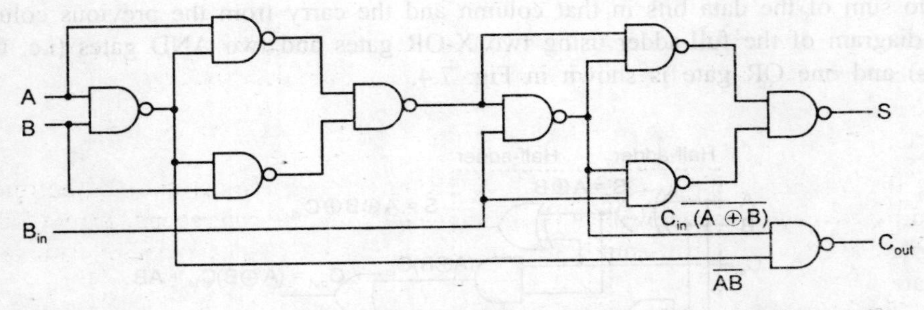

Fig. 7.6 Logic diagram of a full-adder using only 2-input NAND gates.

NOR logic

Let $\quad A \oplus B = X = \overline{\overline{(A + B)} + \overline{\overline{A} + \overline{B}}}$

$$S = A \oplus B \oplus C_{in} = X \oplus C_{in} = \overline{\overline{X + C_{in}} + \overline{\overline{X} + \overline{C_{in}}}}$$

$$C_{out} = AB + C_{in}\,(A \oplus B) = \overline{\overline{\overline{A} + \overline{B}} + \overline{\overline{C_{in}} + \overline{A \oplus B}}}$$

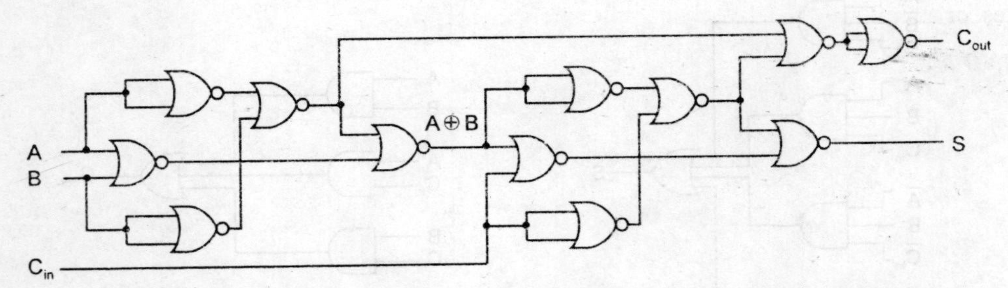

Fig. 7.7 Logic diagram of a full-adder using only 2-input NOR gates.

7.3 THE HALF-SUBTRACTOR

A half-subtractor is an arithmetic circuit that subtracts one bit from the other. It is used to subtract the LSB of the subtrahend from the LSB of the minuend when one binary number is subtracted from the other.

We know that, when a bit B is subtracted from another bit A, a difference bit (d) and a borrow bit (b) result according to the rules given below:

Inputs		Outputs	
A	B	d	b
0	0	0	0
1	0	1	0
1	1	0	0
0	1	1	1

A circuit that produces the correct difference and borrow bits in response to every possible combination of the two 1-bit numbers is, therefore, described by

$$d = A\overline{B} + B\overline{A} = A \oplus B \quad \text{and} \quad b = \overline{A}B$$

That is, the difference bit is obtained by X-ORing the two inputs, and the borrow bit is obtained by ANDing the complement of the minuend with the subtrahend. Figure 7.8 shows two logic diagrams of a half-subtractor—one using an X-OR gate together with one each NOT gate and AND gate and the other using the AOI gates.

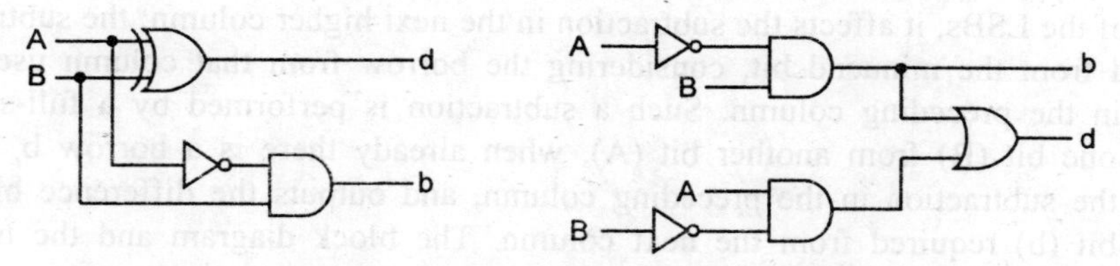

Fig. 7.8 Logic diagrams of a half-subtractor.

A half-subtractor can also be realized using universal logic—either using only NAND gates or using only NOR gates—as explained below.

NAND logic

$$d = A \oplus B = \overline{\overline{A \cdot \overline{AB}} \cdot \overline{B \cdot \overline{AB}}}$$

$$b = \overline{A}B = B(\overline{A} + \overline{B}) = B(\overline{AB}) = \overline{\overline{B \cdot \overline{AB}}}$$

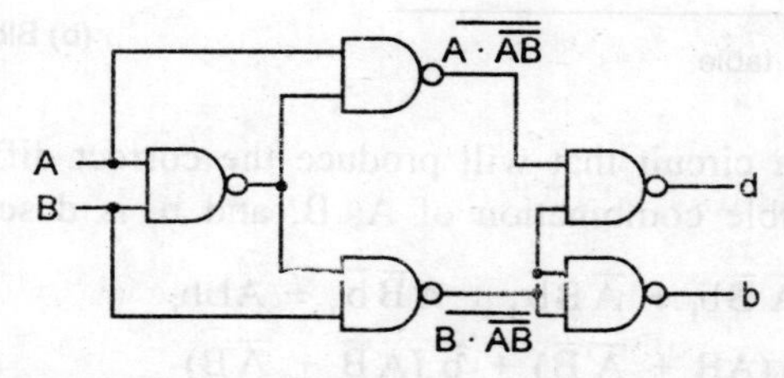

Fig. 7.9 Logic diagram of a half-subtractor using only 2-input NAND gates.

NOR logic

$$d = A \oplus B = A\bar{B} + \bar{A}B = A\bar{B} + B\bar{B} + \bar{A}B + A\bar{A}$$

$$= \bar{B}(A + B) + \bar{A}(A + B) = \overline{B + \overline{A + B}} + \overline{A + \overline{A + B}}$$

$$b = \bar{A}B = \bar{A}(A + B) = \overline{\overline{\bar{A}(A + B)}} = \overline{A + (\overline{A + B})}$$

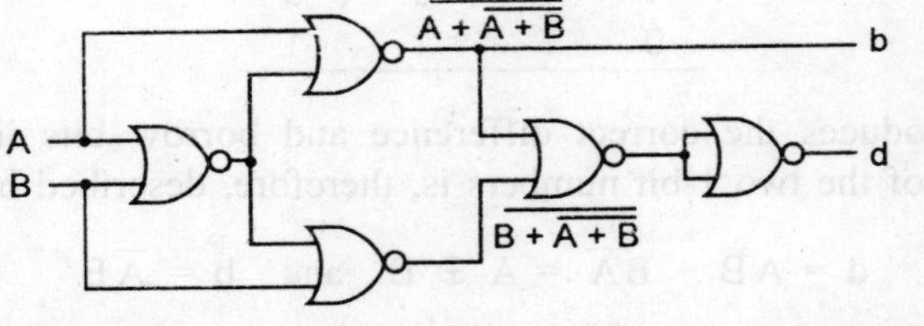

Fig. 7.10 Logic diagram of a half-subtractor using only 2-input NOR gates.

7.4 THE FULL-SUBTRACTOR

The half-subtractor can be used only for LSB subtraction. If there is a borrow during the subtraction of the LSBs, it affects the subtraction in the next higher column; the subtrahend bit is subtracted from the minuend bit, considering the borrow from that column used for the subtraction in the preceding column. Such a subtraction is performed by a full-subtractor. It subtracts one bit (B) from another bit (A), when already there is a borrow b_i from this column for the subtraction in the preceding column, and outputs the difference bit (d) and the borrow bit (b) required from the next column. The block diagram and the truth table of a full-subtractor are shown below.

Inputs			Difference	Borrow
A	B	b_i	d	b
0	0	0	0	0
0	0	1	1	1
0	1	0	1	1
0	1	1	0	1
1	0	0	1	0
1	0	1	0	0
1	1	0	0	0
1	1	1	1	1

(a) Turth table

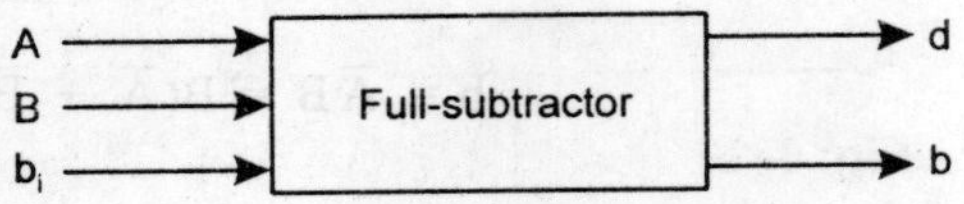

(b) Block diagram

From the truth table, a circuit that will produce the correct difference and borrow bits in response to every possible combination of A, B, and b_i is described by

$$d = \bar{A}\bar{B}b_i + \bar{A}B\bar{b}_i + A\bar{B}\bar{b}_i + ABb_i$$

$$= b_i(AB + \bar{A}\bar{B}) + \bar{b}_i(A\bar{B} + \bar{A}B)$$

$$= b_i(\overline{A \oplus B}) + \bar{b}_i(A \oplus B) = A \oplus B \oplus b_i$$

and

$$b = \bar{A}\bar{B}b_i + \bar{A}B\bar{b}_i + \bar{A}Bb_i + ABb_i = \bar{A}B + \overline{(A \oplus B)}b_i$$

A full-subtractor can, therefore, be realized using X-OR gates and AOI gates as shown in Fig. 7.11.

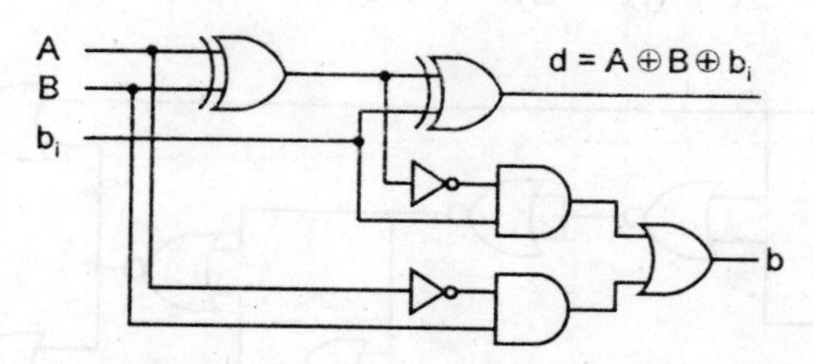

Fig. 7.11 Logic diagram of a full-subtractor.

The full subtractor can also be realized in universal logic using either only NAND gates or only NOR gates as explained below.

NAND logic

$$d = A \oplus B \oplus b_i = \overline{\overline{(A \oplus B)\overline{(A \oplus B)b_i} \cdot b_i\overline{(A \oplus B)b_i}}}$$

$$b = \bar{A}B + b_i\overline{(A \oplus B)} = \overline{\overline{\bar{A}B + b_i\overline{(A \oplus B)}}}$$

$$= \overline{\overline{\bar{A}B} \cdot \overline{b_i\overline{(A \oplus B)}}} = \overline{\overline{B(\bar{A} + \bar{B})} \cdot \overline{b_i[\bar{b}_i + \overline{(A \oplus B)}]}}$$

$$= \overline{\overline{B \cdot \overline{AB}} \cdot \overline{b_i\overline{[b_i \cdot (A \oplus B)]}}}$$

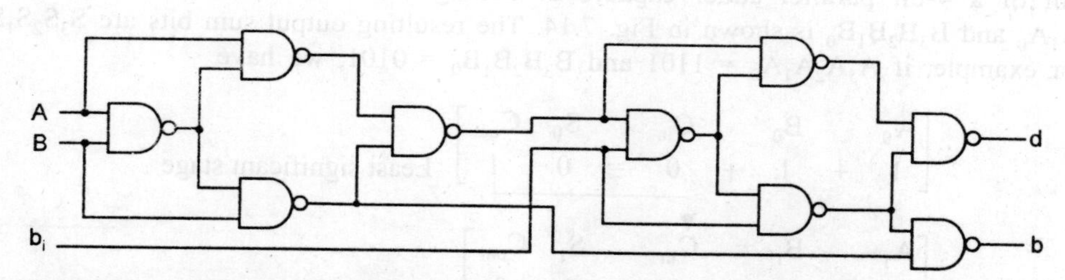

Fig. 7.12 Logic diagram of a full-subtractor using only 2-input NAND gates.

NOR logic

$$d = A \oplus B \oplus b_i$$

$$= \overline{(A \oplus B)b_i + \overline{(A \oplus B)}\bar{b}_i}$$

$$= \overline{[(A \oplus B) + \overline{(A \oplus B)}\,\bar{b}_i]\,[b_i + \overline{(A \oplus B)}\,\bar{b}_i]}$$

$$= \overline{(A \oplus B) + \overline{(A \oplus B) + b_i}} + \overline{b_i + \overline{(A \oplus B) + b_i}}$$

$$= \overline{\overline{(A \oplus B) + \overline{(A \oplus B) + b_i}} + \overline{b_i + \overline{(A \oplus B) + b_i}}}$$

$$b = \overline{A}B + b_i(\overline{A \oplus B})$$

$$= \overline{A}(A + B) + (\overline{A \oplus B})\,[(A \oplus B) + b_i]$$

$$= \overline{\overline{A + \overline{(A + B)}} + \overline{(A \oplus B) + \overline{(A \oplus B) + b_i}}}$$

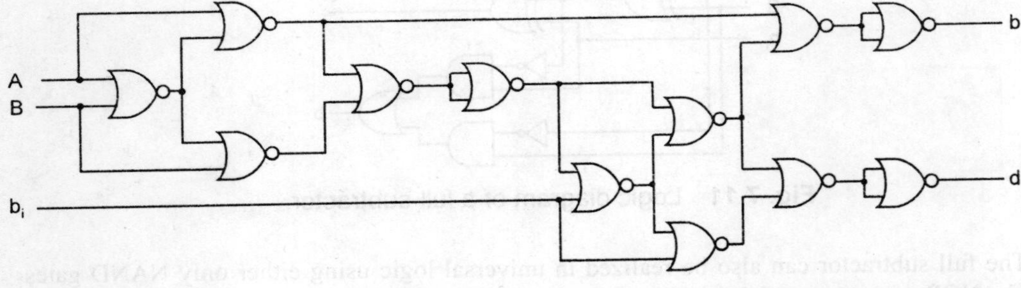

Fig. 7.13 Logic diagram of a full subtractor using only 2-input NOR gates.

7.5 PARALLEL BINARY ADDERS

A full-adder is capable of adding two 1-bit binary numbers and a carry-in. When two *n*-bit binary numbers are to be added, the number of full-adders required will be equal to the number of bits *n* in each number. Of course, the addition of LSBs can be done by using either a half-adder or a full-adder with C_{in} terminal grounded. The carry-out of each full-adder is connected to the carry-in of the next higher order adder. In practical parallel adders, the least significant stage is also a full-adder to facilitate cascading. A parallel adder is used to add two numbers in parallel form and to produce the sum bits as parallel outputs. A block diagram of a 4-bit parallel adder capable of adding two 4-bit numbers designated as $A_3A_2A_1A_0$ and $B_3B_2B_1B_0$ is shown in Fig. 7.14. The resulting output sum bits are $S_3S_2S_1S_0$.

For example, if $A_3A_2A_1A_0$ = 1101 and $B_3B_2B_1B_0$ = 0101, we have

$$\begin{bmatrix} A_0 & & B_0 & & C_{in} & & S_0 & C_{out} \\ 1 & + & 1 & + & 0 & = & 0 & 1 \end{bmatrix} \text{ Least significant stage}$$

$$\begin{bmatrix} A_1 & & B_1 & & C_{in} & & S_1 & C_{out} \\ 0 & + & 0 & + & 1 & = & 1 & 0 \end{bmatrix}$$

$$\begin{bmatrix} A_2 & & B_2 & & C_{in} & & S_2 & C_{out} \\ 1 & + & 1 & + & 0 & = & 0 & 1 \end{bmatrix}$$

$$\begin{bmatrix} A_3 & & B_3 & & C_{in} & & S_3 & C_{out} \\ 1 & + & 0 & + & 1 & = & 0 & 1 \end{bmatrix} \text{ Most significant stage}$$

$$\downarrow$$

$$S_4$$

We see that $S_3S_2S_1S_0 = 0010$. Since the carry-out from the most significant stage is a 1, we have an overflow, i.e. the sum (10010) must be expressed in 5 bits.

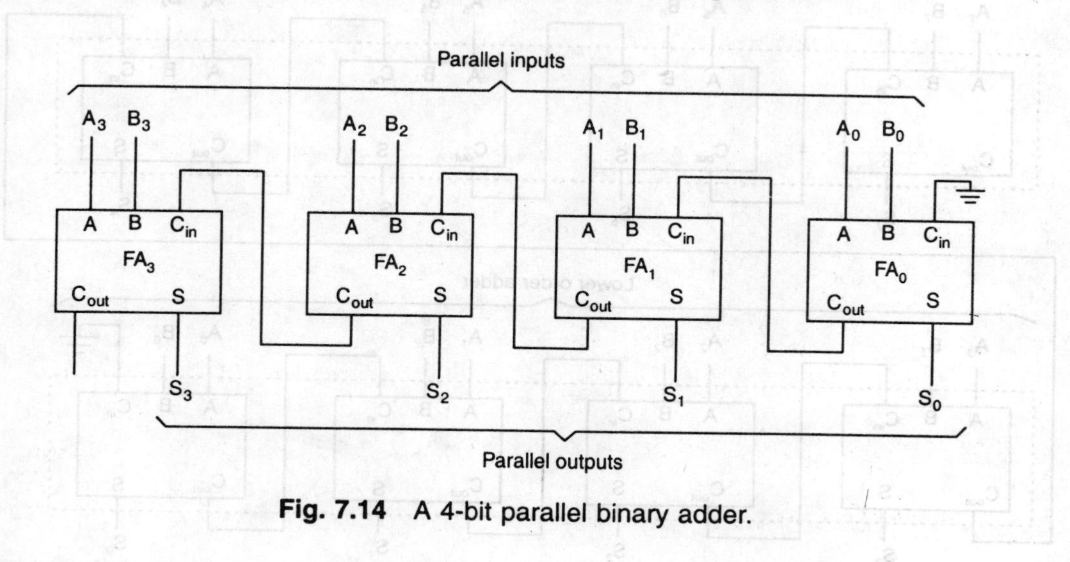

Fig. 7.14 A 4-bit parallel binary adder.

Cascading Parallel Adders

An 8-bit parallel adder can be constructed by cascading two 4-bit parallel adders as shown in Fig. 7.15. One 4-bit adder (the lower order adder) is used to add the 4 LSBs of the inputs and the other (the higher order) adder adds the 4 MSBs. The carry-out from the lower order adder is the carry-in to the least significant stage of the higher order adder. Additional adders can be cascaded to create parallel adders for any number of bits.

The Ripple Carry Adder

In the parallel adders discussed above, the carry-out of each stage is connected to the carry-in of the next stage. The sum and carry-out bits of any stage cannot be produced, until some time after the carry-in of that stage occurs. This is due to the propagation delays in the logic circuitry, which lead to a time delay in the addition process. The carry propagation delay for each full-adder is the time between the application of the carry-in and the occurrence of the carry-out.

Referring to the 4-bit parallel adder in Fig. 7.14, we see that the sum (S_0) and carry-out (C_{out}) bits given by FA_0 are not valid, until after the propagation delay of FA_0. Similarly, S_1 is not valid until after the cumulative propagation delay of two full adders (FA_0 and FA_1), and so on. At each stage, the sum bit is not valid until after the carry bits in all the preceding stages are valid. In effect, carry bits must propagate or ripple through all stages before the most significant sum bit is valid. Thus, the total sum (the parallel output) is not valid until after the cumulative delay of all the adders.

The parallel adder, in which the carry-out of each full-adder is the carry-in to the next most significant adder is called a *ripple carry adder.* The greater the number of bits that a ripple carry adder must add, the greater the time required for it to perform a valid addition.

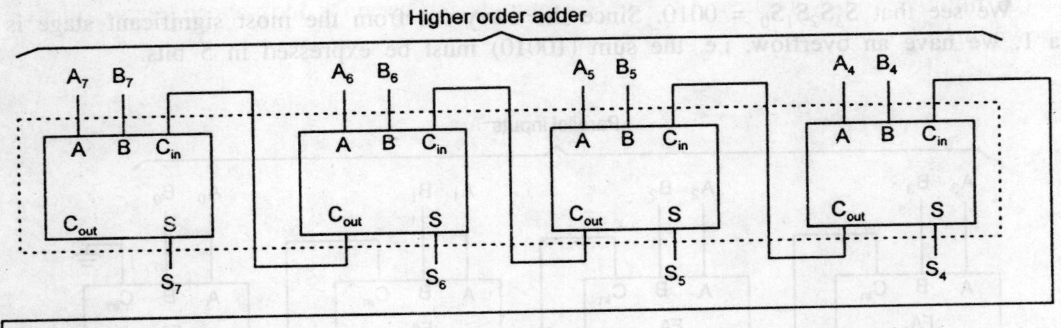

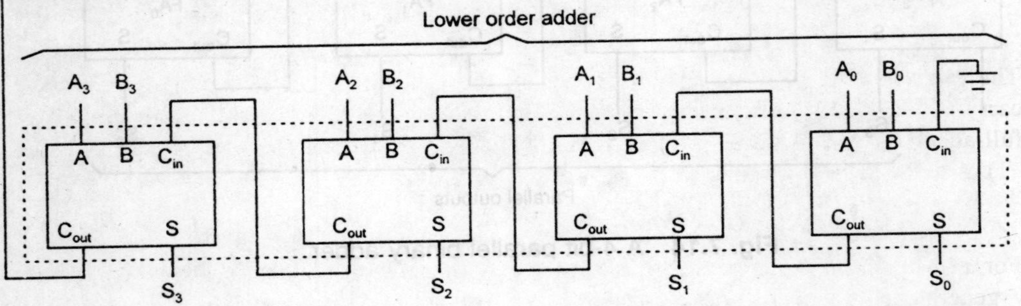

Fig. 7.15 An 8-bit parallel adder obtained by cascading two 4-bit parallel adders.

If two numbers are added such that no carries occur between stages, then the add time is simply the propagation time through a single full-adder.

7.6 THE LOOK-AHEAD CARRY ADDER

In the case of the parallel adder, the speed with which an addition can be performed is governed by the time required for the carries to propagate or ripple through all of the stages of the adder. The look-ahead carry adder speeds up the process by eliminating this ripple carry delay. It examines all the input bits simultaneously and also generates the carry-in bits for all the stages simultaneously.

The method of speeding up the addition process is based on the two additional functions of the full-adder, called the *carry generate* and *carry propagate functions*.

The carry generate (CG) function indicates as to when a carry-out would be generated by the full-adder. A carry-out is generated only when both the input bits are 1. This condition is expressed as the AND function of the two input bits A and B. Thus,

$$CG = A \cdot B$$

A carry-in may be propagated by the full-adder when either or both of the input bits are 1. This condition is expressed as the OR function of the inputs A and B. Thus,

$$CP = A + B$$

A full-adder that produces CG and CP outputs for use in a look-ahead carry adder is shown in Fig. 7.16.

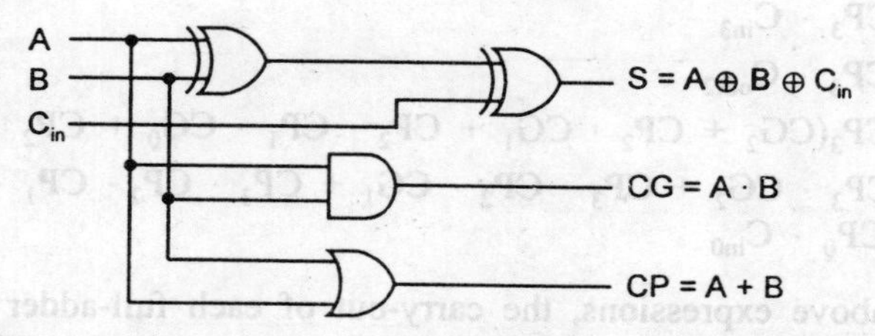

Fig. 7.16 A full-adder that produces CG and CP functions.

The carry-out (C_{out}) of a full-adder is a 1, if the CG is a 1 OR if the CP is a 1 AND the carry-in (C_{in}) is a 1. In other words, we get a carry-out of a 1, if it is generated by the full-adder (A = 1 AND B = 1) or if the adder can propagate the carry-in (A = 1 OR B = 1) AND C_{in} = 1. That is,

$$C_{out} = CG + CP \cdot C_{in}$$

For a 4-bit parallel adder, the carry-out (C_{out}) of each full-adder is dependent on its carry generate (CG), carry propagate (CP), and its carry-in (C_{in}).

The CG and CP functions for each stage are immediately available as soon as the input bits A and B and the carry-in to the LSB adder are applied, because they are dependent only on these bits. The carry input to each stage is the carry output of the previous stage.

Based on these, the expressions for carry-outs of various full-adders are:

Full-adder FA_0:

$$C_{out0} = CG_0 + CP_0 \cdot C_{in0}$$

Full-adder FA_1:

$$C_{in1} = C_{out0}$$

$$\begin{aligned} C_{out1} &= CG_1 + CP_1 \cdot C_{in1} \\ &= CG_1 + CP_1 \cdot C_{out0} \\ &= CG_1 + CP_1(CG_0 + CP_0 \cdot C_{in0}) \\ &= CG_1 + CP_1 \cdot CG_0 + CP_1 \cdot CP_0 \cdot C_{in0} \end{aligned}$$

Full-adder FA_2:

$$C_{in2} = C_{out1}$$

$$C_{out2} = CG_2 + CP_2 \cdot C_{in2}$$

$$\begin{aligned} C_{out2} &= CG_2 + CP_2 \cdot C_{out1} \\ &= CG_2 + CP_2(CG_1 + CP_1 \cdot CG_0 + CP_1 \cdot CP_0 \cdot C_{in0}) \\ &= CG_2 + CP_2 \cdot CG_1 + CP_2 \cdot CP_1 \cdot CG_0 + CP_2 \cdot CP_1 \cdot CP_0 \cdot C_{in0} \end{aligned}$$

Full-adder FA_3:

$$C_{in3} = C_{out2}$$

$$\begin{aligned} C_{out3} &= CG_3 + CP_3 \cdot C_{in3} \\ &= CG_3 + CP_3 \cdot C_{out2} \\ &= CG_3 + CP_3(CG_2 + CP_2 \cdot CG_1 + CP_2 \cdot CP_1 \cdot CG_0 + CP_2 \cdot CP_1 \cdot CP_0 \cdot C_{in0}) \\ &= CG_3 + CP_3 \cdot CG_2 + CP_3 \cdot CP_2 \cdot CG_1 + CP_3 \cdot CP_2 \cdot CP_1 \cdot CG_0 + CP_3 \cdot CP_2 \\ &\quad \cdot CP_1 \cdot CP_0 \cdot C_{in0} \end{aligned}$$

In each of the above expressions, the carry-out of each full-adder stage is dependent only on the initial carry-in, C_{in0}, its CG and CP functions, and the CG and CP functions of the preceding stages. Since each of the CG and CP functions can be expressed in terms of inputs A and B to the full-adders, all of the output carries are immediately available (except for gate delays) and there is no need to wait for a carry to ripple through all of the stages before a final result is achieved. Thus, the look-ahead carry adder speeds up the addition process. The logic diagram of a look-ahead carry adder is shown in Fig. 7.17.

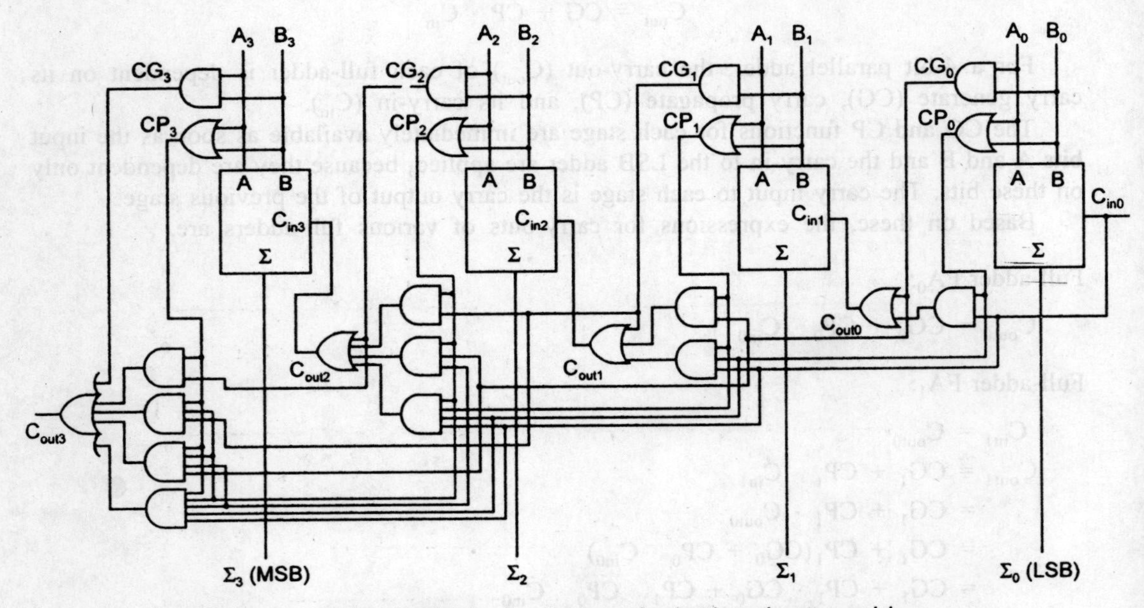

Fig. 7.17 Logic diagram of a 4-stage look-ahead carry adder.

7.7 IC PARALLEL ADDERS

Several parallel adders are available as ICs. The most common one is a 4-bit parallel adder IC that contains four interconnected full-adders and a look-ahead carry circuit needed for high speed operation. The 7483A, the 74LS83A, the 74283, and the 74LS283 are all TTL 4-bit parallel adder chips.

Figure 7.18a shows the functional symbol for the 74LS83 4-bit parallel adder (and its equivalents). The inputs to this IC are two 4-bit numbers, $A_3A_2A_1A_0$ and $B_3B_2B_1B_0$ and the carry C_0, into the LSB position; the outputs are the sum bits $S_3S_2S_1S_0$ and the carry C_4 out of the MSB position. The sum bits are often labelled $\Sigma_3\Sigma_2\Sigma_1\Sigma_0$.

Cascading IC Parallel Adders

The addition of large binary numbers can be accomplished by cascading two or more parallel adder chips. Figure 7.18b shows the cascading of two 74LS83 chips to add two

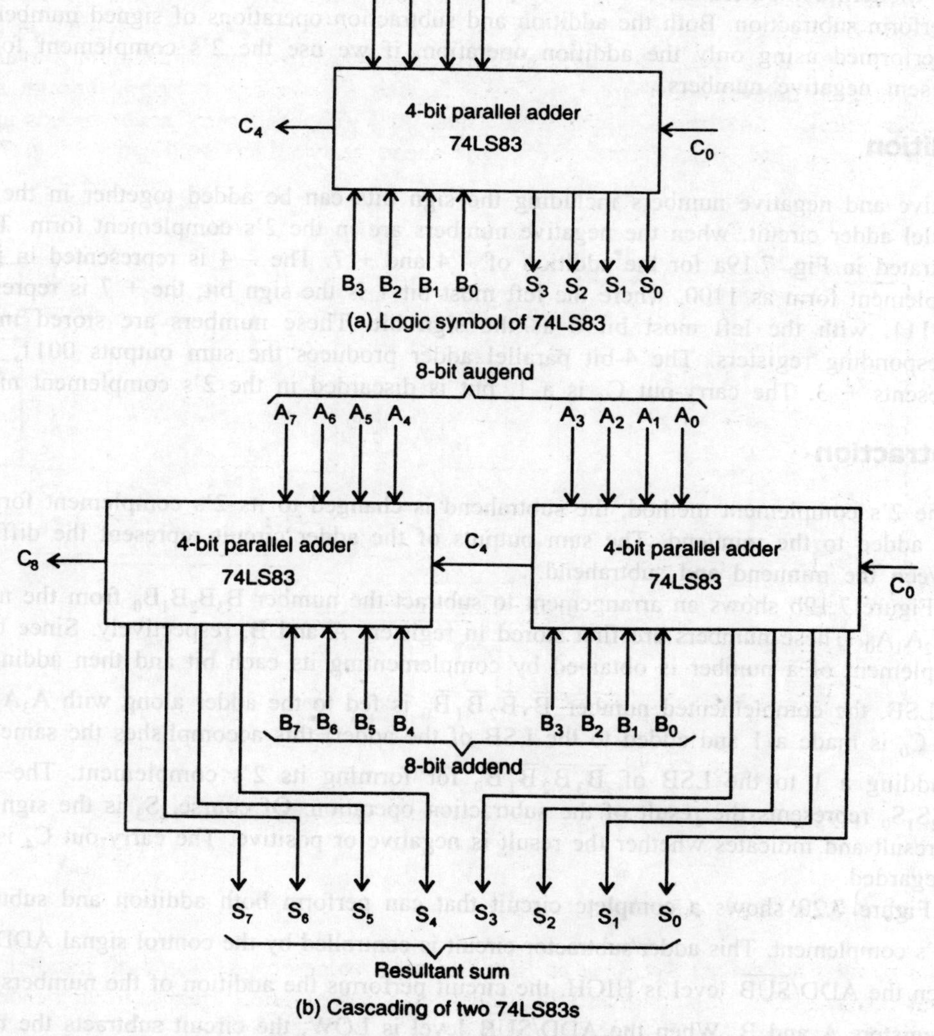

(a) Logic symbol of 74LS83

(b) Cascading of two 74LS83s

Fig. 7.18 IC parallel adders.

8-bit numbers. The adder on the right adds the 4 LSBs of the numbers. The C_4 output of this adder is connected as the input carry to the first position of the second adder which adds the 4 MSBs of the numbers. The eight sum outputs represent the resultant sum of the two 8-bit numbers. The C_8 is the carry-out of the last position (MSB) of the second adder. The C_8 can be used as an overflow bit or as a carry into another adder stage if still larger binary numbers are to be handled.

7.8 TWO'S COMPLEMENT ADDITION AND SUBTRACTION USING PARALLEL ADDERS

Most modern computers use the 2's complement system to represent negative numbers and to perform subtraction. Both the addition and subtraction operations of signed numbers can be performed using only the addition operation, if we use the 2's complement form to represent negative numbers.

Addition

Positive and negative numbers including the sign bits can be added together in the basic parallel adder circuit, when the negative numbers are in the 2's complement form. This is illustrated in Fig. 7.19a for the addition of – 4 and + 7. The – 4 is represented in its 2's complement form as 1100, where the left most bit 1 is the sign bit; the + 7 is represented as 0111, with the left most bit 0 as the sign bit. These numbers are stored in their corresponding registers. The 4-bit parallel adder produces the sum outputs 0011, which represents + 3. The carry-out C_4 is a 1, but is discarded in the 2's complement method.

Subtraction

In the 2's complement method, the subtrahend is changed to its 2's complement form and then added to the minuend. The sum outputs of the adder circuit represent the difference between the minuend and subtrahend.

Figure 7.19b shows an arrangement to subtract the number $B_3B_2B_1B_0$ from the number $A_3A_2A_1A_0$. These numbers are first stored in registers A and B, respectively. Since the 2's complement of a number is obtained by complementing its each bit and then adding 1 to the LSB, the complemented number $\overline{B}_3\overline{B}_2\overline{B}_1\overline{B}_0$ is fed to the adder along with $A_3A_2A_1A_0$. The C_0 is made a 1 and added to the LSB of the adder; this accomplishes the same effect as adding a 1 to the LSB of $\overline{B}_3\overline{B}_2\overline{B}_1\overline{B}_0$ for forming its 2's complement. The output $S_3S_2S_1S_0$ represents the result of the subtraction operation. Of course, S_3 is the sign bit of the result and indicates whether the result is negative or positive. The carry-out C_4 is again disregarded.

Figure 7.20 shows a complete circuit that can perform both addition and subtraction in 2's complement. This adder/subtractor circuit is controlled by the control signal ADD/$\overline{\text{SUB}}$. When the ADD/$\overline{\text{SUB}}$ level is HIGH, the circuit performs the addition of the numbers stored in registers A and B. When the ADD/$\overline{\text{SUB}}$ level is LOW, the circuit subtracts the number in register B from the number in register A. The operation is described as follows:

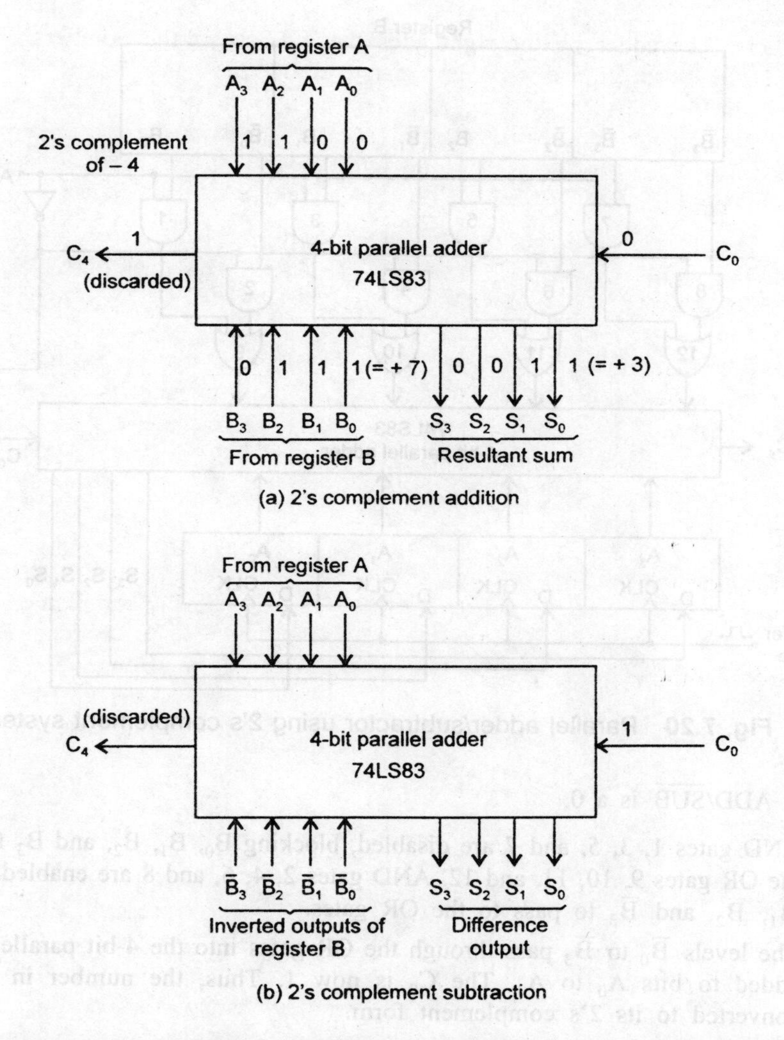

(a) 2's complement addition

(b) 2's complement subtraction

Fig. 7.19 Two's complement addition and subtraction using parallel adders.

When ADD/SUB = 1

1. AND gates 1, 3, 5, and 7 are enabled, allowing B_0, B_1, B_2, and B_3 to pass to the OR gates 9, 10, 11, and 12. AND gates 2, 4, 6, and 8 are disabled, blocking $\overline{B}_0$, $\overline{B}_1$, $\overline{B}_2$, and $\overline{B}_3$ from reaching the OR gates 9, 10, 11, and 12.
2. The levels B_0 to B_3 pass through the OR gates to the 4-bit parallel adder, to be added to the bits A_0 to A_3. The sum appears at the outputs S_0 to S_3.
3. ADD/SUB = 1 causes no carry into the adder.

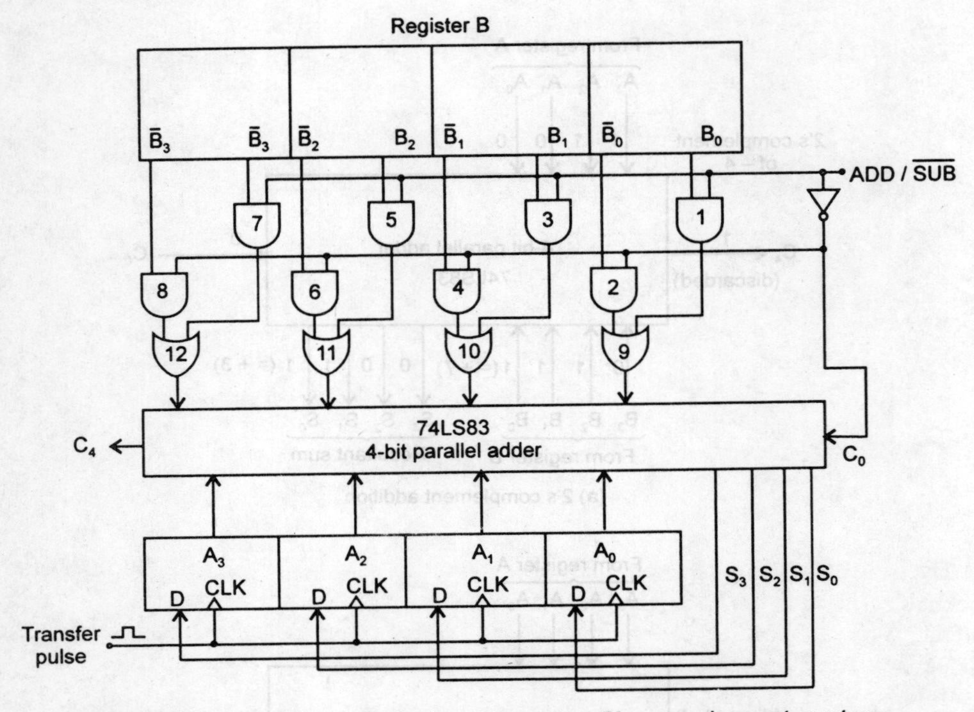

Fig. 7.20 Parallel adder/subtractor using 2's complement system.

When ADD/$\overline{SUB}$ is a 0.

1. AND gates 1, 3, 5, and 7 are disabled, blocking B_0, B_1, B_2, and B_3 from reaching the OR gates 9, 10, 11, and 12. AND gates 2, 4, 6, and 8 are enabled allowing $\overline{B}_0$, $\overline{B}_1$, $\overline{B}_2$, and $\overline{B}_3$ to pass to the OR gates.
2. The levels $\overline{B}_0$ to $\overline{B}_3$ pass through the OR gates into the 4-bit parallel adder, to be added to bits A_0 to A_3. The C_0 is now 1. Thus, the number in register B is converted to its 2's complement form.
3. The difference appears at the outputs S_0 to S_3.

Circuits like the adder/subtractor of Fig. 7.20, are used in computers because they provide a relatively simple means for adding and subtracting signed binary numbers. In most computers, the output is usually transferred into the register A (accumulator), so that the results of the addition or subtraction always end up stored in the register A. This is accomplished by applying a transfer pulse to the CLK inputs of register A.

7.9 SERIAL ADDERS

A serial adder is used to add serial binary numbers. Since the bits are added serially, that is, one pair of bits at a time, only one full-adder is required for this addition, as shown in Fig. 7.21.

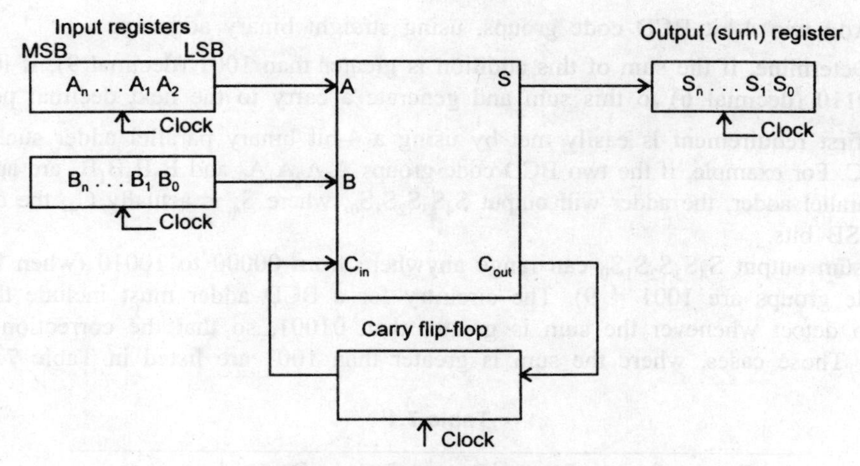

Fig. 7.21 A serial adder.

The serial input bits appear in synchronism, first the LSBs A_0 and B_0, then after one clock pulse A_1 and B_1, after another clock pulse A_2 and B_2, and so on. The carry bit generated in any one of these additions must be saved and added to the next-higher-order pair of input bits. The purpose of the flip-flop in the circuit shown is to store the carry bit C_{out} for the duration of one clock pulse, and then present it as C_{in} when the next pair of input bits is added. The output sum bits are shifted into the output register, as the input bits are shifted out of the input registers as inputs to the full adder. In practice, the output bits are often shifted into one of the input registers behind the data being shifted out. The register, which contains one of the binary numbers to be added before the addition commences and then contains the sum after the addition is completed, is called an ***accumulator***.

Serial adders are slower than the parallel adders, since they require one clock pulse per pair of bits to be added. Serial adders are used where circuit minimization is more important than speed as in pocket calculators.

7.10 BCD ADDER

The BCD addition process has been discussed in Chapter 3. It is briefly reviewed here.

1. Add the 4-bit BCD code groups for each decimal digit position using ordinary binary addition.
2. For those positions where the sum is 9 or less, the sum is in proper BCD form and no correction is needed.
3. When the sum of two digits is greater than 9, a correction of 0110 should be added to that sum, to produce the proper BCD result. This will produce a carry to be added to the next decimal position.

A BCD adder circuit must be able to operate in accordance with the above steps. In other words, the circuit must be able to do the following.

1. Add two 4-bit BCD code groups, using straight binary addition.
2. Determine, if the sum of this addition is greater than 1001 (decimal 9); if it is, add 0110 (decimal 6) to this sum and generate a carry to the next decimal position.

The first requirement is easily met by using a 4-bit binary parallel adder such as the 74LS83 IC. For example, if the two BCD code groups $A_3A_2A_1A_0$ and $B_3B_2B_1B_0$ are applied to a 4-bit parallel adder, the adder will output $S_4S_3S_2S_1S_0$, where S_4 is actually C_4, the carry-out of the MSB bits.

The sum output $S_4S_3S_2S_1S_0$ can range anywhere from 00000 to 10010 (when both the BCD code groups are 1001 = 9). The circuitry for a BCD adder must include the logic needed to detect whenever the sum is greater than 01001, so that the correction can be added in. Those cases, where the sum is greater than 1001 are listed in Table 7.1.

Table 7.1

S_4	S_3	S_2	S_1	S_0	Decimal number
0	1	0	1	0	10
0	1	0	1	1	11
0	1	1	0	0	12
0	1	1	0	1	13
0	1	1	1	0	14
0	1	1	1	1	15
1	0	0	0	0	16
1	0	0	0	1	17
1	0	0	1	0	18

Let's define a logic output X, that will go HIGH only when the sum is greater than 01001 (i.e. for the cases in Table 7.1). If we examine these cases, we see that X will be HIGH for either of the following conditions.

1. Whenever $S_4 = 1$ (sum greater than 15)
2. Whenever $S_3 = 1$ and either S_2 or S_1 or both are 1 (sums 10 to 15)

This condition can be expressed as

$$X = S_4 + S_3(S_2 + S_1)$$

Whenever X = 1, it is necessary to add the correction factor 0110 to the sum bits, and to generate a carry. Figure 7.22 shows the complete circuitry for a BCD adder, including the logic circuit implementation for X.

The circuit consists of three basic parts. The two BCD code groups $A_3A_2A_1A_0$ and $B_3B_2B_1B_0$ are added together in the upper 4-bit adder, to produce the sum $S_4S_3S_2S_1S_0$. The logic gates shown implement the expression for X. The lower 4-bit adder will add the correction 0110 to the sum bits, only when X = 1, producing the final BCD sum output represented by $\Sigma_3\Sigma_2\Sigma_1\Sigma_0$. The X is also the carry-out that is produced when the sum is greater than 01001. Of course, when X = 0, there is no carry and no addition of 0110. In such cases, $\Sigma_3\Sigma_2\Sigma_1\Sigma_0 = S_3S_2S_1S_0$.

Cascading BCD Adders

The circuit of Fig. 7.22 is used for adding two BCD coded decimal digits. When decimal numbers with several digits are to be added, it is necessary to use a separate BCD adder for each digit position.

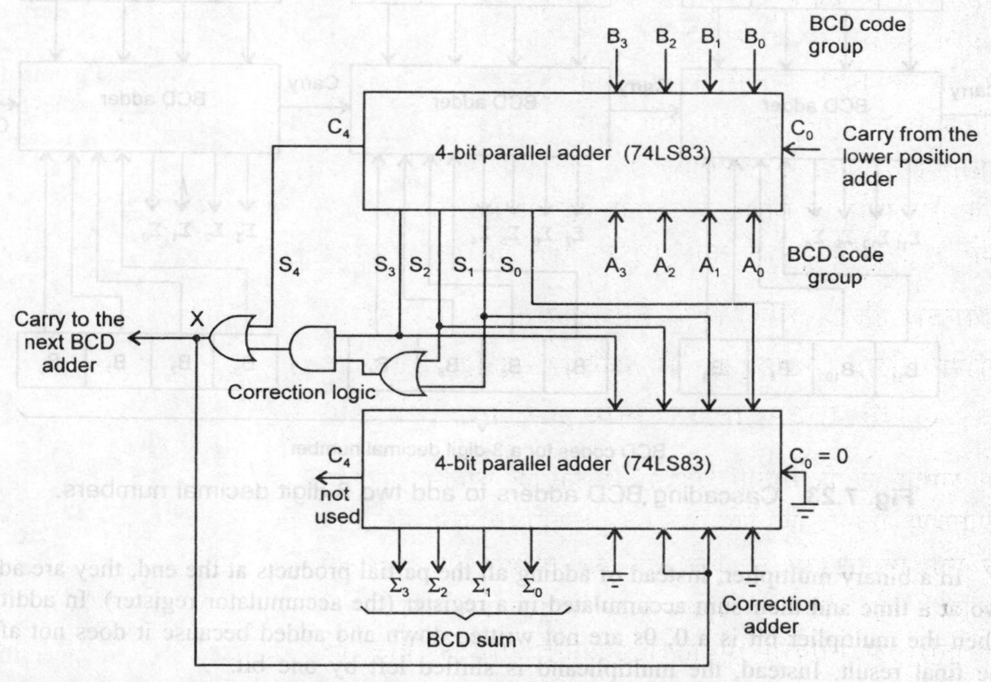

Fig. 7.22 A BCD adder using two 4-bit adders and a correction-detector circuit.

Figure 7.23 is a block diagram of a circuit for the addition of two 3-digit decimal numbers. The register A contains 12 bits, which are the three BCD code groups for one of the 3-digit decimal numbers; similarly the register B contains the BCD representation of the other 3-digit decimal number. The code groups $A_3 - A_0$ and $B_3 - B_0$ representing the least significant digits, are fed to the first BCD adder. Each BCD adder block is assumed to contain the circuitry of Fig. 7.22. This first BCD adder produces the sum output $\Sigma_3\Sigma_2\Sigma_1\Sigma_0$, which is the BCD code for the least significant digit of the sum. It also produces a carry-out, that is sent to the second BCD adder. This adder adds A_7 through A_4 to B_7 through B_4, and produces the sum $\Sigma_7\Sigma_6\Sigma_5\Sigma_4$. This arrangement can, of course, be extended to decimal numbers of any size by simply adding more flip-flops to the registers and including a BCD adder for each digit position.

7.11 BINARY MULTIPLIERS

In Chapter 2 we discussed binary muliplication by the paper and pencil method and by computer method. The paper and pencil method is modified somewhat in digital machines because a binary adder can add only two binary numbers at a time.

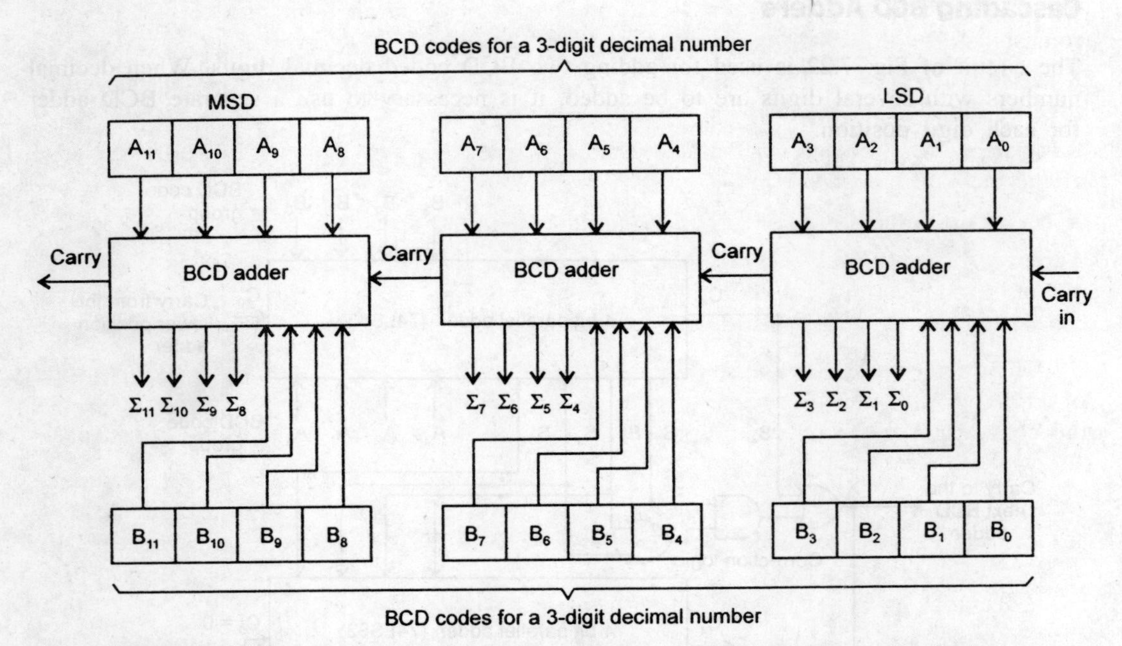

Fig. 7.23 Cascading BCD adders to add two 3-digit decimal numbers.

In a binary multiplier, instead of adding all the partial products at the end, they are added two at a time and their sum accumulated in a register (the accumulator register). In addition, when the multiplier bit is a 0, 0s are not written down and added because it does not affect the final result. Instead, the multiplicand is shifted left by one bit.

The multiplication of 1110 by 1001 using this process is illustrated below

Multiplicand:	1 1 1 0	
Multiplier:	1 0 0 1	
	1 1 1 0	The LSB of the multiplier is a 1; write down the multiplicand; shift the multiplicand one position to the left (1 1 1 0 0).
	1 1 1 0	The second multiplier bit is a 0; write down the previous result 1 1 1 0; shift the multiplicand to the left again (1 1 1 0 0 0).
	1 1 1 0	The third multiplier bit is a 0; write down the previous result 1 1 1 0; shift the multiplicand to the left again (1 1 1 0 0 0 0).
	+ 1 1 1 0 0 0 0	The fourth multiplier bit is a 1; write down the new multiplicand; add it to the first partial product to obtain the final product.
	1 1 1 1 1 1 0	

This multiplication process can be performed by the serial multiplier circuit shown in Fig. 7.24, which multiplies two 4-bit numbers to produce an 8-bit product. The circuit consists of the following elements.

X register. A 4-bit shift register that stores the multiplier—it will shift right on the falling edge of the clock. Note that 0s are shifted in from the left.

B register. An 8-bit register that stores the multiplicand; it will shift left on the falling edge of the clock. Note that 0s are shifted in from the right.

A register. An 8-bit register, i.e. the accumulator that accumulates the partial products.

Adder. An 8-bit parallel adder that produces the sum of A and B registers. The adder outputs S_7 through S_0 are connected to the D inputs of the accumulator so that the sum can be transferred to the accumulator only when a clock pulse gets through the AND gate.

The circuit operation can be described by going through each step in the multiplication of 1110 by 1001. The complete process requires 4 clock cycles. Refer to Fig. 7.24A for the contents of each register and adder outputs as we describe the sequence of steps.

1. *Before the first clock pulse:* Prior to occurrence of the first clock pulse, the register A is loaded with 00000000, the register B with the multiplicand 00001110, and the register X with the multiplier 1001. We can assume that each of these registers is loaded using its asynchronous inputs (i.e. PRESET and CLEAR). The output of the adder will be the sum of A and B, that is, 00001110.
2. *First clock pulse:* Since the LSB of the multiplier (X_0) is a 1, the first clock pulse gets through the AND gate and its positive going transition transfers the sum outputs into the accumulator. The subsequent negative going transition causes the X and B registers to shift right and left, respectively. This, of course, produces a new sum of A and B.
3. *Second clock pulse:* The second bit of the original multiplier is now in X_0. Since this bit is a 0, the second clock pulse is inhibited from reaching the accumulator. Thus, the sum outputs are not transferred into the accumulator and the number in the accumulator does not change. The negative going transition of the clock pulse will again shift the X and B registers.
4. *Third clock pulse:* The third bit of the original multiplier is now in X_0; since this bit is a 0, the third clock pulse is inhibited from reaching the accumulator. Thus, the sum outputs are not transferred into the accumulator and the number in the accumulator does not change. The negative going transition of the clock pulse will again shift the X and B registers.
5. *Fourth clock pulse:* The last bit of the original multiplier is now in X_0, and since it is a 1, the positive going transition of the fourth pulse transfers the sum into the accumulator. The accumulator now holds the final product. The negative going transition of the clock pulse shifts X and B again. Note that, X is now 0000, since all the multiplier bits have been shifted out.

7.12 CODE CONVERTERS

Code converters are logic circuits whose inputs are bit patterns representing numbers (or characters) in one code and whose outputs are the corresponding representations in a different

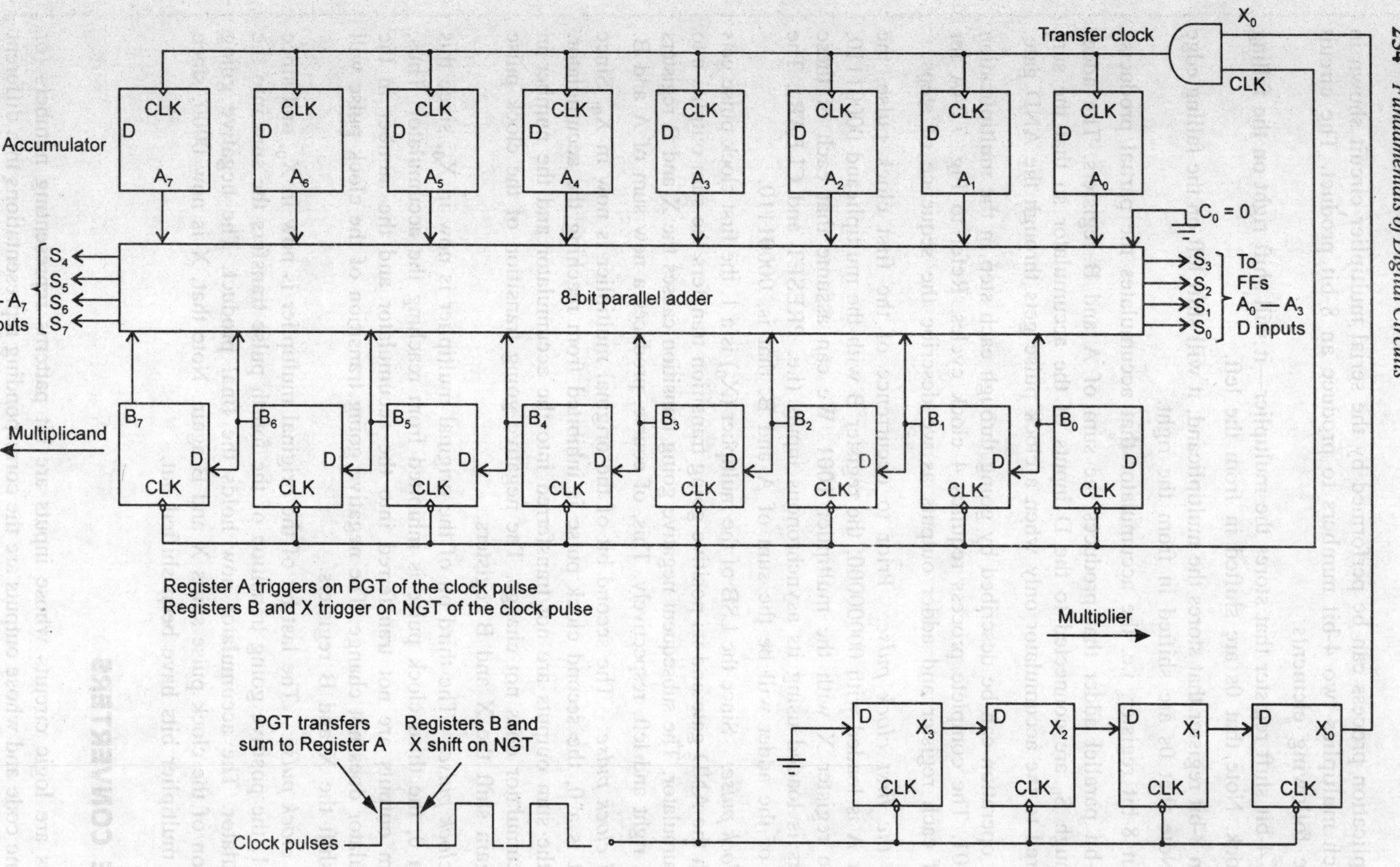

Fig. 7.24 Binary multiplier.

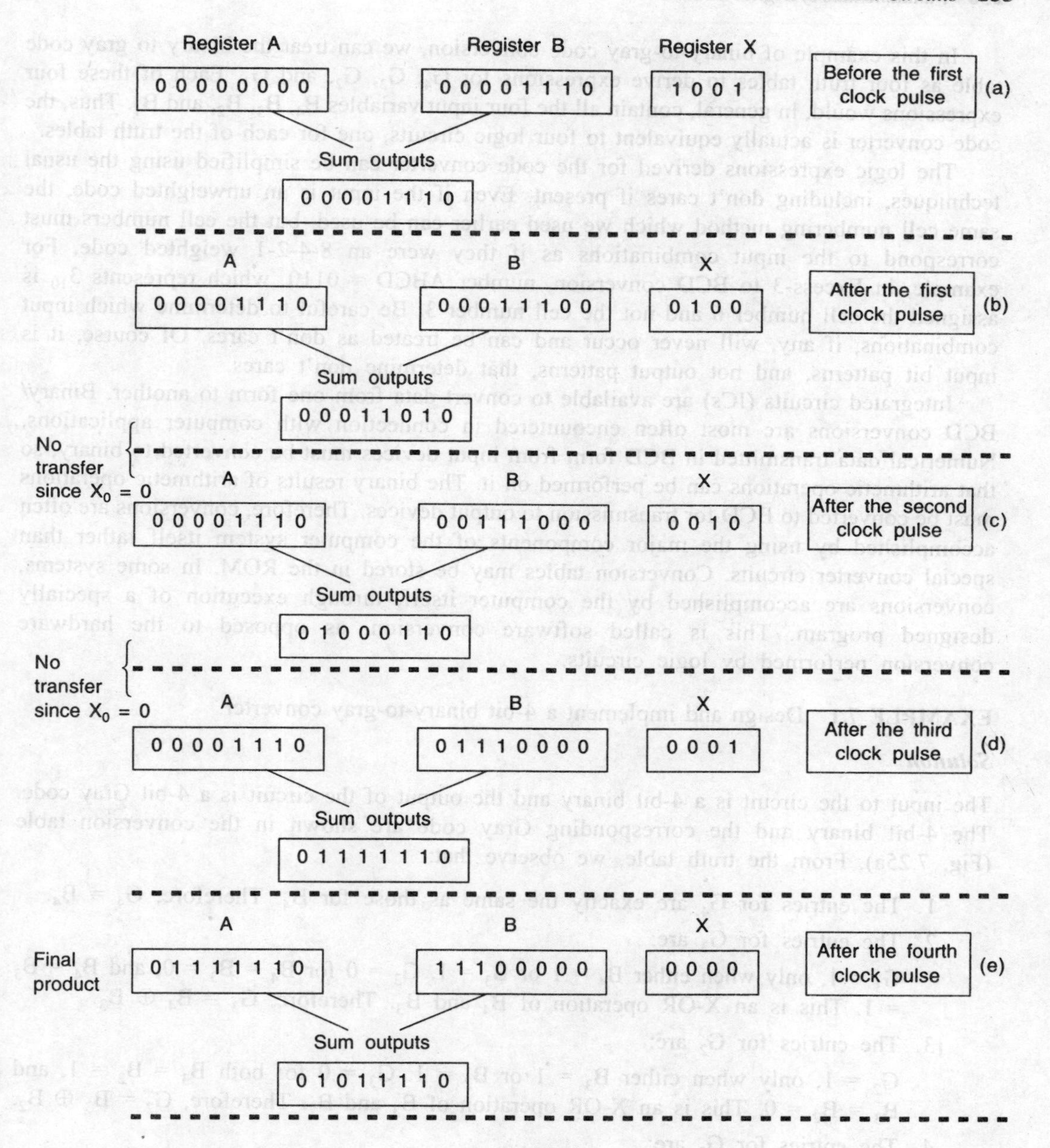

Fig. 7.24A

code. For example, a binary-to-gray code converter has four binary input lines B_4, B_3, B_2, and B_1, and four gray code output lines G_4, G_3, G_2, and G_1. When the input is 0010, for instance, the output should be 0011 and so forth. To design a code converter, we use a code table treating it as a truth table to express each output as a Boolean algebraic function of all the inputs.

In this example of binary-to-gray code conversion, we can treat the binary to gray code table as four truth tables to derive expressions for G_4, G_3, G_2, and G_1. Each of these four expressions would, in general, contain all the four input variables B_4, B_3, B_2, and B_1. Thus, the code converter is actually equivalent to four logic circuits, one for each of the truth tables.

The logic expressions derived for the code converter can be simplified using the usual techniques, including don't cares if present. Even if the input is an unweighted code, the same cell numbering method which we used earlier can be used, but the cell numbers must correspond to the input combinations as if they were an 8-4-2-1 weighted code. For example, in Excess-3 to BCD conversion, number ABCD = 0110, which represents 3_{10} is assigned the cell number 6 and not the cell number 3. Be careful to determine which input combinations, if any, will never occur and can be treated as don't cares. Of course, it is input bit patterns, and not output patterns, that determine don't cares.

Integrated circuits (ICs) are available to convert data from one form to another. Binary/BCD conversions are most often encountered in connection with computer applications. Numerical data transmitted in BCD form from input devices must be converted to binary, so that arithmetic operations can be performed on it. The binary results of arithmetic operations must be converted to BCD for transmission to output devices. Therefore, conversions are often accomplished by using the major components of the computer system itself rather than special converter circuits. Conversion tables may be stored in the ROM. In some systems, conversions are accomplished by the computer itself, through execution of a specially designed program. This is called software conversion, as opposed to the hardware conversion performed by logic circuits.

EXAMPLE 7.1 Design and implement a 4-bit binary-to-gray converter.

Solution

The input to the circuit is a 4-bit binary and the output of the circuit is a 4-bit Gray code. The 4-bit binary and the corresponding Gray code are shown in the conversion table (Fig. 7.25a). From the truth table, we observe that:

1. The entries for G_4 are exactly the same as those for B_4. Therefore, $G_4 = B_4$.
2. The entries for G_3 are:

   $G_3 = 1$, only when either $B_4 = 1$ or $B_3 = 1$. $G_3 = 0$ for $B_4 = B_3 = 0$, and $B_4 = B_3 = 1$. This is an X-OR operation of B_4 and B_3. Therefore, $G_3 = B_4 \oplus B_3$.
3. The entries for G_2 are:

   $G_2 = 1$, only when either $B_3 = 1$ or $B_2 = 1$. $G_3 = 0$ for both $B_3 = B_2 = 1$, and $B_3 = B_2 = 0$. This is an X-OR operation of B_3 and B_2. Therefore, $G_2 = B_3 \oplus B_2$.
4. The entries for G_1 are:

   $G_1 = 1$, only when $B_2 = 1$ or $B_1 = 1$. $G_1 = 0$ for both $B_2 = B_1 = 1$ and $B_2 = B_1 = 0$. This is an X-OR operation of B_2 and B_1. Therefore, $G_1 = B_2 \oplus B_1$.

So, the conversion can be achieved by using three X-OR gates as shown in Fig. 7.25c. The same circuit can be obtained by implementing the minimal expressions for G_4, G_3, G_2, and G_1 in terms of B_4, B_3, B_2, and B_1 obtained by minimizing the K-maps. Thus, the minimal expressions for G_4, G_3, G_2, and G_1 are:

$$G_4 = B_4$$

$$G_3 = \overline{B}_4 B_3 + B_4 \overline{B}_3 = B_4 \oplus B_3$$

$$G_2 = \overline{B}_3 B_2 + B_2 \overline{B}_3 = B_3 \oplus B_2$$

$$G_1 = \overline{B}_2 B_1 + B_1 \overline{B}_2 = B_2 \oplus B_1$$

The K-map for G_4 and its minimization is shown in Fig. 7.25b.

4-bit binary				4-bit gray			
B_4	B_3	B_2	B_1	G_4	G_3	G_2	G_1
0	0	0	0	0	0	0	0
0	0	0	1	0	0	0	1
0	0	1	0	0	0	1	1
0	0	1	1	0	0	1	0
0	1	0	0	0	1	1	0
0	1	0	1	0	1	1	1
0	1	1	0	0	1	0	1
0	1	1	1	0	1	0	0
1	0	0	0	1	1	0	0
1	0	0	1	1	1	0	1
1	0	1	0	1	1	1	1
1	0	1	1	1	1	1	0
1	1	0	0	1	0	1	0
1	1	0	1	1	0	1	1
1	1	1	0	1	0	0	1
1	1	1	1	1	0	0	0

(a) Conversion table

B_4B_3 \ B_2B_1	00	01	11	10
00				
01				
11	1	1	1	1
10	1	1	1	1

$G_4 = B_4$

(b) K- map for G_4

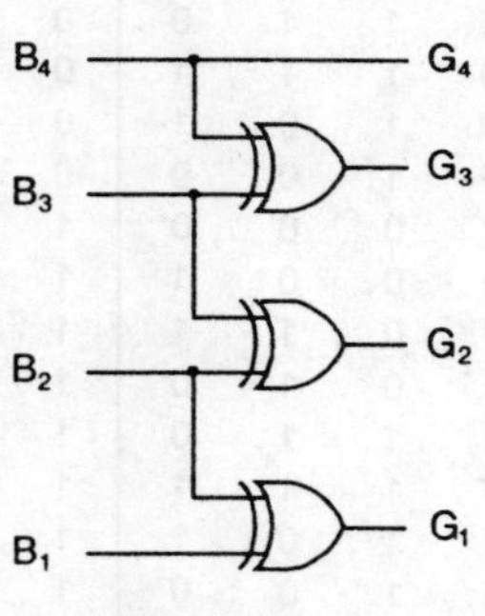

(c) Logic diagram

Fig. 7.25 Example 7.1.

EXAMPLE 7.2 Design and implement a 4-bit gray-to-binary converter.

Solution

The 4-bit input Gray code and the corresponding output binary numbers are shown in the conversion table of Fig. 7.26a. From the table, we observe that:

1. The entries for B_4 are exactly the same as those for G_4. Therefore, $B_4 = G_4$.
2. The entries for B_3 are:

   $B_3 = 1$, only when the number of 1s in G_4 and G_3 is an odd number. Otherwise $B_3 = 0$. So, B_3 is the modulo sum of G_4 and G_3. Therefore, $B_3 = G_4 \oplus G_3$.
3. The entries for B_2 are:

$B_2 = 1$, only when the number of 1s in G_4, G_3, and G_2 is an odd number. Otherwise $B_2 = 0$. So, B_2 is the modulo sum of G_4, G_3, and G_2, i.e. modulo sum of B_3 and G_2. Therefore, $B_2 = B_3 \oplus G_2$.

4. The entries for B_1 are:

   $B_1 = 1$, only when the number of 1s in G_4, G_3, G_2, and G_1 is an odd number. Otherwise $B_1 = 0$. So, B_1 is the modulo sum of G_4, G_3, G_2 and G_1, i.e. modulo sum of B_2 and G_1. Therefore, $B_1 = B_2 \oplus G_1$.

The same expressions can also be obtained using K-maps. Implementing the above expressions, the logic circuit is shown in Fig. 7.26c.

4-bit gray				4-bit binary			
G_4	G_3	G_2	G_1	B_4	B_3	B_2	B_1
0	0	0	0	0	0	0	0
0	0	0	1	0	0	0	1
0	0	1	1	0	0	1	0
0	0	1	0	0	0	1	1
0	1	1	0	0	1	0	0
0	1	1	1	0	1	0	1
0	1	0	1	0	1	1	0
0	1	0	0	0	1	1	1
1	0	0	0	1	0	0	0
1	0	0	1	1	0	0	1
1	0	1	1	1	0	1	0
1	0	1	0	1	0	1	1
1	1	1	0	1	1	0	0
1	1	1	1	1	1	0	1
1	1	0	1	1	1	1	0
1	1	0	0	1	1	1	1

(a) Conversion table

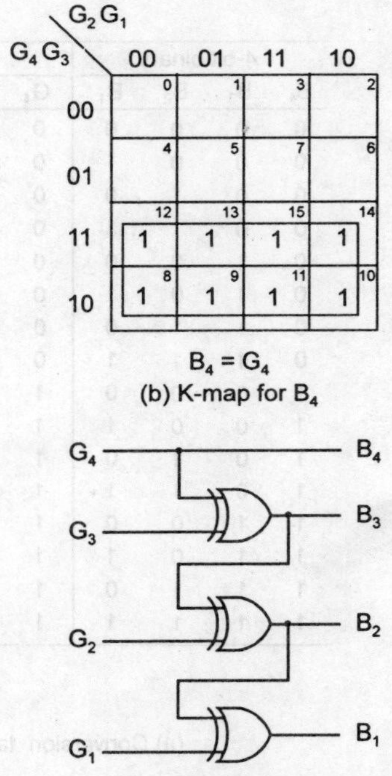

G_4G_3 \ G_2G_1	00	01	11	10
00				
01				
11	1	1	1	1
10	1	1	1	1

$B_4 = G_4$

(b) K-map for B_4

(c) Logic diagram

Fig. 7.26 Example 7.2.

Drawing the K-maps for B_4, B_3, B_2 and B_1 in terms of G_4, G_3, G_2, and G_1 based on the conversion table of Fig. 7.26 and simplifying them, the minimal expressions for B_4, B_3, B_2, and B_1 are obtained as:

$$B_4 = G_4$$

$$B_3 = \overline{G}_4G_3 + G_4\overline{G}_3 = G_4 \oplus G_3$$

$$B_2 = \overline{G}_4G_3\overline{G}_2 + \overline{G}_4\overline{G}_3G_2 + G_4\overline{G}_3\overline{G}_2 + G_4G_3G_2$$

$$= \overline{G}_4(G_3 \oplus G_2) + G_4(\overline{G_3 \oplus G_2}) = G_4 \oplus G_3 \oplus G_2 = B_3 \oplus G_2$$

$$B_1 = \overline{G}_4\overline{G}_3\overline{G}_2G_1 + \overline{G}_4\overline{G}_3G_2\overline{G}_1 + \overline{G}_4G_3\overline{G}_2\overline{G}_1 + \overline{G}_4G_3G_2G_1 + G_4G_3\overline{G}_2G_1 + G_4G_3G_2\overline{G}_1 + G_4\overline{G}_3\overline{G}_2\overline{G}_1 + G_4\overline{G}_3G_2G_1$$

$$= \overline{G}_4\overline{G}_3(G_2 \oplus G_1) + G_4G_3(G_2 \oplus G_1) + \overline{G}_4G_3(\overline{G_2 \oplus G_1}) + G_4\overline{G}_3(\overline{G_2 \oplus G_1})$$

$$= (G_2 \oplus G_1)\,(\overline{G_4 \oplus G_3}) + (\overline{G_2 \oplus G_1})\,(G_4 \oplus G_3)$$

$$= G_4 \oplus G_3 \oplus G_2 \oplus G_1$$

$$= B_2 \oplus G_1$$

The K-map for B_4 and its simplification is shown in Fig. 7.26b.

EXAMPLE 7.3 Design a 4-bit binary-to-BCD converter.

Solution

The input is a 4-bit binary. There are 16 possible combinations of 4-bit inputs (representing 0–15) and all are valid. Hence, the output has to be an 8-bit one; but since the first three bits will all be a 0 for all combinations of inputs, the output can be treated as a 5-bit one. The conversion is shown in Fig. 7.27a.

Drawing the K-maps and minimizing them, the minimal expressions for the BCD outputs A, B, C, D, and E in terms of the 4-bit binary inputs B_4, B_3, B_2, and B_1 are obtained as:

$$A = B_4B_3 + B_4B_2$$

$$B = B_4\overline{B}_3\overline{B}_2$$

$$C = \overline{B}_4B_3 + B_3B_2$$

$$D = B_4B_3\overline{B}_2 + \overline{B}_4B_2$$

$$E = B_1$$

The K-map for A and its minimization is shown in Fig. 7.27c. A logic diagram can be drawn based on the above minimal expressions.

Code Converters Using ICs

Figure 7.28a shows the use of 74184, a ROM device programmed as a BCD-to-binary converter. Figure 7.28b shows the use of 74185, a ROM device programmed as a binary-to-BCD converter.

Figures 7.29 and 7.30 show how these devices can be expanded for conversion involving larger number of bits. Figure 7.31 shows the use of ICs for obtaining 9's and 10's complement of a number.

7.13 PARITY BIT GENERATORS/CHECKERS

Binary data, when transmitted and processed, is susceptible to noise that can alter its 1s to 0s and 0s to 1s. To detect such errors, an additional bit called the parity bit is added to data

Decimal	4-bit binary B_4 B_3 B_2 B_1	BCD output A B C D E
0	0 0 0 0	0 0 0 0 0
1	0 0 0 1	0 0 0 0 1
2	0 0 1 0	0 0 0 1 0
3	0 0 1 1	0 0 0 1 1
4	0 1 0 0	0 0 1 0 0
5	0 1 0 1	0 0 1 0 1
6	0 1 1 0	0 0 1 1 0
7	0 1 1 1	0 0 1 1 1
8	1 0 0 0	0 1 0 0 0
9	1 0 0 1	0 1 0 0 1
10	1 0 1 0	1 0 0 0 0
11	1 0 1 1	1 0 0 0 1
12	1 1 0 0	1 0 0 1 0
13	1 1 0 1	1 0 0 1 1
14	1 1 1 0	1 0 1 0 0
15	1 1 1 1	1 0 1 0 1

(a) Conversion table

B_4, B_3, B_2, B_1 → SOP circuit → A, B, C, D, E

(b) Block diagram

B_4B_3 \ B_2B_1	00	01	11	10
00	0	1	3	2
01	4	5	7	6
11	12: 1	13: 1	15: 1	14: 1
10	8	9	11: 1	10: 1

$A = B_4B_3 + B_4B_2$

(c) K-map for A

Fig. 7.27 Example 7.3.

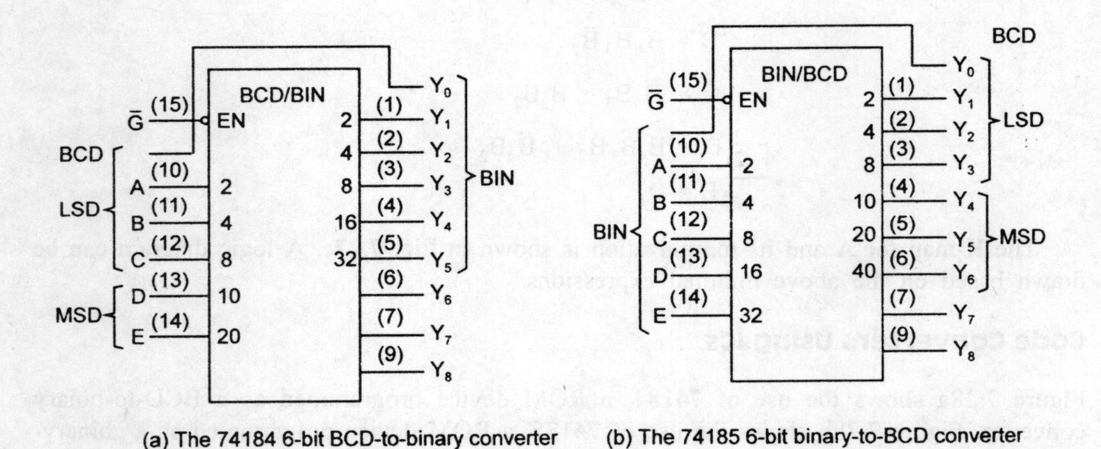

(a) The 74184 6-bit BCD-to-binary converter (b) The 74185 6-bit binary-to-BCD converter

Fig. 7.28 BCD-to-binary and binary-to-BCD conversion using ICs.

bits and the word containing data bits and the parity bit is transmitted. At the receiving end the number of 1s in the word received are counted and the error, if any, is detected. This parity check, however, detects only single bit errors.

A parity bit of a 0 or a 1 is attached to the data bits such that the total number of 1s in the word is even for even parity and odd for odd parity. The parity bit can be attached

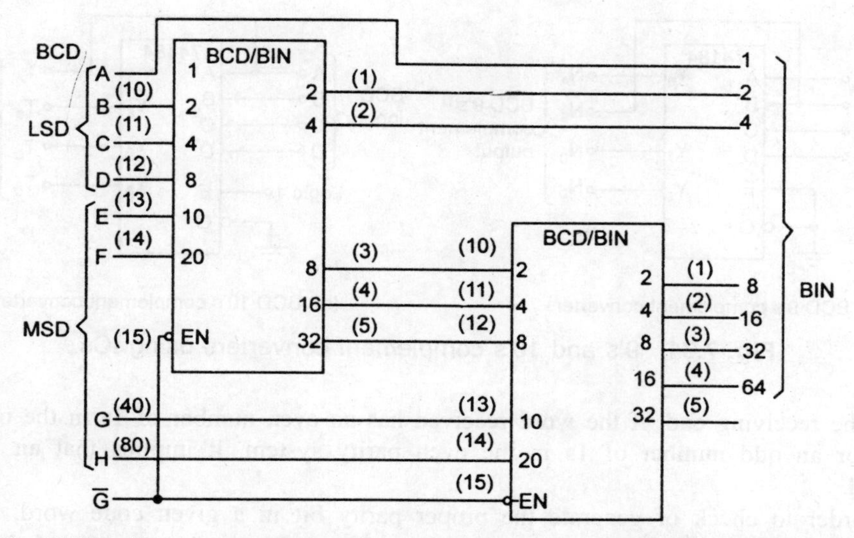

Fig. 7.29 74184s expanded for conversion of two BCD digits to binary.

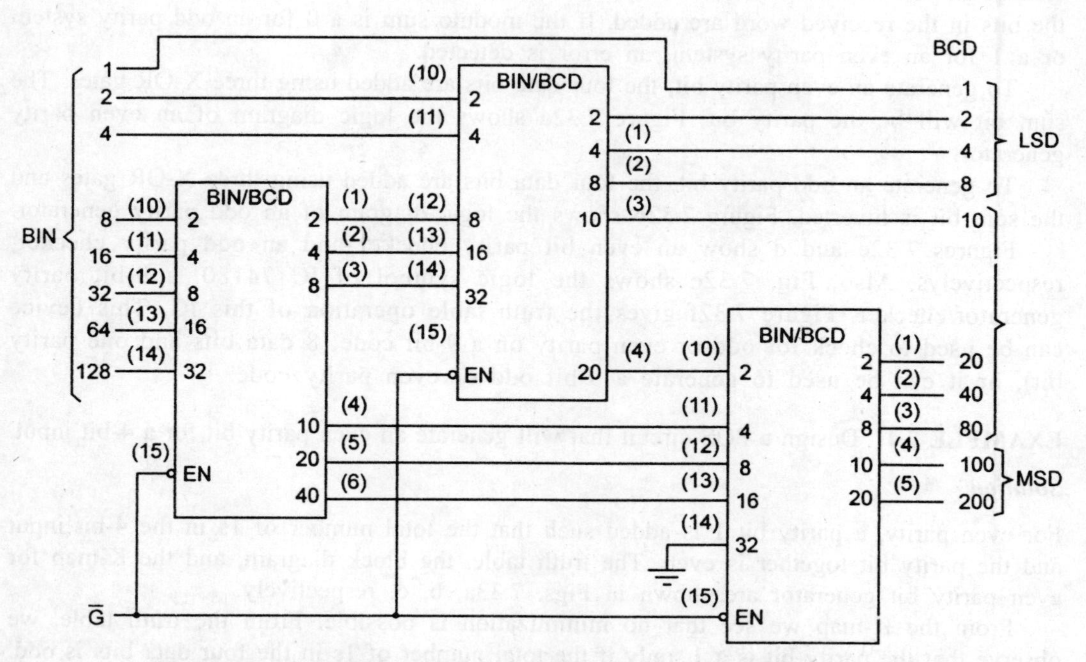

Fig. 7.30 74185s expanded for 8-bit binary to BCD conversion.

to the code group either at the beginning or at the end depending on system design. A given system operates with either even or odd parity but not both. So, a word always contains either an even or an odd number of 1s.

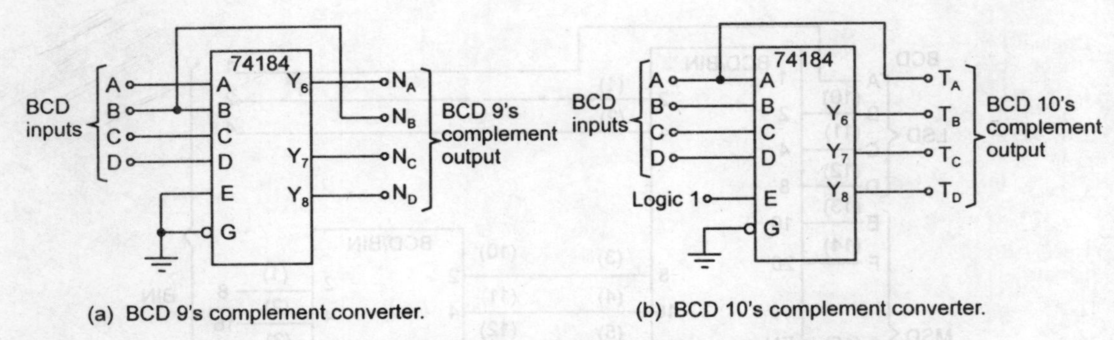

(a) BCD 9's complement converter. (b) BCD 10's complement converter.

Fig. 7.31 9's and 10's complement converters using ICs.

At the receiving end, if the word received has an even number of 1s in the odd parity system or an odd number of 1s in the even parity system, it implies that an error has occurred.

In order to check or generate the proper parity bit in a given code word, the basic principle used is, "the modulo sum of an even number of 1s is always a 0 and the modulo sum of an odd number of 1s is always a 1." Therefore, in order to check for an error, all the bits in the received word are added. If the modulo sum is a 0 for an odd parity system or a 1 for an even parity system, an error is detected.

To generate an even parity bit, the four data bits are added using three X-OR gates. The sum bit will be the parity bit. Figure 7.32a shows the logic diagram of an even parity generator.

To generate an odd parity bit, the four data bits are added using three X-OR gates and the sum bit is inverted. Figure 7.32b shows the logic diagram of an odd parity generator.

Figures 7.32c and d show an even bit parity checker and an odd parity checker, respectivelys. Also, Fig. 7.32e shows the logic symbol of IC 74180, a 9-bit parity generator/checker. Figure 7.32f gives the truth table operation of this IC. This device can be used to check for odd or even parity on a 9-bit code (8 data bits and one parity bit), or it can be used to generate a 9-bit odd or even parity code.

EXAMPLE 7.4 Design a POS circuit that will generate an even parity bit for a 4-bit input.

Solution

For even parity, a parity bit 1 is added such that the total number of 1s in the 4-bit input and the parity bit together is even. The truth table, the block diagram, and the K-map for even parity bit generator are shown in Figs. 7.33a, b, c, respectively.

From the K-map we see that no minimization is possible. From the truth table, we observe that the parity bit is a 1, only if the total number of 1s in the four data bits is odd. Therefore, the parity bit is the modulo sum of the four data bits. Hence,

$$f = A \oplus B \oplus C \oplus D$$

Therefore, three X-OR gates are required to realize the above expression as shown in Fig. 7.33d. It can also be realized by using twelve 2-input NAND gates.

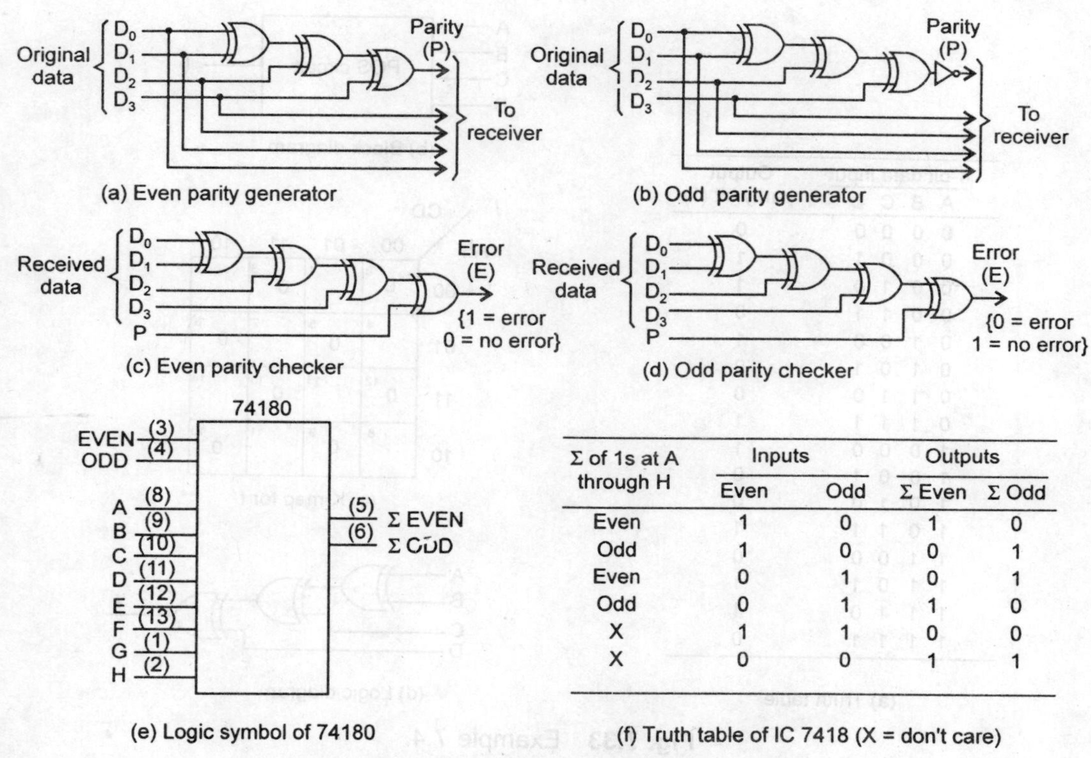

(a) Even parity generator
(b) Odd parity generator
(c) Even parity checker
(d) Odd parity checker
(e) Logic symbol of 74180

Σ of 1s at A through H	Inputs		Outputs	
	Even	Odd	Σ Even	Σ Odd
Even	1	0	1	0
Odd	1	0	0	1
Even	0	1	0	1
Odd	0	1	1	0
X	1	1	0	0
X	0	0	1	1

(f) Truth table of IC 7418 (X = don't care)

Fig. 7.32 Parity bit generator/checker.

From the K-map, the POS output is

$$
\begin{aligned}
f = {} & (A + B + C + D)\,(A + B + \overline{C} + \overline{D})\,(\overline{A} + \overline{B} + C + D)\,(\overline{A} + \overline{B} + \overline{C} + \overline{D}) \\
& (A + \overline{B} + C + \overline{D})\,(A + \overline{B} + \overline{C} + D)\,(\overline{A} + B + C + \overline{D})\,(\overline{A} + B + \overline{C} + D) \\
= {} & (A + B + \overline{CD} \cdot \overline{\overline{C}\,\overline{D}})\,(\overline{A} + \overline{B} + \overline{CD} \cdot \overline{\overline{C}\,\overline{D}})\,(A + \overline{B} + \overline{\overline{C}D} \cdot \overline{C\overline{D}}) \\
& (\overline{A} + B + \overline{\overline{C}D} \cdot \overline{C\overline{D}}) \\
= {} & (\overline{C \cdot D} \cdot \overline{\overline{C}\,\overline{D}} + \overline{A \cdot B} \cdot \overline{\overline{A}\,\overline{B}})\,(\overline{C\overline{D}} \cdot \overline{\overline{C}D} + \overline{A\overline{B}} \cdot \overline{\overline{A}B}) \\
= {} & [(C \oplus D) + (A \oplus B)]\,[\overline{C \oplus D} + \overline{A \oplus B}] \\
= {} & A \oplus B \oplus C \oplus D
\end{aligned}
$$

EXAMPLE 7.5 Design an SOP circuit that will generate an odd parity bit for a 4-bit input.

Solution

An odd parity bit generator outputs a 1, when the number of 1s in the data bits is even, so that the total number of 1s in the data bits and the parity bit together is odd.

A, B, C, D → POS circuit → f

(b) Block diagram

4-bit data input A	B	C	D	Output parity bit (f)
0	0	0	0	0
0	0	0	1	1
0	0	1	0	1
0	0	1	1	0
0	1	0	0	1
0	1	0	1	0
0	1	1	0	0
0	1	1	1	1
1	0	0	0	1
1	0	0	1	0
1	0	1	0	0
1	0	1	1	1
1	1	0	0	0
1	1	0	1	1
1	1	1	0	1
1	1	1	1	0

(a) Truth table

AB \ CD	00	01	11	10
00	0		0	
01		0		0
11	0		0	
10		0		0

(c) K-map for f′

(d) Logic diagram

Fig. 7.33 Example 7.4.

In Example 7.4, we derived an expression for the even parity generator as

$$f = A \oplus B \oplus C \oplus D$$

Since odd parity is the complement of even parity, we get

$$f' = \overline{A \oplus B \oplus C \oplus D}.$$

So, we put an inverter at the output of an even parity bit generator. The same result is directly obtained as follows.

The truth table, the block diagram, and the K-map and the logic diagram for the odd parity generator are shown in Figs. 7.34a, b, and c, respectively.

From the K-map, we see that no minimization is possible. If the expression for the parity bit is implemented as it is, 40 gate inputs are required. To reduce the cost, the expression may be manipulated in terms of X-OR gates and implemented.

$$f = \bar{A}\bar{B}\bar{C}\bar{D} + \bar{A}\bar{B}CD + AB\bar{C}\bar{D} + ABCD + \bar{A}B\bar{C}D + \bar{A}BC\bar{D} + A\bar{B}\bar{C}D + A\bar{B}C\bar{D}$$

$$= \bar{A}\bar{B}\,(\overline{C \oplus D}) + AB\,(\overline{C \oplus D}) + \bar{A}B\,(C \oplus D) + A\bar{B}(C \oplus D)$$

$$= \overline{(C \oplus D)}\;\overline{(A \oplus B)} + (C \oplus D)\,(A \oplus B)$$

$$= \overline{(A \oplus B) \oplus (C \oplus D)}$$

This is nothing but the complement of the output of the even parity bit generator.

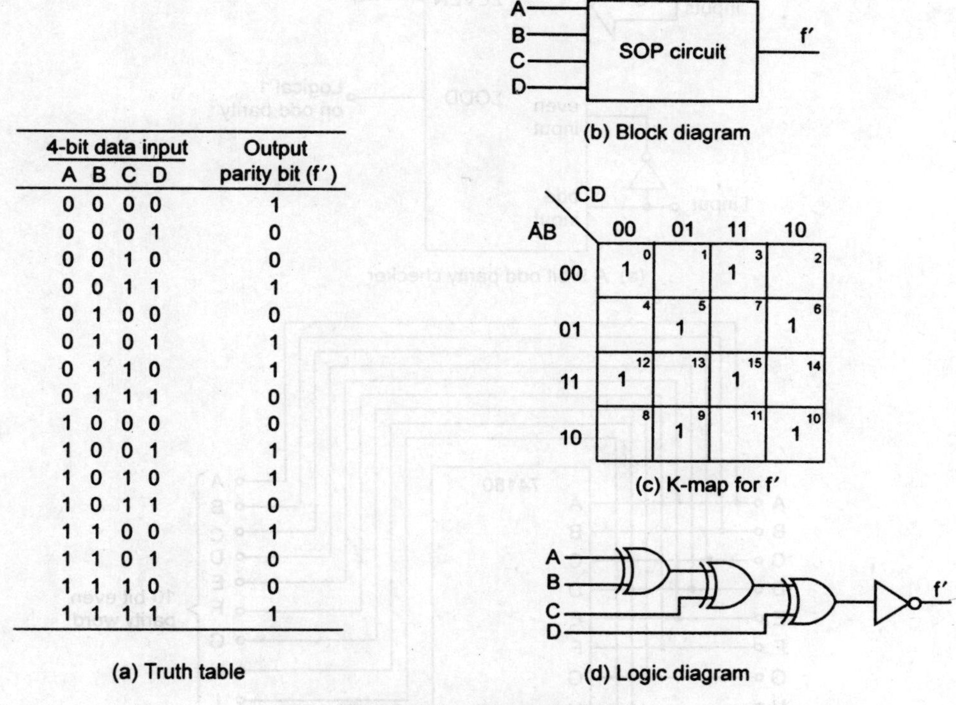

4-bit data input A B C D	Output parity bit (f′)
0 0 0 0	1
0 0 0 1	0
0 0 1 0	0
0 0 1 1	1
0 1 0 0	0
0 1 0 1	1
0 1 1 0	1
0 1 1 1	0
1 0 0 0	0
1 0 0 1	1
1 0 1 0	1
1 0 1 1	0
1 1 0 0	1
1 1 0 1	0
1 1 1 0	0
1 1 1 1	1

(a) Truth table

(b) Block diagram

(c) K-map for f′

(d) Logic diagram

Fig. 7.34 Example 7.5.

EXAMPLE 7.6

(a) Make a 9-bit odd parity checker using a 74180 and an inverter.

(b) Make a 10-bit even parity generator using a 74180 and an inverter.

(c) Make a 16-bit even parity checker using two 74180s.

Solution

(a) 8 of the 9-bits are applied at A-H inputs and the ninth bit, I, is applied to the ODD input. The circuit is shown in Fig. 7.35a.

(b) The 9-bit word consisting of A through I is converted to a 10-bit word with even parity. The circuit is shown in Fig. 7.35b.

(c) The 16-bit even parity checker is shown in Fig. 7.35c.

7.14 COMPARATORS

A comparator is a logic circuit, used to compare the magnitudes of two binary numbers. Depending on the design, it may either simply provide an output that is active (goes HIGH for

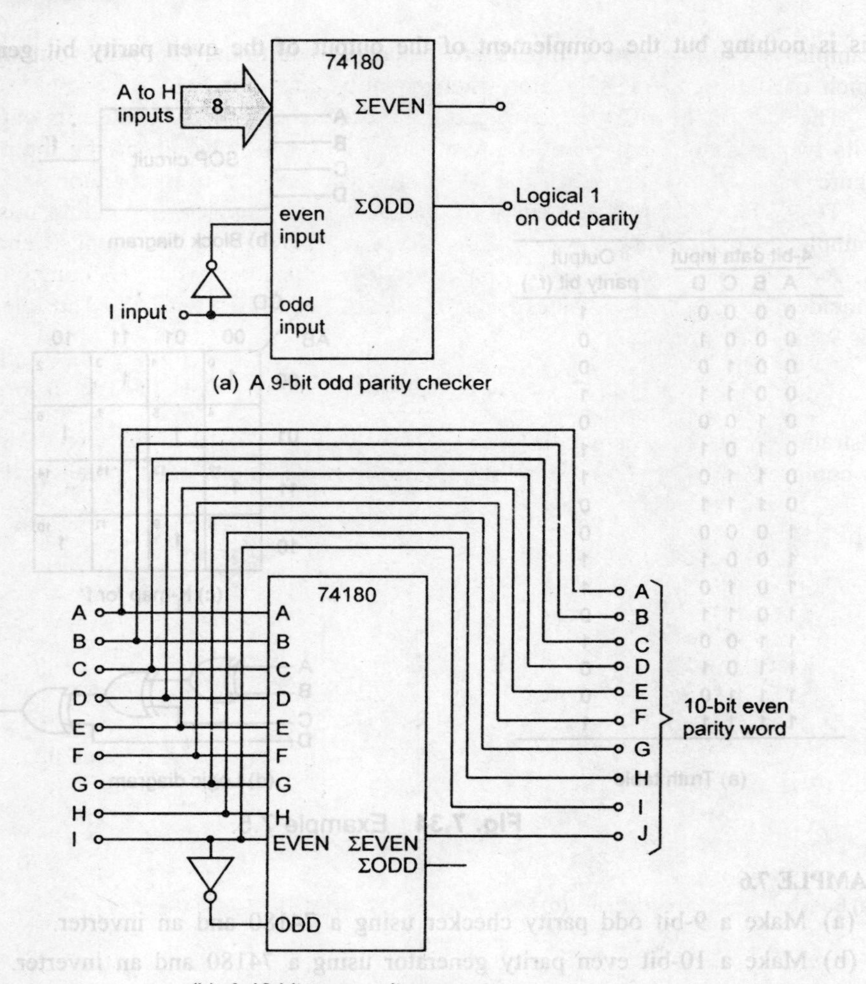

(a) A 9-bit odd parity checker

(b) A 10-bit even parity generator

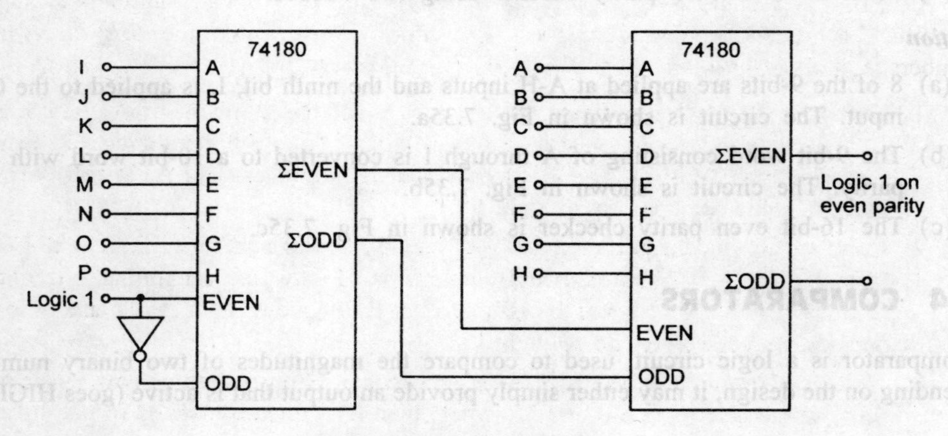

(c) A 16-bit even parity checker

Fig. 7.35 Example 7.6.

example) when the two numbers are equal, or additionally provide outputs that signify which of the numbers is greater when equality does not hold.

The X-NOR gate (coincidence gate) is a basic comparator, because its output is a 1 only if its two input bits are equal, i.e. the output is a 1 if and only if the input bits coincide. Figure 7.36a shows the operation of an X-NOR gate as a comparator.

Two binary numbers are equal, if and only if all their corresponding bits coincide. For example, two 4-bit binary numbers, $A_3A_2A_1A_0$ and $B_3B_2B_1B_0$ are equal, if and only if, $A_3 = B_3$, $A_2 = B_2$, $A_1 = B_1$ and $A_0 = B_0$. Thus, equality holds when A_3 coincides with B_3, A_2 coincides with B_2, A_1 coincides with B_1, and A_0 coincides with B_0. The implementation of this logic,

$$\text{EQUALITY} = (A_3 \odot B_3)\,(A_2 \odot B_2)\,(A_1 \odot B_1)\,(A_0 \odot B_0)$$

is straightforward and is shown in Fig. 7.36b. It is obvious that this circuit can be expanded or compressed to accommodate binary numbers with any other number of bits.

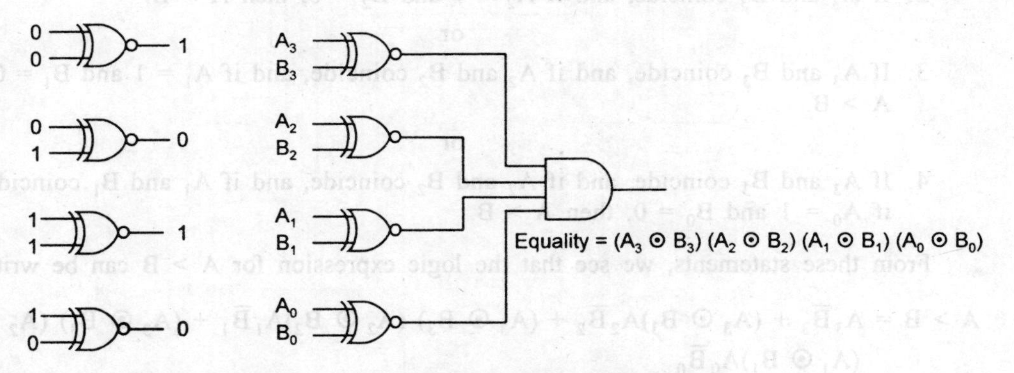

(a) Basic comparator operation (b) Logic diagram of an equality comparator

Fig. 7.36 Comparators.

A comparator capable of signifying which of the two binary numbers has greater magnitude is somewhat more complex. The procedure to determine this for two 4-bit numbers $A = A_3A_2A_1A_0$ and $B = B_3B_2B_1B_0$ is as follows:

1. Compare the most significant bits A_3 and B_3 of the numbers. If A_3 is a 1 and B_3 is a 0, then A is greater than B regardless of how the remaining bits compare. Similarly, if A_3 is a 0 and B_3 is a 1, then B is greater than A. If and only if, A_3 and B_3 coincide, then go to step 2.
2. Compare the next significant bits, A_2 and B_2 of the numbers. If A_2 is a 1 and B_2 is a 0, then A is greater than B regardless of how the remaining bits compare. Similarly, if A_2 is a 0 and B_2 is a 1, then B is greater than A. If and only if, A_2 and B_2 coincide, then go to step 3.
3. If A_1 is a 1 and B_1 is a 0, then A is greater than B regardless of how the remaining bits compare. If A_1 is a 0 and B_1 is a 1, then B is greater than A. If and only if, A_1 and B_1 coincide, then go to step 4.

4. If A_0 is a 1 and B_0 is a 0, then A is greater than B. If A_0 is a 0 and B_0 is a 1, then B is greater than A. If A_0 and B_0 are equal, then the numbers A and B are equal.

When still larger numbers are compared, continue checking in this fashion each pair of less significant bits, until finding a pair that does not coincide. The bit that is a 1, belongs to the greater number. If all pairs of bits coincide, then of course, the numbers are equal.

To illustrate this procedure, let us compare, A = 1110 with B = 1101.

A_3 and B_3 coincide (both bits are 1). So, we check A_2 and B_2. A_2 and B_2 also coincide (both bits are 1). So, we check A_1 and B_1. Since $A_1 = 1$ and $B_1 = 0$, we conclude that A is greater than B, and then we do not check any further pair of bits.

The logic we used above to determine if A is greater than B, can be expressed in a set of statements as follows.

1. If $A_3 = 1$ and $B_3 = 0$, then A > B.

or

2. If A_3 and B_3 coincide, and if $A_2 = 1$ and $B_2 = 0$, then A > B.

or

3. If A_3 and B_3 coincide, and if A_2 and B_2 coincide, and if $A_1 = 1$ and $B_1 = 0$, then A > B.

or

4. If A_3 and B_3 coincide, and if A_2 and B_2 coincide, and if A_1 and B_1 coincide, and if $A_0 = 1$ and $B_0 = 0$, then A > B.

From these statements, we see that the logic expression for A > B can be written as

$$A > B = A_3\overline{B}_3 + (A_3 \odot B_3)A_2\overline{B}_2 + (A_3 \odot B_3)(A_2 \odot B_2)A_1\overline{B}_1 + (A_3 \odot B_3)(A_2 \odot B_2)(A_1 \odot B_1)A_0\overline{B}_0$$

Similarly, the logic expression for B > A can be written as

$$B > A = \overline{A}_3B_3 + (A_3 \odot B_3)\overline{A}_2B_2 + (A_3 \odot B_3)(A_2 \odot B_2)\overline{A}_1B_1 + (A_3 \odot B_3)(A_2 \odot B_2)(A_1 \odot B_1)\overline{A}_0B_0$$

Figure 7.37 shows the logic diagram of a comparator that implements the logic we have described. Note that, it provides three active-HIGH outputs: A > B, A < B, and A = B.

7.15 IC COMPARATOR

Figure 7.38 shows the 7485 4-bit comparator. Pins labelled $(A < B)_{IN}$, $(A = B)_{IN}$, and $(A > B)_{IN}$ are used for cascading. Figure 7.39 shows how two 4-bit comparators are cascaded to perform 8-bit comparisons. The $(A < B)_{OUT}$, $(A = B)_{OUT}$ and $(A > B)_{OUT}$ outputs from the lower order comparator used for the least significant 4 bits, are connected to the $(A < B)_{IN}$, $(A = B)_{IN}$, and $(A > B)_{IN}$ inputs of the higher order comparator. Note that, $(A < B)_{IN}$ input of the lower order comparator is connected to V_{CC}, and $(A = B)_{IN}$ and $(A > B)_{IN}$ inputs of the lower order comparator are connected to ground.

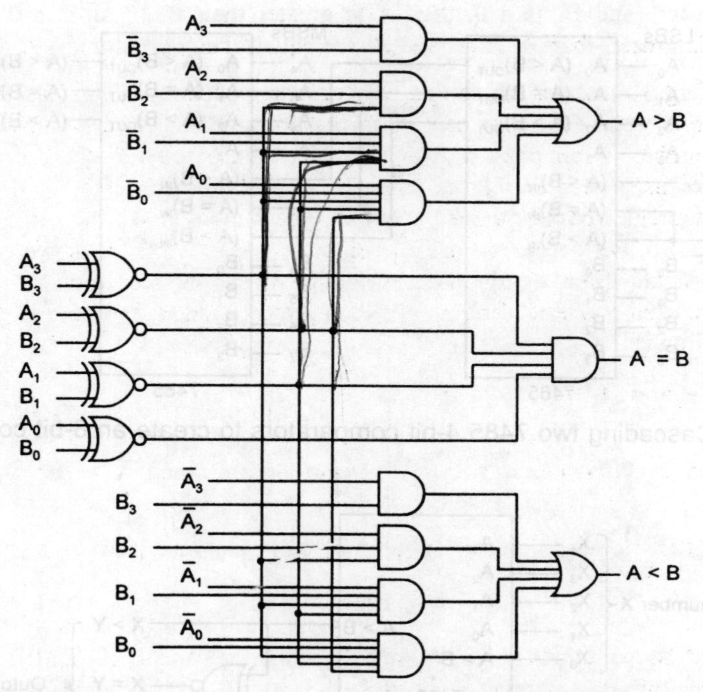

Fig. 7.37 Logic diagram of a 4-bit magnitude comparator.

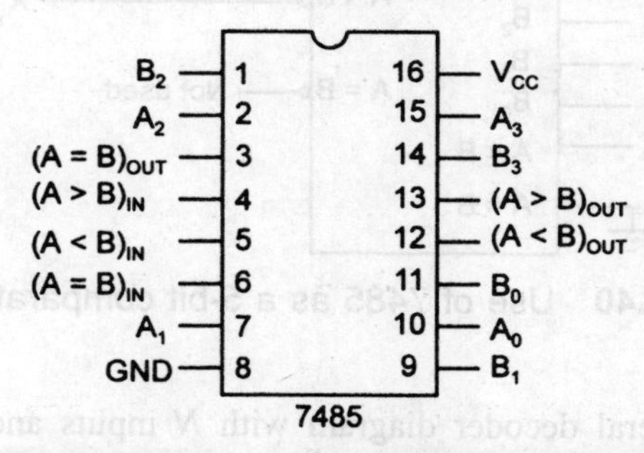

Fig. 7.38 Pin diagram of the 7485 4-bit comparator.

EXAMPLE 7.7 Design a 5-bit comparator using a single 7485 and one gate.

Solution

The circuit is shown in Fig. 7.40. The two 5-bit numbers to be compared are $X_4X_3X_2X_1X_0$ and $Y_4Y_3Y_2Y_1Y_0$.

7.16 DECODERS

A decoder is a logic circuit that converts an *N*-bit binary input code into *M* output lines such that only one output line is activated for each one of the possible combinations of inputs.

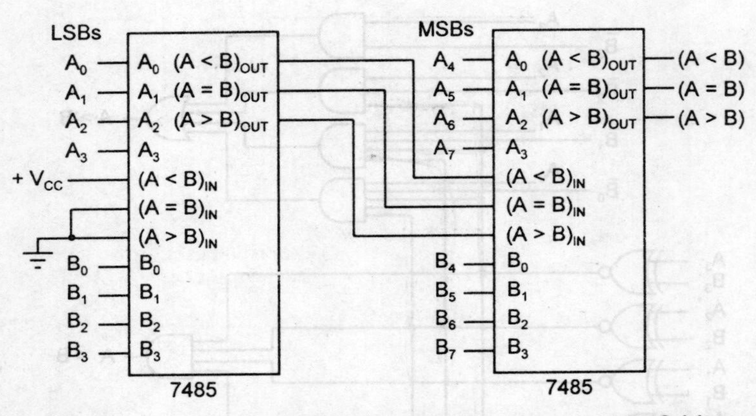

Fig. 7.39 Cascading two 7485 4-bit comparators to create an 8-bit comparator.

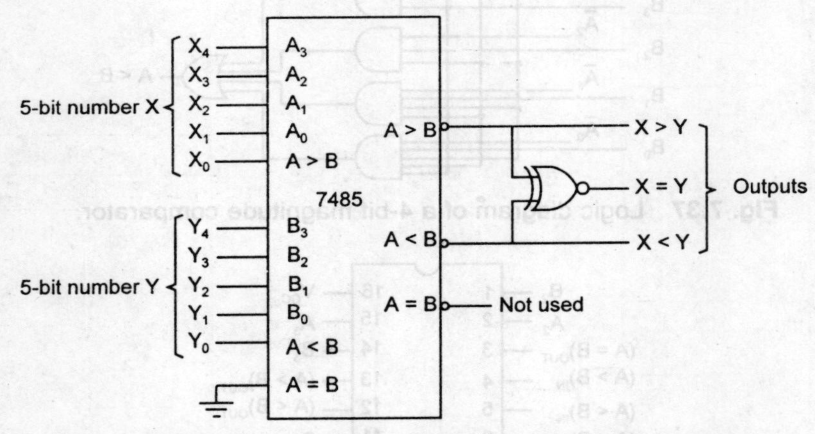

Fig. 7.40 Use of 7485 as a 5-bit comparator.

Figure 7.41 shows the general decoder diagram with N inputs and M outputs. Since each of the N inputs can be a 0 or a 1, there are 2^N possible input combinations or codes. For each of these input combinations, only one of the M outputs will be active (HIGH), all the other outputs will remain inactive (LOW). Some decoders are designed to produce active LOW output, while all the other outputs remain HIGH.

Some decoders do not utilize all of the 2^N possible input codes. For example, a BCD to decimal decoder has a 4-bit input code and 10 output lines that correspond to the 10 BCD code groups 0000 through 1001. Decoders of this type are often designed so that if any of the unused codes are applied to the input, none of the outputs will be activated.

3-Line to 8-Line Decoder

Figure 7.42 shows the circuitry for a decoder with three inputs and eight outputs. It uses all AND gates, and therefore, the outputs are active-HIGH. For active-LOW outputs, NAND

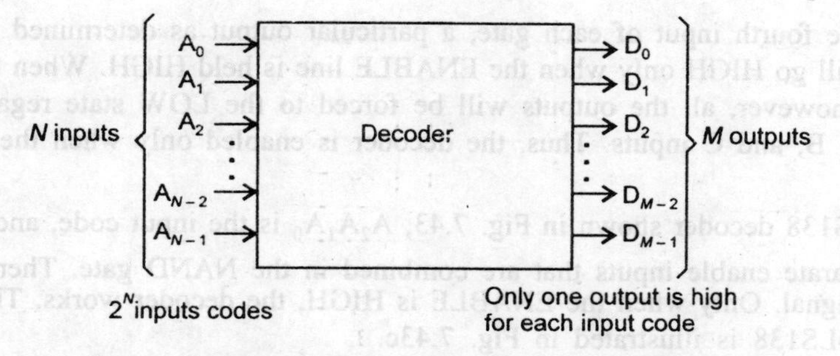

Fig. 7.41 General block diagram of a decoder.

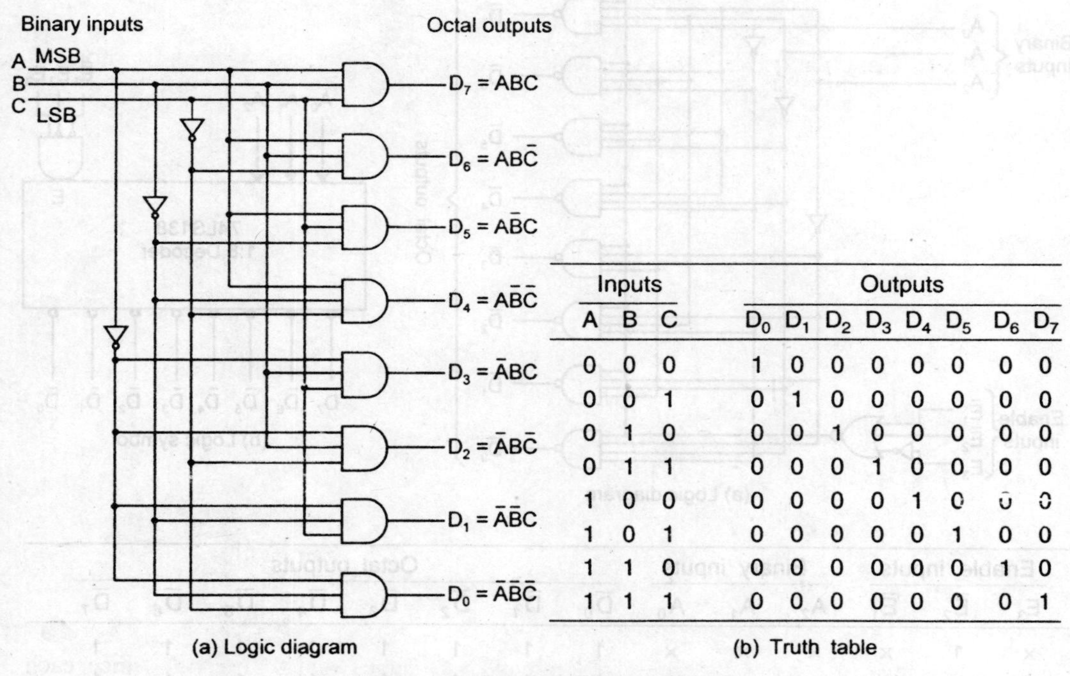

Inputs			Outputs							
A	B	C	D_0	D_1	D_2	D_3	D_4	D_5	D_6	D_7
0	0	0	1	0	0	0	0	0	0	0
0	0	1	0	1	0	0	0	0	0	0
0	1	0	0	0	1	0	0	0	0	0
0	1	1	0	0	0	1	0	0	0	0
1	0	0	0	0	0	0	1	0	0	0
1	0	1	0	0	0	0	0	1	0	0
1	1	0	0	0	0	0	0	0	1	0
1	1	1	0	0	0	0	0	0	0	1

(b) Truth table

Fig. 7.42 A 3-line to 8-line decoder.

gates are used. The truth table of the decoder is shown in Fig. 7.42b. This decoder can be referred to in several ways. It can be called a ***3-line to 8-line decoder*** because it has three input lines and eight output lines. It is also called a ***binary-to-octal decoder*** because it takes a 3-bit binary input code and activates one of the eight (octal) outputs corresponding to that code. It is also referred to as a ***1-of-8 decoder*** because only one of the eight outputs is activated at one time.

Enable Inputs

Some decoders have one or more ENABLE inputs, that are used to control the operation of the decoder. For example, in the 3-line to 8-line decoder, if a common ENABLE line is connected to the fourth input of each gate, a particular output as determined by the A, B, C input code will go HIGH only when the ENABLE line is held HIGH. When the ENABLE is held LOW, however, all the outputs will be forced to the LOW state regardless of the levels at the A, B, and C inputs. Thus, the decoder is enabled only when the ENABLE is HIGH.

In the 74LS138 decoder shown in Fig. 7.43, $A_2A_1A_0$ is the input code, and E_3, $\overline{E}_2$, and $\overline{E}_1$ are the separate enable inputs that are combined in the NAND gate. Then, $E_3\overline{E}_2\overline{E}_1$ is the ENABLE signal. Only when the ENABLE is HIGH, the decoder works. The truth table operation of 74LS138 is illustrated in Fig. 7.43c.

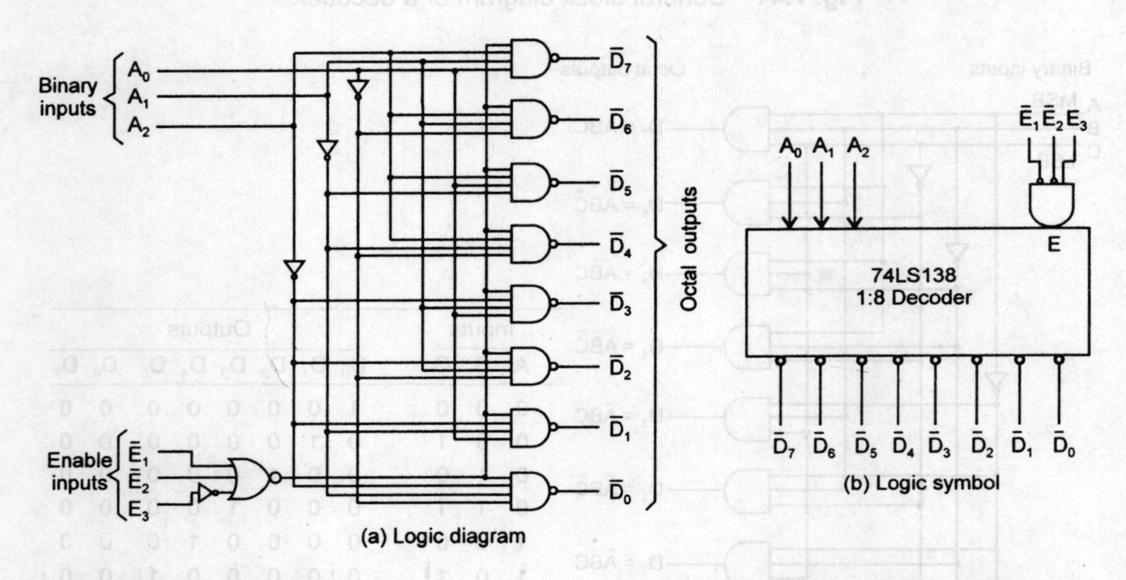

(a) Logic diagram

(b) Logic symbol

Enable inputs			Binary inputs			Octal outputs							
E_3	$\overline{E}_2$	$\overline{E}_1$	A_2	A_1	A_0	$\overline{D}_0$	$\overline{D}_1$	$\overline{D}_2$	$\overline{D}_3$	$\overline{D}_4$	$\overline{D}_5$	$\overline{D}_6$	$\overline{D}_7$
×	1	×	×	×	×	1	1	1	1	1	1	1	1
×	×	1	×	×	×	1	1	1	1	1	1	1	1
0	×	×	×	×	×	1	1	1	1	1	1	1	1
1	0	0	0	0	0	0	1	1	1	1	1	1	1
1	0	0	0	0	1	1	0	1	1	1	1	1	1
1	0	0	0	1	0	1	1	0	1	1	1	1	1
1	0	0	0	1	1	1	1	1	0	1	1	1	1
1	0	0	1	0	0	1	1	1	1	0	1	1	1
1	0	0	1	0	1	1	1	1	1	1	0	1	1
1	0	0	1	1	0	1	1	1	1	1	1	0	1
1	0	0	1	1	1	1	1	1	1	1	1	1	0

(c) Truth table

Fig. 7.43 74LS138, 3-line to 8-line decoder.

8421 BCD-to-Decimal Decoder (4-Line to 10-Line Decoder)

BCD-to-Decimal Decoder (7442)

Figure 7.44a shows the logic diagram for a 7442 BCD-to-decimal decoder. It is also available as a 74LS42 and 74HC42. An output goes LOW only when its corresponding BCD input is applied. For example, $\overline{D}_4$ will go LOW only when the input $A_3A_2A_1A_0 = 0100$, $\overline{D}_7$ will go LOW only when the input $A_3A_2A_1A_0 = 0111$. For input combinations that are invalid for BCD, none of the outputs will be activated. This decoder can also be referred to as a 4-to-10 decoder, or a 1-of-10 decoder. The pin diagram and the truth table for 7442 are shown in Figs. 7.44b and c, respectively.

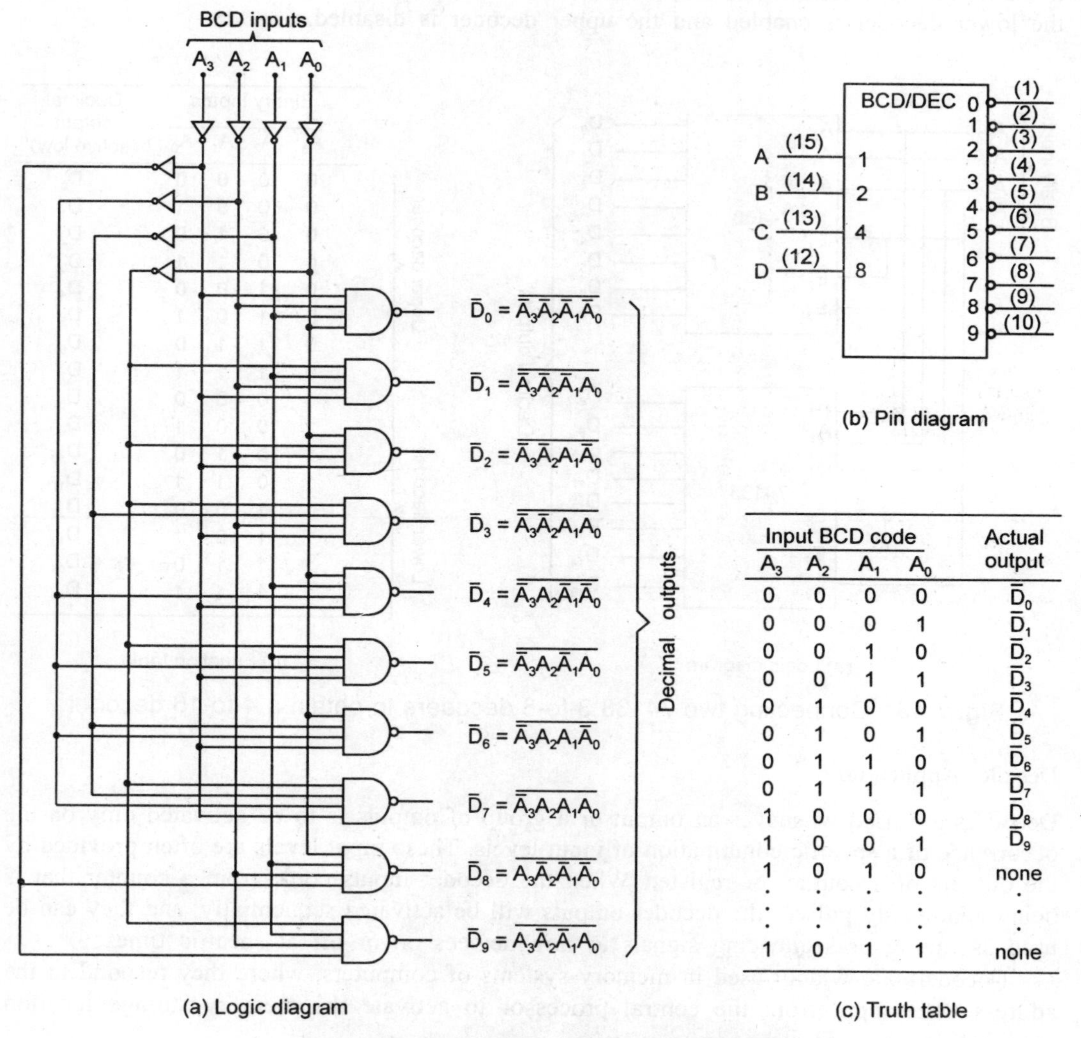

Input BCD code				Actual output
A_3	A_2	A_1	A_0	
0	0	0	0	$\overline{D}_0$
0	0	0	1	$\overline{D}_1$
0	0	1	0	$\overline{D}_2$
0	0	1	1	$\overline{D}_3$
0	1	0	0	$\overline{D}_4$
0	1	0	1	$\overline{D}_5$
0	1	1	0	$\overline{D}_6$
0	1	1	1	$\overline{D}_7$
1	0	0	0	$\overline{D}_8$
1	0	0	1	$\overline{D}_9$
1	0	1	0	none
⋮	⋮	⋮	⋮	⋮
1	0	1	1	none

(c) Truth table

Fig. 7.44 BCD-to-decimal decoder 7442.

BCD-to-Decimal Decoder/Driver

The TTL 7445 IC is a BCD-to-decimal decoder/driver. The term 'driver' is added to its description because this IC has open-collector outputs that can operate at higher current and voltage limits than a normal TTL output. It makes 7445 suitable for directly driving loads such as indicator LEDs or lamps, relays or DC motors.

4-to-16 Decoder from two 3-to-8 Decoders

Figure 7.45 shows the arrangement for using two 74138s, 3-to-8 decoders, to obtain a 4-to-16 decoder. The most significant input bit A_3 is connected to $\overline{E}_1$ on the upper decoder (for D_0 through D_7) and to E_3 on the lower decoder (for D_8 through D_{15}). Thus, when A_3 is LOW, the upper decoder is enabled and the lower decoder is disabled. When A_3 is HIGH, the lower decoder is enabled and the upper decoder is disabled.

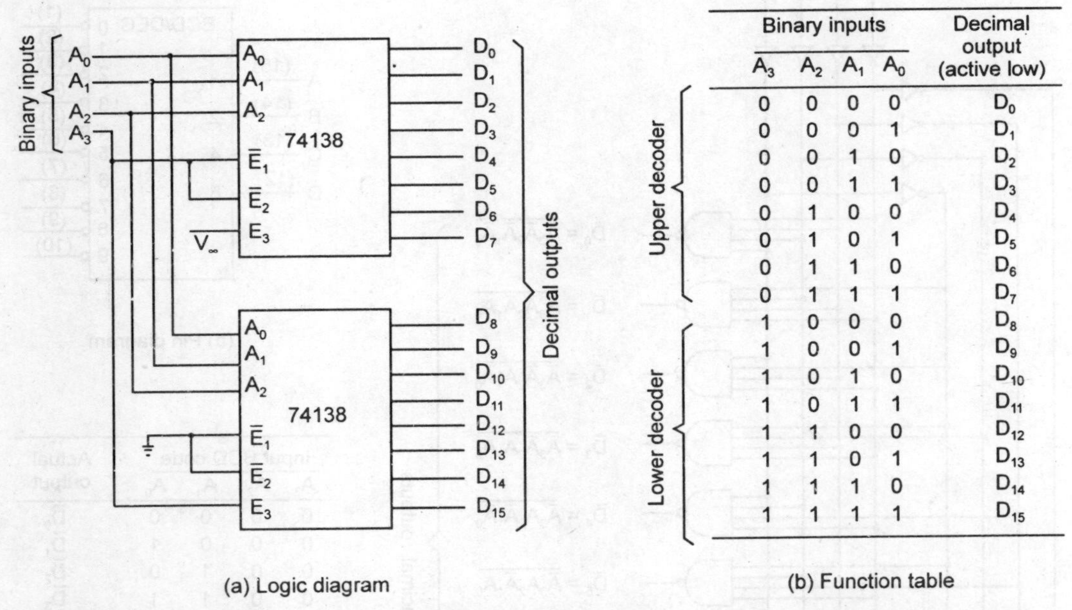

Binary inputs				Decimal output (active low)
A_3	A_2	A_1	A_0	
0	0	0	0	D_0
0	0	0	1	D_1
0	0	1	0	D_2
0	0	1	1	D_3
0	1	0	0	D_4
0	1	0	1	D_5
0	1	1	0	D_6
0	1	1	1	D_7
1	0	0	0	D_8
1	0	0	1	D_9
1	0	1	0	D_{10}
1	0	1	1	D_{11}
1	1	0	0	D_{12}
1	1	0	1	D_{13}
1	1	1	0	D_{14}
1	1	1	1	D_{15}

(a) Logic diagram (b) Function table

Fig. 7.45 Connecting two 74138 3-to-8 decoders to obtain a 4-to-16 decoder.

Decoder Applications

Decoders are used whenever an output or a group of outputs is to be activated only on the occurrence of a specific combination of input levels. These input levels are often provided by the outputs of a counter or register. When the decoder inputs come from a counter that is being continually pulsed, the decoder outputs will be activated sequentially, and they can be used as timing or sequencing signals to turn devices on or off at specific times.

Decoders are widely used in memory systems of computers, where they respond to the address code input from the central processor to activate the memory storage location specified by the address code.

7.17 BCD-TO-SEVEN SEGMENT DECODERS

This type of decoder accepts the BCD code and provides outputs to energize seven segment display devices in order to produce a decimal read out. Sometimes, the hex characters A through F may be produced. Each segment is made up of a material that emits light when current is passed through it. The most commonly used materials include LEDs, ncandescent filaments and LCDs. The LEDs generally provide greater illumination levels but require more power than that by LCDs.

Figure 7.46a shows a seven segment display consisting of seven light emitting segments. The segments are designated by letters a–g as shown in the figure. By illuminating various combinations of segments as shown in Fig. 7.46b, the numbers 0–9 can be displayed. Figures 7.46c and d show two types of LED displays—the common-anode and the common-cathode types. In the common-anode type, a low voltage applied to an LED cathode allows current to flow through the diode, which causes it to emit light. In the common-cathode type, a high voltage applied to an LED anode causes the current to flow and produces the resulting light emission.

An 8-4-2-1 BCD-to-seven segment decoder is a logic circuit as shown in Fig. 7.47a. The function table for such a decoder is shown in Fig. 7.47b. Since a 1 (HIGH) on any output line activates that line, we assume that the display is of the common-cathode type. The K-map used to simplify the logic expression for driving segment b is shown in Fig. 7.47c. Entries 10–15 are don't cares as usual. Since LEDs require considerable power, decoders often contain output drivers capable of supplying sufficient power.

$$
\begin{aligned}
b &= \overline{A}_3\overline{A}_2\overline{A}_1\overline{A}_0 + \overline{A}_3\overline{A}_2\overline{A}_1A_0 + \overline{A}_3\overline{A}_2A_1\overline{A}_0 + \overline{A}_3\overline{A}_2A_1A_0 + \overline{A}_3A_2\overline{A}_1\overline{A}_0 \\
&\quad + \overline{A}_3A_2A_1A_0 + A_3\overline{A}_2\overline{A}_1\overline{A}_0 + A_3\overline{A}_2\overline{A}_1A_0 \\
&= \Sigma\, m(0, 1, 2, 3, 4, 7, 8, 9)
\end{aligned}
$$

Don't cares, $d = \Sigma\, m(10, 11, 12, 13, 14, 15)$

7.18 DISPLAY DEVICES

Display devices provide a visual display of numbers, letters, and symbols in response to electrical input, and serve as constituents of an electronic display system.

Classification of Displays

The commonly used displays in the digital electronic field are as follows:

1. Cathode Ray Tube (CRT)
2. Light Emitting Diode (LED)
3. Liquid Crystal Display (LCD)
4. Gas Discharge Plasma Display (Cold-cathode Display or Nixie)
5. Electro-luminescent (EL) Display
6. Incandescent Display

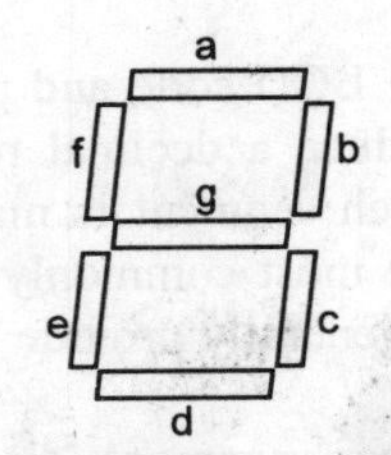

(a) Letters used to designate the segments.

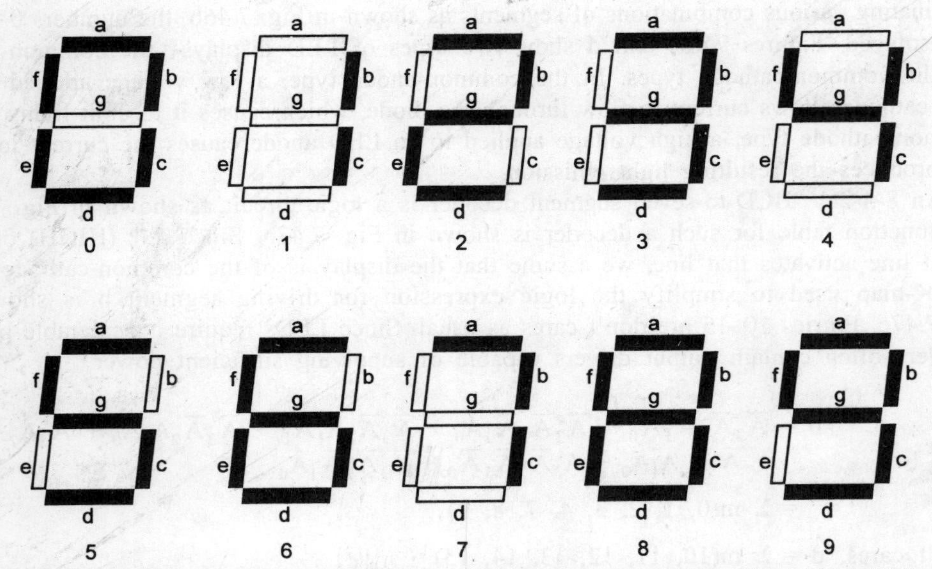

(b) By causing different combinations of the segments to illuminate (shown with solid black), the numerals 0–9 can be displayed.

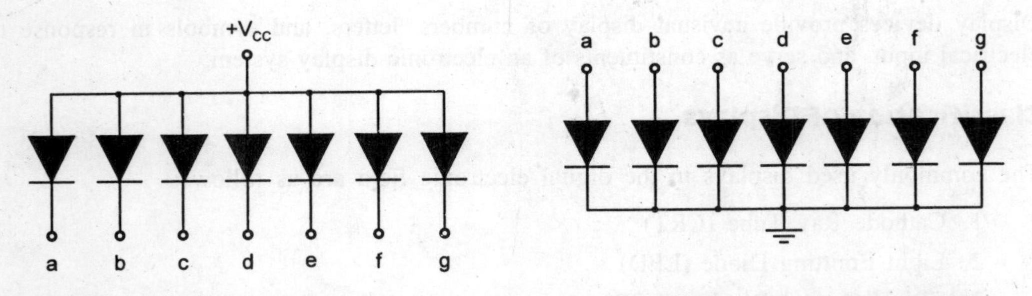

(c) A common-anode LED display. (d) A common-cathode LED display.

Fig. 7.46 The seven segment display.

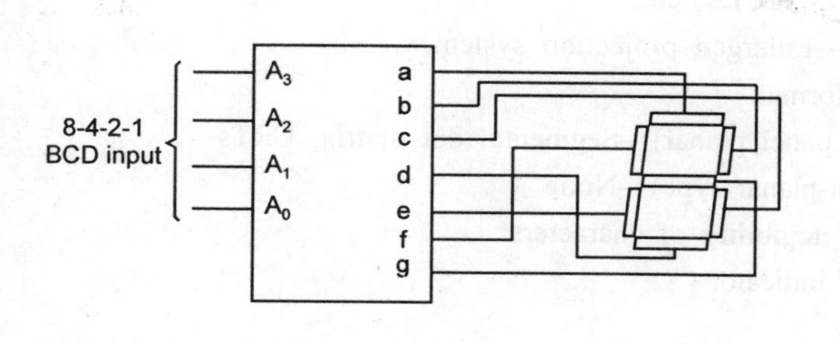

(a) Logic circuit

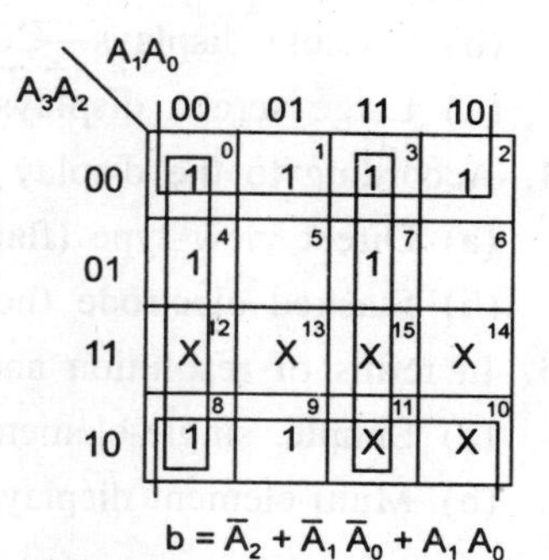

(c) K-map to derive simplified expression for driving segment (b)

Decimal digit	8-4-2-1 BCD				Seven segment code						
	A_3	A_2	A_1	A_0	a	b	c	d	e	f	g
0	0	0	0	0	1	1	1	1	1	1	0
1	0	0	0	1	0	1	1	0	0	0	0
2	0	0	1	0	1	1	0	1	1	0	1
3	0	0	1	1	1	1	1	1	0	0	1
4	0	1	0	0	0	1	1	0	0	1	1
5	0	1	0	1	1	0	1	1	0	1	1
6	0	1	1	0	1	0	1	1	1	1	1
7	0	1	1	1	1	1	1	0	0	0	0
8	1	0	0	0	1	1	1	1	1	1	1
9	1	0	0	1	1	1	1	1	0	1	1

(b) Function table

Fig. 7.47 BCD-to-seven segment decoder.

7. Electrophoretic Image Display (EPID)
8. Liquid Vapour Display (LVD)

In general, displays are classified in a number of ways, such as follows:

1. On the methods of conversion of electrical data to visible light:
   (a) Active displays (light emitters—CRTs, Gas discharge plasma, LEDs, etc.)
   (b) Passive displays (light controllers—LCDs, EPIDs, etc.)
2. On the applications:
   (a) Analog displays—Bargraph displays (CRT)
   (b) Digital displays—Nixies, Alphanumeric, LEDs, etc.
3. According to the display size and physical dimensions:
   (a) Symbolic displays—Alphanumeric, Nixie tubes, LEDs, etc.

(b) Console displays—CRTs, LCDs, etc.

(c) Large screen displays—Enlarged projection system.

4. According to the display format:

(a) Direct view type (flat panel planar)—Segmental dot matrix, CRTs

(b) Stacked electrode (non-planar type)—Nixie

5. In terms of resolution and legibility of characters:

(a) Simple, single-element indicator

(b) Multi-element displays.

The Light Emitting Diode (LED)

In a forward-biased diode, free electrons cross the junction and fall into holes. When they recombine, these free electrons radiate energy as they fall from a higher energy level to a lower one. In a rectifier diode, the energy is dissipated as heat, but in an LED, the energy is radiated as light. By using elements like gallium, arsenic, and phosphorous, a manufacturer can produce LEDs that radiate red, green, yellow, orange, and infrared radiation (invisible). LEDs that produce visible radiation are used in instrument displays, calculators, digital watches, etc.

The infrared (IR) LED finds applications in burglar-alarm systems and other areas requiring invisible radiation. Generally, the infrared emitting LEDs are coated with phosphor so that, by the excitation of phosphor, visible light can be produced.

The advantages of using LEDs in electronic displays are:

1. LEDs are very small and can be considered as point sources of light. They can, therefore, be stacked in a high density matrix to serve as a numeric and alphanumeric display.
2. The light output from an LED is a function of current flowing through it. An LED can, therefore, be controlled smoothly by varying the current. This is particularly useful for operating LED displays under different ambient lighting conditions.
3. LEDs are highly efficient emitters of EM radiation. LEDs with light output of different colours, i.e. red, yellow, amber, and green are commonly available.
4. LEDs are very fast devices, having a turn-on/off time of less than 1 ms.
5. The low supply voltage and current requirements of LEDs make them compatible with TTL ICs.
6. LEDs are manufactured with the same type of technology as that used for transistors and ICs and, therefore, they are economical and have a high degree of reliability.
7. LEDs are rugged and can, therefore, withstand shocks and vibrations. They can be operated over a wide range of temperature say, 0–70°C.

The disadvantage of LEDs compared to LCDs is their high power requirement. Also, LEDs are not suited for large area displays, primarily because of their high cost.

The Liquid Crystal Display (LCD)

A liquid crystal is a state of matter between a solid and a liquid. The characteristic feature of a liquid crystal is its long cylindrical molecules. The alignment of molecules can exist only over a limited temperature range of 0–50°C with most available devices.

Liquid crystal displays (LCDs) are used in similar applications where LEDs are used. These applications are display of numeric and alphanumeric characters in dot matrix and segmental displays.

LCDs are of two types—dynamic scattering type and field effect type. The dynamic scattering liquid crystal cell is constructed by layering the liquid crystal between glass sheets with transparent electrodes deposited on the inside faces. The liquid crystal material may be one of the several organic compounds which exhibit optical properties of a crystal, though they remain in liquid form. When a potential is applied across the cell, charge carriers flowing through a liquid disrupt the molecular alignment and produce turbulence. When the liquid is not activated, it is transparent. When the liquid is activated, the molecular turbulence causes light to be scattered in all directions and the cell appears to be bright. The phenomenon is called *dynamic scattering.*

The construction of a field effect liquid crystal display is similar to that of the dynamic scattering type, with the exception that two thin polarizing optical filters are placed at the inside of each glass sheet. The liquid crystal material in the field effect cell is also of different type from that employed in the dynamic scattering cell. The material used is twisted nematic type and it actually twists the light passing through the cell when the latter is not energized. This allows the light to pass through the optical filters and the cell appears bright. When the cell is energized, no twisting of light takes place and the cell appears dull.

Liquid crystal cells are of two types—transmittive type and reflective type. In the transmittive type cell, both glass sheets are transparent, so that light from a rear source is scattered in the forward direction when the cell is activated. The reflective type cell has a reflective surface on one side of the glass sheets. The incident light on the front surface of the cell is dynamically scattered by an activated cell. Both types of cells appear quite bright when activated even under ambient light conditions.

The liquid crystals are light reflectors or transmitters and, therefore, consume small amounts of energy.

The advantages of LCDs are low power consumption and low cost. The disadvantages are that they occupy a large area and their operating speed is low. LCDs are normally used for seven segment displays.

Operation of Liquid Crystal Displays

LCDs operate by polarizing light so that a non-activated segment reflects incident light and thus appears invisible against its background. An activated segment does not reflect incident light and thus appears dark. LCDs consume much less power than LED displays and are widely used in battery powered devices such as calculators and watches. An LCD does not emit light energy like an LED. So, it cannot be seen in the dark. It requires an external source of light.

Basically, LCDs operate from a low voltage (typically 3–15 V rms), low frequency (20–60 Hz) ac signal and draw very little current. They are often arranged as seven segment displays for numerical read-outs as shown in Fig. 7.48. The ac voltage needed to turn on a segment is applied between the segment and the backplane which is common to all segments. The segment and the backplane form a capacitor, that draws very little current as long as the ac frequency is kept low. The frequency is generally not lower than 25 Hz, because that would produce visible flicker.

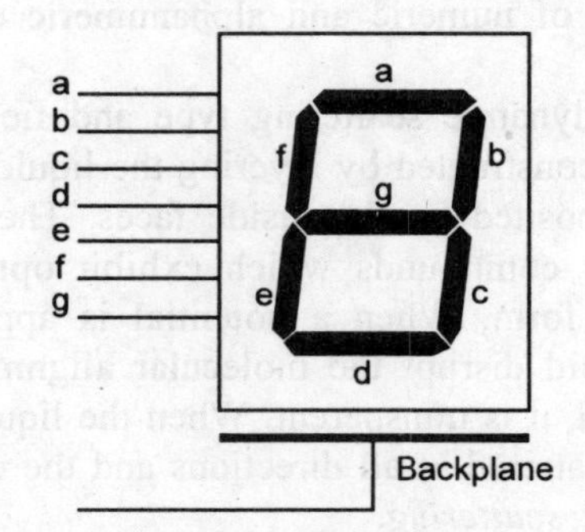

Fig. 7.48 Liquid crystal display.

Driving an LCD

An LCD segment will turn on, when an ac voltage is applied between the segment and the backplane, and will turn off, when there is no voltage between the two. Rather than generating an ac signal, it is common practice to produce the required ac voltage by applying out-of-phase square waves to the segment and the backplane. Figure 7.49 illustrates the driving arrangement for one segment. A 40 Hz square wave is applied to the backplane and also to the input of a CMOS 4070 X-OR gate. The other input to the X-OR is a control input, that controls the ON/OFF state of the segment.

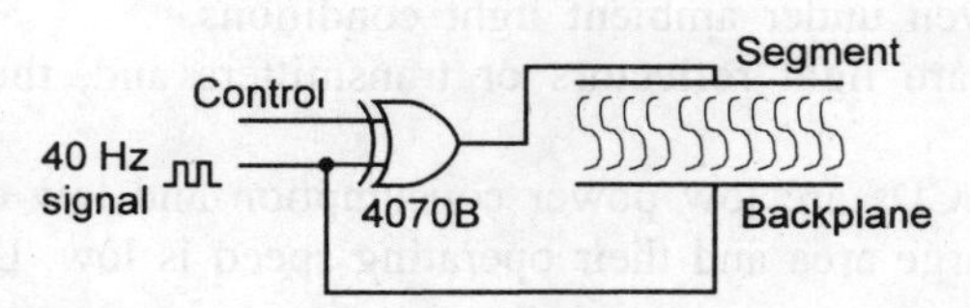

Fig. 7.49 Method for driving the seven segment LCD.

When the CONTROL input is LOW, the X-OR output will be exactly the same as the 40 Hz square wave, so that the signals applied to the segment and the backplane are equal. Since there is no difference in voltage, the segment will be OFF. When the CONTROL input is HIGH, the X-OR output will be the inverse of the 40 Hz square wave, so that the signal applied to the segment is out-of-phase with the signal applied to the backplane. As a result, the segment voltage will alternatively be at + 5 V and – 5 V relative to the backplane. This ac voltage will turn on the segment. The same idea can be extended to a complete seven segment LCD display as shown in Fig. 7.50.

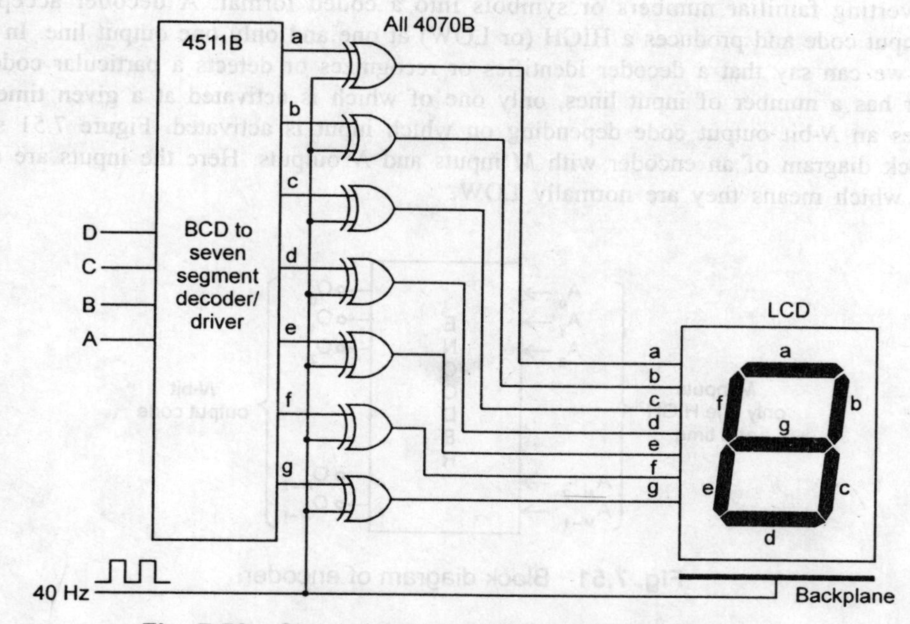

Fig. 7.50 Circuit for driving the seven segment LCD.

In general, CMOS devices are used to drive LCDs for two reasons. First, they require much less power than that by TTL and are more suited to the battery-operated applications where LCDs are used. Second, as the TTL LOW state voltage is not exactly 0 V, and can be as much as 0.4 V, it produces a dc component of voltage between the segment and the backplane which considerably shortens the life of an LCD.

Incandescent Seven Segment Displays

Incandescent displays use light-bulb filaments as the segments. When current is made to flow through these filaments, they become hot and thus glow. They emit a bright white light that is easy to read. They are sometimes covered with coloured filters to provide coloured light. These displays are much less efficient than the LED types because they require more current per segment, and therefore, are not used in battery-operated devices such as calculators and multimeters. They are, however, used in electronic cash registers and other line-operated devices where power consumption is not critical.

7.19 ENCODERS

An encoder is a device whose inputs are decimal digits and/or alphabetic characters and whose outputs are the coded representation of those inputs. In other words, an encoder may be said to be a combinational logic circuit that performs the 'reverse' operation of the decoder. The opposite of the decoding process is called encoding, i.e. encoding is a process of converting familiar numbers or symbols into a coded format. A decoder accepts an *N*-bit input code and produces a HIGH (or LOW) at one and only one output line. In other words, we can say that a decoder identifies or recognizes or detects a particular code. An encoder has a number of input lines, only one of which is activated at a given time, and produces an *N*-bit output code depending on which input is activated. Figure 7.51 shows the block diagram of an encoder with *M* inputs and *N* outputs. Here the inputs are active HIGH, which means they are normally LOW.

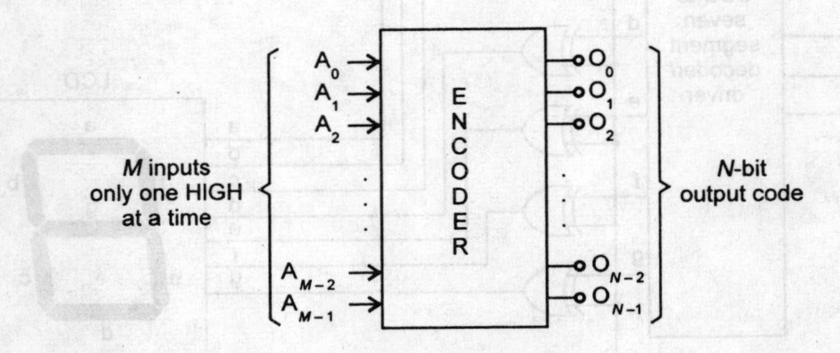

Fig. 7.51 Block diagram of encoder.

Octal-to-Binary Encoder

A binary-to-octal decoder (3-line to 8-line decoder) accepts a 3-bit input code and activates one of 8 output lines corresponding to that code. An octal-to-binary encoder (8-line to 3-line encoder) accepts 8 input lines and produces a 3-bit output code corresponding to the activated input. Figure 7.52 shows the truth table and the logic circuit for an octal-to-binary encoder with active HIGH inputs.

Octal digits		Binary A_2	A_1	A_0
D_0	0	0	0	0
D_1	1	0	0	1
D_2	2	0	1	0
D_3	3	0	1	1
D_4	4	1	0	0
D_5	5	1	0	1
D_6	6	1	1	0
D_7	7	1	1	1

(a) Truth table

Octal inputs
D_1 D_2 D_3 D_4 D_5 D_6 D_7
A_2
A_1
A_0
Binary outputs

(b) Logic diagram

Fig. 7.52 Octal-to-binary encoder.

From the truth table, we see that A_2 is a 1 if any of the digits D_4 or D_5 or D_6 or D_7 is a 1. Therefore,

$$A_2 = D_4 + D_5 + D_6 + D_7$$

Similarly

$$A_1 = D_2 + D_3 + D_6 + D_7$$

and

$$A_0 = D_1 + D_3 + D_5 + D_7$$

We see that D_0 is not present in any of the expressions. So, D_0 is a don't care.

Decimal-to-BCD Encoder

This type of encoder has 10 inputs—one for each decimal digit, and 4 outputs corresponding to the BCD code as shown in Fig. 7.53a. This is a basic 10-line to 4-line encoder. The BCD code is listed in the truth table (Fig. 7.53b) and from this we can determine the relationships between each BCD bit and the decimal digits. There is no explicit input for a decimal 0. The BCD output is 0000 when the decimal inputs 1–9 are all 0.

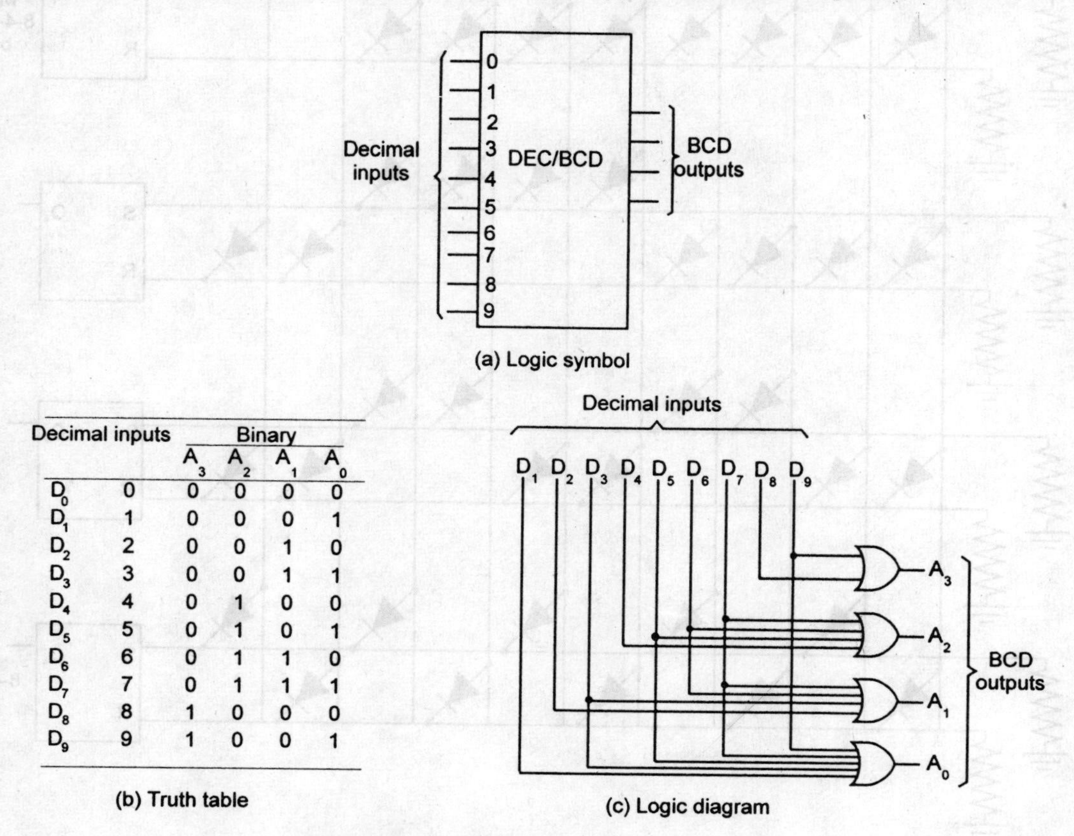

(a) Logic symbol

Decimal inputs		Binary A_3	A_2	A_1	A_0
D_0	0	0	0	0	0
D_1	1	0	0	0	1
D_2	2	0	0	1	0
D_3	3	0	0	1	1
D_4	4	0	1	0	0
D_5	5	0	1	0	1
D_6	6	0	1	1	0
D_7	7	0	1	1	1
D_8	8	1	0	0	0
D_9	9	1	0	0	1

(b) Truth table

(c) Logic diagram

Fig. 7.53 Decimal-to-BCD encoder.

The logic circuit of the decoder is shown in Fig. 7.53c. From the table, we get

$$A_3 = D_8 + D_9$$
$$A_2 = D_4 + D_5 + D_6 + D_7$$
$$A_1 = D_2 + D_3 + D_6 + D_7$$
$$A_0 = D_1 + D_3 + D_5 + D_7 + D_9$$

7.20 KEYBOARD ENCODERS

Figure 7.54 shows a typical keyboard encoder consisting of a diode matrix, used to encode the 10 decimal digits in 8-4-2-1 BCD.

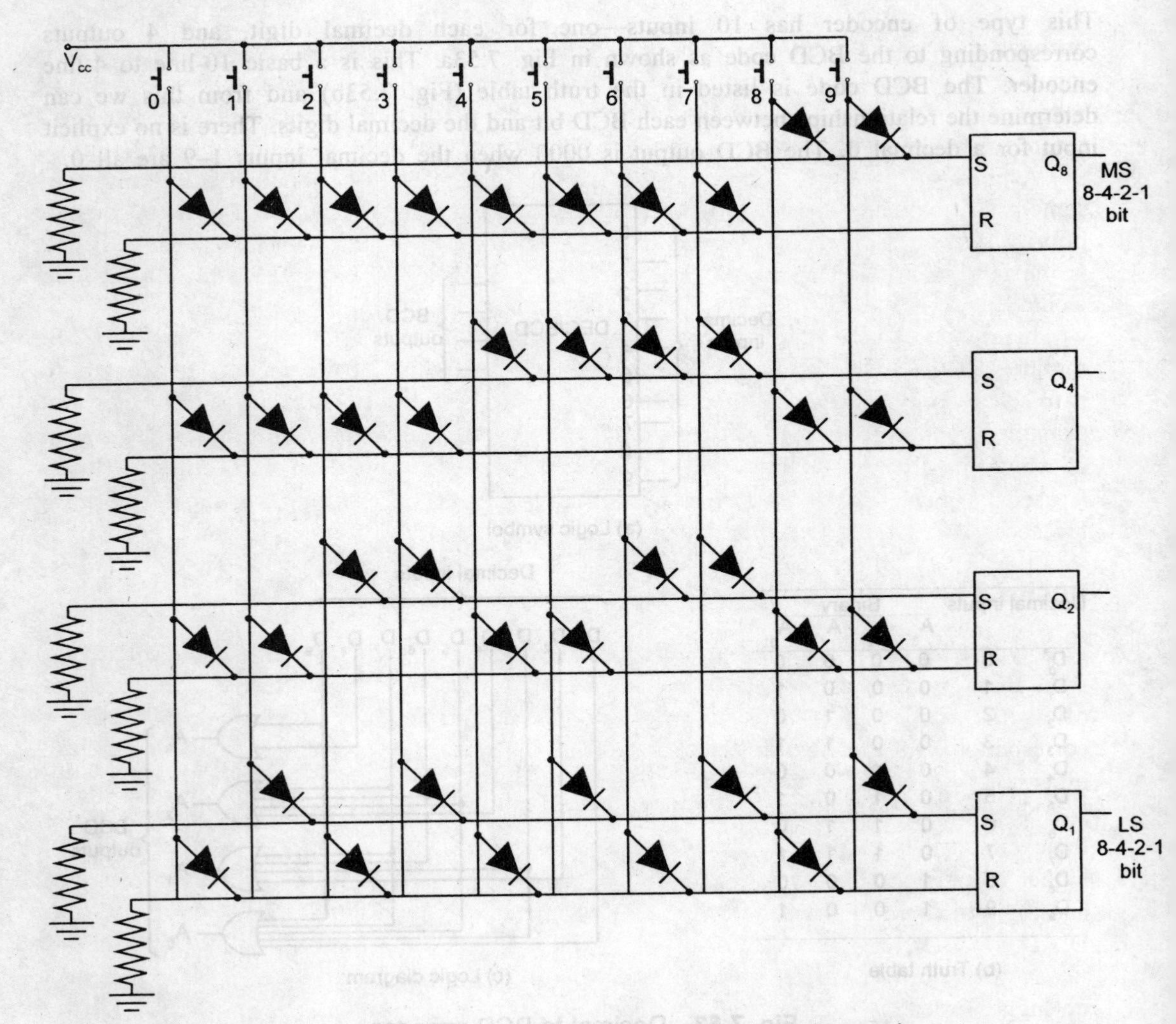

Fig. 7.54 A keyboard encoder employing a diode matrix.

The S-R flip-flops are used to store the BCD output. When a key corresponding to one of the decimal digits is pressed, a positive voltage forward biases the selected diodes connected to the SET(S) and RESET(R) inputs of the flip-flops. The diodes are so arranged that each flip-flop sets or resets, as necessary to produce the 4-bit code corresponding to the decimal digit. For example, when the key 7 is pressed, the diodes connected to the S inputs of Q_4, Q_2, and Q_1 are forward biased, as is that connected to the R input of Q_8. Thus, the output is 0111. Note that, the diode configuration at each S and R input is essentially a diode OR gate. Diode matrix encoders are found on printed circuit boards of many devices having a keyboard as the means of data entry.

7.21 PRIORITY ENCODERS

The encoders discussed so far will operate correctly, provided that one and only one decimal input is HIGH at any given time. In some practical systems, two or more decimal inputs may inadvertently become HIGH at the same time. For example, a person operating a keyboard might press a second key before releasing the first. Let us say he presses key 3 before releasing key 4. In such a case the output will be 7_{10} (0111) instead of being 4_{10} or 3_{10}.

A priority encoder is a logic circuit that responds to just one input in accordance with some priority system, among all those that may be simultaneously HIGH. The most common priority system is based on the relative magnitudes of the inputs; whichever decimal input is the largest, is the one that is encoded. Thus, in the above example, a priority encoder would encode decimal 4 if both 3 and 4 are simultaneously HIGH.

In some practical applications, priority encoders may have several inputs that are routinely HIGH at the same time, and the principal function of the encoder in those cases is to select the input with the highest priority. This function is called ***arbitration***. A common example is found in computer systems, where there are numerous input devices and several of which may attempt to supply data to the computer at the same time. A priority encoder is used to enable that input device which has the highest priority among those competing for access to the computer at the same time.

Decimal-to-BCD Priority Encoder

This type of encoder performs the same basic function of encoding the decimal digits into 4-bit BCD outputs, as that performed by a normal decimal-to-BCD encoder. It, however, offers the additional facility of providing priority. That is, it produces a BCD output corresponding to the highest order decimal digit appearing on the inputs and ignores all others.

Now, let us look at the requirements for the priority detection logic. The purpose of this logic circuitry is to prevent a lower order digit input from disrupting the encoding of a higher order digit. This is accomplished by using inhibit gates. Referring to the truth table and logic diagram of the decimal-to-BCD encoder of Fig. 7.53, note that A_0 is HIGH when D_1, D_3, D_5, D_7, or D_9 is HIGH.

In the priority encoder, input digit 1 will be allowed to activate the A_0 output, only if no higher order digits other than those that also activate A_0 are HIGH, i.e. D_2, D_4, D_6, and

D_8 must be LOW. If any of those are HIGH, then A_0 will be LOW. This can be stated as follows:

A_0 is HIGH if,

D_1 is HIGH and D_2, D_4, D_6, and D_8 are LOW

OR

D_3 is HIGH and D_4, D_6, and D_8 are LOW

OR

D_5 is HIGH and D_6 and D_8 are LOW

OR

D_7 is HIGH and D_8 is LOW

OR

D_9 is HIGH.

Thus,

$$A_0 = D_1\overline{D}_2\overline{D}_4\overline{D}_6\overline{D}_8 + D_3\overline{D}_4\overline{D}_6\overline{D}_8 + D_5\overline{D}_6\overline{D}_8 + D_7\overline{D}_8 + D_9$$

Similarly, A_1 is HIGH when D_2, D_3, D_6, or D_7 is HIGH. So, in the priority encoder A_1 will be HIGH if,

D_2 is HIGH and D_4, D_5, D_8, and D_9 are LOW

OR

D_3 is HIGH and D_4, D_5, D_8, and D_9 are LOW

OR

D_6 is HIGH and D_8 and D_9 are LOW

OR

D_7 is HIGH and D_8 and D_9 are LOW.

Thus,

$$A_1 = D_2\overline{D}_4\overline{D}_5\overline{D}_8\overline{D}_9 + D_3\overline{D}_4\overline{D}_5\overline{D}_8\overline{D}_9 + D_6\overline{D}_8\overline{D}_9 + D_7\overline{D}_8\overline{D}_9$$

Also, A_2 is HIGH when D_4, D_5, D_6, or D_7 is HIGH. So, in the priority encoder A_2 will become HIGH if,

D_4 is HIGH and D_8 and D_9 are LOW

OR

D_5 is HIGH and D_8 and D_9 are LOW

OR

D_6 is HIGH and D_8 and D_9 are LOW

OR

D_7 is HIGH and D_8 and D_9 are LOW.

Thus,

$$A_2 = D_4\overline{D}_8\overline{D}_9 + D_5\overline{D}_8\overline{D}_9 + D_6\overline{D}_8\overline{D}_9 + D_7\overline{D}_8\overline{D}_9$$

Finally, A_3 is HIGH if D_8 is HIGH or if D_9 is HIGH. So, in the priority encoder A_3 will be HIGH if D_8 is HIGH OR D_9 is HIGH.

Thus,

$$A_3 = D_8 + D_9$$

The truth table operation of the priority encoder is shown in Table 7.2. The truth table clearly shows that the magnitudes of the decimal inputs determine their priorities. If any decimal input is active it is encoded, provided all higher value inputs are inactive regardless of the states of all lower value inputs.

Table 7.2 Truth table of a decimal-to-BCD priority encoder

Decimal inputs (× = Don't care)									BCD outputs			
D_1	D_2	D_3	D_4	D_5	D_6	D_7	D_8	D_9	A_3	A_2	A_1	A_0
1	1	1	1	1	1	1	1	1	1	1	1	1
×	×	×	×	×	×	×	×	0	0	1	1	0
×	×	×	×	×	×	×	0	1	0	1	1	1
×	×	×	×	×	×	0	1	1	1	0	0	0
×	×	×	×	×	0	1	1	1	1	0	0	1
×	×	×	×	0	1	1	1	1	1	0	1	0
×	×	×	0	1	1	1	1	1	1	0	1	1
×	×	0	1	1	1	1	1	1	1	1	0	0
×	0	1	1	1	1	1	1	1	1	1	0	1
0	1	1	1	1	1	1	1	1	1	1	1	0

Figure 7.55 shows the logic diagram for the 74147 priority encoder. In this device, the inputs and the outputs are active LOW.

Octal-to-Binary Priority Encoder

The octal code is often used at the inputs of digital circuits that require manual entering of long binary words. Priority encoder 74148 IC has been designed to achieve this operation. Its block diagram is given in Fig. 7.56. This circuit has active LOW inputs and active LOW outputs. The enable input and the Gray outputs which are also active LOW are used to cascade circuits to handle more inputs. A hexadecimal-to-binary encoder which is also a very useful circuit because of the widespread use of the hexadecimal code in computers, microprocessors, etc. can be designed using this facility.

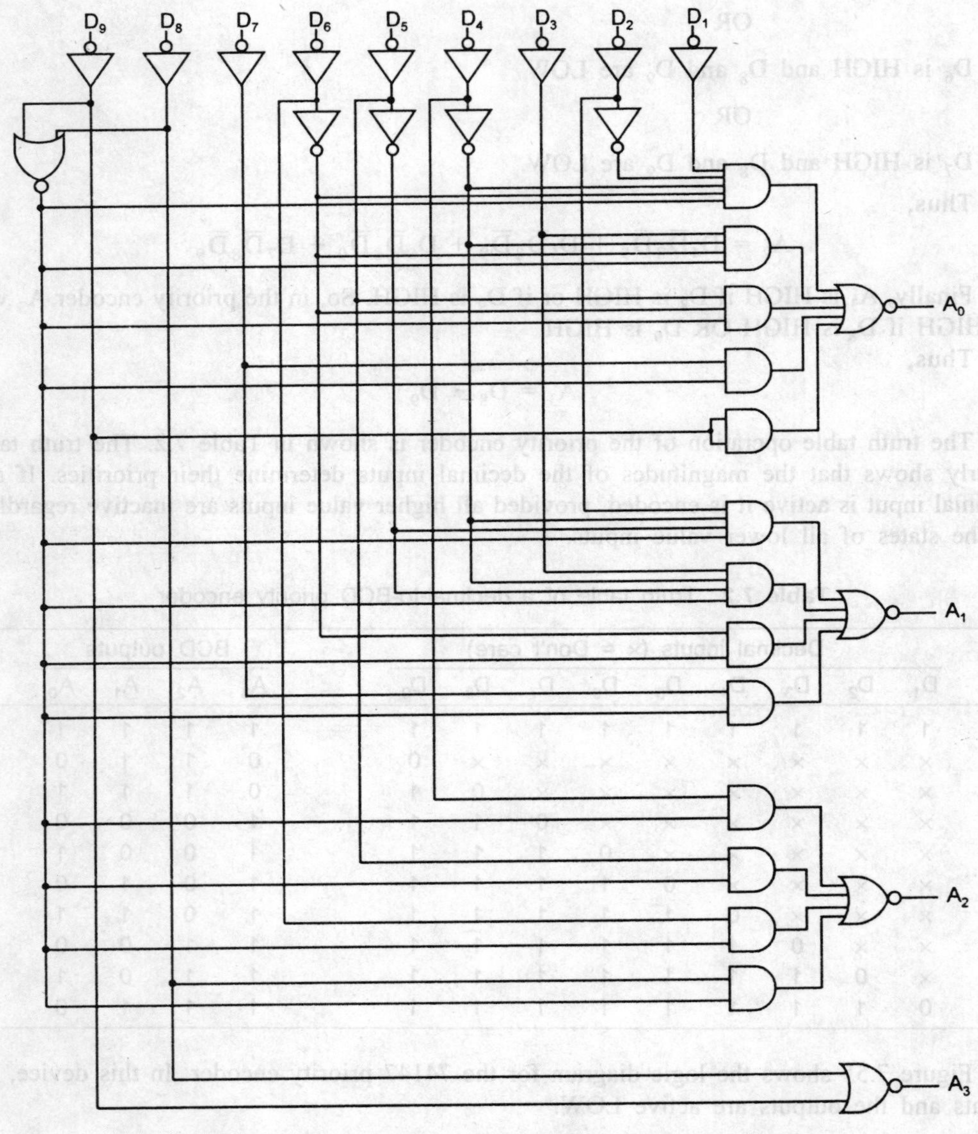

Fig. 7.55 Logic diagram of the 74147 priority encoder.

EXAMPLE 7.8 Design a hexadecimal-to-binary encoder using 74148 encoders and 74157 multiplexer.

Solution

Since there are 16 symbols (0–F) in the hexadecimal number system, two 74148 encoders are required. Hexadecimal inputs 0–7 are applied to IC1 input lines and hexadecimal inputs 8–F to IC2 input lines. Whenever one of the inputs of IC2 is active (LOW), IC1 must be

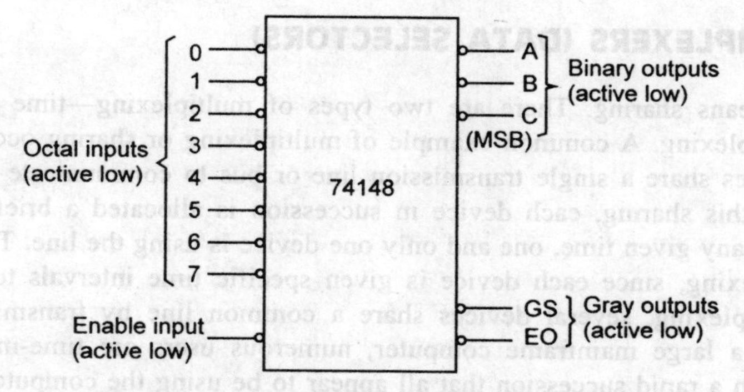

Fig. 7.56 Octal-to-binary priority encoder.

disabled. On the other hand, if all the inputs of IC2 are HIGH, then IC1 must be enabled. This is achieved by connecting the EO line of IC2 to the EI line of IC1. A quad 2:1 multiplexer is required to get the proper 4-bit binary outputs. The complete circuit is shown in Fig. 7.57. The GS output of 74148 goes LOW whenever one of its inputs is active. Therefore, the GS of IC2 is connected to the SELECT input of 74157. The 74157 selects its A inputs if the SELECT input is LOW, otherwise B inputs are selected. The outputs of the multiplexer are the required binary outputs and are active LOW. This circuit is also a priority encoder.

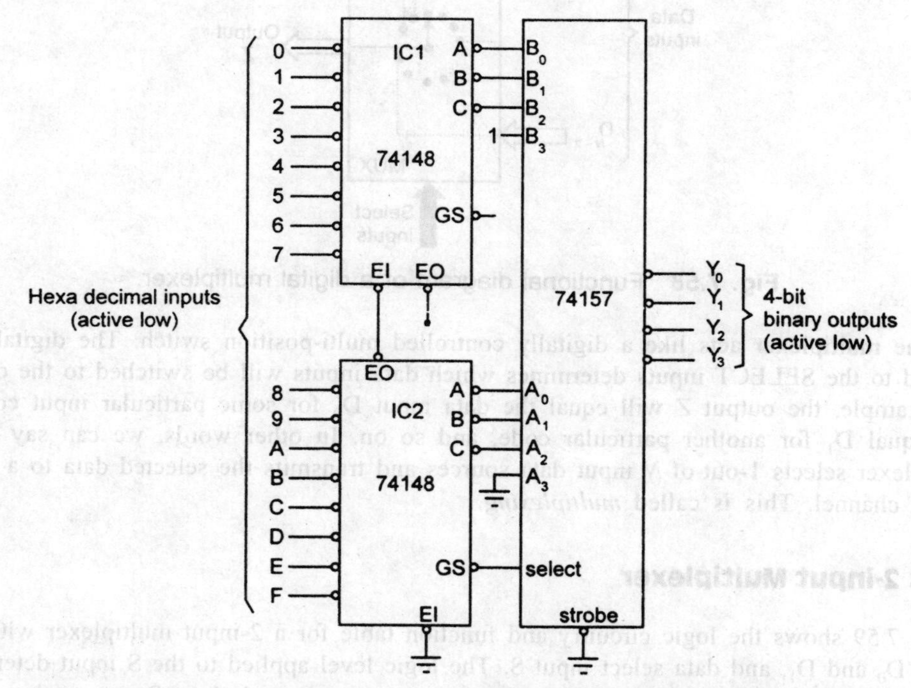

Fig. 7.57 Hexadecimal-to-binary encoder using 74148s and 74157.

7.22 MULTIPLEXERS (DATA SELECTORS)

Multiplexing means sharing. There are two types of multiplexing—time multiplexing and frequency multiplexing. A common example of multiplexing or sharing occurs when several peripheral devices share a single transmission line or bus to communicate with a computer. To accomplish this sharing, each device in succession is allocated a brief time to send or receive data. At any given time, one and only one device is using the line. This is an example of time multiplexing, since each device is given specific time intervals to use the line. In frequency multiplexing, several devices share a common line by transmitting at different frequencies. In a large mainframe computer, numerous users are time-multiplexed to the computer in such a rapid succession that all appear to be using the computer simultaneously.

A multiplexer (MUX) or data selector is a logic circuit that accepts several data inputs and allows only one of them at a time to get through the output. The routing of the desired data input to the output is controlled by SELECT inputs (sometimes referred to as ADDRESS inputs). Figure 7.58 shows the functional diagram of a general multiplexer. In this diagram, the inputs and outputs are drawn as large arrows to indicate that they may constitute one or more signal lines.

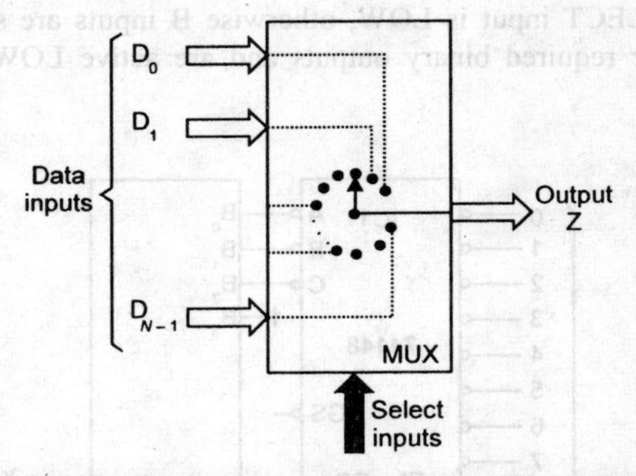

Fig. 7.58 Functional diagram of a digital multiplexer.

The multiplexer acts like a digitally controlled multi-position switch. The digital code applied to the SELECT inputs determines which data inputs will be switched to the output. For example, the output Z will equal the data input D_0 for some particular input code; Z will equal D_1 for another particular code, and so on. In other words, we can say that a multiplexer selects 1-out-of-*N* input data sources and transmits the selected data to a single output channel. This is called *multiplexing*.

Basic 2-input Multiplexer

Figure 7.59 shows the logic circuitry and function table for a 2-input multiplexer with data inputs D_0 and D_1, and data select input S. The logic level applied to the S input determines which AND gate is enabled, so that its data input passes through the OR gate to the output.

The output, $Z = \bar{S}D_0 + SD_1$.

When S = 0, AND gate 1 is enabled and AND gate 2 is disabled. So, $Z = D_0$.

When S = 1, AND gate 1 is disabled and AND gate 2 is enabled. So, $Z = D_1$.

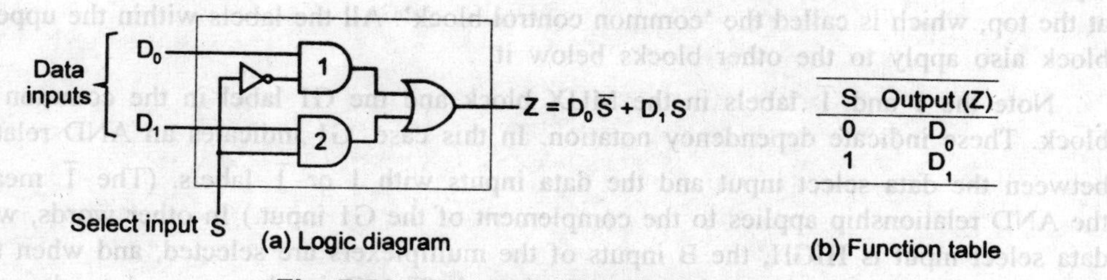

Fig. 7.59 The 2-input multiplexer.

The 74157 Quadruple 2-Input Data Selector/Multiplexer

The 74157 consists of four separate 2-input multiplexers on a single chip. Each of the four multiplexers share a common data select line and a common $\overline{E}$ (Enable) line as shown in Fig. 7.60a. Because there are only two inputs to be selected from each multiplexer, a single data select input is sufficient. A LOW on the $\overline{E}$ input allows the selected input data

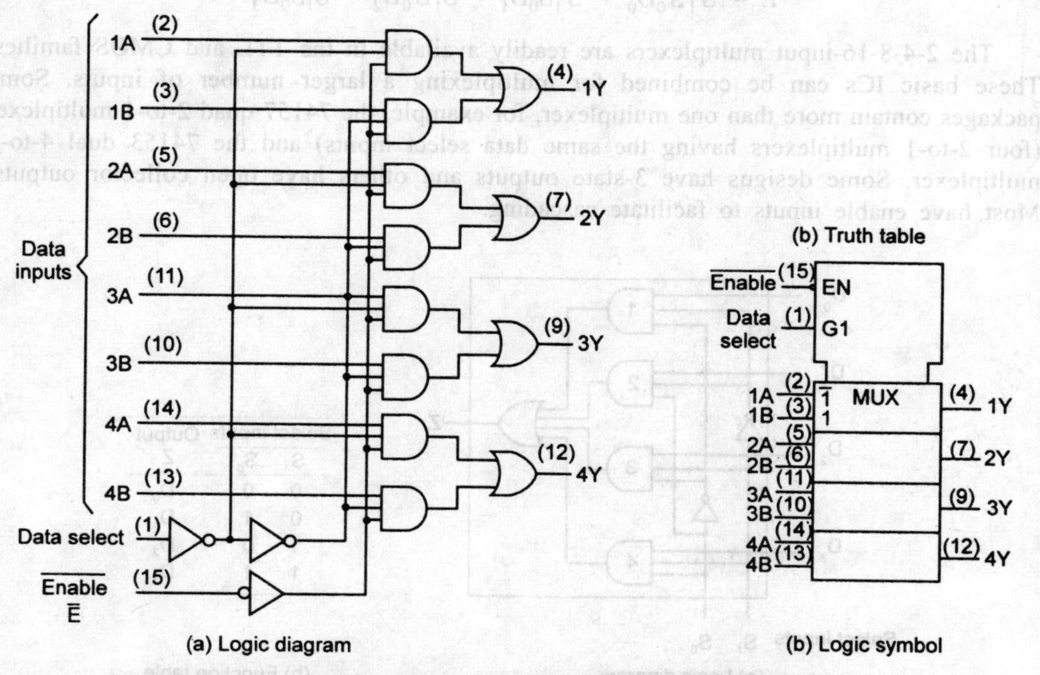

Fig. 7.60 The 74157 quadruple 2-input data selector/multiplexer.

to pass through to the output. A HIGH on the $\overline{E}$ input prevents data from going through to the output, i.e. it disables the multiplexers. The ANSI/IEEE logic symbol for this device is shown in Fig. 7.60b.

Note that the four multiplexers are indicated by the partitioned output line and that the inputs common to all the four multiplexers are indicated as inputs to the 'notched' block at the top, which is called the 'common control block'. All the labels within the upper MUX block also apply to the other blocks below it.

Note the 1 and $\overline{1}$ labels in the MUX block and the G1 label in the common control block. These indicate dependency notation. In this case, G1 indicates an AND relationship between the data select input and the data inputs with 1 or $\overline{1}$ labels. (The $\overline{1}$ means that the AND relationship applies to the complement of the G1 input.) In other words, when the data select input is HIGH, the B inputs of the multiplexers are selected, and when the data select input is LOW, the A inputs are selected. The G is always used to denote AND dependency.

The 4-input Multiplexer

Figure 7.61a shows the logic circuitry for a 4-input multiplexer with data inputs D_0, D_1, D_2, and D_3, and data select inputs S_0 and S_1. The logic levels applied to the S_0 and S_1 inputs determine which AND gate is enabled, so that its data input passes through the OR gate to the output. The function table in Fig. 7.61b gives the output for the input select codes as

$$Z = \overline{S}_1\overline{S}_0D_0 + \overline{S}_1S_0D_1 + S_1\overline{S}_0D_2 + S_1S_0D_3$$

The 2-4-8-16-input multiplexers are readily available in the TTL and CMOS families. These basic ICs can be combined for multiplexing a larger number of inputs. Some packages contain more than one multiplexer, for example, the 74157 quad 2-to-1 multiplexer (four 2-to-1 multiplexers having the same data select inputs) and the 74153 dual 4-to-1 multiplexer. Some designs have 3-state outputs and others have open collector outputs. Most have enable inputs to facilitate cascading.

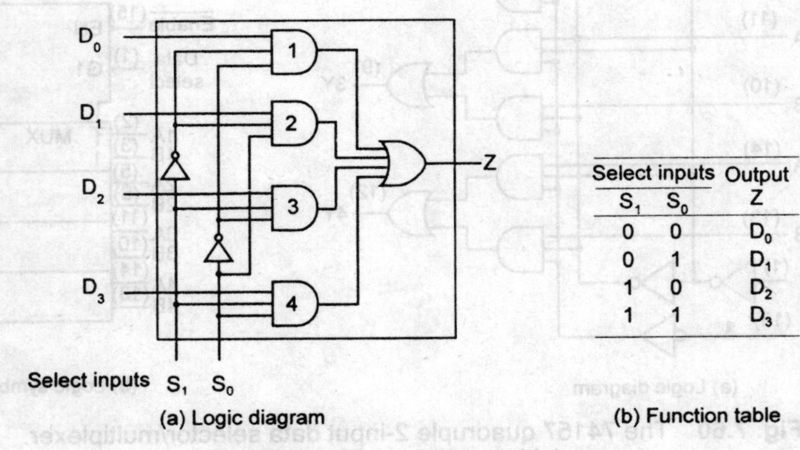

(a) Logic diagram

Select inputs		Output
S_1	S_0	Z
0	0	D_0
0	1	D_1
1	0	D_2
1	1	D_3

(b) Function table

Fig. 7.61 The 4-input multiplexer.

The 74151A 8-input Data Selector/Multiplexer

Figure 7.62a shows the logic diagram for an 8-input multiplexer. This multiplexer has 8 data inputs, 3 data select inputs and an enable input $\overline{E}$ (Enable). A LOW on the $\overline{E}$ input allows the selected input data to pass through to the output. When $\overline{E}$ = 1, the multiplexer is disabled, so that Z = 0, regardless of the select input code. Note that the output as well as its complement are available. The operation of the MUX is illustrated in the truth table of Fig. 7.62b.

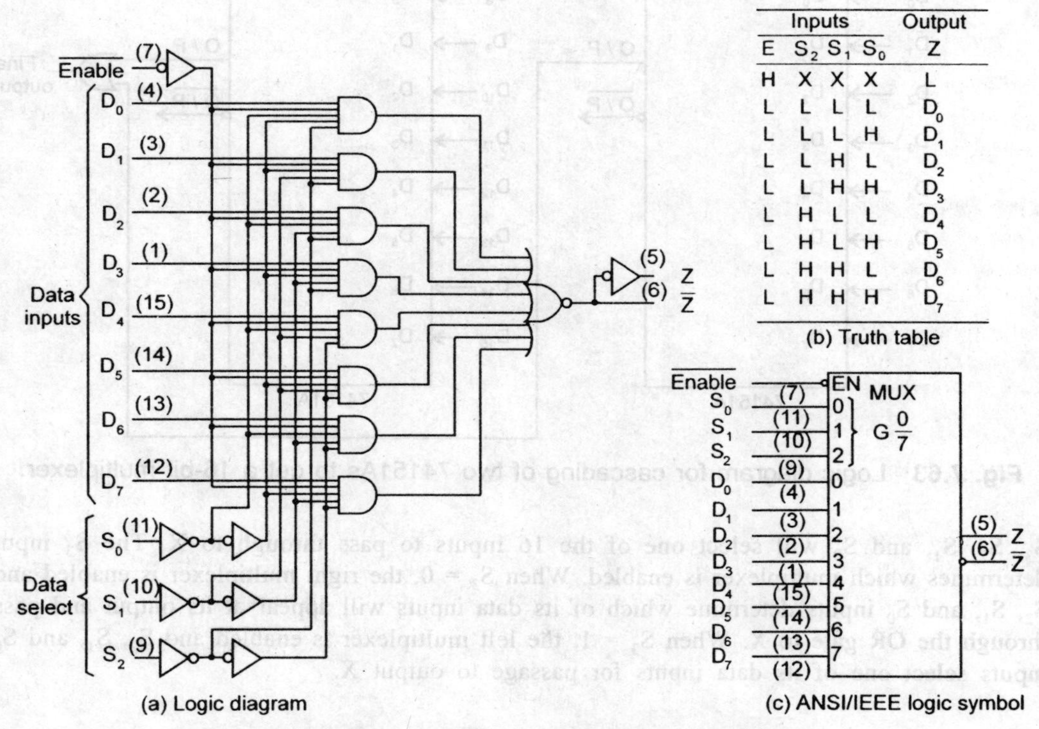

Inputs				Output
E	S_2	S_1	S_0	Z
H	X	X	X	L
L	L	L	L	D_0
L	L	L	H	D_1
L	L	H	L	D_2
L	L	H	H	D_3
L	H	L	L	D_4
L	H	L	H	D_5
L	H	H	L	D_6
L	H	H	H	D_7

(b) Truth table

Fig. 7.62 The 74151A 8-input data selector/multiplexer.

The ANSI/IEEE logic symbol for 74151A is shown in Fig. 7.62c. In this case, there is no need for a common control block on the logic symbol, because there is only one multiplexer to be controlled, not four as in the 74157. The G $\frac{0}{7}$ label within the logic symbol indicates the AND relationship between the data select input and each of the data inputs 0–7.

The 16-input Multiplexer from Two 8-input Multiplexers

Figure 7.63 shows an arrangement to use two 8-input multiplexers (74151A) to get a 16-input multiplexer. One OR gate and one inverter are also required. The four select inputs

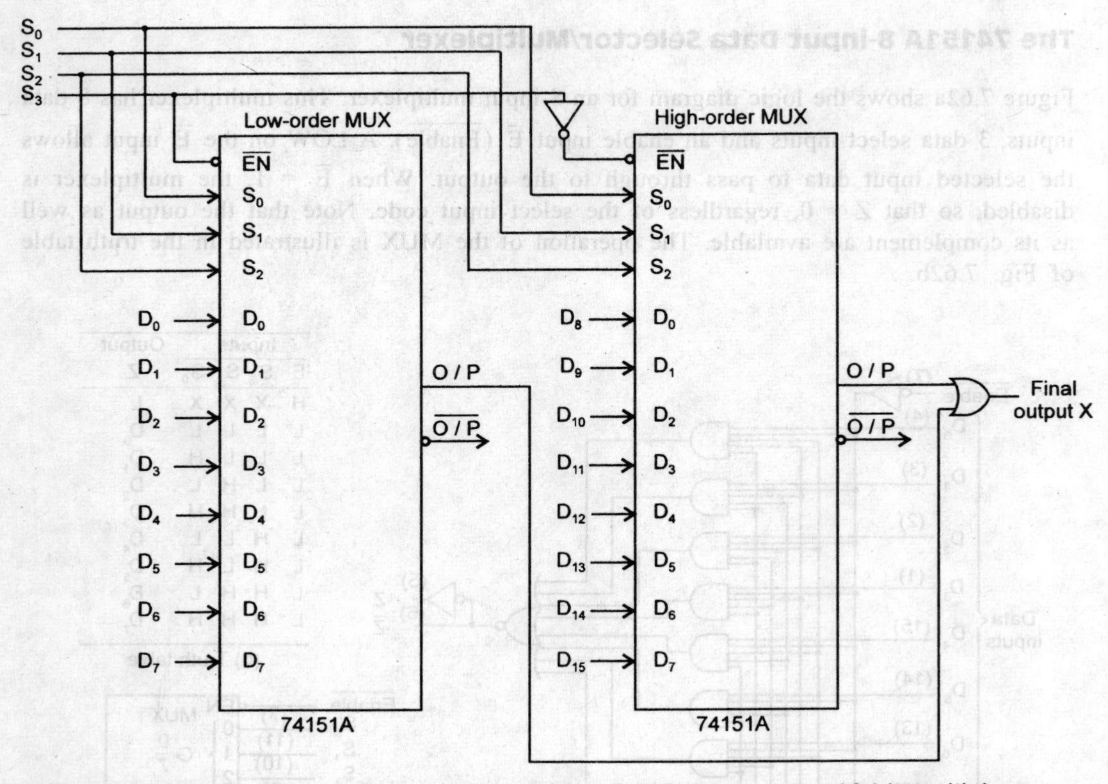

Fig. 7.63 Logic diagram for cascading of two 74151As to get a 16-bit multiplexer.

S_3, S_2, S_1, and S_0 will select one of the 16 inputs to pass through to X. The S_3 input determines which multiplexer is enabled. When $S_3 = 0$, the right multiplexer is enabled and S_2, S_1, and S_0 inputs determine which of its data inputs will appear at its output and pass through the OR gate to X. When $S_3 = 1$, the left multiplexer is enabled and S_2, S_1, and S_0 inputs select one of its data inputs for passage to output X.

7.23 APPLICATIONS OF MULTIPLEXERS

Multiplexers find numerous and varied applications in digital systems of all types. These applications include data selection, data routing, operation sequencing, parallel-to-serial conversion, waveform generation, and logic function generation.

Logic Function Generator

A multiplexer can be used in place of logic gates to implement a logic expression. It can be so connected that it duplicates the logic of any truth table, i.e. it can generate any Boolean algebraic function of a set of input variables. In such applications, the multiplexer can be viewed as a function generator, because we can easily set or change the logic function it implements. One advantage of using a multiplexer in place of logic gates is that, a single integrated circuit can perform a function that might otherwise require numerous integrated circuits. Morever, it is very easy to change the logic function implemented, if and when redesign of a system becomes necessary.

The first step in the design of a function generator using a multiplexer is to construct a truth table for the function to be implemented. Then, connect logical 1 to each data input of the multiplexer corresponding to each combination of the input variables, for which the truth table shows the function to be equal to 1. Logical 0 is connected to the remaining data inputs. The variables themselves are connected to the data select inputs of the multiplexer. For example, suppose the truth table specifies that the function F equals 1 for the input combination 110. So, when $S_2S_1S_0$ = 110, the data input 6, which is connected to logical 1, will be selected. This will route a 1 to the output of the multiplexer.

EXAMPLE 7.9 Use a multiplexer to implement the logic function $F = A \oplus B \oplus C$.

Solution

The truth table for F and the logic diagram to implement F are shown in Fig. 7.64. Since there are three input variables, we can use a multiplexer with three data select inputs (an 8-to-1 MUX). The truth table shows the use of data select inputs S_2, S_1, and S_0 for input variables A, B, and C respectively. Since F = 1 when ABC = 001, 010, 100, and 111, we

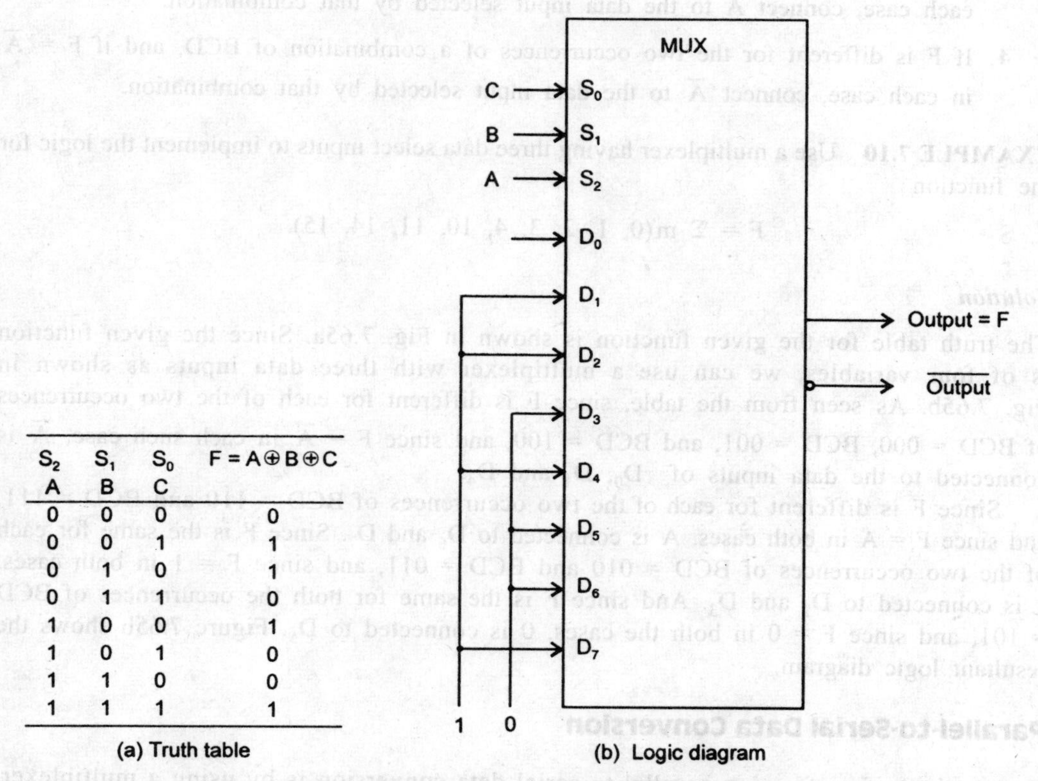

S_2 A	S_1 B	S_0 C	$F = A \oplus B \oplus C$
0	0	0	0
0	0	1	1
0	1	0	1
0	1	1	0
1	0	0	1
1	0	1	0
1	1	0	0
1	1	1	1

(a) Truth table

(b) Logic diagram

Fig. 7.64 Use of 74151A to implement the logic function $F = A \oplus B \oplus C$.

connect logical 1 to data inputs D_1, D_2, D_4, and D_7. Logical 0 is connected to other data inputs D_0, D_3, D_5, and D_6. When the data select inputs are any of the combinations for which F = 1, the output will be 1, and when the data select inputs are any of the combinations for which F = 0, the output will be 0. Thus, the multiplexer behaves in exactly the same way that a set of logic gates implementing the function F would behave.

In general, a multiplexer with *n*-data select inputs can implement any function of $n + 1$ variables. The key to this design is to use the most significant input variable and its complement to drive some of the data inputs. Suppose, we wish to implement a 4-variable logic function using a multiplexer with three data select inputs. Let the input variables be A, B, C, and D; A is the MSB. A truth table for the function F (A, B, C, D) is constructed. In the truth table, we note that BCD progresses twice through the sequence 000, 001,..., 111; once with A = 0 and again with A = 1. The following rules are used to determine the connections that should be made to the data inputs of the multiplexer.

1. If F = 0 both times when the same combination of BCD occurs, connect logical 0 to the data input selected by that combination.
2. If F = 1 both times when the same combination of BCD occurs, connect logical 1 to the data input selected by that combination.
3. If F is different for the two occurrences of a combination of BCD, and if F = A in each case, connect A to the data input selected by that combination.
4. If F is different for the two occurrences of a combination of BCD, and if $F = \overline{A}$ in each case, connect $\overline{A}$ to the data input selected by that combination.

EXAMPLE 7.10 Use a multiplexer having three data select inputs to implement the logic for the function

$$F = \Sigma\, m(0, 1, 2, 3, 4, 10, 11, 14, 15).$$

Solution

The truth table for the given function is shown in Fig. 7.65a. Since the given function is of four variables, we can use a multiplexer with three data inputs as shown in Fig. 7.65b. As seen from the table, since F is different for each of the two occurrences of BCD = 000, BCD = 001, and BCD = 100, and since $F = \overline{A}$ in each such case, $\overline{A}$ is connected to the data inputs of D_0, D_1 and D_4.

Since F is different for each of the two occurrences of BCD = 110 and BCD = 111, and since F = A in both cases, A is connected to D_6 and D_7. Since F is the same for each of the two occurrences of BCD = 010 and BCD = 011, and since F = 1 in both cases, 1 is connected to D_2 and D_3. And since F is the same for both the occurrences of BCD = 101, and since F = 0 in both the cases, 0 is connected to D_5. Figure 7.65b shows the resultant logic diagram.

Parallel-to-Serial Data Conversion

One method of performing parallel-to-serial data conversion is by using a multiplexer. Figure 7.66a shows the logic diagram to convert an 8-bit parallel data into serial form using

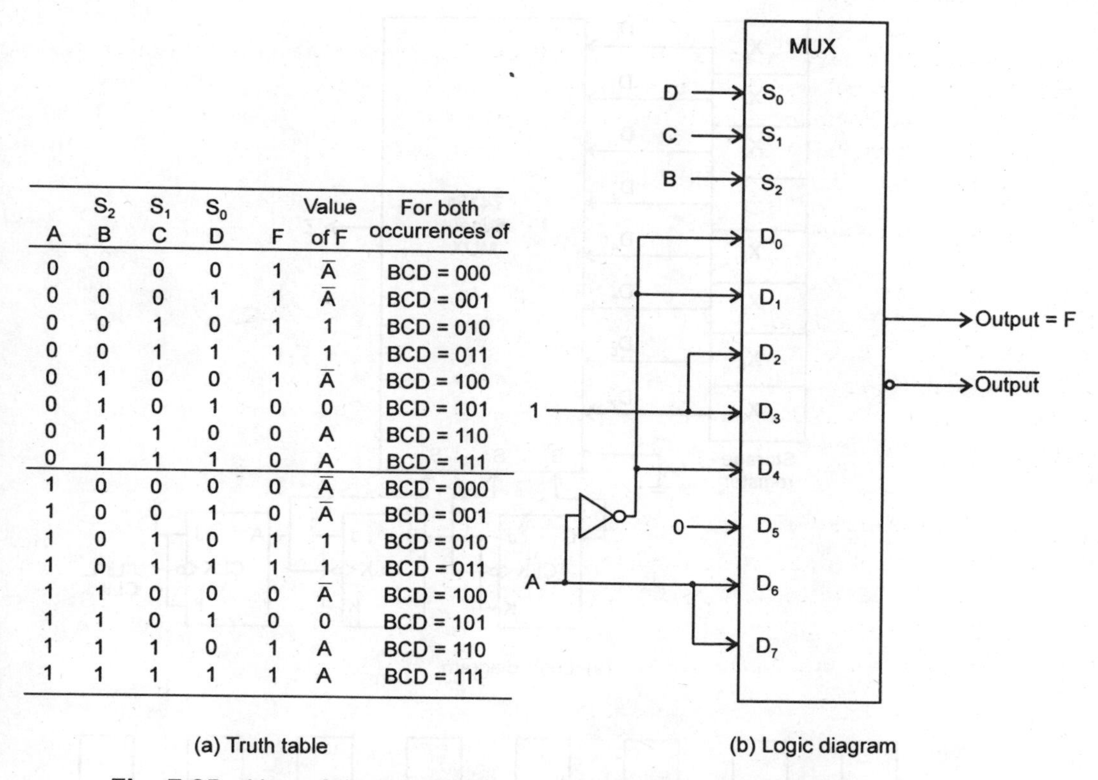

A	S_2 B	S_1 C	S_0 D	F	Value of F	For both occurrences of
0	0	0	0	1	$\overline{A}$	BCD = 000
0	0	0	1	1	$\overline{A}$	BCD = 001
0	0	1	0	1	1	BCD = 010
0	0	1	1	1	1	BCD = 011
0	1	0	0	1	$\overline{A}$	BCD = 100
0	1	0	1	0	0	BCD = 101
0	1	1	0	0	A	BCD = 110
0	1	1	1	0	A	BCD = 111
1	0	0	0	0	$\overline{A}$	BCD = 000
1	0	0	1	0	$\overline{A}$	BCD = 001
1	0	1	0	1	1	BCD = 010
1	0	1	1	1	1	BCD = 011
1	1	0	0	0	$\overline{A}$	BCD = 100
1	1	0	1	0	0	BCD = 101
1	1	1	0	1	A	BCD = 110
1	1	1	1	1	A	BCD = 111

(a) Truth table

(b) Logic diagram

Fig. 7.65 Use of the multiplexer to implement the logic for the function $F = \Sigma\, m(0, 1, 2, 3, 4, 10, 11, 14, 15)$.

an 8-input multiplexer. The data are present in parallel form at the outputs of the register X and are fed to the 8-input multiplexer. A 3-bit (mod-8) counter is used to provide the select code bits $S_2S_1S_0$, so that they cycle through from 000 to 111 as clock pulses are applied. In this way, the output of the multiplexer will be X_0 during the first clock period, X_1 during the second clock period, and so on. The output Z is a waveform, which is a serial representation of the parallel input data. The waveforms in Fig. 7.66b are for the case $X_7X_6X_5X_4X_3X_2X_1X_0 = 11001010$. This conversion process takes a total of eight clock cycles. Note that X_0 (the LSB) is transmitted first and X_7 (MSB) is transmitted last.

Multiplexing Seven Segment Displays

Optical output displays such as seven segment displays, typically consume considerable power. In applications where power consumption is a major concern, such as pocket calculators, and where several displays must be illuminated simultaneously, multiplexers are used to reduce power consumption. Instead of illuminating all the displays simultaneously, a multiplexer selects each display in turn. If the rate at which the displays are illuminated is fast enough (usually around 30 times per second), the human eye will not be able to detect any flicker, and it will appear that all displays are illuminated simultaneously. The power consumption will be no greater than that of a single display illuminated continuously.

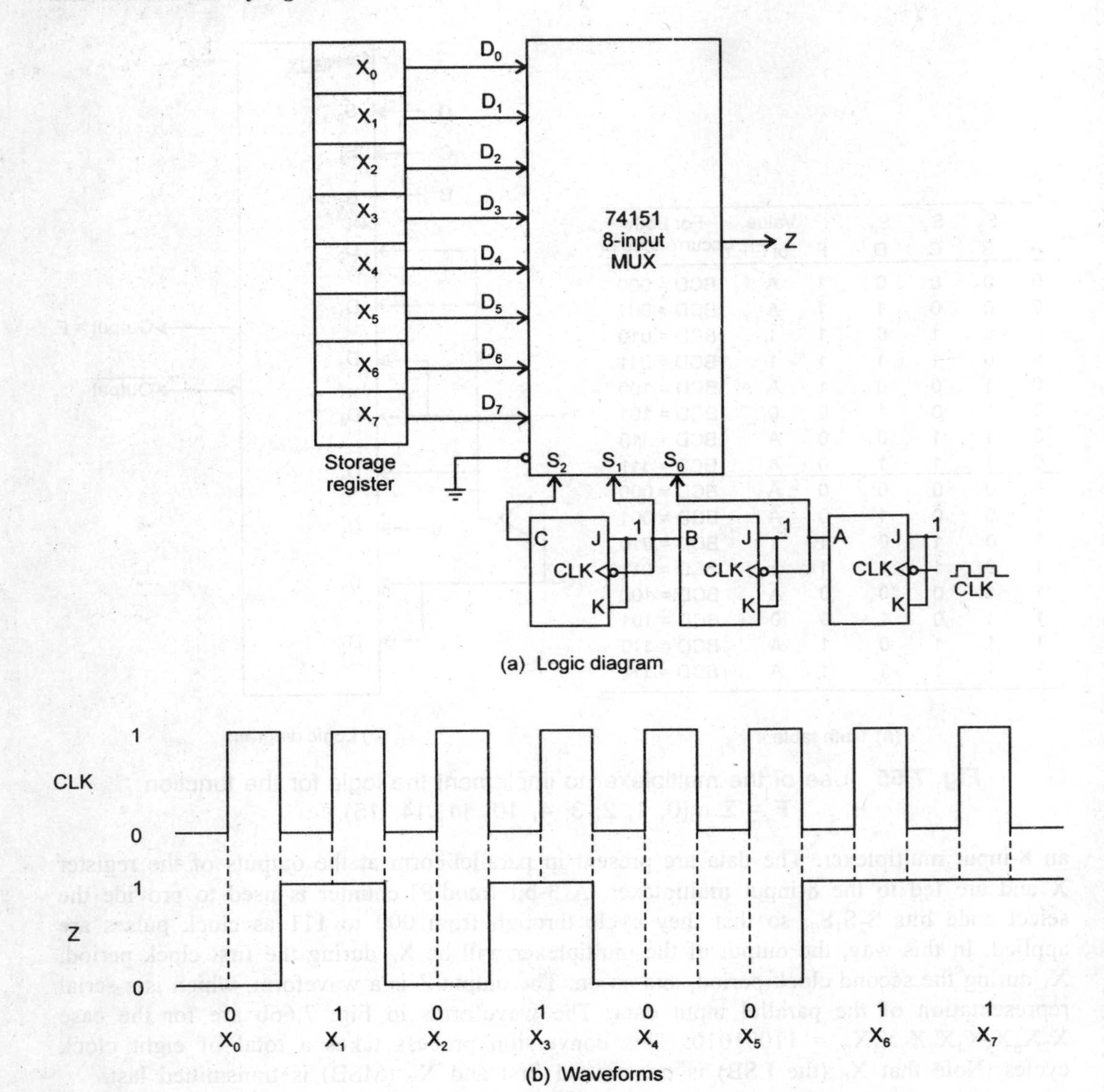

Fig. 7.66 Parallel-to-serial data converter.

Figure 7.67 shows a simplified method of multiplexing BCD numbers to a seven segment display. In this example, 2-digit numbers are displayed on the segment read-out using a single BCD to seven segment decoder. This basic method of multiplexing can be extended to displays with any number of inputs. The basic operation is as follows:

Two BCD digits (A and B) are applied to the multiplexer inputs. A square wave is applied to the data select line and when it is LOW, the A bits ($A_3A_2A_1A_0$) are routed to the inputs of the 7449 BCD-to-seven segment decoder. The LOW on the data select also puts a LOW on the 1 input of the 74139, 2-line to 4-line decoder, thus activating its 0 output

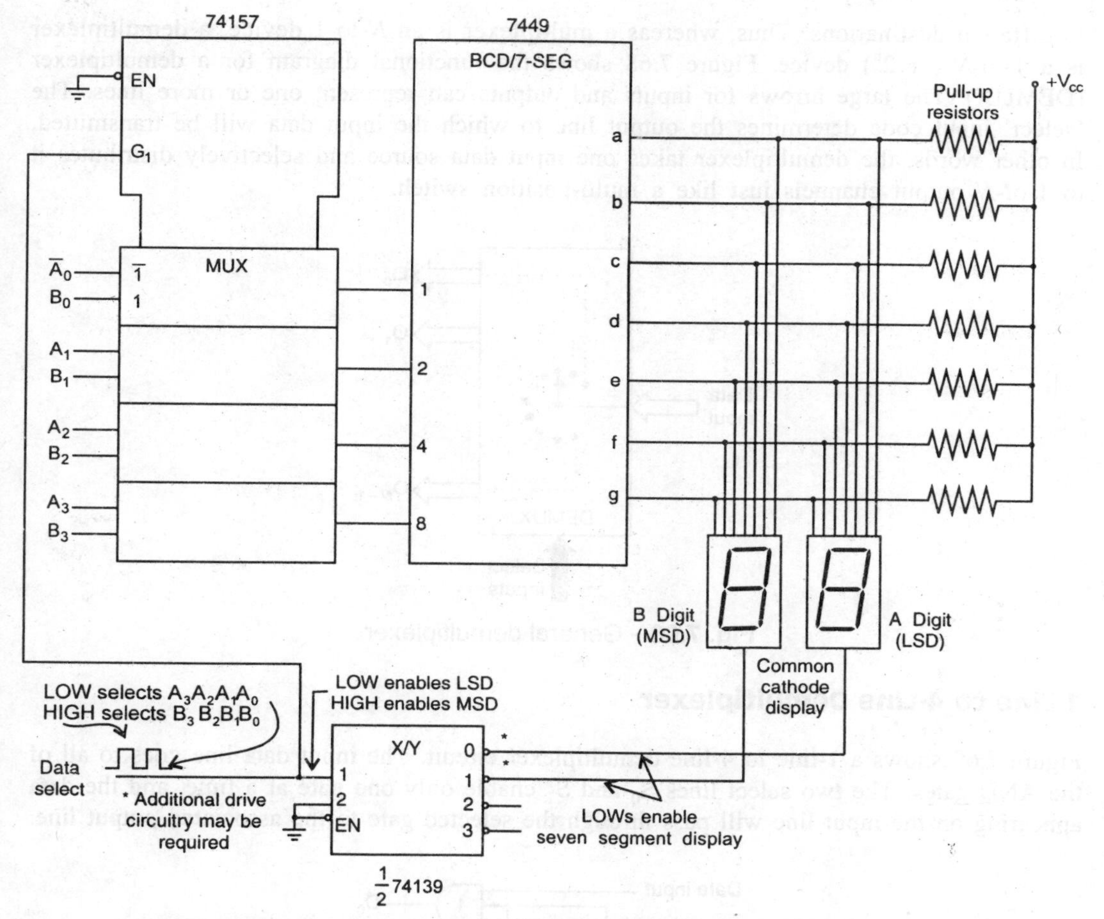

Fig. 7.67 Simplified seven segment display multiplexing logic.

and enabling the A-digit display by effectively connecting its common terminal to ground. The A-digit is now ON and the B-digit is OFF.

When the data select line goes HIGH, the B bits ($B_3B_2B_1B_0$) are routed to the inputs of the BCD-to-seven segment decoder. Also, the 74139 decoder's 1 output is activated, thus, enabling the B-digit display. The B-digit is now ON and the A-digit is OFF. The cycle repeats at the frequency of the data select square wave. The frequency must be high enough to prevent visual flicker as the digit displays are multiplexed.

7.24 DEMULTIPLEXERS (DATA DISTRIBUTORS)

A multiplexer takes several inputs and transmits one of them to the output. A demultiplexer performs the reverse operation; it takes a single input and distributes it over several outputs. So, a demultiplexer can be thought of as a 'distributor', since it transmits the same data to different destinations. Thus, whereas a multiplexer is an N-to-1 device, a demultiplexer is a 1-to-N (or 2^n) device. Figure 7.68 shows the functional diagram for a demultiplexer (DEMUX). The large arrows for inputs and outputs can represent one or more lines. The 'select' input code determines the output line to which the input data will be transmitted. In other words, the demultiplexer takes one input data source and selectively distributes it to 1-of-N output channels just like a multi-position switch.

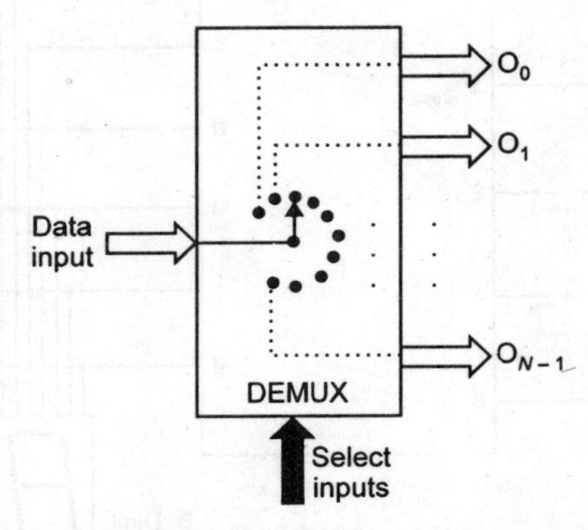

Fig. 7.68 General demultiplexer.

1-Line to 4-Line Demultiplexer

Figure 7.69 shows a 1-line to 4-line demultiplexer circuit. The input data line goes to all of the AND gates. The two select lines S_0 and S_1 enable only one gate at a time, and the data appearing on the input line will pass through the selected gate to the associated output line.

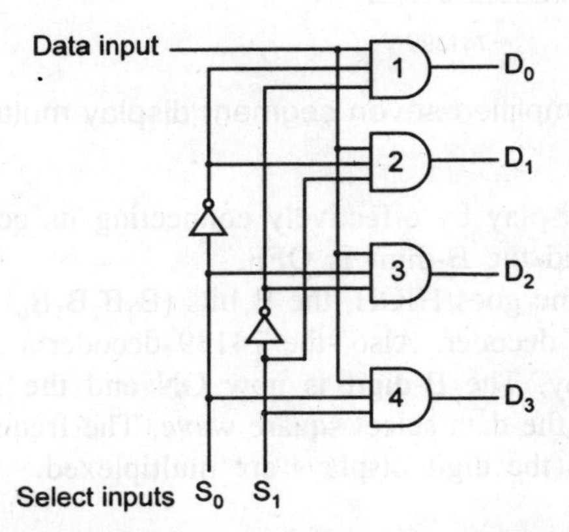

Fig. 7.69 Logic diagram of a 1-line to 4-line demultiplexer.

1-Line to 8-Line Demultiplexer

Figure 7.70a shows the logic diagram for a demultiplexer that distributes one input line to eight output lines. The single data input line D is connected to all eight AND gates,

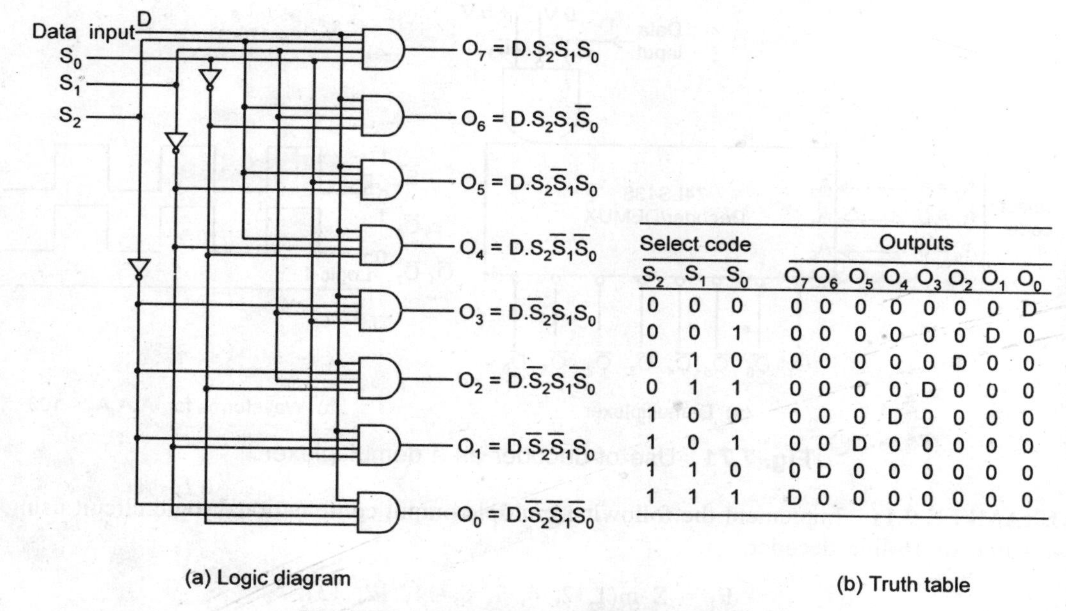

Select code			Outputs							
S_2	S_1	S_0	O_7	O_6	O_5	O_4	O_3	O_2	O_1	O_0
0	0	0	0	0	0	0	0	0	0	D
0	0	1	0	0	0	0	0	0	D	0
0	1	0	0	0	0	0	0	D	0	0
0	1	1	0	0	0	0	D	0	0	0
1	0	0	0	0	0	D	0	0	0	0
1	0	1	0	0	D	0	0	0	0	0
1	1	0	0	D	0	0	0	0	0	0
1	1	1	D	0	0	0	0	0	0	0

(a) Logic diagram (b) Truth table

Fig. 7.70 1-line to 8-line demultiplexer.

but only one of these gates will be enabled by the select input lines. For example, with $S_2S_1S_0$ = 000, only the AND gate O_0 will be enabled, and the data input D will appear at output O_0. Other select codes cause input D to reach the other outputs. The truth table in Figure 7.70b summarizes the operation.

The demultiplexer circuit of Fig. 7.70a is very similar to the 3-line to 8-line decoder circuit of Fig. 7.42a, except that a fourth input D has been added to each gate. The inputs ABC of Fig. 7.42a are here labelled $S_2S_1S_0$ and become the data select inputs.

In the 3-to-8 IC decoder, there are three input lines and eight output lines. The enable input $\overline{E}$ is used to enable or disable the decoding process. This 3-to-8 decoder can be used as a 1-to-8 demultiplexer as follows.

The enable input $\overline{E}$ is used as the data input D, and the binary code inputs are used as the select inputs. Depending on the select inputs, the data input will be routed to a particular output. For this reason, the IC manufacturers often call this type of device a decoder/demultiplexer.

The 74LS138 decoder can be used as a demultiplexer by using $\overline{E}_1$ as the data input D, holding the other two enable inputs in their active states and using the $A_2A_1A_0$ inputs as the select code as shown in Fig. 7.71a.

Figure 7.71b shows the typical waveforms for the case when $A_2A_1A_0$ = 100, and selects the output $\overline{O}_4$. For this case, the data signal applied to $\overline{E}_1$ will be transmitted to $\overline{O}_4$ and all other outputs will remain in their inactive HIGH states.

Demultiplexers are used as clock demultiplexers in synchronous data transmission systems in the receivers, and in security monitoring systems, etc.

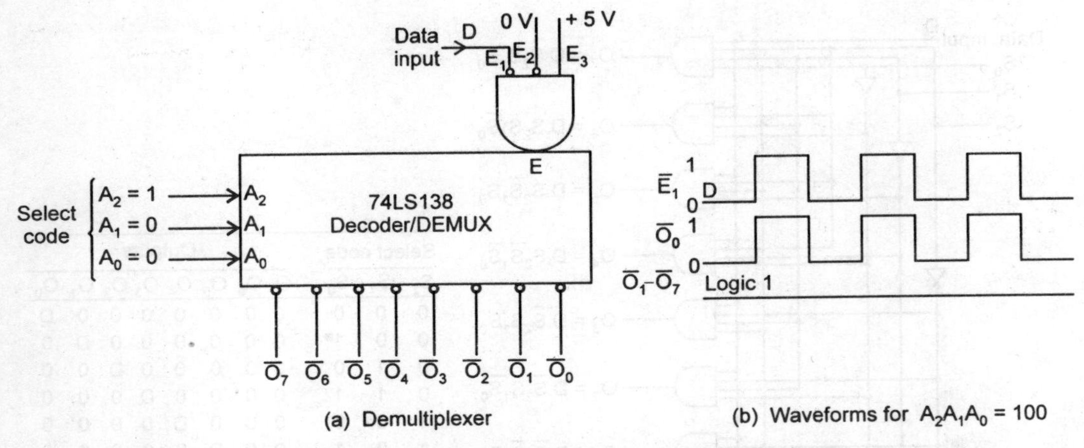

Fig. 7.71 Use of decoder as a demultiplexer.

EXAMPLE 7.11 Implement the following multiple output combinational logic circuit using a 4-line to 16-line decoder.

$$F_1 = \Sigma\ m(1, 2, 4, 7, 8, 11, 12, 13)$$

$$F_2 = \Sigma\ m(2, 3, 9, 11)$$

$$F_3 = \Sigma\ m(10, 12, 13, 14)$$

$$F_4 = \Sigma\ m(2, 4, 8)$$

Solution

The realization is shown in Fig. 7.72. The decoder's outputs are active LOW; therefore, a NAND gate is required for every output of the combinational circuit. In combinational logic

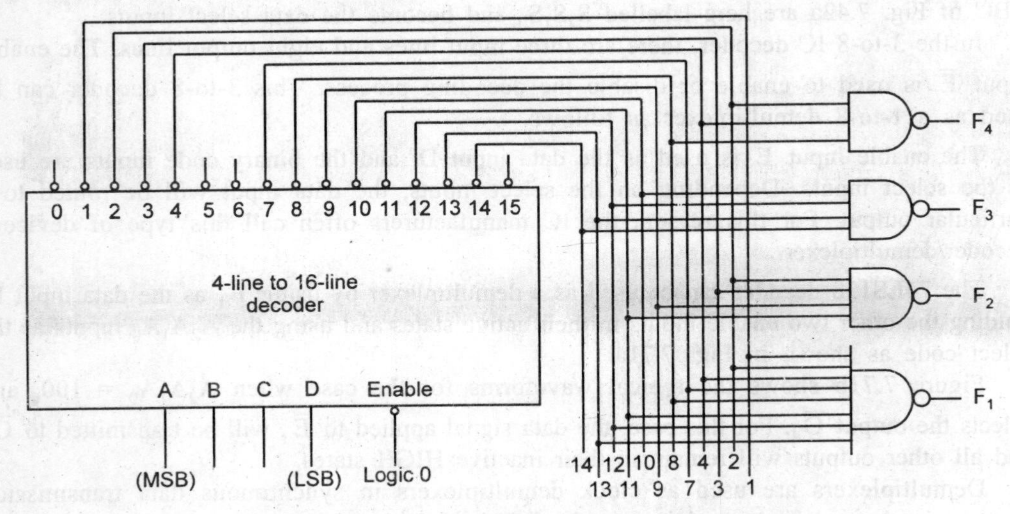

Fig. 7.72 Logic diagram (Example 7.11).

design using a multiplexer, additional gates are not required, whereas the design using a demultiplexer requires additional gates. However, even with this disadvantage, the decoder is more economical in cases where non-trivial, multiple-output expressions of the same input variables are required. In such cases, one multiplexer is required for each output, whereas it is likely that only one decoder supported with a few gates would be required. Therefore, using a decoder could have advantages over using a multiplexer.

7.25 ANSI/IEEE SYMBOLS

Adders

The qualifying symbol for an adder is the Greek letter sigma, Σ. Figure 7.73a shows the logic symbol for the 74AS283 4-bit adder. Binary grouping symbols show that all the possible combinations of the 4-bit inputs, P and Q, produce all combinations of the output sum Σ. The input labelled CI is carry-in and the output labelled CO is carry-out.

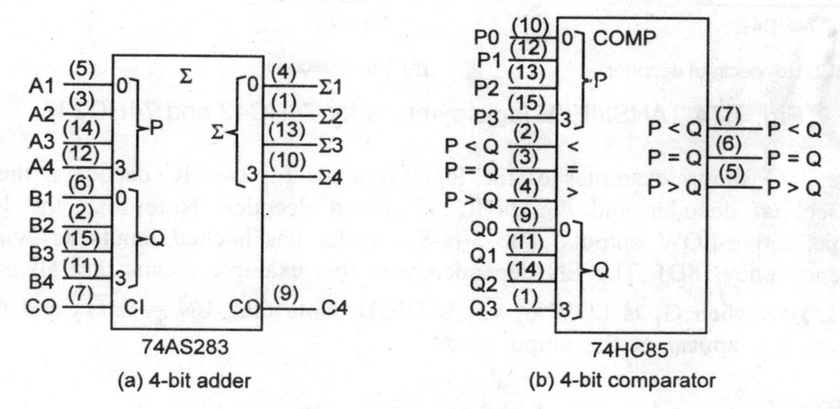

(a) 4-bit adder (b) 4-bit comparator

Fig. 7.73 ANSI/IEEE logic symbols.

Comparators

The qualifying symbol for a comparator is COMP. Figure 7.73b shows the logic symbol for the 74HC85 4-bit comparator. The comparator outputs are indicated by P < Q, P = Q, and P > Q. The logic inputs labelled <, =, and > are used for cascading.

Decoders, Encoders, and Code Converters

The qualifying symbol for a decoder, encoder, or code converter is X/Y, where X represents the input code and Y represents the output code. The letters X and Y can be used in the symbol, or one of them can be replaced by characters that identify the codes they represent such as BIN, DEC, etc. The ANSI/IEEE standards permit input and output lines to be identified in several different ways. If the input code is binary, the lines can be identified by the weights of the bit positions they represent, i.e. 1, 2, 4, etc. Alternatively, the binary grouping symbol} can be used. In cases where the input or the output code is apparent from the way in which the lines are identified, the letters X and Y in the qualifying symbol need not be replaced by characters representing the code. The output lines can be numbered consecutively inside the outline and identified outside the outline by characters that refer to a function table defining the code.

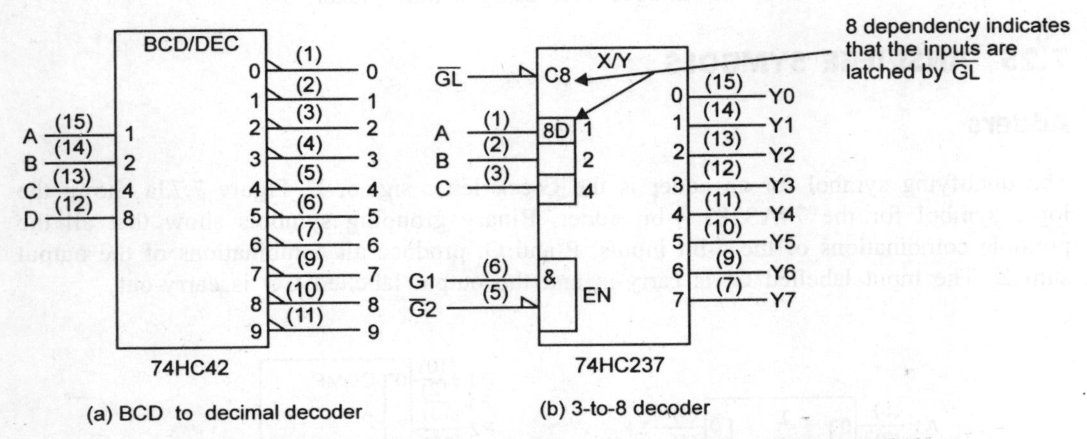

Fig. 7.74 ANSI/IEEE logic symbols for 74HC42 and 74HC237.

Figure 7.74 shows examples of the logic symbols for two IC decoders, the 74HC42 BCD-to-Decimal decoder and the 74HC237 3-to-8 decoder. Note that, the BCD/DEC decoder has active-LOW outputs. The 3-to-8 decoder has latched inputs as evident from the C8 dependency (8D). The EN dependency in this example means that all outputs are inactive (LOW) when G_1 is LOW or $\overline{G}_2$ is HIGH. Note that, EN = $G_1\overline{G}_2$ and the 3-state symbol does not appear at the output lines.

Multiplexers

The qualifying symbol for a multiplexer is MUX. Figure 7.75 shows the logic symbol for the 74S251 8-to-1 multiplexer with 3-state output and bus driving capability. The right pointing triangle in the qualifying symbol signifies the bus driving capability. The output and its complement have the 3-state symbols and are enabled by the input $\overline{G}$. The binary grouping symbol shows that the output depends on the eight possible binary combinations of the data-select inputs, A, B, and C.

Demultiplexers

The qualifying symbol for a demultiplexer is DMUX. Figure 7.76 shows the 74HC38 1-to-8 demultiplexer. The data input in this case is $G1 \cdot \overline{G}2A \cdot \overline{G}2B$. So, a single active-HIGH input (G1) can be used by connecting $\overline{G}2A$ and $\overline{G}2B$ both LOW, or active-LOW inputs can be used with G1 connected HIGH. A demultiplexer can be regarded as a decoder. Figure 7.76b shows an equivalent logic symbol for the 74HC38 when its function is interpreted as a

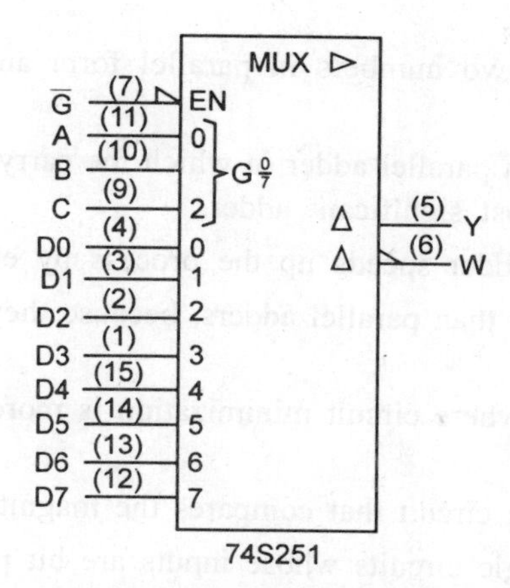

Fig. 7.75 ANSI/IEEE logic symbol for 74S251 8-to-1 multiplexer with 3-state output.

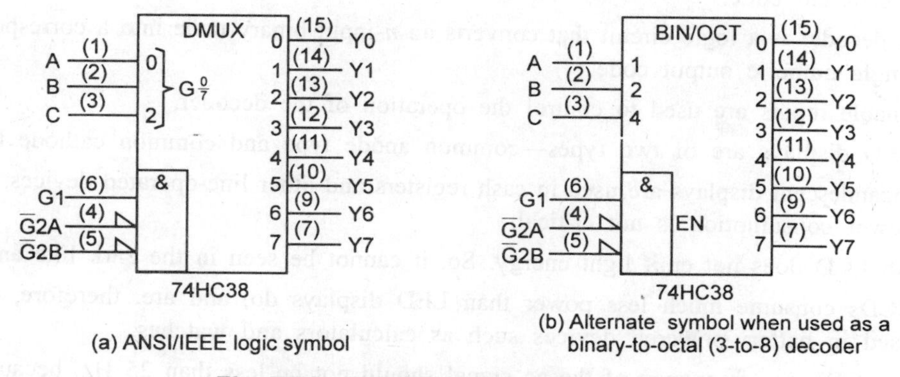

Fig. 7.76 74HC38 1-to-8 demultiplexer.

3-to-8 (BIN/OCT) decoder. In this interpretation, the inputs G1, G2A, and G2B create an enable (EN) dependency rather than serving as data inputs. Any combination that makes

$$G1 \cdot \overline{G}2A \cdot \overline{G}2B = 0$$

causes all outputs to be inactive (LOW).

SUMMARY

- A half-adder is an arithmetic circuit that adds two binary digits.
- A half-subtractor is an arithmetic circuit that subtracts one binary digit from another.
- A full-adder is an arithmetic circuit that adds two binary digits and a carry, i.e. 3 bits.
- A full-subtractor is an arithmetic circuit that subtracts one binary digit from another considering a borrow.

- A parallel adder adds two numbers in parallel form and produces the sum bits in parallel form.
- A ripple carry adder is a parallel adder in which the carry-out of each full-adder is the carry-in to the next most significant adder.
- The look-ahead carry adder speeds up the process by eliminating the ripple carry.
- Serial adders are slower than parallel adders, because they require one clock pulse for each pair of bits added.
- Serial adders are used where circuit minimization is more important than speed, as in pocket calculators.
- A comparator is a logic circuit that compares the magnitudes of two binary numbers.
- Code converters are logic circuits whose inputs are bit patterns representing numbers or characters in one code and whose outputs are the corresponding representations in a different code.
- A decoder is a logic circuit that converts an *n*-input binary code into a corresponding single numeric output code.
- Enable inputs are used to control the operation of the decoder.
- LED displays are of two types—common anode type and common cathode type.
- Incandescent displays are used in cash registers and other line-operated devices, where power consumption is not critical.
- An LCD does not emit light energy. So, it cannot be seen in the dark like an LED.
- LCDs consume much less power than LED displays do, and are, therefore, widely used in battery-powered devices such as calculators and watches.
- In LCDs, the frequency of the ac signal should not be less than 25 Hz, because this would produce visible flicker.
- An encoder is a device whose inputs are decimal digits and/or alphabetic characters and whose outputs are the coded representations of those inputs.
- A priority encoder is a logic circuit that responds to just one input, in accordance with some priority system, among those that may be simultaneously HIGH.
- A multiplexer is a logic circuit that accepts several data inputs and allows only one of them at a time to get through to the output.
- Multiplexers are used for data selection, data routing, operation sequencing, parallel-to-serial conversion, waveform generation, and logic function generation, etc.
- DEMUX is a logic circuit that, depending on the status of its select inputs, channels its data input to one of several data outputs.
- The term 'driver' is a technical term, sometimes added to an IC's description to indicate that its outputs can operate with higher current and/or voltage limits than those of a normal standard IC.

QUESTIONS

1. Describe the operations performed by the following arithmetic circuits:
   (a) Half-adder (b) Full-adder (c) Half-subtractor
   (d) Full-subtractor
2. Briefly describe the following:
   (a) Parallel adder (b) Serial adder (c) Ripple carry adder
   (d) Look-ahead carry adder.
3. How does the look-ahead carry adder speed up the addition process?
4. When is a carry generated and when is a carry propagated?
5. What is the disadvantage of serial adders? For which applications are they preferred?
6. How do you compare serial and parallel adders?
7. What do you mean by cascading of parallel adders? Why is it required?
8. What is a parity bit generator?
9. What is (a) an odd parity generator, and (b) an even parity generator?
10. What is a code converter? List some of the code converters.
11. What is the use of the enable input in a decoder?
12. Describe the operations performed by the following logic circuits:
    (a) Comparator (b) Decoder (c) Encoder
13. What does the term 'driver' mean in a decoder/driver?
14. What is the number of inputs and outputs of a decoder that accepts 64 different input combinaions?
15. Why are CMOS devices preferred over TTL devices for driving LCDs?
16. What are the two types of LED displays?
17. What is the drawback of incandescent seven segment displays? Where are they used?
18. Which LED segments will be ON for a decoder/driver input of 1001?
19. What type of displays are used in calculators and watches?
20. How does a priority encoder differ from an ordinary encoder?
21. How is ac voltage produced in LCDs? Why does an LCD require an external source of light? What is the minimum frequency required in LCDs and why?
22. Explain the terms: (a) multiplexing, and (b) demultiplexing.
23. What is a keyboard encoder?
24. List some of the applications of multiplexers and demultiplexers.
25. Which of the following statements refer to LCD displays and which to LED displays?
    (a) Emit light
    (b) Reflect ambient light
    (c) Are best for low power applications

(d) Require an ac voltage
(e) Require current limiting resistors

26. Explain the difference between a MUX and a DEMUX.

27. Which of the following statements refer to a decoder and which to an encoder?

(a) Has more inputs than outputs.
(b) Is used to convert key actuations to a binary code.
(c) Only one output can be activated at one time.
(d) Can be used to interface a BCD input to an LED display.

29. Which of the following statements refer to a decoder, an encoder, a MUX, or a DEMUX?

(a) Has more inputs than outputs.
(b) Uses SELECT inputs.
(c) Can be used in parallel-to-serial conversion.
(d) Produces a binary code at its output.
(e) Only one of its outputs can be active at one time.
(f) Can be used to route an input signal to one of several possible outputs.
(g) Can be used to generate arbitrary logic functions.

30. Indicate true or false:

(a) When a MUX is used to implement a logic function, the logic variables are applied to the MUX's data inputs.
(b) The circuit for a DEMUX is basically the same as that for a decoder.

PROBLEMS

7.1 Design an 8421-to-2421 BCD code converter and draw its logic diagram.

7.2 Find the simplest possible logic expressions for a 2421-to-51111 BCD code converter.

7.3 Draw a logic diagram for an Excess-3-to-decimal decoder. Inputs and outputs should be active HIGH.

7.4 Draw a logic diagram for a 2421-to-decimal decoder.

7.5 Draw a logic circuit that generates an even parity bit for the 2421 BCD code.

7.6 Draw a logic circuit that generates an odd parity bit for the 3321 BCD code.

7.7 Design the following code converters: (a) 5211 to 2421 (b) 4-bit binary to excess-3 (c) 4-bit BCD to gray.

7.8 Design a logic circuit to generate (i) an even parity bit, and (ii) an odd parity bit for a 3-bit binary input.

Chapter 8

FLIP-FLOPS AND TIMING CIRCUITS

8.1 INTRODUCTION

Basically, switching circuits may be combinational switching circuits or sequential switching circuits. The switching circuits considered so far have been combinational switching circuits. Combinational switching circuits are those whose output levels at any instant of time are dependent only on the levels present at the inputs at that time. Any prior input level conditions have no effect on the present outputs, because combinational logic circuits have no memory. On the other hand, sequential switching circuits are those whose output levels at any instant of time are dependent not only on the levels present at the inputs at that time, but also on the prior input level conditions. It means that sequential switching circuits have memory. Sequential circuits are thus made of combinational circuits and memory elements.

The most important memory element is the flip-flop, which is made up of an assembly of logic gates. Even though a logic gate by itself has no storage capability, several logic gates can be connected together in ways that permit information to be stored. There are several different gate arrangements that are used to construct flip-flops in a wide variety of ways. Each type of flip-flop has special features or characteristics necessary for particular applications.

A flip-flop (FF), known more formally as a bistable multivibrator, has two stable states. It can remain in either of the states indefinitely. Its state can be changed by applying the proper triggering signal.

Figure 8.1 shows the general type of symbol used for a flip-flop. The flip-flop has two outputs, labelled Q and $\overline{Q}$. Actually any letter can be used to represent the output, but Q is the one most often used. The Q output is the normal output of the flip-flop and $\overline{Q}$ is the

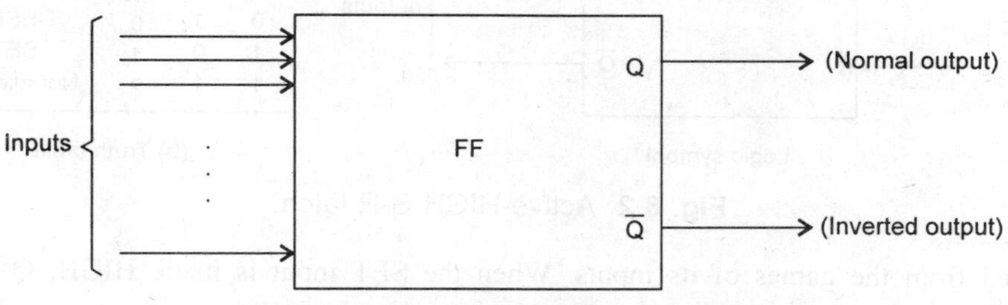

Fig. 8.1 General flip-flop symbol.

inverted output. The state of the flip-flop always refers to the state of the normal output Q. The inverted output $\overline{Q}$ is in the opposite state. A flip-flop is said to be in HIGH state or logic 1 state or SET state when Q = 1, and in LOW state or logic 0 state or RESET state or CLEAR state when Q = 0.

As the symbol in Fig. 8.1 implies, a flip-flop can have one or more inputs. These inputs are used to cause the flip-flop to switch back and forth (i.e. 'flip-flop') between its possible output states. A flip-flop input has to be pulsed momentarily to cause a change in the flip-flop output, and the output will remain in that new state even after the input pulse has been removed. This is the flip-flop's memory characteristic.

There are a number of applications of flip-flops. As such, the flip-flop serves as a storage device. It stores a 1 when its Q output is a 1, and stores a 0 when its Q output is a 0. Flip-flops are the fundamental components of shift registers and counters.

The term 'latch' is used for certain flip-flops. It refers to non-clocked flip-flops, because these flip-flops 'latch on' to a 1 or a 0 immediately upon receiving the input pulse called SET or RESET. Gated latches are latches which respond to the inputs and latch on to a 1 or a 0 only when they are enabled, i.e. only when the input ENABLE or gating signal is HIGH. In the absence of ENABLE or gating signal, the latch does not respond to the changes in its inputs.

A latch may be an active-HIGH input latch or an active-LOW input latch. Active-HIGH means that the SET and RESET inputs are normally resting in the LOW state and one of them will be pulsed HIGH whenever we want to change the latch outputs. Active-LOW means that the SET and RESET inputs are normally resting in the HIGH state and one of them will be pulsed LOW whenever we want to change the latch outputs.

8.2 THE S-R LATCH

The simplest type of flip-flop is called an S-R latch. It has two outputs labelled Q and $\overline{Q}$ and two inputs labelled S and R. The state of the latch corresponds to the level of Q (HIGH or LOW, 1 or 0) and $\overline{Q}$ is, of course, the complement of that state. The output as well as its complement are available for each flip-flop.

Figure 8.2 shows the logic symbol and truth table of an S-R latch. Q_0 represents the state of the flip-flop before applying the inputs. The name of the latch, S-R or SET-RESET,

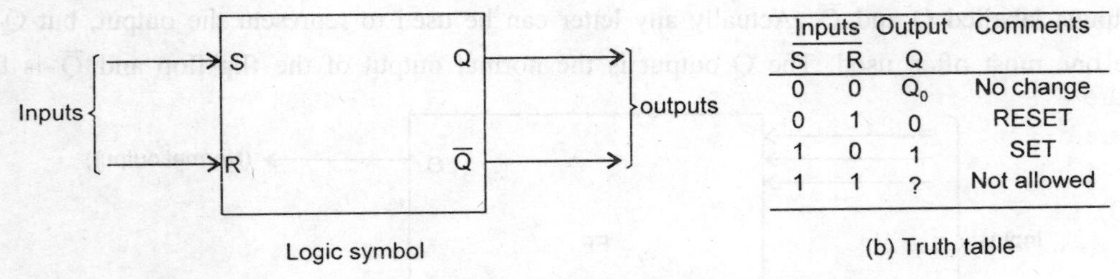

Inputs		Output	Comments
S	R	Q	
0	0	Q_0	No change
0	1	0	RESET
1	0	1	SET
1	1	?	Not allowed

(b) Truth table

Fig. 8.2 Active-HIGH S-R latch.

is derived from the names of its inputs. When the SET input is made HIGH, Q becomes 1 (and $\overline{Q}$ equals 0). When the RESET input is made HIGH, Q becomes 0 (and $\overline{Q}$ equals 1.) If both the inputs S and R are made LOW, there is no change in the state of the latch.

It means that the latch remains in the same state in which it was, prior to the application of inputs. If both the inputs are made HIGH, the output is unpredictable, i.e. both Q and $\overline{Q}$ may be HIGH, or both may be LOW or any one of them may be HIGH and the other LOW. This condition is described as not-allowed, unpredictable, invalid, or indeterminate. The S-R latch is also called R-S latch or S-C (SET-CLEAR) latch. Resetting is also called clearing because we CLEAR out the 1 in the output by resetting to 0. In more complex flip-flops, called gated latches, the change of state does not take place immediately after the application of the inputs. The change of state takes place only after applying a gate pulse.

The NOR Gate S-R Latch

An S-R latch can be constructed using two cross-coupled NOR gates or NAND gates. Using two NOR gates, an active-HIGH S-R latch can be constructed and using two NAND gates an active-LOW S-R latch can be constructed. Figure 8.3 shows the logic diagram of an active-HIGH S-R latch composed of two cross-coupled NOR gates. Note that the output of each gate is connected to one of the inputs of the other gate.

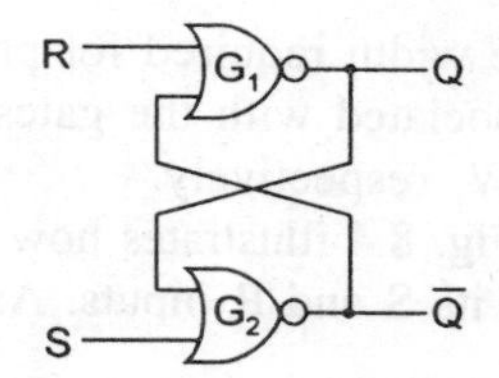

Fig. 8.3 Logic diagram of an active-HIGH S-R latch using NOR gates.

Let us assume that the latch is initially SET, i.e. Q = 1 and $\overline{Q}$ = 0. If the inputs are S = 0 and R = 0, the inputs to G_1 are a 0 (R) and a 0 ($\overline{Q}$) and so its output is a 1, i.e. Q remains as a 1. The inputs to G_2 are a 0 (S) and a 1 (Q) and so its output is a 0, i.e. $\overline{Q}$ remains as a 0. That is, S = 0 and R = 0 do not result in a change of state. Similarly, if Q = 0 and $\overline{Q}$ = 1 initially, and if S = 0 and R = 0 are the inputs applied, the inputs to G_2 are a 0 (Q) and a 0 (S) and so its output is a 1, i.e. $\overline{Q}$ remains as a 1. The inputs to G_1 are a 0 (R) and a 1 ($\overline{Q}$) and so its output is a 0, i.e. Q remains as a 0. This implies that the latch remains in the same state, when S = 0, R = 0 is applied.

If Q = 1 and $\overline{Q}$ = 0 initially, and if inputs S = 1 and R = 0 are applied, the inputs to G_2 are a 1 (S) and a 1 (Q), and so its output is a 0, i.e. $\overline{Q}$ remains as a 0. The inputs to G_1 are a 0 (R) and a 0 ($\overline{Q}$), and so its output is a 1, i.e. Q remains as a 1. If Q = 0 and $\overline{Q}$ = 1 initially and if inputs S = 1 and R = 0 are applied, the inputs to G_2 are a 1 (S) and a 0 (Q), and so its output is a 0, i.e. $\overline{Q}$ goes to a 0. The inputs to G_1 are a 0 (R) and a 0 ($\overline{Q}$), and so its output is a 1, i.e. Q goes to a 1. This implies that irrespective of the present state, the output of the S-R latch goes to SET state, i.e. state 1 after application of the input, S = 1 and R = 0.

If Q = 1 and $\overline{Q}$ = 0 initially, and if inputs S = 0 and R = 1 are applied, the inputs to G_1 are a 1 (R) and a 0 ($\overline{Q}$), and so its output is a 0, i.e. Q goes to a 0. The inputs to G_2 are a 0 (S) and a 0 (Q), and so its output is a 1, i.e. $\overline{Q}$ goes to a 1. If Q = 0 and $\overline{Q}$ = 1 initially, and if inputs S = 0 and R = 1 are applied, the inputs to G_1 are a 1 (R) and a 1 ($\overline{Q}$), and so the output of G_1 is a 0, i.e. Q remains as a 0. The inputs to G_2 are a 0 (S) and a 0 (Q), and so the output of G_2 is a 1, i.e. $\overline{Q}$ remains as a 1. This implies that whatever may be its present state, when S = 0, R = 1 are applied, the flip-flop goes to RESET state, i.e. state 0.

When both the inputs S and R are a 1, the corresponding outputs will be Q = 0 and $\overline{Q}$ = 0, which is invalid.

It is thus necessary only to pulse a SET or RESET input to change the state of the latch. For example, if the latch is initially RESET, a pulse applied to its SET is the same as making S momentarily a 1, followed by a 0. The 1 sets the latch, after which R and S are once again a 0, the no-change condition. Since a pulse must remain HIGH long enough for NOR gates to change states, the minimum pulse width is the sum of the propagation delays through the gates. One gate must change from LOW to HIGH and the other from HIGH to LOW. Thus,

$$PW_{min} = t_{PLH} + t_{PHL}$$

where PW_{min} is the minimum pulse width required for proper operation of the gate, t_{PLH} and t_{PHL} are the propagation delays associated with the gates when the output is changing from LOW to HIGH and HIGH to LOW, respectively.

The timing diagram shown in Fig. 8.4 illustrates how the active-HIGH S-R latch responds to arbitrarily selected waveforms at its S and R inputs. Assume that initially the latch is SET,

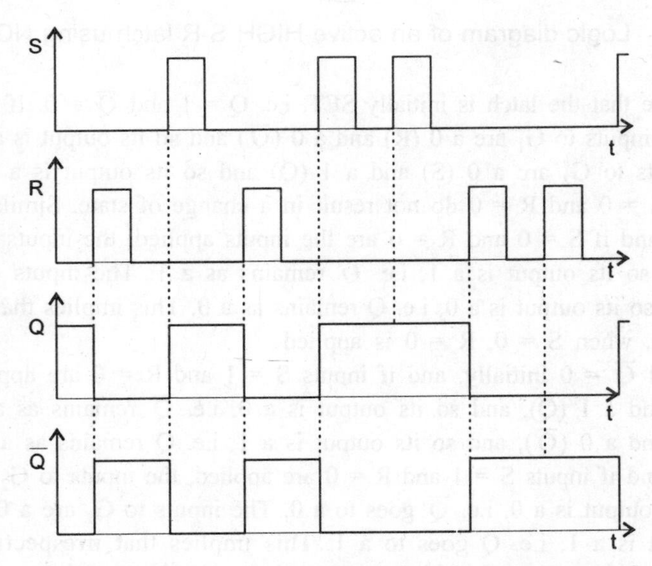

Fig. 8.4. Active-HIGH S-R latch—timing diagram.

i.e. Q = 1 and $\overline{Q}$ = 0. The latch remains in that state till the RESET pulse arrives. At the positive going edge of the RESET pulse the latch resets, i.e. Q becomes a 0 and $\overline{Q}$ becomes a 1. The latch remains in that state (i.e. $Q = 0$ and $\overline{Q} = 1$) till the SET pulse arrives. At the positive going edge of the SET pulse, Q becomes a 1 and $\overline{Q}$ a 0. The latch remains in that state till the next RESET pulse arrives. Observe that when the latch is already in SET state, the arrival of a SET pulse does not affect the state of the latch and similarly, when the latch is already in RESET state, the arrival of a RESET pulse does not have any effect on the state of the latch.

The analysis of the operation of the active-HIGH NOR latch can be summarized as follows.

1. SET = 0, RESET = 0: This is the normal resting state of the NOR latch and it has no effect on the output state. Q and $\overline{Q}$ will remain in whatever state they were prior to the occurrence of this input condition.
2. SET = 1, RESET = 0: This will always set Q = 1, where it will remain even after SET returns to 0.
3. SET = 0, RESET = 1: This will always reset Q = 0, where it will remain even after RESET returns to 0.
4. SET = 1, RESET = 1: This condition tries to SET and RESET the latch at the same time, and it produces $Q = \overline{Q} = 0$. If the inputs are returned to zero simultaneously, the resulting output state is unpredictable. This input condition should not be used.

The SET and RESET inputs are normally in the LOW state and one of them will be pulsed HIGH, whenever we want to change the latch outputs.

The NAND Gate S-R Latch

An active-LOW S-R latch can be constructed using two cross-coupled NAND gates. Figures 8.5a and b show the logic diagram and truth table of an active-LOW S-R latch. Since the NAND gate is equivalent to an active-LOW OR gate, an active-LOW S-R latch using OR gates may also be represented as shown in Fig. 8.5c.

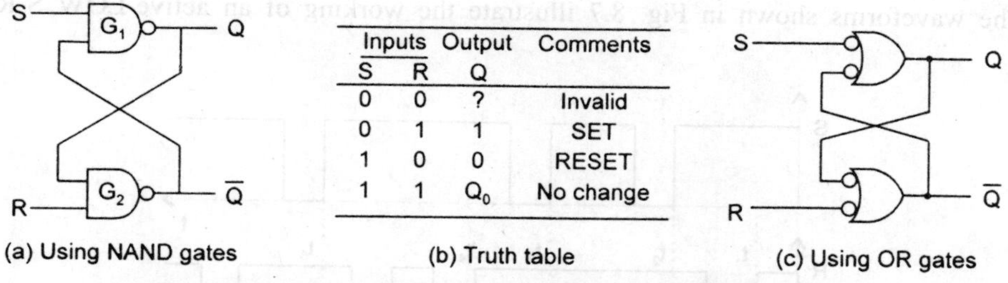

Inputs S	R	Output Q	Comments
0	0	?	Invalid
0	1	1	SET
1	0	0	RESET
1	1	Q_0	No change

(b) Truth table

Fig. 8.5 Logic diagram of an active-LOW S-R latch.

Suppose initially, $Q = 1$ and $\overline{Q} = 0$. If inputs are $S = 0$ and $R = 0$, G_1 inputs are a 0 (S) and a 0 ($\overline{Q}$). So, its output $Q = 1$. G_2 inputs are a 0 (R) and a 1 (Q) and so its output $\overline{Q} = 1$. Now, suppose $Q = 0$ and $\overline{Q} = 1$; then if inputs are $S = 0$ and $R = 0$, G_1 inputs are a 0 (S) and a 1 ($\overline{Q}$). So, its output $Q = 1$. G_2 inputs are a 0 (R) and a 1 (Q) and so its output $\overline{Q} = 1$. That is, whatever may be the initial state, when the inputs are $S = 0$ and $R = 0$, both Q and $\overline{Q}$ will be equal to 1.

Similarly, we can show that, when

(a) S = 1 and R = 0, whatever may be the initial state, Q will go to 0 and $\overline{Q}$ to 1. That is, the FF resets.

(b) S = 0 and R = 1, whatever may be the initial state, Q will go to 1 and $\overline{Q}$ to 0. That is, the FF sets.

(c) S = 1 and R = 1, no change of state takes place, i.e., the output (Q or $\overline{Q}$) remains the same as it was before.

The operation of this latch is the reverse of the NOR gate latch discussed earlier. That is why it is called an active-LOW S-R latch. If the 0s are replaced by 1s and 1s by 0s in Fig. 8.5b, we get the same truth table as that of the NOR gate latch shown in Fig. 8.2b.

The SET and RESET inputs are normally resting in the HIGH state and one of them will be pulsed LOW, whenever we want to change the latch outputs.

An active-LOW NAND latch can be converted into an active-HIGH NAND latch by inserting the inverters at the S and R inputs. Figure 8.6 shows the logic diagram and truth table of an active-HIGH NAND latch.

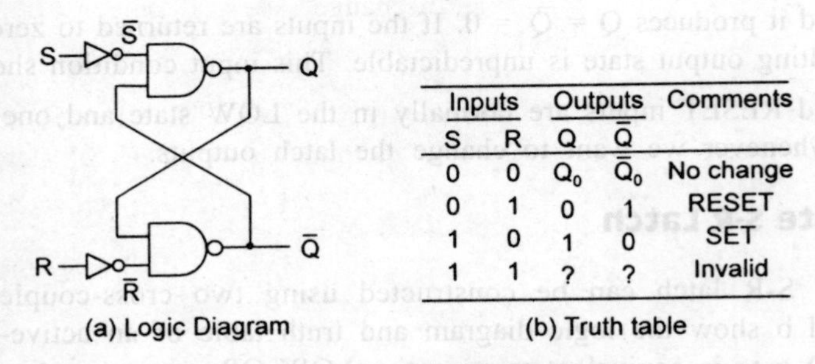

Inputs		Outputs		Comments
S	R	Q	$\overline{Q}$	
0	0	Q_0	$\overline{Q}_0$	No change
0	1	0	1	RESET
1	0	1	0	SET
1	1	?	?	Invalid

Fig. 8.6 An active-HIGH NAND latch.

The waveforms shown in Fig. 8.7 illustrate the working of an active-LOW S-R latch.

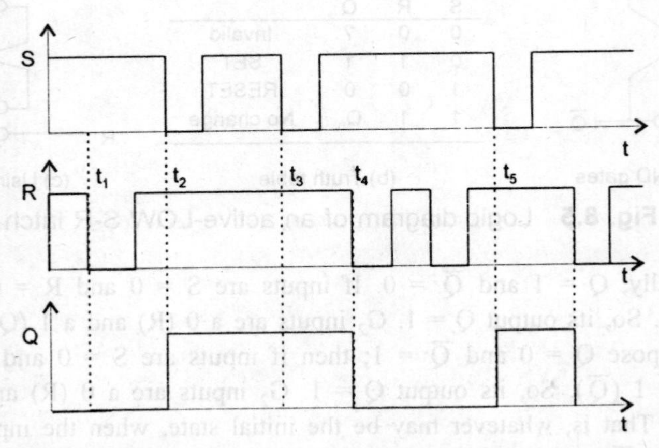

Fig. 8.7 Active-LOW S-R latch—timing diagram.

Initially, Q = 0 and S and R are in normal resting states, i.e. S = 1 and R = 1. Since Q is already in a 0 state, a negative pulse applied at R at time t_1 cannot produce any effect and Q remains as 0. But a negative pulse applied to S at time t_2 brings Q to a 1 state. Since Q is already equal to 1, a negative pulse applied to S at time t_3 cannot produce any effect and Q remains at a 1. The negative pulse applied to R at time t_4, however, changes the state of the latch and Q becomes a 0.

This example shows that the latch output 'remembers' the last input that was activated and will not change states until the opposite input is activated.

When power is applied to a circuit, it is not possible to predict the starting state of a flip-flop output, whether its SET and RESET inputs are in their inactive states (i.e. S = R = 1 for a NAND latch, and S = R = 0 for a NOR latch). There is just as much chance that the starting state will be a Q = 0 as Q = 1. It will depend on things like internal propagation delays, parasitic capacitance and external loading.

8.3 GATED LATCHES

The Gated S-R Latch

In the latches described earlier, the output can change state any time the input conditions are changed. So, they are called *asynchronous* latches. A gated S-R latch requires an ENABLE (EN) input. Its S and R inputs will control the state of the flip-flop only when the ENABLE is HIGH. When the ENABLE is LOW, the inputs become ineffective and no change of state can take place. The ENABLE input may be a clock. So, a gated S-R latch is also called clocked S-R latch or synchronous S-R latch. Since this type of flip-flop responds to the changes in inputs only as long as the clock is HIGH, these types of flip-flops are also called *level triggered flip-flops*. The logic diagram, the logic symbol and the truth table for a gated S-R latch are shown in Fig. 8.8. In this circuit, the invalid state occurs when both S and R are simultaneously HIGH.

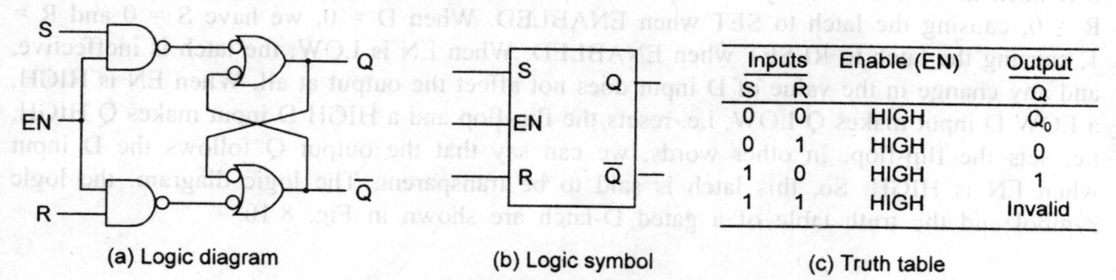

(a) Logic diagram

(b) Logic symbol

Inputs S	Inputs R	Enable (EN)	Output Q
0	0	HIGH	Q_0
0	1	HIGH	0
1	0	HIGH	1
1	1	HIGH	Invalid

(c) Truth table

Fig. 8.8 A gated S-R latch.

EXAMPLE 8.1 Determine the output waveform Q if the inputs shown in Fig. 8.9a are applied to a gated S-R latch shown in Fig. 8.9b, that was initially SET.

Solution

The output waveform Q shown in Fig. 8.9c is drawn as follows:

Prior to t_0, Q is HIGH. Even though R goes HIGH prior to t_0, Q will not change because EN is LOW. Similarly, even though S goes HIGH prior to t_1, Q will not change because EN is LOW. Any time S is HIGH and R is LOW, a HIGH on the EN sets the latch, and any time S is LOW and R is HIGH, a HIGH on the EN resets the latch.

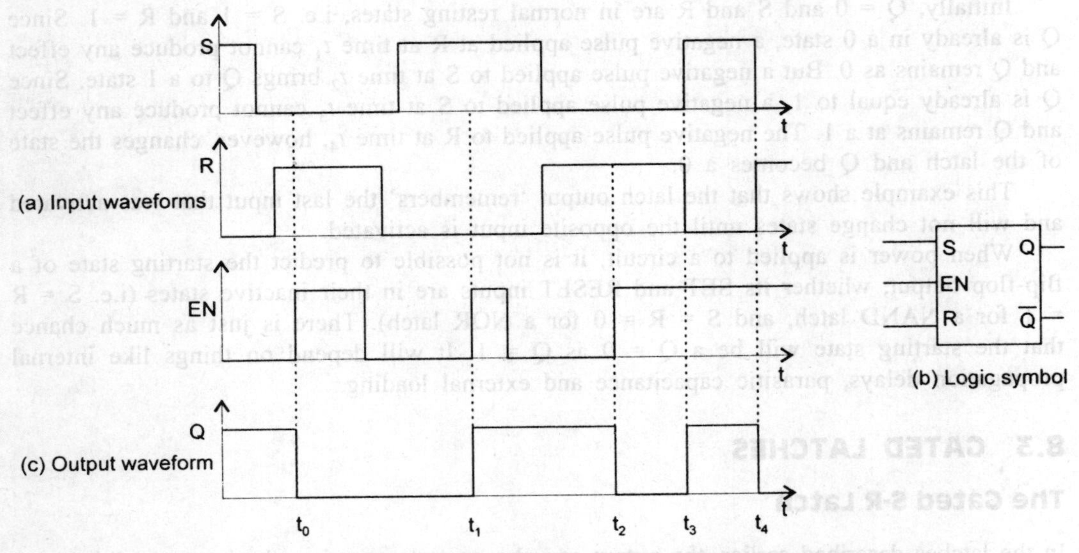

Fig. 8.9 Waveforms—the gated S-R latch (Example 8.1).

The Gated D-latch

In many applications, it is not necessary to have separate S and R inputs to a latch. If the input combinations S = R = 0 and S = R = 1 are never needed, the S and R are always the complement of each other. So, we can construct a latch with a single input (S) and obtain the R input by inverting it. This single input is labelled D (for data) and the device is called a D-latch. So, another type of gated latch is the gated D-latch. It differs from the S-R latch in that, it has only one input in addition to EN. When D = 1, we have S = 1 and R = 0, causing the latch to SET when ENABLED. When D = 0, we have S = 0 and R = 1, causing the latch to RESET when ENABLED. When EN is LOW, the latch is ineffective, and any change in the value of D input does not affect the output at all. When EN is HIGH, a LOW D input makes Q LOW, i.e. resets the flip-flop and a HIGH D input makes Q HIGH, i.e. sets the flip-flop. In other words, we can say that the output Q follows the D input when EN is HIGH. So, this latch is said to be transparent. The logic diagram, the logic symbol and the truth table of a gated D-latch are shown in Fig. 8.10.

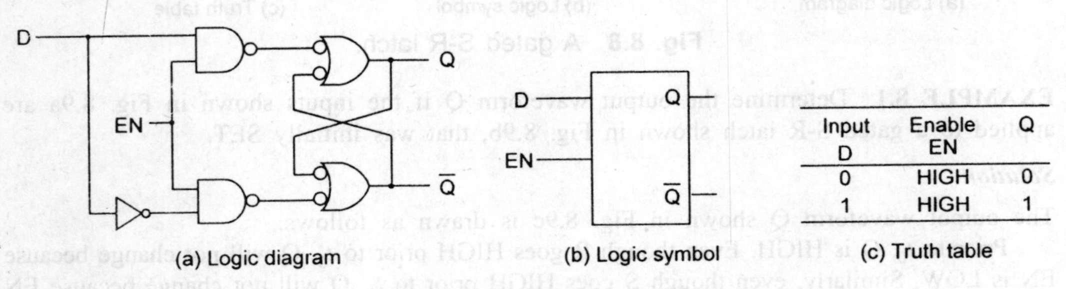

Input D	Enable EN	Q
0	HIGH	0
1	HIGH	1

(c) Truth table

Fig. 8.10 A gated D-latch.

EXAMPLE 8.2 Determine the Q output waveform if the inputs shown in Fig. 8.11a are applied to the gated D-latch shown in Fig. 8.11b, which is initially RESET.

Solution

The Q output waveform is shown in Fig. 8.11c. The D input controls the Q output only when EN is HIGH.

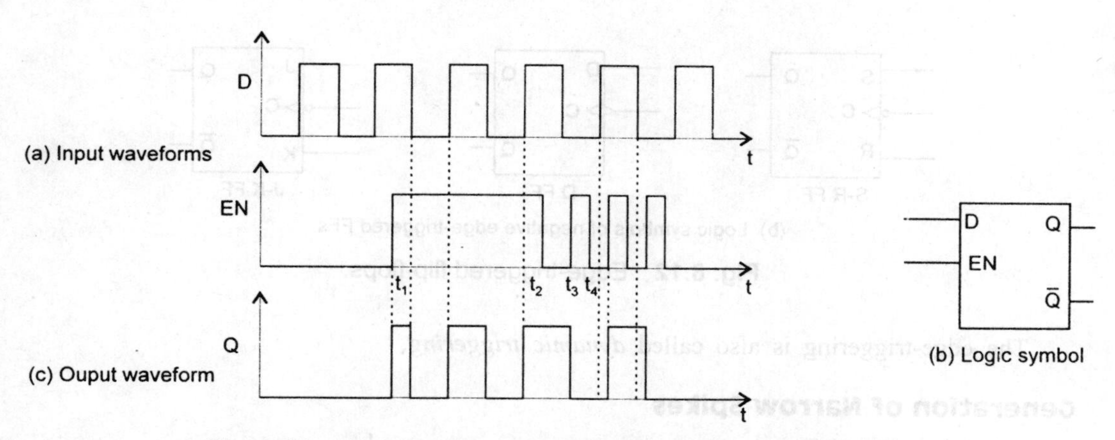

Fig. 8.11 Waveforms—the gated D-latch (Example 8.2).

8.4 EDGE-TRIGGERED FLIP-FLOPS

Digital systems can operate either synchronously or asynchronously. In asynchronous systems, the outputs of logic circuits can change state any time, when one or more of the inputs change. An asynchronous system is difficult to design and troubleshoot. In synchronous systems, the exact times at which any output can change states are determined by a signal commonly called the *clock*. The flip-flops using the clock signal are called the *clocked flip-flops*. Control signals are effective only if they are applied in synchronization with the clock signal. The clock signal is distributed to all parts of the system and most of the system outputs can change state only when the clock makes a transition. Clocked flip-flops may be positive edge-triggered or negative edge-triggered. Positive edge-triggered flip-flops are those in which 'state transitions' take place only at the positive-going (0 to 1, or LOW to HIGH) edge of the clock pulse and negative edge-triggered flip-flops are those in which 'state transitions' take place only at the negative-going (1 to 0, or HIGH to LOW) edge of the clock signal. Positive-edge triggering is indicated by a 'triangle' at the clock terminal of the flip-flop. Negative-edge triggering is indicated by a 'triangle' with a bubble at the clock terminal of the flip-flop. Thus edge-triggered flip-flops are sensitive to their inputs only at the transition of the clock. There are three basic types of edge-triggered flip-flops: S-R, J-K, and D (Fig. 8.12). Of these, D and J-K flip-flops are the most widely used ones and readily available in the IC form than is the S-R type. But the S-R flip-flop is also covered here because it is a good base upon which to build both the D and the J-K flip-flops, having been derived from the S-R flip-flop.

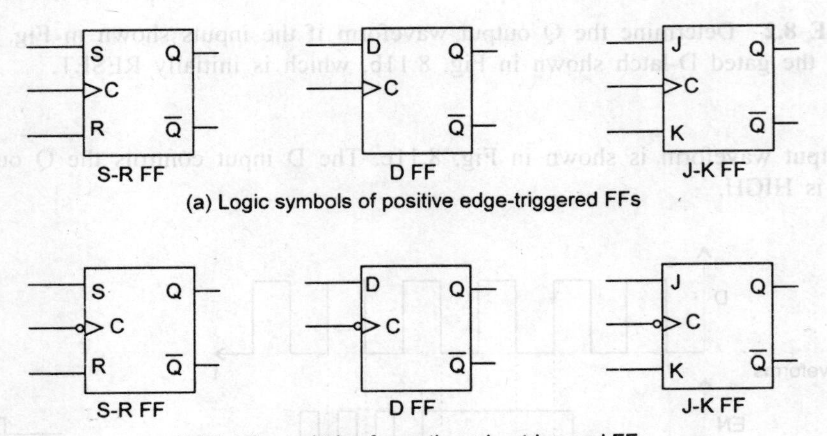

Fig. 8.12 Edge-triggered flip-flops.

The edge-triggering is also called *dynamic triggering.*

Generation of Narrow Spikes

A narrow positive spike is generated at the rising edge of the clock using an inverter and an AND gate as shown in Fig. 8.13a. The inverter produces a delay of a few nanoseconds. The AND gate produces an output spike that is HIGH only for a few nanoseconds, when CLK and $\overline{\text{CLK}}$ are both HIGH. This results in a narrow pulse at the output of the AND gate which occurs at the positive-going transition of the clock signal.

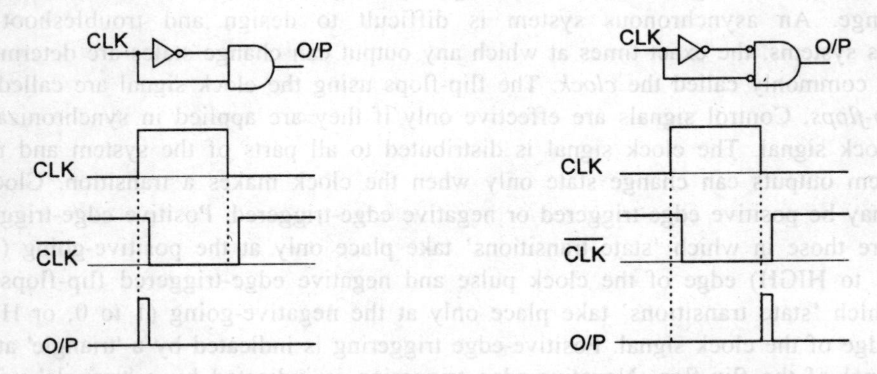

Fig. 8.13 Generation of narrow spikes using edge detector.

Similarly, a narrow positive spike is generated at the falling edge of the clock by using an inverter and an active-LOW AND gate as shown in Fig. 8.13b. The inverter produces a delay of a few nanoseconds. The active-LOW AND gate produces an output spike that is HIGH only for a few nanoseconds, when CLK and $\overline{CLK}$ are both LOW. This results in a narrow pulse at the output of the AND gate which occurs at the negative-going transition of the clock.

The Edge-triggered S-R Flip-flop

Figure 8.14 shows the logic symbol and the truth table for a positive edge-triggered S-R flip-flop. The S and R inputs of the S-R flip-flop are called the *synchronous*

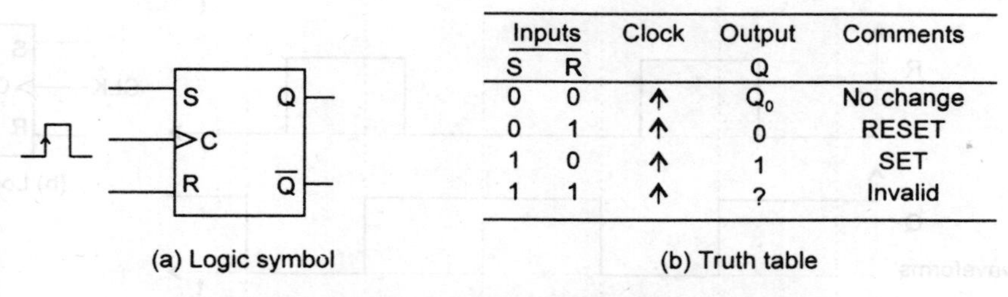

Inputs		Clock	Output	Comments
S	R		Q	
0	0	↑	Q_0	No change
0	1	↑	0	RESET
1	0	↑	1	SET
1	1	↑	?	Invalid

(a) Logic symbol (b) Truth table

Fig. 8.14 Positive edge-triggered S-R flip-flop.

control inputs because data on these inputs affect the flip-flop's output only on the triggering (positive going) edge of the clock pulse. Without a clock pulse, the S and R inputs cannot affect the output. When S is HIGH and R is LOW, the Q output goes HIGH on the positive-going edge of the clock pulse and the flip-flop is SET. (If it is already in SET state, it remains SET.) When S is LOW and R is HIGH, the Q output goes LOW on the positive-going edge of the clock pulse and the flip-flop is RESET, i.e. cleared. (If it is already in RESET state, it remains RESET.) When both S and R are LOW, the output does not change from its prior state. (If it is in SET state, it remains SET and if it is in RESET state, it remains RESET.) When both S and R are HIGH simultaneously, an invalid condition exists. The basic operation described above is illustrated in Fig. 8.14b.

Figure 8.15 shows the logic symbol and the truth table of a negative edge-triggered S-R flip-flop. This flip-flop will trigger only when the clock input goes from 1 to 0.

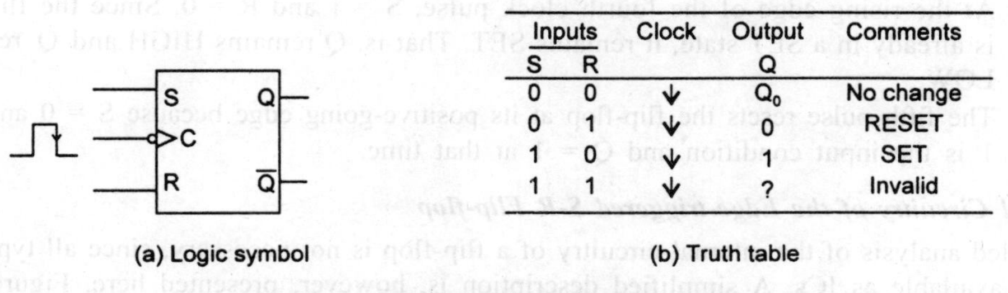

Inputs		Clock	Output	Comments
S	R		Q	
0	0	↓	Q_0	No change
0	1	↓	0	RESET
1	0	↓	1	SET
1	1	↓	?	Invalid

(a) Logic symbol (b) Truth table

Fig. 8.15 Negative edge-triggered S-R flip-flop.

EXAMPLE 8.3 The waveforms shown in Fig. 8.16a are applied to the positive edge-triggered S-R flip-flop shown in Fig. 8.16b. Sketch the output waveforms.

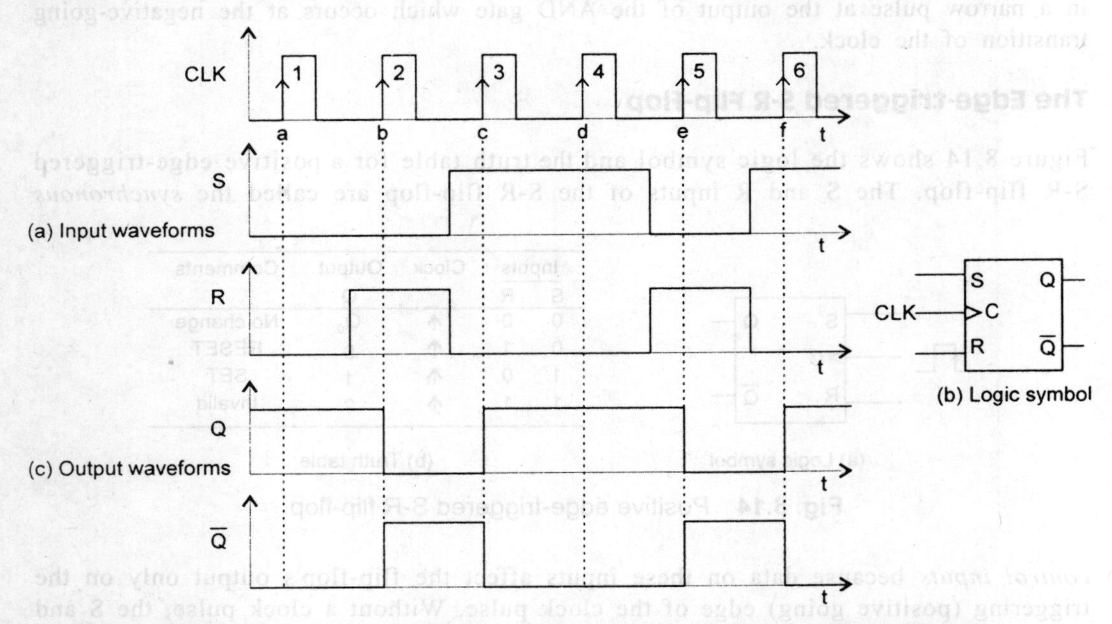

Fig. 8.16 Waveforms—positive edge-triggered S-R flip-flop (Example 8.3).

Solution

The output waveform is drawn as shown in Fig. 8.16c after going through the following steps.

1. Initially, S = 0 and R = 0 and Q is assumed to be HIGH.
2. At the positive-going transition of the first clock pulse (i.e. at a), both S and R are LOW. So, no change of state takes place. Q remains HIGH and $\overline{Q}$ remains LOW.
3. At the leading edge of the second clock pulse (i.e. at b), S = 0 and R = 1. So, the flip-flop resets. Hence, Q goes LOW and $\overline{Q}$ goes HIGH.
4. At the positive-going edge of the third clock pulse (i.e. at c), S = 1 and R = 0. So, the flip-flop sets. Hence, Q goes HIGH and $\overline{Q}$ goes LOW.
5. At the rising edge of the fourth clock pulse, S = 1 and R = 0. Since the flip-flop is already in a SET state, it remains SET. That is, Q remains HIGH and $\overline{Q}$ remains LOW.
6. The fifth pulse resets the flip-flop at its positive-going edge because S = 0 and R = 1 is the input condition and Q = 1 at that time.

Internal Circuitry of the Edge-triggered S-R Flip-flop

A detailed analysis of the internal circuitry of a flip-flop is not necessary, since all types are readily available as ICs. A simplified description is, however, presented here. Figure 8.17 shows the simplified circuitry of the edge-triggered S-R flip-flop.

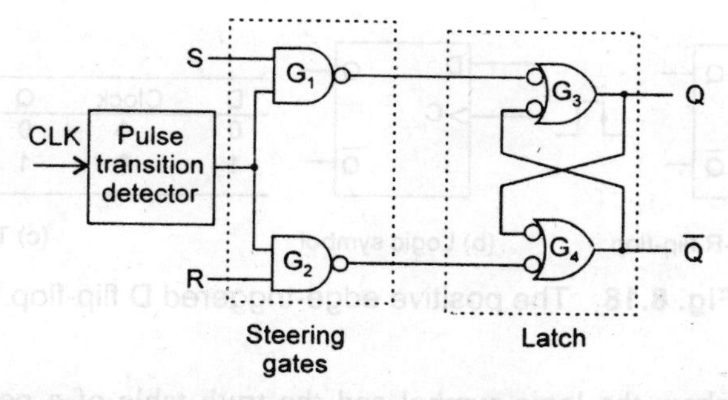

Fig. 8.17 Simplified circuitry of the edge-triggered S-R flip-flop.

It contains three sections.

1. A basic NAND gate latch formed by NAND gates (active-LOW OR gates) G_3 and G_4.
2. A pulse steering circuit formed by NAND gates G_1 and G_2.
3. An edge (pulse transition) detector circuit.

The edge detector generates a positive spike at the positive-going or negative-going edge of the clock pulse. The steering gates 'direct' or 'steer' the narrow spike either to G_3 or to G_4 depending on the state of the S and R inputs.

Let us say, initially Q = 0 and $\overline{Q}$ = 1. If S, R, and CLK are LOW, the outputs of G_1 and G_2 will be HIGH. Now since Q = 0 and $\overline{Q}$ = 1, inputs to G_3 are a 1 and a 1 ($\overline{Q}$). So, its output is a 0, i.e. Q remains a 0. Inputs to G_4 will be a 1 and a 0 (Q). So, its output will be a 1, i.e. $\overline{Q}$ remains a 1. So, no change of state takes place. If a clock pulse is applied (keeping S = 0 and R = 0) when Q is HIGH and $\overline{Q}$ is LOW, the outputs of G_1 and G_2 will be HIGH. So, the outputs of G_3 and G_4, i.e. Q and $\overline{Q}$ remain the same. Therefore, no change of state takes place. Thus, S = 0, R = 0 is the normal resting state.

Suppose, now S is made HIGH, and R kept LOW and a clock pulse is applied. Since the S input to G_1 is HIGH and the spike generated is also HIGH, the output of G_1 goes LOW for a very short duration, when the clock goes HIGH, causing the Q output to go HIGH. Both inputs to G_4 are now HIGH (G_2 output is HIGH because R is LOW) forcing $\overline{Q}$ to go LOW. This LOW on $\overline{Q}$ is coupled into one input of gate G_3, ensuring that the Q will remain HIGH. The flip-flop is now in the SET state.

Now, suppose S is made LOW and R HIGH and the clock pulse is applied. Since R is now HIGH, the positive-going edge of the clock pulse produces a negative spike at the output of G_2 forcing $\overline{Q}$ to go HIGH. Since both the inputs to G_3 are now HIGH, its output Q will be LOW. Since Q is coupled to one input of G_4, $\overline{Q}$ remains HIGH. The flip-flop is now in the RESET state.

As in the case of the gated S-R latch, S = 1, R = 1 is an invalid state.

The Edge-triggered D Flip-flop

The edge-triggered D flip-flop has only one input terminal. The D flip-flop may be obtained from an S-R flip-flop by just putting one inverter between the S and R terminals (Fig. 8.18a).

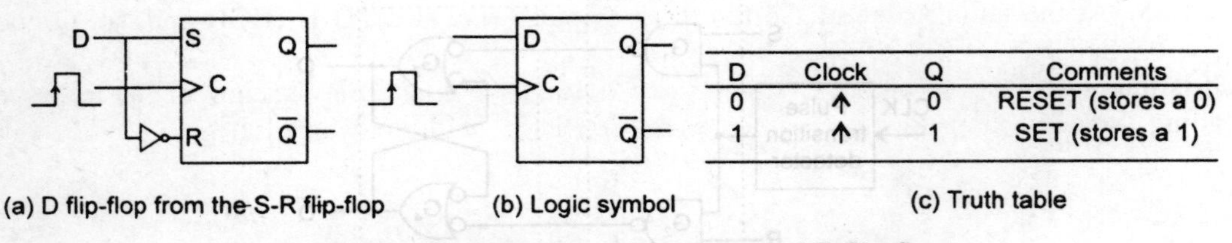

D	Clock	Q	Comments
0	↑	0	RESET (stores a 0)
1	↑	1	SET (stores a 1)

Fig. 8.18 The positive edge-triggered D flip-flop.

Figures 8.18b and c show the logic symbol and the truth table of a positive edge-triggered D flip-flop. Note that, this flip-flop has only one synchronous control input in addition to the clock. This is called the D (data) input. The operation of the D flip-flop is very simple. The output Q will go to the same state that is present on the D input at the positive-going transition of the clock pulse. In other words, the level present at D will be stored in the flip-flop at the instant the positive-going transition occurs. That is, if D is a 1 and the clock is applied, Q goes to a 1 and $\overline{Q}$ to a 0 at the rising edge of the pulse and thereafter remain so. If D is a 0 and the clock is applied, Q goes to a 0 and $\overline{Q}$ to a 1 at the rising edge of the clock pulse and thereafter remain so.

The negative edge-triggered D flip-flop operates in the same way as a positive edge-triggered D flip-flop except that the change of state takes place at the negative-going edge of the clock pulse. Figures 8.19a, b and c show the flip-flop, its logic symbol and the truth table, respectively.

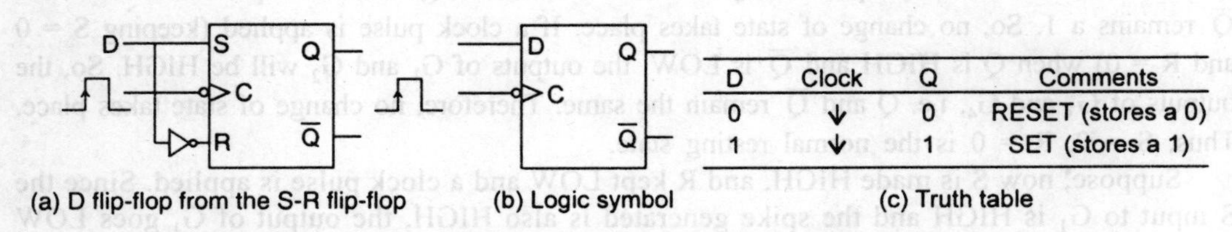

D	Clock	Q	Comments
0	↓	0	RESET (stores a 0)
1	↓	1	SET (stores a 1)

Fig. 8.19 The negative edge-triggered D flip-flop.

EXAMPLE 8.4 The waveforms shown in Fig. 8.20a are applied to the D flip-flop shown in Fig. 8.20b. Draw the output waveform.

Solution

The output waveform shown in Fig. 8.21c is drawn as explained below:

1. Initially Q is assumed to be HIGH, D LOW, and the clock LOW.
2. At the negative-going edge of the first clock pulse (i.e. at a), D is LOW and, so, Q goes LOW.
3. At the negative transition edge of the second clock pulse (i.e. at b), D is HIGH and, so, Q goes HIGH.
4. At the trailing edge of the third clock pulse (i.e. at c), D is LOW and, so, Q goes LOW.

5. At the falling edge of the fourth clock pulse (i.e. at d), D is HIGH and, so, Q goes HIGH.

It is thus seen that the output Q assumes the state of the input D at the time of the negative-going (negative transition or trailing or falling) edge of the clock pulse.

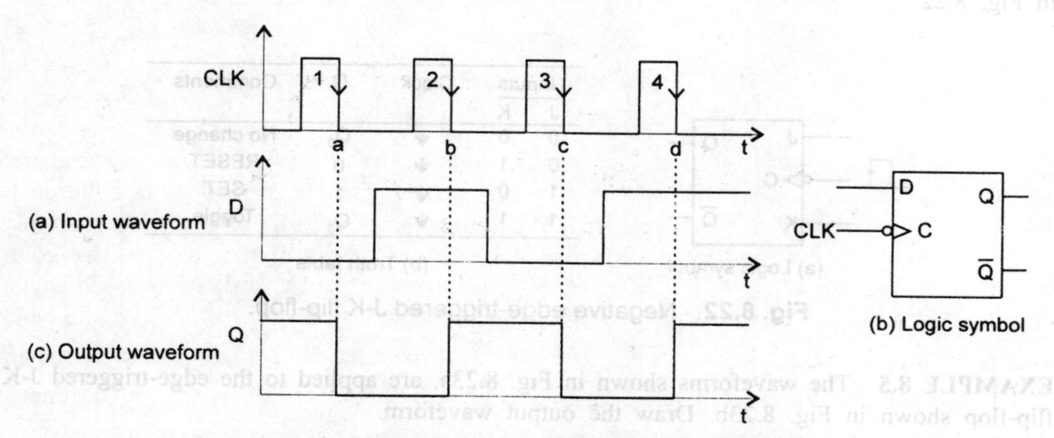

Fig. 8.20 Waveforms—D flip-flop (Example 8.4).

The Edge-triggered J-K Flip-flop

The J-K flip-flop is very versatile and also the most widely used. The J and K designations for the synchronous control inputs have no known significance.

The functioning of the J-K flip-flop is identical to that of the S-R flip-flop, except that it has no invalid state like that of the S-R flip-flop. The logic symbol and the truth table for a positive edge-triggered J-K flip-flop are shown in Fig. 8.21.

J, C, K, Q, $\overline{Q}$

(a) Logic symbol

Inputs		Clock	Q	Comments
J	K			
0	0	↑	Q_0	No change
0	1	↑	0	RESET
1	0	↑	1	SET
1	1	↑	$\overline{Q}_0$	Toggle

(b) Truth table

Fig. 8.21 Positive edge-triggered J-K flip-flop.

When J = 0 and K = 0, no change of state takes place even if a clock pulse is applied.

When J = 0 and K = 1, the flip-flop resets at the positive-going edge of the clock pulse.

When J = 1 and K = 0, the flip-flop sets at the positive-going edge of the clock pulse.

When J = 1 and K = 1, the flip-flop toggles, i.e. goes to the opposite state at the positive-going edge of the clock pulse. In this mode, the flip-flop toggles or changes state for each occurrence of the positive-going edge of the clock pulse.

The logic symbol and the truth table of a negative edge-triggered J-K flip-flop are shown in Fig. 8.22.

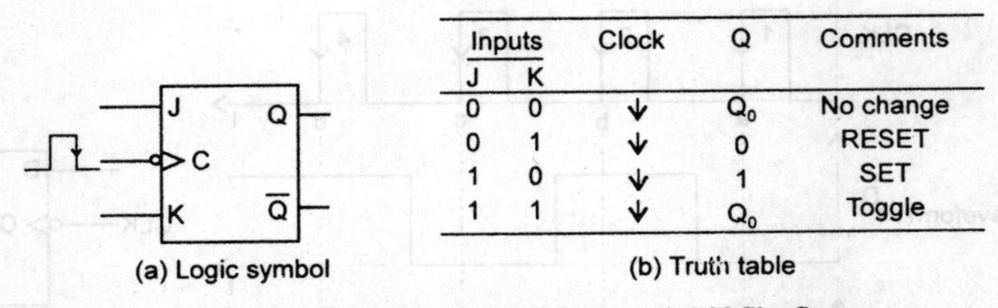

Inputs J	Inputs K	Clock	Q	Comments
0	0	↓	Q_0	No change
0	1	↓	0	RESET
1	0	↓	1	SET
1	1	↓	$\overline{Q}_0$	Toggle

(b) Truth table

Fig. 8.22 Negative edge-triggered J-K flip-flop.

EXAMPLE 8.5 The waveforms shown in Fig. 8.23a, are applied to the edge-triggered J-K flip-flop shown in Fig. 8.23b. Draw the output waveform.

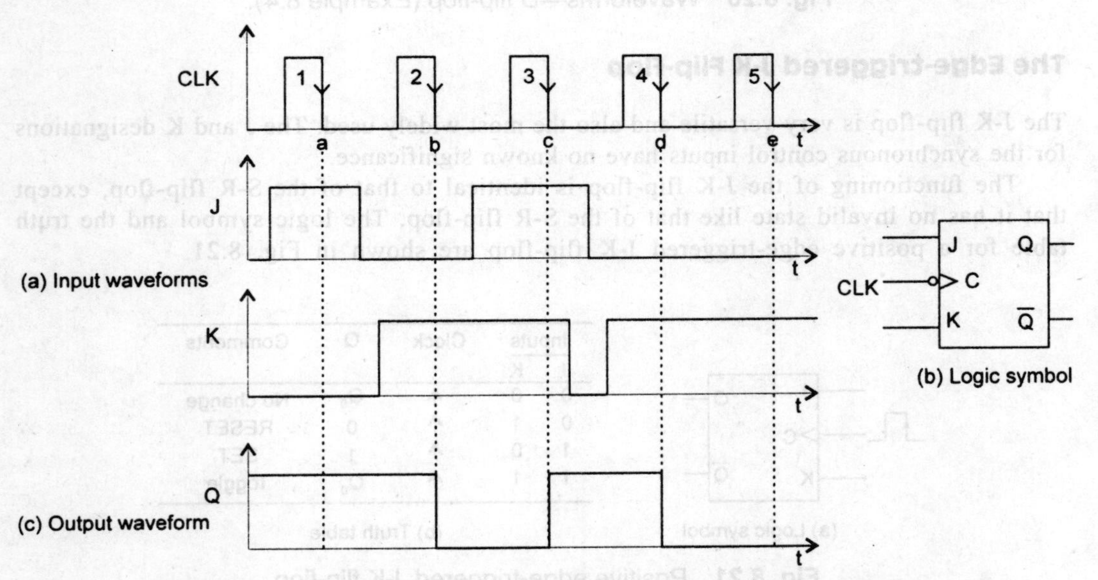

Fig. 8.23 Waveforms—edge-triggered J-K flip-flop (Example 8.5).

Solution

The output waveform shown in Fig. 8.23c is drawn as explained below:

1. Initially J = 0, K = 0 and CLK = 0. Assume that the initial state of the flip-flop is a 1, i.e. Q = 1 initially.

2. At the negative-going edge of the first clock pulse (i.e. at a), J = 1 and K = 0. So, Q remains as a 1 and, therefore, $\overline{Q}$ as a 0.
3. At the trailing edge of the second clock pulse (i.e. at b), J = 0 and K = 1. So, the flip-flop resets. That is, Q goes to a 0 and $\overline{Q}$ to a 1.
4. At the falling edge of the third clock pulse (i.e. at c), both J and K are a 1. So, the flip-flop toggles. That is, Q changes from a 0 to a 1 and $\overline{Q}$ from a 1 to a 0.
5. At the negative-going transition of the fourth clock pulse (i.e. at d), J = 0 and K = 1. So, the flip-flop RESETS, i.e. Q goes to a 0 and $\overline{Q}$ to a 1.

Internal Circuitry of the Edge-triggered J-K Flip-flop

A simplified version of the internal circuitry of the edge-triggered J-K flip-flop is shown in Fig. 8.24. It contains the same three sections as those of the edge-triggered S-R flip-flop. In fact, the only difference between the two circuits is that, the Q and $\overline{Q}$ outputs are fed back to the pulse steering NAND gates. This feedback connection is what gives the J-K flip-flop its toggle operation for J = K = 1 condition.

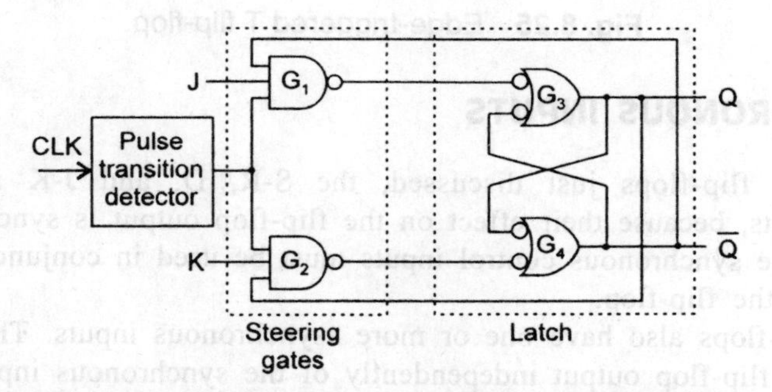

Fig. 8.24 Simplified circuit diagram of the edge-triggered J-K flip-flop.

The toggling operation may be explained as given below.

1. Suppose Q = 0, $\overline{Q}$ = 1 and J = K = 1. When a clock pulse is applied, the narrow positive pulse of the edge detector is inverted by gate G_1, i.e. G_1 steers the spike inverted (because its other inputs J and $\overline{Q}$ are both a 1) to gate G_3. So, the output of G_3, i.e. Q goes HIGH. Since Q is connected as one input of G_2, the output of G_2 remains HIGH (because initially Q was a 0). So, G_4 has both inputs as a 1. Thus, its output $\overline{Q}$ will be a 0.
2. Suppose, Q = 1, $\overline{Q}$ = 0, and J = K = 1. When a clock pulse is applied, the narrow positive spike at G_2 is steered (inverted) to the input of G_4. So, the output of G_4, i.e. $\overline{Q}$ goes HIGH. Since the output of G_1 is HIGH (because initially $\overline{Q}$ was a 0), both the inputs to G_3 are HIGH. So, the output of G_3, i.e. Q goes LOW.

So, we can say that, if a clock pulse is applied when J = 1 and K = 1, the flip-flop toggles, i.e. it changes its state.

The Edge-triggered T Flip-flop

A T flip-flop has a single control input, labelled T for toggle. When T is HIGH, the flip-flop toggles on every new clock pulse. When T is LOW, the flip-flop remains in whatever state it was before. Although T flip-flops are not widely available commercially, it is easy to convert a J-K flip-flop to the functional equivalent of a T flip-flop by just connecting J and K together and labelling the common connection as T. Thus, when T = 1, we have J = K = 1, and the flip-flop toggles. When T = 0, we have J = K = 0, and there is no change. The logic symbol and the truth table of a T flip-flop are shown in Fig. 8.25.

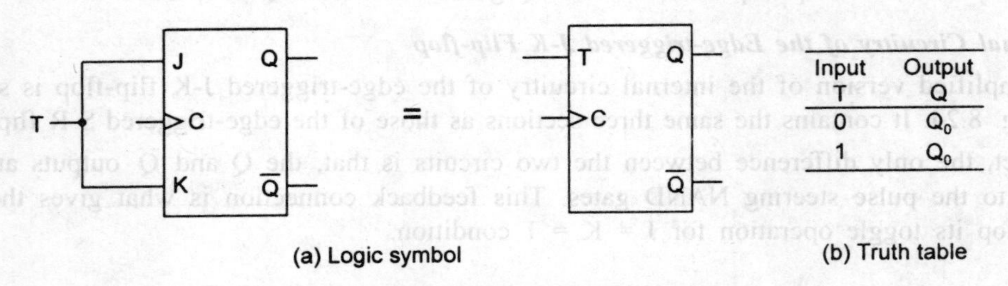

Input T	Output Q
0	Q_0
1	$\overline{Q_0}$

Fig. 8.25 Edge-triggered T flip-flop.

8.5 ASYNCHRONOUS INPUTS

For the clocked flip-flops just discussed, the S-R, D, and J-K inputs are called synchronous inputs, because their effect on the flip-flop output is synchronized with the clock input. These synchronous control inputs must be used in conjunction with a clock signal to trigger the flip-flop.

Most IC flip-flops also have one or more asynchronous inputs. These asynchronous inputs affect the flip-flop output independently of the synchronous inputs and the clock input. These asynchronous inputs can be used to SET the flip-flop to the 1 state or RESET the flip-flop to the 0 state at any time regardless of the conditions at the other inputs. In other words, we can say that the asynchronous inputs are the override inputs, which can be used to override all the other inputs in order to place the flip-flop in one state or the other. They are normally labelled PRESET (PRE) or direct SET (S_D) or DC SET, and CLEAR (CLR) or direct RESET (R_D) or DC CLEAR. An active level on the PRESET input will SET the flip-flop and an active level on the CLEAR input will RESET it. If the asynchronous inputs are active-LOW, the same is indicated by a small bubble at the input terminals. The inputs in that case are labelled as $\overline{\text{PRE}}$ and $\overline{\text{CLR}}$ or $\overline{S}_D$ and $\overline{R}_D$. Most IC flip-flops have both DC SET and DC CLEAR inputs. Some have only DC CLEAR. Some have active-HIGH inputs, some have active-LOW inputs.

It is important to realize that these asynchronous inputs respond to the DC levels. That means, in the case of active-HIGH (LOW) inputs, if a constant 1 (0) is held on the PRE ($\overline{\text{PRE}}$) input, the flip-flop will remain in the Q = 1 state regardless of what is occurring at the other inputs. Similarly, if a constant 1 (0) is held on the CLR ($\overline{\text{CLR}}$) input, the flip-flop will remain in the Q = 0 state regardless of what is occurring at the other inputs. Most often, however, the asynchronous inputs are used to SET or CLEAR the flip-flop to the desired state by the application of a momentary pulse. When DC SET and DC RESET conditions are not used in any application, they must be held at their inactive levels.

The logic symbol and the truth table of a J-K flip-flop with active-LOW PRESET and CLEAR inputs are shown in Fig. 8.26. In this case, both PRESET and CLEAR inputs must be kept HIGH for synchronous operation.

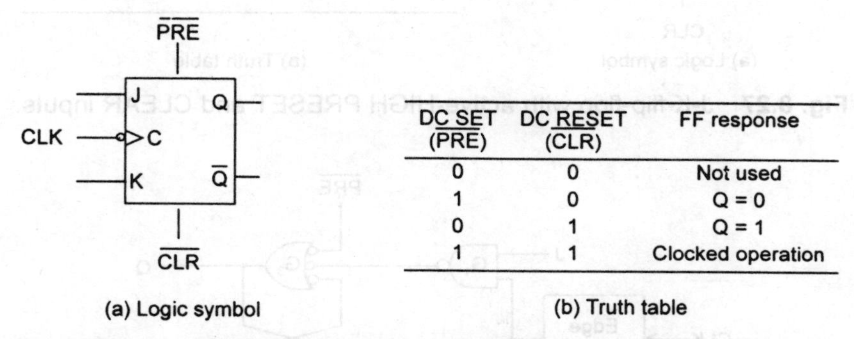

DC SET ($\overline{PRE}$)	DC RESET ($\overline{CLR}$)	FF response
0	0	Not used
1	0	Q = 0
0	1	Q = 1
1	1	Clocked operation

(a) Logic symbol (b) Truth table

Fig. 8.26 J-K flip-flop with active-LOW PRESET and CLEAR inputs.

The operation is discussed below.

1. $\overline{PRE}$ = 1, $\overline{CLR}$ = 1, i.e. DC SET = 1 and DC CLEAR = 1. The asynchronous inputs are inactive and the flip-flop responds freely to J, K and CLK inputs in the normal way. In other words, the clocked operation can take place.
2. $\overline{PRE}$ = 0, $\overline{CLR}$ = 1, the DC SET is activated and Q is immediately SET to a 1, no matter what conditions are present at the J, K and CLK inputs. The CLK input cannot affect the flip-flop while DC SET = 0.
3. $\overline{PRE}$ = 1, $\overline{CLR}$ = 0. The DC CLEAR is activated and Q is immediately cleared to a 0 independent of the conditions on the J, K or CLK inputs. The CLK input has no effect when DC CLEAR = 0.
4. DC SET = DC CLEAR = 0. This condition should not be used, since it can result in an invalid state.

Figure 8.27 shows the logic symbol and the truth table of a J-K flip-flop with active-HIGH PRESET and CLEAR. Figure 8.28 shows the logic diagram for an edge-triggered J-K flip-flop with PRESET ($\overline{PRE}$) and CLEAR ($\overline{CLR}$) inputs. As shown in the figure, these inputs are connected directly into the latch portion of the flip-flop so that they override the effect of the synchronous inputs J, K, and the CLK.

EXAMPLE 8.6 The waveforms shown in Fig. 8.29a are applied to the J-K flip-flop shown in Fig. 8.29b. Draw the output waveform.

Solution

The output waveform shown in Fig. 8.29c is drawn as explained below:

1. Initially $\overline{PRE}$ and $\overline{CLR}$ are both a 1, and Q is LOW.

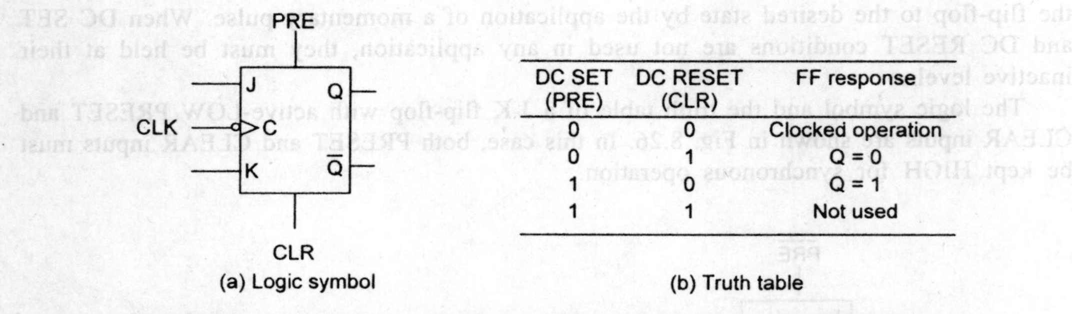

(a) Logic symbol

DC SET (PRE)	DC RESET (CLR)	FF response
0	0	Clocked operation
0	1	Q = 0
1	0	Q = 1
1	1	Not used

(b) Truth table

Fig. 8.27 J-K flip-flop with active-HIGH PRESET and CLEAR inputs.

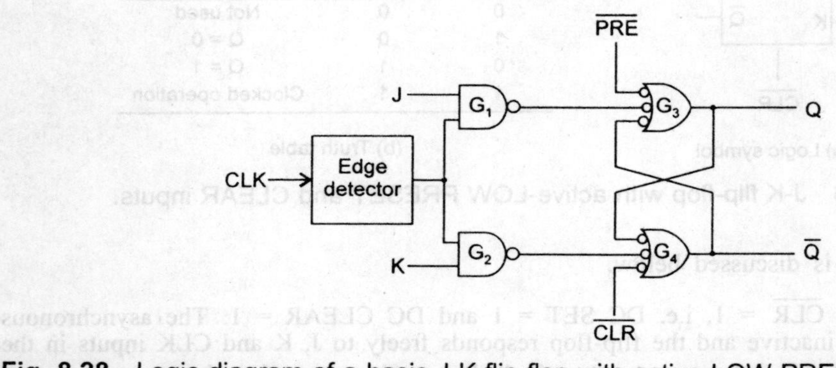

Fig. 8.28 Logic diagram of a basic J-K flip-flop with active-LOW PRESET and CLEAR.

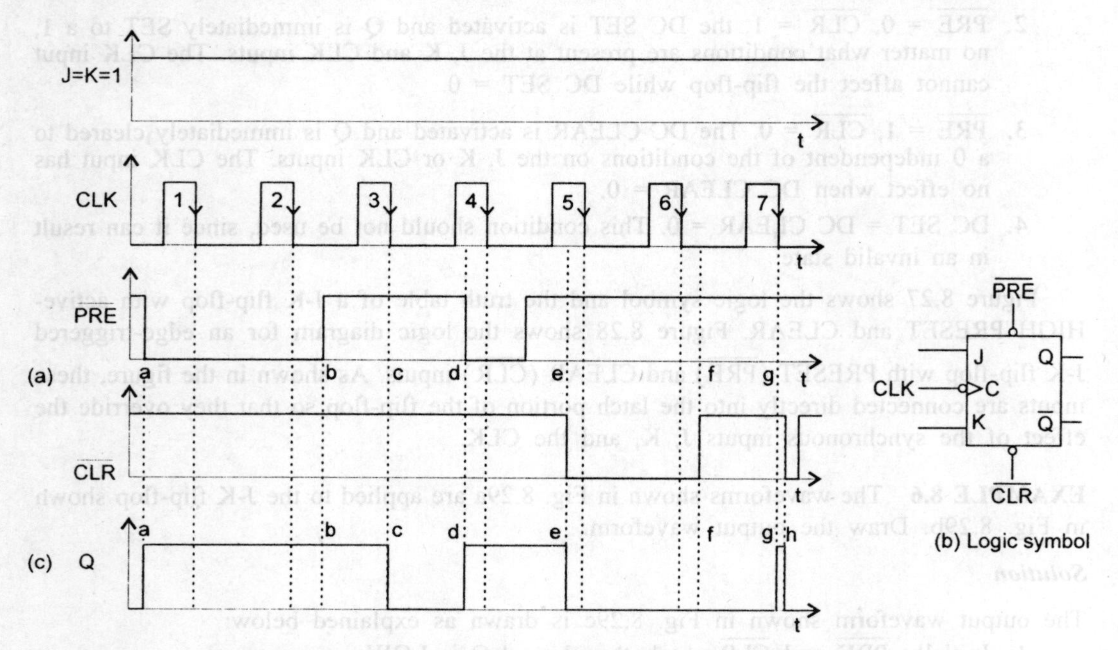

Fig. 8.29 Waveforms—J-K flip-flop (Example 8.6).

2. At the instant a, $\overline{PRE}$ goes LOW. So, Q is SET to a 1, and remains SET up to b, because $\overline{PRE}$ is kept LOW up to b. From b to c it also remains at a 1, because both $\overline{PRE}$ and $\overline{CLR}$ are a 1.
3. Since the flip-flop is in the clocked mode (i.e. $\overline{PRE}$ = 1 and $\overline{CLR}$ = 1) and since J and K are both a 1, the flip-flop toggles and goes to a 0 at the negative-going edge of the third clock pulse at c.
4. At d, $\overline{PRE}$ goes LOW. So, Q is SET to a 1 and remains SET till e.
5. At e, $\overline{CLR}$ goes LOW. So, Q is RESET to a 0 and remains RESET till f.
6. After f, Q toggles and goes to a 1 at g at the negative-going edge of the seventh clock pulse.
7. At h, $\overline{CLR}$ goes LOW. So, Q also goes LOW.

8.6 FLIP-FLOP OPERATING CHARACTERISTICS

Manufacturers of IC flip-flops specify several important characteristics and timing parameters that must be considered before a flip-flop is used in any circuit application. They are typically found in the data sheets for ICs, and they are applicable to all flip-flops regardless of the particular form of the circuit.

Propagation Delay Time

The output of a flip-flop will not change state immediately after the application of the clock signal or asynchronous inputs. The time interval between the time of application of the triggering edge or asynchronous inputs and the time at which the output actually makes a transition is called the *propagation delay time* of the flip-flop. It is usually in the range of a few ns to 1 μs. Several categories of propagation delay are important in the operation of a flip-flop. The propagation delays that occur in response to a positive transition on the clock input are illustrated in Fig. 8.30. They are:

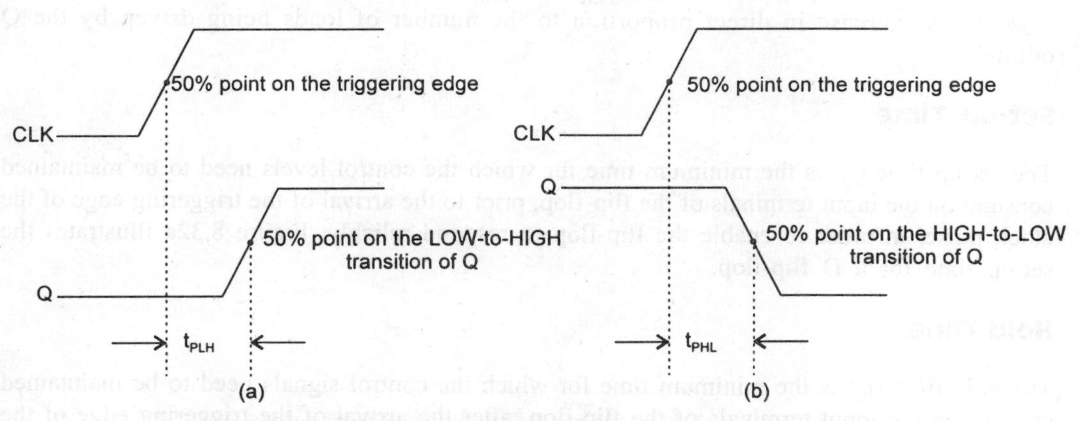

Fig. 8.30 Propagation delays t_{PLH} and t_{PHL} w.r.t. CLK.

2. Propagation delay t_{PLH} measured from the triggering of the clock pulse to the LOW-to-HIGH transition of the output (shown in Fig. 8.30a).
3. Propagation delay t_{PHL} measured from the triggering of the clock pulse to the HIGH-to-LOW transition of the output (shown in Fig. 8.30b).

The propagation delays that occur in response to signals on a flip-flop's asynchronous inputs (PRESET and CLEAR) are illustrated in Fig. 8.31. They are:

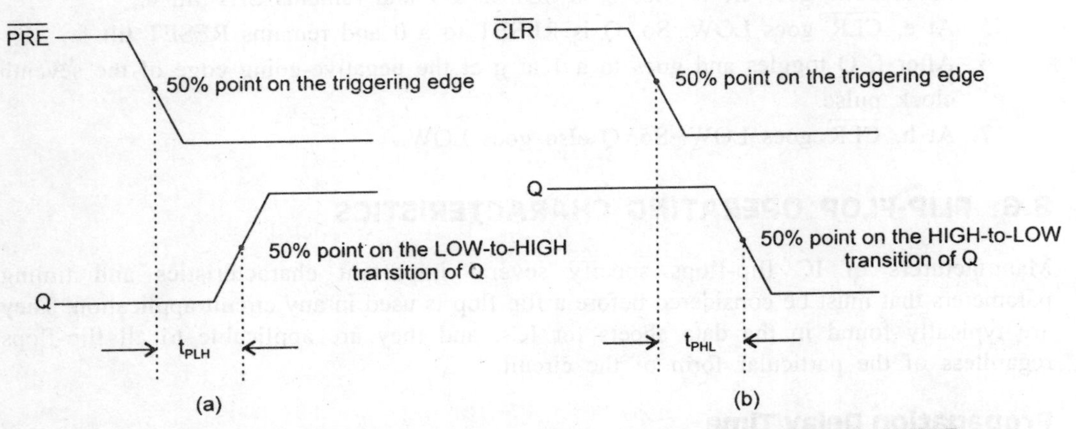

Fig. 8.31 Propagation delays t_{PLH} and t_{PHL} w.r.t. PRESET and CLEAR.

1. Propagation delay t_{PLH} measured from the PRESET input to the LOW-to-HIGH transition of the output. Figure 8.31a. illustrates this delay for active-LOW PRESET.
2. Propagation delay t_{PHL} measured from the CLEAR input to the HIGH-to-LOW transition of the output. Figure 8.31b illustrates this delay for active-LOW CLEAR.

Note that these delays are measured between the 50% points on the input and output waveforms. The propagation delays t_{PLH} and t_{PHL} are usually in the range of a few ns to 1 μs. They increase in direct proportion to the number of loads being driven by the Q output.

Set-up Time

The set-up time (t_s) is the minimum time for which the control levels need to be maintained constant on the input terminals of the flip-flop, prior to the arrival of the triggering edge of the clock pulse, in order to enable the flip-flop to respond reliably. Figure 8.32a illustrates the set-up time for a D flip-flop.

Hold Time

The hold time (t_h) is the minimum time for which the control signals need to be maintained constant at the input terminals of the flip-flop, after the arrival of the triggering edge of the clock pulse, in order to enable the flip-flop to respond reliably. Figure 8.32b illustrates the hold time for a D flip-flop.

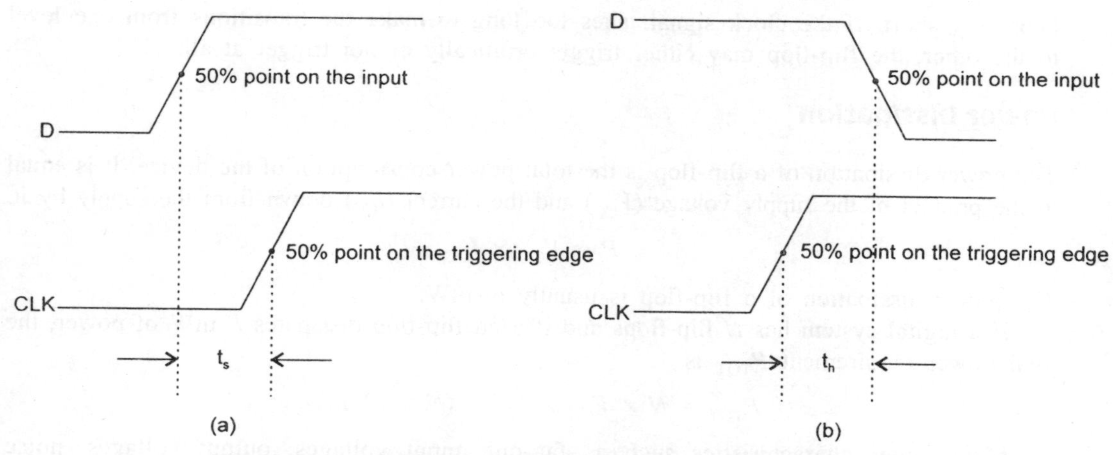

Fig. 8.32 Set-up time and hold time for a D flip-flop.

Maximum Clock Frequency

The maximum clock frequency (f_{MAX}) is the highest frequency at which a flip-flop can be reliably triggered. If the clock frequency is above this maximum, the flip-flop would be unable to respond quickly enough and its operation will be unreliable. The f_{MAX} limit will vary from one flip-flop to another.

Pulse Widths

The manufacturer usually specifies the minimum pulse widths for the clock and asynchronous inputs. For the clock signal, the minimum HIGH time $t_W(H)$ and the minimum LOW time $t_W(L)$ are specified and for asynchronous inputs, i.e. PRESET and CLEAR, the minimum active state time is specified. Failure to meet these minimum time requirements can result in unreliable operation. Figure 8.33 shows pulse widths for CLK and asynchronous inputs.

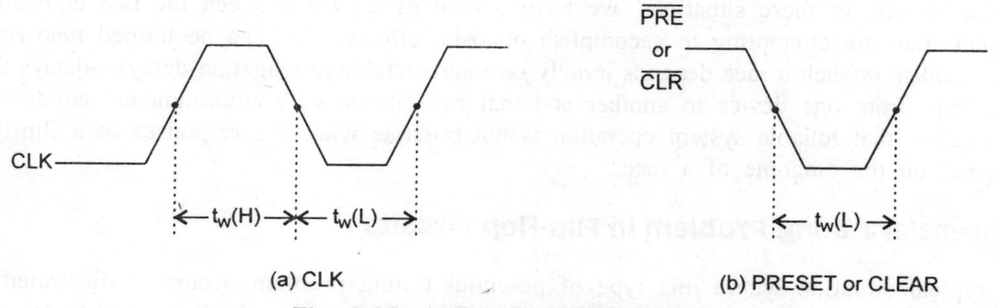

Fig. 8.33 Minimum pulse widths.

Clock Transition Times

For reliable triggering, the clock waveform transition times (rise and fall times) should be kept very short. If the clock signal takes too long to make the transitions from one level to the other, the flip-flop may either trigger erratically or not trigger at all.

Power Dissipation

The power dissipation of a flip-flop is the total power consumption of the device. It is equal to the product of the supply voltage (V_{CC}) and the current (I_{CC}) drawn from the supply by it.

$$P = V_{CC} \times I_{CC}$$

The power dissipation of a flip-flop is usually in mW.

If a digital system has N flip-flops and if each flip-flop dissipates P mW of power, the total power requirement P_{TOT} is

$$P_{TOT} = N \times V_{CC} \times I_{CC} = (N \times P) \text{ mW}$$

Many other characteristics such as fan-out, input voltages, output voltages, noise margin, etc., will be discussed in Chapter 11 in relation to logic gates.

Clock Skew and Time Race

One of the most common timing problems in synchronous circuits is clock skew. In many digital circuits, the output of one flip-flop is connected either directly or through logic gates to the input of another flip-flop, and both flip-flops are triggered by the same clock signal. The propagation delay of a flip-flop and/or the delays of the intervening gates make it difficult to predict precisely when the changing state of one flip-flop will be experienced at the input of another.

The clock signal which is applied simultaneously to all flip-flops in a synchronous system may undergo varying degrees of delay caused by wiring between components, and arrive at the CLK inputs of different flip-flops at different times. This delay is called *clock skew*. If the clock skew is minimal, a flip-flop may get clocked before it receives a new input (derived from the output of another clocked flip-flop). On the other hand, if the clock pulse is delayed significantly, the inputs to a flip-flop may have changed before the clock pulse arrives. In these situations, we have a kind of a *race* between the two competing signals that are attempting to accomplish opposite effects. This can be termed *time race*. The *winner* in such a race depends largely on unpredictable propagation delays—delays that can vary from one device to another and that can change with environmental conditions. It is clear that reliable system operation is not possible when the responses of a flip-flop depend on the outcome of a race.

Potential Timing Problem in Flip-flop Circuits

A typical situation where this type of potential timing problem occurs is illustrated in Fig. 8.34, where the output of the first flip-flop Q_1 is connected to the S input of the second flip-flop and both the flip-flops are clocked by the same signal at their CLK inputs.

The potential timing problem is like this: Since Q_1 will change on the positive-going transition of the clock pulse, the S_2 input of the second flip-flop will be in a changing state as it receives the same positive-going transition. This could lead to an unpredictable response at Q_2.

Let us assume that, initially Q_1 = 1 and Q_2 = 0. Thus, FF_1 has S_1 = 0, R_1 = 1 and FF_2 has S_2 = 1, R_2 = 0 prior to the positive-going transition of the clock pulse. When the positive-going transition occurs, Q_1 will go to the LOW state, but cannot actually go LOW until after the propagation delay t_{PHL}. The same positive-going transition will reliably clock FF_2 to the HIGH state, provided that t_{PHL} is greater than the FF_2's hold time requirement t_h. If this condition is not met, the response of FF_2 will be unpredictable.

Fortunately, all modern edge-triggered flip-flops have hold time requirements that are 5 ns or less; most have t_h = 0 which means that they have no hold time requirements. So, we can say that:

"The flip-flop output will go to a state determined by the logic levels present at its synchronous control inputs just prior to the active clock transition."

If we apply this rule to Fig. 8.34, it says that Q_2 of FF_2 will go to a state determined by the S_2 = 1 and R_2 = 0, a condition that is present just prior to the positive-going transition of the clock pulse. The fact that S_2 is changing in response to the same positive-going transition has no effect on Q_2.

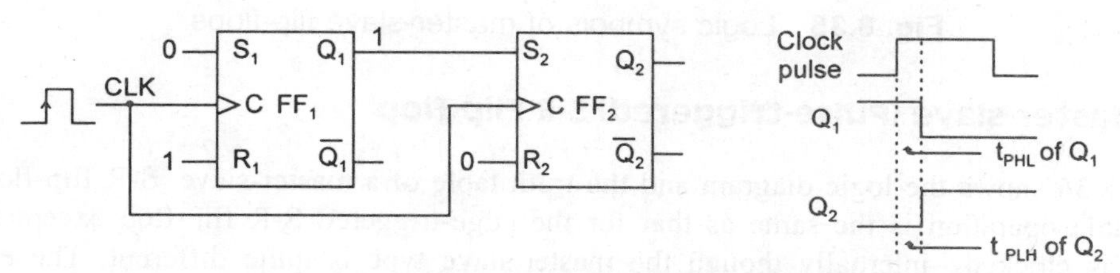

Fig. 8.34 Illustration of timing problem.

8.7 MASTER-SLAVE (PULSE-TRIGGERED) FLIP-FLOPS

Before the development of edge-triggered flip-flops with little or no hold time requirements, the timing problems such as those shown in Fig. 8.34 were often handled by a class of flip-flops called the master-slave flip-flops.

The master-slave flip-flop was developed to make the synchronous operation more predictable, that is, to avoid the problems of logic race in clocked flip-flops. This improvement is achieved by introducing a known time delay (equal to the width of one clock pulse) between the time that the flip-flop responds to a clock pulse and the time the response appears at its output. A master-slave flip-flop is also called a *pulse-triggered flip-flop* because the length of the time required for its output to change state equals the width of one clock pulse.

The master-slave or pulse-triggered flip-flop actually contains two flip-flops—a master flip-flop and a slave flip-flop. The control inputs are applied to the master flip-flop and maintained constant for the set-up time t_s prior to the application of the clock pulse. On the rising edge of the clock pulse, the levels on the control inputs are used to determine the output of the master. On the falling edge of the clock pulse, the state of the master is transferred to the slave, whose outputs are Q and $\overline{Q}$. Thus, the actual outputs of the flip-flop, i.e. Q and $\overline{Q}$ change just after the negative-going transition of the clock. These master-slave flip-flops function very much like the negative edge-triggered flip-flops except for one major disadvantage. The control inputs must be held stable while CLK is HIGH, otherwise an unpredictable operation may occur. This problem with the master-slave flip-flop is overcome with an improved master-slave version called the *master-slave with data lock-out.*

There are three basic types of master-slave flip-flops—S-R, D, and J-K. The J-K is by far the most commonly available in IC form. Figure 8.35 shows the logic symbols. The key to identifying a master-slave flip-flop by its logic symbol is the postponed output symbol ˥ at the outputs. Note that there is no dynamic input indicator at the clock input.

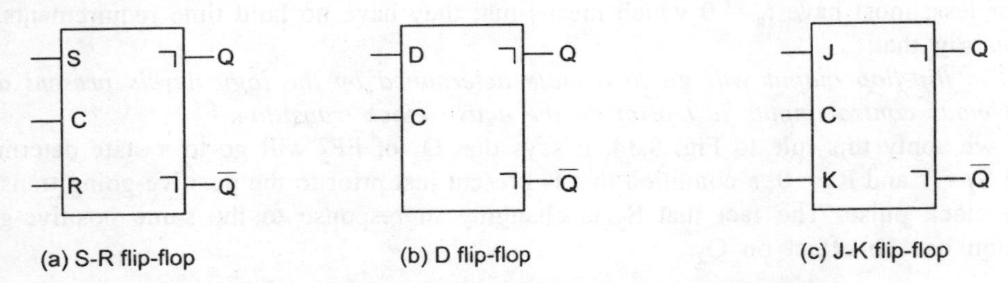

(a) S-R flip-flop (b) D flip-flop (c) J-K flip-flop

Fig. 8.35 Logic symbols of master-slave flip-flops.

The Master-slave (Pulse-triggered) S-R Flip-flop

Figure 8.36 shows the logic diagram and the truth table of a master-slave, S-R flip-flop. The truth table operation is the same as that for the edge-triggered S-R flip-flop except for the way it is clocked—internally though the master-slave type is quite different. The external control inputs S and R are applied to the master section. The master section is basically a gated S-R latch, and responds to the external S-R inputs applied to it at the positive-going edge of the clock signal. The slave section is the same as the master section except that it is clocked on the inverted clock pulse and thus responds to its control inputs (which are nothing but the outputs of the master flip-flop) at the negative-going edge of the clock pulse. Thus, the master section assumes the state determined by the S and R inputs at the positive-going edge of the clock pulse and the slave section copies the state of the master section at the negative-going edge of the clock pulse. The state of the slave then immediately appears on its Q and $\overline{Q}$ outputs.

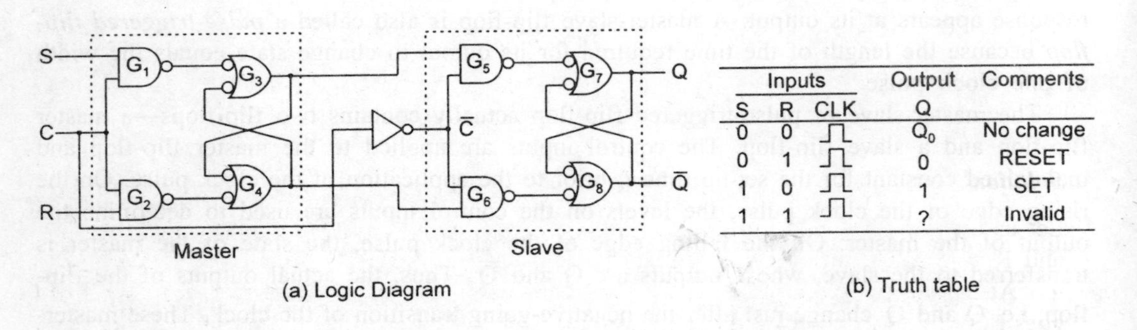

Inputs			Output	Comments
S	R	CLK	Q	
0	0	⎍	Q_0	No change
0	1	⎍	0	RESET
1	0	⎍	1	SET
1	1	⎍	?	Invalid

(a) Logic Diagram (b) Truth table

Fig. 8.36 The master-slave S-R flip-flop.

EXAMPLE 8.7 The waveforms shown in Fig. 8.37a are applied to the master-slave flip-flop shown in Fig. 8.36. Draw the output waveform.

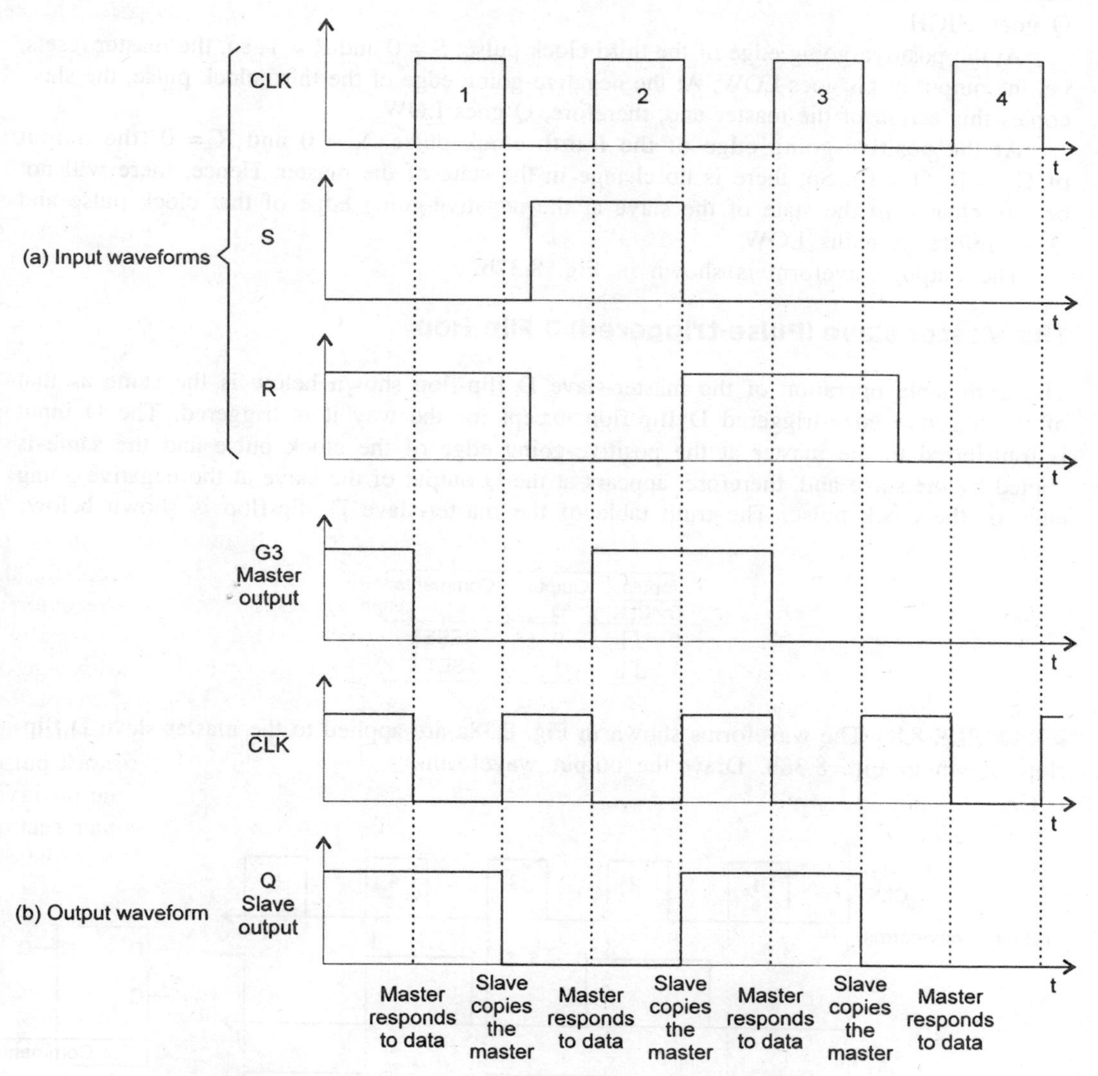

Fig. 8.37 Waveforms—master-slave flip-flop (Example 8.7).

Solution

Let us assume that, initially the flip-flop is SET, i.e. Q = 1 and the control inputs are S = 0 and R = 1 and the output of master is a 1.

At the positive-going edge of the first clock pulse, the master resets, i.e. the output of G_3 goes LOW. At the negative-going edge of the first clock pulse, the slave copies it. So, Q goes LOW. The inputs S and R now change when the clock is LOW; so, it does not affect the operation. At the positive-going edge of the second clock pulse, S = 1 and R = 0 (and G_3 = 0, Q = 0). So, the master sets, i.e. the output of G_3 goes HIGH. At the negative-going edge of the second clock pulse, the slave copies this action of the master and, therefore, Q goes HIGH.

At the positive-going edge of the third clock pulse, S = 0 and R = 1, so, the master resets, i.e. the output of G_3 goes LOW. At the negative-going edge of the third clock pulse, the slave copies this action of the master and, therefore, Q goes LOW.

At the positive-going edge of the fourth clock pulse, S = 0 and R = 0 (the output of G_3 = 0, Q = 0). So, there is no change in the state of the master. Hence, there will not be any change in the state of the slave at the negative-going edge of that clock pulse and Q, therefore, remains LOW.

The output waveform is shown in Fig. 8.37b.

The Master-slave (Pulse-triggered) D Flip-flop

The truth table operation of the master-slave D flip-flop shown below is the same as that of the negative edge-triggered D flip-flop except for the way it is triggered. The D input is transferred to the master at the positive-going edge of the clock pulse and the same is copied by the slave and, therefore, appears at the Q output of the slave at the negative-going edge of the clock pulse. The truth table of the master-slave D flip-flop is shown below:

Inputs		Output	Comments
D	CLK	Q	
0	⎍	0	RESET
1	⎍	1	SET

EXAMPLE 8.8 The waveforms shown in Fig. 8.38a are applied to the master-slave D flip-flop shown in Fig. 8.38b. Draw the output waveform.

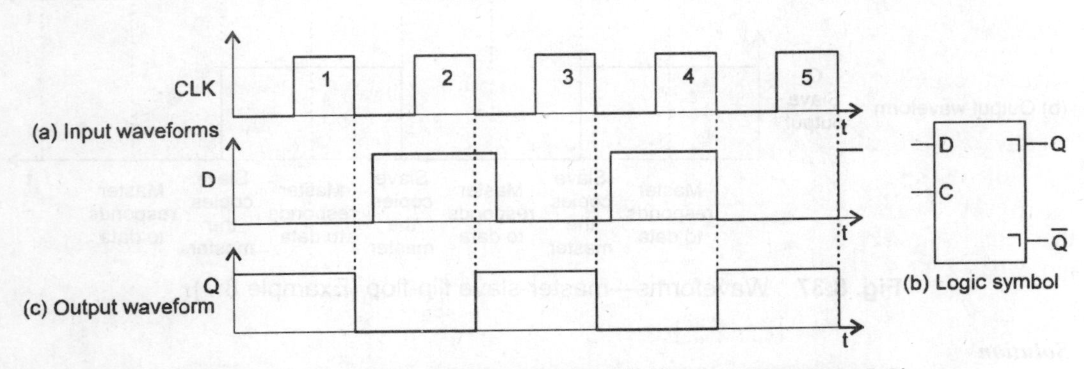

Fig. 8.38 Waveforms—master-slave D flip-flop (Example 8.8).

Solution

The output waveform shown in Fig. 8.38c is drawn as explained below:

Assume that initially Q = 1. At the positive-going edge of the first clock pulse, D is LOW. So, Q goes LOW at the negative-going edge of the first clock pulse.

At the positive-going edge of the second clock pulse, D is HIGH. So, Q goes HIGH at the negative-going edge of the second clock pulse.

D is LOW at the positive-going edge of the third clock pulse. So, Q goes LOW at the negative-going edge of the third clock pulse, and so on.

A major restriction on the use of a master-slave flip-flop is that, the input data must be held constant while the clock is HIGH. If the data changes states while the clock pulse is HIGH, the master responds immediately to the new data, thus, loosing the data which was available at the preceding positive-going edge of the clock pulse. The slave then responds to this new state of the master at the next negative-going edge of the clock pulse. Thus, the flip-flop provides an output corresponding to the last input level seen while the clock pulse was HIGH. This is illustrated by the example given below.

EXAMPLE 8.9 The waveforms shown in Fig. 8.39a are applied to the master-slave D flip-flop shown in Fig. 8.39b. Draw the output waveform.

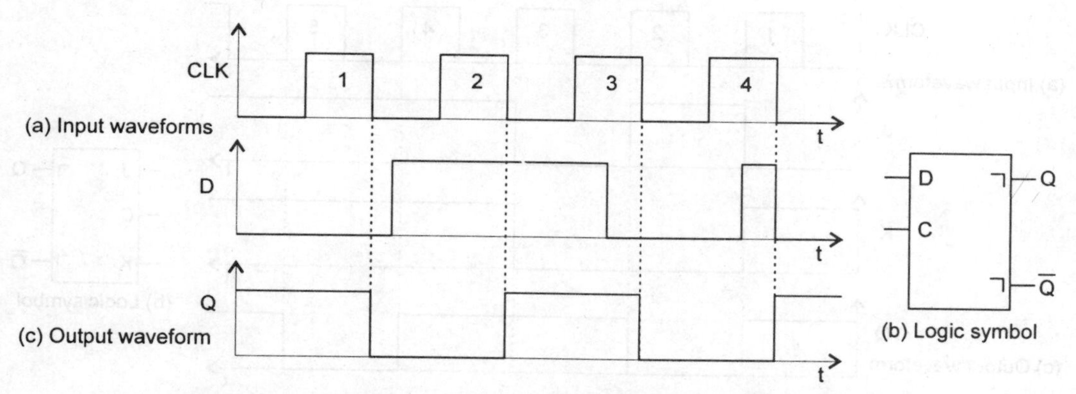

Fig. 8.39 Waveforms—master-slave D flip-flop (Example 8.9).

Solution

The output waveform shown in Fig. 8.39c is drawn as explained below:

At the positive-going edge of the third clock pulse, D is HIGH, but Q goes LOW at the negative-going edge of this clock pulse because D went LOW when the third clock pulse was still HIGH.

At the positive-going edge of the fourth clock pulse, D is LOW, but Q goes HIGH at the negative-going edge of this clock pulse, because D went HIGH when the fourth clock pulse was still HIGH.

The Master-slave (Pulse-triggered) J-K Flip-flop

Figure 8.40 shows the logic diagram and the truth table of a master-slave J-K flip-flop. The truth table operation is the same as that of a negative edge-triggered J-K flip-flop except for the way in which it is triggered. The logic diagram of the master-slave J-K flip-flop is similar to that of the master-slave S-R flip-flop. The difference is that the Q output is connected back to the input of G_2 and the $\overline{Q}$ output is connected back to the input of G_1 and the external inputs are designated as J and K.

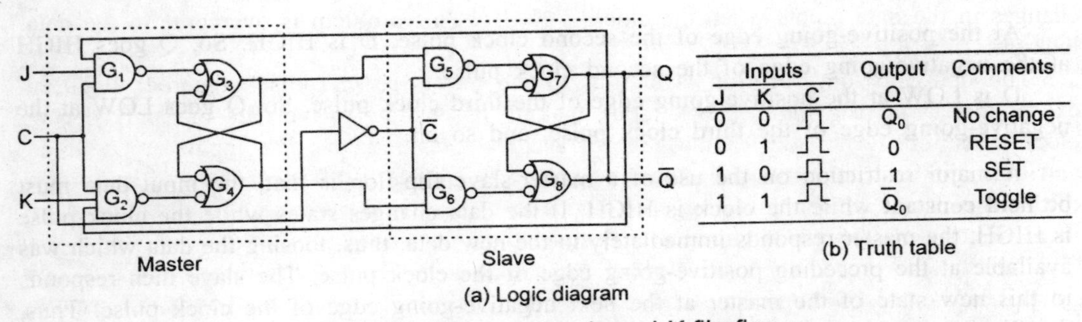

(b) Truth table

Inputs			Output	Comments
J	K	C	Q	
0	0	⎍	Q_0	No change
0	1	⎍	0	RESET
1	0	⎍	1	SET
1	1	⎍	$\bar{Q}_0$	Toggle

Fig. 8.40 The master-slave J-K flip-flop.

EXAMPLE 8.10 The waveforms shown in Fig. 8.41a are applied to the master-slave J-K flip-flop shown in Fig. 8.41b. Draw the output waveform.

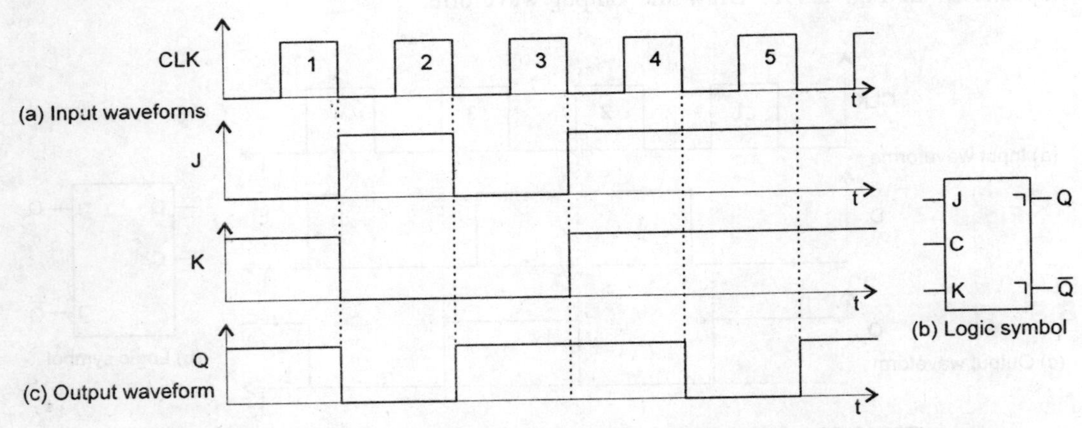

Fig. 8.41 Waveforms—master-slave J-K flip-flop (Example 8.10).

Solution

The output waveform shown in Fig. 8.41c is drawn as explained below:

Initially, J = 0 and K = 1 and the flip-flop is assumed to be in SET state, i.e. Q = 1.

At the positive-going edge of the first clock pulse, J = 0 and K = 1. So, the flip-flop resets, and Q goes LOW at the negative-going edge of this clock pulse.

At the positive-going edge of the second clock pulse, J = 1 and K = 0. So, the flip-flop sets, and Q goes HIGH at the negative-going edge of this clock pulse.

At the positive-going edge of the third clock pulse, J = 0 and K = 0. So, there will not be any change in the state of the flip-flop at the negative-going edge. Thus, the flip-flop remains SET. Hence Q remains HIGH. There afterwards, both J and K remain HIGH. So, the flip-flop will be in toggle mode. Hence, Q goes to the opposite state at the negative-going edge of each of the subsequent clock pulses.

The Data Lock-out Flip-flop

Earlier it was mentioned that a severe limitation of the master-slave flip-flop is that the data inputs must be held constant while the clock is HIGH, because it responds to any changes in the data inputs when the clock is HIGH. This problem is overcome in the data lock-out flip-flop.

The data lock-out flip-flop is similar to the master-slave (pulse-triggered) flip-flop except that it has a dynamic clock input, making it sensitive to the data bits only during clock transitions. After the leading edge of a clock transition, the data inputs are disabled and thus not held constant while the clock pulse is HIGH. In essence, the master portion of this flip-flop is like an edge-triggered device and the slave portion performs like the slave in a master-slave device to produce a postponed output.

Figure 8.42 shows the logic symbol for a data lock-out J-K flip-flop. Note that this symbol has both the dynamic input indicator for the clock, and the postponed output indicators. This type of flip-flop is classified by most manufacturers as a master-slave with a special lock-out feature. The master-slave flip-flop has now become obsolete although we may encounter it in older equipment.

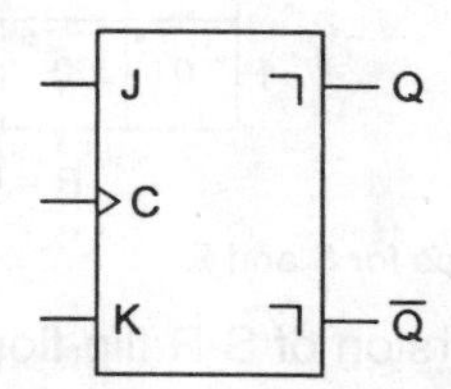

Fig. 8.42 Logic symbol of the master-slave J-K flip-flop with data lock-output.

8.8 CONVERSION OF FLIP-FLOPS

To convert one type of flip-flop into another type, a combinational circuit is designed such that if the inputs of the required flip-flop are fed as inputs to the combinational circuit and the output of the combinational circuit is connected to the inputs of the actual flip-flop, then the output of the actual flip-flop is the output of the required flip-flop.

S-R Flip-flop to J-K Flip-flop

Here the external inputs to the already available S-R flip-flop will be J and K. S and R are the outputs of the combinational circuit, which are also the actual inputs to the S-R flip-flop. We write a truth table with J, K, Q_n, Q_{n+1}, S, and R, where Q_n is the present state of the flip-flop and Q_{n+1} is the next state obtained when the particular J and K inputs are applied.

J, K and Q_n can have eight combinations. For each combination of J, K, and Q_n, find the corresponding Q_{n+1}, i.e. determine to which next state (Q_{n+1}) the J-K flip-flop will go from the present state Q_n if the present inputs J and K are applied. Now complete the table by writing the values of S and R required to get each Q_{n+1} from the corresponding Q_n, i.e. write what values of S and R are required to change the state of the flip-flop from Q_n to Q_{n+1}.

The conversion table, the K-maps for S and R in terms of J, K, and Q_n and the logic diagram showing the conversion from S-R to J-K are shown in Fig. 8.43.

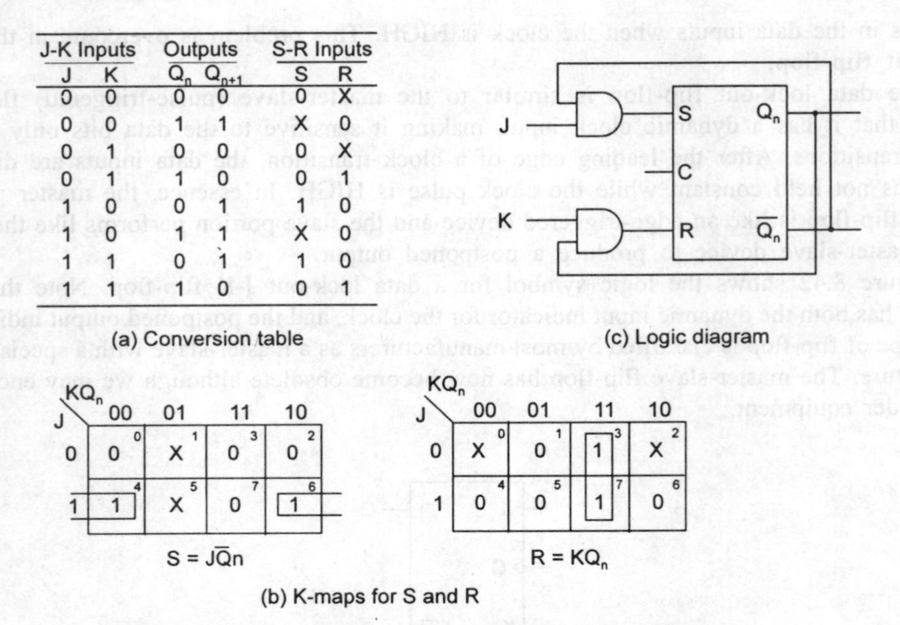

J-K Inputs		Outputs		S-R Inputs	
J	K	Q_n	Q_{n+1}	S	R
0	0	0	0	0	X
0	0	1	1	X	0
0	1	0	0	0	X
0	1	1	0	0	1
1	0	0	1	1	0
1	0	1	1	X	0
1	1	0	1	1	0
1	1	1	0	0	1

Fig. 8.43 Conversion of S-R flip-flop to J-K flip-flop.

J-K Flip-flop to S-R Flip-flop

Here, S and R become the external inputs, and the outputs of the combinational circuit are the actual inputs to the J-K flip-flop. So, we have to get the values of J and K in terms of S, R and Q_n. Thus, write a table using S, R, Q_n, Q_{n+1}, J, and K. The external inputs S and R and the output Q_n can make eight possible combinations. For each combination, find the corresponding Q_{n+1}. In the S-R flip-flop, the combination S = and R = 1 is not permitted. So, the corresponding output is invalid and, therefore, the corresponding J and K are don't cares. Complete the table by writing the values of J and K required to get each Q_{n+1} from the corresponding Q_n.

The conversion table, the K-maps for J and K in terms of S, R, and Q_n and the logic diagram showing the conversion from J-K to S-R are shown in Fig. 8.44.

S-R Flip-flop to D Flip-flop

D is the external input to the flip-flop, S and R are the actual inputs to the flip-flop. Express S and R in terms of D and Q_n.

The conversion table, the K-maps for S and R in terms of D and Q_n, and the logic diagram showing the conversion from S-R to D are shown in Fig. 8.45.

D Flip-flop to S-R Flip-flop

S and R are the external inputs and D is the actual input to the flip-flop. S, R, and Q_n make eight possible combinations, but S = R = 1 is an invalid combination. So, the corresponding entries for Q_{n+1} and D are the don't cares. Express D in terms of S, R, and Q_n.

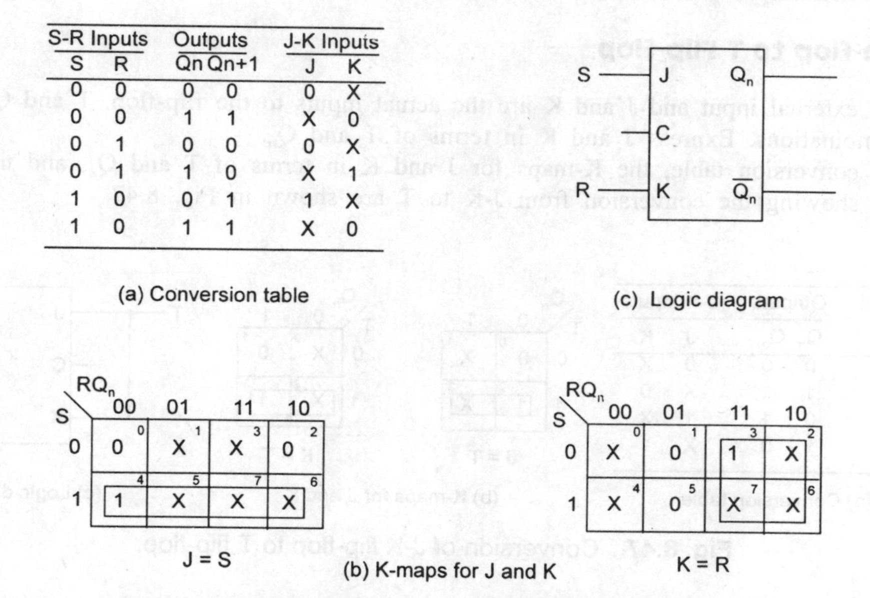

S-R Inputs		Outputs		J-K Inputs	
S	R	Qn	Qn+1	J	K
0	0	0	0	0	X
0	0	1	1	X	0
0	1	0	0	0	X
0	1	1	0	X	1
1	0	0	1	1	X
1	0	1	1	X	0

Fig. 8.44 Conversion of J-K flip-flop to S-R flip-flop.

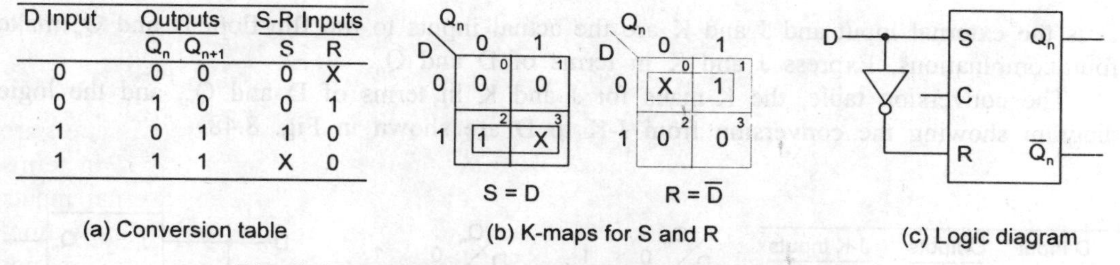

D Input	Outputs		S-R Inputs	
	Q_n	Q_{n+1}	S	R
0	0	0	0	X
0	1	0	0	1
1	0	1	1	0
1	1	1	X	0

Fig. 8.45 Conversion of S-R flip-flop to D flip-flop.

The conversion table, the K-map for D in terms of S, R, and Q_n, and the logic diagram showing the conversion from D to S-R are shown in Fig. 8.46.

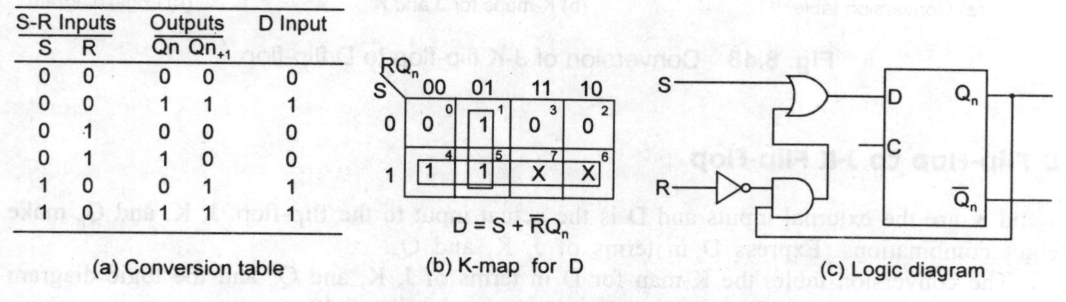

S-R Inputs		Outputs		D Input
S	R	Qn	Qn_{+1}	
0	0	0	0	0
0	0	1	1	1
0	1	0	0	0
0	1	1	0	0
1	0	0	1	1
1	0	1	1	1

Fig. 8.46 Conversion of D flip-flop to S-R flip-flop.

J-K Flip-flop to T Flip-flop

T is the external input and J and K are the actual inputs to the flip-flop. T and Q_n make four combinations. Express J and K in terms of T and Q_n.

The conversion table, the K-maps for J and K in terms of T and Q_n, and the logic diagram showing the conversion from J-K to T are shown in Fig. 8.47.

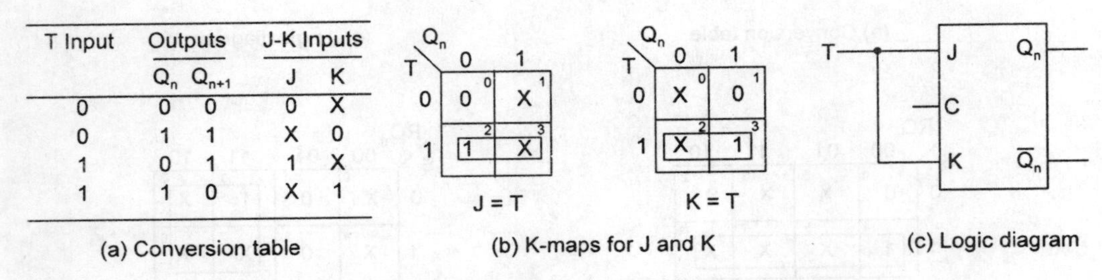

T Input	Outputs Q_n	Outputs Q_{n+1}	J-K Inputs J	J-K Inputs K
0	0	0	0	X
0	1	1	X	0
1	0	1	1	X
1	1	0	X	1

(a) Conversion table (b) K-maps for J and K (c) Logic diagram

Fig. 8.47 Conversion of J-K flip-flop to T flip-flop.

J-K Flip-flop to D Flip-flop

D is the external input and J and K are the actual inputs to the flip-flop. D and Q_n make four combinations. Express J and K in terms of D and Q_n.

The conversion table, the K-maps for J and K in terms of D and Q_n, and the logic diagram showing the conversion from J-K to D are shown in Fig. 8.48.

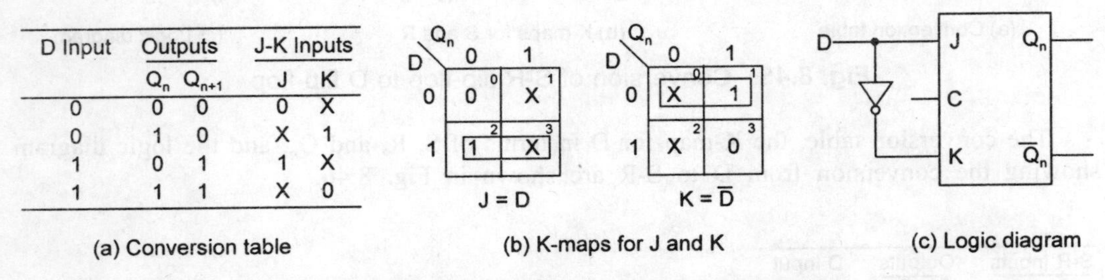

D Input	Outputs Q_n	Outputs Q_{n+1}	J-K Inputs J	J-K Inputs K
0	0	0	0	X
0	1	0	X	1
1	0	1	1	X
1	1	1	X	0

(a) Conversion table (b) K-maps for J and K (c) Logic diagram

Fig. 8.48 Conversion of J-K flip-flop to D flip-flop.

D Flip-flop to J-K Flip-flop

J and K are the external inputs and D is the actual input to the flip-flop. J, K, and Q_n make eight combinations. Express D in terms of J, K, and Q_n.

The conversion table, the K-map for D in terms of J, K, and Q_n and the logic diagram showing the conversion from D to J-K are shown in Fig. 8.49.

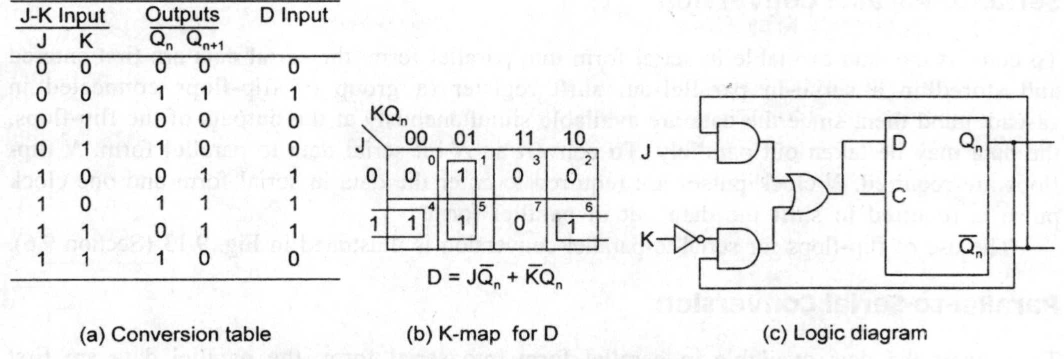

J-K Input		Outputs		D Input
J	K	Q_n	Q_{n+1}	
0	0	0	0	0
0	0	1	1	1
0	1	0	0	0
0	1	1	0	0
1	0	0	1	1
1	0	1	1	1
1	1	0	1	1
1	1	1	0	0

(a) Conversion table (b) K-map for D (c) Logic diagram

Fig. 8.49 Conversion of D flip-flop to J-K flip-flop.

8.9 APPLICATIONS OF FLIP-FLOPS

There are a large number of applications of flip-flops. Some of the basic applications are parallel data storage, serial data storage, transfer of data, frequency division, counting, parallel to serial conversion, serial to parallel conversion, synchronizing the effect of asynchronous data, detection of an input sequence, etc. These are discussed in detail as under.

Parallel Data Storage

A group of flip-flops is called a register. To store a data of *N* bits, *N* flip-flops are required. Since the data is available in parallel form, i.e. all bits are present at a time, these bits may be made available at the D input terminals of the flip-flops and when a clock pulse is applied to all the flip-flops simultaneously, these bits will be transferred to the Q outputs of the flip-flops and the flip-flops (register) then store the data. The use of the flip-flops for parallel data storage is illustrated in Fig. 9.17 (see Section 9.8).

Serial Data Storage

To store a data of *N* bits available in serial form, *N* number of D flip-flops are connected in cascade. The clock signal is connected to all the flip-flops. The serial data is applied to the D input terminal of the first flip-flop. Each clock pulse transfers the D input to its Q output. So, after *N* clock pulses the register (group of flip-flops) contains the data and then stores it. The use of flip-flops for serial data storage is illustrated in Fig. 9.10 (Section 9.5) and Fig. 9.13 (Section 9.6).

Transfer of Data

Data stored in flip-flops may be transferred out in a serial fashion, i.e. bit-by-bit from the output of one flip-flop or may be transferred out in parallel form, i.e. all bits at a time from the Q outputs of each of the flip-flops.

Serial-to-Parallel Conversion

To convert the data available in serial form into parallel form the serial data are first entered and stored in a serial-in parallel-out shift register (a group of flip-flops connected in cascade) and then, since the data are available simultaneously at the outputs of the flip-flops, the data may be taken out parallely. To convert an *N*-bit serial data to parallel form, *N* flip-flops are required. *N* clock pulses are required to enter the data in serial form and one clock pulse is required to shift the data out in parallel form.

The use of flip-flops for serial-to-parallel conversion is illustrated in Fig. 9.13 (Section 9.6).

Parallel-to-Serial Conversion

To convert the data available in parallel form into serial form, the parallel data are first entered in the parallel-in serial-out shift register in parallel form, i.e. all bits at a time and then that data is shifted out of the register serially, i.e. bit-by-bit by the application of clock pulses. To convert an *N*-bit parallel data to serial form, *N* flip-flops are required. One clock pulse is required to shift the parallel data into the register and *N* clock pulses are required to shift the data out of the register serially.

The use of flip-flops for parallel-to-serial conversion is illustrated in Fig. 9.15 (Section 9.7).

Counting

A number of flip-flops may be connected in a particular fashion to count the pulses electronically. One flip-flop can count up to 2 pulses, two flip-flops can count up to $2^2 = 4$ pulses. In general, *N* flip-flops can count up to 2^N pulses. In a simple counter, all the flip-flops are connected in toggle mode. The clock pulses are applied to the first flip-flop and the clock terminal of each subsequent flip-flop is connected to the Q output of the previous flip-flop. Feedback may be provided if the maximum count required is not 2^N. Flip-flops may be used to count up or down or up/down. Figures 10.1, 10.2, 10.3, 10.7, 10.8, and 10.9 illustrate the use of flip-flops for counting.

Frequency Division

Flip-flops may be used to divide the input signal frequency by any number. A single flip-flop may be used to divide the input frequency by 2. Two flip-flops may be used to divide the input frequency by 4. In general, *N* flip-flops may be used to divide the input frequency by 2^N. If *N* flip-flops are connected as a ripple counter (a counter in which the external signal is applied to the clock terminal of the first flip-flop and the Q output of each flip-flop is connected to the clock input of next flip-flop) and if the input signal of frequency f is fed to the first flip-flop, the output of this flip-flop will be of frequency $f/2$, the output of the second flip-flop will be of frequency $f/4$, and so on.

Figure 10.7a and the waveforms in Fig. 10.7b (also Figs. 10.1, 10.2, 10.4, 10.5) illustrate the use of flip-flops for frequency division.

8.10 ANSI/IEEE SYMBOLS

The symbols which we have used till now for the latches and the flip-flops are called the traditional symbols. We now introduce the ANSI/IEEE symbols for the latches and flip-flops.

Figure 8.50a shows the ANSI/IEEE logic symbol for a single D latch. It uses the letter C to denote the ENABLE input. In fact, the ANSI/IEEE symbology uses C for any input that controls the effect of other inputs on the output. The logic level applied to the ENABLE input controls the D input when the same is applied to effect a change in Q and $\overline{Q}$. Note that the outputs Q and $\overline{Q}$ are labelled outside the block. The right triangle on $\overline{Q}$ indicates that it is an inverted output.

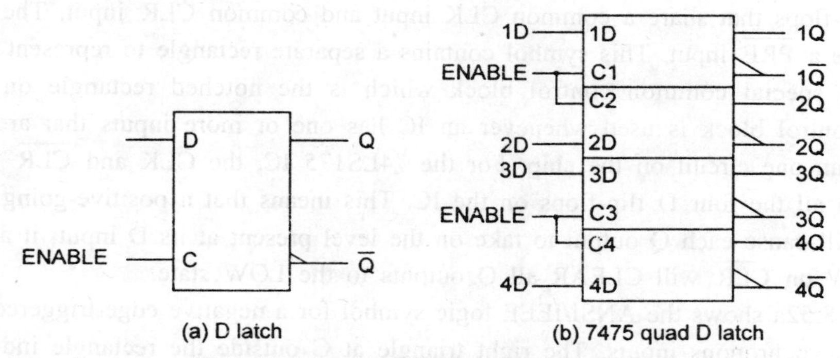

Fig. 8.50 ANSI/IEEE logic symbols.

Figure 8.50b shows the ANSI/ IEEE logic symbol for the IC—TTL 7475 quad latch. This IC contains four D latches that operate individually in a manner described earlier. This symbol also applies to the corresponding ICs in the /other TTL and CMOS series, for example, 74LS75, 74L75, 74C75, and 74HC75.

The overall symbol outline for the 7475 IC contains four smaller rectangles that represent the individual latches. Note how the inputs and outputs to each latch are labelled. For example, the input to the top latch is labelled 1D, its ENABLE input is labelled C1 and its outputs are 1Q and 1$\overline{Q}$. The top two latches have a common ENABLE input, that is, C1 and C2 are connected together internally and brought out to a single pin on the IC package. Likewise, the bottom two latches share a common ENABLE input.

Figure 8.51a shows the ANSI/IEEE logic symbol for a positive edge-triggered D flip-flop with asynchronous inputs. The clock input is labelled C inside the symbol's rectangular

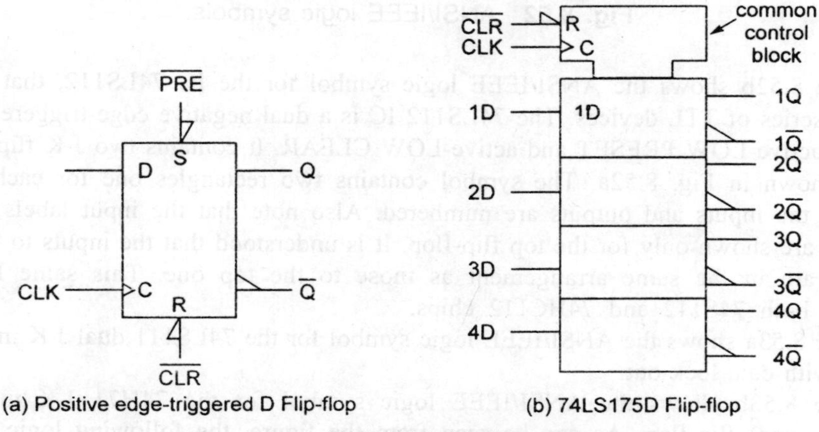

Fig. 8.51 ANSI/IEEE logic symbols.

outline. The triangle at C indicates that the flip-flop is positive edge-triggered. The $\overline{PRE}$ and $\overline{CLR}$ inputs are active-LOW as symbolized by the right triangles. The labels S and R inside the rectangle indicate asynchronous SET and RESET which are equivalent to PRESET and CLEAR, respectively.

Figure 8.51b shows the ANSI/IEEE logic symbol for a 74LS175 IC, which contains four D flip-flops that share a common CLK input and common $\overline{CLR}$ input. The flip-flops do not have a $\overline{PRE}$ input. This symbol contains a separate rectangle to represent each flip-flop and a special common control block which is the notched rectangle on top. The common control block is used whenever an IC has one or more inputs that are common to more than one circuit on the chip. For the 74LS175 IC, the CLK and $\overline{CLR}$ inputs are common to all the four D flip-flops on the IC. This means that a positive-going transition on CLK will cause each Q output to take on the level present at its D input; it also means that a LOW on $\overline{CLR}$ will CLEAR all Q outputs to the LOW state.

Figure 8.52a shows the ANSI/IEEE logic symbol for a negative edge-triggered J-K flip-flop with asynchronous inputs. The right triangle at C outside the rectangle indicates that the flip-flop is triggered at the negative-going transition of the clock.

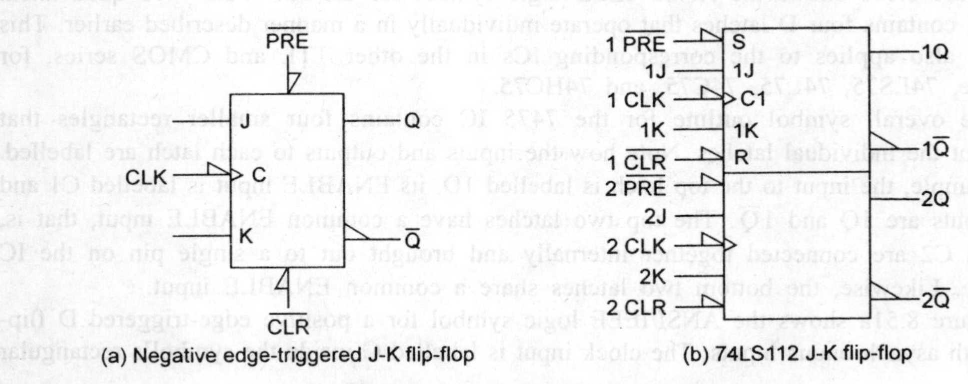

(a) Negative edge-triggered J-K flip-flop (b) 74LS112 J-K flip-flop

Fig. 8.52 ANSI/IEEE logic symbols.

Figure 8.52b shows the ANSI/IEEE logic symbol for the IC 74LS112, that is part of the 74LS series of TTL devices. The 74LS112 IC is a dual negative edge-triggered J-K flip-flop with active-LOW PRESET and active-LOW CLEAR. It contains two J-K flip-flops like the one shown in Fig. 8.52a. The symbol contains two rectangles one for each flip-flop. Note how the inputs and outputs are numbered. Also note that the input labels inside the rectangles are shown only for the top flip-flop. It is understood that the inputs to the bottom flip-flops are in the same arrangement as those to the top one. This same IC symbol represents both 74S112 and 74HC112 chips.

Figure 8.53a shows the ANSI/IEEE logic symbol for the 74LS111 dual J-K master-slave flip-flop with data lock-out.

Figure 8.53b shows the ANSI/IEEE logic symbol for the 74H71 J-K master-slave (pulse-triggered) flip-flop. As can be seen from the figure, the following logic operations are performed on the inputs to generate J and K.

$$1J = (J1A \cdot J1B) + (J2A \cdot J2B)$$

$$1K = (K1A \cdot K1B) + (K2A \cdot K2B)$$

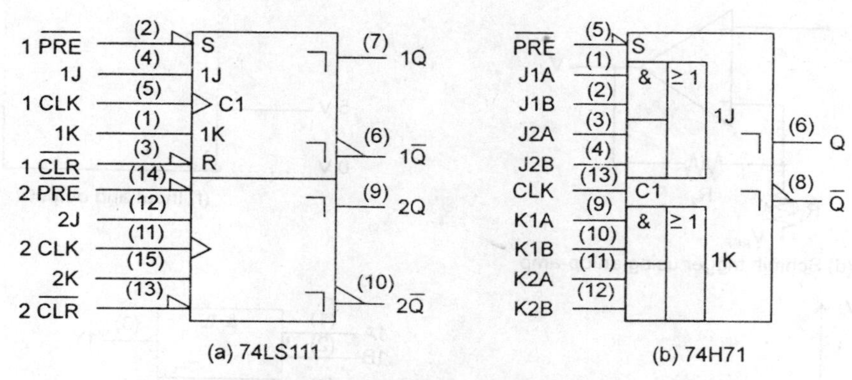

Fig. 8.53 ANSI/IEEE logic symbols.

The 74H71 flip-flop does not have data lock-out, since the clock input is not dynamic.

8.11 SCHMITT TRIGGER

A Schmitt trigger is not classified as a flip-flop, but it does exhibit a type of memory characteristic that makes it useful in certain special situations.

A Schmitt trigger inverter accepts slow changing signals and produces an output that has oscillation-free transitions. Figures 8.54a and f show a Schmitt trigger inverter and its response to a slow changing input. From the waveform of Fig. 8.54f, it can be noticed that the output does not change from HIGH to LOW until the input exceeds the positive-going threshold voltage V_{T+} (also called UTL) and then remains LOW even if the input falls below V_{T+} and changes from LOW to HIGH only when the input falls below the negative-going threshold voltage V_{T-} (also called LTL). Logic designers use ICs with Schmitt trigger inputs to convert slow-changing signals to clean, fast-changing signals that can drive standard IC inputs.

There are several ICs available with Schmitt trigger inputs. The 7414, 74LS14, and 74HC14 are hex inverter ICs with Schmitt trigger inputs. The 7413, 74LS13, and 74HC13 are dual 4-input NANDs with Schmitt trigger inputs. The 74132 contains four 2-input NAND gates with built-in Schmitt triggers.

The logic symbol for these devices shown in Figs. 8.54a, b, and c contain a box-like symbol called the *hysterisis loop*, which represents the transfer characteristic (output voltage vs input voltage) of a device having hysterisis. The lower and upper trigger levels of TTL Schmitt triggers are fixed by design, therefore, hysterisis is not adjustable. Typical values are, V_{T+} = 1.7 V and V_{T-} = 0.9 V, giving a hysterisis of 0.8 V.

Figure 8.54d shows a Schmitt trigger inverter using an op-amp. Figure 8.54e shows the hysterisis loop.

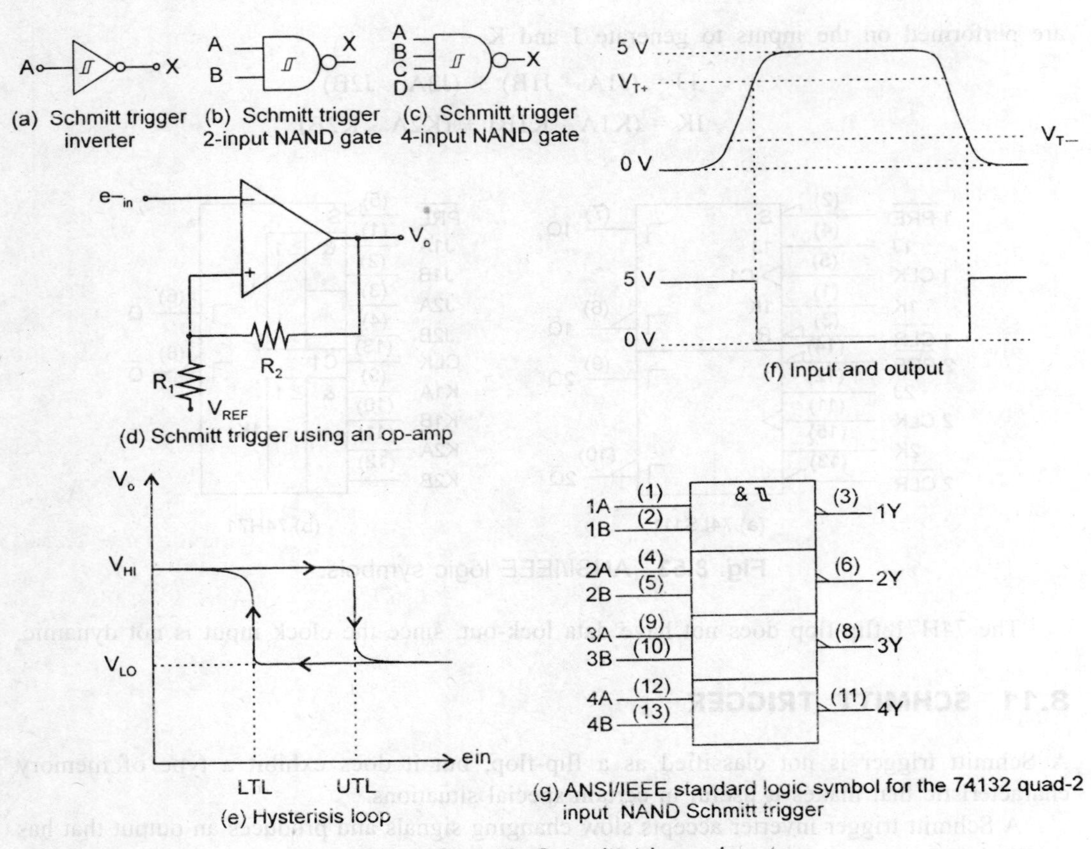

Fig. 8.54 ICs with Schmitt trigger inputs.

Figure 8.54g shows the ANSI/IEEE logic symbol for 74132 quad 2-input NAND Schmitt trigger.

8.12 MONOSTABLE MULTIVIBRATOR (ONE-SHOT)

There are three types of multivibrators: (1) Bistable multivibrator, commonly known as flip-flop, (2) monostable multivibrator, commonly known as one-shot or single-shot or simply monostable, and (3) astable multivibrator usually called free-running multivibrator.

A bistable multivibrator (flip-flop) has two stable states. A stable state is a state in which the multivibrator can remain indefinitely. A triggering signal is required to change the state of the bistable multivibrator. It flip-flops (i.e. changes back and forth) between its two stable states when triggering pulses are applied.

A monostable multivibrator, as the name itself indicates, has only one stable state. The other state is quasi-stable. When triggered, it changes from its stable state (LOW state) to its quasi-stable state (HIGH state) and remains there for a specified length of time before returning automatically to its stable state, i.e. it produces a pulse of predetermined width in response to a trigger input. The trigger itself may be a pulse, whose LOW-to-HIGH or HIGH-to-LOW transition (depending on design) initiates the output pulse. The width of the output pulse is usually determined by the resistance and capacitance values in an RC network, called the timing circuit, connected to the device. A monostable multivibrator is often called simply a monostable, a one-shot or a single-shot because it produces a **single pulse** in response to a trigger input. The output of the monostable is labelled Q. Many monostables also have a $\overline{Q}$ output, the complement of Q which goes LOW when Q goes HIGH and vice versa.

Figure 8.55a shows a basic one-shot circuit composed of a NOR gate and an inverter. In the normal stable state, the input is LOW and Q output is also LOW. So, the output of the NOR gate (G_1) is HIGH. When a positive triggering pulse is applied to G_1, the output of G_1 goes LOW. This HIGH-to-LOW transition is coupled through the capacitor to the input to inverter (G_2). The apparent LOW on G_2 makes its output go HIGH. This HIGH is connected back into G_1 keeping its output LOW. So, a trigger pulse has caused the Q output to go HIGH.

The capacitor immediately begins to charge through R towards the HIGH voltage level. The rate at which it charges is determined by the RC time constant. When the capacitor charges to a certain level which appears as a HIGH to G_2, the output goes back to LOW. So, a pulse of fixed time duration is generated at the Q terminal of the monostable circuit.

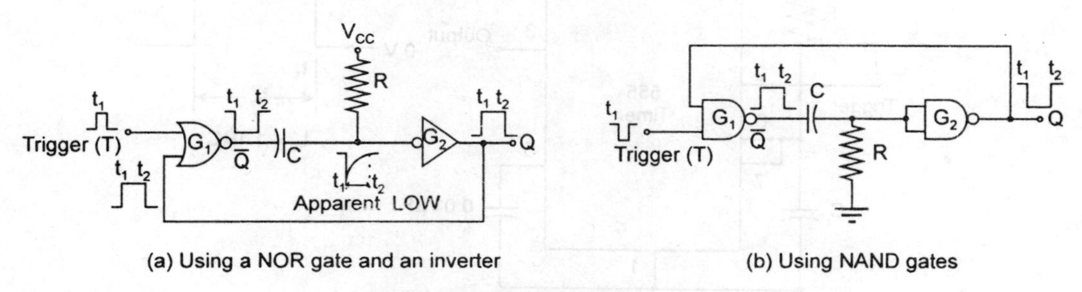

Fig. 8.55 Monostable circuits.

Figure 8.55b shows a monostable using NAND gates. G_1 is a 2-input NAND gate. G_2 is used as an inverter. Under the resting condition, voltage across R is zero which is the input to G_2. So, the output of G_2 is HIGH. As both the inputs to G_1 are HIGH, its output is LOW.

When a negative trigger is applied at G_1, the output of G_1 goes HIGH. Since the voltage across C cannot change instantaneously, this appears as a HIGH at the input to G_2, and, therefore, its output goes LOW. As the capacitor charges, the voltage across R and, therefore, the input to G_2 decreases with a time constant RC. When this voltage falls below a certain level, it appears as a LOW to G_2, and, therefore, its output goes back to HIGH. So, a triggering pulse causes the output of the circuit to go LOW for a specific time. Hence, the circuit acts as a monostable multivibrator.

Figure 8.56 shows the various logic symbols used for monostable multivibrators.

Figure 8.57 shows a monostable multivibrator using 555 timer. The 555 timer is TTL compatible. The output at pin 3 is normally in LOW state. When a negative triggering pulse is applied at pin 2, the output goes HIGH for a specific time ($t_w = 1.1RC$) and then comes back to its normal LOW state. So, a positive pulse of width $t_w = 1.1RC$ is generated. This is a non-retriggerable one-shot. Application of new trigger pulses during the timing cycle

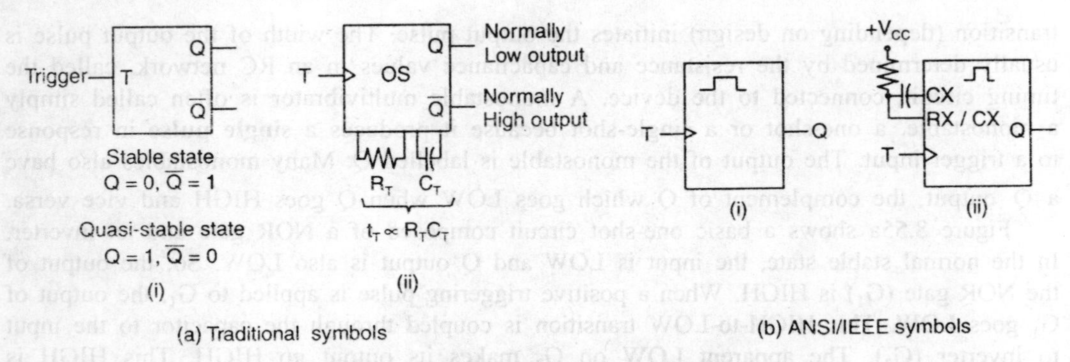

Fig. 8.56 Monostable multivibrators.

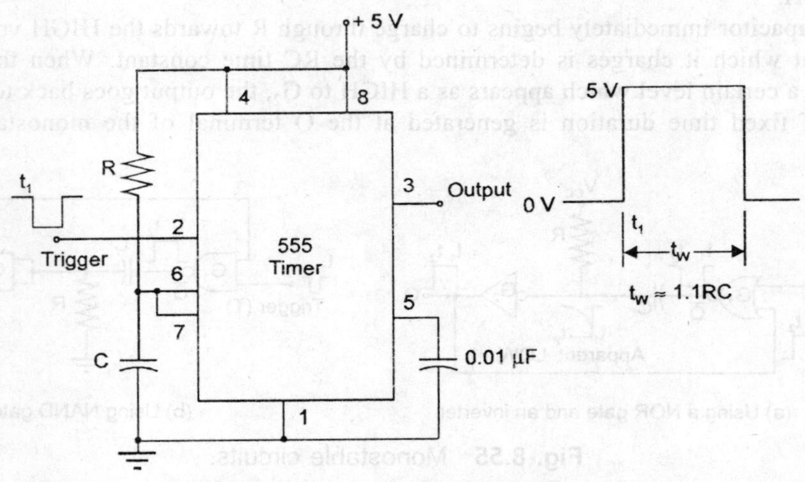

Fig. 8.57 Monostable using 555 timer.

has no effect. However, the RESET input at pin 4 can be used to terminate an output pulse during the timing cycle, if desired.

Integrated Circuit One-shots

Two types of one-shots are available in IC form—retriggerable type and non-retriggerable type. A retriggerable one-shot is one that accepts a new trigger input while the output pulse produced by the previous trigger is still in progress. The new trigger initiates a new timing cycle. So, the output pulse is extended, beginning from where the new trigger occurred, for a length of time equal to the monostable's full output pulse width. In other words, regardless of how long an output pulse had been HIGH, a new trigger input effectively restarts time and superimposes a new pulse beginning from where the trigger occurred. A non-retriggerable one-shot simply ignores any new trigger that occurs while a pulse output is in progress.

Figure 8.58a shows the waveforms for a non-retriggerable monostable multivibrator. From the waveforms, observe that if t_p is the pulse width of the one-shot, the output of the one-shot goes HIGH only for t_p whenever a positive going transition of the trigger occurs. Also, the duration of the triggering pulse is of no consequence. Also, a triggering pulse applied when the output is already HIGH, does not affect the output at all.

Figure 8.58b shows the waveforms for a retriggerable one-shot. Note that the output remains HIGH for a time t_p after the application of the trigger at any instant.

Figure 8.58c shows the comparison of waveforms for the outputs of retriggerable and non-retriggerable one-shots.

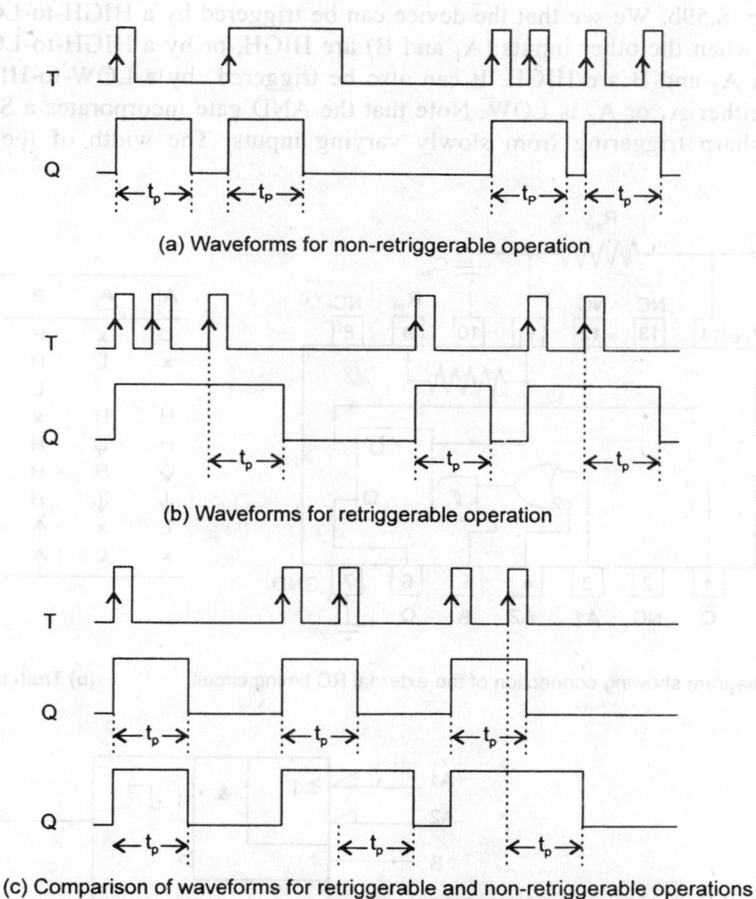

Fig. 8.58 Waveforms of non-retriggerable and retriggerable monoshots.

Actual Devices

There are several one-shot ICs available in both the retriggerable and non-retriggerable versions. The 74121 and 74L121 are single non-retriggerable one-shot ICs. The 74221,

74LS221, and 74HC221 are dual non-retriggerable one-shot ICs; the 74122 and 74LS122 are single retriggerable one-shots. The 74123, 74LS123, and 74HC123 are dual retriggerable one-shot ICs.

The 74121 is a widely used non-retriggerable one-shot IC. It is constructed with logic gates at its inputs. Depending on how the external signal inputs are connected to these gates, the monostable can be triggered by a LOW-to-HIGH or by a HIGH-to-LOW level transition.

Figure 8.59a shows the wiring diagram of the 74121 with connections to the external *RC* timing circuit. The truth table governing the operation and triggering of this device is shown in Fig. 8.59b. We see that the device can be triggered by a HIGH-to-LOW transition on input A_2 when the other inputs (A_1 and B) are HIGH, or by a HIGH-to-LOW transition on A_1, when A_2 and B are HIGH. It can also be triggered by a LOW-to-HIGH transition on B when either A_1 or A_2 is LOW. Note that the AND gate incorporates a Schmitt trigger to provide sharp triggering from slowly varying inputs. The width of the output pulse

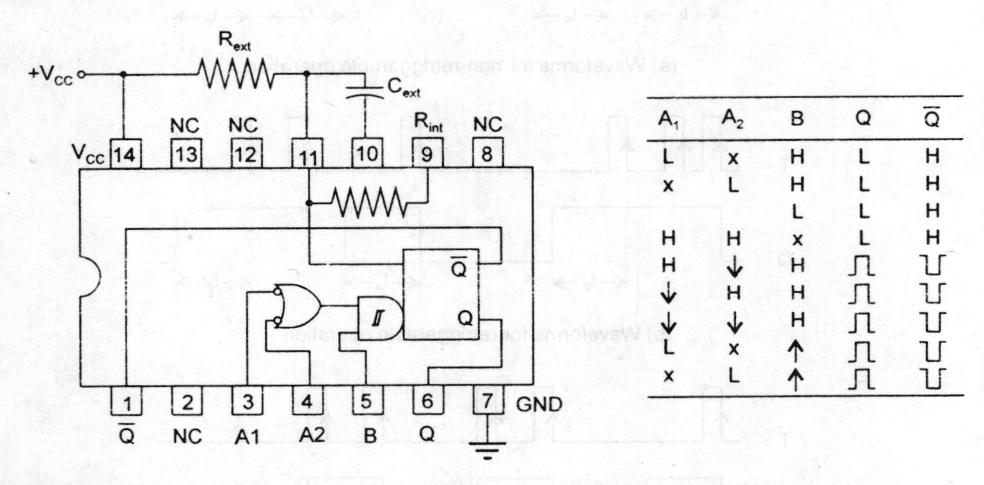

A_1	A_2	B	Q	$\bar{Q}$
L	x	H	L	H
x	L	H	L	H
		L	L	H
H	H	x	L	H
H	↓	H	⊓	⊔
↓	H	H	⊓	⊔
↓	↓	H	⊓	⊔
L	x	↑	⊓	⊔
x	L	↑	⊓	⊔

(a) Wiring diagram showing connection of the external RC timing circuit

(b) Truth table

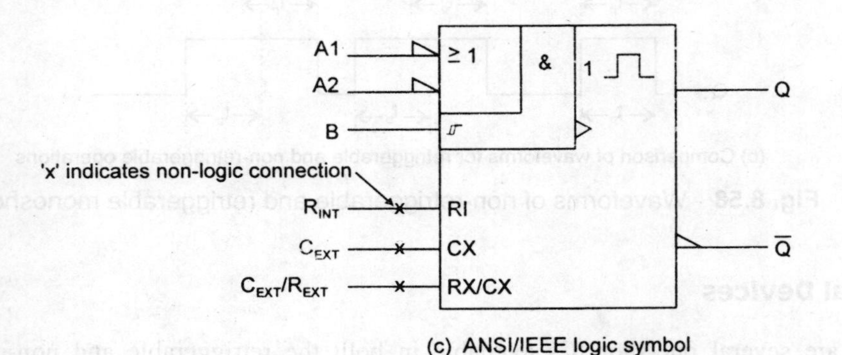

(c) ANSI/IEEE logic symbol

Fig. 8.59 The 74121 non-triggerable one-shot IC.

produced by the 74121 is given by

$$PW = (\ln 2)R_{ext}C_{ext} = 0.69R_{ext}C_{ext}$$

Figure 8.59c shows the ANSI/IEEE logic symbol for the 74121 non-retriggerable IC one-shot. The qualifying symbol for a non-retriggerable monostable is a 1 followed by a single pulse. The qualifying symbol for a retriggerable monostable is a pulse.

Figure 8.60 shows the wiring diagram and the truth table for one-half of the dual 74123 retriggerable monostable multivibrator. Note that this version has a CLEAR input (CLR) which, when made LOW will cause an output pulse, already in progress, to be terminated.

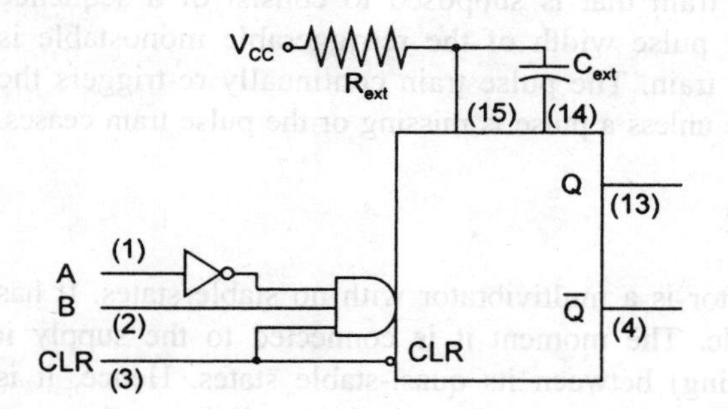

CLR	A	B	Q	$\overline{Q}$
L	x	x	L	H
x	H	x	L	H
x	x	L	L	H
H	L	↑	⎍	⊔
H	↓	H	⎍	⊔
↓	L	H	⎍	⊔

(a) Wiring diagram for one-half of the 74123 dual monostable. (b) Truth table

Fig. 8.60 Dual 74123 retriggerable one-shot.

When $C_{ext} > 1000$ pF, the output pulse width is approximately given by

$$t_w \approx 0.28R_{ext}C_{ext}(1 + 0.7/R_{ext})$$

If C_{ext} is an electrolytic capacitor or if the CLR function is used, the manufacturer recommends that a diode be inserted between R_{ext} and pin 15 (cathode to pin 15), and the coefficient 0.28 in the equation be changed to 0.25.

Applications of Monostable Multivibrators

Monostable multivibrators can be used for gating, creating time delays, generating a sequence of pulses, and for detection of missing pulses, etc.

Gating. In many digital systems, it is necessary to enable (disable) a logic gate to permit (stop) the passage of digital signals to another part of the system for a prescribed period of time. A monostable can be used to enable (or inhibit) logic gates for the necessary periods of time. The Q ($\overline{Q}$) output of the monostable is connected as one input to an AND gate, whose other input is the clock signal. When a triggering signal is applied to the monostable, its Q output goes HIGH for a specific period of time enabling (disabling) the AND gate and thereby permitting (stopping) the passage of the clock signal.

Time delays. Monostables are widely used to deliver a pulse for a certain duration after the occurrence of another pulse, i.e. to create a prescribed time delay in delivering a pulse. For example, to delay a pulse of 1 μs by 1 ms, the 1 μs pulse is applied as the triggering signal to the first one-shot (of t_p = 1 ms) and the $\overline{Q}$ output of the first one-shot is applied as the triggering input to the second one-shot (of t_p = 1 μs). The output of the second one-shot is then a 1 μs pulse delayed by 1 ms.

Synchronizing. Digital computer operations are often synchronized by sequences of pulses that occur on different control lines at different times. A sequence of *N* pulses can be generated by using *N* one-shots.

Detection of missing pulses. A retriggerable monostable can be used to detect a missing pulse or the cessation of pulses in a pulse train that is supposed to consist of a sequence of regularly recurring pulses. For this, the pulse width of the retriggerable monostable is set to between 1 and 2 periods of the pulse train. The pulse train continually re-triggers the monostable which never goes to the OFF state unless a pulse is missing or the pulse train ceases.

8.13 ASTABLE MULTIVIBRATOR

As the name indicates, an astable multivibrator is a multivibrator with no stable states. It has two states and both of them are quasi-stable. The moment it is connected to the supply it keeps on switching back and forth (oscillating) between its quasi-stable states. Hence, it is also called a free-running multivibrator. It is useful for providing clock signals to synchronous digital circuits.

There are several types of astable multivibrators in common use. Some of them are presented here.

Astable Multivibrator Using Schmitt Trigger

Figure 8.61a shows how a Schmitt trigger INVERTER can be connected as an oscillator. The signal at V_{OUT} is an approximate square wave (Fig. 8.61b) that depends on the values of *R* and *C*. The relation between the frequency of oscillation and the *RC* product is shown in Fig. 8.61c for three different Schmitt trigger ICs. The maximum permitted value of *R* in each case is also given. The circuit will fail to oscillate if *R* is not kept below these limits.

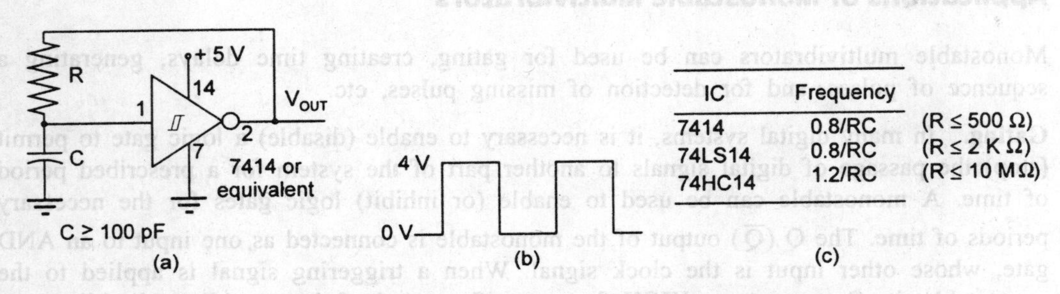

IC	Frequency	
7414	0.8/RC	(R ≤ 500 Ω)
74LS14	0.8/RC	(R ≤ 2 K Ω)
74HC14	1.2/RC	(R ≤ 10 MΩ)

Fig. 8.61 Astable multivibrator using Schmitt trigger.

Astable Multivibrator Using 555 Timer

The 555 timer is a TTL compatible device that can operate in several different modes. Figure 8.62 shows an astable multivibrator using the 555 timer. Its output is a repetitive rectangular waveform that switches between two logic levels; the time interval for each logic level is determined by the values of R and C. The formulas for these time intervals t_1 and t_2 and the overall period of the oscillations and the limiting values of the components are shown in Fig. 8.62. The capacitor C charges from V_{CC} through R_A and R_B with a time constant $(R_A + R_B)C$. When the voltage across C reaches $(2/3)V_{CC}$, the output goes LOW, and the capacitor starts discharging through R_B, with a time constant

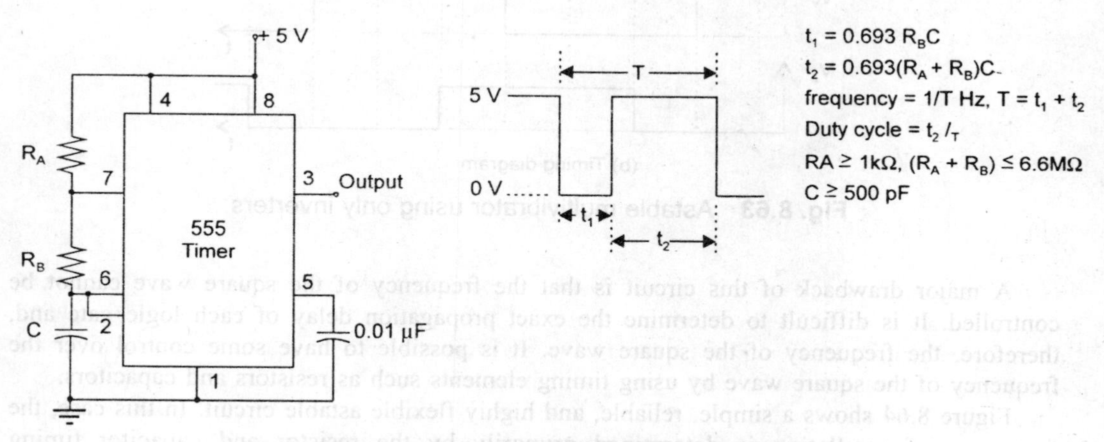

Fig. 8.62 Astable multivibrator using 555 timer.

R_BC. When it discharges to $(1/3)V_{CC}$, the output goes HIGH. So, the output is LOW for t_1 and HIGH for t_2 and then this cycle repeats itself. With this arrangement, we can only get a square wave with more than 50 per cent duty cycle. R_B can be made very large compared to R_A to get an approximate square wave (50 per cent duty cycle). A diode can be connected across R_B (with anode at pin 7 and cathode at pin 6) to get a perfect square wave output. *Even a square wave with less than 50 per cent duty cycle can be obtained when R_B is shunted by a diode.*

Astable Multivibrator Using Inverters

A very simple astable multivibrator can be constructed using three inverters. The inverters are connected in cascade and the output of the third inverter is connected as the input to the first inverter as shown in Fig. 8.63a. Hence the circuit is called a *ring oscillator*.

We know that the output of an inverter changes to the opposite state after a propagation delay when its input is changed. So, the signal at any point in this circuit changes state after a time equal to the sum of the propagation delays of the preceding gates and, therefore, a square wave is generated.

If at time t_0, V_x goes from LOW to HIGH, V_y will go from HIGH to LOW after a propagation delay of one gate, V_z will go from LOW to HIGH after a propagation delay of two gates, and V_o, i.e. V_x will go from HIGH to LOW after a propagation delay of three

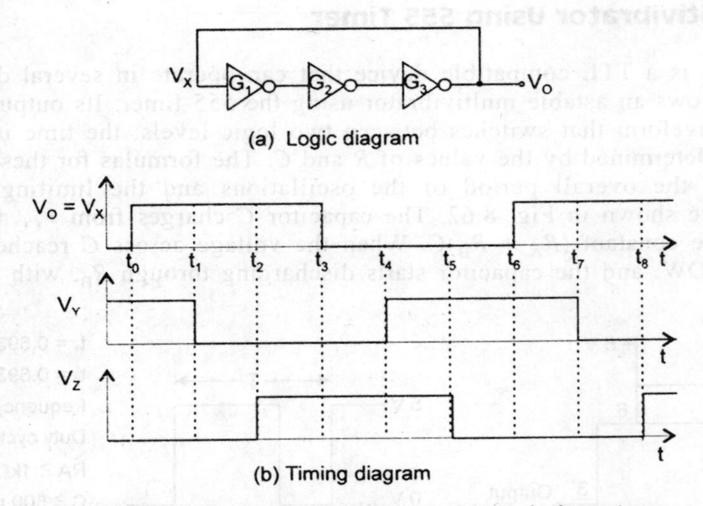

Fig. 8.63 Astable multivibrator using only inverters.

A major drawback of this circuit is that the frequency of the square wave cannot be controlled. It is difficult to determine the exact propagation delay of each logic gate and, therefore, the frequency of the square wave. It is possible to have some control over the frequency of the square wave by using timing elements such as resistors and capacitors.

Figure 8.64 shows a simple, reliable, and highly flexible astable circuit. In this case, the frequency of oscillation is determined primarily by the resistor and capacitor timing components. Hence it is called an *RC* oscillator. The exact relationship between the oscillation frequency and the *R* and *C* components depends in part on the electrical characteristics of the logic gates. For the standard TTL family of components, a resistance of approximately 400 Ω produces the relationship

$$f = \frac{1}{0.001C}$$

where C is measured in μF and f in Hz.

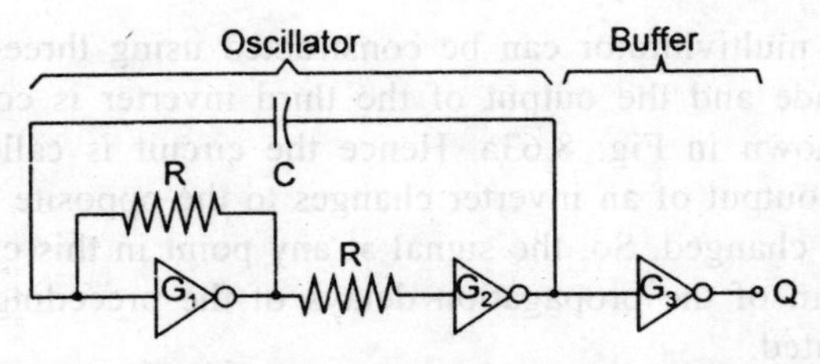

Fig. 8.64 Astable multivibrator using *R* and *C* as timing components.

Figure 8.65 shows an astable multivibrator using an AND gate and an inverter. Both the inputs to the AND gate are shorted. So, it just provides a time delay equal to the propagation delay time of the gate and acts as a buffer. When the supply is switched on

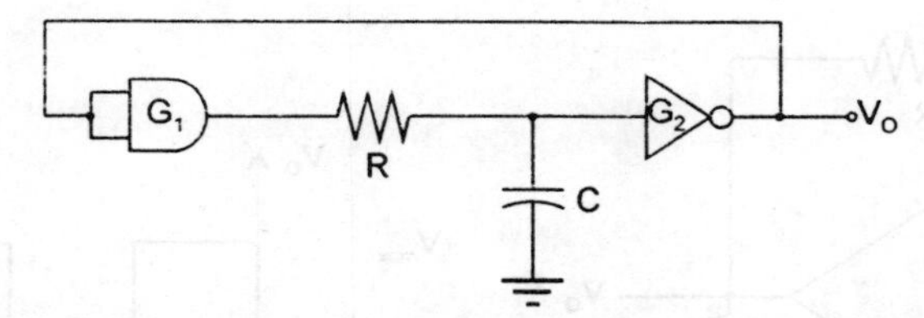

Fig. 8.65 Astable multivibrator using an AND gate and an inverter.

and as the capacitor is uncharged, the input to the inverter is 0 V. Therefore, its output is HIGH. Hence the output of the AND gate also goes HIGH. The capacitor now starts charging and when the voltage across it rises to a level which appears as a logic 1 to the inverter, the output goes LOW. Thus, the output of the AND gate also goes LOW. Now the capacitor starts discharging. When the voltage across it falls to a level which appears as a logic 0 to the inverter, the output of the inverter goes HIGH again. This cycle of events repeats itself, generating a square wave.

Figure 8.66a shows an astable multivibrator using two inverters. The output of each inverter is coupled to the input of the other through a capacitor. The capacitive coupling networks prevent either inverter from having a stable state. If designed properly, the circuit will start oscillating on its own and require no initial input trigger.

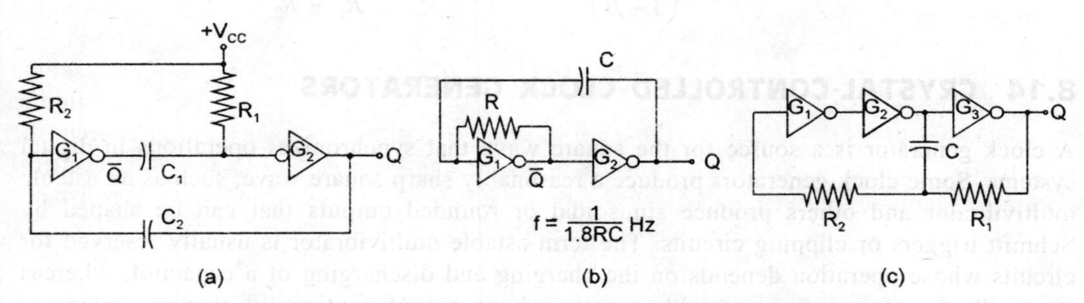

Fig. 8.66 Astable multivibrator using CMOS inverters.

Figure 8.66b shows another astable multivibrator circuit using inverters. The device used is IC 74HC04, CMOS inverter.

Figure 8.66c uses the IC 74HC04. Although this circuit can operate reliably, its frequency is not easily predictable.

The advantage of using CMOS devices in astable multivibrators is that, they have a much higher input impedance than that of their TTL counterparts. This characteristic makes the performance of CMOS multivibrators more predictable than that of TTL designs—in the sense that CMOS multivibrators are less sensitive to variations in device characteristics.

Astable Multivibrator Using Op-amps

Figure 8.67 shows an astable multivibrator using an op-amp. Assuming that the maximum positive and negative outputs of the comparator are $\pm V_{max}$, capacitor C continually

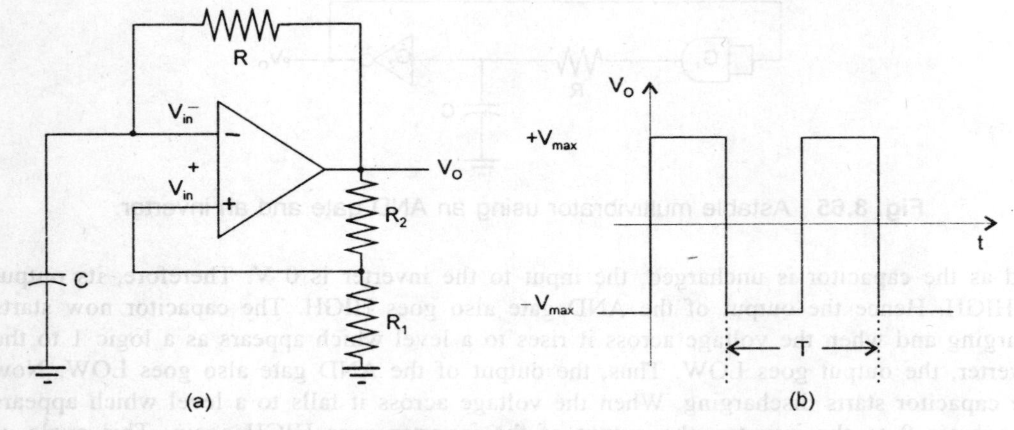

Fig. 8.67 Astable multivibrator using op-amp.

charges and discharges towards those values, and so, the output of the circuit will be a square wave with a period

$$T = 2RC \ln\left(\frac{1+\beta}{1-\beta}\right) \text{ s}, \qquad \text{where } \beta = \frac{R_1}{R_1 + R_2}$$

8.14 CRYSTAL-CONTROLLED CLOCK GENERATORS

A clock generator is a source for the square wave that synchronizes operations in digital systems. Some clock generators produce a reasonably sharp square wave, such as an astable multivibrator and others produce sinusoidal or rounded outputs that can be shaped by Schmitt triggers or clipping circuits. The term astable multivibrator is usually reserved for circuits whose operation depends on the charging and discharging of a capacitor, whereas an oscillator refers to any unstable system whose output continually changes value.

Many clock generators employ a crystal as the frequency sensitive component. Crystals are available in a wide range of frequencies and are more stable than inductor/capacitor networks. Crystal-controlled oscillators have good frequency stability, i.e. they generate signals whose frequency is less likely to drift. Also, since crystal-controlled oscillators produce signals whose frequencies equal their crystal frequency, they are predictable. The frequency stability is particularly important in applications where the frequency serves as a time reference. For example, in digital watches.

Figure 8.68a shows a simple quartz crystal oscillator using TTL 7404 inverters. The capacitor in parallel with the crystal is a trimming capacitor which allows very slight frequency adjustments.

Figure 8.68b shows a popular crystal-controlled clock generator constructed by using TTL 7404 inverters. Although the oscillations produced by this design are not perfectly square, the output is generally adequate for TTL synchronizing and triggering functions. The 7414 hex Schmitt trigger inverter can also be used. The value of R in the circuit controls the gain and is usually between 300 Ω and 1.5 kΩ. The optimum value of R

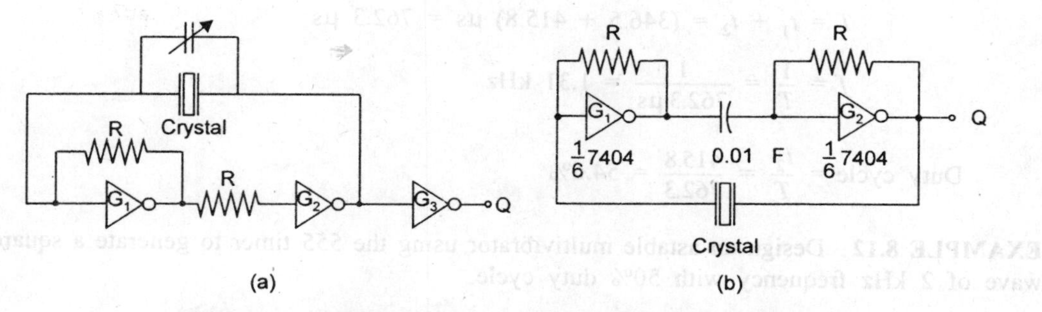

Fig. 8.68 Crystal clock generator using TTL inverters.

depends on the type of the crystal used and its frequency. If R is too low or too high, the generator may oscillate at a harmonic of the crystal frequency and thus have a smaller amplitude. This design has been used to produce clock frequencies from about 1 MHz to 20 MHz.

Figure 8.69 shows two crystal-controlled oscillators constructed with inverters from 74HC04 CMOS hex inverters. As in the TTL circuit of Fig. 8.68, the oscillation frequencies may be sensitive to the values of R used. A typical value for R in Fig. 8.69a is 100 kΩ but it may have to be specially selected to prevent oscillations at a harmonic frequency. The 100 pF capacitor suppresses the spurious high frequency oscillations in the 30 MHz to 50 MHz range. The resistor in Fig. 8.69b is of the order of 1–5 MΩ and the circuit will oscillate for crystal frequencies up to about 9 MHz.

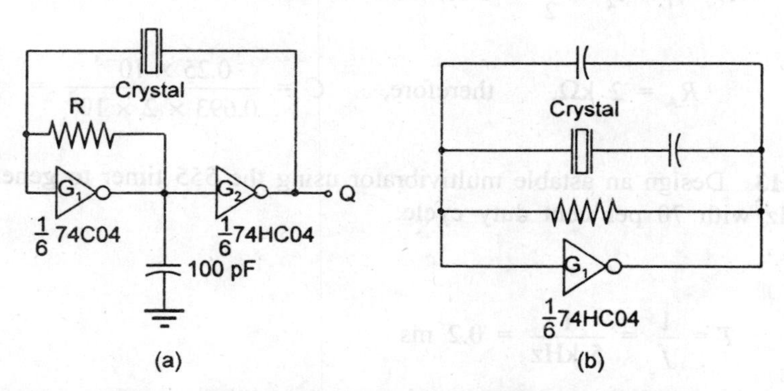

Fig. 8.69 Crystal clock generator using CMOS inverters.

EXAMPLE 8.11 Calculate the frequency and duty cycle of the 555 astable multivibrator output for

$$C = 0.01\ \mu\text{F}, \quad R_A = 10\ \text{k}\Omega \quad \text{and} \quad R_B = 50\ \text{k}\Omega.$$

Solution

$$t_1 = 0.693 R_B C = 0.693 \times 50 \times 10^3 \times 0.01 \times 10^{-6} = 346.5\ \mu\text{s}$$

$$t_2 = 0.693(R_A + R_B)C = 0.693 \times 60 \times 10^3 \times 0.01 \times 10^{-6} = 415.8\ \mu\text{s}$$

$$T = t_1 + t_2 = (346.5 + 415.8)\ \mu s = 762.3\ \mu s$$

$$f = \frac{1}{T} = \frac{1}{762.3\ \mu s} = 1.31\ \text{kHz}$$

$$\text{Duty cycle} = \frac{t_2}{T} = \frac{415.8}{762.3} = 54.6\%$$

EXAMPLE 8.12 Design an astable multivibrator using the 555 timer to generate a square wave of 2 kHz frequency with 50% duty cycle.

Solution

To obtain a square wave using the 555 timer, a diode is connected across R_B (see Fig. 8.62), such that it conducts and shorts R_B when C is charging and opens when C is discharging. Therefore,

$$t_1 = 0.693 R_B C$$

$$t_2 = 0.693 R_A C$$

As $t_1 = t_2$, therefore, $R_A = R_B$

As $f = 2$ kHz, therefore, $T = \dfrac{1}{2\ \text{kHz}} = 0.5$ ms

Hence, $t_1 = t_2 = \dfrac{T}{2} = 0.25$ ms

Let $R_A = 2\ \text{k}\Omega$, therefore, $C = \dfrac{0.25 \times 10^{-3}}{0.693 \times 2 \times 10^3} = 0.18\ \mu\text{F}$

EXAMPLE 8.13 Design an astable multivibrator using the 555 timer to generate a square wave of 5 kHz with 70 per cent duty cycle.

Solution

$$T = \frac{1}{f} = \frac{1}{5\ \text{kHz}} = 0.2\ \text{ms}$$

$$t_1 = 0.693 R_B C = 0.3 \times 0.2 = 0.06\ \text{ms}$$

$$t_2 = 0.693(R_A + R_B)C = 0.7 \times 0.2 = 0.14\ \text{ms}$$

Let $C = 5000$ pF, therefore, $R_B = \dfrac{0.06 \times 10^{-3}}{0.693 \times 5000 \times 10^{-12}} = 17.25\ \text{k}\Omega$

$$R_A + R_B = \frac{0.14 \times 10^{-3}}{0.693 \times 5000 \times 10^{-12}} = 40.4\ \text{k}\Omega$$

Therefore, $R_A = 40.4 - 17.25 = 23.15\ \text{k}\Omega$

EXAMPLE 8.14 For the astable multivibrator using the op-amp shown in Fig. 8.67,

$$R = 10\ \text{k}\Omega, \quad C = 0.01\ \mu\text{F} \quad \text{and} \quad R_1 = 5\ \text{k}\Omega.$$

It is desired to make the frequency of the output square wave adjustable from 10 kHz through 100 kHz by making R_2 adjustable. Through what range of values should R_2 be made adjustable to obtain the required frequency range?

Solution

The period of the output square wave must range from

$$T = \frac{1}{10\ \text{kHz}} = 0.1\ \text{ms} \quad \text{through} \quad T = \frac{1}{100\ \text{kHz}} = 0.01\ \text{ms}$$

We know that, $T = 2RC\ \ln\left(\frac{1+\beta}{1-\beta}\right)$ s, where $\beta = \frac{R_1}{R_1 + R_2}$.

Therefore,

$$T = 10^{-4}\ \text{s} = 2 \times 10 \times 10^3 \times 0.01 \times 10^{-6}\ \ln\left(\frac{1+\beta}{1-\beta}\right)$$

or $$\ln\left(\frac{1+\beta}{1-\beta}\right) = \frac{10^{-4}}{2 \times 10^4 \times 0.01 \times 10^{-6}} = 0.5$$

or $$\frac{1+\beta}{1-\beta} = e^{0.5} = 1.6487$$

or $$\beta = 0.25 = \frac{R_1}{R_1 + R_2} = \frac{5\ \text{k}\Omega}{5\ \text{k}\Omega + R_2}$$

or $$R_2 = 15\ \text{k}\Omega$$

Again,

$$T = 10^{-5}\ \text{s} = 2 \times 10 \times 10^3 \times 0.01 \times 10^{-6}\ \ln\left(\frac{1+\beta}{1-\beta}\right)$$

or $$\ln\left(\frac{1+\beta}{1-\beta}\right) = \frac{10^{-5}}{2 \times 10^4 \times 0.01 \times 10^{-6}} = 0.05$$

or $$\frac{1+\beta}{1-\beta} = e^{0.05} = 1.05$$

or $$\beta = 0.025 = \frac{R_1}{R_1 + R_2} = \frac{5\ \text{k}\Omega}{5\ \text{k}\Omega + R_2}$$

or $$R_2 = 192\ \text{k}\Omega$$

Hence the range of R_2 must be 15 kΩ through 192 kΩ.

SUMMARY

- A flip-flop is the basic memory element; it can store a 0 or a 1.
- A flip-flop is known more formally as a bistable multivibrator. It has two stable states.
- Flip-flops are used for data storage, counting, frequency division, parallel-to-serial and serial-to-parallel data conversion, etc.
- An unclocked flip-flop is called a latch, because the output of the flip-flop latches on to a 1 or a 0 immediately after the input is applied.
- Non-gated latches are called asynchronous latches and clocked (but not edge-triggered) latches are called synchronous latches.
- A latch is constructed using two cross-coupled NAND gates or NOR gates.
- A latch may be an active-HIGH latch or an active-LOW latch.
- The NOR gate S-R latch is an active-HIGH S-R latch.
- The NAND gate S-R latch is an active-LOW S-R latch.
- The clocked D latch is called a transparent D latch, because its output follows the input when the clock is HIGH.
- Flip-flops may be level-triggered or edge-triggered.
- The level-triggered flip-flops are those which respond to changes in inputs when the clock is HIGH. The edge-triggered flip-flops are those which respond only to inputs present at the transition of the clock pulse.
- Edge triggering is also called dynamic triggering.
- The J-K flip-flop is the most versatile and most widely used of all the flip-flops.
- T flip-flops are not widely available as commercial items.
- PRESET and CLEAR inputs may be active-LOW or active-HIGH.
- PRESET and CLEAR inputs are asynchronous inputs. They override all other inputs. They are also called DC SET, and DC RESET or DC CLEAR, or Direct SET (S_D) and Direct RESET (R_D).
- A triangle at the clock input terminal of a flip-flop indicates that it is a positive edge-triggered flip-flop. A bubble and a triangle at the clock input terminal of a flip-flop indicates that it is a negative edge-triggered flip-flop.
- No symbol at the clock input terminal of a flip-flop indicates that it is level-triggered.
- For an S-R flip-flop, S = 0, R = 0 is no change, S = 1, R = 0 is SET, S = 0, R = 1 is RESET, and S = 1, R = 1 is invalid.
- For a J-K flip-flop, J = 0, K = 0 is no change, J = 1, K = 0 is SET, J = 0, K = 1 is RESET, and J = 1, K = 1 is the toggle mode.
- The T flip-flop toggles, i.e. it goes to the opposite state after each clock pulse.
- In a D flip-flop, the input is transferred to the output when a clock is applied. D = 0 is RESET and D = 1 is SET.
- Due to wiring between components, a clock pulse may undergo varying degrees of delay before it arrives at the flip-flops. It is called *clock skew*.

- The master-slave flip-flop is made up of two flip-flops—a master and a slave. It was developed to make synchronous operation of flip-flops more predictable by overcoming the problem of *race* in clocked flip-flops. The master-slave flip-flop is now obsolete.
- Master-slave flip-flops are called pulse-triggered flip-flops, because the length of time required for its output to change state equals the width of one (clock) pulse.
- The inputs to a master-slave flip-flop must not change when the clock is HIGH. If they change, the flip-flop gives an output corresponding to the last combination of inputs present before the clock terminates.
- Data lock-out flip-flops are nothing but master-slave flip-flops in which the master is an edge-triggered flip-flop.
- A master-slave flip-flop is recognized by the postponed output symbol ⏋ at its Q and $\overline{Q}$ terminals. No dynamic indicator symbol is provided at the clock terminal.
- A data lock-out type flip-flop will have postponed output symbol ⏋ at its Q and $\overline{Q}$ terminals and a dynamic indicator symbol at the clock terminal.
- The time interval between the application of a triggering edge or asynchronous inputs and the output actually making a transition is called the propagation delay of a flip-flop.
- The minimum time for which the control levels need to be maintained constant on the input terminals of a flip-flop, prior to the arrival of the triggering edge of the clock pulse, for the reliable operation of the flip-flop is called the *set-up* time.
- The minimum time for which the control signals need to be maintained constant at the input terminals of a flip-flop, after the arrival of the triggering edge of the clock pulse, in order to enable the flip-flop to respond reliably is called the *hold time*.
- The highest frequency at which a flip-flop can be reliably triggered is called the maximum clock fréquency.
- Any one type of flip-flop can be converted to any other type by providing a suitable combinational circuit.
- A Schmitt trigger inverter is used to convert a slowly varying waveform to a square wave.
- A monostable multivibrator has only one stable state. It is also called *one-shot* or *single-shot* or *monostable*.
- A monostable can be used for gating, providing time delays, generating a sequence of pulses, and for detection of missing pulses.
- There are two basic types of IC one-shots—retriggerable and non-retriggerable.
- An astable multivibrator has no stable states. It does not require any triggering signal for its operation. It is also called a free-running multivibrator.
- An astable multivibrator can be used as a master oscillator to provide clock pulses.

QUESTIONS

1. Distinguish between combinational and sequential switching circuits.
2. What is meant by stable state?
3. What do you mean by (a) a latch, and (b) a gated latch?
4. How do you build a latch using universal gates?

5. What is an active-HIGH latch and an active-LOW latch?
6. Distinguish between synchronous and asynchronous latches.
7. What is the normal resting state of SET and CLEAR inputs in a NAND gate S-R latch? What is the active state of each input?
8. Name the two types of inputs which a clocked flip-flop has.
9. Why is a gated D latch called the 'transparent' latch?
10. List the different types of latches and flip-flops. Name the applications of each type.
11. How does a J-K flip-flop differ from an S-R flip-flop in its basic operation? What is its advantage over an S-R flip-flop?
12. Describe the main difference between a gated S-R latch and an edge-triggered S-R flip-flop.
13. What do you mean by a level-triggered flip-flop? How does it differ from an edge-triggered flip-flop?
14. What J, K condition will always set Q upon the occurrence of the active CLK transition?
15. Explain why the S and R inputs of an edge-triggered S-R flip-flop affect Q only during the active transition of CLK.
16. Describe the differences between the pulse-triggered and edge-triggered flip-flops.
17. Suppose that the D input of a flip-flop changes from LOW to HIGH in the middle of a clock pulse. Describe what would happen if the flip-flop is a positive edge-triggered type, a pulse-triggered type.
18. Can a D flip-flop respond to its D and CLK inputs while $\overline{PRE} = 1$?
19. What do you mean by toggling?
20. Which is the most versatile flip-flop? Which flip-flop is preferred for data transfer?
21. Why are asynchronous inputs called overriding inputs?
22. What is a master-slave flip-flop? Discuss its working.
23. What is the major restriction when operating a pulse-triggered flip-flop?
24. What do you mean by (a) clock skew, and (b) time race?
25. Symbolically, how is a data lock-out flip-flop distinguished from a pulse-triggered flip-flop?
26. What do you mean by a flip-flop with data lock-out? How does it differ from a normal master-slave flip-flop?
27. Typically, a manufacturer's data sheet specifies four different propagation delay times associated with a flip-flop. Name and describe each of them.
28. Define the following terms as applied to flip-flops.
    (a) set-up time (b) hold time
    (c) propagation delay (d) maximum clock frequency
    (e) power dissipation
29. How must a J-K flip-flop be connected to function as a divide-by-two element? How many flip-flops are required to produce a divide-by-64 device?
30. What is meant by dynamic triggering?

31. How do you convert one type of flip-flop into another type?

32. What is the difference between the retriggerable and non-retriggerable one-shots?

33. Describe the operation and applications of monostable multivibrators, astable multivibrators, bistable multivibrators, and Schmitt trigger inverter?

34. Explain the difference in operation of a monostable and a astable multivibrator.

35. What are the advantages of the crystal clock generator?

36. State TRUE or FALSE:

(a) The CLK input will affect the flip-flop output only when the active transition of the control input occurs.

(b) The SET input can never be used to make Q a 0.

(c) A J-K flip-flop can be used as an S-C flip-flop, but an S-C flip-flop cannot be used as a J-K flip-flop.

(d) The Q output of a D flip-flop will equal the level at the D input at all times.

(e) A D latch is in its transparent mode when EN is a 0.

(f) Synchronous data transfer requires less circuitry than that required by asynchronous transfer.

(g) Why are MOS inverters preferred over TTL inverters in making astable multivibrators?

PROBLEMS

8.1 If the waveforms shown in Fig. 8.70 are applied to an active-HIGH S-R latch which is in the RESET state, draw the output waveform.

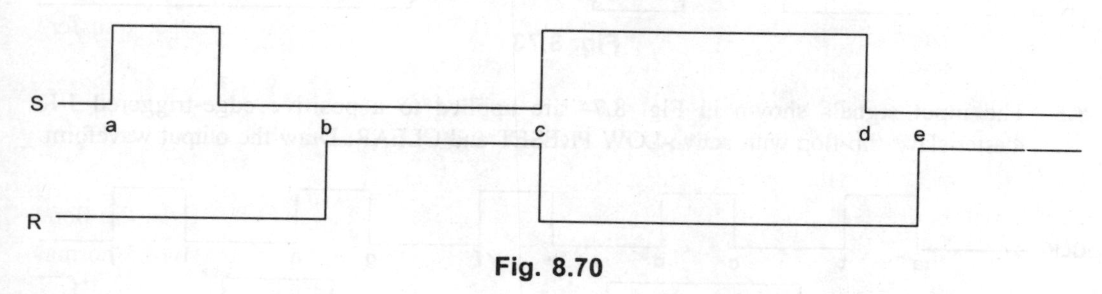

Fig. 8.70

8.2 If the waveforms shown in Fig. 8.71 are applied to an active-LOW S-R latch which is in the RESET state, draw the output waveform.

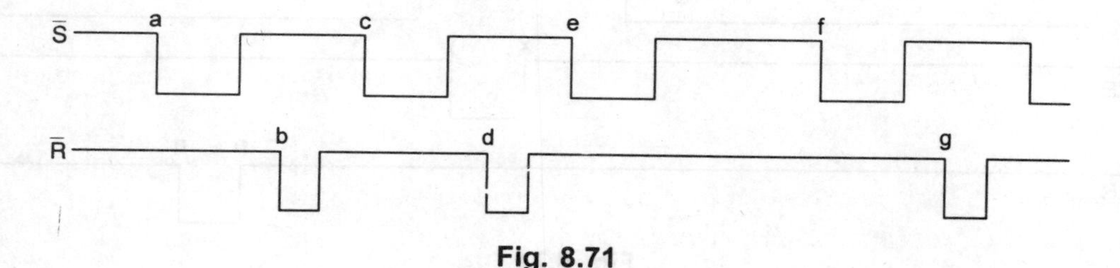

Fig. 8.71

8.3 The waveforms shown in Fig. 8.72 are applied to (a) a positive edge-triggered S-R flip-flop, (b) a negative edge-triggered S-R flip-flop, and (c) a master-slave S-R flip-flop. Draw the output waveform in each case.

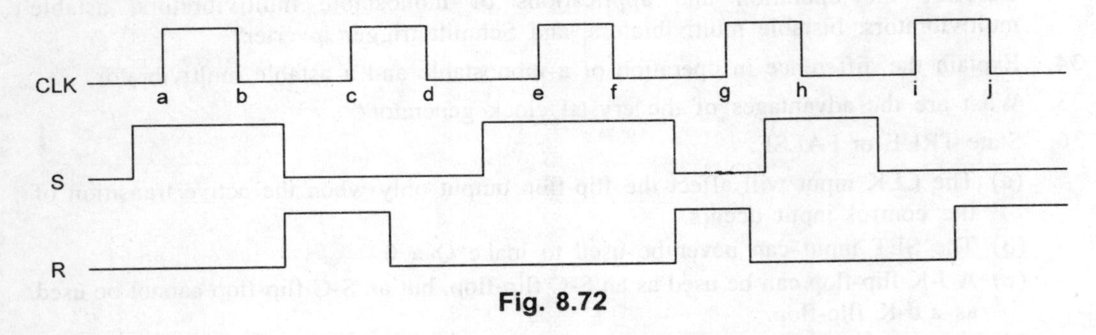

Fig. 8.72

8.4 The waveforms shown in Fig. 8.73 are applied to (a) a positive-edge-triggered J-K flip-flop, (b) a negative edge-triggered J-K flip-flop, and (c) a master-slave J-K flip-flop. Draw the output waveform in each case.

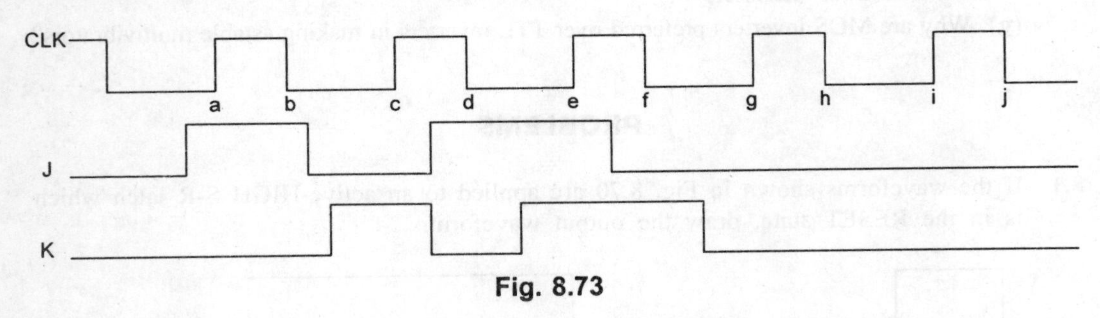

Fig. 8.73

8.5 The input signals shown in Fig. 8.74 are applied to a positive edge-triggered J-K master-slave flip-flop with active-LOW PRESET and CLEAR. Draw the output waveform.

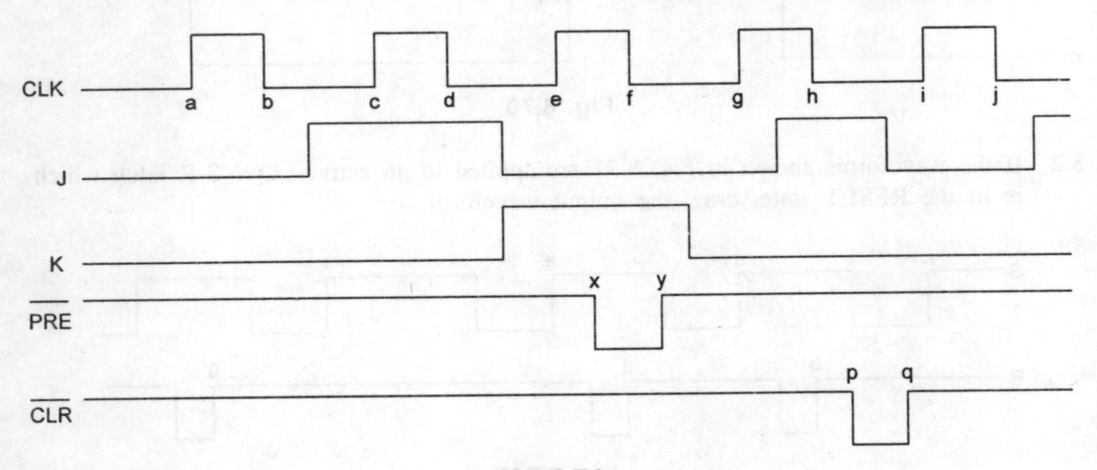

Fig. 8.74

8.6 The waveforms shown in Fig. 8.75 are applied to a negative edge-triggered S-R flip-flop with active-HIGH PRESET and CLEAR. Draw the output waveform.

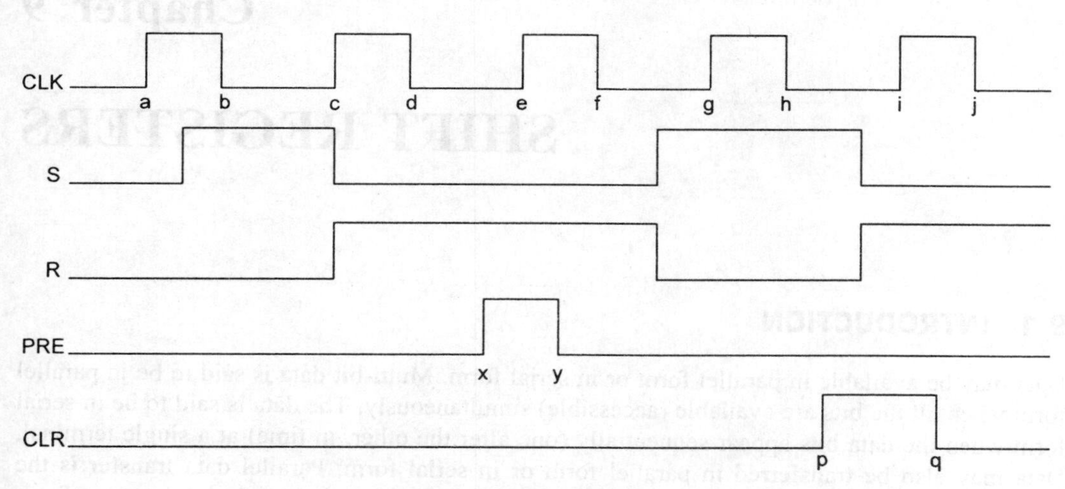

Fig. 8.75

8.7 The following serial data are applied to the flip-flop shown in Fig. 8.76. Determine the resulting serial data that appears on the Q output. There is one clock pulse for each bit time. Assume that, Q is initially 0. The rightmost bits are applied first.

J_1 = 10110110 $\qquad$ J_2 = 11011001

K_1 = 10010110 $\qquad$ K_2 = 11011011

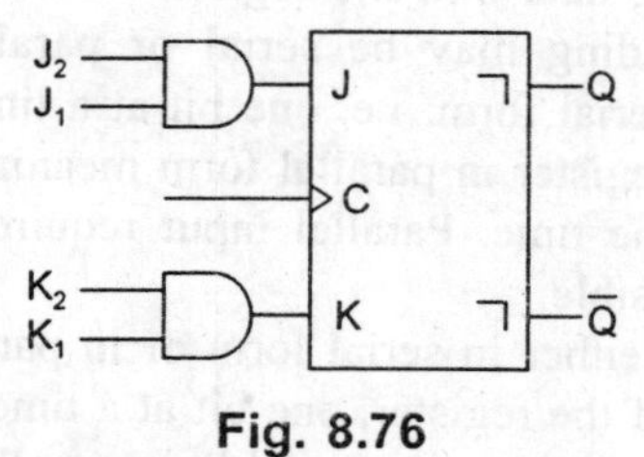

Fig. 8.76

8.8 Design an astable multivibrator using the 555 timer to generate a square wave of 5 kHz with 40 per cent duty cycle.

8.9 For the astable multivibrator using an op-amp as shown in Fig. 8.67, R = 20 kΩ, C = 0.05 μF and R_1 = 10 kΩ. It is desired to make the frequency of the output square wave adjustable from 15 kHz through 150 kHz by making R_2 adjustable. Through what range of values should R_2 be made adjustable to obtain the required frequency range?

8.10 Determine the values of R_{ext} and C_{ext} that will produce a pulse width of 2 μs when connected to a 74123 as shown in Fig. 8.60.

8.11 Design a one-shot using the 555 timer to generate a pulse of width 1 ms.

Chapter 9

SHIFT REGISTERS

9.1 INTRODUCTION

Data may be available in parallel form or in serial form. Multi-bit data is said to be in parallel form when all the bits are available (accessible) simultaneously. The data is said to be in serial form when the data bits appear sequentially (one after the other, in time) at a single terminal. Data may also be transferred in parallel form or in serial form. Parallel data transfer is the simultaneous transmission of all bits of data from one device to another. Serial data transfer is the transmission of one bit of data at a time from one device to another. Serial data must be transmitted under the synchronization of a clock, since the clock provides the means to specify the time at which each new bit is sampled.

As a flip-flop (FF) can store only one bit of data, a 0 or a 1, it is referred to as a single-bit register. When more bits of data are to be stored, a number of FFs are used. A register is a set of FFs used to store binary data. The storage capacity of a register is the number of bits (1s and 0s) of digital data it can retain. Loading a register means setting or resetting the individual FFs, i.e. inputting data into the register so that their states correspond to the bits of data to be stored. Loading may be serial or parallel. In serial loading, data is transferred into the register in serial form, i.e. one bit at a time, whereas in parallel loading, the data is transferred into the register in parallel form meaning that all the FFs are triggered into their new states at the same time. Parallel input requires that the SET and/or RESET controls of every FF be accessible.

A register may output data either in serial form or in parallel form. Serial output means that the data is transferred out of the register, one bit at a time serially. Parallel output means that the entire data stored in the register is available in parallel form, and can be transferred out at the same time.

Shift registers are a type of logic circuits closely related to counters. They are used basically for the storage and transfer of digital data. The basic difference between a shift register and a counter is that, a shift register has no specified sequence of states except in certain very specialized applications, whereas a counter has a specified sequence of states.

A shift register is a very important digital building block. It has innumerable applications. Registers are often used to momentarily store binary information appearing at the output of an encoding matrix. A register might be used to accept input data from an alphanumeric keyboard and then present the data at the input of a microprocessor chip. Similarly, shift registers are often used to momentarily store binary data at the output of a decoder. A shift register also forms the basis for some very important arithmetic operations. For example, the operations of complementation, multiplication, and division are frequently implemented by means of a register. A shift register can also be connected to form a number of different types of counters. These counters offer some very distinct advantages.

9.2 BUFFER REGISTER

Some registers do nothing more than storing a binary word. The buffer register is the simplest of registers. It simply stores the binary word. The buffer may be a controlled buffer. Most of the buffer registers use D flip-flops.

Figure 9.1 shows a 4-bit buffer register. The binary word to be stored is applied to the data terminals. On the application of clock pulse, the output word becomes the same as the word applied at the input terminals, i.e. the input word is loaded into the register by the application of clock pulse.

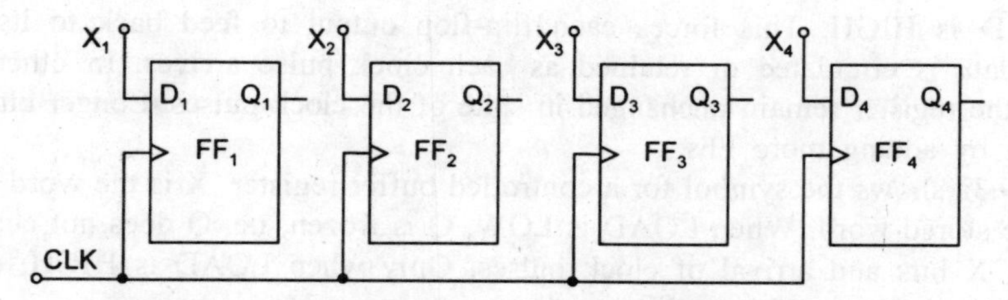

Fig. 9.1 Logic diagram of a 4-bit buffer register.

When the positive clock edge arrives, the stored word becomes

$$Q_4\ Q_3\ Q_2\ Q_1 = X_4\ X_3\ X_2\ X_1$$

or

$$Q = X$$

This circuit is too primitive to be of any use. What it needs is some control over the X bits, i.e. some way of holding them off until we are ready to store them.

9.3 CONTROLLED BUFFER REGISTER

Figure 9.2 shows a controlled buffer register. If $\overline{CLR}$ goes LOW, all the FFs are RESET and the output becomes, Q = 0000.

When $\overline{CLR}$ is HIGH, the register is ready for action. LOAD is the control input. When LOAD is **HIGH**, the data bits X can reach the D inputs of FFs. At the positive-going edge of the next **clock** pulse, the register is loaded, i.e.

$$Q_4\ Q_3\ Q_2\ Q_1 = X_4\ X_3\ X_2\ X_1 \quad \text{or} \quad Q = X$$

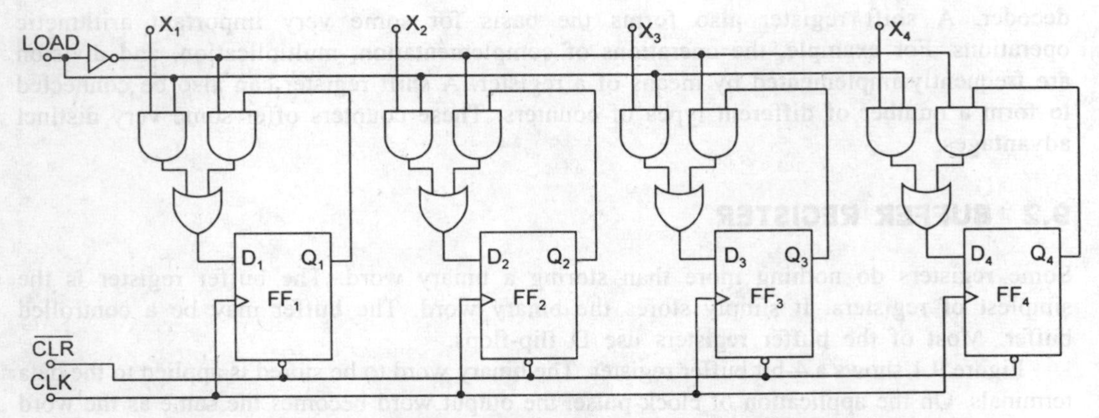

Fig. 9.2 Logic diagram of a 4-bit controlled buffer register.

When LOAD is LOW, the X bits cannot reach the FFs. At the same time, the inverted signal $\overline{\text{LOAD}}$ is HIGH. This forces each flip-flop output to feed back to its data input. Therefore, data is circulated or retained as each clock pulse arrives. In other words, the contents of the register remain unchanged in spite of the clock pulses. Longer buffer registers can be built by adding more FFs.

Figure 9.3a shows the symbol for a controlled buffer register. X is the word to be loaded and Q is the stored word. When LOAD is LOW, Q is frozen, i.e. Q does not change in spite of changing X bits and arrival of clock pulses. Only when LOAD is HIGH, can the next positive clock edge load X into the register.

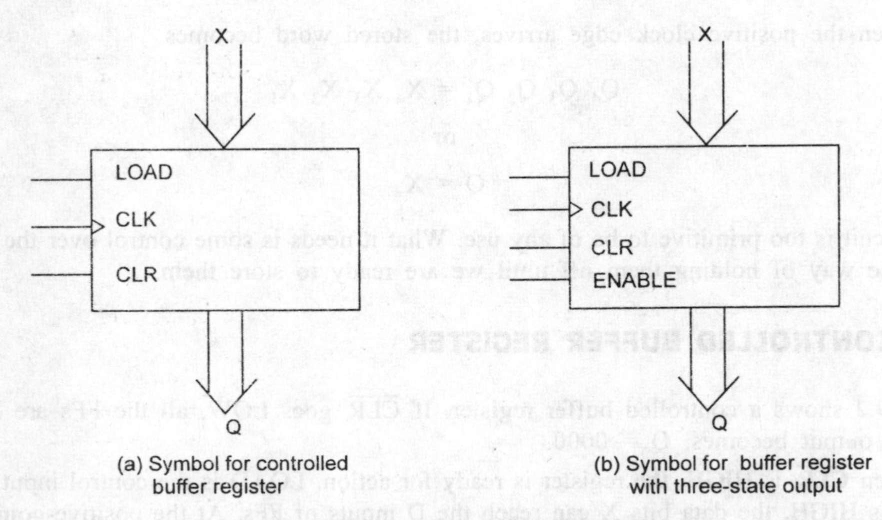

(a) Symbol for controlled buffer register

(b) Symbol for buffer register with three-state output

Fig. 9.3 Logic symbols of buffer registers.

A register with three-state output is required, if the buffer output is to be connected to a common bus. Here the output of the register can be floating or can have the word available at the output terminals. Figure 9.3b shows the symbol of a buffer register with three-state output. Let the ENABLE signal be an active-HIGH signal. Then, when the ENABLE is HIGH, the word stored by the register is available at the output terminals and when the ENABLE is LOW, the output terminals are floating.

9.4 DATA TRANSMISSION IN SHIFT REGISTERS

A number of FFs connected together such that data may be shifted into and shifted out of them is called a shift register. Data may be shifted into or out of the register either in serial form or in parallel form. So, there are four basic types of shift registers: serial-in, serial-out; serial-in, parallel-out; parallel-in, serial-out; and parallel-in, parallel-out. The process of data shifting in these registers is illustrated in Fig. 9.4. All of these

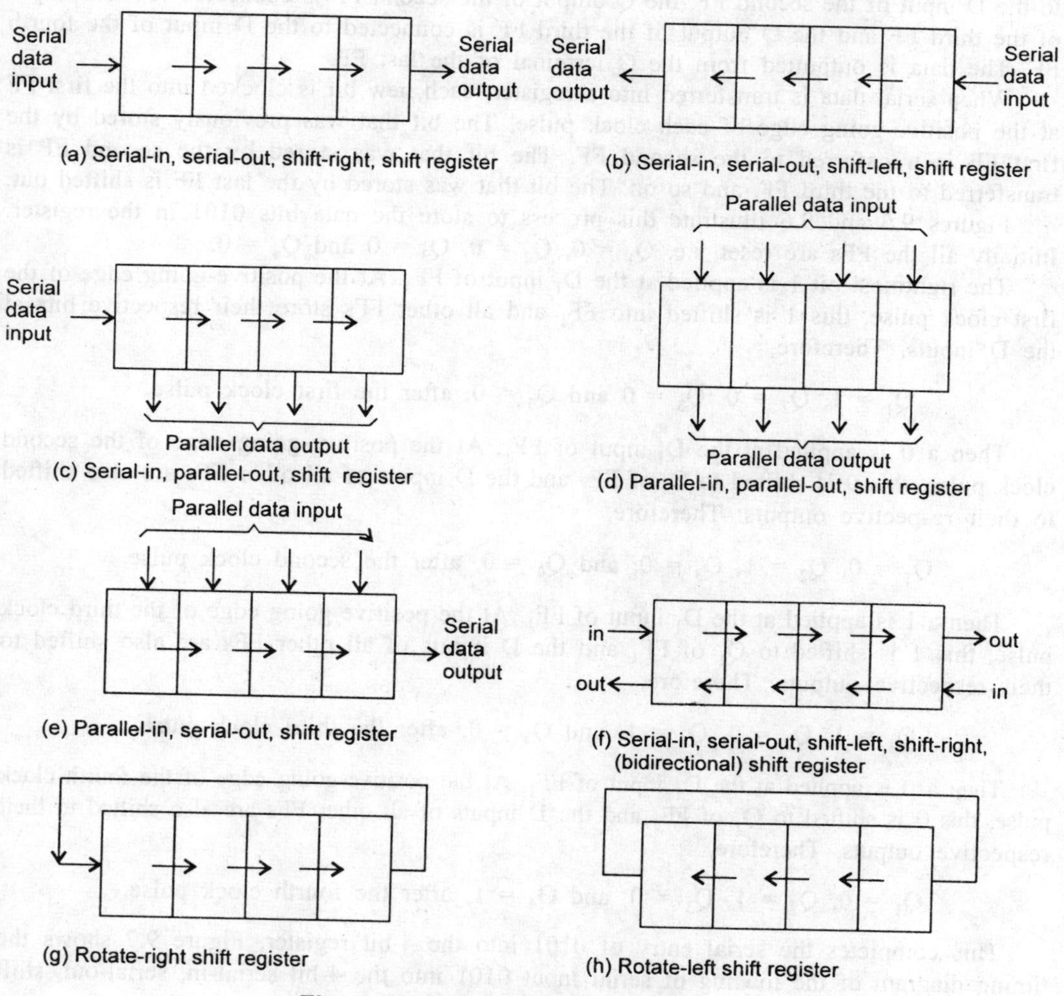

Fig. 9.4 Data transfer in registers.

configurations are commercially available as TTL MSI/LSI circuits. Data may be rotated left or right. Data may be shifted from left to right or right to left at will, i.e. in a bidirectional way. Also, data may be shifted in serially (in either way) or in parallel and shifted out serially (in either way) or in parallel.

9.5 SERIAL-IN, SERIAL-OUT, SHIFT REGISTER

This type of shift register accepts data serially, i.e. one bit at a time, and also outputs data serially.

The logic diagram of a 4-bit serial-in, serial-out, shift-right, shift register is shown in Fig. 9.5a. With four stages, i.e. four FFs, the register can store up to four bits of data. Serial data is applied at the D input of the first FF. The Q output of the first FF is connected to the D input of the second FF, the Q output of the second FF is connected to the D input of the third FF and the Q output of the third FF is connected to the D input of the fourth FF. The data is outputted from the Q terminal of the last FF.

When serial data is transferred into a register, each new bit is clocked into the first FF at the positive-going edge of each clock pulse. The bit that was previously stored by the first FF is transferred to the second FF. The bit that was stored by the second FF is transferred to the third FF, and so on. The bit that was stored by the last FF is shifted out.

Figures 9.5 and 9.6 illustrate this process to store the data bits 0101 in the register. Initially all the FFs are reset, i.e. $Q_1 = 0$, $Q_2 = 0$, $Q_3 = 0$ and $Q_4 = 0$.

The rightmost bit 1 is applied at the D_1 input of FF_1. At the positive-going edge of the first clock pulse, this 1 is shifted into FF_1 and all other FFs store their respective bits at the D inputs. Therefore,

$$Q_1 = 1,\ Q_2 = 0,\ Q_3 = 0 \text{ and } Q_4 = 0, \text{ after the first clock pulse.}$$

Then a 0 is applied at the D_1 input of FF_1. At the positive-going edge of the second clock pulse, this 0 is shifted to Q_1 of FF_1 and the D inputs of all other FFs are also shifted to their respective outputs. Therefore,

$$Q_1 = 0,\ Q_2 = 1,\ Q_3 = 0, \text{ and } Q_4 = 0, \text{ after the second clock pulse.}$$

Then a 1 is applied at the D_1 input of FF_1. At the positive-going edge of the third clock pulse, this 1 is shifted to Q_1 of FF_1 and the D inputs of all other FFs are also shifted to their respective outputs. Therefore,

$$Q_1 = 1,\ Q_2 = 0,\ Q_3 = 1, \text{ and } Q_4 = 0, \text{ after the third clock pulse.}$$

Then a 0 is applied at the D_1 input of FF_1. At the positive-going edge of the fourth clock pulse, this 0 is shifted to Q_1 of FF_1 and the D inputs of all other FFs are also shifted to their respective outputs. Therefore,

$$Q_1 = 0,\ Q_2 = 1,\ Q_3 = 0, \text{ and } Q_4 = 1, \text{ after the fourth clock pulse.}$$

This completes the serial entry of 0101 into the 4-bit register. Figure 9.7 shows the timing diagram of the loading of serial input 0101 into the 4-bit serial-in, serial-out, shift register.

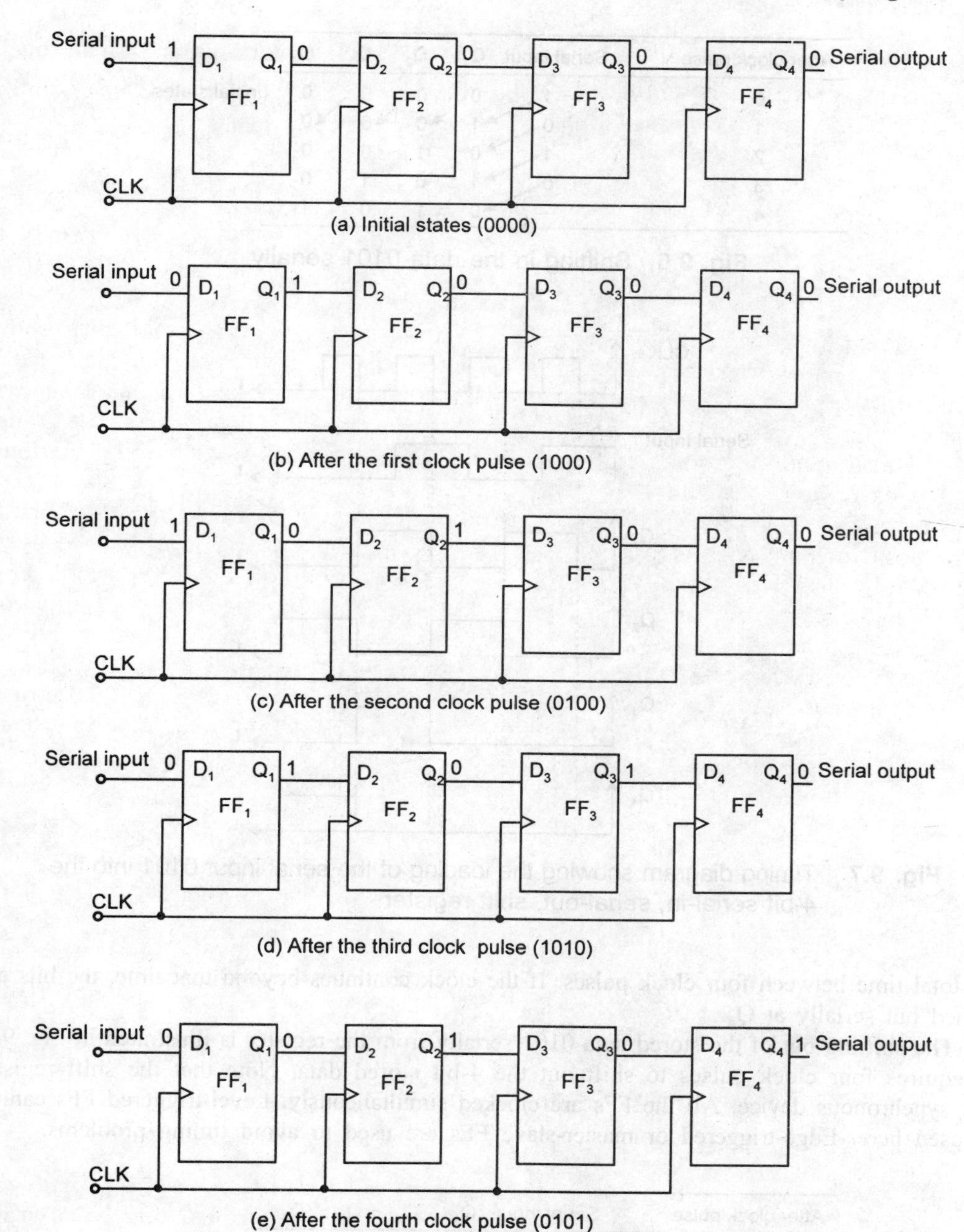

Fig. 9.5 Loading of the 4-bit serial-in, serial-out, shift register.

In this example, we have assumed that initially all the FFs are reset, although the initial states have no effect on the loading of new data. The initial states represent old data, whose bits are effectively shifted out of the register in serial form at Q_4 as the new bits are shifted in at Q_1. Since four bits must be shifted into the register, the time required to load it equals

After clock pulse	Serial input	Q_1	Q_2	Q_3	Q_4	
0	1	0	0	0	0	(initial states
1	0	1	0	0	0	
2	1	0	1	0	0	
3	0	1	0	1	0	
4	—	0	1	0	1	

Fig. 9.6 Shifting in the data 0101 serially.

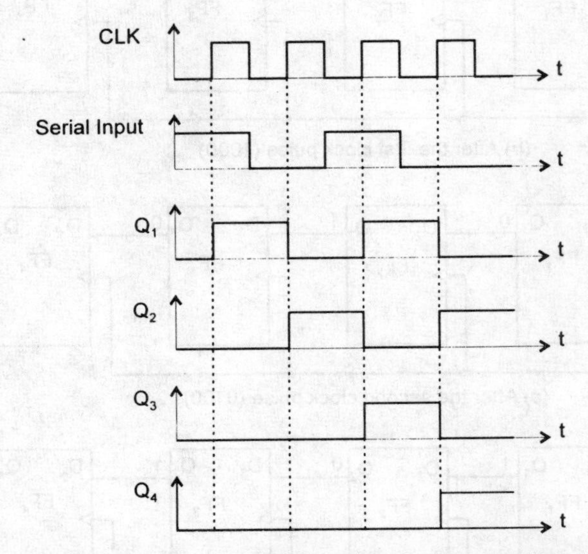

Fig. 9.7 Timing diagram showing the loading of the serial input 0101 into the 4-bit serial-in, serial-out, shift register.

the total time between four clock pulses. If the clock continues beyond that time, the bits are shifted out serially at Q_4.

The shifting out of the stored data 0101 serially from the register is illustrated in Fig. 9.8. It requires four clock pulses to shift out the 4-bit stored data. Note that the shift register is a synchronous device. All the FFs are clocked simultaneously. Level-triggered FFs cannot be used here. Edge-triggered or master-slave FFs are used to avoid timing problems.

After clock pulse	Serial input	Q_1	Q_2	Q_3	Q_4	
0	0	0	1	0	1	(initial states)
1	0	0	0	1	0	
2	0	0	0	0	1	
3	0	0	0	0	0	
4	—	0	0	0	0	(final states)

Fig. 9.8 Shifting out the data 0101 serially.

A shift register can also be constructed using J-K FFs or S-R FFs as shown in Figs. 9.9a and 9.9b, respectively. The data is applied at the J (S) input of the first FF. The complement of this is fed to the K (R) terminal of the first FF. The Q output of the first FF is connected to J (S) input of the second FF, the Q output of the second FF to J (S) input of the third FF, and so on. Also, $\overline{Q}_1$ is connected to K_2 (R_2), $\overline{Q}_2$ is connected to K_3 (R_3), and so on.

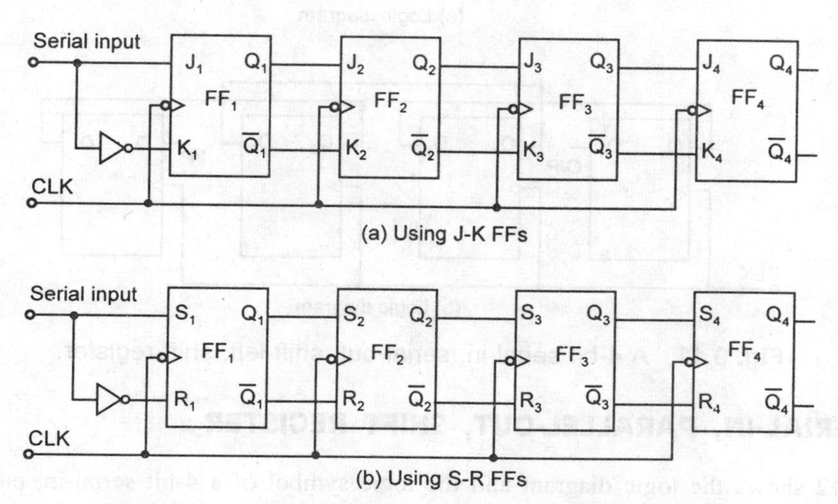

Fig. 9.9 A 4-bit serial-in, serial-out, shift register.

Figure 9.10 shows the functional diagram and the logic symbol of an 8-bit shift register, which uses master-slave S-R FFs. Note that the register has two inputs labelled A and B. If the serial input is connected to A, it will be loaded into the register only if B is high, and vice versa. This gating allows one of the inputs to serve as a control. When it is HIGH, the serial input on the other line is enabled and when it is LOW, the serial input is inhibited.

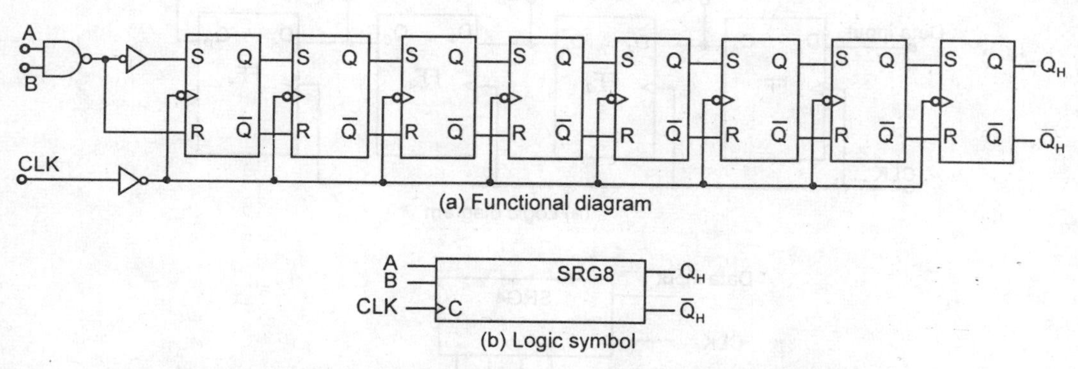

Fig. 9.10 The 7491A 8-bit serial-in, serial-out, shift register.

Figure 9.11 shows the logic diagrams of a 4-bit serial-in, serial-out, shift-left, shift register.

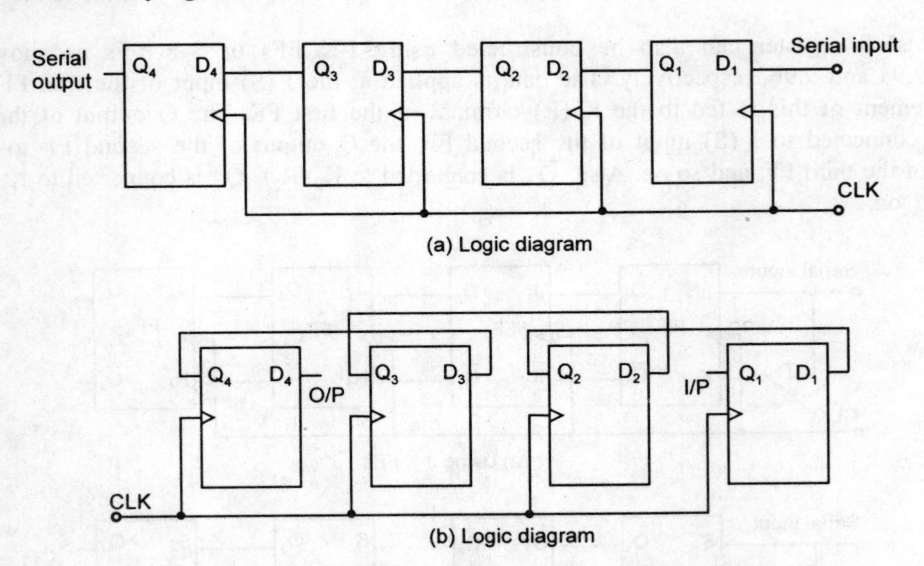

Fig. 9.11 A 4-bit serial-in, serial-out, shift-left, shift register.

9.6 SERIAL-IN, PARALLEL-OUT, SHIFT REGISTER

Figure 9.12 shows the logic diagram and the logic symbol of a 4-bit serial-in, parallel-out, shift register. In this type of register, the data bits are entered into the register serially, but the data stored in the register is shifted out in parallel form.

Once the data bits are stored, each bit appears on its respective output line and all bits are available simultaneously, rather than on a bit-by-bit basis as with the serial output. The serial-in, parallel-out, shift register can be used as a serial-in, serial-out, shift register if the output is taken from the Q terminal of the last FF.

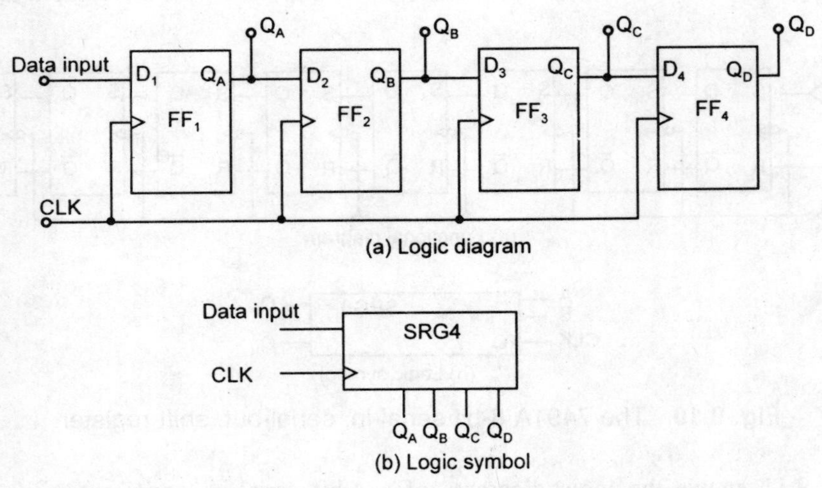

Fig. 9.12 A 4-bit serial-in, parallel-out, shift register.

8-Bit Serial-in, Parallel-out, Shift Register

Figure 9.13 shows the logic diagram and the logic symbol of an 8-bit serial-in, parallel-out, shift register. In the logic diagram, A and B are the gated serial inputs. Suppose that the serial data is connected to A, then B is used as a control line. When B is held HIGH, the NAND gate is enabled and the serial input data after passing through the NAND gate is inverted. The input data is shifted serially into the register. When B is held LOW, the NAND gate output is forced HIGH, the input data stream is inhibited and the next positive clock transition will shift a 0 into the first FF. Each succeeding positive clock transition will shift another 0 into the register. After eight clock pulses, the register will be full of 0s. This register also has an active-LOW asynchronous clear input that can be used to reset every FF in the register.

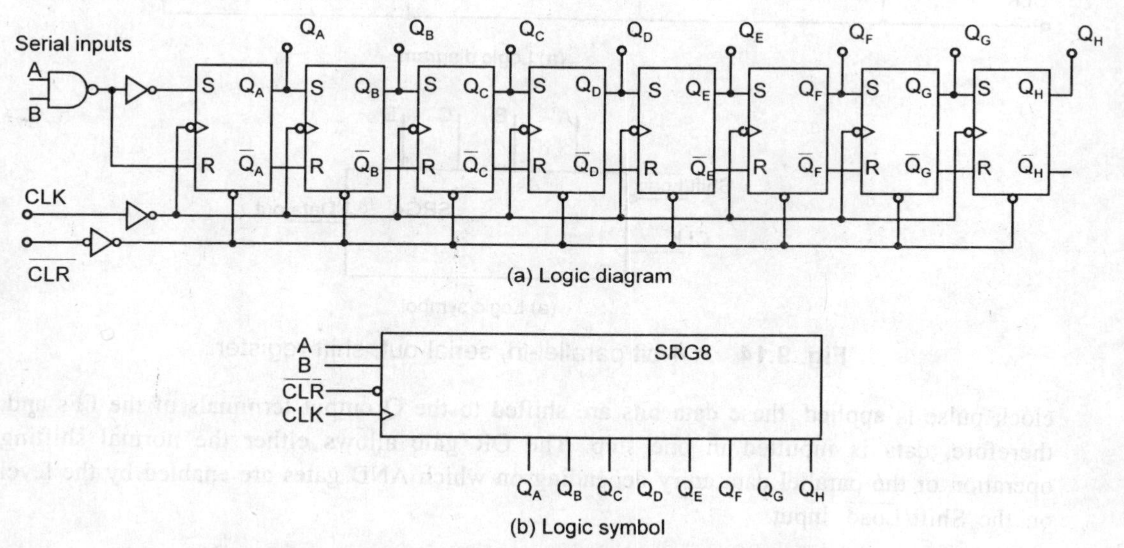

(a) Logic diagram

(b) Logic symbol

Fig. 9.13 An 8-bit serial-in, parallel-out, shift register.

9.7 PARALLEL-IN, SERIAL-OUT, SHIFT REGISTER

For a parallel-in, serial-out, shift register, the data bits are entered simultaneously into their respective stages on parallel lines, rather than on a bit-by-bit basis on one line as with serial data inputs, but the data bits are transferred out of the register serially, i.e. on a bit-by-bit basis over a single line.

Figure 9.14 illustrates a 4-bit parallel-in, serial-out, shift register using D FFs. There are four data lines A, B, C, and D through which the data is entered into the register in parallel form. The signal Shift/$\overline{\text{Load}}$ allows (a) the data to be entered in parallel form into the register and (b) the data to be shifted out serially from terminal Q_4.

When Shift/$\overline{\text{Load}}$ line is HIGH, gates G_1, G_2, and G_3 are disabled, but gates G_4, G_5, and G_6 are enabled allowing the data bits to shift-right from one stage to the next. When Shift/$\overline{\text{Load}}$ line is LOW, gates G_4, G_5, and G_6 are disabled, whereas gates G_1, G_2, and G_3 are enabled allowing the data input to appear at the D inputs of the respective FFs. When a

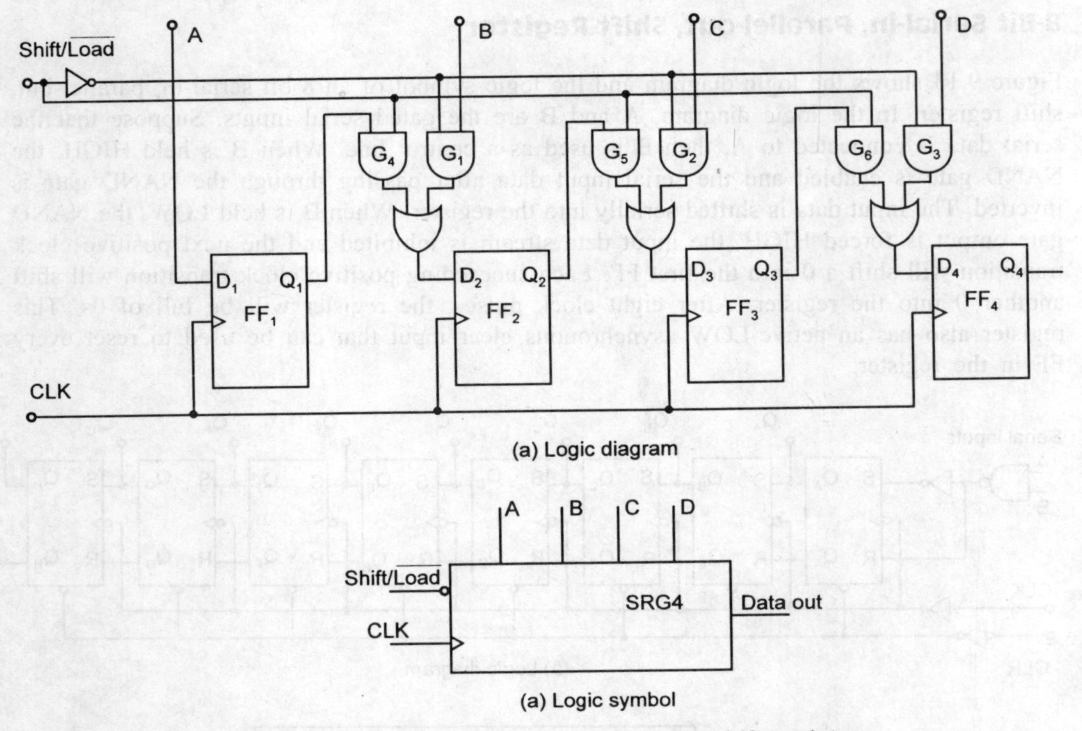

Fig. 9.14 A 4-bit parallel-in, serial-out, shift register.

clock pulse is applied, these data bits are shifted to the Q output terminals of the FFs and, therefore, data is inputted in one step. The OR gate allows either the normal shifting operation or the parallel data entry depending on which AND gates are enabled by the level on the Shift/$\overline{\text{Load}}$ input.

The 74165 8-Bit Parallel-in, Serial-out, Shift Register

Figure 9.15 shows the internal logic diagram and the logic symbol of the IC74165 shift register. The 74165 is an example of an IC shift register, that has parallel-in, serial-out operation. It can also be operated as a serial-in, serial-out, shift register.

A LOW on the Shift/Load (SH/$\overline{\text{LD}}$) input enables all the NAND gates for parallel loading. When an input data bit is a 1, the FF is asynchronously SET by a LOW output of the upper gate. When an input data bit is a 0, the FF is asynchronously RESET by a LOW output of the lower gate. The clock is inhibited during parallel loading. A HIGH on the SH/$\overline{\text{LD}}$ input enables the clock, causing the data on the register to shift-right. Data can be entered serially on the SER input. Also, the clock can be inhibited any time with a HIGH on the CLK INH input. The serial data outputs of the register are Q_H and its complement $\overline{Q}_H$. To load a shift register in parallel, it is necessary that at least one control input (SET or RESET) from each FF be accessible externally. When loading data, the SET and RESET functions are always complement of each other. So, most parallel-in registers use only a single control input for each FF and use an internal inversion to generate the other.

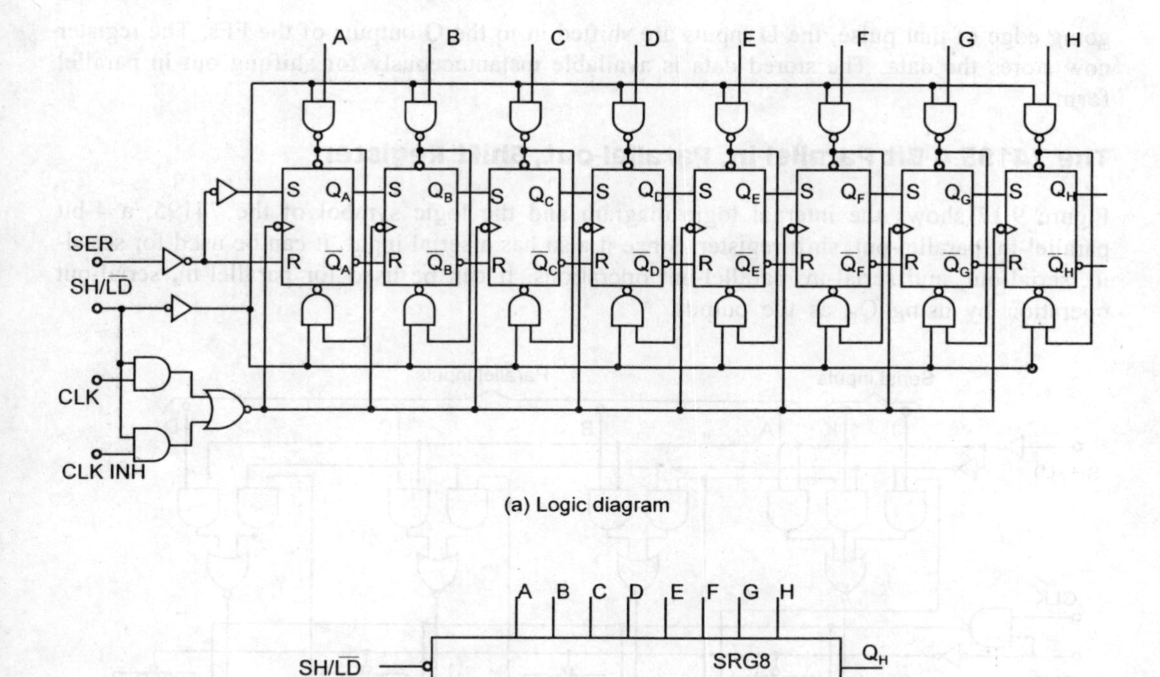

(a) Logic diagram

(b) Logic symbol

Fig. 9.15 The 74165 8-bit parallel-in, serial-out, shift register.

9.8 PARALLEL-IN, PARALLEL-OUT, SHIFT REGISTER

In a parallel-in, parallel-out, shift register, the data is entered into the register in parallel form, and also the data is taken out of the register in parallel form. Immediately following the simultaneous entry of all data bits, the bits appear on the parallel outputs.

Figure 9.16 shows a 4-bit parallel-in, parallel-out, shift register using D FFs. Data is applied to the D input terminals of the FFs. When a clock pulse is applied, at the positive-

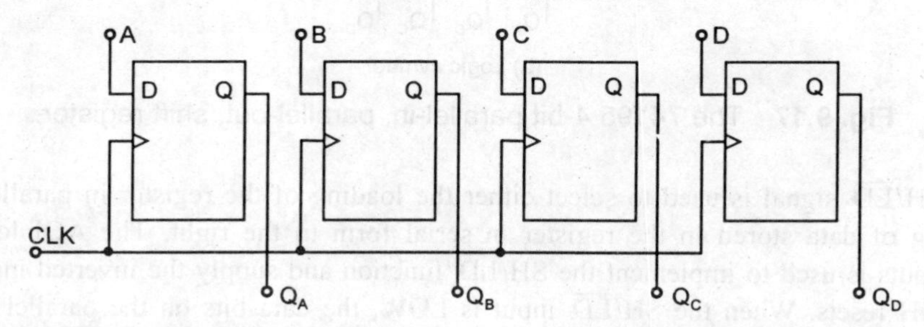

Fig. 9.16 Logic diagram of a 4-bit parallel-in, parallel-out, shift register.

going edge of that pulse, the D inputs are shifted in to the Q outputs of the FFs. The register now stores the data. The stored data is available instantaneously for shifting out in parallel form.

The 74195 4-Bit Parallel-in, Parallel-out, Shift Register

Figure 9.17 shows the internal logic diagram and the logic symbol of the 74195, a 4-bit parallel-in, parallel-out, shift register. Since it also has a serial input, it can be used for serial-in, serial-out, and serial-in, parallel-out operations. It can be used for parallel-in, serial-out operation by using Q_D as the output.

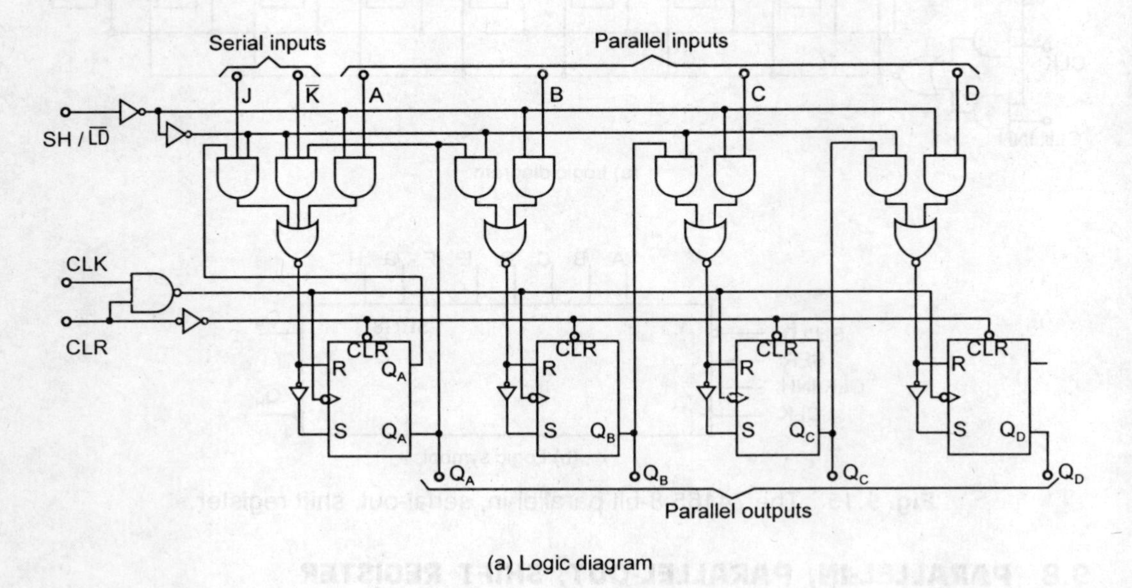

(a) Logic diagram

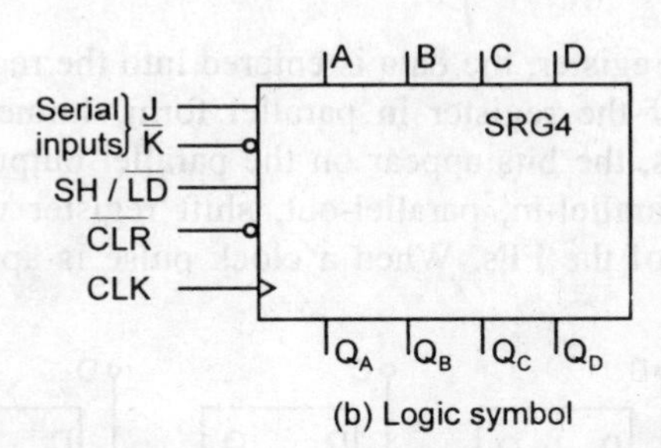

(b) Logic symbol

Fig. 9.17 The 74195 4-bit parallel-in, parallel-out, shift register.

The SH/$\overline{\text{LD}}$ signal is used to select either the loading of the register in parallel form or the shifting of data stored in the register in serial form to the right. The AOI logic at the parallel inputs is used to implement the SH/$\overline{\text{LD}}$ function and supply the inverted inputs to the active-HIGH resets. When the SH/$\overline{\text{LD}}$ input is LOW, the data bits on the parallel inputs are entered synchronously on the positive transition of the clock. When the SH/$\overline{\text{LD}}$ input is HIGH, stored data will shift right synchronously with the clock. The J and $\overline{K}$ are the serial data inputs to the first stage of the register (Q_A). The Q_D can be used for the serial data output. The active-LOW CLR is asynchronous. These can be used to control the state of the first (Q_A) FF in the same way that J and K control a clocked J-K FF, except that K is active-LOW.

9.9 BIDIRECTIONAL SHIFT REGISTER

A bidirectional shift register is one in which the data bits can be shifted from left to right or from right to left.

Figure 9.18 shows the logic diagram of a 4-bit serial-in, serial-out, bidirectional (shift-left, shift-right) shift register. Right/$\overline{\text{Left}}$ is the mode signal. When Right/$\overline{\text{Left}}$ is a 1, the logic circuit works as a shift-right shift register. When Right/$\overline{\text{Left}}$ is a 0, it works as a shift-left shift register. The bidirectional operation is achieved by using the mode signal and two AND gates and one OR gate for each stage as shown in Fig. 9.18.

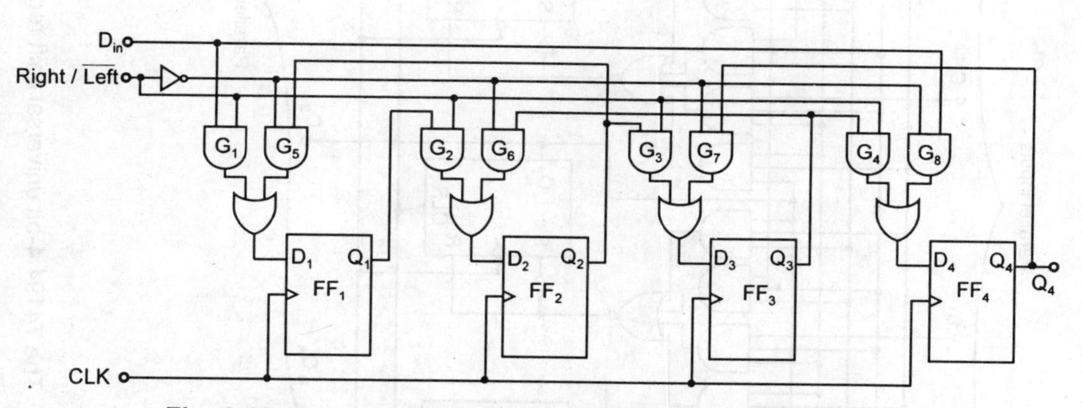

Fig. 9.18 Logic diagram of a 4-bit bidirectional shift register.

A HIGH on the Right/$\overline{\text{Left}}$ control input enables the AND gates G_1, G_2, G_3, and G_4 and disables the AND gates G_5, G_6, G_7, and G_8, and the state of Q output of each FF is passed through the gate to the D input of the following FF. When a clock pulse occurs, the data bits are then effectively shifted one place to the right. A LOW on the Right/$\overline{\text{Left}}$ control input enables the AND gates G_5, G_6, G_7, and G_8 and disables the AND gates G_1, G_2, G_3, and G_4, and the Q output of each FF is passed to the D input of the preceding FF. When a clock pulse occurs, the data bits are then effectively shifted one place to the left. Hence, the circuit works as a bidirectional shift register.

9.10 UNIVERSAL SHIFT REGISTERS

A universal shift register is a bidirectional register, whose input can be either in serial form or in parallel form and whose output also can be either in serial form or in parallel form.

Figure 9.19a shows the logic diagram of the 74194 4-bit universal shift register. Note that the output of each FF is routed through AOI logic to the stage on its right and to the stage

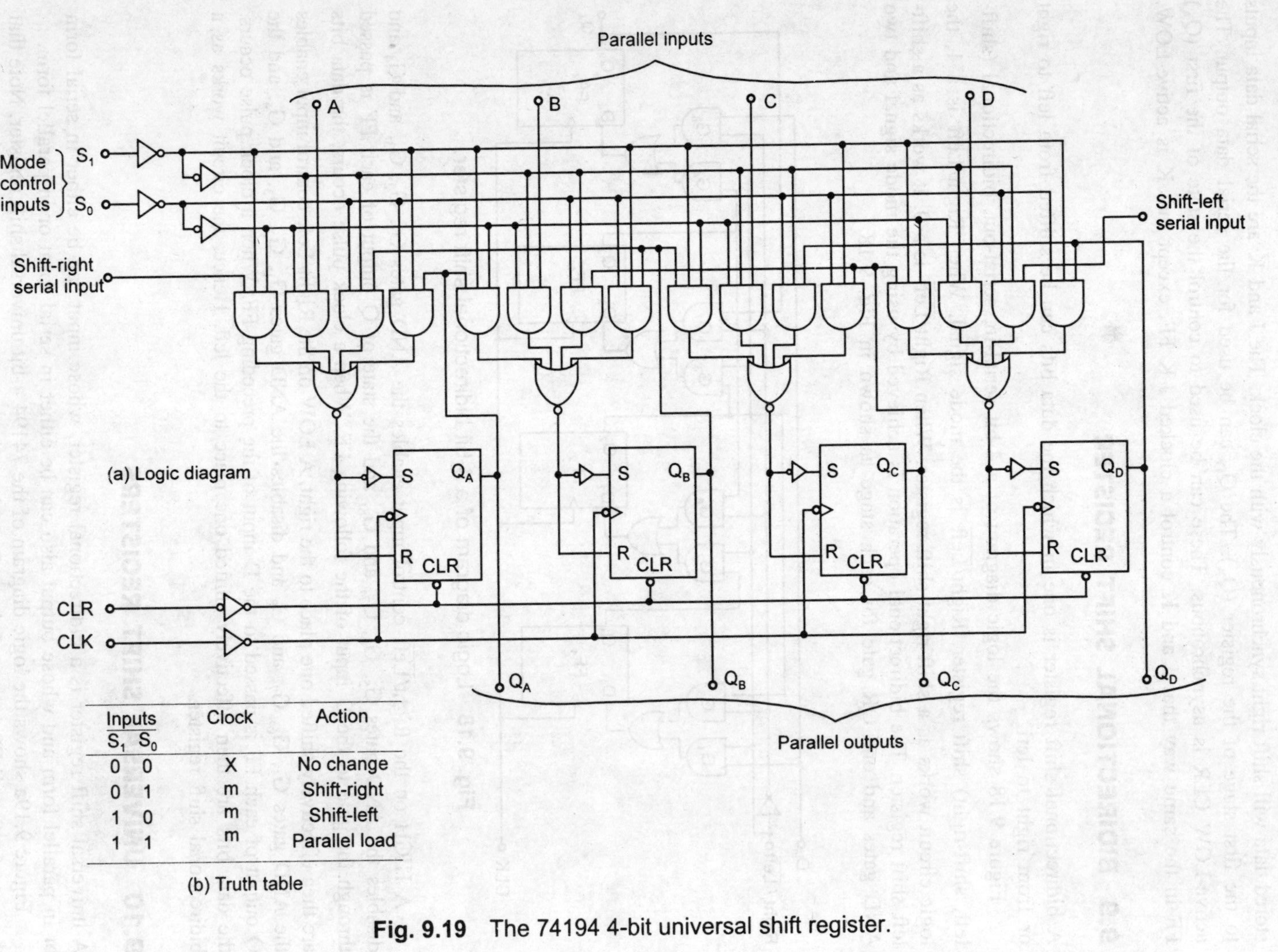

Inputs S_1	S_0	Clock	Action
0	0	X	No change
0	1	m	Shift-right
1	0	m	Shift-left
1	1	m	Parallel load

(b) Truth table

Fig. 9.19 The 74194 4-bit universal shift register.

on its left. The mode control inputs S_0 and S_1 are used to enable the left-to-right connections when it is desired to shift-right, and the right-to-left connections when it is desired to shift-left.

The truth table (Fig. 9.19b) shows that no shifting occurs when S_0 and S_1 are both LOW or both HIGH. When $S_0 = S_1 = 0$, there is no change in the contents of the register, and when $S_0 = S_1 = 1$, the parallel input data A, B, C and D are loaded into the register on the rising edge of the clock pulse. The combination $S_0 = S_1 = 0$, is said to inhibit the loading of serial or parallel data, since the register contents cannot change under that condition. The register has an asynchronous active-LOW clear input, which can be used to reset all the FFs irrespective of the clock and any serial or parallel inputs.

9.11 DYNAMIC SHIFT REGISTERS

All the shift registers we have discussed till now are called *static shift registers*, because each one of the memory elements (i.e. FFs) used to build the register can retain the data bit indefinitely. So, once loaded, the contents of each element of the register remain the same. In a dynamic shift register, storage is accomplished by continually shifting the bits from one stage to the next and re-circulating the output of the last stage into the first stage. The data continually circulates through the register under the control of a clock. To obtain output, a serial output terminal must be accessed at a specific clock pulse, otherwise, the sequence of bits will not correspond to the data stored.

For example, if a 32-bit word is circulating through a 32-bit register, serial output must be given at multiples of 32 clock pulses. To store new data in such a register, the re-circulation path between the last stage and the first stage is intercepted and the new data is loaded serially into the first stage.

Since each stage of a dynamic shift register needs to retain a bit only for a time equal to one clock period, it is not necessary that each stage of the shift register be a FF. In particular, dynamic shift registers are constructed using dynamic inverters. The clock pulses cause bits to be transferred from one inverter stage to the next, by transferring the charge stored on the inherent capacitance of the MOS devices. This design requires the use of a clock, having certain minimum frequency to ensure that the capacitance does not fully discharge between the 'refresh' cycles. The main advantages of dynamic MOS registers are their small power consumption and simplicity, which permits a very large number of stages to be fabricated on a single IC. Their disadvantage is that all data transfer must be in serial form, which is much slower than parallel data transfer.

Dynamic MOS registers are widely used as memory devices in digital systems that operate on serial data. Because of their small power consumption and the inherent slowness of the serial systems, they are used in applications where power consumption and physical size are more important considerations than speed, such as in pocket calculators. In the context of memory applications, loading a register is called *writing* into it and taking an output from it is called *reading*.

Figure 9.20 shows the logic diagram and the truth table for the 2401 dual 1024-bit dynamic N-MOS shift register. It has a write/re-circulate ($W/\overline{R}$) control, that is used to govern whether new serial data is written (loaded) into the register or whether the existing data is re-circulated (stored) by the register. When $W/\overline{R}$ is HIGH, AND gate 3 is enabled

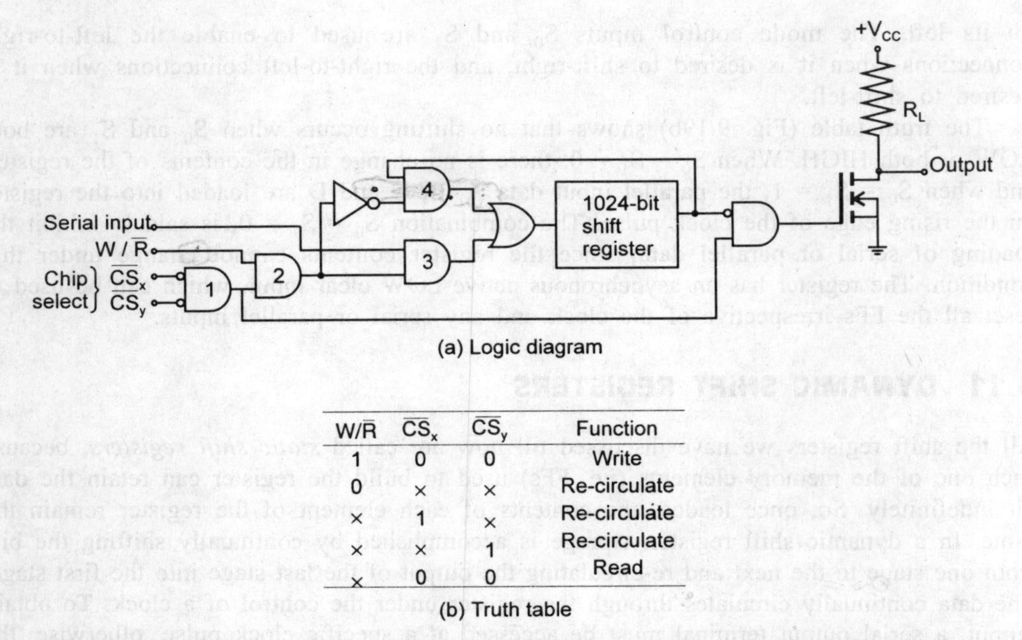

(a) Logic diagram

$W/\overline{R}$	$\overline{CS}_x$	$\overline{CS}_y$	Function
1	0	0	Write
0	×	×	Re-circulate
×	1	×	Re-circulate
×	×	1	Re-circulate
×	0	0	Read

(b) Truth table

Fig. 9.20 One-half of the 2401 dual 1024-bit dynamic NMOS shift register.

and the serial input is transferred through it to the register. When $W/\overline{R}$ is LOW, AND gate 4 is enabled and a re-circulation path from output to input is completed. Serial data appears at the output, regardless of the state of $W/\overline{R}$. The circuit also has two active-LOW chip-select inputs, labelled $\overline{CS}_x$ and $\overline{CS}_y$. Both $\overline{CS}_x$ and $\overline{CS}_y$ must be LOW in order to read or write data. Note, however, that re-circulation is independent of $\overline{CS}_x$ and $\overline{CS}_y$, and re-circulation is also independent of $W/\overline{R}$ if at least one of the chip-selects is HIGH.

9.12 APPLICATIONS OF SHIFT REGISTERS

Time Delays

In many digital systems, it is necessary to delay the transfer of data until such time as operations on other data have been completed, or to synchronize the arrival of data at a subsystem where it is processed with other data. A shift register can be used to delay the arrival of serial data by a specific number of clock pulses, since the number of stages corresponds to the number of clock pulses required to shift each bit completely through the register. The total time delay can be controlled by adjusting the clock frequency and by prescribing the number of stages in the register. In practice, the clock frequency is fixed and the total delay can be adjusted only by controlling the number of stages through which the data is passed. By using a serial-in, parallel-out register and by taking the serial output at any one of the intermediate stages, we have the flexibility to delay the output by any number of clock pulses equal to or less than the number of stages in the register. The arrangement shown in Fig. 9.12 can be used to delay the data by 4 clock pulses.

Serial/Parallel Data Conversion

We know that data can be available either in serial form or in parallel form. Transfer of data in parallel form is much faster than that in serial form. Similarly, the processing of data is much faster when all the data bits are available simultaneously. For this reason, digital systems in which speed is an important consideration, are designed to operate on data in parallel form. When large data is to be transmitted over long distances, transmitting data on parallel lines is costly and impracticable. It is convenient and economical to transmit data in serial form, since serial data transmission requires only one line. Shift registers are used for converting serial data to parallel form, so that a serial input can be processed by a parallel system and for converting parallel data to serial form, so that parallel data can be transmitted serially.

A serial-in, serial-out, shift register can be used to perform serial-to-parallel conversion, and a parallel-in, serial-out, shift register can be used to perform parallel-to-serial conversion. A universal shift register can be used to perform both the serial-to-parallel and parallel-to-serial data conversions. A bidirectional shift register can be used to reverse the order of data. The arrangement shown in Fig. 9.12 can be used for serial-to-parallel conversion of a 4-bit data. The arrangement shown in Fig. 9.14 can be used for parallel-to-serial conversion of a 4-bit data.

Ring Counters

Ring counters are constructed by modifying the serial-in, serial-out, shift registers. There are two types of ring counters—basic ring counter and Johnson counter. The basic ring counter can be obtained from a serial-in, serial-out, shift register by connecting the Q output of the last FF to the D input of the first FF. The Johnson counter can be obtained from a serial-in, serial-out, shift register by connecting the $\overline{Q}$ output of the last FF to the D input of the first FF. Ring counter outputs can be used as a sequence of synchronizing pulses. The ring counter is a decimal counter. It is a divide-by-*N* counter, where *N* is the number of stages. The keyboard encoder is an example of the application of a shift register used as a ring counter in conjunction with other devices.

Ring counters are dealt within detail in Chapter 10. Figure 10.86 (Section 10.11) illustrates the use of the shift register as a ring counter. Figure 10.90 (Section 10.11) illustrates the use of the shift register as a twisted ring counter.

Universal Asynchronous Receiver Transmitter (UART)

Computers and microprocessor-based systems often send and receive data in a parallel format. Frequently these systems must communicate with external devices that send and/or receive serial data. An interfacing device used to accomplish these conversions is the UART.

A UART is a specially designed integrated circuit that contains all the registers and synchronizing circuitry necessary to receive data in serial form and to convert and transmit it in parallel form and vice versa. Figure 9.21 shows the use of UART as an interfacing device.

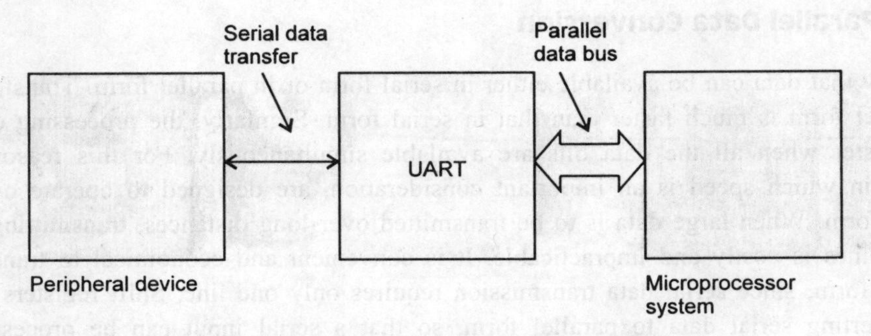

Fig. 9.21 UART as an interfacing device.

9.13 ANSI/IEEE STANDARD SYMBOLS

The qualifying symbol that identifies a shift register is SRG*n*, where *n* is the number of bits. For example, SRG4 and SRG8 identify 4-bit and 8-bit shift registers, respectively. Figure 9.22a shows the symbol for an 8-bit serial-in, parallel-out, shift register, i.e. 74HC164. The arrow indicates that shifting occurs from left-to-right, i.e. from Q_A towards Q_H. The $\overline{CLR}$ and CLK are the inputs to the control block. All eight bits are clocked simultaneously. The R at the $\overline{CLR}$ input indicates that clearing is independent of the clock. If the clearing is clock-dependent, it is indicated by 1R at the $\overline{CLR}$ input. Inputs A and B are ANDed to create the clock-dependent data input ID.

In ANSI/IEEE notation, the letter D is always used to represent input data to a storage element. The noninverting symbol at the CLK terminal indicates that the bits are shifted synchronously on the leading edge of the clock.

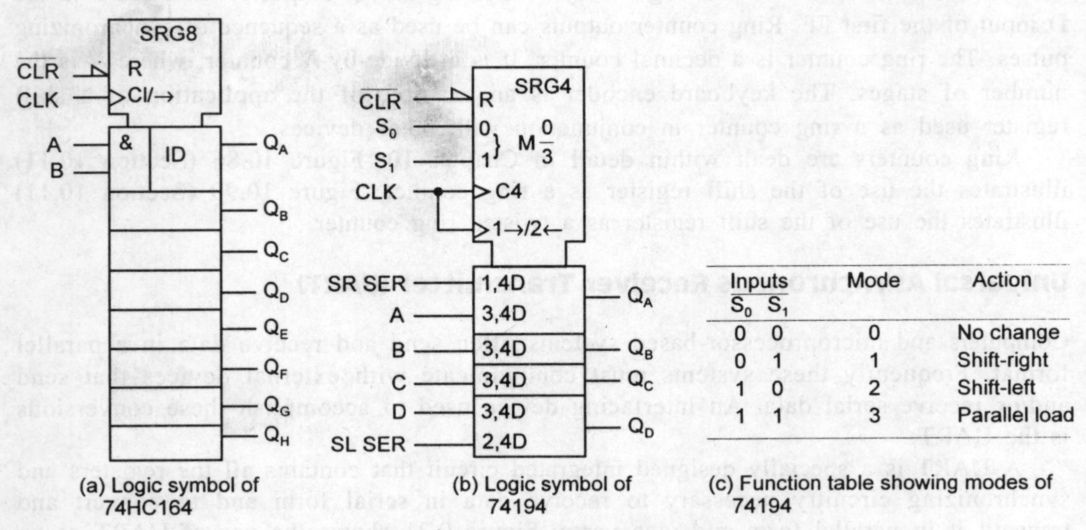

Inputs S_0	S_1	Mode	Action
0	0	0	No change
0	1	1	Shift-right
1	0	2	Shift-left
1	1	3	Parallel load

(a) Logic symbol of 74HC164 (b) Logic symbol of 74194 (c) Function table showing modes of 74194

Fig. 9.22 IEEE/ANSI symbols.

Mode Dependency

Shift registers like other digital devices are often designed so that they can be operated in one of several modes, under the control of one or more binary inputs. The S_0 and S_1 inputs to the 74194 universal shift register shown in Fig. 9.22b, determine whether it is shifted right, shifted left, or loaded in parallel as given by the function table of Fig. 9.22c. $M\frac{0}{3}$ indicates that there are four modes indicated by 0, 1, 2, and 3. The numeral 4 is used to represent clock dependency. The R at the $\overline{CLR}$ terminal indicates that $\overline{CLR}$ is asynchronous. If $\overline{CLR}$ is synchronous, it is indicated by 4R. Shift-right serial input (SR SER) has mode dependency 1,4D, meaning that the input data is accepted at that point when in mode 1 and that the data (D) is clock dependent (synchronous), i.e. loaded synchronously.

Similarly, shift-right serial input (SL SER) shift-left serial input data has mode dependency 2 and clock dependency 4, meaning that the input data is accepted at that point when in mode 2 and that the data is clock dependent. Parallel inputs A, B, C, and D have mode-dependency 3 and clock-dependency 4, meaning that parallel data is inputted at A, B, C, and D synchronously. The symbol $1 \rightarrow /2 \leftarrow$ at the clock terminal indicates that right shifting occurs in mode 1 and left shifting occurs in mode 2. The two CLK lines are used in the symbol only to separate the clock and mode dependencies.

SUMMARY

- Data is said to be in serial form if the bits are available sequentially.
- Data is said to be in parallel form if the bits are available simultaneously.
- A flip-flop can store only one bit of data, a 0 or a 1. So, a flip-flop is called a single-bit register.
- A register is a set of FFs used to store binary data.
- The registers which are used to simply store the data are called buffer registers.
- A register in which shifting of data takes place is called a shift register.
- A shift register has no specified sequence of states except in some very specialized applications.
- Transfer of data into and out of shift registers may take place either in serial form or in parallel form.
- In a serial-in, serial-out, shift register, data is fed in serially, that is one bit at a time on a single line and data is also shifted out serially.
- In a serial-in, parallel-out, shift register, data is fed in serially but data is shifted out in parallel form, that is all bits at a time.
- In a parallel-in, serial-out, shift register, data is fed in, in parallel form but shifted out in serial form.
- In a parallel-in, parallel-out, shift register, data is both fed in and shifted out in parallel form.

- In a bidirectional shift register, data can be shifted from left-to-right or right-to-left.
- In a universal shift register, data can be shifted from left-to-right or right-to-left and also data can be shifted in or shifted out in serial form or in parallel form.
- A shift register using FFs is called a static shift register.
- In a static shift register, data stored is stationary.
- Dynamic shift registers are made up of MOS inverters.
- In a dynamic shift register, storage is accomplished by continually shifting data from one stage to the next and re-circulating the output of the last stage into the first stage.
- Dynamic MOS registers are used in applications where power consumption and space are more important than speed as in pocket calculators.
- Shift registers are used in digital systems to provide time delays. They are also used for serial/parallel data conversion and in construction of ring counters.

QUESTIONS

1. What do you mean by (a) serial data and (b) parallel data?
2. Why is a FF called a single-bit register? What does the 'state' of a FF indicate?
3. What is a register? What is a shift register?
4. What are the various types of shift registers?
5. What is a serial-in, serial-out, shift register?
6. What is a parallel-in, serial-out, shift register?
7. What is a serial-in, parallel-out, shift register?
8. What is a parallel-in, parallel-out, shift register?
9. What is a bidirectional shift register?
10. What is a universal shift register?
11. What are the applications of shift registers?
12. What is the basic difference between a counter and a shift register?
13. What do you understand by the following?

    (a) Buffer register (b) Controlled buffer register
    (c) Dynamic shift register (d) Static shift register
14. What is mode dependency? What is clock dependency?
15. Where are dynamic MOS registers used?
16. With neat diagrams, explain the working of the following types of shift registers:

    (a) Serial-in, serial-out, (b) Serial-in, parallel-out
    (c) Parallel-in, serial-out (d) Parallel-in, parallel-out
    (e) Bidirectional shift registers.

Chapter 10

COUNTERS

10.1 INTRODUCTION

A digital counter is a set of flip-flops (FFs) whose states change in response to pulses applied at the input to the counter. The FFs are interconnected such that their combined state at any time is the binary equivalent of the total number of pulses that have occurred up to that time. Thus, as its name implies, a counter is used to count pulses. A counter can also be used as a frequency divider to obtain waveforms with frequencies that are specific fractions of the clock frequency. They are also used to perform the timing function as in digital watches, to create time delays, to produce non-sequential binary counts, to generate pulse trains, and to act as frequency counters, etc.

Counters may be *asynchronous* counters or *synchronous* counters. Asynchronous counters are also called *ripple counters*. The ripple counter is the simplest type of counter, the easiest to design and requires the least amount of hardware. In ripple counters, the FFs within the counter are not made to change the states at exactly the same time. This is because the FFs are not triggered simultaneously. The clock does not directly control the time at which every stage changes state. An asynchronous counter uses T FFs to perform a counting function. The actual hardware used is usually J-K FFs connected in toggle mode, i.e. with Js and Ks connected to logic 1. Even D FFs may be used here.

The asynchronous counter has a disadvantage, insofar as the unwanted spikes are concerned. This limitation is overcome in parallel counters. The asynchronous counter is called ripple counter because when the counter, for example, goes from 1111 to 0000, the first stage causes the second to flip, the second causes the third to flip, and the third causes the fourth to flip, and so on. In other words, the transition of the first stage ripples through to the last stage. In doing so, many intermediate stages are briefly entered. If there is a gate that will AND during any state, a brief spike will be seen at the gate output every time the counter goes from 1111 to 0000. Ripple counters are also called *serial* or *series counters*. Synchronous counters are clocked such that each FF in the counter is triggered at the same time. This is accomplished by connecting the clock line to each stage of the counter. Synchronous counters are faster than asynchronous counters, because the propagation delay involved is less.

A counter may be an *up-counter* or a *down-counter*. An up-counter is a counter which counts in the upward direction, i.e. 0, 1, 2, 3,…, N. A down-counter is a counter which counts in the downward direction, i.e. N, $N-1$, $N-2$, $N-3$, …, 1, 0. Each of the counts of the counter is called the *state* of the counter. The number of states through which the counter passes before returning to the starting state is called the *modulus* of the counter. Hence, the modulus of a counter is equal to the total number of distinct states (counts) including zero that a counter can store. In other words, the number of input pulses that causes the counter to reset to its initial count is called the modulus of the counter. Since a 2-bit counter has 4 states, it is called a mod-4 counter. It divides the input clock signal frequency by 4, therefore, it is also called a divide-by-4 counter. It requires two FFs. Similarly, a 3-bit counter uses 3 FFs and has $2^3 = 8$ states. It divides the input clock frequency by 2^3, i.e. 8. In general, an *n*-bit counter will have *n* FFs and 2^n states, and divides the input frequency by 2^n. Hence, it is a divide-by-2^n counter.

A counter may have a shortened modulus. This type of counter does not utilize all the possible states. Some of the states are unutilized, i.e. invalid. The number of FFs required to construct a mod-*N* counter equals the smallest *n* for which $N \leq 2^n$. A mod-*N* counter divides the input frequency by *N*, hence, it is called a divide-by-*N* counter. In an asynchronous counter, the invalid states are bypassed by providing a suitable feedback. In a synchronous counter, the invalid states are taken care of by treating the corresponding excitations as don't cares. The least significant bit (LSB) of any counter is that bit which changes most often. In ripple counters, the LSB is the Q output of the FF to which the external clock is applied.

A counter which goes through all the possible states before restarting is called the *full modulus counter*. A counter in which the maximum number of states can be changed is called the *variable modulus counter*. The final state of the counter sequence is called the *terminal count*.

Lock-out

In shortened-modulus counters, there may occur the problem of *lock-out*. Sometimes when the counter is switched on, or any time during counting, because of noise spikes, the counter may find itself in some unused (invalid) state. Subsequent clock pulses may cause the counter to move from one unused state to another unused state and the counter may never come to a valid state. So, the counter becomes useless. A counter whose unused states have this feature is said to suffer from the problem of lock-out. To ensure that, at 'start-up' the counter is in its initial state, external logic circuitry is provided which properly resets each FF. The logic circuitry for presetting the counter to its initial state can be provided either by obtaining an expression for reset/preset for the FFs or by modifying the design such that the counter goes from each invalid state to the initial state after the clock pulse. So, no don't cares are permitted in this design.

Combination of Modulo Counters

A single FF is a mod-2 counter. We can have a counter of any modulus by choosing an appropriate number of FFs and providing proper feedback. Counters of different mods can be combined to get another mod counter. For example, a mod-2 counter and a mod-5 counter can be combined to get a mod-10 counter; a mod-5 counter and a mod-4 counter can be combined to get a mod-20 counter, and so on. The connection between the individual counters may be a ripple connection, or the counters may be operated in synchronism with one another independently of whether the individual counters are ripple or synchronous. Further, we are at liberty to choose the order of the individual counters in a chain of counters. Such permutations will not change the modulus of the composite counter but may well make a substantive difference in the code in which the counter state is to be read.

10.2 ASYNCHRONOUS COUNTERS

Two-bit Ripple Up-counter Using Negative Edge-triggered Flip-flops

The 2-bit up-counter counts in the order 0, 1, 2, 3, 0, 1,..., i.e. 00, 01, 10, 11, 00, 01,...,etc. Figure 10.1 shows a 2-bit ripple up-counter, using negative edge-triggered J-K

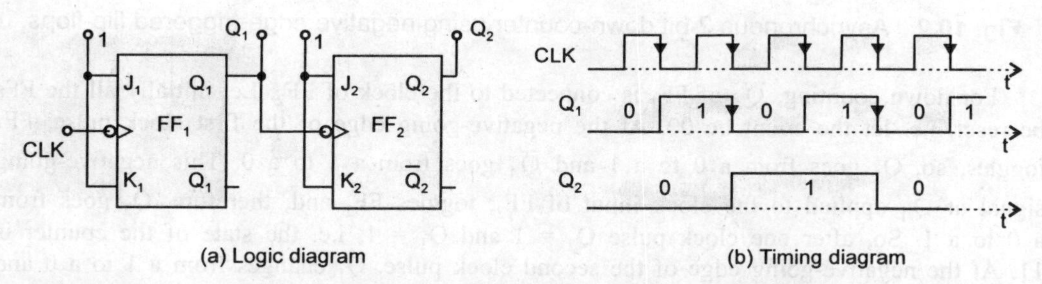

Fig. 10.1 Asynchronous 2-bit up-counter using negative edge-triggered flip-flops.

FFs, and its timing diagram. The counter is initially reset to 00. When the first clock pulse is applied, FF_1 toggles at the negative-going edge of this pulse, therefore, Q_1 goes from LOW to HIGH. This becomes a positive-going signal at the clock input of FF_2. So, FF_2 is not affected, and hence, the state of the counter after one clock pulse is $Q_1 = 1$ and $Q_2 = 0$, i.e. 01. At the negative-going edge of the second clock pulse, FF_1 toggles. So, Q_1 changes from HIGH to LOW and this negative-going signal applied to CLK of FF_2 activates FF_2, and hence, Q_2 goes from LOW to HIGH. Therefore, $Q_1 = 0$ and $Q_2 = 1$, i.e. 10 is the state of the counter after the second clock pulse. At the negative-going edge of the third clock pulse, FF_1 toggles, so, Q_1 changes from a 0 to a 1. This becomes a positive-going signal to FF_2, hence, FF_2 is not affected. Therefore, $Q_2 = 1$ and $Q_1 = 1$, i.e. 11 is the state of the counter after the third clock pulse. At the negative-going edge of the fourth clock pulse, FF_1 toggles. So, Q_1 goes from a 1 to a 0. This negative-going signal at Q_1 toggles FF_2, hence, Q_2 also changes from a 1 to a 0. Therefore, $Q_2 = 0$ and $Q_1 = 0$, i.e. 00 is the state of the counter after the fourth clock pulse. For subsequent clock pulses, the counter goes through the same sequence of states. So, it acts as a mod-4 counter with Q_1 as the LSB and Q_2 as the MSB. The counting sequence is thus 00, 01, 10, 11, 00, 01,...,etc.

Two-bit Ripple Down-counter Using Negative Edge-triggered Flip-flops

A 2-bit down-counter counts in the order 0, 3, 2, 1, 0, 3,..., i.e. 00, 11, 10, 01, 00, 11,

..., etc. Figure 10.2 shows a 2-bit ripple down-counter, using negative-edge triggered J-K FFs, and its timing diagram.

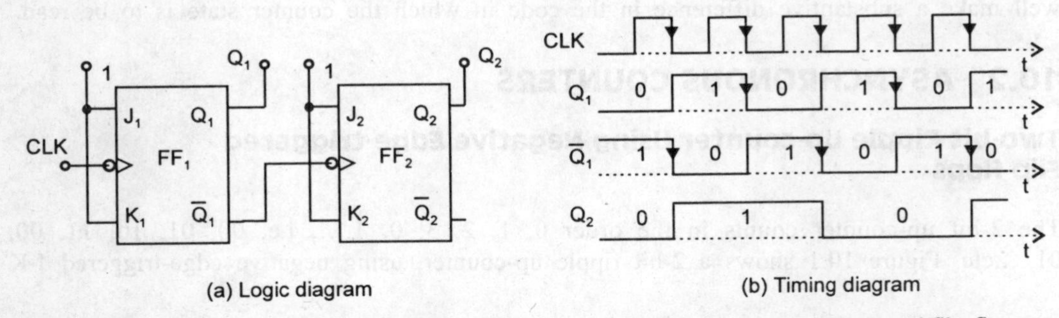

Fig. 10.2 Asynchronous 2-bit down-counter using negative edge-triggered flip-flops.

For down counting, $\overline{Q}_1$ of FF_1 is connected to the clock of FF_2. Let initially all the FFs be reset, i.e. let the count be 00. At the negative-going edge of the first clock pulse, FF_1 toggles, so, Q_1 goes from a 0 to a 1 and $\overline{Q}_1$ goes from a 1 to a 0. This negative-going signal at $\overline{Q}_1$ applied to the clock input of FF_2, toggles FF_2 and, therefore, Q_2 goes from a 0 to a 1. So, after one clock pulse $Q_2 = 1$ and $Q_1 = 1$, i.e. the state of the counter is 11. At the negative-going edge of the second clock pulse, Q_1 changes from a 1 to a 0 and $\overline{Q}_1$ from a 0 to a 1. This positive-going signal at $\overline{Q}_1$ does not affect FF_2 and, therefore, Q_2 remains at a 1. Hence, the state of the counter after the second clock pulse is 10. At the negative-going edge of the third clock pulse, FF_1 toggles. So, Q_1 goes from a 0 to a 1 and $\overline{Q}_1$ from a 1 to a 0. This negative-going signal at $\overline{Q}_1$ toggles FF_2 and, so, Q_2 changes from a 1 to a 0. Hence, the state of the counter after the third clock pulse is 01. At the negative-going edge of the fourth clock pulse, FF_1 toggles. So, Q_1 goes from a 1 to a 0 and $\overline{Q}_1$ from a 0 to a 1. This positive-going signal at $\overline{Q}_1$ does not affect FF_2. So, Q_2 remains at a 0. Hence, the state of the counter after the fourth clock pulse is 00. For subsequent clock pulses the counter goes through the same sequence of states, i.e. the counter counts in the order 00, 11, 10, 01, 00, and 11...

Two-bit Ripple Up/Down Counter Using Negative Edge-triggered Flip-flops

As the name indicates, an up/down counter is a counter which can count both in upward and downward directions (Fig. 10.3). An up/down counter is also called a forward/backward counter or a bidirectional counter. So, a control signal or a mode signal M is required to choose the direction of count. When M = 1 for up counting, Q_1 is transmitted to clock of FF_2 and when M = 0 for down counting, $\overline{Q}_1$ is transmitted to clock of FF_2. This is achieved by using two AND gates and one OR gate as shown in Fig. 10.3. The external clock signal is applied to FF_1.

$$\text{Clock signal to } FF_2 = (Q_1 \cdot \text{Up}) + (\overline{Q}_1 \cdot \text{Down}) = Q_1M + \overline{Q}_1\overline{M}$$

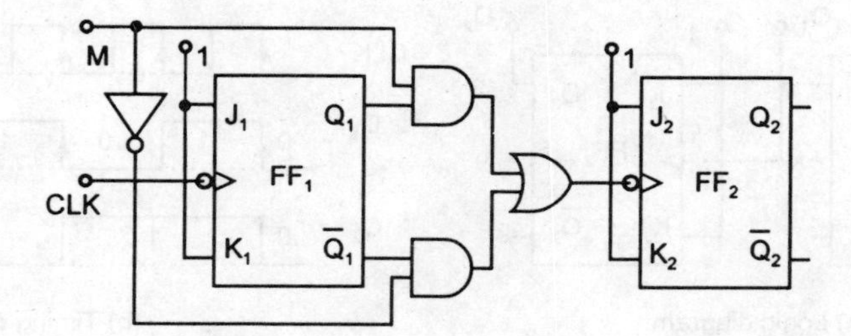

Fig. 10.3 Logic diagram of a 2-bit ripple up/down counter using negative edge-triggered flip-flops.

Two-bit Ripple Up-counter Using Positive Edge-triggered Flip-flops

A 2-bit ripple up-counter, using positive edge-triggered J-K FFs, and its timing diagram are shown in Fig. 10.4. The $\overline{Q}_1$ output of the first FF is connected to the clock of FF_2. The external clock signal is applied to the first flip-flop FF_1. The FF_1 toggles at the positive-going edge of each clock pulse and FF_2 toggles whenever $\overline{Q}_1$ changes from a 0 to a 1. State transitions occur at the positive-going edges of the clock pulses. The counting sequence is 00, 01, 10, 11, 00, 01, ..., etc.

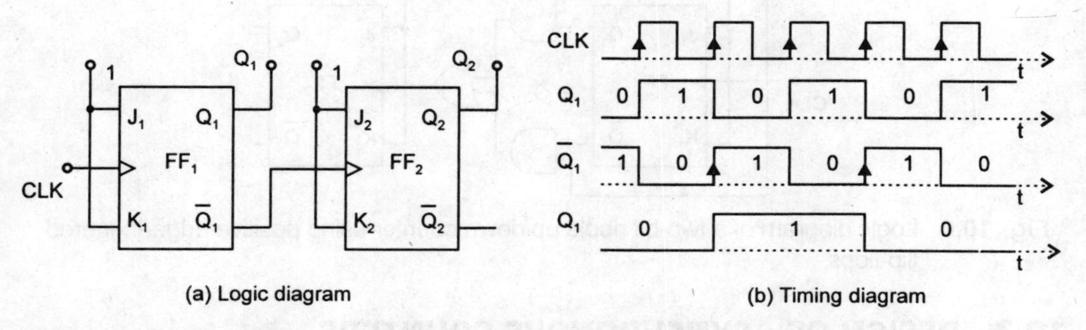

Fig. 10.4 Asynchronous 2-bit up-counter using positive-edge triggered flip-flops.

Two-bit Ripple Down-counter Using Positive Edge-triggered Flip-flops

A 2-bit ripple down-counter, using positive edge-triggered J-K FFs, and its timing diagram are shown in Fig. 10.5. The Q_1 output of the first FF is connected to the clock of FF_2. The external clock signal is applied to FF_1. The FF_1 toggles at the positive-going edge of each clock pulse. The FF_2 toggles whenever Q_1 changes from a 0 to a 1. The counting sequence is 00, 11, 10, 01, 00, 11,... etc.

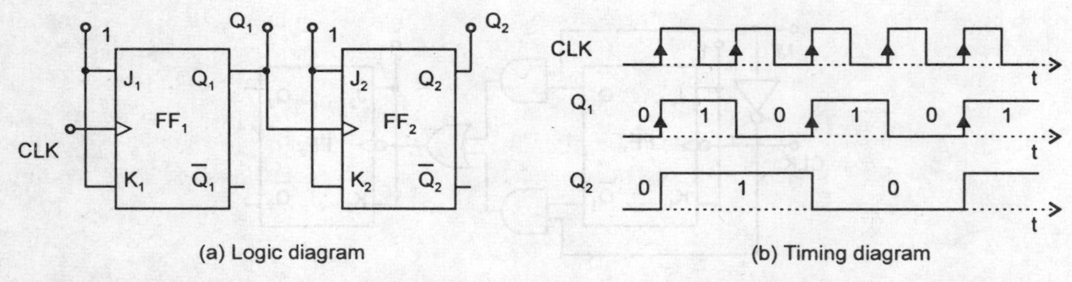

Fig. 10.5 Asynchronous 2-bit down-counter using positive edge-triggered flip-flops.

Two-bit Ripple Up/Down Counter Using Positive Edge-triggered Flip-flops

Figure 10.6 shows a 2-bit ripple up/down counter using positive edge-triggered J-K FFs. When M = 1 for up counting, $\overline{Q}_1$ is transmitted to the clock of FF_2 and when M = 0 for down counting, Q_1 is transmitted to the clock of FF_2. This is achieved by using two AND gates and one OR gate as shown in Fig. 10.6.

Clock signal to $FF_2 = (\overline{Q}_1 \cdot \text{Up}) + (Q_1 \cdot \text{Down}) = \overline{Q}_1 M + Q_1 \overline{M}$

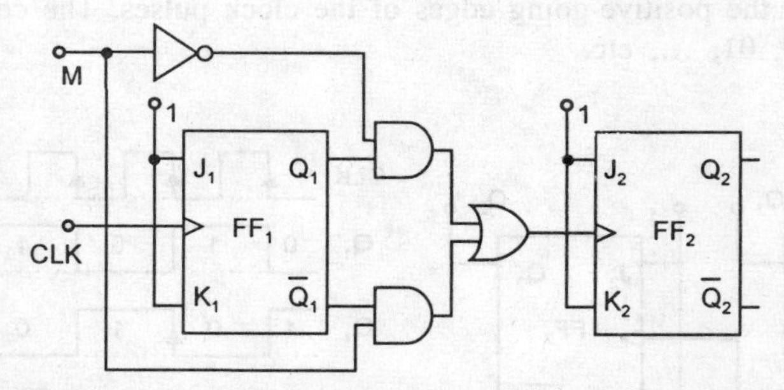

Fig. 10.6 Logic diagram of a two-bit ripple up/down counter using positive edge-triggered flip-flops.

10.3 DESIGN OF ASYNCHRONOUS COUNTERS

To design an asynchronous counter, first write the counting sequence, then tabulate the values of reset signal R for various states of the counter and obtain the minimal expression for R or $\overline{R}$ using K-map or any other method. Provide a feedback such that R or $\overline{R}$ resets all the FFs after the desired count.

EXAMPLE 10.1 Design and implement a mod-6 asynchronous counter using T FFs.

Solution

A mod-6 counter has six stable states 000, 001, 010, 011, 100, and 101. When the sixth clock pulse is applied, the counter temporarily goes to 110 state, but immediately resets to

000 because of the feedback provided. It is a 'divide-by-6 counter', in the sense that it divides the input clock frequency by 6. It requires three FFs, because the smallest value of n satisfying the condition $N \leq 2^n$ is $n = 3$; three FFs can have eight possible states, out of which only six are utilized and the remaining two states 110 and 111, are invalid. If, initially the counter is in 000 state, then after the first clock pulse it goes to 001, after the second clock pulse it goes to 010, and so on. After the sixth clock pulse it goes to 000.

For the design, write a truth table [Fig. 10.7(c)] with the present state outputs Q_3, Q_2, and Q_1 as the variables, and reset R as the output and obtain an expression for R in terms

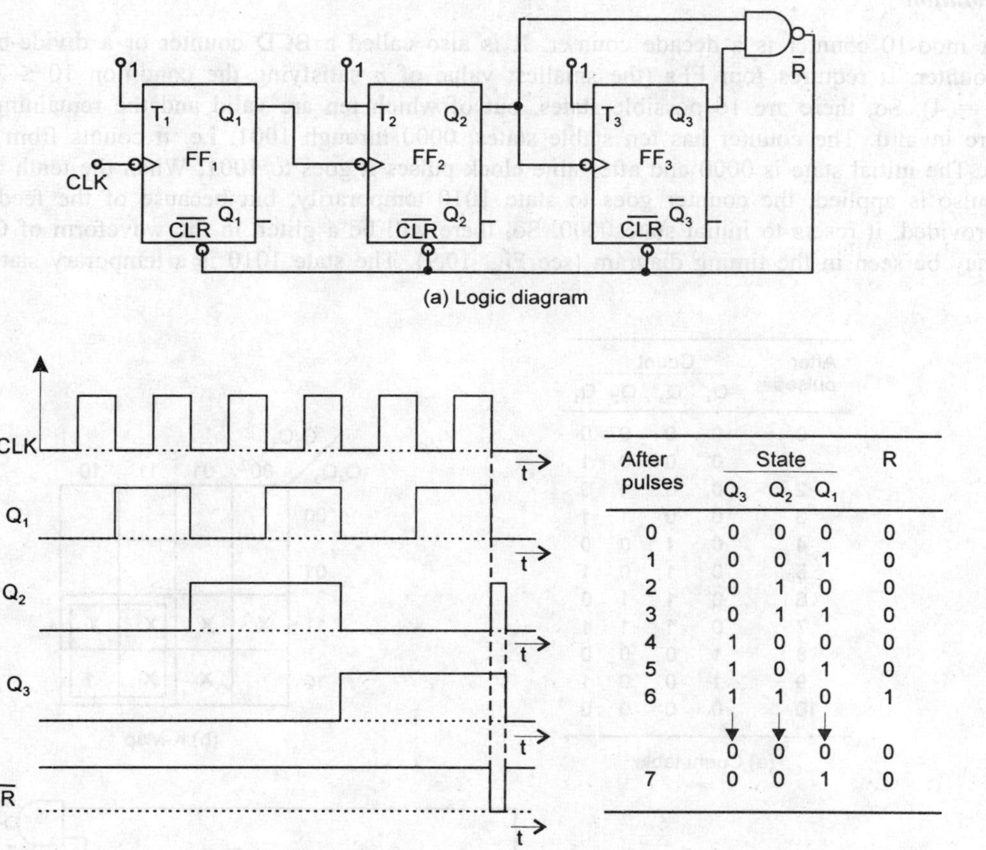

After pulses	State Q_3	Q_2	Q_1	R
0	0	0	0	0
1	0	0	1	0
2	0	1	0	0
3	0	1	1	0
4	1	0	0	0
5	1	0	1	0
6	1	1	0	1
	↓ 0	↓ 0	↓ 0	0
7	0	0	1	0

(c) Table for R

Fig. 10.7 Asynchronous mod-6 counter using T flip-flops (Example 10.1).

of Q_3, Q_2, and Q_1. That decides the feedback to be provided. From the truth table, $R = Q_3Q_2$. For active-LOW reset, $\overline{R}$ is used. The reset pulse is of very short duration, of the order of ns and it is equal to the propagation delay time of the NAND gate used. The expression for R can also be determined as follows.

$$R = 0 \text{ for } 000 \text{ to } 101,\ R = 1 \text{ for } 110, \text{ and } R = X \text{ for } 111$$

Therefore,

$$R = Q_3Q_2\overline{Q}_1 + Q_3Q_2Q_1 = Q_3Q_2$$

The logic diagram, the timing diagram, and the table for R of a mod-6 counter are all shown in Fig. 10.7. From the timing diagram it is seen that a glitch appears in the waveform of Q_2.

EXAMPLE 10.2 Design and implement a mod-10 asynchronous counter using T FFs.

Solution

A mod-10 counter is a decade counter. It is also called a BCD counter or a divide-by-10 counter. It requires four FFs (the smallest value of n satisfying the condition $10 \le 2^n$, is $n = 4$). So, there are 16 possible states, out of which ten are valid and the remaining six are invalid. The counter has ten stable states, 0000 through 1001, i.e. it counts from 0 to 9. The initial state is 0000 and after nine clock pulses it goes to 1001. When the tenth clock pulse is applied, the counter goes to state 1010 temporarily, but because of the feedback provided, it resets to initial state 0000. So, there will be a glitch in the waveform of Q_2 as may be seen in the timing diagram (see Fig. 10.9). The state 1010 is a temporary state for

After pulses	Count			
	Q_4	Q_3	Q_2	Q_1
0	0	0	0	0
1	0	0	0	1
2	0	0	1	0
3	0	0	1	1
4	0	1	0	0
5	0	1	0	1
6	0	1	1	0
7	0	1	1	1
8	1	0	0	0
9	1	0	0	1
10	0	0	0	0

(a) Count table

Q_4Q_3 \ Q_2Q_1	00	01	11	10
00				
01				
11	X	X	X	X
10		X	X	1

(b) K-Map

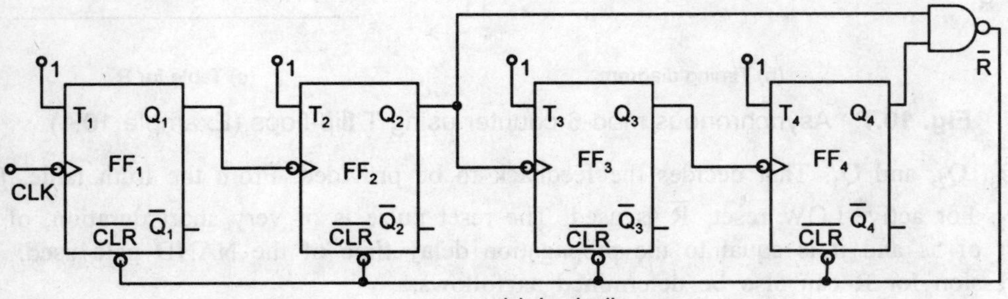

(c) Logic diagram

Fig. 10.8 Asynchronous mod-10 counter using T flip-flops (Example 10.2).

which the reset signal R = 1, R = 0 for 0000 to 1001, and R = X (don't care) for 1011 to 1111.

The count table and the K-map for reset are shown in Fig.10.8(a) and 10.8(b), respectively. From the K-map, $R = Q_4Q_2$. So, feedback is provided from second and fourth FFs. For active-HIGH reset, Q_4Q_2 is applied to the CLEAR terminal. For active-LOW reset, $\overline{Q_4Q_2}$ is connected to $\overline{\text{CLR}}$ of all the FFs. The logic diagram of the decade counter is shown in Fig. 10.8(c) and its timing diagram without considering propagation delay is shown in Fig. 10.9.

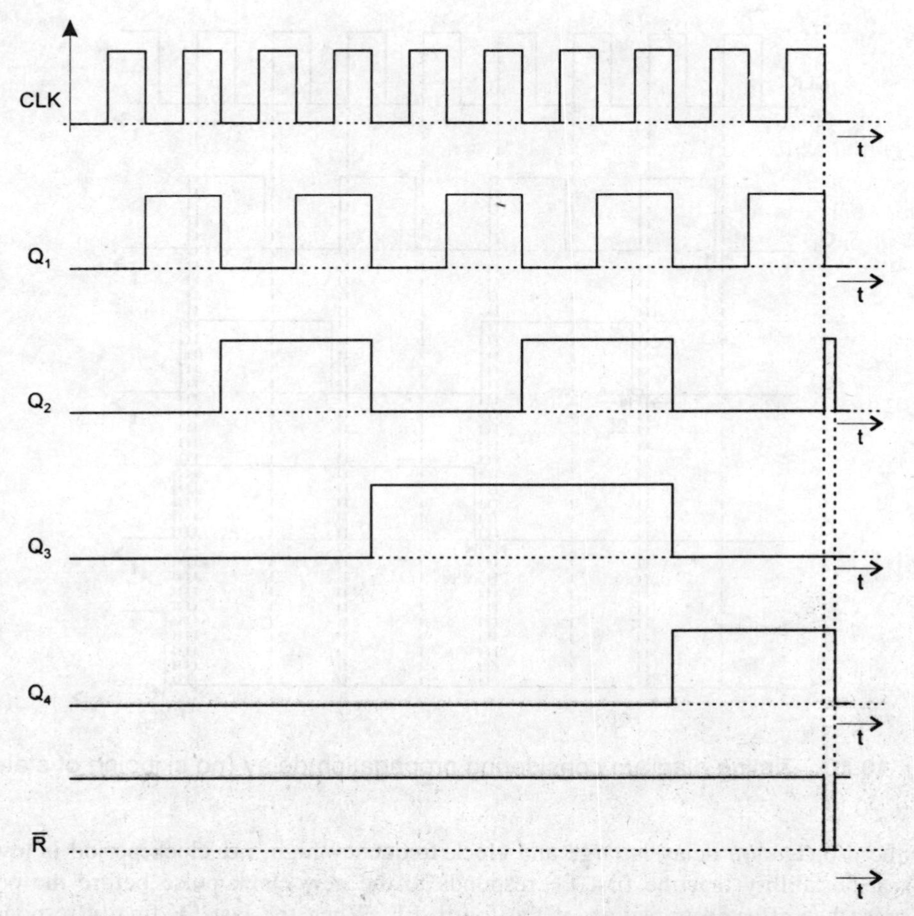

Fig. 10.9 Timing diagram of a mod-10 counter without considering propagation delay (Example 10.2).

10.4 EFFECTS OF PROPAGATION DELAY IN RIPPLE COUNTERS

In ripple counters, each FF is toggled by the changing state of the preceding FF. So, no FF can change state until after the propagation delay of the FF that precedes it, i.e. until all preceding FFs complete their transitions. This delay accumulates as we proceed through additional stages. This will have a detrimental effect on the operation of the ripple counter. For a decade counter or a 4-bit counter, all the FFs change states when the count goes from 0111 to 1000. So, after the eighth clock pulse is applied, Q_1 will not change state instantaneously, but goes from a 1 to a 0 after a propagation delay of t_{pd}. Q_2 changes state after another propagation delay time of $2t_{pd}$. Q_3 changes state after $3t_{pd}$ and Q_4 changes state after $4t_{pd}$. The timing diagram considering propagation delays is shown in Fig. 10.10.

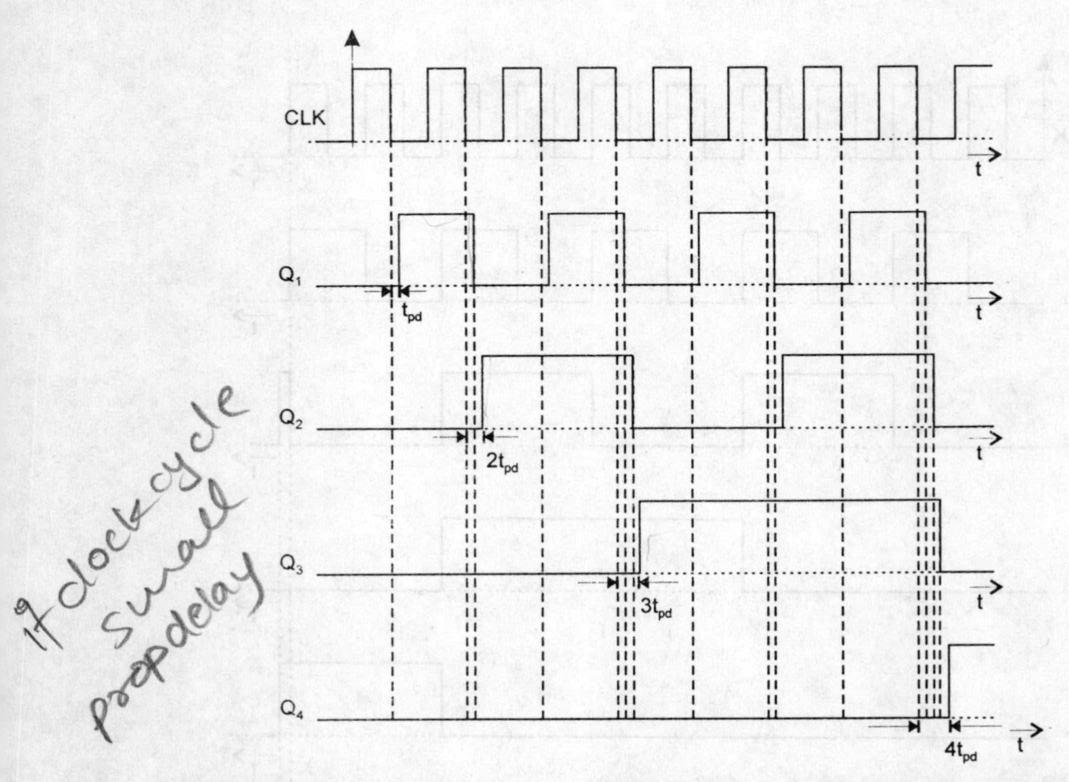

Fig. 10.10 Timing diagram considering propagation delay (no skipping of states).

If the propagation delay is large and clock frequency high, i.e. clock period is low, then there is a possibility that the first FF responds to the new clock pulse before the previous clock pulse has effected transition of the fourth FF. When the last FF finally responds, the counter may read 1001 having skipped 1000. Thus the count goes directly from 0111 to 1001. If the clock frequency is so high that it is possible for the clock pulse to change the state of the first stage, before the state changes caused by the previous clock pulse have rippled through to the last stage, then a count will be skipped. Thus it is obvious that propagation delays in the FFs of a ripple counter impose a limit on the frequency at which the counter can be clocked.

If T_C is the period of the clock pulse, n the number of stages and t_{pd} the propagation delay in each stage, then the clock frequency f_C is constrained by

$$\frac{1}{T_C} = f_C < \frac{1}{nt_{pd}}$$

Suppose T_C = 0.1 μs, i.e. f_C = 1/(0.1 μs) = 10 MHz and suppose t_{pd} = 0.05 μs, the timing diagram would then be as shown in Fig. 10.11. Here skipping of states occur. The count 1000 is not reached at all as shown in Fig. 10.11.

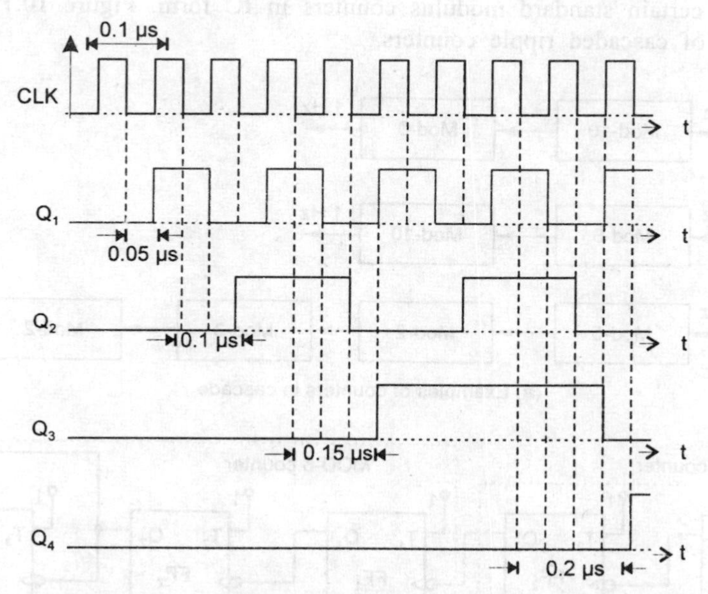

Fig. 10.11 Timing diagram of a ripple counter showing skipping of states.

EXAMPLE 10.3 Implement a 3-bit ripple counter using D FFs.

Solution

For ripple counters, the FFs used must be in toggle mode. The D FFs may be used in toggle mode by connecting the $\overline{Q}$ of each FF to its D terminal. The 3-bit ripple counter using D FFs is shown in Fig. 10.12.

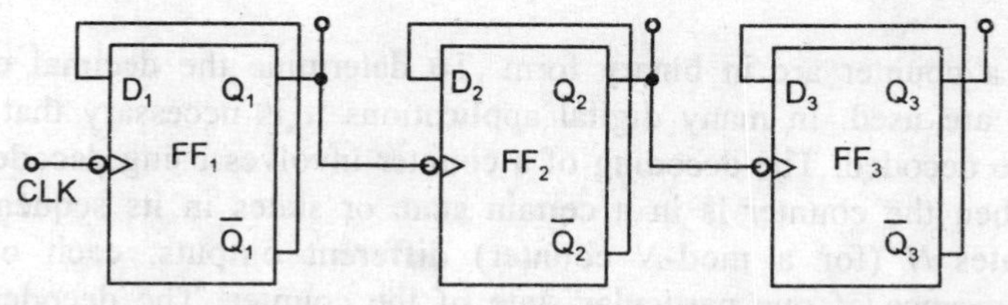

Fig. 10.12 Logic diagram of a 3-bit asynchronous counter using D flip-flops (Example 10.3).

Cascading of Ripple Counters

Ripple counters can be connected in cascade to increase the modulus of the counter. A mod-M and a mod-N counter in cascade give a mod-MN counter. While cascading, the most significant stage of the first counter is connected to the toggling stage of the second counter. Some examples of counters in cascade are shown in Fig. 10.13(a).

The order of cascading does not affect the frequency division, however, the duty cycle of the most significant output may depend on the order in which the counters are cascaded. Mod counters are often constructed by cascading lower modulus counters because of the availability of certain standard modulus counters in IC form. Figure 10.13(b) shows the logic diagram of cascaded ripple counters.

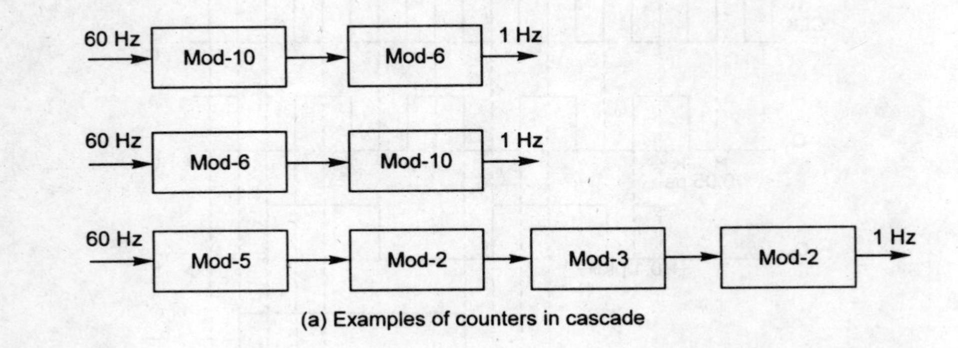

(a) Examples of counters in cascade

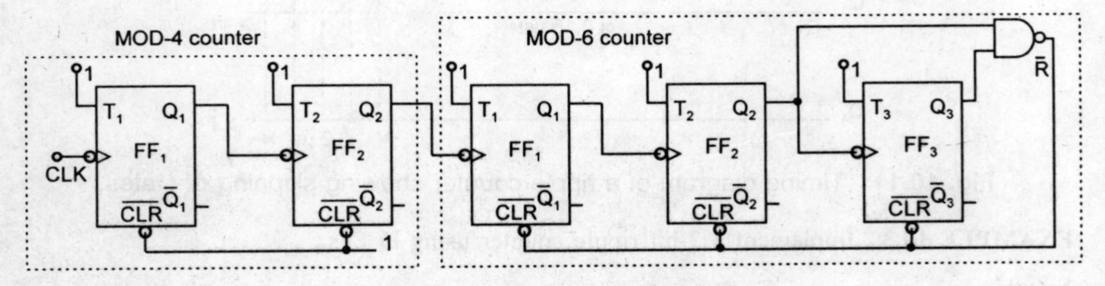

(b) Logic diagram of cascaded ripple counters

Fig. 10.13 Ripple counters.

10.5 DECODING OF RIPPLE COUNTERS

The outputs of a counter are in binary form. To determine the decimal equivalent of the count, decoders are used. In many digital applications it is necessary that some or all the counter states be decoded. The decoding of a counter involves using decoders or logic gates to determine when the counter is in a certain state or states in its sequence. The decoder network generates N (for a mod-N counter) different outputs, each of which detects (decodes) the presence of one particular state of the counter. The decoder outputs can be designed to produce either a HIGH or a LOW level when the detection occurs. An active-

HIGH decoder produces HIGH outputs to indicate detection. An active-LOW decoder produces LOW outputs to indicate detection.

Figure 10.14(a) shows a binary-to-decimal decoder used with a 2-bit ripple counter. The four AND gates produce high output in succession for each of the four successive counts

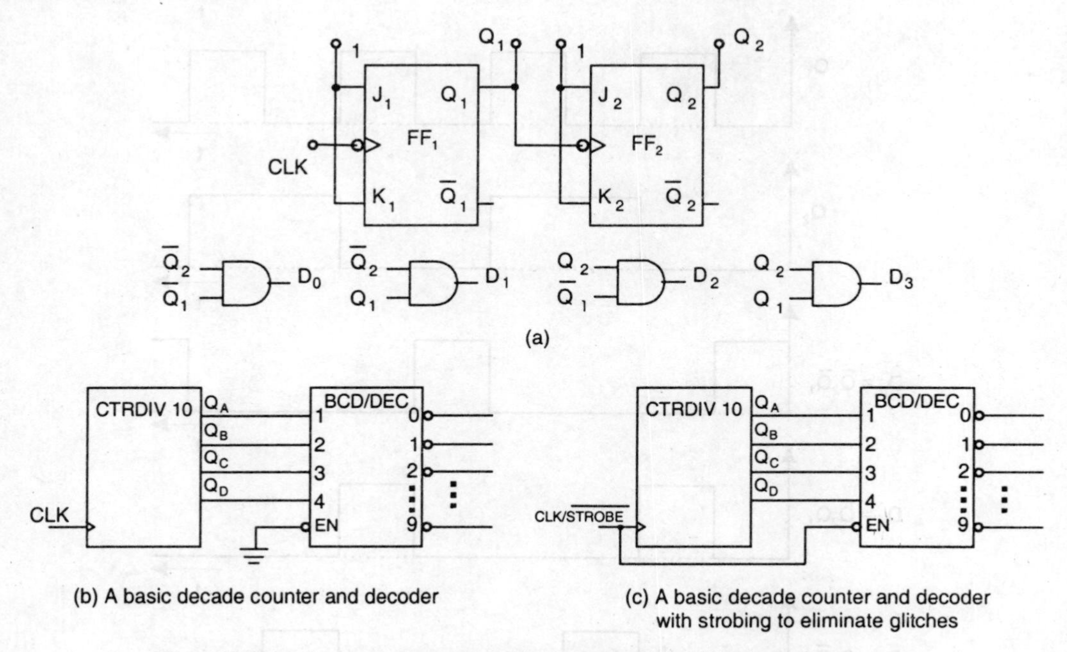

Fig. 10.14 Asynchronous 2-bit up counter with decoder gates.

0, 1, 2, 3 corresponding to states 00 ($\overline{Q}_2\overline{Q}_1$), 01 ($\overline{Q}_2Q_1$), 10 ($Q_2\overline{Q}_1$), and 11 ($Q_2Q_1$). This is active-HIGH decoding. For active-LOW decoding, NAND gates are used in place of AND gates. The waveforms in Fig. 10.15 indicate that one and only one decoder output is high at any given time. These waveforms correspond to active-HIGH decoding. These waveforms are ideal in the sense that they ignore the effects of propagation delay.

Figure 10.16 shows the actual decoder waveforms that result when the propagation delays of the 2-bit ripple counter are taken into account. The propagation delays of the AND gates are neglected. The propagation delays owing to the ripple effect in asynchronous counters create transitional states in which the counter outputs are changing at slightly different times. These transitional states produce undesired voltage spikes of short duration (glitches on the outputs of a decoder). We see that glitches occur in the D_0 and D_2 waveforms because of intermittent combinations of states that activate the corresponding AND gates. The width of these glitches equals the width of the propagation delay t_{pd}. The practical consequences of the glitches depend on the application.

If the decoder outputs drive indicator lamps, the glitches may be too narrow to create any response. On the other hand, if the decoder outputs drive edge-triggered devices, the glitches may be responsible for unsatisfactory performance. In those cases, the outputs can be 'deglitched' by ANDing them with the clock. However, this remedy reduces the width

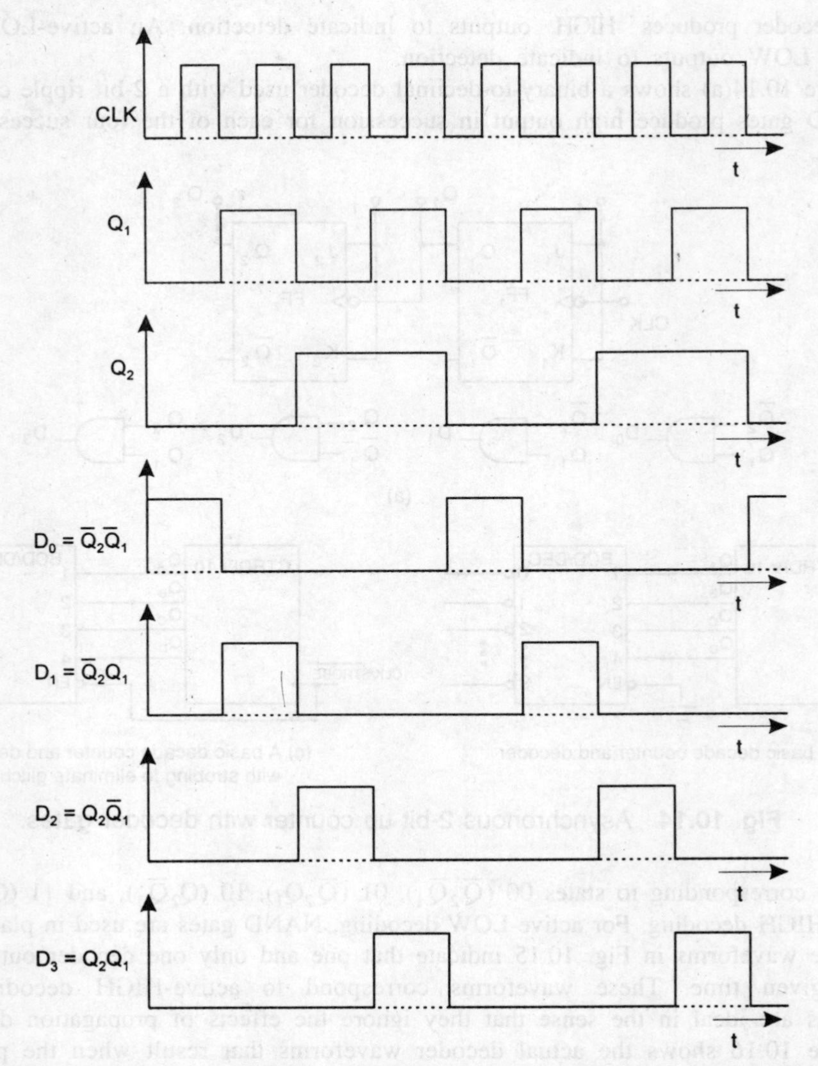

Fig. 10.15 Decoder waveforms without considering the propagation delays.

of a deglitched output to the width of a clock pulse, which causes gaps to appear between decoder outputs. One way to eliminate the glitches, is to enable the decoded outputs at a time after the glitches have had time to disappear. This method is known as *strobing* and can be accomplished by using the LOW level of the clock to enable the decoder. Figure 10.14(b) shows the logic diagram of a basic decade counter and decoder. Figure 10.14(c) shows the logic diagram of a basic decade counter and decoder with strobing to eliminate glitches.

EXAMPLE 10.4 A binary ripple counter is required to count up to $16{,}383_{10}$. How many FFs are required? If the clock frequency is 8.192 MHz, what is the frequency at the output of the MSB?

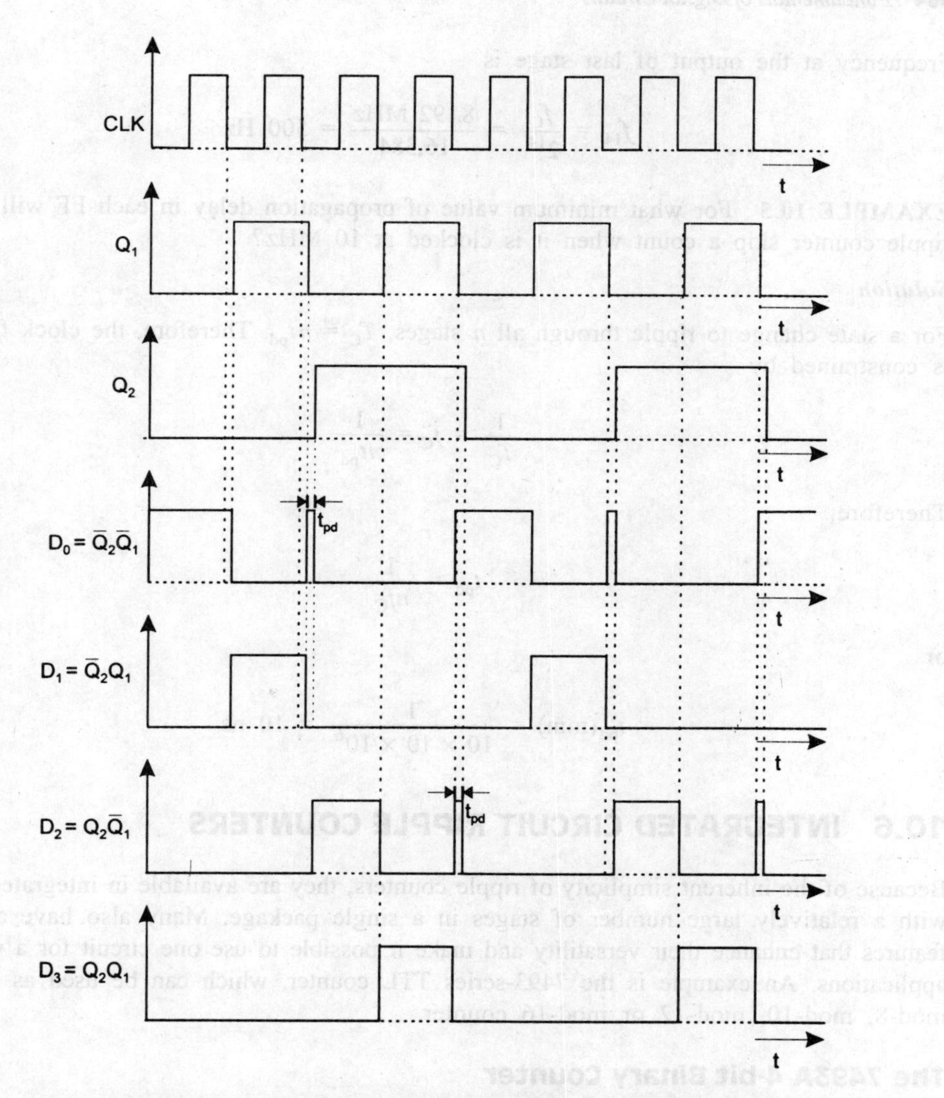

Fig. 10.16 Decoder waveform considering the propagation delays.

Solution

The number of FFs n is to be selected such that the number of states $N \le 2^n$. With n FFs, the largest count possible is $2^n - 1$. Therefore,

$$2^n - 1 = 16{,}383$$

or

$$n = \log_2 16{,}384 = 14$$

So, the number of FFs required is 14.

Frequency at the output of last stage is

$$f_{14} = \frac{f_C}{2^{14}} = \frac{8.192 \text{ MHz}}{16,384} = 500 \text{ Hz}$$

EXAMPLE 10.5 For what minimum value of propagation delay in each FF will a 10-bit ripple counter skip a count when it is clocked at 10 MHz?

Solution

For a state change to ripple through all n stages, $T_C = nt_{pd}$. Therefore, the clock frequency is constrained by

$$\frac{1}{T_C} = f_C = \frac{1}{nt_{pd}}$$

Therefore,

$$t_{pd} = \frac{1}{nf_C}$$

or

$$t_{pd}(\text{min}) = \frac{1}{10 \times 10 \times 10^6} = 10 \text{ ns}$$

10.6 INTEGRATED CIRCUIT RIPPLE COUNTERS

Because of the inherent simplicity of ripple counters, they are available in integrated circuits with a relatively large number of stages in a single package. Many also have additional features that enhance their versatility and make it possible to use one circuit for a variety of applications. An example is the 7493-series TTL counter, which can be used as a mod-2, mod-8, mod-10, mod-12 or mod-16 counter.

The 7493A 4-bit Binary Counter

As the logic diagram in Fig. 10.17 shows, this device actually consists of a single FF and a 3-bit asynchronous counter. This arrangement is for flexibility. It can be used as a divide-by-2 device using only the single FF; or it can be used as a mod-8 counter using only the 3-bit counter portion. It can also be used as a mod-10, mod-12, or mod-16 counter by making proper connections. This device also provides gated reset inputs, R_{01}, R_{02}. When both of these inputs are HIGH, the counter is reset to the 0000 state by $\overline{\text{CLR}}$.

To use the 7493A as a mod-2 counter, apply the clock signal to CLKA (pin 14) and take the output from $Q_{A(LSB)}$ (pin 12). The R_{01} and R_{02} are grounded.

To use the 7493A as a mod-8 counter, apply the clock signal to CLKB (pin 1) and take outputs from Q_B, Q_C, and $Q_{D(MSB)}$. The R_{01} and R_{02} are grounded.

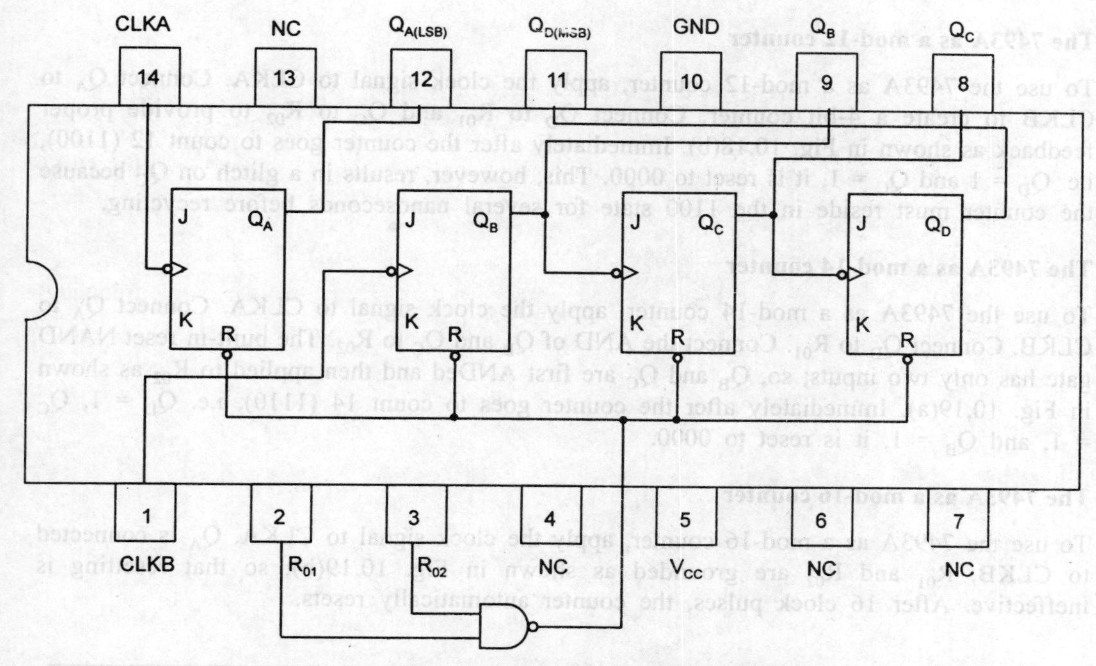

Fig. 10.17 The 7493A 4-bit ripple counter.

The 7493A as a mod-10 counter

To use the 7493A as a mod-10 counter, apply the clock signal to CLKA, connect Q_A to CLKB (since all the FFs are to be used), connect Q_B to R_{02}, and Q_D to R_{01} because feedback is provided from the 8's place and 2's place ($10_{10} = 1010_2$) as shown in Fig. 10.18(a). As soon as the count goes to 10 (1010), i.e. $Q_D = 1$ and $Q_B = 1$, all the FFs will be reset.

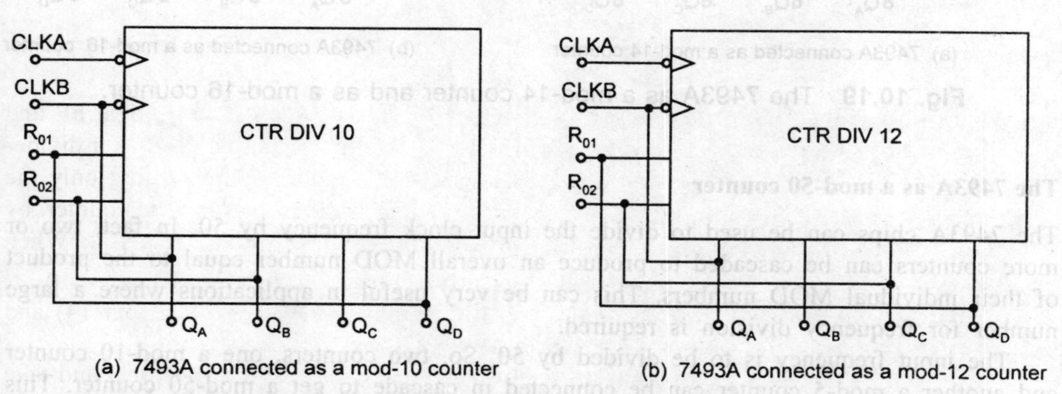

Fig. 10.18 The 7493A as a mod-10 counter and as a mod-12 counter.

The 7493A as a mod-12 counter

To use the 7493A as a mod-12 counter, apply the clock signal to CLKA. Connect Q_A to CLKB to create a 4-bit counter. Connect Q_D to R_{01} and Q_C to R_{02} to provide proper feedback as shown in Fig. 10.18(b). Immediately after the counter goes to count 12 (1100), i.e. $Q_D = 1$ and $Q_C = 1$, it is reset to 0000. This, however, results in a glitch on Q_C because the counter must reside in the 1100 state for several nanoseconds before recycling.

The 7493A as a mod-14 counter

To use the 7493A as a mod-14 counter, apply the clock signal to CLKA. Connect Q_A to CLKB. Connect Q_D to R_{01}. Connect the AND of Q_B and Q_C to R_{02}. The built-in reset NAND gate has only two inputs; so, Q_B and Q_C are first ANDed and then applied to R_{02} as shown in Fig. 10.19(a). Immediately after the counter goes to count 14 (1110), i.e. $Q_D = 1$, $Q_C = 1$, and $Q_B = 1$, it is reset to 0000.

The 7493A as a mod-16 counter

To use the 7493A as a mod-16 counter, apply the clock signal to CLKA. Q_A is connected to CLKB, R_{01} and R_{02} are grounded as shown in Fig. 10.19(b), so that resetting is ineffective. After 16 clock pulses, the counter automatically resets.

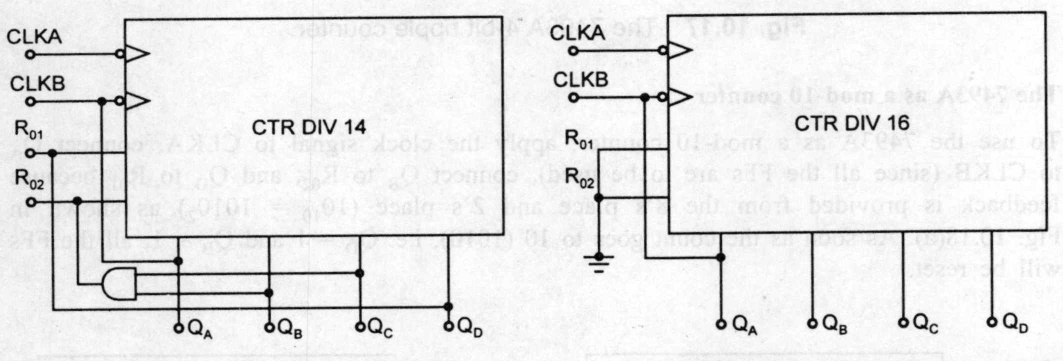

(a) 7493A connected as a mod-14 counter

(b) 7493A connected as a mod-16 counter

Fig. 10.19 The 7493A as a mod-14 counter and as a mod-16 counter.

The 7493A as a mod-50 counter

The 7493A chips can be used to divide the input clock frequency by 50. In fact, two or more counters can be cascaded to produce an overall MOD number equal to the product of their individual MOD numbers. This can be very useful in applications where a large number for frequency division is required.

The input frequency is to be divided by 50. So, two counters, one a mod-10 counter and another a mod-5 counter can be connected in cascade to get a mod-50 counter. This can be achieved by using two 7493A chips as shown in Fig. 10.20. The first counter divides f_{in} by 10; so, Q_D of the first counter has a frequency of $f_{in}/10$. This Q_D is connected to

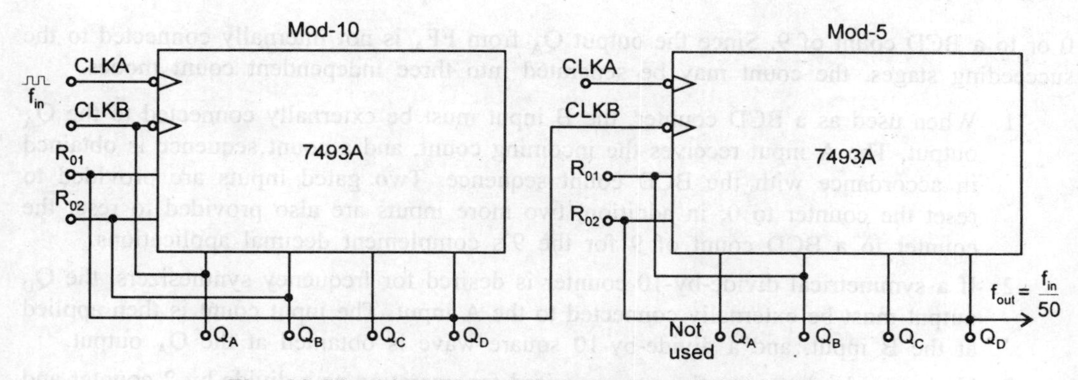

Fig. 10.20 Mod-50 counter using two 7493A chips.

CLKB. The second 7493A divides this $f_{in}/10$ by 5 so that the output signal at Q_D of the second 7493A is of frequency $(f_{in}/10)/5 = f_{in}/50$.

The 7490 Decade Counter

The 7490 is a decade counter (mod-10) which consists of four master-slave flip-flops internally connected to provide a divide-by-2 counter and a divide-by-5 counter as shown in Fig. 10.21. Gated direct-reset lines are provided to inhibit count inputs to either a logic

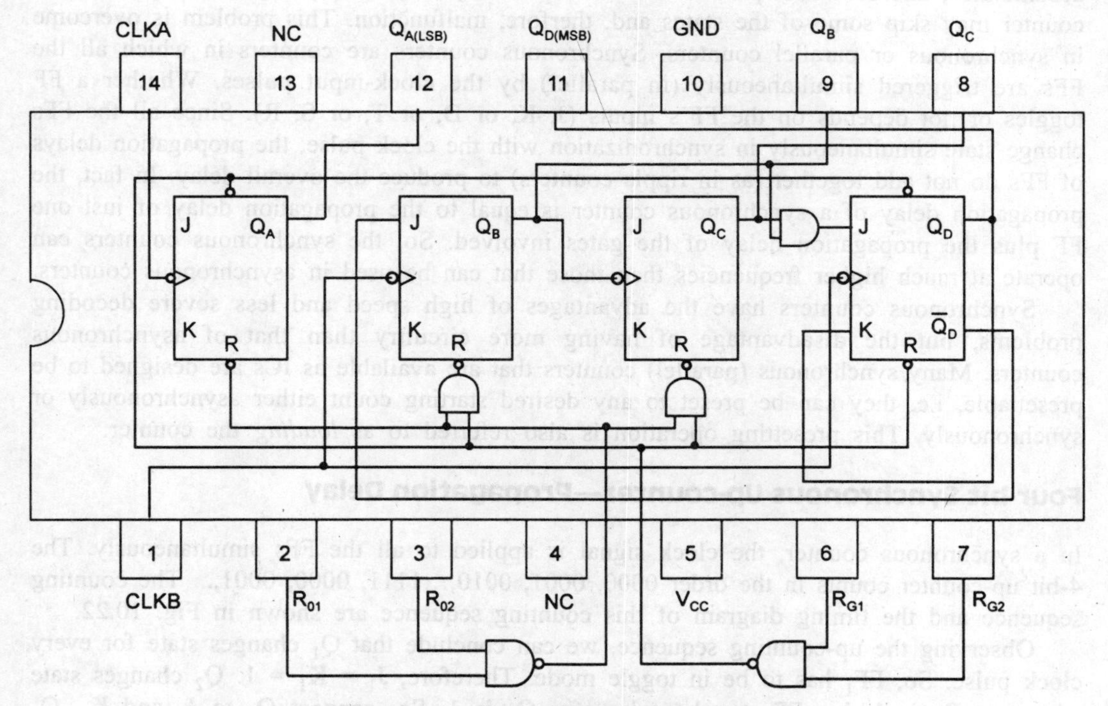

Fig. 10.21 The 7490 decade counter.

0 or to a BCD count of 9. Since the output Q_A from FF_A is not internally connected to the succeeding stages, the count may be separated into three independent count modes.

1. When used as a BCD counter, the B input must be externally connected to the Q_A output. The A input receives the incoming count, and a count sequence is obtained in accordance with the BCD count sequence. Two gated inputs are provided to reset the counter to 0; in addition, two more inputs are also provided to reset the counter to a BCD count of 9 for the 9's complement decimal applications.
2. If a symmetrical divide-by-10 counter is desired for frequency synthesizers, the Q_D output must be externally connected to the A input. The input count is then applied at the B input, and a divide-by-10 square wave is obtained at the Q_A output.
3. No external interconnections are required for operation as a divide-by-2 counter and a divide-by-5 counter. The FF_A is used as a binary element for the divide-by-2 function. The B input is used to obtain binary divide-by-5 operation at the Q_B, Q_C, and Q_D outputs. In this mode, the two counters operate independently. However, all the four FFs are reset simultaneously.

10.7 SYNCHRONOUS COUNTERS

Asynchronous counters are serial counters. They are slow because each FF can change state only if all the preceding FFs have changed their state. The propagation delay thus gets accumulated, and so causes problems. If the clock frequency is very high, the asynchronous counter may skip some of the states and, therfore, malfunction. This problem is overcome in synchronous or parallel counters. Synchronous counters are counters in which all the FFs are triggered simultaneously (in parallel) by the clock-input pulses. Whether a FF toggles or not depends on the FF's inputs (J, K, or D, or T, or S, R). Since all the FFs change state simultaneously in synchronization with the clock pulse, the propagation delays of FFs do not add together (as in ripple counters) to produce the overall delay. In fact, the propagation delay of a synchronous counter is equal to the propagation delay of just one FF plus the propagation delay of the gates involved. So, the synchronous counters can operate at much higher frequencies than those that can be used in asynchronous counters.

Synchronous counters have the advantages of high speed and less severe decoding problems, but the disadvantage of having more circuitry than that of asynchronous counters. Many synchronous (parallel) counters that are available as ICs are designed to be presettable, i.e. they can be preset to any desired starting count either asynchronously or synchronously. This presetting operation is also referred to as *loading* the counter.

Four-bit Synchronous Up-counter—Propagation Delay

In a synchronous counter, the clock signal is applied to all the FFs simultaneously. The 4-bit up-counter counts in the order 0000, 0001, 0010,...1111, 0000, 0001,... The counting sequence and the timing diagram of this counting sequence are shown in Fig. 10.22.

Observing the up-counting sequence, we can conclude that Q_1 changes state for every clock pulse. So, FF_1 has to be in toggle mode. Therefore, $J_1 = K_1 = 1$. Q_2 changes state whenever Q_1 is 1, i.e. FF_2 toggles whenever Q_1 is 1. So, connect Q_1 to J_2 and K_2. Q_3

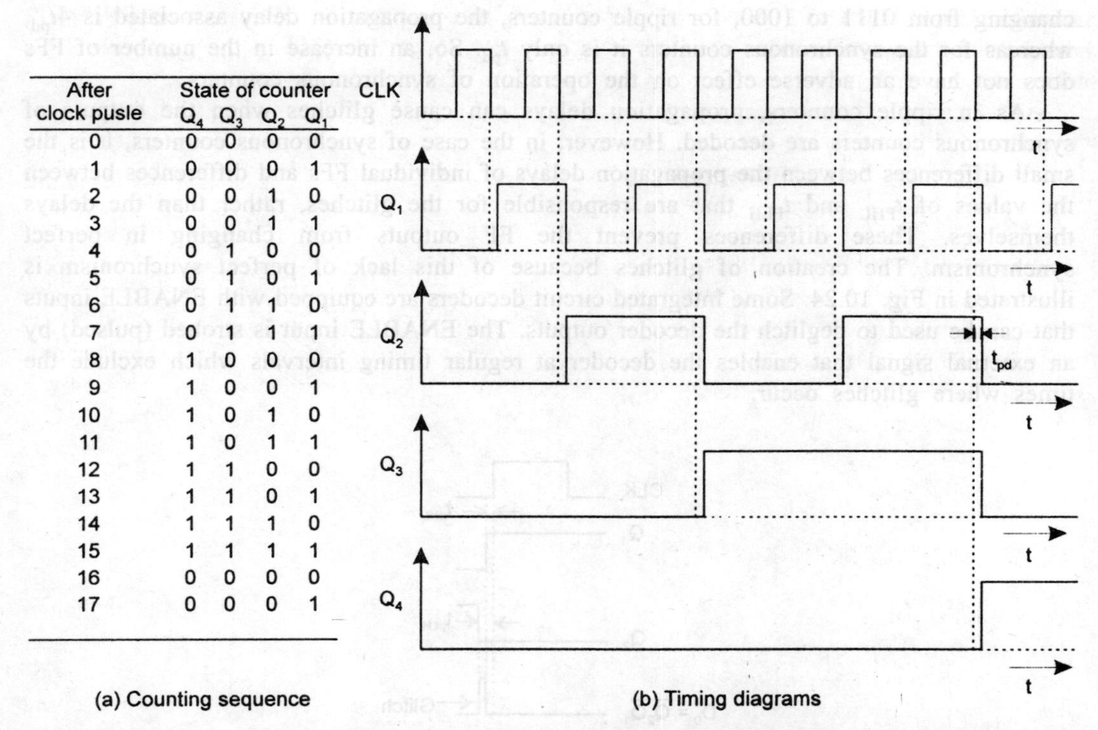

After clock pusle	Q_4	Q_3	Q_2	Q_1
0	0	0	0	0
1	0	0	0	1
2	0	0	1	0
3	0	0	1	1
4	0	1	0	0
5	0	1	0	1
6	0	1	1	0
7	0	1	1	1
8	1	0	0	0
9	1	0	0	1
10	1	0	1	0
11	1	0	1	1
12	1	1	0	0
13	1	1	0	1
14	1	1	1	0
15	1	1	1	1
16	0	0	0	0
17	0	0	0	1

Fig. 10.22 Four-bit synchronous up-counter considering the effect of propagation delay.

changes state whenever $Q_2 = 1$ and $Q_1 = 1$; that means, FF_3 toggles whenever $Q_1Q_2 = 1$. So, Q_1Q_2 is connected to J_3 and K_3. Q_4 changes state whenever $Q_1 = 1$, $Q_2 = 1$, and $Q_3 = 1$, i.e. FF_4 toggles whenever $Q_1Q_2Q_3 = 1$. So, $Q_1Q_2Q_3$ is connected to J_4 and K_4. Hence, the logic diagram will be as shown in Fig. 10.23.

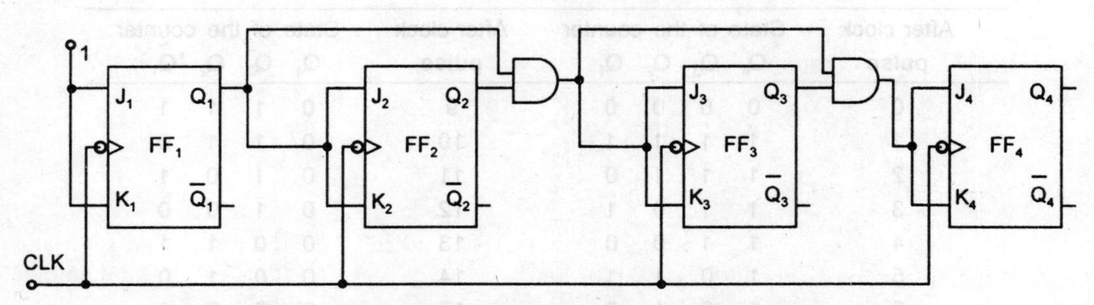

Fig. 10.23 Logic diagram of a 4-bit synchronous up-counter.

As may be seen in the timing diagram of Fig. 10.22, the propagation delay does not get accumulated in synchronous counters as it does in ripple counters. When the count is changing from 0111 to 1000, for ripple counters, the propagation delay associated is $4t_{pd}$, whereas for the synchronous counters it is only t_{pd}. So, an increase in the number of FFs does not have an adverse effect on the operation of synchronous counters.

As in ripple counters, propagation delays can cause glitches when the outputs of synchronous counters are decoded. However, in the case of synchronous counters, it is the small differences between the propagation delays of individual FFs and differences between the values of t_{PHL} and t_{PLH} that are responsible for the glitches, rather than the delays themselves. These differences prevent the FF outputs from changing in perfect synchronism. The creation of glitches because of this lack of perfect synchronism is illustrated in Fig. 10.24. Some integrated circuit decoders are equipped with ENABLE inputs that can be used to deglitch the decoder outputs. The ENABLE input is strobed (pulsed) by an external signal that enables the decoder at regular timing intervals which exclude the times where glitches occur.

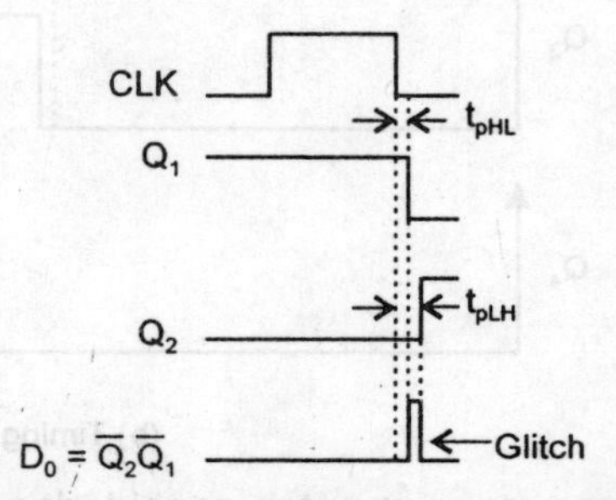

Fig. 10.24 Glitch in synchronous counter.

Four-bit Synchronous Down-counter

A 4-bit synchronous down-counter counts in the order 0000, 1111, 1110, ..., 0000, 1111. The counting sequence is shown in Table 10.1.

Table 10.1 Counting sequence of a 4-bit synchronous down-counter

After clock pulse	Q_4	Q_3	Q_2	Q_1	After clock pulse	Q_4	Q_3	Q_2	Q_1
0	0	0	0	0	9	0	1	1	1
1	1	1	1	1	10	0	1	1	0
2	1	1	1	0	11	0	1	0	1
3	1	1	0	1	12	0	1	0	0
4	1	1	0	0	13	0	0	1	1
5	1	0	1	1	14	0	0	1	0
6	1	0	1	0	15	0	0	0	1
7	1	0	0	1	16	0	0	0	0
8	1	0	0	0	17	1	1	1	1

(Header: State of the counter spans Q_4–Q_1 columns.)

Observing the down-counting sequence, we can conclude that:

Q_1 changes state after each clock pulse. So, FF_1 is in toggle mode. Therefore, $J_1 = 1$ and $K_1 = 1$. Q_2 changes state only when $Q_1 = 0$, i.e. $\overline{Q}_1 = 1$, i.e. FF_2 toggles only when $\overline{Q}_1 = 1$. So, connect $\overline{Q}_1$ to J_2 and K_2.

Q_3 changes state only when $Q_1 = 0$ and $Q_2 = 0$, i.e. $\overline{Q}_1 = 1$ and $\overline{Q}_2 = 1$, i.e. FF_3 toggles only when $\overline{Q}_1\overline{Q}_2 = 1$. So, connect $\overline{Q}_1\overline{Q}_2$ to J_3 and K_3.

Q_4 changes state only when $Q_1 = 0$, $Q_2 = 0$, and $Q_3 = 0$, i.e. only when $\overline{Q}_1 = 1$, $\overline{Q}_2 = 1$ and $\overline{Q}_3 = 1$, i.e. FF_4 toggles only when $\overline{Q}_1\overline{Q}_2\overline{Q}_3 = 1$. So, connect $\overline{Q}_1\overline{Q}_2\overline{Q}_3$ to J_4 and K_4.

Hence, the logic diagram for a 4-bit down-counter is as shown in Fig. 10.25.

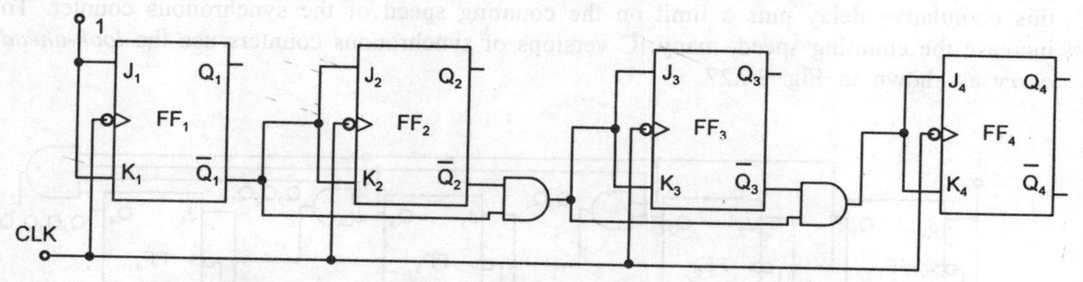

Fig. 10.25 Logic diagram of a 4-bit synchronous down-counter.

Four-bit Synchronous Up/Down (Bidirectional) Counter

A 4-bit synchronous up/down counter can be obtained by combining the up-counting and down-counting operations in a single counter using control or mode signal. Let us say, we want the counter to count up when mode signal M = 1 and count down when mode signal M = 0. We can obtain the expressions for excitations of an up/down counter by combining the excitations of up- and down-counters using the mode signal. Therefore, the design equations for an up/down counter are:

$$J_1 = K_1 = 1$$

$$J_2 = K_2 = (Q_1 \cdot \text{Up}) + (\overline{Q}_1 \cdot \text{Down}) = Q_1M + \overline{Q}_1\overline{M}$$

$$J_3 = K_3 = (Q_1 \cdot Q_2 \cdot \text{Up}) + (\overline{Q}_1 \cdot \overline{Q}_2 \cdot \text{Down}) = Q_1Q_2M + \overline{Q}_1\overline{Q}_2\overline{M}$$

$$J_4 = K_4 = (Q_1 \cdot Q_2 \cdot Q_3 \cdot \text{Up}) + (\overline{Q}_1 \cdot \overline{Q}_2 \cdot \overline{Q}_3 \cdot \text{Down}) = Q_1Q_2Q_3M + \overline{Q}_1\overline{Q}_2\overline{Q}_3\overline{M}$$

The logic diagram of a 4-bit synchronous up/down counter is shown in Fig. 10.26. Most up/down counters can be reversed at any point in the sequence.

Look-ahead Carry

In the 4-bit up-counter shown in Fig. 10.23, the term Q_1Q_2, ... is called a *carry*, that is, it is brought forward to each stage. The carry must ripple through the successive AND gates, i.e. the output of the rightmost AND gate is not valid until the outputs of all the

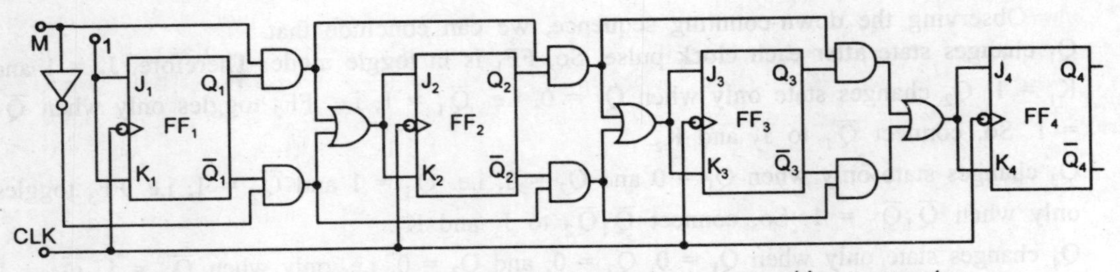

Fig. 10.26 Logic diagram of a 4-bit synchronous up/down counter.

preceding AND gates are valid. The propagation delays of the AND gates accumulate and this cumulative delay puts a limit on the counting speed of the synchronous counter. To increase the counting speed, many IC versions of synchronous counters use the *look-ahead carry* as shown in Fig. 10.27.

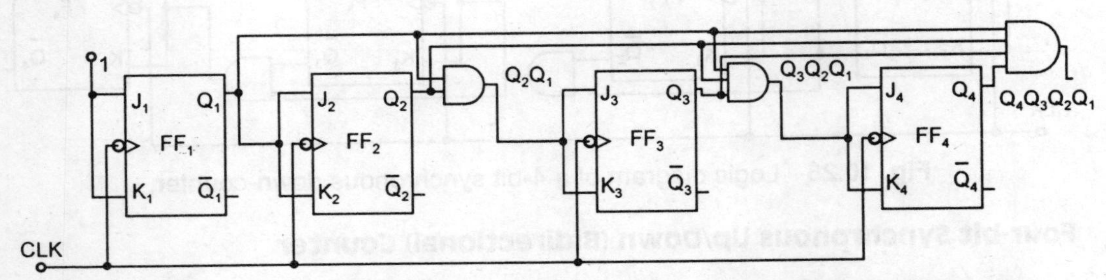

Fig. 10.27 Logic diagram of a 4-bit synchronous counter with look-ahead carry.

The J and K control logic is the same as that in the ripple-carry circuit, but the accumulation of propagation delay is eliminated. The total delay at each stage in this case is equal to the propagation delay of just one AND gate.

10.8 HYBRID COUNTERS

A hybrid counter is a counter in which the output of a synchronous counter drives the clock input of another counter to get a divide-by-N operation. The block diagram of a hybrid counter is shown in Fig. 10.28.

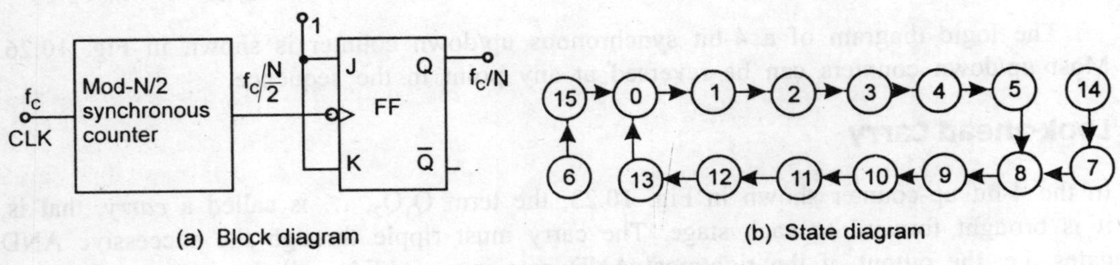

Fig. 10.28 Hybrid counter.

Hybrid counters can be used to obtain a symmetrical divide-by-N output. For example, when N is any number divisible by 2, we can obtain a symmetrical divide-by-N counter, by making the output of a synchronous mod-$N/2$ counter drive a mod-2 counter. The output of the mod-2 counter has a frequency of f_C/N. The logic diagram of a mod-12 synchronous hybrid counter obtained by using a mod-6 counter and a mod-2 counter is shown in Fig. 10.29. The timing diagram shown in Fig. 10.30 indicates that the sequence of the states of the counter is 0, 1, 2, 3, 4, 5, 8, 9, 10, 11, 12, 13, 0, 1, 2, etc. The counter is self-starting as seen from the state diagram of Fig. 10.38. Whenever it goes to an invalid state, it comes back to the valid state in 1 or 2 clock pulses.

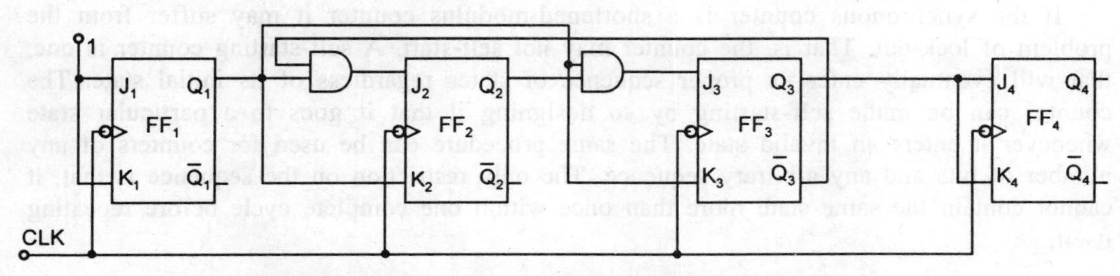

Fig. 10.29 Logic diagram of a hybrid mod-12 counter.

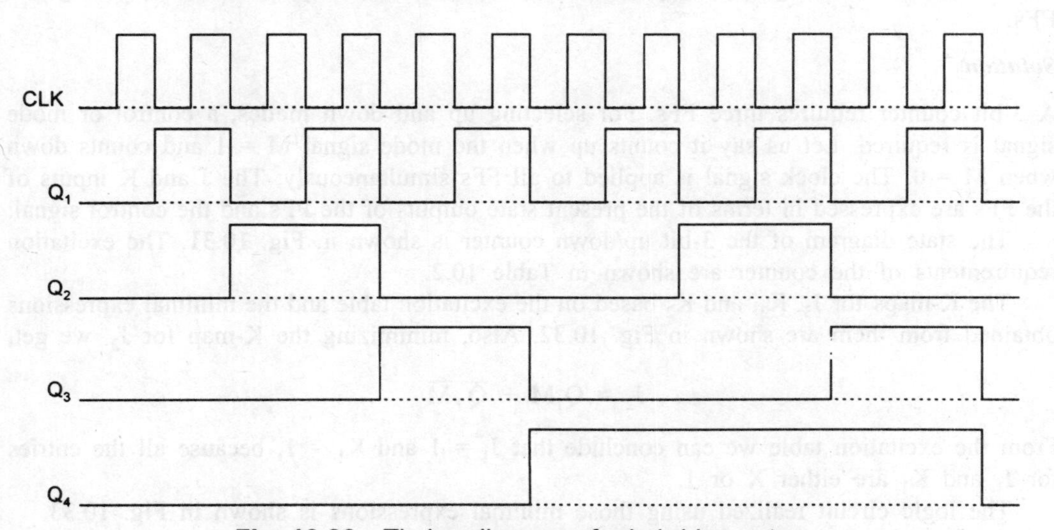

Fig. 10.30 Timing diagram of a hybrid counter.

10.9 DESIGN OF SYNCHRONOUS COUNTERS

For a systematic design of synchronous counters, the following procedure is used:

Step 1 Based on the description of the problem, determine the required number n of the FFs—the smallest value of n is such that the number of states $N \leq 2^n$—and the desired counting sequence.

Step 2 Draw the state diagram showing all the possible states. A state diagram, which can also be called the transition diagram, is a graphical means of depicting the sequence of states through which the counter progresses. In case the counter goes to a particular state from the invalid states on the next clock pulse, the same can also be included in the state diagram.

Step 3 Write the excitation table that lists the present state (PS), the next state (NS) and the required excitation.

Step 4 Obtain the minimal expressions for the excitations of the FFs using K-maps.

Step 5 Implement the minimal expressions to get the logic circuit.

If the synchronous counter is a shortened-modulus counter it may suffer from the problem of lock-out. That is, the counter may not self-start. A self-starting counter is one, that will eventually enter its proper sequence of states regardless of its initial state. The counter can be made self-starting by so designing it that it goes to a particular state whenever it enters an invalid state. The same procedure can be used for counters of any number of bits and any arbitrary sequence. The only restriction on the sequence is that, it cannot contain the same state more than once within one complete cycle before repeating itself.

EXAMPLE 10.6 Design and implement a synchronous 3-bit up/down counter using J-K FFs.

Solution

A 3-bit counter requires three FFs. For selecting up and down modes, a control or mode signal is required. Let us say it counts up when the mode signal M = 1 and counts down when M = 0. The clock signal is applied to all FFs simultaneously. The J and K inputs of the FFs are expressed in terms of the present state outputs of the FFs and the control signal.

The state diagram of the 3-bit up/down counter is shown in Fig. 10.31. The excitation requirements of the counter are shown in Table 10.2.

The K-maps for J_3, K_3, and K_2 based on the excitation table and the minimal expressions obtained from them are shown in Fig. 10.32. Also, minimizing the K-map for J_2, we get,

$$J_2 = Q_1M + \overline{Q}_1\overline{M}$$

From the excitation table we can conclude that $J_1 = 1$ and $K_1 = 1$, because all the entries for J_1 and K_1 are either X or 1.

The logic circuit realized using those minimal expressions is shown in Fig. 10.33.

Second method

A 3-bit up/down counter can also be realized by designing the up-counter and the down-counter separately and then combining them using a mode signal and additional gates.

Three-bit up-counter. The state diagram of a 3-bit up-counter is shown in Fig. 10.34(a). The excitation requirements when JK FFs are chosen as memory elements are shown in Table 10.3.

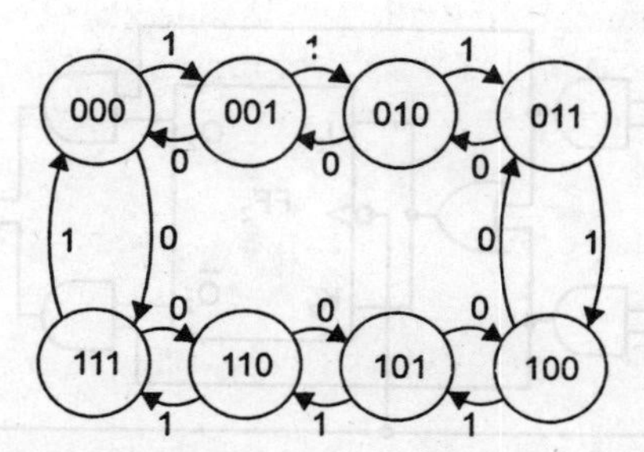

Fig. 10.31 State diagram of the 3-bit up/down counter (Example 10.6).

Table 10.2 Excitation table

PS			Mode	NS			Required excitations					
Q_3	Q_2	Q_1	M	Q_3	Q_2	Q_1	J_3	K_3	J_2	K_2	J_1	K_1
0	0	0	0	1	1	1	1	X	1	X	1	X
0	0	0	1	0	0	1	0	X	0	X	1	X
0	0	1	0	0	0	0	0	X	0	X	X	1
0	0	1	1	0	1	0	0	X	1	X	X	1
0	1	0	0	0	0	1	0	X	X	1	1	X
0	1	0	1	0	1	1	0	X	X	0	1	X
0	1	1	0	0	1	0	0	X	X	0	X	1
0	1	1	1	1	0	0	1	X	X	1	X	1
1	0	0	0	0	1	1	X	1	1	X	1	X
1	0	0	1	1	0	1	X	0	0	X	1	X
1	0	1	0	1	0	0	X	0	0	X	X	1
1	0	1	1	1	1	0	X	0	1	X	X	1
1	1	0	0	1	0	1	X	0	X	1	1	X
1	1	0	1	1	1	1	X	0	X	0	1	X
1	1	1	0	1	1	0	X	0	X	0	X	1
1	1	1	1	0	0	0	X	1	X	1	X	1

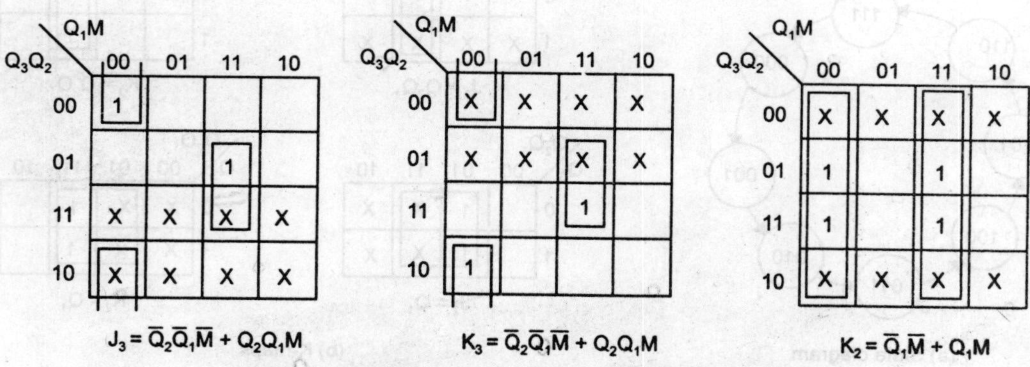

Fig. 10.32 Karnaugh maps (Example 10.6).

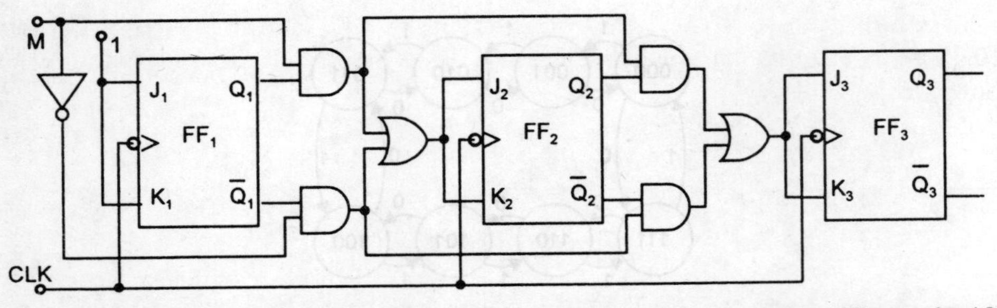

Fig. 10.33 Logic diagram of the synchronous 3-bit up/down counter (Example 10.6).

Table 10.3 Excitation table of a 3-bit up-counter (Example 10.6)

PS			NS			Required excitation					
Q_3	Q_2	Q_1	Q_3	Q_2	Q_1	J_3	K_3	J_2	K_2	J_1	K_1
0	0	0	0	0	1	0	X	0	X	1	X
0	0	1	0	1	0	0	X	1	X	X	1
0	1	0	0	1	1	0	X	X	0	1	X
0	1	1	1	0	0	1	X	X	1	X	1
1	0	0	1	0	1	X	0	0	X	1	X
1	0	1	1	1	0	X	0	1	X	X	1
1	1	0	1	1	1	X	0	X	0	1	X
1	1	1	0	0	0	X	1	X	1	X	1

The K-maps for excitations based on the excitation table and the minimal expressions for excitations J_3, K_3, J_2, and K_2 in terms of the present outputs Q_3, Q_2, and Q_1 obtained by minimizing the K-maps are shown in Fig. 10.34(b). From the excitation table it is seen that, $J_1 = K_1 = 1$, because all the entries for J_1 and K_1 are either a 1 or an X.

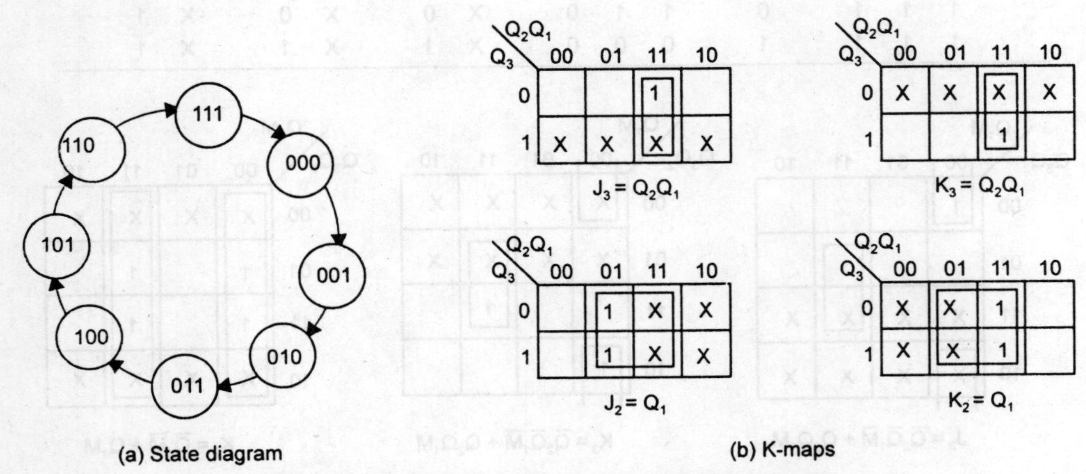

Fig. 10.34 Three-bit up-counter (Example 10.6).

Three-bit down-counter. The state diagram of a 3-bit down-counter is shown in Fig. 10.35(a). The excitation requirements when J-K FFs are selected as memory elements are shown in Table 10.4.

Table 10.4 Excitation table of a 3-bit down-counter (Example 10.6)

PS			NS			Required excitation					
Q_3	Q_2	Q_1	Q_3	Q_2	Q_1	J_3	K_3	J_2	K_2	J_1	K1
0	0	0	1	1	1	1	X	1	X	1	X
0	0	1	0	0	0	0	X	0	X	X	1
0	1	0	0	0	1	0	X	X	1	1	X
0	1	1	0	1	0	0	X	X	0	X	1
1	0	0	0	1	1	X	1	1	X	1	X
1	0	1	1	0	0	X	0	0	X	X	1
1	1	0	1	0	1	X	0	X	1	1	X
1	1	1	1	1	0	X	0	X	0	X	1

The K-maps and the minimal expressions for excitations J_3, K_3, J_2, and K_2 obtained by simplifying them are shown in Fig. 10.35(b).

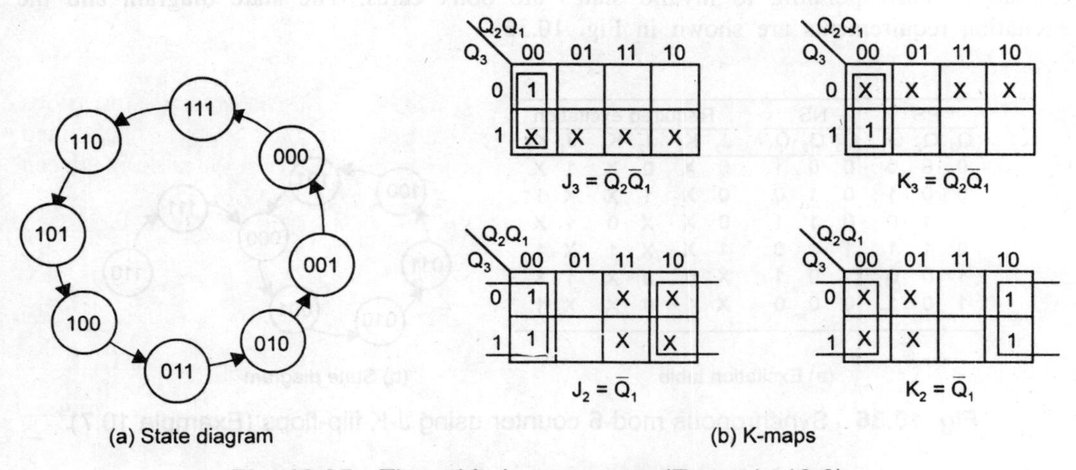

Fig. 10.35 Three-bit down counter (Example 10.6).

From the design equations of the up-counter and the down-counter we see that, in both cases $J_1 = 1$ and $K_1 = 1$. Therefore, for the up/down counter too, $J_1 = 1$ and $K_1 = 1$. For the up-counter $J_2 = K_2 = Q_1$ and for the down-counter, $J_2 = K_2 = \overline{Q}_1$. So, for the up/down counter, $J_2 = K_2 = (Q_1 \cdot \text{Up}) + (\overline{Q}_1 \cdot \text{Down})$.

Similarly, for the up-counter, $J_3 = K_3 = Q_2Q_1$, and for the down-counter, $J_3 = K_3 = \overline{Q}_2\overline{Q}_1$. So, for an up/down counter

$$J_3 = K_3 = (Q_1 \quad Q_2 \cdot \text{Up}) + (\overline{Q}_1 \cdot \overline{Q}_2 \cdot \text{Down})$$

By using a control signal M, and taking M = 1 for the up-mode and M = 0 for the down-mode and combining the above equations, the design equations of an up/down counter are

$$J_1 = K_1 = 1$$

$$J_2 = K_2 = (Q_1 \cdot \text{Up}) + (\overline{Q}_1 \cdot \text{Down}) = Q_1M + \overline{Q}_1\overline{M}$$

$$J_3 = K_3 = (Q_1 \cdot Q_2 \cdot \text{Up}) + (\overline{Q}_1 \cdot \overline{Q}_2 \cdot \text{Down}) = Q_1Q_2M + \overline{Q}_1\overline{Q}_2\overline{M}$$

We see that these equations are the same as the equations obtained by the direct design of the up/down counter. So, the circuit will be the same as that shown in Fig. 10.33.

EXAMPLE 10.7 Design a mod-6 counter using J-K FFs with separate logic circuitry for each J and K input. Construct timing diagrams and determine the duty cycle of the output of the most significant stage. Construct a state diagram to determine whether the counter is self-starting or not.

Solution

We know that the counting sequence for a mod-6 counter is 000, 001, 010, 011, 100, 101, and 000... It requires three FFs and has two invalid states, 110 and 111. The entries for excitations corresponding to invalid states are don't cares. The state diagram and the excitation requirements are shown in Fig. 10.36.

PS			NS			Required excitation					
Q_3	Q_2	Q_1	Q_3	Q_2	Q_1	J_3	K_3	J_2	K_2	J_1	K_1
0	0	0	0	0	1	0	X	0	X	1	X
0	0	1	0	1	0	0	X	1	X	X	1
0	1	0	0	1	1	0	X	X	0	1	X
0	1	1	1	0	0	1	X	X	1	X	1
1	0	0	1	0	1	X	0	0	X	1	X
1	0	1	0	0	0	X	1	0	X	X	1

(a) Excitation table

(b) State diagram

Fig. 10.36 Synchronous mod-6 counter using J-K flip-flops (Example 10.7).

The Karnaugh maps for J_3, K_3, J_2, K_2, J_1, and K_1 in terms of Q_3, Q_2, and Q_1, and the minimal expressions for excitations obtained from them are shown in Fig. 10.37. The logic diagram based on those minimal expressions is shown in Fig. 10.38. The timing diagram is shown in Fig. 10.39. The timing diagram should be studied carefully to verify how each waveform ensures that proper FFs change states to produce the desired sequence. When constructing a timing diagram for a synchronous counter it is important to show the waveforms that control the J and K inputs. For example, in this design it is necessary to show $Q_1\overline{Q}_3$ which controls J_2 and Q_2Q_1 which controls J_3. We see that the duty cycle of the most significant output is $2T/6T = 1/3$. Note also that the frequency of the output is one-sixth of the clock frequency as expected in a mod-6 counter.

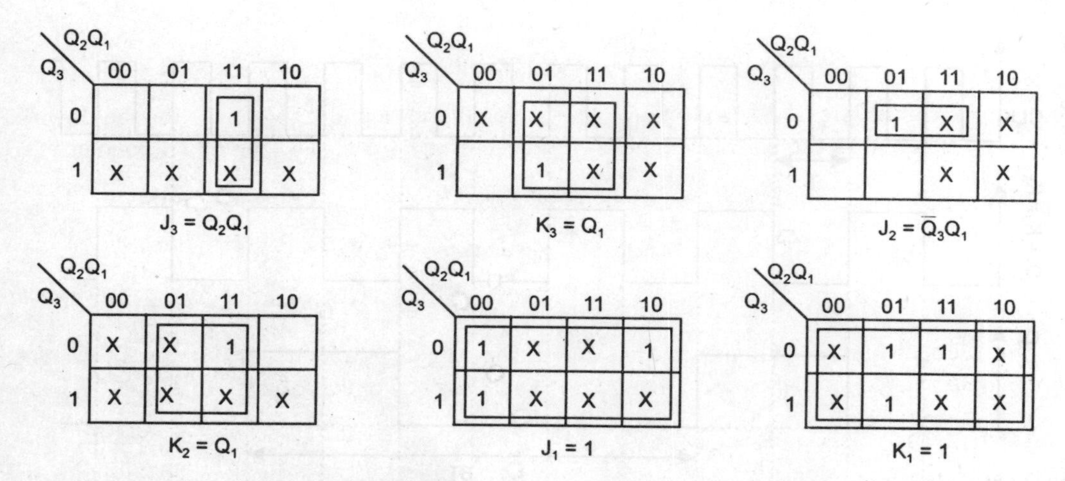

Fig. 10.37 K-maps for excitations of synchronous mod-6 counter using J-K flip-flops (Example 10.7).

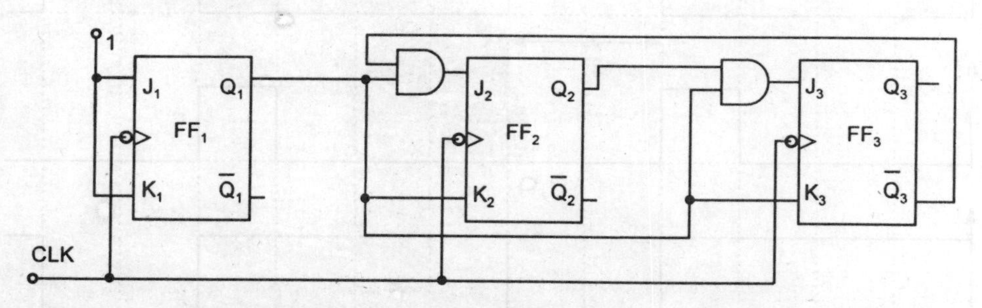

Fig. 10.38 Logic diagram of the synchronous mod-6 counter using J-K flip-flops (Example 10.7).

For this counter, 110 and 111 are the invalid states. We can determine the counter's sequence when its initial present state is 110 or 111. Given each present state, we can determine the J and K inputs from the logic diagram. With these inputs applied to the counter, which is in the present state, we can determine each of the next states of the counter. From Table 10.5 we see that, if the present state ($Q_3Q_2Q_1$) is 110, the excitations are $J_3 = 0$, $K_3 = 0$, $J_2 = 0$, $K_2 = 0$, $J_1 = 1$, and $K_1 = 1$. With these excitations the counter changes from 110 to 111. If the present state is 111, the excitations are $J_3 = 1$, $K_3 = 1$, $J_2 = 0$, $K_2 = 1$, $J_1 = 1$, and $K_1 = 1$. With these excitations the counter changes from 111

Table 10.5 Table to check for lock-out (Example 10.7)

PS			Present inputs						NS		
Q_3	Q_2	Q_1	J_3	K_3	J_2	K_2	J_1	K_1	Q_3	Q_2	Q_1
1	1	0	0	0	0	0	1	1	1	1	1
1	1	1	1	1	0	1	1	1	0	0	0

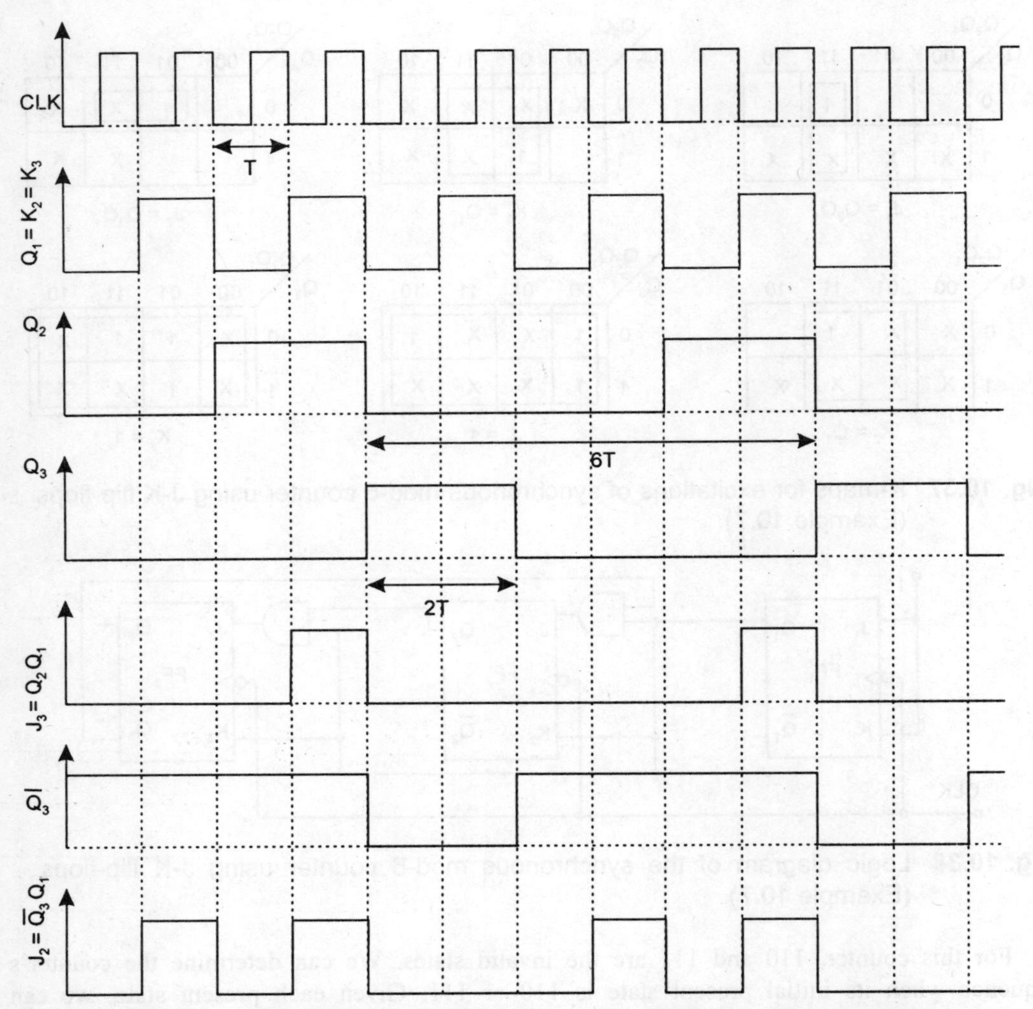

Fig. 10.39 Timing diagram for the mod-6 synchronous counter (Example 10.7).

to 000. From the table, to check for lock-out, we see that the counter is self-starting because if the counter initially finds itself in state 111, it goes to 000 after one clock pulse and if the counter is initially in state 110 then it goes to 111 after one clock pulse and goes to 000 after two clock pulses.

EXAMPLE 10.8 Design and implement a BCD counter using J-K FFs. Is the counter self-starting?

Solution

A BCD counter is nothing but a mod-10 counter. It requires four FFs. States 0000 through 1001 are stable. After the tenth clock pulse, the counter resets. States 1010 through 1111 are invalid. The excitation requirements are shown in Table 10.6. The K-maps for J_4, K_4,

J_3, K_3, J_2, K_2, J_1, and K_1 based on the excitation table, and the minimal expressions obtained from those K-maps are shown in Fig. 10.40. The logic diagram based on those minimal

Table 10.6 Excitation requirements

PS				NS				Required excitation							
Q_4	Q_3	Q_2	Q_1	Q_4	Q_3	Q_2	Q_1	J_4	K_4	J_3	K_3	J_2	K_2	J_1	K_1
0	0	0	0	0	0	0	1	0	X	0	X	0	X	1	X
0	0	0	1	0	0	1	0	0	X	0	X	1	X	X	1
0	0	1	0	0	0	1	1	0	X	0	X	X	0	1	X
0	0	1	1	0	1	0	0	0	X	1	X	X	1	X	1
0	1	0	0	0	1	0	1	0	X	X	0	0	X	1	X
0	1	0	1	0	1	1	0	0	X	X	0	1	X	X	1
0	1	1	0	0	1	1	1	0	X	X	0	X	0	1	X
0	1	1	1	1	0	0	0	1	X	X	1	X	1	X	1
1	0	0	0	1	0	0	1	X	0	0	X	0	X	1	X
1	0	0	1	0	0	0	0	X	1	0	X	0	X	X	1

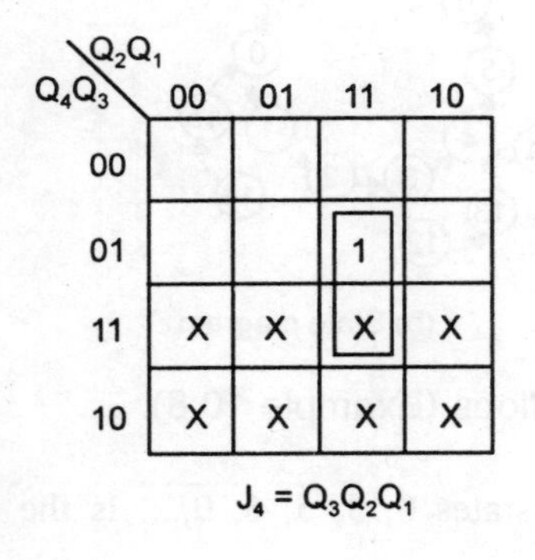

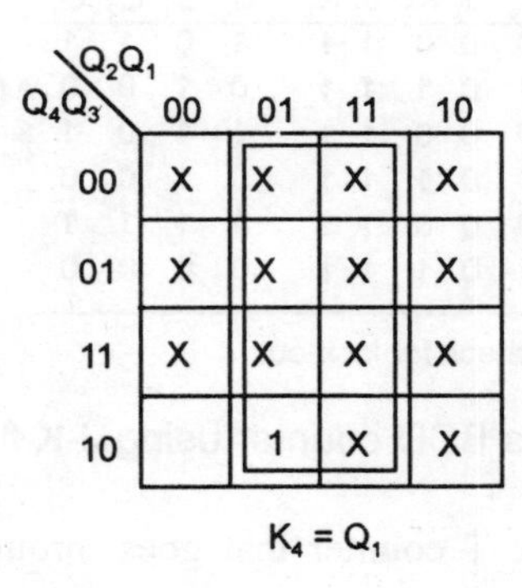

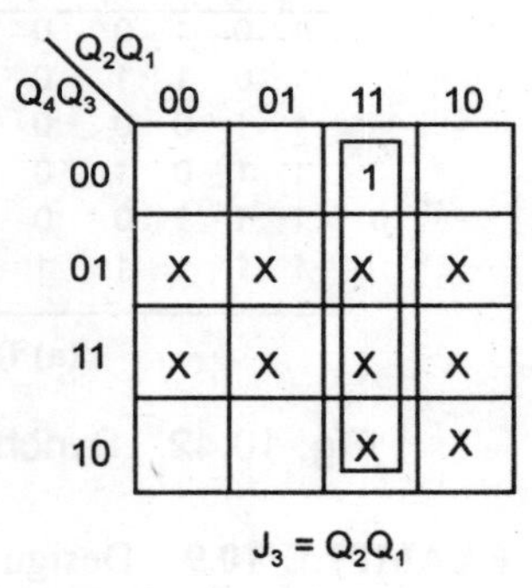

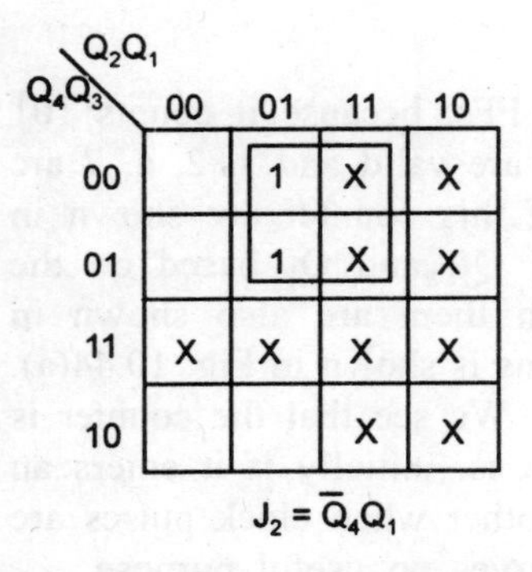

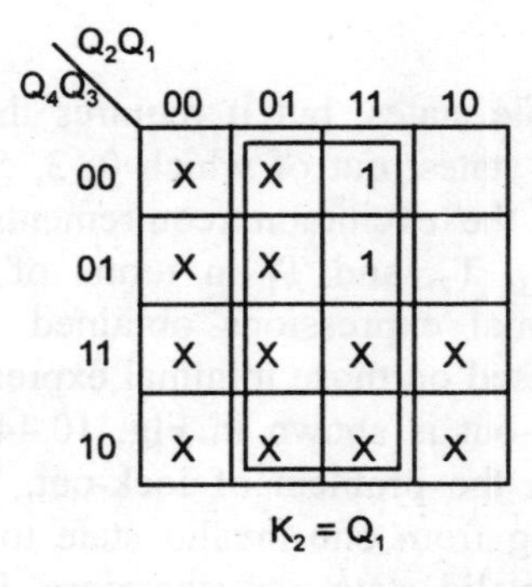

Q_4Q_3 \ Q_2Q_1	00	01	11	10
00	X	X	X	X
01			1	
11	X	X	X	X
10	X	X	X	X

$K_3 = Q_2Q_1$

Fig. 10.40 K-maps for excitations of the BCD counter using J-K flip-flops (Example 10.8).

expressions is shown in Fig. 10.41. The state diagram and the table to check for lock-out are shown in Fig. 10.42. The table to check for lock-out shows that, if the counter finds itself in an invalid state initially, it moves to a valid state after one or two clock pulses and then counts in the normal way. Therefore, the counter is self-starting.

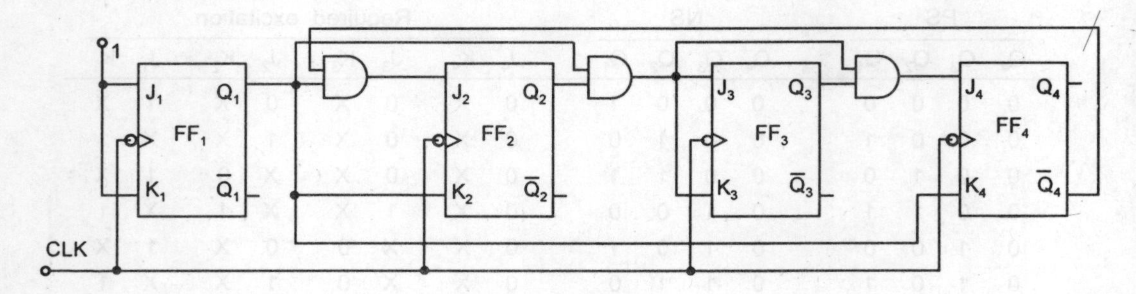

Fig. 10.41 Logic diagram of the synchronous BCD counter using J-K flip-flops (Example 10.8).

PS				Present inputs								NS			
Q_4	Q_3	Q_2	Q_1	J_4	K_4	J_3	K_3	J_2	K_2	J_1	K_1	Q_4	Q_3	Q_2	Q_1
1	0	1	0	0	0	0	0	0	0	1	1	1	0	1	1
1	0	1	1	0	1	1	1	0	1	1	1	0	1	0	0
1	1	0	0	0	0	0	0	0	0	1	1	1	1	0	1
1	1	0	1	0	1	0	0	0	1	1	1	0	1	0	0
1	1	1	0	0	0	0	0	0	0	1	1	1	1	1	1
1	1	1	1	1	1	1	1	0	1	1	1	0	0	0	0

(a) Table to check for lock-out

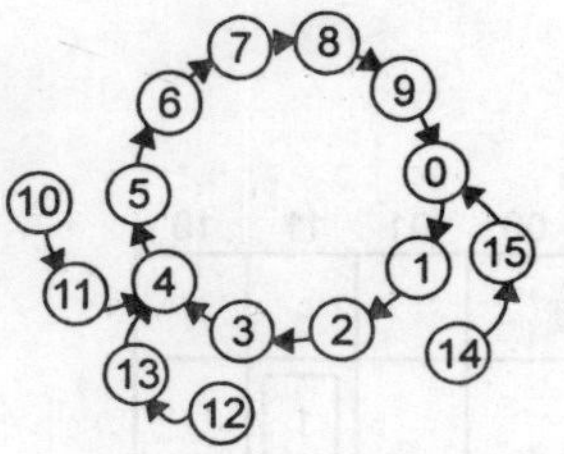

(b) State diagram

Fig. 10.42 Synchronous BCD counter using J-K flip-flops (Example 10.8).

EXAMPLE 10.9 Design a type T counter that goes through states 0, 3, 5, 6, 0,.... Is the counter self-starting?

Solution

This counter has only four stable states, but it requires three FFs, because it counts 101 and 110. Three FFs can have 8 states, out of which 0, 3, 5, 6 are valid and 1, 2, 4, 7 are invalid. The state diagram and the excitation requirements of this counter are shown in Fig. 10.43. The K-maps for T_3, T_2, and T_1 in terms of Q_3, Q_2, and Q_1 based on the excitation table, and the minimal expressions obtained from them are also shown in Fig. 10.43. The logic diagram based on those minimal expressions is shown in Fig. 10.44(a).

The table to check for lock-out is shown in Fig. 10.44(b). We see that the counter is not self-starting. It suffers from the problem of lock-out. That is, initially if it enters an invalid state, it keeps on moving from one invalid state to another when clock pulses are applied and never returns to a valid state and, therefore, it serves no useful purpose.

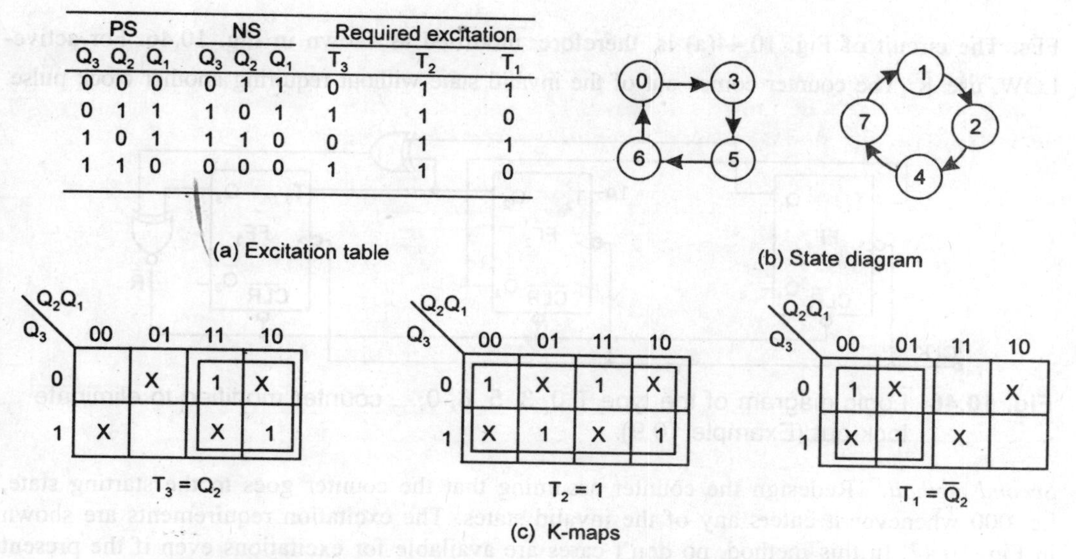

PS			NS			Required excitation		
Q_3	Q_2	Q_1	Q_3	Q_2	Q_1	T_3	T_2	T_1
0	0	0	0	1	1	0	1	1
0	1	1	1	0	1	1	1	0
1	0	1	1	1	0	0	1	1
1	1	0	0	0	0	1	1	0

Fig. 10.43 T type, 0, 3, 5, 6, 0... counter (Example 10.9).

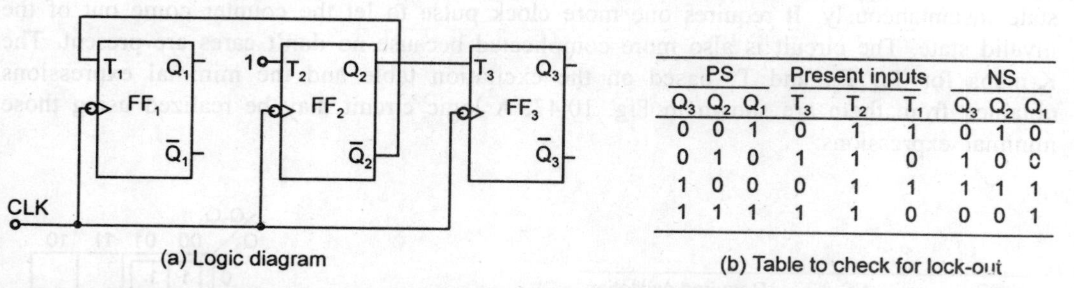

PS			Present inputs			NS		
Q_3	Q_2	Q_1	T_3	T_2	T_1	Q_3	Q_2	Q_1
0	0	1	0	1	1	0	1	0
0	1	0	1	1	0	1	0	0
1	0	0	0	1	1	1	1	1
1	1	1	1	1	0	0	0	1

(b) Table to check for lock-out

Fig. 10.44 T type counter that goes through states 0, 3, 5, 6, 0,...(Example 10.9).

Elimination of lock-out. There are two methods to eliminate lock-out.

First method. Obtain a reset pulse as shown in Fig. 10.45 such that, whenever the counter goes to an invalid state, it automatically resets to the initial state 000.

PS			Reset
Q_3	Q_2	Q_1	R
0	0	1	1
0	1	0	1
1	0	0	1
1	1	1	1

Q_3 \ Q_2Q_1	00	01	11	10
0		1		1
1	1		1	

$$
\begin{aligned}
R &= \overline{Q}_3\overline{Q}_2Q_1 + \overline{Q}_3Q_2\overline{Q}_1 + Q_3\overline{Q}_2\overline{Q}_1 + Q_3Q_2Q_1 \\
&= \overline{Q}_3(Q_2 \oplus Q_1) + Q_3(\overline{Q_2 \oplus Q_1}) \\
&= Q_3 \oplus Q_2 \oplus Q_1
\end{aligned}
$$

Fig. 10.45 Table for R, K-map, and the minimal expression for R (Example 10.9).

Obtain this reset pulse from the counter already designed (taking the excitations corresponding to the invalid states as don't cares) and feed it to the CLR terminals of the

FFs. The circuit of Fig. 10.44(a) is, therefore, modified as shown in Fig. 10.46. For active-LOW, use $\overline{R}$. The counter comes out of the invalid state without requiring another clock pulse.

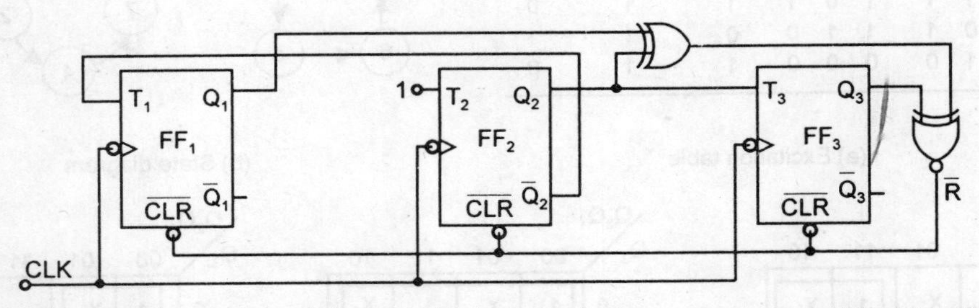

Fig. 10.46 Logic diagram of the type T 0, 3, 5, 6, 0,... counter modified to eliminate lock-out (Example 10.9).

Second method. Redesign the counter assuming that the counter goes to the starting state, i.e. 000 whenever it enters any of the invalid states. The excitation requirements are shown in Fig. 10.47. In this method, no don't cares are available for excitations even if the present state is an invalid one. The counter designed by this method cannot come out of the invalid state instantaneously. It requires one more clock pulse to let the counter come out of the invalid state. The circuit is also more complicated because no don't cares are present. The K-maps for T_3, T_2, and T_1 based on the excitation table and the minimal expressions obtained from them are shown in Fig. 10.47. A logic circuit may be realized using those minimal expressions.

PS			NS			Required excitation		
Q_3	Q_2	Q_1	Q_3	Q_2	Q_1	T_3	T_2	T_1
0	0	0	0	1	1	0	1	1
0	0	1	0	0	0	0	0	1
0	1	0	0	0	0	0	1	0
0	1	1	1	0	1	1	1	0
1	0	0	0	0	0	1	0	0
1	0	1	1	1	0	0	1	1
1	1	0	0	0	0	1	1	0
1	1	1	0	0	0	1	1	1

(a) Excitation table

$Q_3 \backslash Q_2Q_1$	00	01	11	10
0			1	
1	1		1	1

$T_3 = Q_3\overline{Q}_1 + Q_2Q_1$

$Q_3 \backslash Q_2Q_1$	00	01	11	10
0	1	1		
1		1	1	

$T_1 = \overline{Q}_3\overline{Q}_2 + Q_3Q_1$

$Q_3 \backslash Q_2Q_1$	00	01	11	10
0	1		1	1
1		1	1	1

$T_2 = Q_2 + \overline{Q}_3\overline{Q}_2 + Q_3Q_1$

(b) K-maps

Fig. 10.47 Excitation table and K-maps (Example 10.9).

EXAMPLE 10.10 Design a type D counter that goes through states 0, 1, 2, 4, 0,... The undesired (unused) states must always go to zero (000) on the next clock pulse.

Solution

This counter has only four stable states, but it requires three FFs because it counts 100 as well. Three FFs can have eight states, so, four of them are invalid. Since the counter must go to 000 on the next clock pulse, the second method of Example 10.9 is applied. The state diagram and the excitation requirements are shown in Fig. 10.48.

PS			NS			Required excitation		
Q_3	Q_2	Q_1	Q_3	Q_2	Q_1	D_3	D_2	D_1
0	0	0	0	0	1	0	0	1
0	0	1	0	1	0	0	1	0
0	1	0	1	0	0	1	0	0
0	1	1	0	0	0	0	0	0
1	0	0	0	0	0	0	0	0
1	0	1	0	0	0	0	0	0
1	1	0	0	0	0	0	0	0
1	1	1	0	0	0	0	0	0

(a) Excitation table

(a) State diagram

Fig. 10.48 Excitation table and state diagram (Example 10.10).

From the excitation table we can see that no minimization is possible. The expressions for excitations are

$$D_3 = \overline{Q}_3 Q_2 \overline{Q}_1; \qquad D_2 = \overline{Q}_3 \overline{Q}_2 Q_1; \qquad D_1 = \overline{Q}_3 \overline{Q}_2 \overline{Q}_1$$

The logic diagram based on these expressions is shown in Fig. 10.49. The counter is self-starting, i.e. it does not suffer from the problem of lock-out as may be seen from the state diagram.

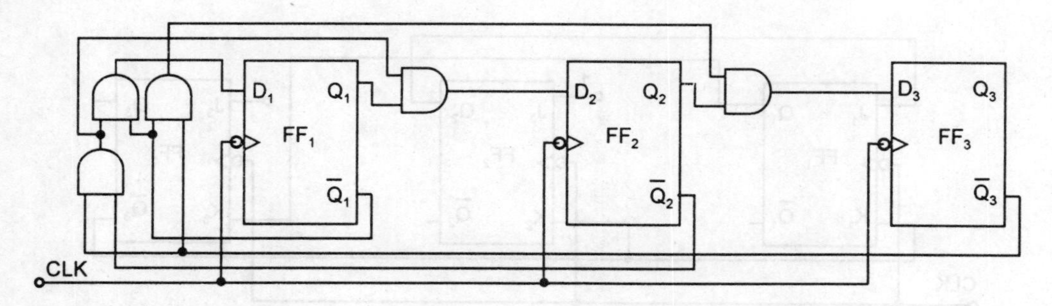

Fig. 10.49 Logic diagram of J-K type D counter that goes through states 0, 1, 2, 4, 0,... (Example 10.10).

EXAMPLE 10.11 Design a J-K counter that goes through states 3, 4, 6, 7, and 3... . Is the counter self-starting? Modify the circuit such that whenever it goes to an invalid state it comes back to state 3.

Solution

This counter requires three FFs. There are four invalid states. The corresponding excitations will be don't cares. The state diagram and the excitation requirements are shown in Fig. 10.50.

PS			NS			Required excitation					
Q_3	Q_2	Q_1	Q_3	Q_2	Q_1	J_3	K_3	J_2	K_2	J_1	K_1
0	1	1	1	0	0	1	X	X	1	X	1
1	0	0	1	1	0	X	0	1	X	0	X
1	1	0	1	1	1	X	0	X	0	1	X
1	1	1	0	1	1	X	1	X	0	X	0

(a) Excitation table

(b) State diagram

Fig. 10.50 State diagram and excitation table of type J-K 3, 4, 6, 7, 3,... counter (Example 10.11).

The K-maps for K_3, K_2, J_1, and K_1 based on the excitation table and the minimal expressions obtained from them are shown in Fig. 10.51. From the excitation table, J_3 = 1 and J_2 = 1, since all the entries for J_3 and J_2 are either a 1 or a X. The logic diagram for the counter based on those minimal expressions is shown in Fig. 10.52.

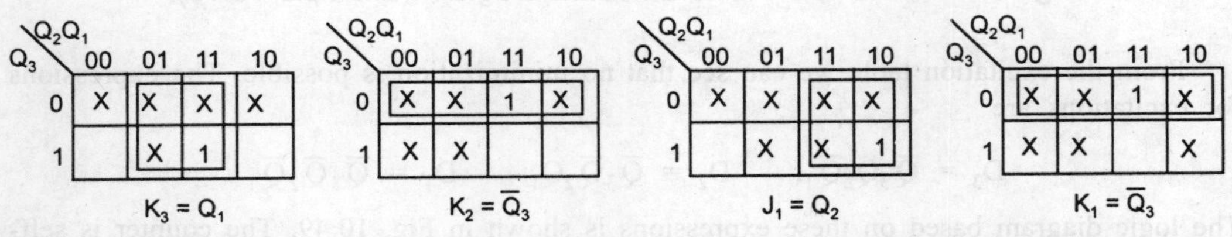

Fig. 10.51 K-maps for excitations of type 3, 4, 6, 7, 3, ... counter using J-K flip-flops (Example 10.11).

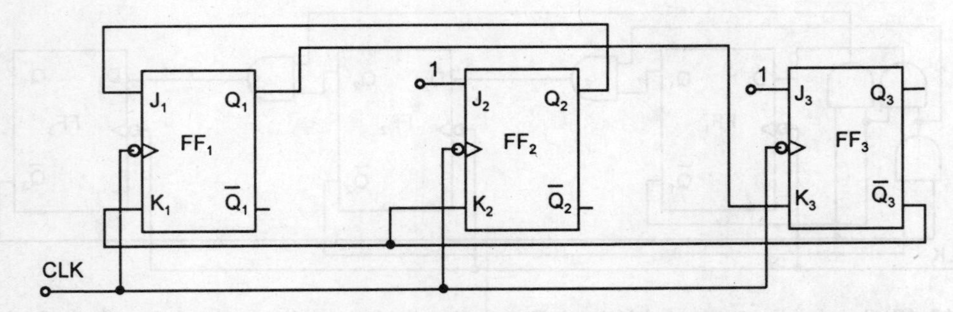

Fig. 10.52 Logic diagram of the J-K counter that goes through states 3, 4, 6, 7, 3,... (Example 10.11).

Test for lock-out. The NS entries of Table 10.7 to check for lock-out show that there is no problem of lock-out and the counter is self-starting, because any time it goes into an invalid state, it comes out and goes into a useful state after one clock pulse. The state diagram of the counter is shown in Fig. 10.50b.

To see that the counter goes to state 3(011) whenever it enters any of the invalid states, obtain an expression for reset R or set S as shown in Fig. 10.53 and modify the circuit accordingly such that the counter goes to 011 after the next clock pulse.

Table 10.7 Table to check for lock-out (Example 10.11)

PS			Present inputs						NS		
Q_3	Q_2	Q_1	J_3	K_3	J_2	K_2	J_1	K_1	Q_3	Q_2	Q_1
0	0	0	1	0	1	1	0	1	1	1	0
0	0	1	1	1	1	1	1	1	1	1	0
0	1	0	1	0	1	1	0	1	1	0	0
1	0	1	1	1	1	0	1	0	0	1	1

Since we do not want the counter to reset to 000, and instead go to state 011, we apply $\overline{R}$ to the clear terminal of FF_3 and to the preset terminals of FF_2 and FF_1 as shown in the modified circuit of Fig. 10.54.

PS			Reset
Q_3	Q_2	Q_1	R
0	0	0	1
0	0	1	1
0	1	0	1
1	0	1	1

Q_3 \ Q_2Q_1	00	01	11	10
0	1	1		1
1		1		

$R = \overline{Q}_2Q_1 + \overline{Q}_3\overline{Q}_1$

Fig. 10.53 Table and K-map for R of type J-K 3, 4, 6, 7, 3,... counter (Example 10.11).

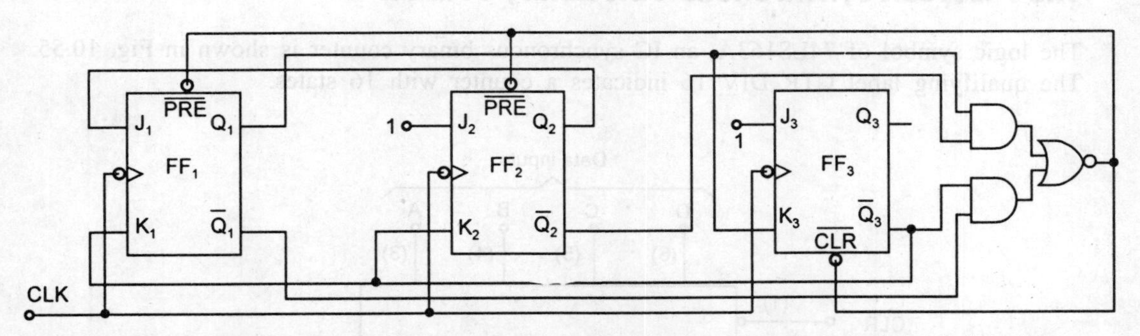

Fig. 10.54 Logic diagram of the J-K counter modified to go to state 3 from invalid states (Example 10.11).

EXAMPLE 10.12 Compare f_{max} of a 4-bit ripple counter with that of a 4-bit synchronous counter using J-K FFs. The t_{pd} for each FF is 50 ns and the t_{pd} for each AND gate is 20 ns. What needs to be done to convert these counters to mod-32? Determine f_{max} for the mod-32 ripple and parallel counters.

Solution

(a) The total delay that must be allowed between the input clock pulses of a synchronous counter is equal to

$$t_{pd}(\text{FF}) + t_{pd}(\text{AND})$$

Thus, $T_{clock} \geq 50 + 20 = 70$ ns. And so, the parallel counter has, $f_{max} = 1/(70 \text{ ns}) = 14.3$ MHZ

(b) A mod-16 ripple counter uses four FFs with $t_{pd} = 50$ ns. Thus, f_{max} for the ripple counter is

$$f_{max} = \frac{1}{(4 \times 50) \text{ ns}} = 5 \text{ MHz}$$

(c) A fifth FF must be added, since $2^5 = 32$.

(d) The f_{max} of the synchronous counter remains the same regardless of the number of FFs. Thus, its f_{max} is still 14.3 MHz. The f_{max} of the 5-bit ripple counter will change to

$$f_{max} = \frac{1}{(5 \times 50) \text{ ns}}$$

$$= 4 \text{ MHz}$$

10.10 IC SYNCHRONOUS COUNTERS

The 74LS163A Synchronous 4-bit Binary Counter

The logic symbol of 74LS163A, an IC synchronous binary counter is shown in Fig. 10.55. The qualifying label CTR DIV 16 indicates a counter with 16 states.

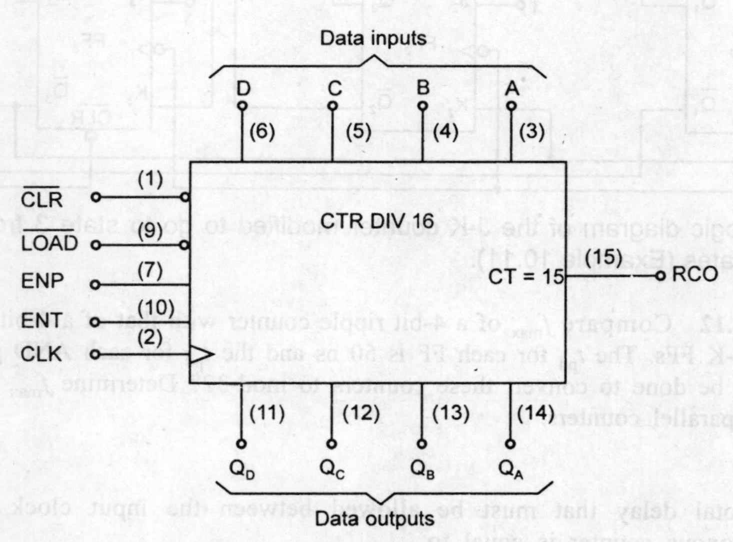

Fig. 10.55 Logic symbol of 74LS163A.

First the counter can be synchronously preset to any 4-bit binary number by applying the proper levels to the data inputs. When a LOW is applied to the $\overline{\text{LOAD}}$ input, the counter will assume the state of the data inputs on the next clock pulse. This, of course, allows the counter sequence to start with any 4-bit binary number.

The active-LOW clear input $\overline{\text{CLR}}$, synchronously resets all the four FFs in the counter on the leading edge of the next clock pulse. The ENP and ENT are the enable inputs. These inputs must both be HIGH for the counter to sequence through its binary states. When at least one is LOW, the counter is disabled.

The ripple carry output RCO goes HIGH, when the counter reaches the last state in the sequence, 15 (1111). This output, in conjunction with the enable inputs, allows these counters to be cascaded for higher count sequences. The A and Q_A are the least significant input and output bits, respectively. For example, if the initial state is to be 1010, to begin with, the LOW-level pulse on the $\overline{\text{CLR}}$ input causes all the outputs (Q_A, Q_B, Q_C, and Q_D) to go LOW. Next the LOW-level pulse on the $\overline{\text{LOAD}}$ input enters the data (1010) on the data inputs (D, C, B, and A) into the counter. These data appear on the Q outputs at the time of the first positive-going clock edge after the $\overline{\text{LOAD}}$ input goes LOW. This is the preset operation, in which the output $Q_DQ_CQ_BQ_A$ is set to 1010_{10}.

The counter now advances through binary counts 11, 12, 13, 14, 15 on the next five positive-going clock edges. It then recycles to 0, 1, 2, 3, and 4... on the following clock pulses. Both ENP and ENT inputs are HIGH during the count sequence. If the count is to be terminated on let us say 6, the ENP is made to go LOW, the count is inhibited, and the counter remains in the binary 6 state.

The 74LS160A Synchronous Decade Counter

This device has the same inputs and outputs as the 74LS163A binary counter as shown in Fig. 10.56. It may be preset to any BCD count using the data inputs and a LOW on the

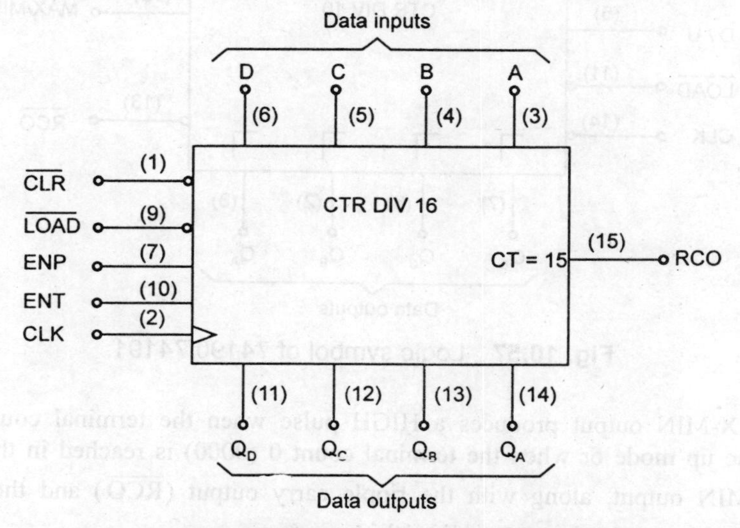

Fig. 10.56 Logic symbol of 74LS160A.

$\overline{\text{LOAD}}$ input. A LOW on the $\overline{\text{CLR}}$ will reset the counter. The count enable inputs ENP and ENT must both be HIGH for the counter to advance through its sequence of states in response to a positive transition on the CLK input. As in the 74LS163A chip, the enable inputs in conjunction with the carry-out (RCO: terminal count of 1001) provide for cascading several decade counters.

The pin diagram is the same for the following 74160–74163 series counters:

74160—Decade counter, asynchronous clear
74161—Binary counter, asynchronous clear
74162—Decade counter, synchronous clear
74163—Binary counter, synchronous clear

The 74190/74191 Series Up/Down Counters

The 74190 is a decade up/down counter and the 74191 is a 4-bit binary up/down counter. The pin diagram is the same for both.

Figure 10.57 shows the logic symbol for the 74190 IC decade up/down counter. The direction of the count is determined by the level of the up/down count ($D/\overline{U}$). When this input is HIGH, the counter counts down; and when it is LOW, the counter counts up. Also, this device can be preset to any desired BCD digit as determined by the states of the data inputs when the load input is LOW.

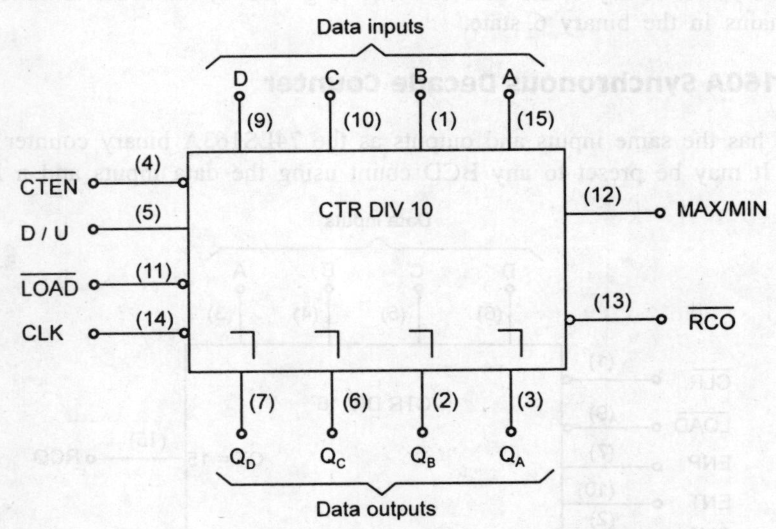

Fig. 10.57 Logic symbol of 74190/74191.

The MAX-MIN output produces a HIGH pulse when the terminal count 9 (1001) is reached in the up mode or when the terminal count 0 (0000) is reached in the down mode. This MAX-MIN output, along with the ripple carry output ($\overline{\text{RCO}}$) and the count enable input ($\overline{\text{CTEN}}$), is used when cascading the counters.

The 74ALS560A/74ALS561A Counters

The 74ALS560A/74ALS561A 4-bit synchronous counters are similar to the 74160–74163 series, but have considerably more versatility. The 74ALS560A is a decade counter and the 74ALS561A is a binary counter. Both have internal look-ahead carry, both are programmable, and both have enable and carry-out functions to facilitate cascading. In addition, both have 3-state outputs. Furthermore, these counters can be cleared either synchronously or asynchronously and can be loaded either synchronously or asynchronously. The pin diagram is shown in Fig. 10.58.

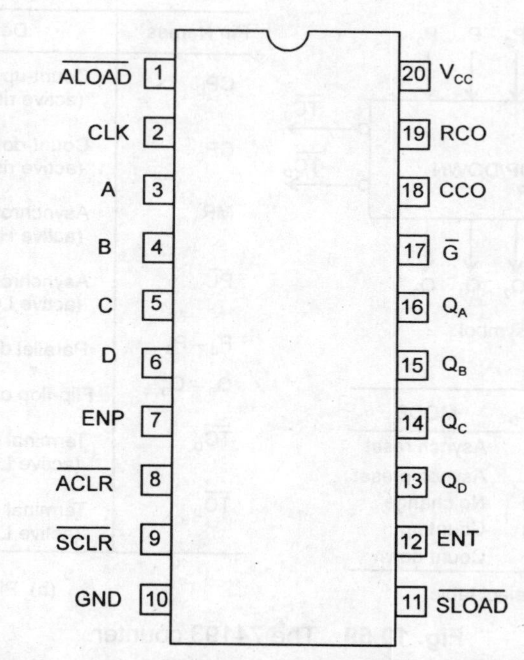

Fig. 10.58 Pin diagram of the 74ALS 560A/561A counter.

Programmable Counters

Most synchronous counters, available in IC form, can be pre-loaded with a binary number in parallel form (as with parallel-in shift registers) prior to initiation of counting. This pre-loading capability makes it possible to begin a count sequence from 0 or any other number. Such counters are said to be programmable. In a counter with asynchronous loading, the initial state is loaded using direct SET and direct CLEAR inputs on the FFs. Thus, loading occurs irrespective of the clock. In a counter with synchronous loading, the initial state is loaded on the occurrence of the clock edge, using the J-K inputs of the FFs. Synchronous loading generally requires the use of a rather elaborate logic circuitry connected to the J and K inputs, since these inputs are also needed for control of the normal counting sequence. In either case, the circuit will have a separate LOAD control input which must be made active to achieve loading. Some circuits can be loaded either synchronously or asynchronously and have separate control inputs such as SLOAD and ALOAD, to enable one type of loading or the other to take place.

10.11 THE 74193 (LS 193/HC 193) COUNTER

Figure 10.59 shows the logic symbol and the input/output description for the 74193 counter. This counter can be described as a mod-16 presettable up/down counter with synchronous counting, asynchronous preset and asynchronous master reset.

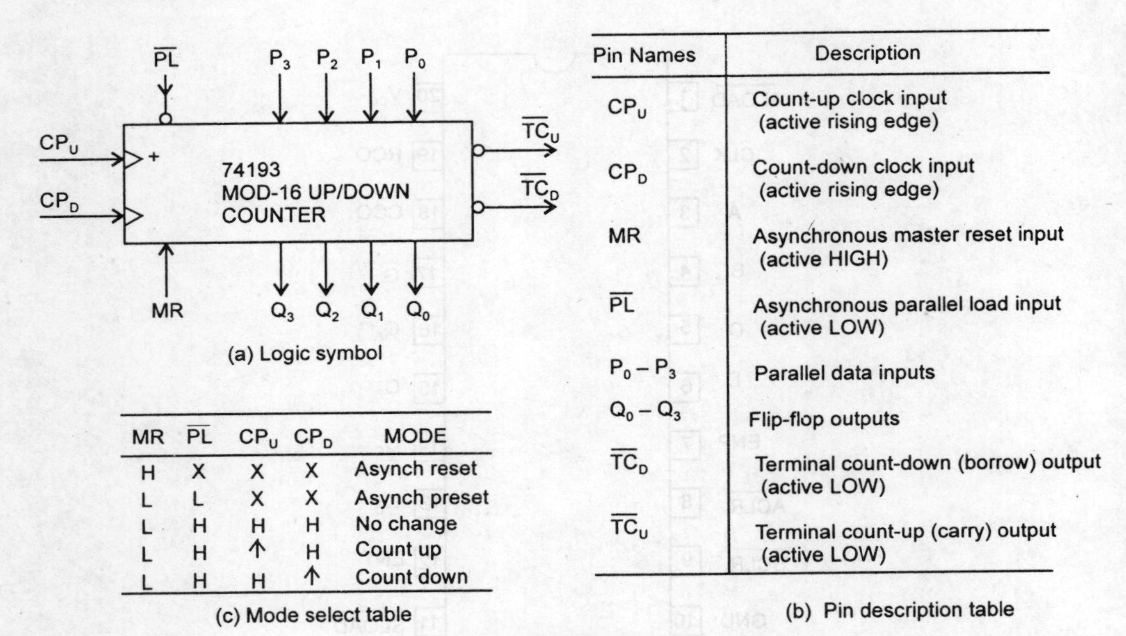

(a) Logic symbol

MR	$\overline{PL}$	CP_U	CP_D	MODE
H	X	X	X	Asynch reset
L	L	X	X	Asynch preset
L	H	H	H	No change
L	H	↑	H	Count up
L	H	H	↑	Count down

(c) Mode select table

Pin Names	Description
CP_U	Count-up clock input (active rising edge)
CP_D	Count-down clock input (active rising edge)
MR	Asynchronous master reset input (active HIGH)
$\overline{PL}$	Asynchronous parallel load input (active LOW)
$P_0 - P_3$	Parallel data inputs
$Q_0 - Q_3$	Flip-flop outputs
$\overline{TC}_D$	Terminal count-down (borrow) output (active LOW)
$\overline{TC}_U$	Terminal count-up (carry) output (active LOW)

(b) Pin description table

Fig. 10.59 The 74193 counter.

Clock Inputs CP_U and CP_D

The counter will respond to the positive-going transitions at one of two clock inputs. The CP_U is the count-up clock input. When pulses are applied to this input, the counter will increment (count up) on each PGT to a maximum count of 1111. Then it recycles to 0000 and starts all over. The CP_D is the count-down clock input. When pulses are applied to this input, the counter will decrement (count down) on each PGT to a minimum count of 0000; then it recycles to 1111 and starts all over. Thus, one clock input or the other will be used for counting while the other clock input is inactive (kept HIGH).

Master Reset (MR)

The master reset is an active-HIGH asynchronous input that resets the counter to the 0000 state. The MR is a DC RESET, and so, it will hold the counter at 0000 as long as MR = 1. It also overrides all other inputs.

Preset Inputs

The counter FFs can be preset to the logic levels present on the parallel data inputs P_3–P_0 by momentarily pulsing the parallel input $\overline{PL}$ from HIGH to LOW. This is an asynchronous preset that overrides the counting operation. The $\overline{PL}$ will have no effect, however, if the MR input is in its active-HIGH state.

Count Outputs

The current count is always present at the FF outputs Q_3–Q_0, where Q_0 is the LSB and Q_3 the MSB.

Terminal Count Outputs

The terminal count outputs are used when two or more 74193s are connected as a multi-stage counter to produce a larger mod number. In the count-up mode, the $\overline{TC}_U$ output of the lower-order counter is connected to the CP_U input of the next higher-order counter. In the count-down mode, the $\overline{TC}_D$ output of the lower-order counter is connected to the CP_D input of the next higher-order counter.

The $\overline{TC}_U$ is generated on the 74193 chip using the logic shown in Fig. 10.60a. Clearly, $\overline{TC}_U$ will be LOW only when the counter is in the 1111 state and CP_U is LOW. Thus, $\overline{TC}_U$ will remain HIGH as the counter counts up from 0000 to 1110. On the next PGT of CP_U, the count goes to 1111, but $\overline{TC}_U$ does not go LOW until CP_U returns LOW. The next PGT at CP_U recycles the count to 0000 and also causes $\overline{TC}_U$ to return HIGH. This PGT at CP_U occurs when the counter recycles from 1111 to 0000, and can be used to clock a second 74193 up-counter to its next higher count.

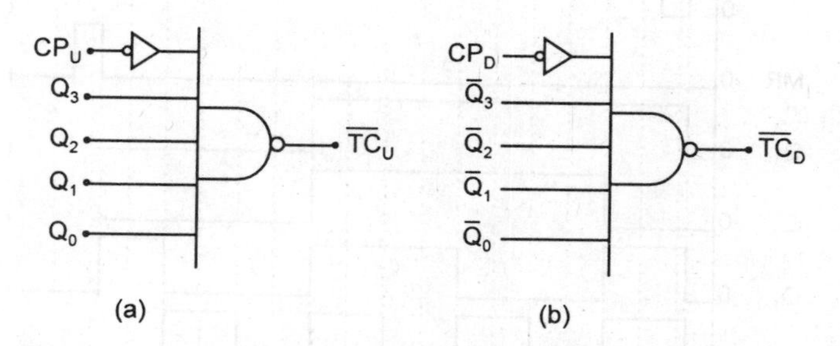

Fig. 10.60 Logic for generating $\overline{TC}_U$ and $\overline{TC}_D$.

The $\overline{TC}_D$ is generated as shown in Fig. 10.60b. It is normally HIGH and does not go LOW until the counter has counted down to the 0000 state and CP_D is LOW. When the next PGT at CP_D recycles the counter to 1111, it causes $\overline{TC}_D$ to return HIGH. This PGT at $\overline{TC}_D$ can be used to clock a second 74193 down-counter to its next lower count.

EXAMPLE 10.13 Figure 10.61a shows a 74193 wired as an up-counter. The parallel data inputs are permanently connected as 1011, and the CP_U, $\overline{PL}$, and MR input waveforms are shown in Fig. 10.61b. Assuming that the counter is initially in the 0000 state, determine the counter output waveforms.

Solution

Initially (at t_0), the counter FFs are LOW. This causes $\overline{TC}_U$ to be HIGH. Just prior to time t_1, the input is pulsed LOW. This immediately loads the counter with 1011 to produce $Q_3 = 1$, $Q_2 = 0$, $Q_1 = 1$, and $Q_0 = 1$. At time t_1, the CP_U makes a PGT, but the counter cannot respond to this because $\overline{PL}$ is still active at that time. At times t_2, t_3, t_4, and t_5 the counter

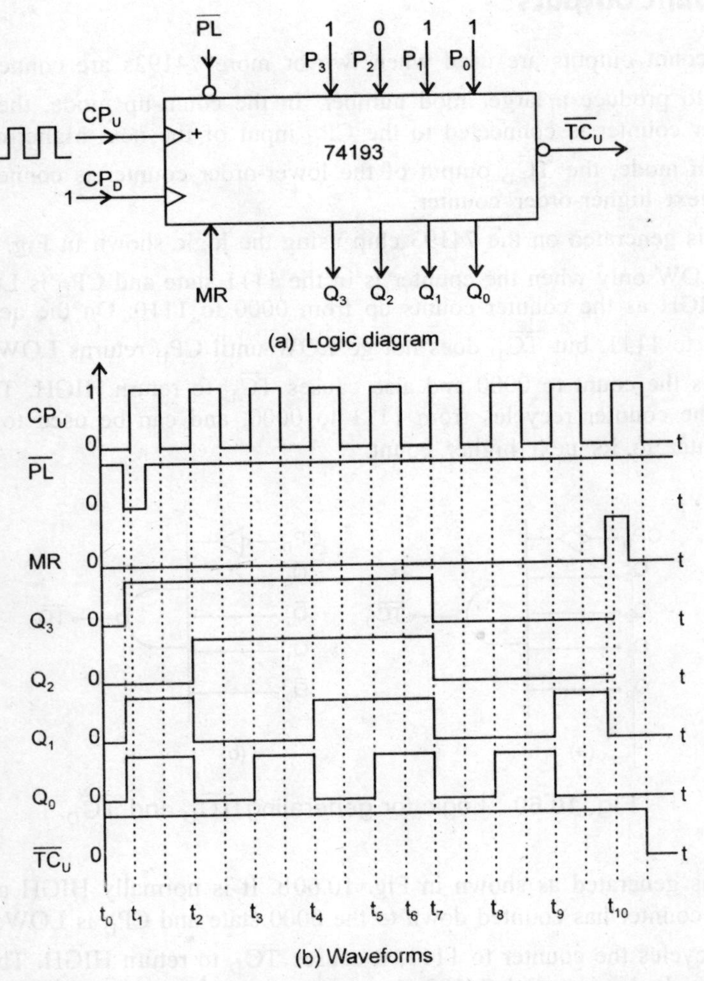

Fig. 10.61 The 74193 wired as an up-counter (Example 10.13).

counts up on each PGT at CP_U. After the PGT at t_5, the counter is in the 1111 state, but $\overline{TC}_U$ does not go LOW at t_5. When the next PGT occurs at t_7, the counter recycles to 0000 and $\overline{TC}_U$ returns HIGH.

The counter will count up in response to the PGTs at t_8 and t_9. For the PGT at t_{10}, $\overline{TC}_U$ has no effect because the MR goes HIGH prior to t_{10} and remains active at t_{10}. This will reset all FFs to 0 and override the CP_U signal.

EXAMPLE 10.14 Figure 10.62a shows the 74193 wired as a down-counter. The parallel data inputs are permanently wired as 0111, and the CP_D and $\overline{PL}$ waveforms are shown in Fig. 10.62b. Assuming that the counter is initially in the 0000 state, determine the output waveforms.

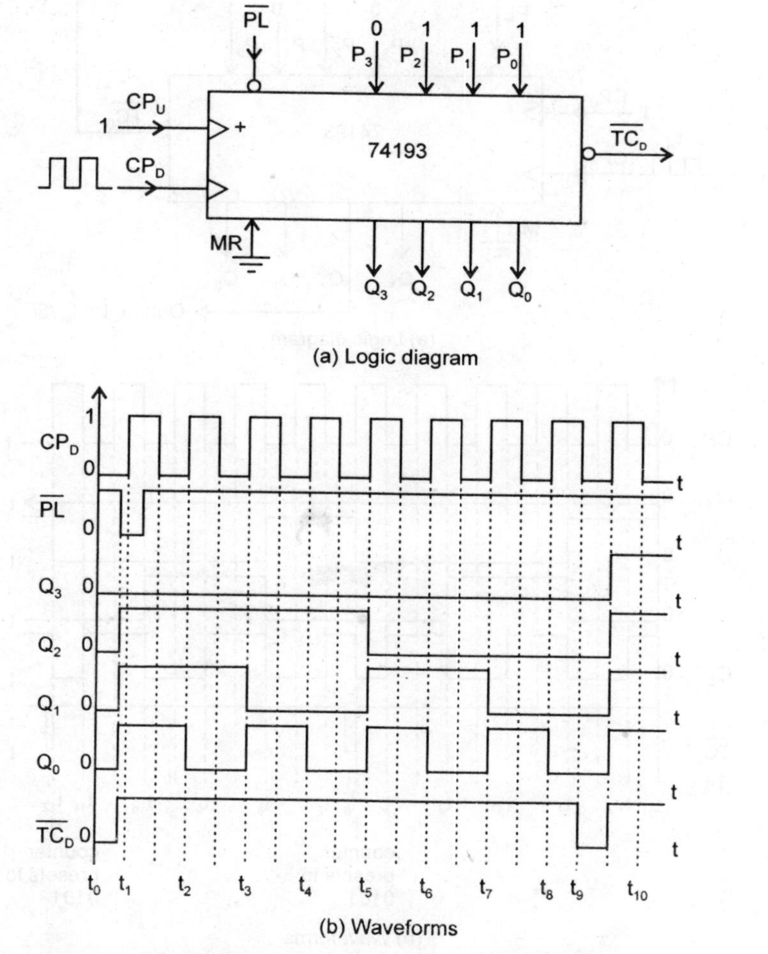

Fig. 10.62 The 74193 as a down-counter (Example 10.14).

Solution

At t_0, all the FF outputs are LOW and CP_D is LOW. These are the conditions that produce $\overline{TC}_D = 0$. Prior to t_1, the $\overline{PL}$ input is pulsed LOW. This immediately presets the counter to 0111 and therefore causes $\overline{TC}_D$ to go HIGH. The PGT of CP_D at t_1 will have no effect, since $\overline{PL}$ is still active. The counter will respond to the PGTs at t_2 to t_8 and counts down to 0000 at t_8. The $\overline{TC}_D$ does not go LOW until t_9 when CP_D goes LOW. At t_{10}, the PGT of CP_D causes the counter to recycle to 1111 and also drives $\overline{TC}_D$ back to HIGH.

Variable Mod Number Using the 74193 Counter

Presettable counters can be easily designed for different mod numbers without the need for

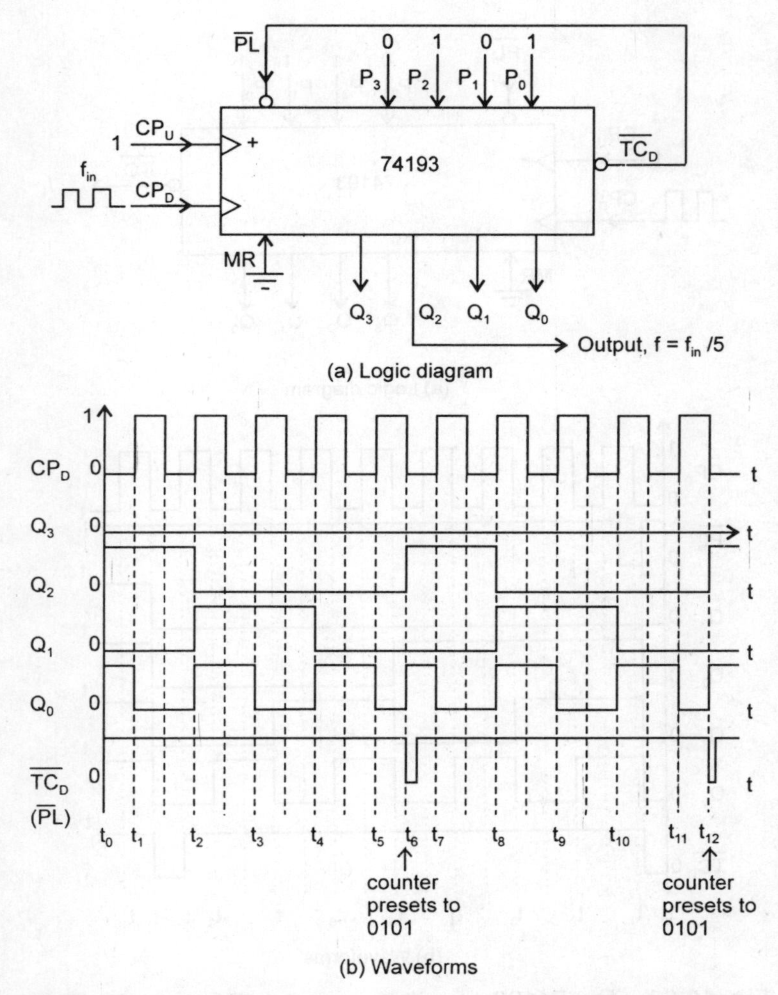

Fig. 10.63 The 74193 wired as a mod-5 counter and its waveforms.

additional logic circuitry. We will demonstrate this for the 74193 using the circuit of Fig. 10.63a. Here the 74193 is used as a down-counter with its parallel load inputs permanently connected at 0101 (5_{10}). Note that the $\overline{TC}_D$ output is connected back to the $\overline{PL}$ input.

We begin our analysis by assuming that the counter is counting down and is in the 0101 state at time t_0. Refer to Fig. 10.63b for the counter waveforms.

The counter will decrement (count down) on the PGTs of CP_D at t_1 to t_5. At t_5, the counter is in the 0000 state. When CP_D goes LOW at t_6, it drives $\overline{TC}_D$ LOW. This immediately activates the $\overline{PL}$ input and presets the counter back to the 0101 state. Note that $\overline{TC}_D$ stays LOW only for a short interval because, once the counter outputs go to 0101 in response to $\overline{PL} = 0$, the condition needed to keep $\overline{TC}_D = 0$ is removed. Thus, there is only a narrow glitch at $\overline{TC}_D$.

The same sequence is repeated at t_7 to t_{12} and at equal intervals thereafter. If we examine the Q_2 waveform, we can see that it goes through one complete cycle for every five cycles of CP_D. For example, there are five clock cycles between the PGT of Q_2 at t_6 and the PGT of Q_2 at t_{12}. Thus, the frequency of the Q_2 waveform is 1/5th of the clock frequency.

The frequency division ratio (5) is the same as the number applied to the parallel data inputs (0101 = 5). In fact, we can vary the frequency division ratio by changing the logic levels applied to the parallel data inputs.

A variable-frequency-divider circuit can be easily implemented by connecting switches to the parallel data inputs of the circuit in Fig. 10.63. The switches can be set to a value equal to the desired frequency-division ratio.

The 74192 Counter

The 74192 is an up/down decade counter. Like the 74193, it is also a dual clock up/down counter with asynchronous parallel load, asynchronous overriding master reset, and internal terminal count logic which allows the counters to be cascaded without any additional logic. It can be reset, preset, and can count up or down. Also, like the 74193, it has four M-S flip-flops plus steering, terminal count decoding, and preset logic. The operation of 74192 is very much the same as that of 74193. The terminal count-up output is LOW while the up-clock input is LOW and the counter is in its highest state 9. Similarly, the terminal count-down output is LOW while the down-clock input is LOW and the counter is in state 0. The logic equations for terminal counter are:

$$\overline{TC}_U = Q_0 \cdot \overline{Q}_1 \cdot \overline{Q}_2 \cdot Q_3 \cdot \overline{CP}_U; \quad \overline{TC}_D = \overline{Q}_0 \cdot \overline{Q}_1 \cdot \overline{Q}_2 \cdot \overline{Q}_3 \cdot \overline{CP}_D$$

10.12 SHIFT REGISTER COUNTERS

One of the applications of shift registers is that they can be arranged to form several types of counters. Shift register counters are obtained from serial-in, serial-out shift registers by providing feedback from the output of the last FF to the input of the first FF. These devices are called counters because they exhibit a specified sequence of states. The most widely used shift register counter is the ring counter (also called the basic ring counter or the simple ring counter) as well as the twisted ring counter (also called the Johnson counter or the switch-tail ring counter).

Ring Counter

This is the simplest shift register counter. The basic ring counter using D FFs is shown in Fig. 10.64. The realization of this counter using J-K FFs is shown in Fig. 10.65. Its state diagram and the sequence table are shown in Fig. 10.66. Its timing diagram is shown in Fig. 10.67. The FFs are arranged as in a normal shift register, i.e. the Q output of each stage is connected to the D input of the next stage, but the Q output of the last FF is connected back to the D input of the first FF such that the array of FFs is arranged in a ring and, therefore, the name *ring counter*.

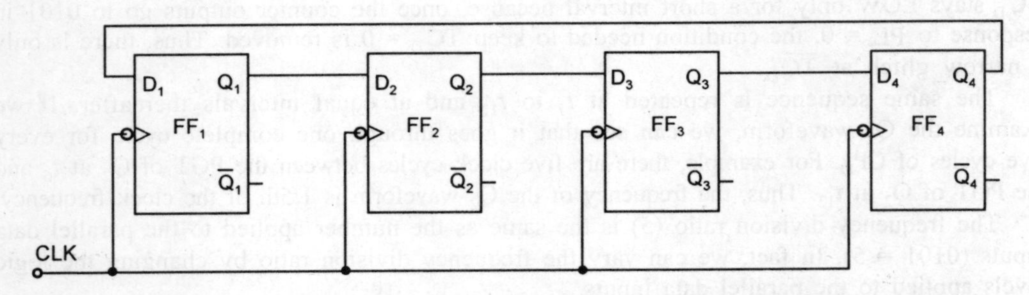

Fig. 10.64 Logic diagram of a 4-bit ring counter using D flip-flips.

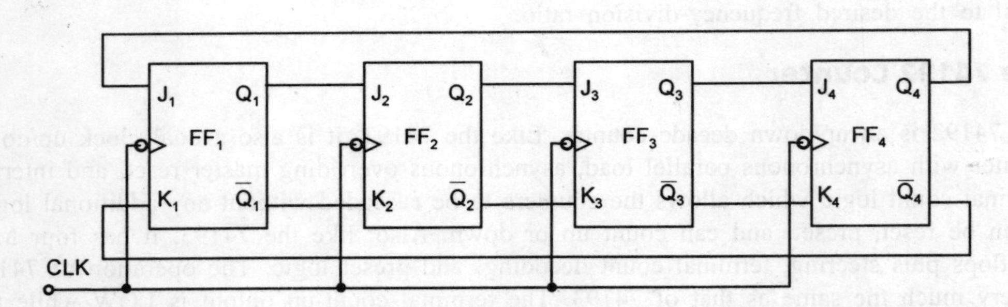

Fig. 10.65 Logic diagram of a 4-bit ring counter using J-K flip-flops.

Q_1	Q_2	Q_3	Q_4	After clock pulse
1	0	0	0	0
0	1	0	0	1
0	0	1	0	2
0	0	0	1	3
1	0	0	0	4
0	1	0	0	5
0	0	1	0	6
0	0	0	1	7

(a) Sequence table

1000
0100
0010
0001

(b) State diagram

Fig. 10.66 Sequence table and state diagram of a 4-bit ring counter.

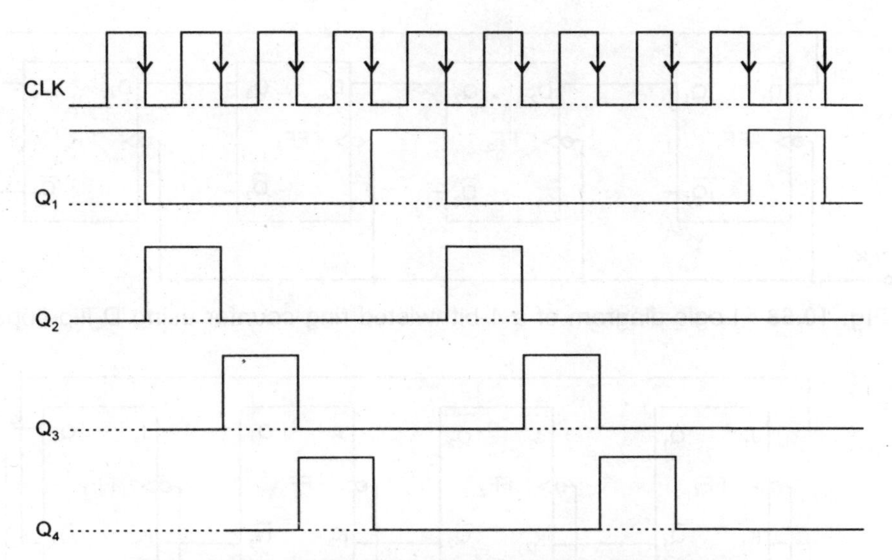

Fig. 10.67 Timing diagram of a 4-bit ring counter.

In most instances, only a single 1 is in the register and is made to circulate around the register as long as clock pulses are applied. Initially, the first FF is preset to a 1. So, the initial state is 1000, i.e. $Q_1 = 1$, $Q_2 = 0$, $Q_3 = 0$, and $Q_4 = 0$. After each clock pulse, the contents of the register are shifted to the right by one bit and Q_4 is shifted to Q_1. The sequence repeats after four clock pulses. The number of distinct states in the ring counter, i.e. the mod of the ring counter is equal to the number of FFs used in the counter. An n-bit ring counter can count only n bits whereas an n-bit ripple counter can count 2^n bits. So, the ring counter is uneconomical compared to a ripple counter, but has the advantage of requiring no decoder, since we can read the count by simply noting which FF is set. Since it is entirely a synchronous operation and requires no gates external to FFs, it has the further advantage of being very fast.

Twisted Ring Counter (Johnson Counter)

This counter is obtained from a serial-in, serial-out shift register by providing a feedback from the inverted output of the last FF to the D input of the first FF. The Q output of each stage is connected to the D input of next stage but the $\overline{Q}$ output of the last stage is connected to the D input of first stage, therefore, the name *twisted ring counter*. This feedback arrangement produces a unique sequence of states.

The logic diagram of a 4-bit Johnson counter using D FFs is shown in Fig. 10.68. The realization of the same using J-K FFs is shown in Fig. 10.69. The state diagram and the sequence table are shown in Fig. 10.70. The timing diagram of a Johnson counter is shown in Fig. 10.71.

Let initially all the FFs be reset, i.e. the state of the counter be 0000. After each clock pulse, the level of Q_1 is shifted to Q_2, the level of Q_2 to Q_3, Q_3 to Q_4 and the level of $\overline{Q}_4$ to Q_1 and the sequence given in Fig. 10.70a is obtained. This sequence is repeated after every eight clock pulses.

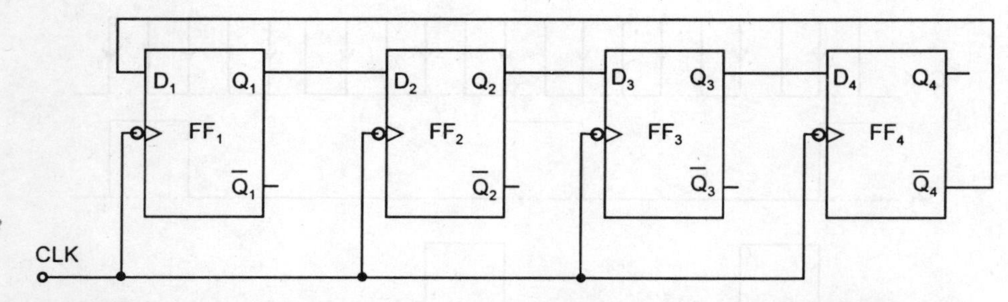

Fig. 10.68 Logic diagram of a 4-bit twisted ring counter using D flip-flops.

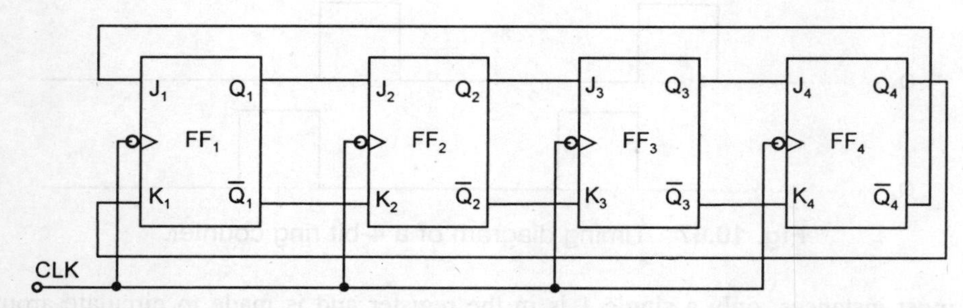

Fig. 10.69 Logic diagram of a 4-bit twisted ring counter using J-K flip-flops.

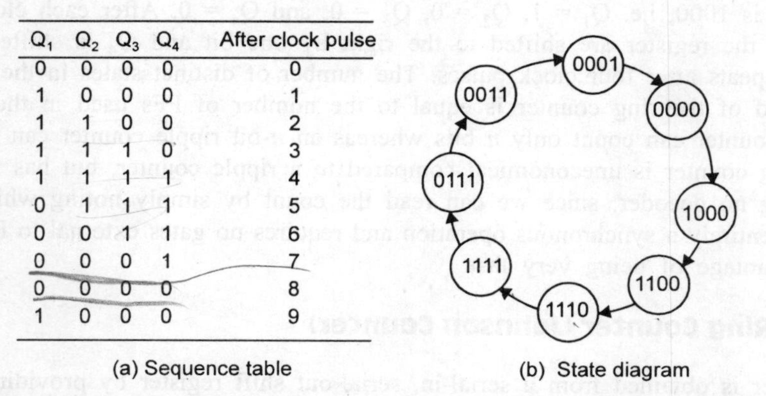

Q_1	Q_2	Q_3	Q_4	After clock pulse
0	0	0	0	0
1	0	0	0	1
1	1	0	0	2
1	1	1	0	3
1	1	1	1	4
0	1	1	1	5
0	0	1	1	6
0	0	0	1	7
0	0	0	0	8
1	0	0	0	9

(a) Sequence table (b) State diagram

Fig. 10.70 Sequence table and state diagram of a twisted ring counter.

An *n* FF Johnson counter can have 2*n* unique states and can count up to 2*n* pulses. So, it is a mod-2*n* counter. It is more economical than the normal ring counter, but less economical than the ripple counter. It requires two input gates for decoding regardless of the size of the counter. Thus, it requires more decoding circuitry than that by the normal ring counter, but less than that by the ripple counter. It represents a middle ground between the ring counter and the ripple counter.

Both types of ring counters suffer from the problem of lock-out, i.e. if the counter finds itself in an unused state, it will persist in moving from one unused state to another and will never find its way to a used state. This difficulty can be corrected by adding a gate. With this addition, if the counter finds itself initially in an unused state, then after a number of

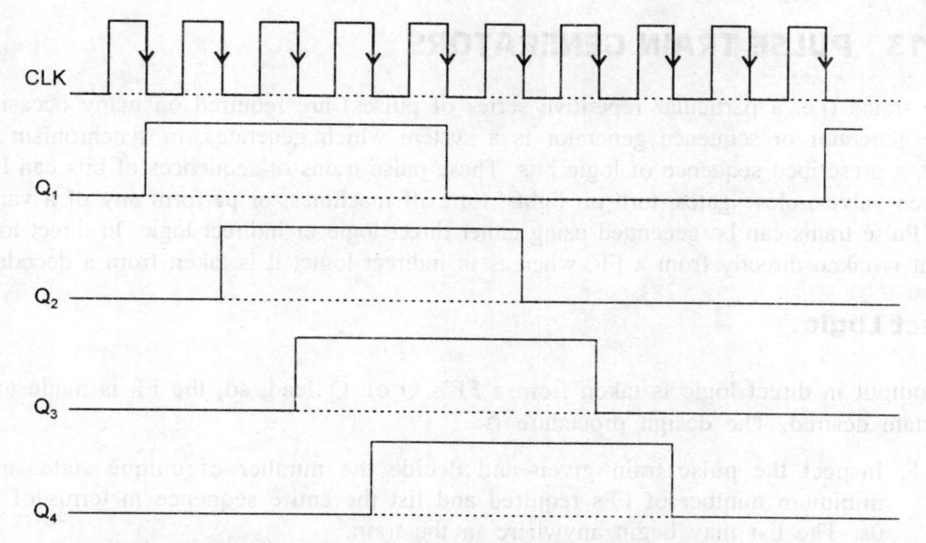

Fig. 10.71 Timing diagram of a 4-bit twisted ring counter.

clock pulses, depending on the state, the counter will find its way to a used state and thereafter, follow the desired sequence. A Johnson counter designed to prevent lock-out is shown in Fig. 10.72. A self-starting ring counter (whatever may be the initial state, a single 1 will eventually circulate) is shown in Fig. 10.73.

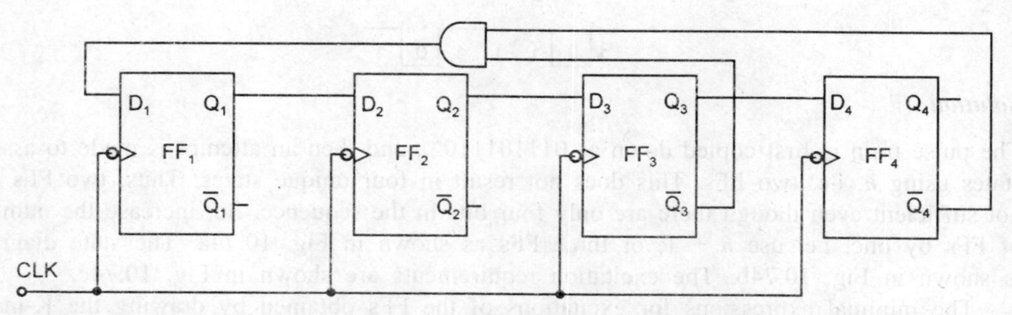

Fig. 10.72 Logic diagram of a 4-bit Johnson counter designed to prevent lock-out.

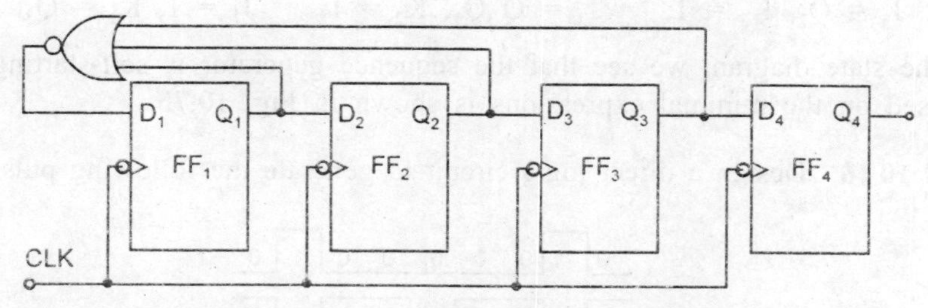

Fig. 10.73 Logic diagram of a self-starting ring counter.

10.13 PULSE TRAIN GENERATORS

Pulse trains (i.e. a particular repetitive series of pulses) are required on many occasions. A pulse generator or sequence generator is a system which generates, in synchronism with a clock, a prescribed sequence of logic bits. These pulse trains or sequences of bits can be used to open valves, close gates, turn on lights, turn off machines, or perform any of a variety of jobs. Pulse trains can be generated using either direct logic or indirect logic. In direct logic the output is taken directly from a FF, whereas in indirect logic, it is taken from a decoder gate.

Direct Logic

The output in direct logic is taken from a FF's Q or $\overline{Q}$ lead; so, the FF is made to go to the state desired. The design procedure is:

1. Inspect the pulse train given and decide the number of unique states and the minimum number of FFs required and list the entire sequence in terms of 1s and 0s. The list may begin anywhere in the train.
2. Taking that these 1s and 0s will form the least significant bits of the state assignment, assign unique states to other FFs. If unique states are not possible with the least number of FFs n, such that the number of states $N \leq 2^n$, increase the number of FFs by one or more to get the unique states. (This is the disadvantage of direct logic.)
3. Design the counter as usual. The output is at the Q or $\overline{Q}$ lead of the LSB FF.

EXAMPLE 10.15 Design a pulse train generator for the waveform shown below.

1 0 1 1 1 0 1

Solution

The pulse train is first copied down as 011101110..., and then an attempt is made to assign states using n, i.e. two FFs. This does not result in four unique states. Thus, two FFs are not sufficient even though there are only four bits in the sequence. So, increase the number of FFs by one, i.e. use $n + 1$, or three FFs as shown in Fig. 10.74a. The state diagram is shown in Fig. 10.74b. The excitation requirements are shown in Fig. 10.74c.

The minimal expressions for excitations of the FFs obtained by drawing the K-maps based on the excitation table are:

$$J_3 = Q_2,\ K_3 = 1; \quad J_2 = \overline{Q}_3 Q_1,\ K_2 = 1; \quad J_1 = 1,\ K_1 = Q_3.$$

From the state diagram we see that the sequence generator is self-starting. The logic diagram based on the minimal expressions is shown in Fig. 10.75.

EXAMPLE 10.16 Design a direct logic circuit to generate the following pulse trains.

(1) 0 1 0 0 0 0 0 1 0

(2) 1 0 1 1 0 1 1 0 1

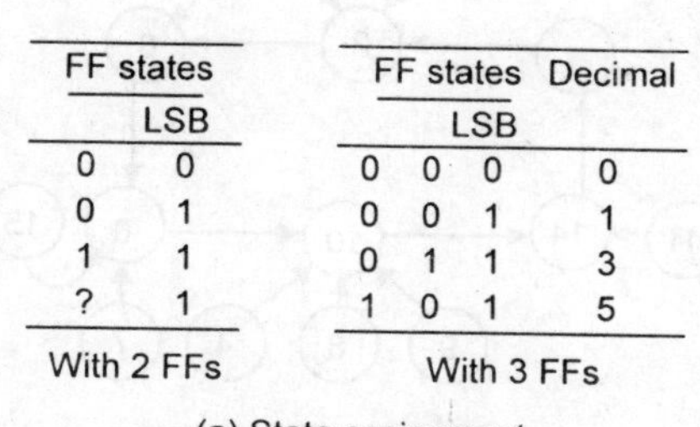

FF states	
	LSB
0	0
0	1
1	1
?	1

With 2 FFs

FF states			Decimal
		LSB	
0	0	0	0
0	0	1	1
0	1	1	3
1	0	1	5

With 3 FFs

(a) State assignment

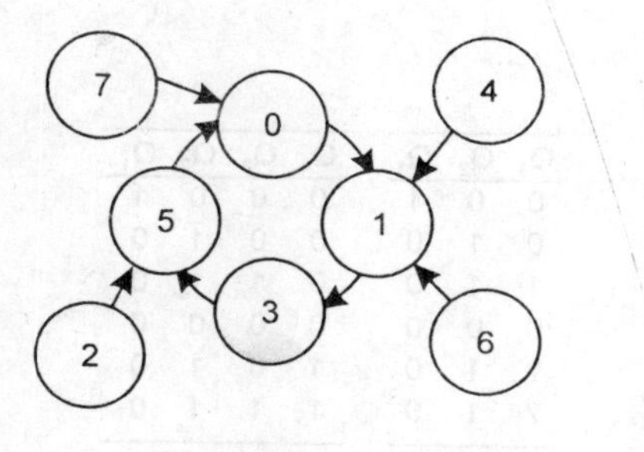

(b) State diagram

PS			NS			Required excitation					
Q_3	Q_2	Q_1	Q_3	Q_2	Q_1	J_3	K_3	J_2	K_2	J_1	K_1
0	0	0	0	0	1	0	X	0	X	1	X
0	0	1	0	1	1	0	X	1	X	X	0
0	1	1	1	0	1	1	X	X	1	X	0
1	0	1	0	0	0	X	1	0	X	X	1

(c) Excitation table

Fig. 10.74 Pulse train generator (Example 10.15).

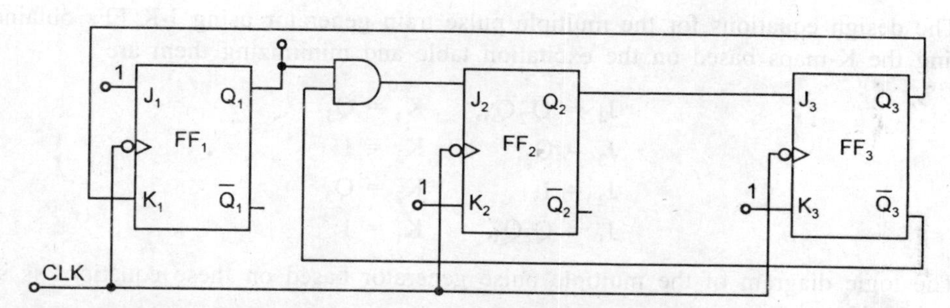

Fig. 10.75 Logic diagram of the pulse train generator (Example 10.15).

Solution

The period of each pulse train is six time frames. The sequences are therefore 6-bits long. As there are 6 states, the minimum number of FFs required is three. Try to assign the states assuming that pulse train (1) is to be available at Q output of FF_1 and pulse train (2) at the Q output of FF_2.

It is not possible to assign the states using only three FFs as seen from the state assignment table. So, try with four FFs. It is now possible to assign the states. The state assignment and the state diagram are shown in Fig. 10.76a and b, respectively. The excitation requirements are shown in Fig. 10.76c.

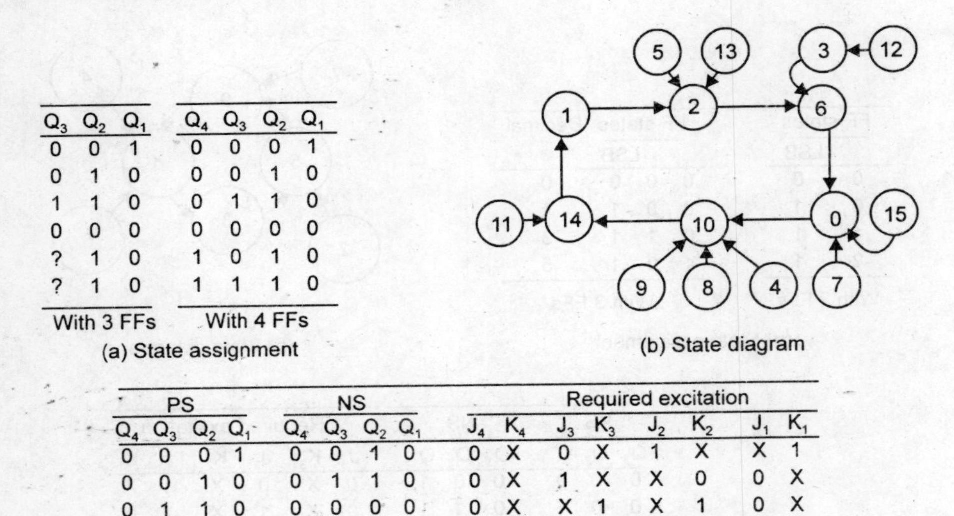

Q_3	Q_2	Q_1	Q_4	Q_3	Q_2	Q_1
0	0	1	0	0	0	1
0	1	0	0	0	1	0
1	1	0	0	1	1	0
0	0	0	0	0	0	0
?	1	0	1	0	1	0
?	1	0	1	1	1	0
With 3 FFs			With 4 FFs			

(a) State assignment

(b) State diagram

PS				NS				Required excitation							
Q_4	Q_3	Q_2	Q_1	Q_4	Q_3	Q_2	Q_1	J_4	K_4	J_3	K_3	J_2	K_2	J_1	K_1
0	0	0	1	0	0	1	0	0	X	0	X	1	X	X	1
0	0	1	0	0	1	1	0	0	X	1	X	X	0	0	X
0	1	1	0	0	0	0	0	0	X	X	1	X	1	0	X
0	0	0	0	1	0	1	0	1	X	0	X	1	X	0	X
1	0	1	0	1	1	1	0	X	0	1	X	X	0	0	X
1	1	1	0	0	0	0	0	X	1	X	1	X	1	1	X

(c) Excitation table

Fig. 10.76 Multiple pulse train generator (Example 10.16).

The design equations for the multiple pulse train generator using J-K FFs obtained by drawing the K-maps based on the excitation table and minimizing them are

$$J_4 = \overline{Q}_2\overline{Q}_1, \quad K_4 = Q_3$$
$$J_3 = Q_2, \quad K_3 = 1$$
$$J_2 = 1, \quad K_2 = Q_3$$
$$J_1 = Q_4Q_3, \quad K_1 = 1$$

The logic diagram of the multiple pulse generator based on these equations is shown in Fig. 10.77.

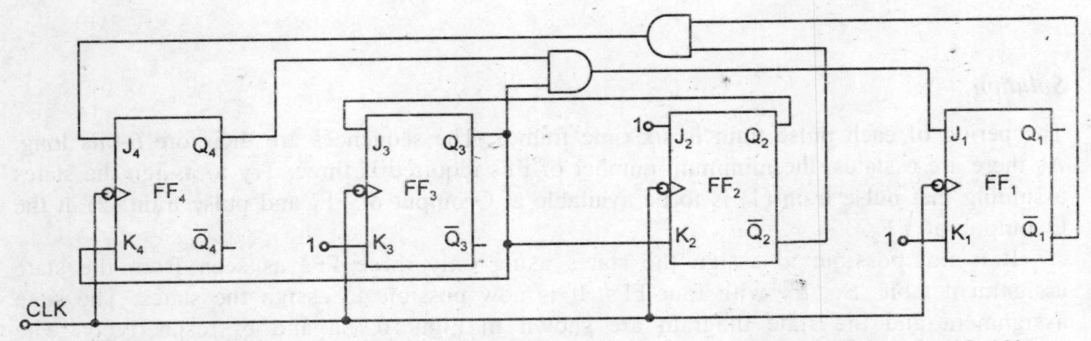

Fig. 10.77 Logic diagram of the multiple pulse train generator (Example 10.16).

Indirect Logic

The block diagram of indirect logic is shown in Fig. 10.78. Generation of pulse trains using indirect logic, has the advantage that any counter (ripple or synchronous) with the correct number of states can form the generator. It is always assured that the number of FFs required, n, is such that $N \leq 2^n$. The gates can be used to detect the proper states when a 1 must be outputted. The same approach can be used for multiple outputs and multiple-output minimization can be used to reduce the logic.

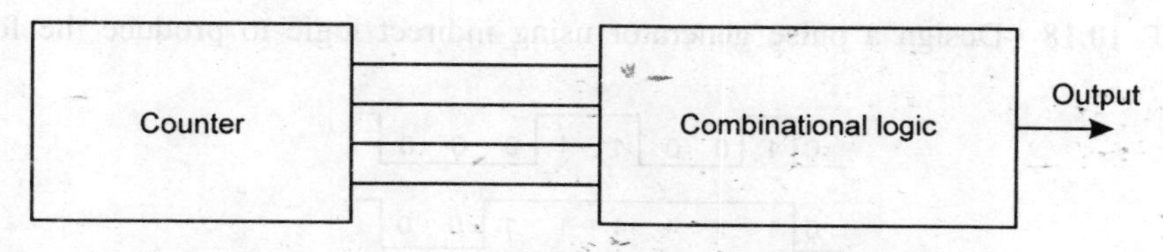

Fig. 10.78 Block diagram of indirect logic.

EXAMPLE 10.17 Generate the following pulse train using indirect logic.

1 0 1 1 0 1

Solution

The given sequence is 10110. We need five states to generate the above pulse train. So, any mod-5 counter can be used. For simplicity, we use a ripple counter. It goes through states 0, 1, 2, 3, 4, and 0..., etc. The truth table and the K-map for output f are shown in Fig. 10.79. The logic diagram based on that value of f is shown in Fig. 10.80. While at state 0, it outputs a 1, i.e. the first bit of the sequence. While at state 1, it outputs a 0, i.e. the second bit of the sequence, and so on. States 5, 6, 7 are invalid, so the corresponding outputs are don't cares.

Q_3	Q_2	Q_1	Output (f)
0	0	0	1
0	0	1	0
0	1	0	1
0	1	1	1
1	0	0	0
1	0	1	X
1	1	0	X
1	1	1	X

(a) Truth table

Q_3 \ Q_2Q_1	00	01	11	10
0	1		1	1
1			X	X

Output, $f = Q_2 + \overline{Q}_3\overline{Q}_1$

(b) K-map

Fig. 10.79 Pulse train generator (Example 10.17).

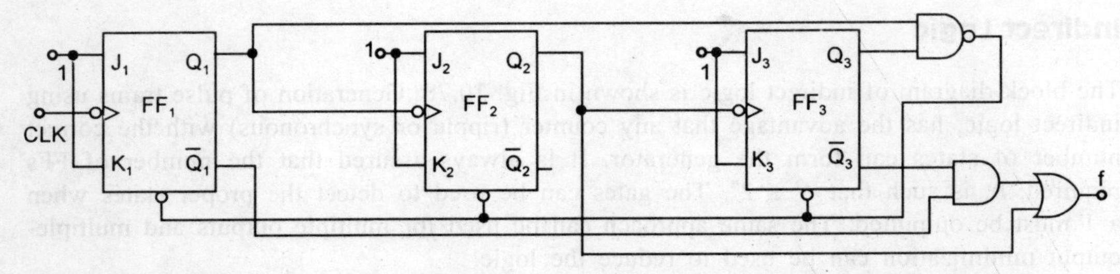

Fig. 10.80 Logic diagram of the pulse train generator (Example 10.17).

EXAMPLE 10.18 Design a pulse generator using indirect logic to produce the following waveforms.

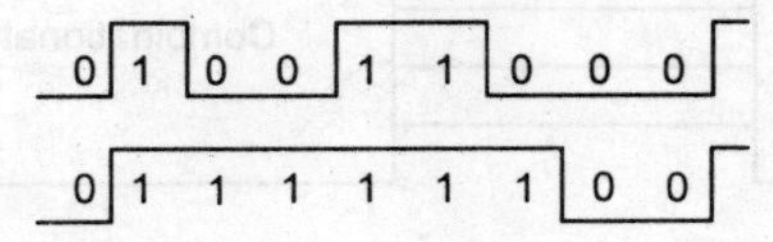

Solution

The pulse trains to be generated are: (a) 10011000 and (b) 1111100. These are both eight bits long and, therefore, a mod-8, i.e. a 3-bit ripple counter can be used. Let f_1 and f_2 be the outputs of the combinational circuits. The state assignment is shown in the truth table. The truth table and the K-maps for outputs f_1 and f_2 are shown in Fig. 10.81. The logic diagram based on the minimal expressions for f_1 and f_2 obtained from K-maps is shown in Fig. 10.82.

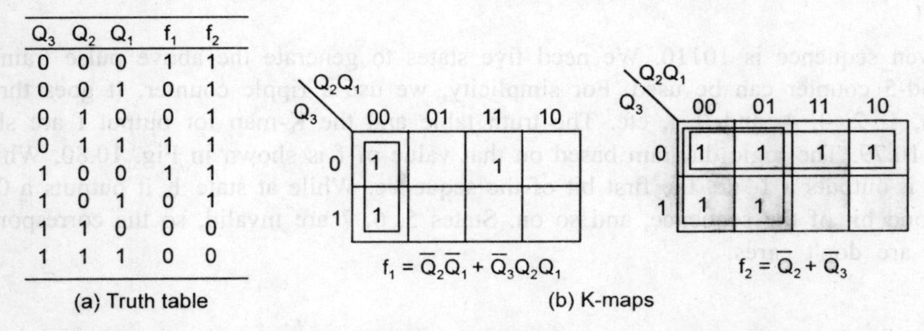

Q_3	Q_2	Q_1	f_1	f_2
0	0	0	1	1
0	0	1	0	1
0	1	0	0	1
0	1	1	1	1
1	0	0	1	1
1	0	1	0	1
1	1	0	0	0
1	1	1	0	0

(a) Truth table

(b) K-maps

Fig. 10.81 Truth table and K-maps for f_1 and f_2 (Example 10.18).

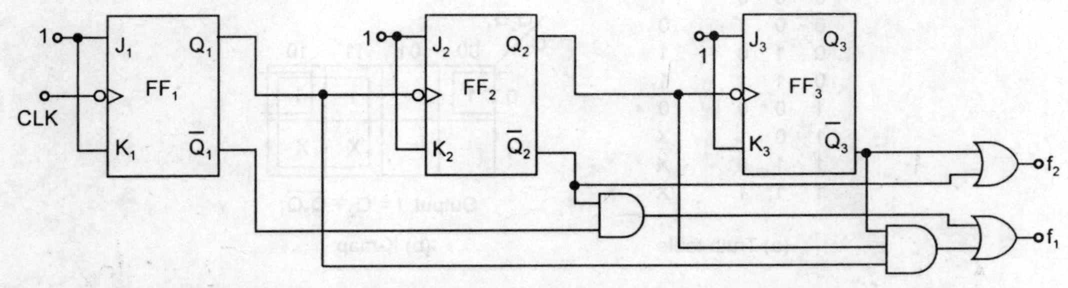

Fig. 10.82 Logic diagram of the pulse train generator (Example 10.18).

10.14 PULSE GENERATORS USING SHIFT REGISTERS

Shift registers can be used to generate single pulse trains. In general, they cannot be used for multiple pulse trains without being derailed. Shift registers can be adopted to serve as sequence generators. The length of a sequence is the number of bits in the sequence before the pattern of bits repeats itself. The sequence length is analogous to the period of a periodic analog waveform. In many applications, the length of the sequence is more important than the exact pattern of the repeated bits. Since the generator repeats the pattern over and over again, the sequence can be read in a number of ways depending on the starting point. If the sequence available at the complementary outputs of the FFs is read, a different pattern may be obtained.

A shift register is quite restrictive in the sense that it cannot go from any one state to any other state of our choice. For example, it cannot go from 0110 to 0101 directly; it can only go to 0011 or 1100 upon receipt of the next clock pulse. So, the pulse train is first examined to see if it can be generated by shifting.

The basic structure of a sequence generator is shown in Fig. 10.83. Here n FFs are cascaded in the usual manner of a shift register. Type D FFs have been used and the clock

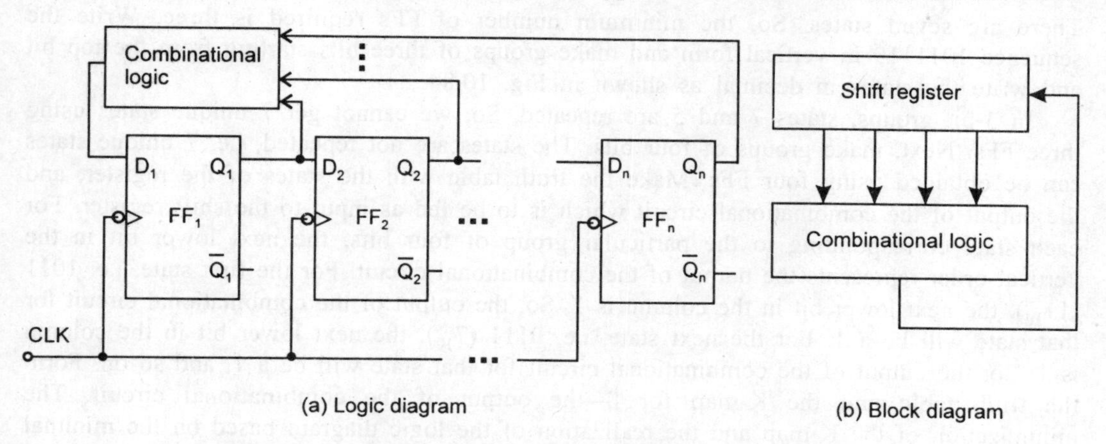

Fig. 10.83 Basic structure of a sequence generator.

signal is applied to all FFs. It is similar to a shift register counter but the difference is that in a ring counter, only the output of the last FF is given as data input to the first FF, whereas in a sequence generator, the logic level of D_1 is determined not only by Q_n but also by the output of other FFs in the cascade. That is,

$$D_1 = f(Q_1, Q_2, \ldots, Q_n)$$

The prescribed sequence appears at the output of each of the FFs. Of course, as we go from FF_1 to FF_2, etc. we find successive delays by one clock interval. In order to build a sequence generator capable of generating a sequence of length S, it is necessary to use at least n FFs, where n satisfies the condition $S \leq 2^n - 1$. If the order of 1s and 0s in the sequence is prescribed, generally it will not be possible to generate a length S in which the minimum number of FFs is used. On the other hand, for any n, there is at least one sequence for which the sequence length S is a maximum, i.e. $2^n - 1$. The design procedure is as follows:

From the information of the waveform (or the sequence of bits, if given), decide the number of FFs required for pulse generation. The minimum number of FFs required for N states is n such that $N \le 2^n$. Convert the waveform to 1s and 0s. Write the bits in vertical order. Form groups of n bits starting from the top bit and convert them to state numbers (decimal) moving down one bit at a time. The procedure is repeated until the cycle is completed. If the entire period of the train can be examined without repeating a state, the job can be done by n FFs. If there is a repetition of states, try $(n + 1)$ FFs. If repeated states still occur, try $(n + 2)$ FFs, and so on.

EXAMPLE 10.19 Design a pulse train generator using a shift register to generate the following waveform.

0 1 0 1 1 1 1 0

Solution

There are seven states. So, the minimum number of FFs required is three. Write the sequence 1011110 in vertical form and make groups of three bits starting from the top bit and write the states in decimal as shown in Fig. 10.84.

In 3-bit groups, states 7 and 5 are repeated. So, we cannot get 7 unique states using three FFs. Next, make groups of four bits. The states are not repeated, i.e. 7 unique states can be obtained using four FFs. Make the truth table with the states of the register, and the output of the combinational circuit which is to be fed as input to the shift register. For each state corresponding to the particular group of four bits, the next lower bit in the vertical order represents the output of the combinational circuit. For the first state, i.e. 1011 (11_{10}), the next lower bit in the column is 1. So, the output of the combinational circuit for that state will be a 1. For the next state, i.e. 0111 (7_{10}), the next lower bit in the column is 1. So, the output of the combinational circuit for that state will be a 1, and so on. Form the truth table and the K-map for f—the output of the combinational circuit. The minimization of the K-map and the realization of the logic diagram based on the minimal expression obtained are both shown in Fig. 10.85.

EXAMPLE 10.20 Design a pulse generator using a shift register to generate the waveform shown below:

0 1 0 1 1 1 0 1

Solution

The given pulse train is 1011101. There are seven states. So, the minimum number of FFs required is three. Make groups of three bits and form the states as shown in Fig. 10.86.

As seen from Fig. 10.86, it is not possible to generate seven unique states using three or four or even five FFs. So, the required number of FFs is six. The corresponding truth table is shown in Fig. 10.87a.

The minimal expression for f obtained by minimizing the K-map is

$$f = \overline{Q}_6 + \overline{Q}_5 Q_4 + Q_6 \overline{Q}_1$$

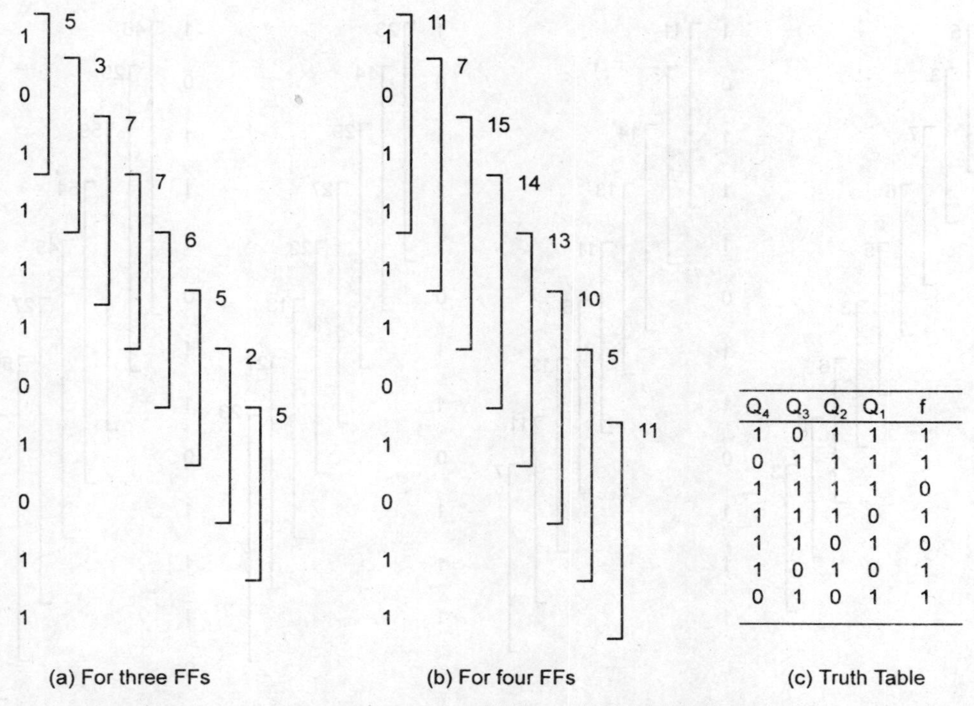

Q_4	Q_3	Q_2	Q_1	f
1	0	1	1	1
0	1	1	1	1
1	1	1	1	0
1	1	1	0	1
1	1	0	1	0
1	0	1	0	1
0	1	0	1	1

Fig. 10.84 State assignment for the pulse train generator (Example 10.19).

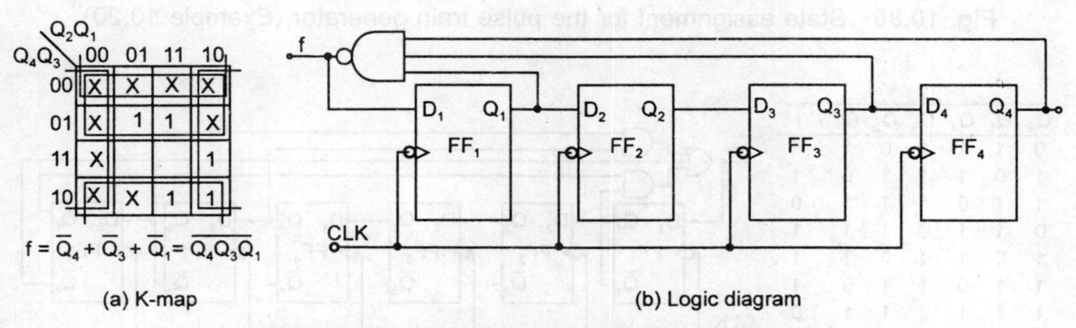

Fig. 10.85 Pulse train generator (Example 10.19).

The logic diagram of the pulse generator based on the minimal expression obtained is shown in Fig. 10.87b.

Linear Sequence Generator

One of the disadvantages of using a shift register as a state generating device is, that it seems very difficult to generate very many unique states. There is one type of circuit called the *linear sequence generator* that can generate $2^n - 1$ unique states for an n-FF shift register. It will generate all states except the state number (decimal) 0. The sequence of

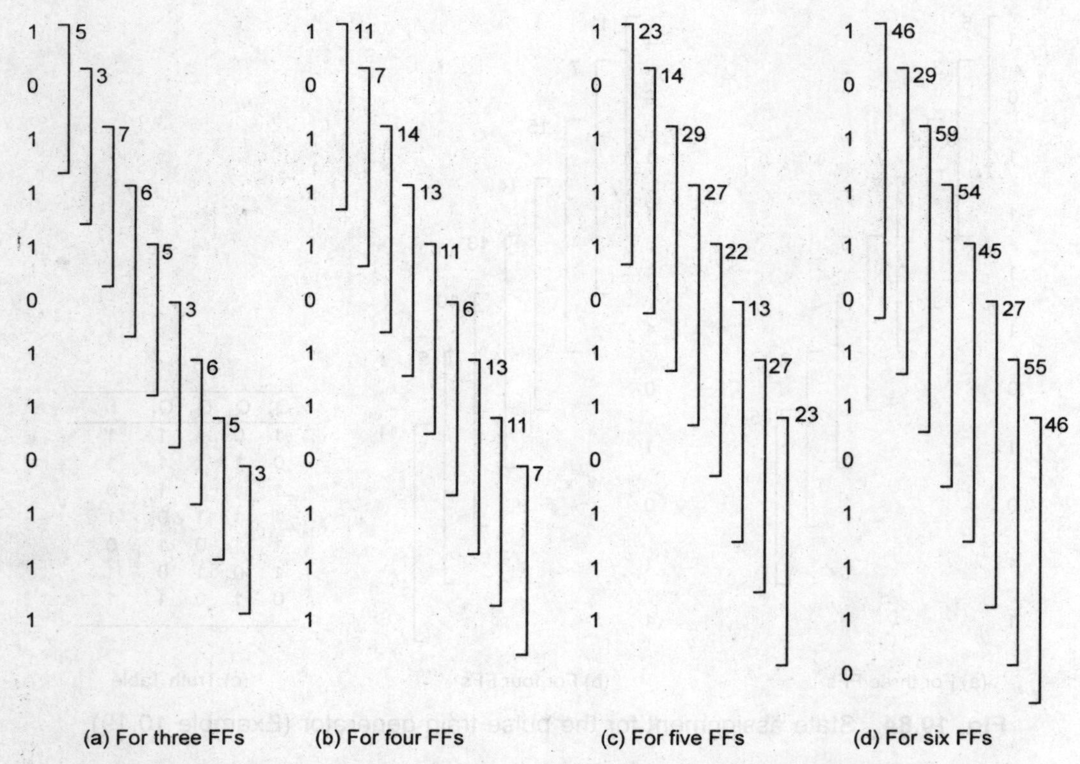

Fig. 10.86 State assignment for the pulse train generator (Example 10.20).

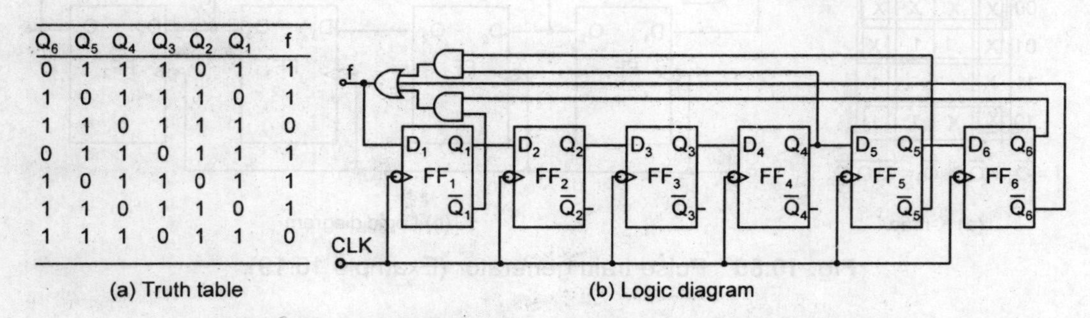

Q_6	Q_5	Q_4	Q_3	Q_2	Q_1	f
0	1	1	1	0	1	1
1	0	1	1	1	0	1
1	1	0	1	1	1	0
0	1	1	0	1	1	1
1	0	1	1	0	1	1
1	1	0	1	1	0	1
1	1	1	0	1	1	0

Fig. 10.87 Pulse train generator (Example 10.20).

states of a 3-FF shift register as a sequence generator is shown in Fig. 10.88. It can generate a 7-bit linear sequence 0100111. This is accomplished by exclusively ORing A and C outputs of the shift register and feeding this XOR output as input to flip-flop C. There is no guarantee that any other 7-bit sequence can be generated by this 3-bit shift register.

The expressions for feedback signals for particular lengths of FFs are shown in Table 10.8.

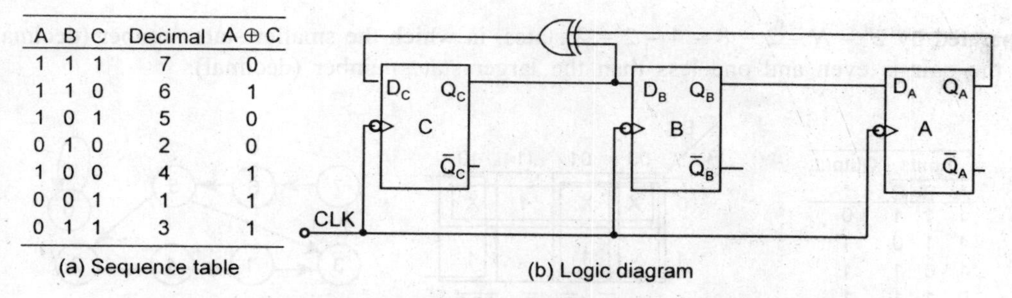

A	B	C	Decimal	A ⊕ C
1	1	1	7	0
1	1	0	6	1
1	0	1	5	0
0	1	0	2	0
1	0	0	4	1
0	0	1	1	1
0	1	1	3	1

(a) Sequence table

(b) Logic diagram

Fig. 10.88 Sequence generator using three type D flip-flops.

Table 10.8 Expressions for feedback signals for particular lengths of flip-flops

Shift register length	Expression for feedback	Shift register length	Expression for feedback
1	A	11	A ⊕ C
2	A ⊕ B	12	A ⊕ B ⊕ C ⊕ K
3	A ⊕ C	13	A ⊕ B ⊕ C ⊕ M
4	A ⊕ B	14	A ⊕ B ⊕ C ⊕ M
5	A ⊕ C	15	A ⊕ B
6	A ⊕ B	16	A ⊕ C ⊕ D ⊕ F
7	A ⊕ B	17	A ⊕ D
8	A ⊕ E ⊕ F ⊕ G	18	A ⊕ H
9	A ⊕ E	19	A ⊕ B ⊕ C ⊕ F
10	A ⊕ D	20	A ⊕ D

A linear sequence can be modified to reduce the number of states to any size by using the following rules:

1. Knowing the number of desired states, N, find the smallest number of FFs, n, needed for that number of states, such that $N \leq 2^n$.
2. Draw the state diagram for a linear sequence generator using n FFs.
3. Examine all pairs of states separated by $2^n - N - 2$ states and locate that pair in which the smaller state number (decimal) is an even number and one less than the larger state number (decimal).
4. Find the state number (decimal) that precedes the smaller state number of the above pair. Draw an arrow from that state number to the larger state number of the pair. This modified state diagram will be the final state diagram of the sequence.

EXAMPLE 10.21 Modify a 3-bit linear sequence generator to output 4 states.

Solution

It requires three FFs because the state number 0 is not permitted. So, draw a state diagram for a 3-bit linear sequence generator (Fig. 10.89c) and locate that pair of state numbers, separated by $2^n - N - 2 = 8 - 4 - 2 = 2$ states, in which the smaller state number (decimal) of the pair is even and one less than the larger state number (decimal).

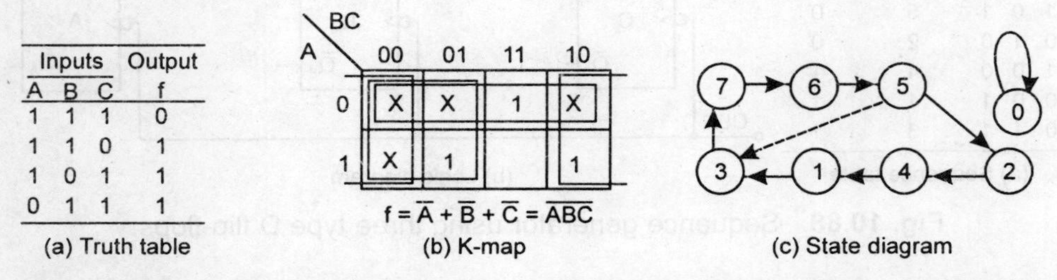

Fig. 10.89 Modification of the linear sequence generator (Example 10.21).

The following pairs of state numbers separated by two states in between are examined to find the difference between their decimal values.

States 7–2 differ by more than 1.
States 6–4 differ by more than 1.
States 5–1 differ by more than 1.
States 2–3 differ by 1. So, this pair fits.

Draw an arrow from the state before 2, that is state 5 to state 3. The required states are, therefore, 7→6→5→3. The truth table is shown in Fig. 10.89a. The K-map and its minimization are shown in Fig. 10.89b. The value of f, i.e. the output of the combinational circuit for each state is taken as the next lower bit in the sequence as shown below.

1] 7
1] 6
1] 5
0] 3
1
1
1

The modified logic diagram based on the value of f is shown in Fig. 10.90.

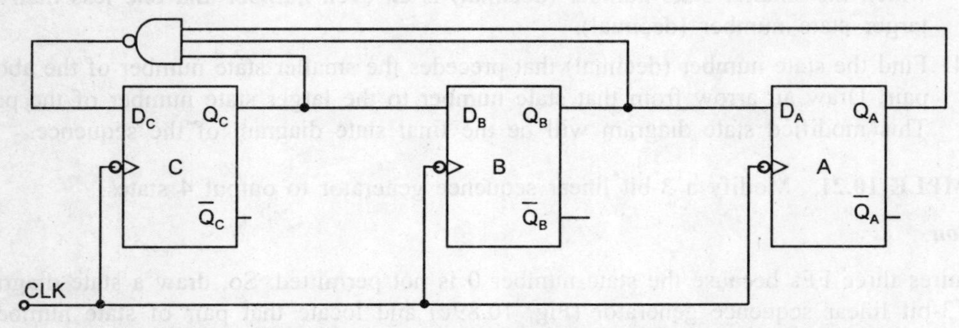

Fig. 10.90 Modified 3-bit linear sequence generator (Example 10.21).

10.15 CASCADING OF SYNCHRONOUS COUNTERS

We have talked about cascading of ripple counters earlier. In cascaded counters, the output of the last stage of one counter drives the input of the next counter.

When operating synchronous counters in a cascaded configuration, it is necessary to use the count enable and the terminal count functions to achieve higher modulus operation on some devices. The count enable is labelled simply CTEN or some other similar designation, and terminal count (TC) is analogous to ripple clock or ripple carry out (RCO) on some IC counters.

Figure 10.91 shows a mod-10 counter and a mod-8 counter connected in cascade. The terminal count (TC) output of counter 1 is connected to the count enable (CTEN) input of

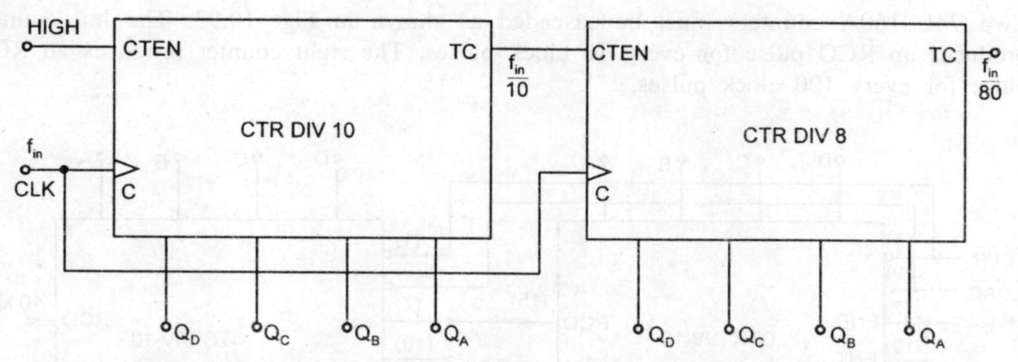

Fig. 10.91 Mod-80 cascaded counter using mod-10 and mod-8 counters.

counter 2. The counter 2 is inhibited by the LOW on its CTEN input until counter 1 reaches its last or terminal state when its terminal count output goes HIGH. This HIGH now enables counter 2, so that when the first clock pulse after counter 1 reaches its last or terminal count (CLK10) is applied, counter 2 goes from its initial state to its second state. Upon completion of the entire second cycle of counter 1 (when counter 1 reaches TC the second time), counter 2 is again enabled and advances to its next state. This sequence continues. Since the first one is a decade counter, it must go through 10 complete cycles before counter 2 completes its first cycle. In other words, for every 10 cycles of counter 1, counter 2 goes through one cycle. Since the second counter is a mod-8 counter, it will complete one cycle only after 80 clock pulses. The overall modulus of these two cascaded counters is

$$10 \times 8 = 80$$

Figure 10.92 illustrates how to obtain a 1 Hz signal from a 1 MHz signal using decade counters.

EXAMPLE 10.22 Draw the logic diagram using 74LS160A counters to obtain a 10 kHz waveform from a 1 MHz clock.

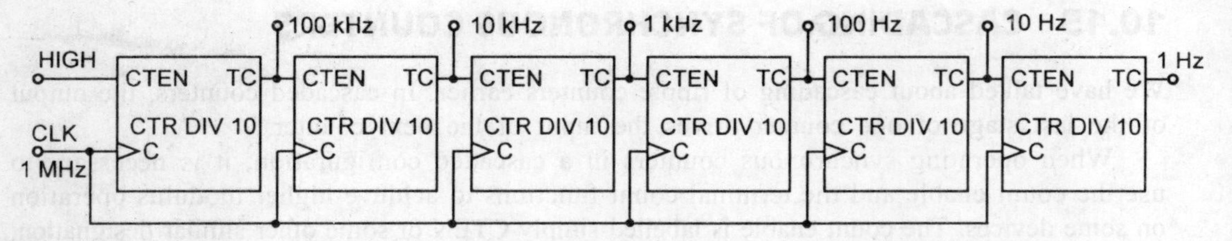

Fig. 10.92 Logic diagram to obtain a 1 Hz signal from a 1 MHz signal using decade counters.

Solution

To obtain a 10 kHz waveform from a 1 MHz clock, a division factor of 100 is required. Two 74LS160A counters must be cascaded as shown in Fig. 10.93. The left counter produces an RCO pulse for every 10 clock pulses. The right counter produces an RCO pulse for every 100 clock pulses.

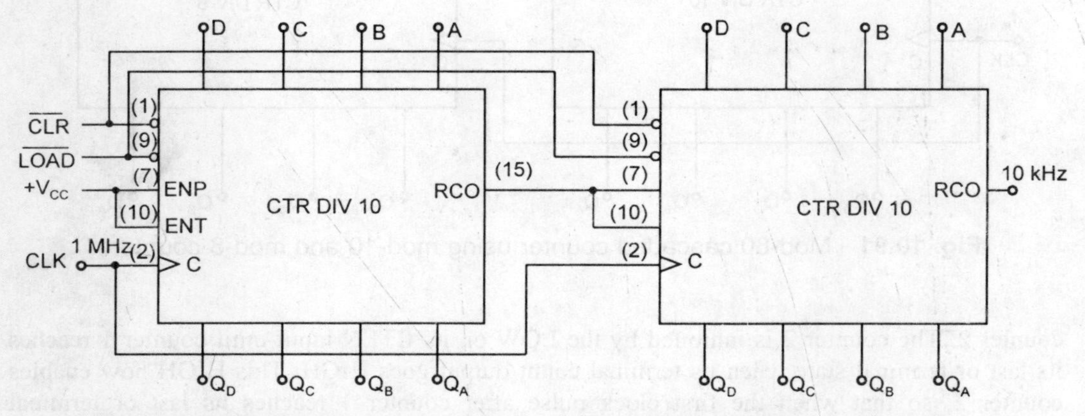

Fig. 10.93 Divide-by-100 counter using two 74LS160A decade counters (Example 10.22).

Cascaded IC Counters with Truncated Sequences

If the overall modulus (divide-by-factor) is the product of the individual moduli of all the cascaded counters, the cascading is called *full modulus cascading*. Often an application requires that the overall modulus be less than that which can be achieved by full modulus cascading. That is, when a truncated sequence must be implemented with cascaded counters.

Figure 10.94 shows how to use three 74LS161A 4-bit binary counters to implement a truncated sequence. If these three counters work cascaded in a full modulus arrangement, the modulus would be $2^{12} = 4096$.

Let us assume that a certain application requires a divide-by-2992 counter (mod-2992). The difference between 4096 and 2992 is 1104, which is the number of states that must be deleted from the full modulus sequence. The technique used in the circuit of

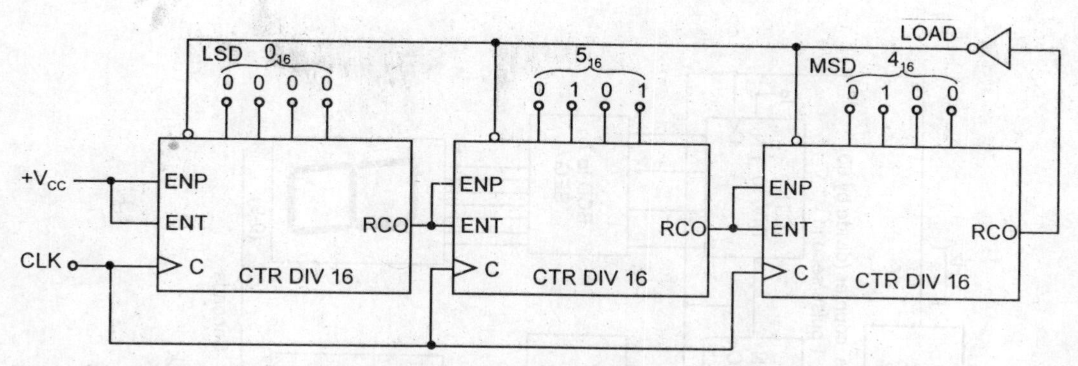

Fig. 10.94 A divide-by-2992 counter using 74LS161A 4-bit binary counters.

Fig. 10.94 is to preset the cascaded counter to 1104_{10} (450_{16}) each time it recycles, so that it counts from 1104 to 4096 on each full cycle. Therefore, each full cycle of the counter consists of 2992 states.

Notice that the RCO output of the rightmost counter is inverted and applied to the $\overline{\text{LOAD}}$ input of each 4-bit counter. Each time the count reaches its terminal value of 4096, RCO goes HIGH and causes the number on the data inputs (450_{16}) to be preset into the counter. Thus, there is one RCO pulse from the rightmost 4-bit counter for every 2992 clock pulses. With this technique, any modulus can be achieved by simply presetting the counter to the appropriate initial state on completion of each cycle.

Counter Applications

Digital clock

One of the most common applications of counters is the digital clock—a time clock which displays the time of the day in hours, minutes and sometimes seconds. In order to construct an accurate digital clock, a very closely controlled basic clock frequency is required. For battery operated digital clocks (or watches) the basic frequency is normally obtained from a quartz crystal oscillator. Digital clocks operated from the ac power line can use the 50 Hz power frequency as the basic clock frequency. In either case, the basic frequency is divided down to a frequency of 1 Hz or one pulse per second. Figure 10.95 shows the basic block diagram for a digital clock operating from 50 Hz.

The 50 Hz signal is sent through a shaping circuit to produce square pulses at the rate of 50 pps. The 50 pps waveform is fed into a mod-50 counter which is used to divide the 50 pps down to 1 pps. The 1 pps signal is fed into the SECONDS section, which is used to count and display seconds from 0 to 59. The BCD counter advances one count per second. After 9 seconds the BCD counter recycles to 0, which triggers the mod-6 counter and causes it to advance one count. This continues for 59 seconds, at which point the mod-6 counter is at the 101(5) count and the BCD counter is at the 1001(9) count, so that the display reads 59 s. The next pulse recycles the BCD counter to 0, which in turn, recycles the mod-6 counter to 0.

The output of the mod-6 counter in the SECONDS section has a frequency of one pulse per minute (the mod-6 recycles every 60 s). This signal is fed to the MINUTES section,

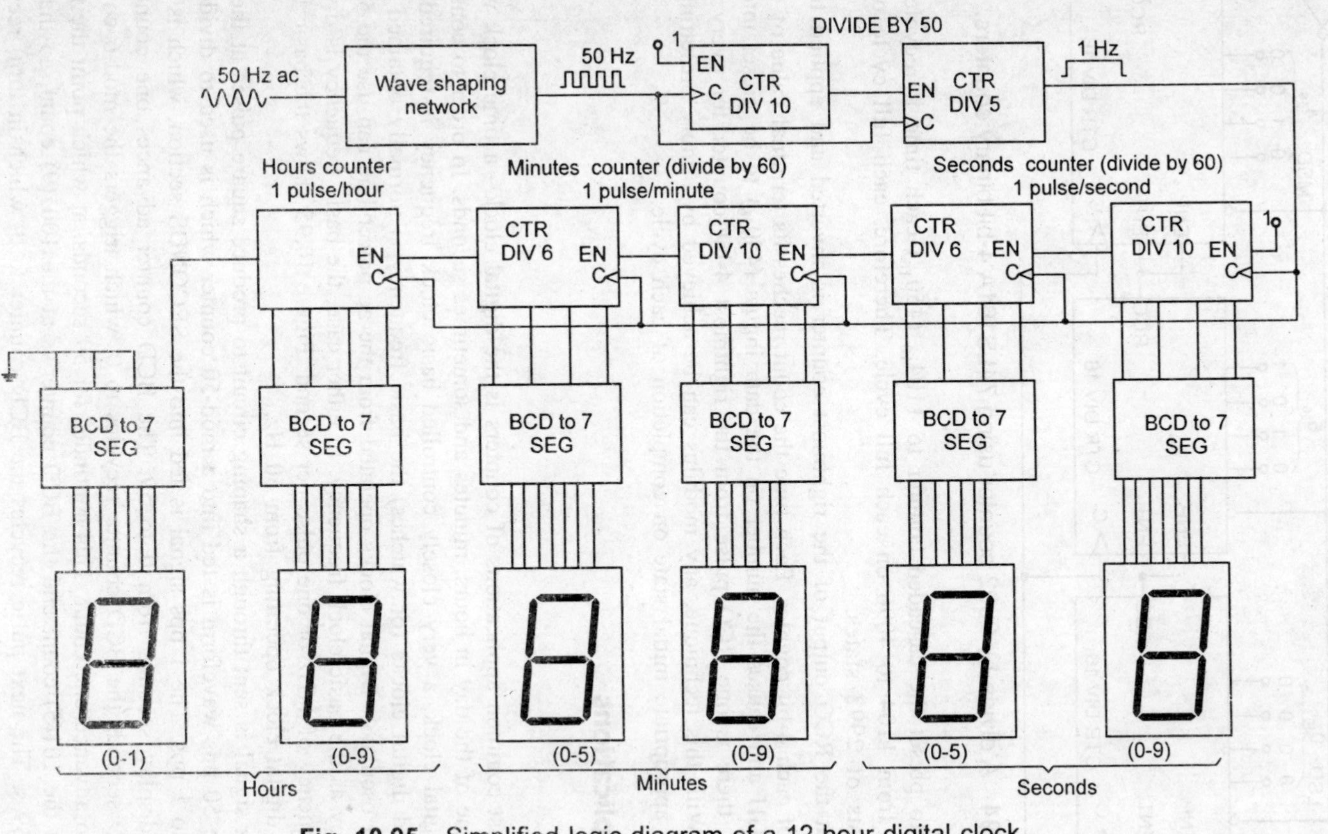

Fig. 10.95 Simplified logic diagram of a 12-hour digital clock.

which counts and displays minutes from 0 through 59. The MINUTES section is identical to the SECONDS section and operates in exactly the same manner.

The output of the mod-6 counter in the MINUTES section has a frequency of one pulse per hour (the mod-6 recycles every 60 min). This signal is fed to the HOURS section which counts and displays hours from 1 through 12. The HOURS section is different from the SECONDS and MINUTES section in that it never goes to the 0 state. Figure 10.96 shows

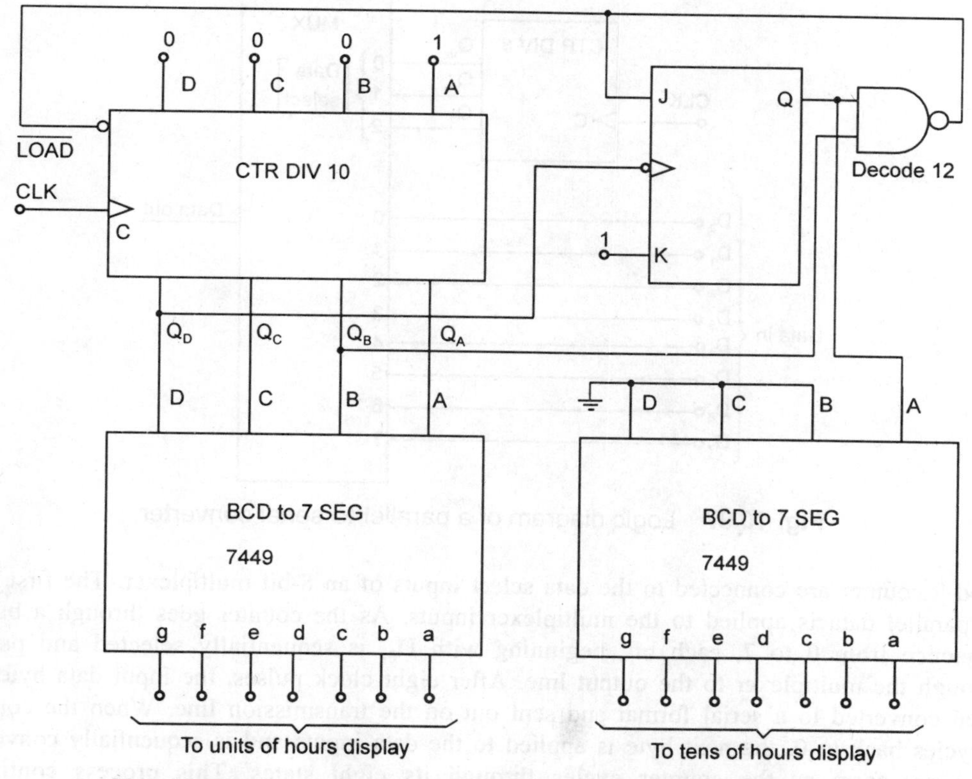

Fig. 10.96 Logic diagram for the hours counter and decoder.

the detailed circuitry in the HOURS section. It includes a BCD counter to count units of hours, and a single FF (mod-2) to count tens of hours. Consider that initially both the decade counter and the FF are RESET and the decode-12 gate output is HIGH. The decade counter advances from 0 to 9, and as it recycles from 9 back to 0, the FF is toggled to the SET state by the HIGH-to-LOW transition of Q_D. This illuminates a 1 on the tens-of-hours display. The total count is now 10_{10} (the decade counter is in the 0 state and the FF is SET).

Next, the total count advances to 11_{10} and then to 12. In state 12, the Q_B output of the decade counter is HIGH, the FF is still SET, and thus the decode gate output is LOW. This activates the LOAD input of the decade counter on the next clock pulse, the decade counter is preset to state 1 by the data inputs, and the FF is reset ($J = 0$, $K = 1$). This logic always causes the counter to recycle from 12 back to 1 rather than back to 0.

Parallel-to-Serial Data Conversions

A group of bits appearing simultaneously on parallel lines is called *parallel data*. A group of bits appearing on a single line in time sequence is called *serial data*. Parallel-to-serial conversions, normally accomplished using a counter, provide a binary sequence for the data select inputs of a data selector/multiplexer as illustrated in Fig. 10.97. The Q outputs of the

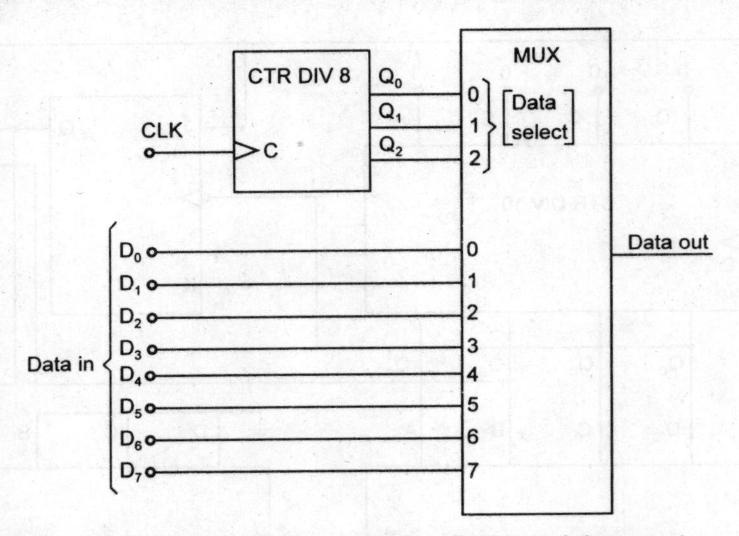

Fig. 10.97 Logic diagram of a parallel-to-serial converter.

mod-8 counter are connected to the data select inputs of an 8-bit multiplexer. The first byte of parallel data is applied to the multiplexer inputs. As the counter goes through a binary sequence from 0 to 7, each bit, beginning with D_0, is sequentially selected and passed through the multiplexer to the output line. After eight clock pulses, the input data byte has been converted to a serial format and sent out on the transmission line. When the counter recycles back to 0, the next byte is applied to the data inputs and is sequentially converted to serial form as the counter cycles through its eight states. This process continues repeatedly to convert each parallel byte to a serial byte.

Frequency counter

A frequency counter is a circuit that can measure and display the frequency of a signal. A simplified form for constructing a frequency counter is shown in Fig. 10.98a. It contains a counter with its associated decoder/display circuitry and an AND gate. The AND gate inputs include the pulses of an unknown frequency f_x and a SAMPLE pulse that controls how long the unknown pulses are allowed to pass through the AND gate into the counter. The counter is usually made up of cascaded BCD counters. A decoder/display unit converts the BCD outputs into a decimal display for easy monitoring.

The waveforms in Fig. 10.98b show that a CLEAR pulse is applied to the counter at t_0 to start the counter at zero. Prior to t_1, the sample pulse waveform is LOW and, so, the AND output Z is LOW and the counter will not be counting. The sample pulse goes HIGH

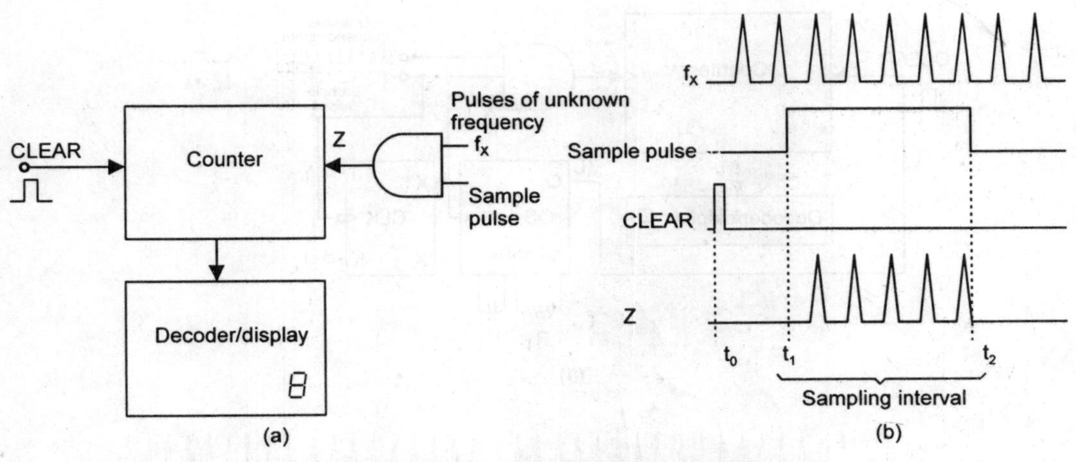

Fig. 10.98 Basic frequency counting method.

from t_1 to t_2. This is called the *sampling interval*. During this sampling interval, the pulses of unknown frequency will pass through the AND gate and get counted by the counter. After t_2, the AND output returns LOW and the counter stops counting. Thus, the counter counts the number of pulses that occur during the sampling interval, and the resulting contents of the counter exhibit the frequency of the pulse waveform.

The accuracy of this method depends almost entirely on the duration of the sampling interval, which must be accurately controlled.

Complete frequency counter

Figure 10.99a shows a complete frequency counter circuit. The circuit contains a one-shot and a J-K FF operating in the toggle mode, in addition to the counter, decoder/display and an AND gate of the simplified circuit discussed above. The AND gate has three inputs, one of which is the FF output, X. The sample pulses are connected to the AND gate and also to the CLK input of the FF.

Figure 10.99b shows the waveforms. Assume that X is in the 0 state initially. Since X is connected to one input of the AND gate, the AND gate is disabled. So, the output of the AND gate is LOW and, therefore, no pulses are fed to the counter even when the first sample pulse occurs between t_1 and t_2. At t_2, the negative-going edge of the first sample pulse toggles the FF to the HIGH state. This positive transition at X triggers the monostable multivibrator, which generates a narrow positive pulse to clear the counter and the counter displays 0. At the positive-going edge of the second sample (at t_3), the AND gate is enabled (since X is now a 1) and allows the unknown frequency to be counted, into the counter until t_4. At the end of the second sample pulse (at t_4), the FF toggles because of the negative edge of the sample pulse. So, the FF output X goes LOW and disables the AND gate. Therefore, the counter stops counting. The third sample pulse does not enable the AND gate because X is LOW. So, till the end of the next sample pulse (i.e. from t_4 to t_6), the counter holds and displays the count that it had reached at t_4. At the negative-going edge of the third sample pulse (i.e. at t_6), the FF toggles and its output X goes HIGH and the operation follows the same sequence of events that had begun earlier at t_2.

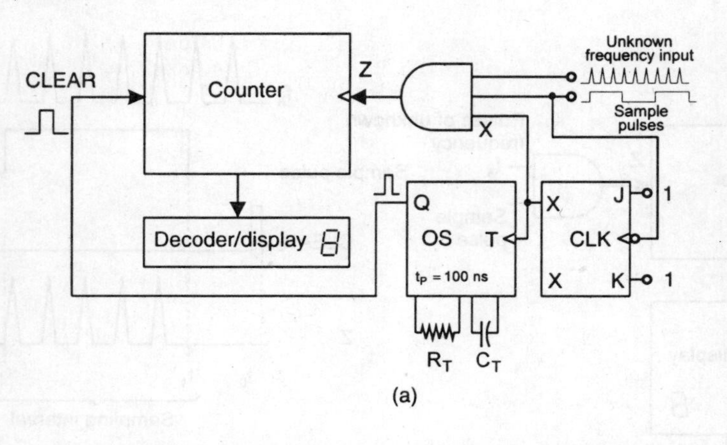

(a)

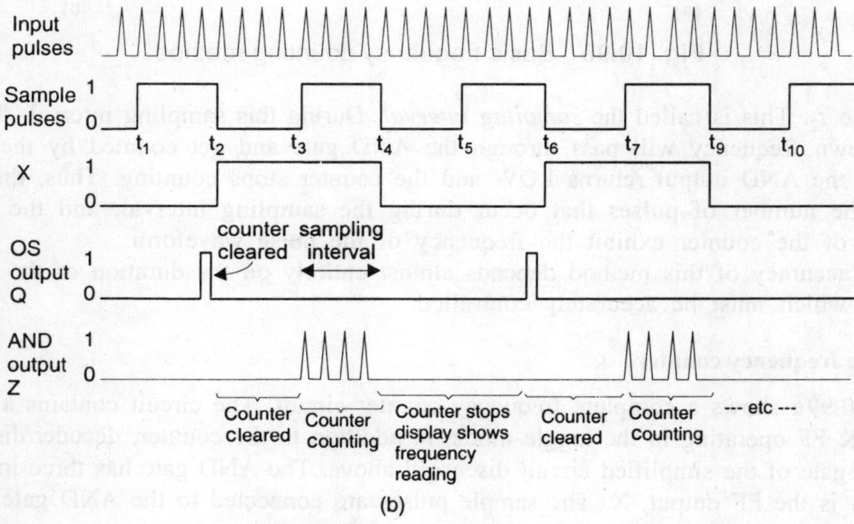

(b)

Fig. 10.99 Frequency counter.

This frequency counter then goes through a repetitive sequence of clearing to 0, counting, holding for display, clearing to 0, counting, and so on.

Since the display is connected directly to the counter outputs, the display will show the clearing and counting action of the counter. This makes it very difficult to read the display to determine the unknown frequency except at very slow sample intervals. This problem can be solved by inserting a buffer register between the counter and the decoder/display unit.

SUMMARY

- A digital counter is a set of FFs whose states change in response to pulses applied at the input to the counter.

- Counters may be asynchronous counters or synchronous counters.
- Asynchronous counters are also called ripple counters.
- In asynchronous counters, all the FFs do not change states simultaneously. They are serial counters.
- In synchronous counters, all the FFs change states simultaneously. They are parallel counters.
- A counter may be an up-counter or a down-counter or an up/down counter.
- The number of states through which the counter passes before returning to the starting state is called the modulus of the counter.
- If the overall modulus is the product of the individual moduli of all the cascaded counters, such cascading is called *full modulus cascading*.
- The LSB of any counter is the bit that changes most often.
- A mod-M and a mod-N counter in cascade give a mod-MN counter.
- In asynchronous counters if the clock frequency is very high owing to accumulation of propagation delay, skipping of states may occur.
- In synchronous counters, the propagation delays of individuals FFs do not add together.
- Synchronous counters have the advantages of high speed and less severe decoding problems but the disadvantage of more circuitry than that of asynchronous counters.
- A counter is said to suffer from the problem of lock-out or is said to be of not self-starting type, if it keeps going from one invalid state to another invalid state after successive clock pulses and never returns to a normal state.
- Shortened modulus counters may suffer from the problem of lock-out.
- A counter is said to be of self-starting type, if it returns to a valid state and counts normally after one or more clock pulses, even if it enters an invalid state.
- A shift register is quite restrictive in the sense that it cannot go from any one state to any other state of our choice.
- Shift registers can be arranged as counters or sequence generators.
- A ring counter is also called a basic ring counter or a simple ring counter.
- A twisted ring counter is also called a Johnson counter or a switch-tail ring counter.
- A synchronous counter in which the output of one counter drives the clock input of another counter is called a hybrid counter.
- Pulse trains can be generated using either direct logic or indirect logic or shift registers.
- In direct logic, the output is taken directly from a FF, whereas in indirect logic it is taken from a decoder gate.
- Generation of pulse trains using indirect logic has the advantage that any counter (ripple or synchronous) with the correct number of states can form the generator.

QUESTIONS

1. What do you mean by the following?
   (a) Counter
   (b) Ripple counter
   (c) Asynchronous counter
   (d) Synchronous counter
   (e) BCD counter
   (f) Decade counter
   (g) Up/down counter
   (h) Serial counter
   (i) Parallel counter
2. Explain the following terms:
   (a) Modulus of a counter
   (b) Maximum modulus
   (c) Truncated sequence
   (d) Presetting a counter
   (e) Resetting a counter
   (f) Enable inputs of a counter
3. What is the terminal count of a 4-bit binary counter in the up-mode? In the down-mode? What is the next state after the terminal count in the up-mode? In the down-mode?
4. A 4-bit up/down counter is in the down mode and in the 1100 state. On the next clock pulse, to what state does the counter go?
5. A binary counter is in the 1010 state. (a) What is its next state? (b) What condition must exist on each FF input to ensure that it goes to the proper next state on the clock pulse?
6. How many decade counters are required to convert a clock of 10 MHz to 100 Hz?
7. Four counters, mod-10, mod-10, mod-8 and mod-6, are connected in cascade. What is the modulus of the cascaded counter?
8. What do you mean by full modulus cascading?
9. With general block diagrams, show how each of the following counters can be obtained using a FF, a decade counter and a 4-bit binary counter or any combination of these?
   (a) Divide-by-40 counter
   (b) Divide-by-32 counter
   (c) Divide-by-160 counter
   (d) Divide-by-320 counter
   (e) Divide-by-10,000 counter
   (f) Divide-by-1000 counter
10. What do you mean by the following?
    (a) Decoding a counter
    (b) Decoding glitches
    (c) Active-LOW decoding
    (d) Active-HIGH decoding
11. How are decoding glitches caused?
12. What is strobing?

13. What is the maximum modulus of a counter with each of the following number of flip-flops?

(a) 2 (b) 5 (c) 6
(d) 8 (e) 10

14. Determine the number of FFs in each of the following counters.

(a) mod-3 (b) mod-8 (c) mod-14
(d) mod-20 (e) mod-32 (f) mod-150

15. A 4-bit binary up/down counter is in the binary state of 0. What is its next state in the up-mode? In the down-mode?

16. What are the invalid states in the decade counter? In a mod-5 counter?

17. With general block diagrams, show how to obtain the following frequencies from a 1 MHz clock using single flip-flops, mod-5 counters and decade counters.

(a) 500 kHz (b) 250 kHz (c) 125 kHz
(d) 40 kHz (e) 10 kHz (f) 1 kHz

18. A 4-bit ripple counter consists of FFs each of which has a clock-to-Q propagation delay of 10 ns. How long will the counter take to recycle from 1111 to 0000 after the triggering edge of the clock pulse?

19. What do you mean by 'dependency notation'?

20. Assume that a 4-bit ripple counter is holding the count 0100. What will be the count after 29 clock pulses?

21. What flip-flop outputs should be connected to a clearing NAND gate to form a mod-11 counter?

22. All BCD counters are decade counters. Yes or No?

23. What is the difference between the counting sequence of an up-counter and a down-counter?

24. Describe how an asynchronous down-counter circuit differs from an up-counter circuit.

25. Explain why does a ripple counter's maximum frequency limitation decrease as more FFs are added to the counter?

26. A certain J-K FF has t_{pd} = 10 ns. What is the largest mod counter that can be constructed from these FFs and which still operate up to 10 MHz?

27. What is the advantage of a synchronous counter over an asynchronous counter? What is its disadvantage?

28. What do you mean by a pre-settable counter?

29. Describe the difference between asynchronous and synchronous presetting.

30. Describe some applications of counters.

31. Draw the logic diagram and the timing diagram of a 3-bit binary ripple—up-counter and down-counter—using (i) positive-edge triggered FFs, (ii) negative-edge triggered FFs.

32. Draw the gates necessary to decode all the stages of a mod-16 counter using active-LOW outputs.

33. What is the difference between a ring counter and a Johnson counter?

34. Write the sequence of stages for a 5-bit Johnson counter.

35. How many FFs will be complemented in an 8-bit ripple counter to reach the next count after the following count?

(a) 10110111 (b) 01101111

36. How many states does a mod-*n* counter have?

37. Draw the logic diagram of a 4-bit ripple counter using D flip-flops.

38. Why do the decoding gates for an asynchronous counter have glitches on their outputs?

39. How does strobing eliminate decoding glitches?

40. Which shift-register counter requires the most number of FFs for a given mod number?

41. Enumerate the advantages and disadvantages of a ring counter.

42. How many states each can an *n*-bit ring counter and an *n*-bit Johnson counter have?

43. What do you mean by the following?

(a) Pulse train generator (b) Indirect logic

(c) Direct logic (d) Sequence generator

(e) Length of a sequence (f) Linear sequence generator

44. What is the advantage of designing a pulse generator using indirect logic?

45. Why is a shift register restrictive?

46. Name the basic blocks of a digital clock.

47. Write the steps in the design of a pulse generator.

48. What do you mean by lock-out of a counter?

49. How do you test for the problem of lock-out of a counter? How do you eliminate this problem?

50. What do you mean by the self-starting type counter?

PROBLEMS

10.1 Design the following counters:

(a) A mod-7 asynchronous counter using J-K FFs

(b) A mod-12 asynchronous counter using J-K FFs

(c) A mod-9 synchronous counter using J-K FFs

(d) A mod-7 synchronous counter using S-R FFs

(e) A mod-12 synchronous counter using T FFs

(f) A mod-14 synchronous counter using D FFs

10.2 Design a J-K counter that goes through states 2, 4, 5, 7, 2, 4,... Is the counter self-starting? Modify the circuit such that whenever it goes to any invalid state it comes back to state 2.

10.3 Design the following types of counters:

(a) A type-D counter that goes through states 0, 2, 4, 6, 0,... . The undesired states must always go to a 0 on the next clock pulse.

(b) A type-T counter that goes through states 6, 3, 7, 8, 2, 9, 1, 12, 14, 0, 6, 3,... . Is the counter self-starting?

(c) A type-T counter that goes through states 0, 5, 4, 2, 0,... . Is the counter self-starting?

(d) A type T counter that must go through states 0, 2, 4, 6, 0,... if the control line is HIGH, and through states 0, 4, 2, 7, 0,... if the control line is LOW.

10.4 Design the following types of generators:

(a) A sequence generator to generate the sequence 1011010...

(b) A pulse train generator to generate the sequence 1100010...

(c) A pulse generator using indirect logic to produce the pulse trains 10000111 and 10011010.

(d) A shift register pulse train generator to generate the pulse train 101110.

(e) A shift register pulse train generator to generate the pulse train 1110010.

10.5 Design a direct logic circuit to generate the sequences 100111 and 011101 simultaneously.

10.6 Generate the pulse train 100110 using indirect logic.

10.7 Design a BCD up/down counter using S-R FFs.

10.8 Design an up/down counter using D FFs to count 0, 3, 2, 6, 4, 0,... .

Chapter 11

LOGIC FAMILIES

11.1 INTRODUCTION

Because of the advances in microelectronics, the digital IC technology in less than four decades has rapidly advanced from small scale integration (SSI), through medium scale integration (MSI), large scale integration (LSI), very large scale integration (VLSI), to ultra large scale integration (ULSI). The technology is now entering giant scale integration (GSI) in which millions of gate equivalent circuits are integrated on a single chip. The use of ICs has thus reduced the overall size of a digital system drastically. Consequently, the cost of digital systems has also reduced. The reliability has improved as well, because the number of external interconnections from one device to another has reduced. The power consumption of digital systems has also reduced greatly, because the miniature circuitry requires much less power.

ICs have certain limitations too. ICs cannot handle very large voltages or currents and also electrical devices like precision resistors, inductors, transformers, and large capacitors cannot be implemented on chips. So, ICs are principally used to perform low power circuit operations. The operations that require high power levels or devices that cannot be integrated are still handled by discrete components.

ICs are fabricated using various technologies such as TTL, ECL, and IIL which use bipolar transistors, whereas the MOS and CMOS technologies use unipolar MOSFETs.

11.2 DIGITAL IC SPECIFICATION TERMINOLOGY

The digital IC nomenclature and terminology is fairly standardized. The most useful specification terms are defined below.

Threshold Voltage

The threshold voltage is defined as that voltage at the input of a gate which causes a change in the state of the output from one logic level to the other.

Propagation Delay

A pulse through a gate takes a certain amount of time to propagate from input to output. This interval of time is known as the *propagation delay* of the gate. It is the average transition delay time t_{pd}, expressed by

$$t_{pd} = \frac{t_{PLH} + t_{PHL}}{2}$$

where t_{PLH} is the signal delay time when the output goes from a logic 0 to a logic 1 state and t_{PHL} is the signal delay time when the output goes from a logic 1 to a logic 0 state (Fig. 11.1).

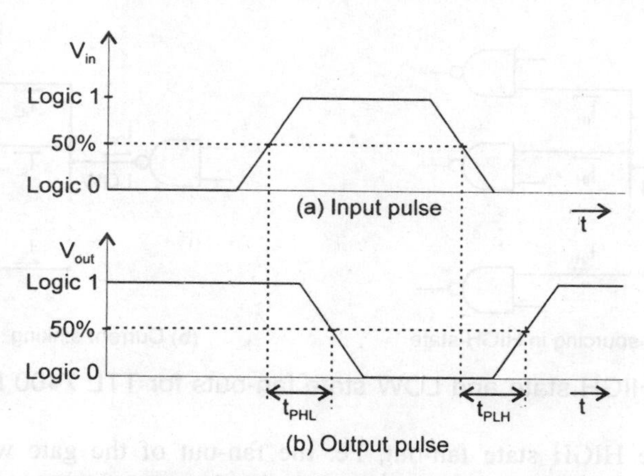

Fig. 11.1 Propagation delay in an inverter.

Power Dissipation

Every logic gate draws some current from the supply for its operation. The current drawn in HIGH state is different from that drawn in LOW state. The *power dissipation*, P_D, of a logic gate is the power required by the gate to operate with 50% duty cycle at a specified frequency and is expressed in milliwatts. This means that 1 and 0 periods of the output are equal. The power dissipation of a gate is given by

$$P_D = V_{CC} \times I_{CC}(\text{avg})/n$$

where V_{CC} is the gate supply voltage, $I_{CC}(\text{avg})$ is the average current drawn from the supply by the entire IC and n is the number of gates in the IC. Now

$$I_{CC}(\text{avg}) = \frac{I_{CCH} + I_{CCL}}{2}$$

where I_{CCH} is the current drawn by the IC when all the gates in the IC are in HIGH state, and I_{CCL} is the current drawn by the IC when all the gates in the IC are in LOW state. The total power consumed by an IC is equal to the product of the power dissipated by each gate and the number of gates in that IC.

Fan-in

The *fan-in* of a logic gate is defined as the number of inputs that the gate is designed to handle.

Fan-out

The *fan-out* (also called the *loading factor*) of a logic gate is defined as the maximum number of standard loads that the output of the gate can drive without impairing its normal operation. A standard load is usually specified as the amount of current needed by an input of another gate of the same IC family. If a gate is made to drive more than this number of gate inputs, the performance of the gate is not guaranteed. The gate may malfunction.

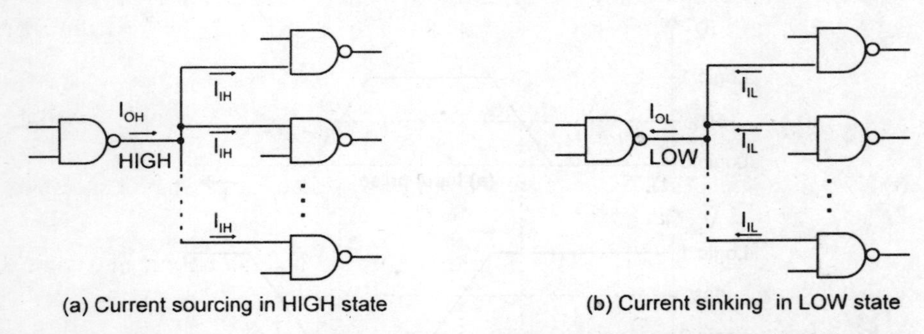

Fig. 11.2 HIGH state and LOW state fan-outs for TTL 7400 NAND gate.

Fan-out may be HIGH state fan-out, i.e. the fan-out of the gate when its output is a logic 1, or it may be LOW state fan-output, i.e. the fan-out of the gate when its output is a logic 0. The smaller of these two numbers is taken as the actual fan-out. The fan-out of a gate affects the propagation delay time as well as saturation. The driving gate sinks current when it is in LOW state and sources current when it is in HIGH state.

$$\text{HIGH state fan-out} = \frac{I_{OH}(\max)}{I_{IH}}$$

where $I_{OH}(\max)$ is the maximum current that the driver gate can source when it is in a 1 state and I_{IH} is the current drawn by each driven gate from the driver gate.

$$\text{LOW state fan-out} = \frac{I_{OL}(\max)}{I_{IL}}$$

where $I_{OL}(\max)$ is the maximum current that the driver gate can sink when its output is a logic 0 and I_{IL} is the current drawn from each driven gate by the driver gate. Figure 11.2 depicts the current sourcing and current sinking actions for the TTL 7400 NAND gate.

Voltage and Current Parameters

The definitions of voltage and current levels corresponding to the logic 0 and logic 1 states are as follows:

V_{IH}(min) (HIGH level input voltage). It is the minimum voltage level required at the input of a gate for that input to be treated as a logic 1. Any voltage below this level will not be accepted as a logic 1 by the logic circuit.

V_{OH}(min) (HIGH level output voltage). It is the minimum voltage level required at the output of a gate for that output to be treated as a logic 1. Any voltage below this level will not be accepted as a logic 1 output.

V_{IL}(max) (LOW level input voltage). It is the maximum voltage level that can be treated as a logic 0 at the input of the gate. Any voltage above this level will not be treated as a logic 0 input by the logic gate.

V_{OL}(max) (LOW level output voltage). It is the maximum voltage level that can be treated as a logic 0 at the output of the gate. Any voltage above this level will not be treated as a logic 0 output.

I_{IH} (HIGH level input current). The current that flows into an input when a specified HIGH level voltage is applied to that input.

I_{IL} (LOW level input current). The current that flows into an input when a specified LOW level voltage is applied to that input.

I_{OH} (HIGH level output current). The current that flows from an output in a logic 1 state under specified load conditions.

I_{OL} (LOW level output current). The current that flows from an output in a logic 0 state under specified load conditions.

Figure 11.3 illustrates the currents and voltages in the HIGH and LOW states.

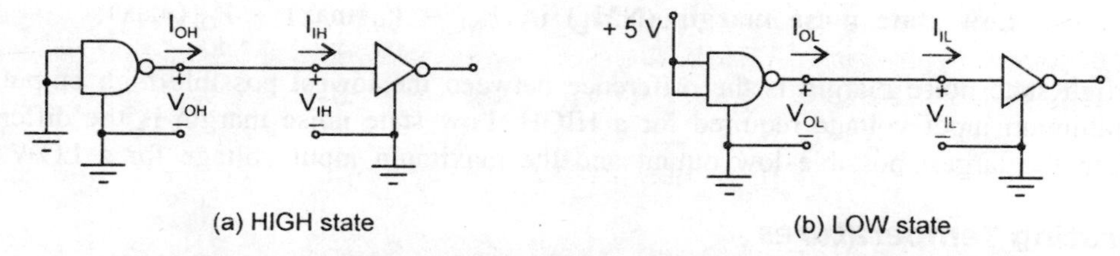

Fig. 11.3 Currents and voltages in the HIGH and LOW states.

Noise Margin

When the digital circuits operate in noisy environment the gates may malfunction if the noise is beyond certain limits. The *noise immunity* of a logic circuit refers to the circuit's ability to tolerate noise voltages at its inputs. A quantitative measure of noise immunity is called *noise margin*. Unwanted spurious signals are called *noise*. Noise may be ac noise or dc noise. A drift in the voltage levels of signals is called dc noise. The ac noise is a random pulse caused by other switching signals. Noise margin is expressed in volts and represents the maximum noise signal that can be added to the input signal of a digital circuit without causing an undesirable change in the circuit output. This is an important criterion for the selection of a logic family for certain applications where environment noise may be high.

Figure 11.4a shows the range of output voltages that can occur in a logic circuit. Voltages greater than V_{OH}(min) are considered as a logic 1 and voltages lower than V_{OL}(max) are considered as a logic 0. Voltages in the disallowed range should not appear at a logic circuit output under normal conditions. Figure 11.4b shows the input voltage require-ments of a logic circuit. The logic circuit will respond to any input greater than V_{IH}(min) as a logic 1 and to any input lower than V_{IL}(max) as a logic 0. Voltages in the indeterminate range will produce an unpredictable response and should not be used.

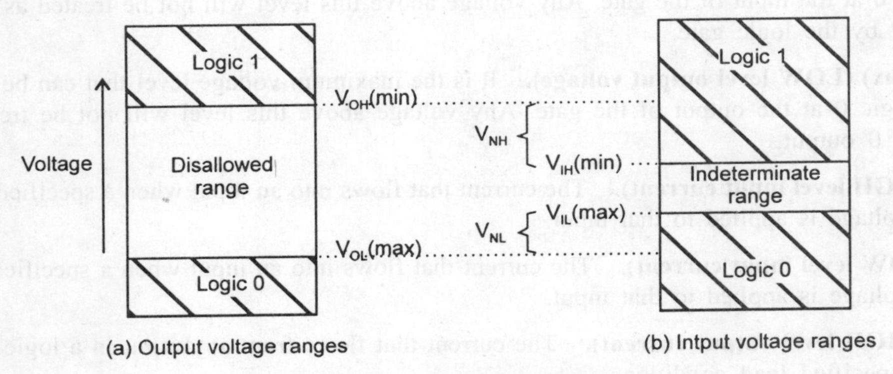

Fig. 11.4 DC noise margins.

Noise margin may be High state noise margin or Low state noise margin

High state noise margin (NM_H) is, $V_{NH} = V_{OH}(\text{min}) - V_{IH}(\text{min})$

Low state noise margin (NM_L) is, $V_{NL} = V_{IL}(\text{max}) - V_{OL}(\text{max})$

High state noise margin is the difference between the lowest possible high output and the minimum input voltage required for a HIGH. Low state noise margin is the difference between the largest possible low output and the maximum input voltage for a LOW.

Operating Temperatures

The IC gates and other circuits are temperature sensitive being semiconductor devices. However, they are designed to operate satisfactorily over a specified range of temperatures. The range specified for commercial applications is 0 to 70°C, for industrial it is 0 to 85°C, and for military applications it is – 55°C to 125°C.

Speed Power Product

A common means for measuring and comparing the overall performance of an IC family is the *speed power product*, which is obtained by multiplying the gate propagation delay by the gate power dissipation. A low value of speed power product is desirable. The smaller the product, the better the overall performance. The speed power product has the units of energy and is expressed in picojoules. It is the *figure of merit* of an IC family. Suppose an IC family has an average propagation delay of 10 ns and an average power dissipation of 5 mW, the speed power product is

$$10 \text{ ns} \times 5 \text{ mW} = 50 \times 10^{-12} \text{ watt-seconds} = 50 \text{ picojoules (pJ)}$$

11.3 LOGIC FAMILIES

Many logic families have been developed. They are: Resistor Transistor Logic (RTL), Direct Coupled Transistor Logic (DCTL), Diode Transistor Logic (DTL), High Threshold Logic (HTL), Transistor Transistor Logic (TTL), Emitter Coupled Logic (ECL), Integrated Injection Logic (IIL), Metal-oxide Semiconductor Logic (MOS), and Complementary Metal-oxide Semiconductor Logic (CMOS). Out of these, RTL, DCTL, DTL, and HTL are obsolete. The logic families TTL, ECL, IIL, MOS, and CMOS are currently in use. The basic function of any type of gate is always the same regardless of the circuit technology used. The TTL and CMOS are suitable for SSI and MSI. The MOS and CMOS are particularly suitable for LSI. The IIL is mainly suitable for VLSI and ULSI. The ECL is mainly used in superfast computers. The logic families currently in use are compared in Table 11.1 in terms of the commonly used specification parameters.

Table 11.1 Comparison of logic families

Logic family	Propagation delay time (ns)	Power dissipation per gate (mW)	Noise margin (V)	Fan-in	Fan-out	Cost
TTL	9	10	0.4	8	10	Low
ECL	1	50	0.25	5	10	High
MOS	50	0.1	1.5		10	Low
CMOS	< 50	0.01	5	10	50	Low
IIL	1	0.1	0.35	5	8	Very low

11.4 TRANSISTOR TRANSISTOR LOGIC (TTL)

The TTL or T^2L family is so named because of its dependence on transistors alone to perform basic logic operations. It is the most popular logic family. It is also the most widely used bipolar digital IC family. The TTL uses transistors operating in saturated mode. It is the fastest of the saturated logic families. The basic TTL logic circuit is the NAND gate. Good speed, low manufacturing cost, wide range of circuits, and the availability in SSI and MSI are its merits. Tight V_{CC} tolerance, relatively high power consumption, moderate packing density, generation of noise spikes and susceptibility to power transients are its demerits.

The TTL logic family consists of several subfamilies or series such as: **Standard TTL, High Speed TTL, Low power TTL, Schottky TTL, Low power Schottky TTL, Advanced Schottky TTL, Advanced low power Schottky TTL and F(fast)TTL.**

The differences between the various TTL subfamilies are in their electrical characteristics such as delay time, power dissipation, switching speed, fan-out, fan-in, noise margin, etc. For standard TTL, propagation delay time = 9 ns, power dissipation per gate = 10 mW, noise margin = 0.4 mV, fan-in = 8, and fan-out = 10.

For standard TTL, 0 V to 0.8 V is treated as a logic 0 and 2 V to 5 V is treated as a logic 1. Signals in 0.8 V to 2 V range should not be applied as input as the corresponding response will be indeterminate. If a terminal is left open in TTL, it is equivalent to connecting it to HIGH, i.e. + 5 V. But such a practice is not recommended, since the floating TTL is extremely susceptible to picking up noise signals that can adversely affect the operation of the device.

Two-input TTL NAND Gate

In the circuit of the two-input TTL NAND gate shown in Fig. 11.5 the input transistor Q_1 is a multiple emitter transistor. Transistor Q_2 is called the phase splitter. Transistor Q_3 'sits above' Q_4 and, therefore, Q_3 and Q_4 make a *totem pole* arrangement. Diodes D_1 and D_2 protect Q_1 from being damaged by the negative spikes of voltages at the inputs. When negative spikes appear at the input terminals, the diodes conduct and bypass the spikes to ground. Diode D ensures that Q_3 and Q_4 do not conduct simultaneously. Transistor Q_3 acts as an emitter follower.

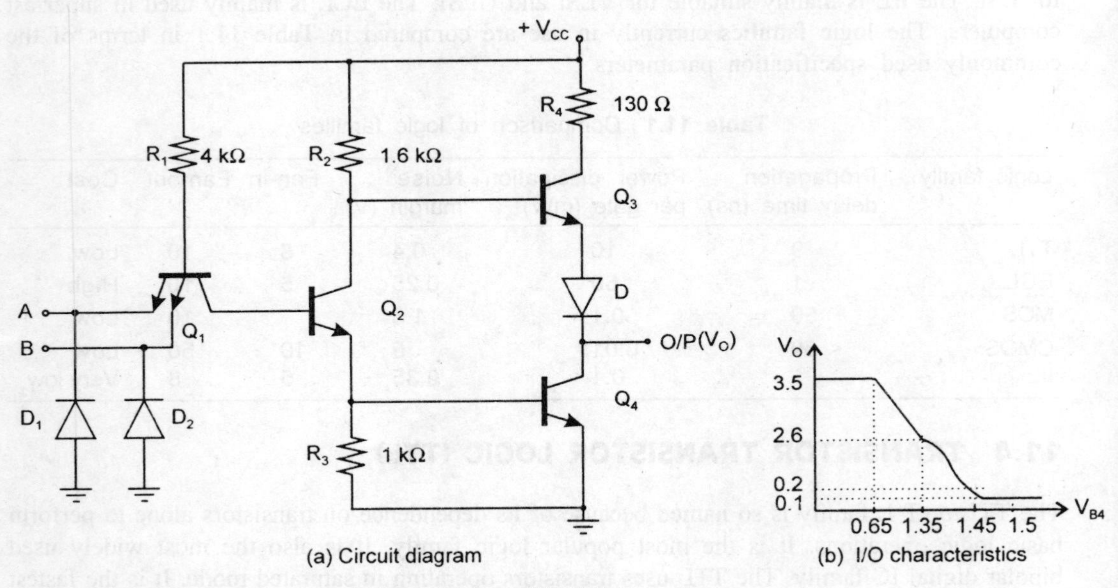

Fig. 11.5 TTL NAND gate.

When both the inputs A and B are HIGH (+ 5 V), both the base-emitter junctions of Q_1 are reverse biased. So, no current flows to the emitters of Q_1. However, the collector-base junction of Q_1 is forward biased. So, a current flows through R_1 to the base of Q_2, and Q_2 turns on. Current from Q_2's emitter flows into the base of Q_4. So, Q_4 is turned on. The collector current of Q_2 flows through R_2 and, so, produces a drop across it thereby reducing the voltage at the collector of Q_2. Therefore, Q_3 is OFF. Since Q_4 is ON, V_O is at its low level (V_{CE} (sat)). So, the output is a logic 0, When either A or B or both are LOW, the corresponding base-emitter junction(s) is (are) forward biased and the collector-base junction of Q_1 is reverse biased. So, the current flows to ground through the emitters of Q_1. Therefore, the base of Q_1 is at 0.7 V, which cannot forward bias the base-emitter junction of Q_2. So, Q_2 is OFF. With Q_2 OFF, Q_4 does not get the required base drive. So, Q_4 is also OFF. Transistor Q_3 gets enough base drive because Q_2 is OFF, i.e. since no current flows into the collector of Q_2, all the current flows into the base of Q_3, and therefore, Q_3 is ON. The output voltage, $V_O = V_{CC} - V_{R2} - V_{BE3} - V_D \approx$ 3.4–3.8 V, which is a logic HIGH level. So, the circuit acts as a two-input NAND gate. When Q_4 is OFF, no current flows through it, but the stray and output capacitances between the output terminal, i.e. the collector of Q_4, and ground get charged to this voltage of 3.4–3.8 V. The I/O characteristics of a TTL NAND gate are shown in Fig. 11.5b.

Totem-pole Output

In the circuit diagram of the two-input TTL NAND gate, transistor Q_3 sits above transistor Q_4. As shown in Fig. 11.6, Q_3 and Q_4 are connected in 'totem pole' fashion. At any time, only one of them will be conducting. Both cannot be ON or OFF simultaneously. Diode D

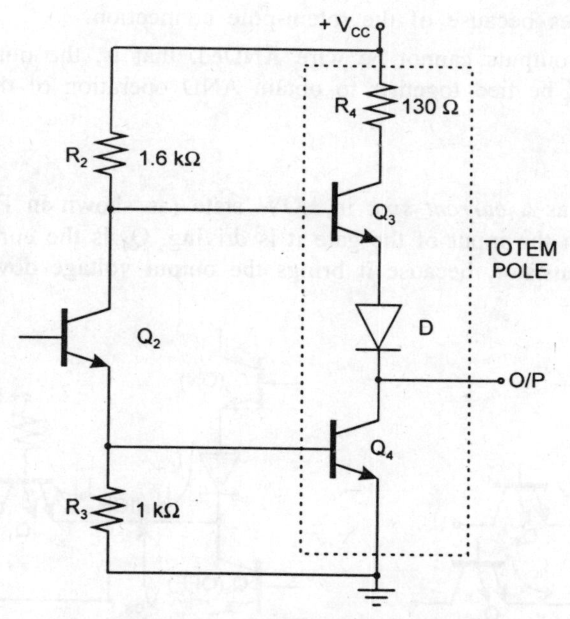

Fig. 11.6 Totem-pole output.

ensures this. If Q_4 is ON, its base is at 0.7 V w.r.t. ground. Q_4 gets base drive from Q_2. So, when Q_4 is ON, Q_2 has to be ON. Therefore, its collector-to-emitter voltage is V_{CE} (sat) ≈ 0.3 V. Hence, $V_{B3} = V_{C2} \approx 0.7 \text{ V} + 0.3 \text{ V} \approx 1 \text{ V}$. For Q_3 to be ON, its base-emitter junction must be forward biased. When Q_4 is ON, D has to be ON for Q_3 to be ON simultaneously. So, the base voltage of Q_3 must be $V_{B3} = V_{CE4}(\text{sat}) + V_D + V_{BE3} \approx 0.7 + 0.3 + 0.7 \approx 1.7$ V, for it to be ON. Since V_{B3} is only 1 V when Q_4 is ON, Q_3 cannot be ON. Hence, it can be concluded that Q_3 and Q_4 do not be conduct simultaneously.

Advantages of totem-pole

1. Even though the circuit can work with Q_3 and D removed and R_4 connected directly to the collector of Q_4, with Q_3 in the circuit, there is no current through R_4 in the output LOW state. So, the inclusion of Q_3 and D keeps the circuit power dissipation low.

2. In the output HIGH state, Q_3 acts as an emitter follower with its associated low output impedance. This low output impedance provides a small time constant for charging up any capacitive load on the output. This action is commonly referred to as active pull-up and it provides very fast rise time waveforms at TTL output.

Disadvantages of totem-pole

1. During transition of the output from LOW to HIGH, Q_4 turns off more slowly than Q_3 turns on, and so, there is a period of a few nanoseconds during which both Q_3 and Q_4 are conducting and, therefore, relatively large currents will be drawn from the supply. So, TTL circuits suffer from internally generated current transients or current spikes because of the totem-pole connection.
2. Totem-pole outputs cannot be wire ANDed, that is, the outputs of a number of gates cannot be tied together to obtain AND operation of those outputs.

Current Sinking

A TTL circuit acts as a *current sink* in LOW state (as shown in Fig. 11.7a), in that, it receives current from the input of the gate it is driving. Q_4 is the current-sinking transistor or the pull-down transistor, because it brings the output voltage down to its LOW state.

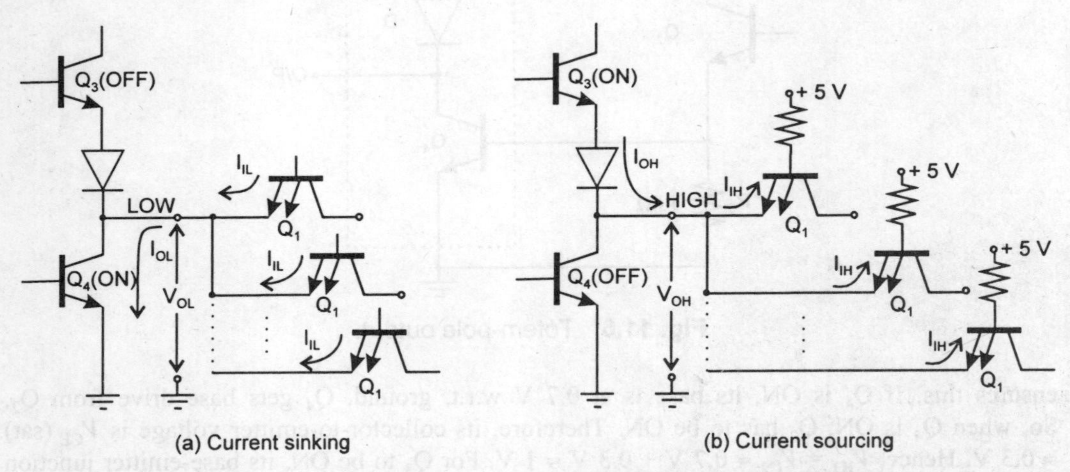

Fig. 11.7 Current sinking and current sourcing in a NAND gate.

Current Sourcing

A TTL circuit acts as a *current source* in the HIGH state (as shown in Fig. 11.7b), in that, it supplies current to the gate it is driving. Q_3 is the current-sourcing transistor or the pull-up transistor, because it pulls up the output voltage to its HIGH state.

TTL Loading and Fan-out

The TTL output has a limit, I_{OL}(max), on how much current it can sink in LOW state and a limit, I_{OH}(max), on how much current it can source in HIGH state. To determine the fan-out, we should know the drive capabilities of the output, i.e. I_{OL}(max) and I_{OH}(max) and the current requirements of each input, i.e. I_{IL} and I_{IH}.

$$\text{HIGH state fan-out} = \frac{I_{OH}\,(\text{max})}{I_{IH}}$$

$$\text{LOW state fan-out} = \frac{I_{OL}\,(\text{max})}{I_{IL}}$$

The smaller of these two numbers is the actual fan-out capability of the gate.

LOW state fan-out

Figure 11.8 shows a standard TTL output in LOW state connected to drive several standard TTL inputs. Transistor Q_4 is ON and is acting as a current sink for an amount of current I_{OL}, that is, the sum of the currents from each input. In its ON state, Q_4's collector-emitter resistance is very small but not zero. So, the current will produce a voltage drop V_{OL}. This voltage drop must not exceed the V_{OL}(max) limit of the IC. This limits the maximum value of I_{OL} and, thus, the number of loads that can be driven, that is, low state fan-out.

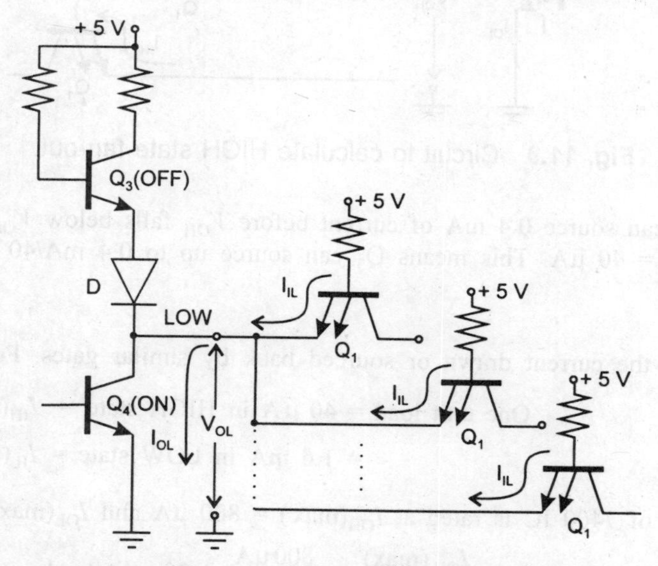

Fig. 11.8 Circuit to calculate LOW state fan-out.

Suppose, Q_4 can sink up to 16 mA before its output reaches V_{OL}(max) = 0.4 V. Suppose I_{IL} = 1.6 mA. This means that Q_4 can sink the current from up to 16/1.6 = 10 loads. If it is connected to more than 10 loads, V_{OL} increases above 0.4 V, and so, the noise margin is reduced and V_{OL} may even go to the indeterminate state.

HIGH state fan-out

When the TTL output is in a HIGH state, Q_3 is acting as an emitter follower, that is, sourcing a total current I_{OH}, which is the sum of the I_{IH} currents of the different TTL inputs as shown in Fig. 11.9. If too many loads are being driven, the current I_{OH} will become large enough to cause a large voltage drop across R_2, to bring V_{OH} $(= V_{CC} - V_{R2} - V_{BE3} - V_D)$ below V_{OH}(min). This is undesirable because it reduces the HIGH state noise margin and could even cause V_{OH} to go into the indeterminate range.

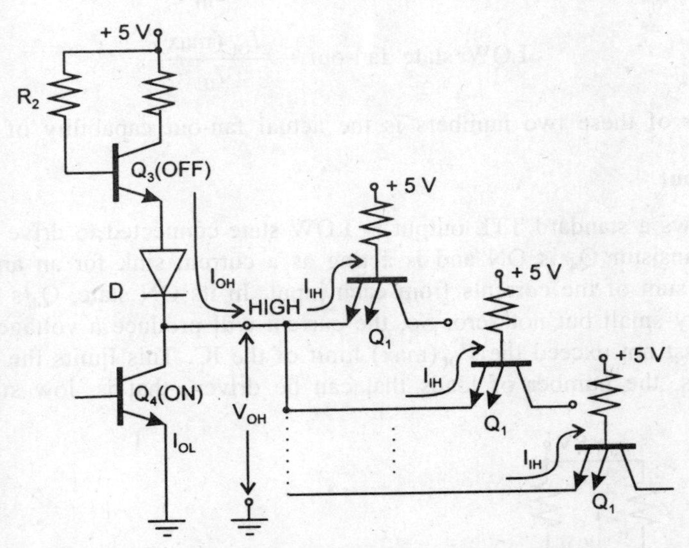

Fig. 11.9 Circuit to calculate HIGH state fan-out.

Suppose Q_3 can source 0.4 mA of current before V_{OH} falls below V_{OH}(min), and each load receives I_{IH} = 40 μA. This means Q_3 can source up to 0.4 mA/40 μA = 10 loads.

Unit load

Unit load means the current drawn or sourced back by similar gates. For 7400,

$$\text{One unit load} = 40\ \mu\text{A in HIGH state} = I_{IH}(\text{max})$$

$$= 1.6\ \text{mA in LOW state} = I_{IL}(\text{max})$$

If the output of 7400 IC is rated at I_{OH}(max) = 800 μA and I_{OL}(max) = 48 mA, then

$$\text{HIGH state fan-out} = \frac{I_{OH}(\text{max})}{I_{IH}} = \frac{800\ \mu\text{A}}{40\ \mu\text{A}} = 20 \text{ unit loads}$$

$$\text{LOW state fan-out} = \frac{I_{OL}(\text{max})}{I_{IL}} = \frac{48\ \text{mA}}{1.6\ \text{mA}} = 30 \text{ unit loads}$$

Therefore, the actual fan-out is equal to the smaller of the above two, i.e. 20 unit loads.

EXAMPLE 11.1 In Fig. 11.5, the 7400 NAND gate has V_{CC} = + 5 V and a 5 kΩ load connected to its output. Find the output voltage (a) when both the inputs are + 5 V and (b) when both the inputs are 0 V.

Solution

When both the inputs are HIGH, i.e. + 5 V, Q_2 and Q_4 are in saturation. When Q_4 is in saturation, its output is $V_{CE}(sat)$ = LOW = 0.3 V.

When both the inputs are LOW, i.e. 0 V, the output is HIGH (Q_3 is ON but Q_2 and Q_4 are OFF.) Therefore,

$$V_{OH} = V_{CC} - I_L(R_4) - V_{CE}(sat) - V_D$$
$$\approx [5 - (I_{OH} \times 130) - 0.1 - 0.7] \text{ V}$$
$$\approx (4.2 - 130 I_L) \text{ V}$$

The load current, $I_{OH} = \dfrac{V_{OH}}{5\ \text{k}\Omega}$ Therefore,

$$V_{OH} = \left[4.2 - 130 \times \frac{V_{OH}}{5000}\right] \text{ V}$$

or

$$V_{OH} \approx \frac{4.2}{1 + \dfrac{13}{500}} \text{ V} = 4.09 \text{ V}$$

EXAMPLE 11.2 In Fig. 11.5, what is the minimum value of the load resistance that can be used if the HIGH state output voltage is not be less than 3.5 V?

Solution

$$V_{OH} = 3.5 \text{ V} = V_{CC} - 130 I_{OH} - V_{CE}(sat) - V_D$$

or

$$3.5 = (5 - 130 I_{OH} - 0.1 - 0.7) \text{ V}$$
$$= (4.2 - 130 I_{OH}) \text{ V}$$
$$= \left[4.2 - 130 \times \frac{3.5}{R_L}\right] \text{ V}$$

Therefore,

$$R_L = \frac{130 \times 3.5}{0.7} = 650\ \Omega$$

EXAMPLE 11.3 Determine the fan-out of the circuit of Fig. 11.10. Also, find its noise margin.

Solution

The given circuit is a standard TTL inverter. From the data sheets, $I_{OL}(\text{max})$ = 16 mA and $I_{IL}(\text{max})$ = 1.6 mA. Therefore,

$$\text{Fan-out} = \frac{I_{OL}(\text{max})}{I_{IL}(\text{max})} = \frac{16 \text{ mA}}{1.6 \text{ mA}} = 10$$

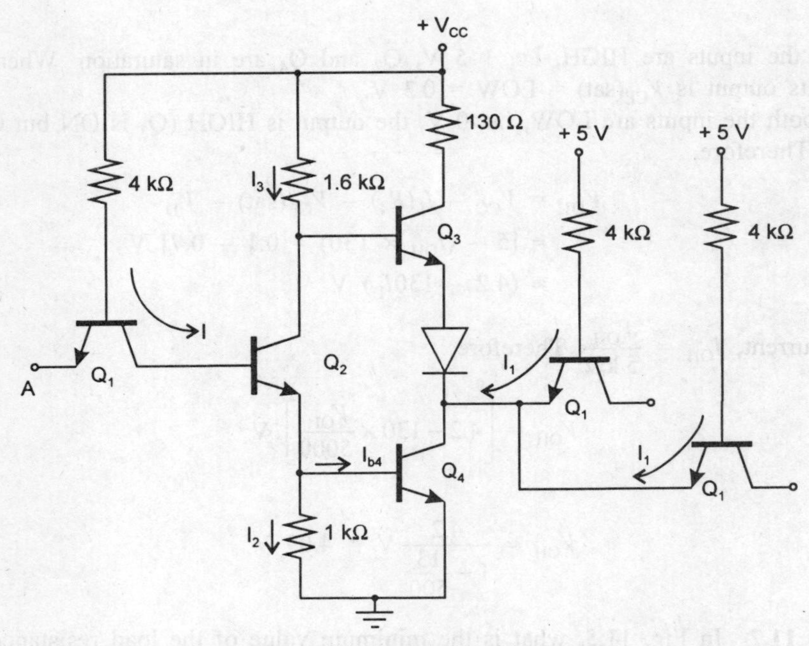

Fig. 11.10 Circuit for Example 11.3.

When the output is LOW, Q_4 is in saturation. Therefore,

$$V_{OL} = V_{CE}(\text{sat}) \approx 0.3 \text{ V}$$

But $V_{IL}(\text{max}) = 0.8$ V. Therefore,
Low level noise margin $= V_{HL} = V_{IL}(\text{max}) - V_{OL} = 0.8 \text{ V} - 0.3 \text{ V} = 0.5 \text{ V}$

Another way of determining fan-out

The fan-out is limited by the amount of current Q_4 can sink, when it is in saturation. Let $V_{OL}(\text{max}) = 0.4$ V (the limiting value). Let I_1 be the current sunk from each load gate. Therefore,

$$I_1 = \frac{V_{CC} - V_{BE}(\text{sat}) - V_{OL}(\text{max})}{4 \text{ k}\Omega} \approx \frac{(5 - 0.75 - 0.4) \text{ V}}{4 \text{ k}\Omega} \approx 1 \text{ mA}$$

The ability of the gate to sink current while keeping Q_4 in saturation is severely limited at its lowest operating temperature, – 55°C. This is about 30 mA. So, fan-out can be taken as 30 mA /1 mA = 30, but to keep V_{OL} well below the 0.4 V limit, $I_{OL}(\text{max})$ is limited to 16 mA. So, the fan-out should be less than 16 mA /1 mA = 16. For safety, the fan-out is taken as 10.

The approximate fan-out may also be calculated as follows:

Current drawn from each driven gate, $I_1 = \dfrac{V_{CC} - V_{BE1} - V_{OL}(\max)}{4\ \text{k}\Omega}$

$$= \frac{(5 - 0.75 - 0.4)\ \text{V}}{4\ \text{k}\Omega} \approx 1\ \text{mA}$$

Calculate the collector current of Q_4. In Fig. 11.10,

$$I = \frac{V_{CC} - V_{BE4} - V_{BE2} - V_{BC1}}{4\ \text{k}\Omega}$$

$$\approx \frac{(5 - 0.75 - 0.75 - 0.7)\ \text{V}}{4\ \text{k}\Omega}$$

$$\approx \frac{2.8\ \text{V}}{4\ \text{k}\Omega} \approx 0.7\ \text{mA}$$

$$I_2 = \frac{V_{BE4}}{1\ \text{k}\Omega} \approx \frac{0.7\ \text{V}}{1\ \text{k}\Omega} \approx 0.7\ \text{mA}$$

$$I_3 = \frac{V_{CC} - V_{B3}}{1.6\ \text{k}\Omega} = \frac{V_{CC} - V_{BE4} - V_{CE2}}{1.6\ \text{k}\Omega}$$

$$\approx \frac{(5 - 0.7 - 0.3)\ \text{V}}{1.6\ \text{k}\Omega} \approx \frac{4\ \text{V}}{1.6\ \text{k}\Omega} \approx 2.5\ \text{mA}$$

Therefore,

$$I_{b4} = I + I_3 - I_2 = (0.75 + 2.5 - 0.7)\ \text{mA} = 2.55\ \text{mA}$$

The transistor Q_4 is in saturation. Therefore, I_{C4} is also saturated. Let I_{C4} be about 4 to 5 times I_{b4} (worst case), i.e. about 10 to 15 mA. Hence, the fan-out is equal to 10 (the worst case).

11.5 OPEN-COLLECTOR GATES

The TTL gates may have totem-pole output or open-collector output. In open-collector TTL, the output is at the collector of Q_4 with nothing connected to it, (i.e. pull-up transistor Q_3 and diode D of the totem-pole output are omitted), therefore, the name *open collector*. The open-collector inverter circuit is shown in Fig. 11.11a. In order to get the proper HIGH and LOW logic levels out of the circuit, an external pull-up resistor is connected to V_{CC} from the collector of Q_4 as shown in Fig. 11.11b. When Q_4 is OFF, the output is pulled to V_{CC} through the external resistor R. When Q_4 is ON, the output is connected to near ground through the saturated transistor. The value of R must be so chosen that when one gate output goes LOW while the others are HIGH, the sink current through the LOW output does not exceed the $I_{OL}(\max)$ limit. Since the output is pulled to logic HIGH level through a

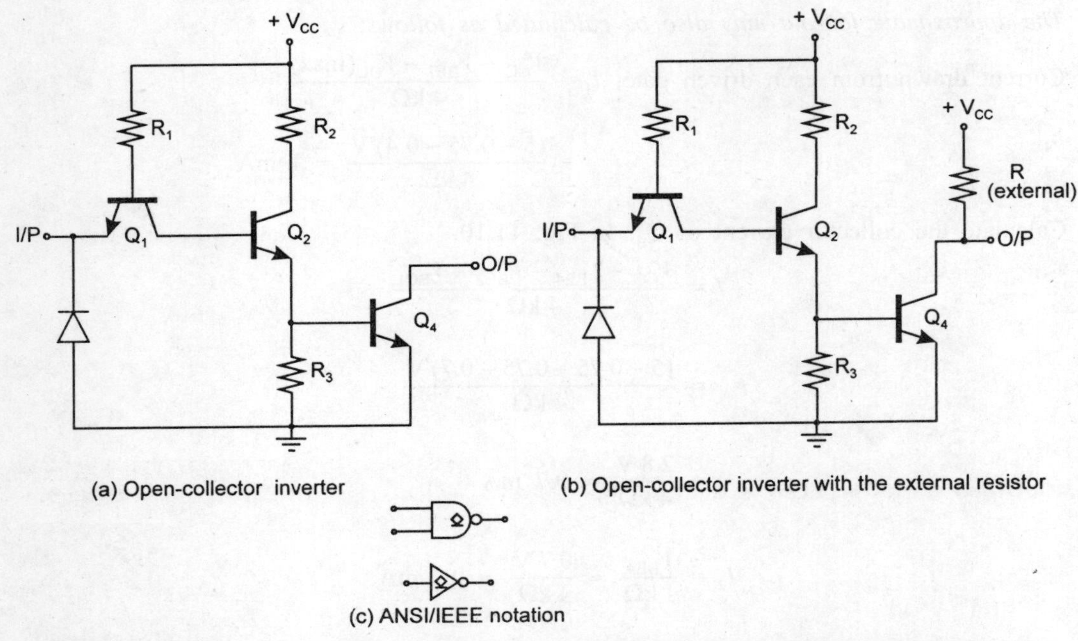

Fig. 11.11 Circuit diagram and logic symbol of open-collector inverter.

resistor, it is called the *passive pull-up*. The open-collector arrangement is much slower than the totem-pole arrangement, because the time constant with which the load capacitance charges in this case is considerably larger. (In the case of totem-pole, it is active pull-up, i.e. pull-up is through transistor Q_3. The R_{ON} of Q_3 is very small; so, the charging time constant is low and the output rises fast.) The speed can be increased only a little bit by choosing a smaller resistance. For this reason, the open-collector circuits should not be used in applications where switching speed is a principal consideration.

The traditional symbols for logic circuits with open-collector outputs are the same as those for totem-pole outputs, but the ANSI/IEEE symbology uses a distinctive notation to identify open-collector outputs. Figure 11.11c shows the ANSI/IEEE designation for an open-collector output. It is an 'underlined diamond'.

Wired AND Operation

Sometimes the outputs of a number of NAND gates may have to be ANDed. This can be achieved by using two more NAND gates (4 and 5) as shown in Fig. 11.12a. The same logic operation can be performed by simply tying the outputs of NAND gates 1, 2, and 3 as shown in Fig. 11.12b. This is called *wired AND operation*, because the AND operation is obtained by simply connecting the output wires together. With this, when any of the gate outputs go to a LOW state, the common output point also goes LOW as a result of its shorting to ground through the ON transistor. The common output will be HIGH only when all the gate outputs are in a HIGH state. This arrangement has the advantage of needing fewer gates compared to the conventional arrangement.

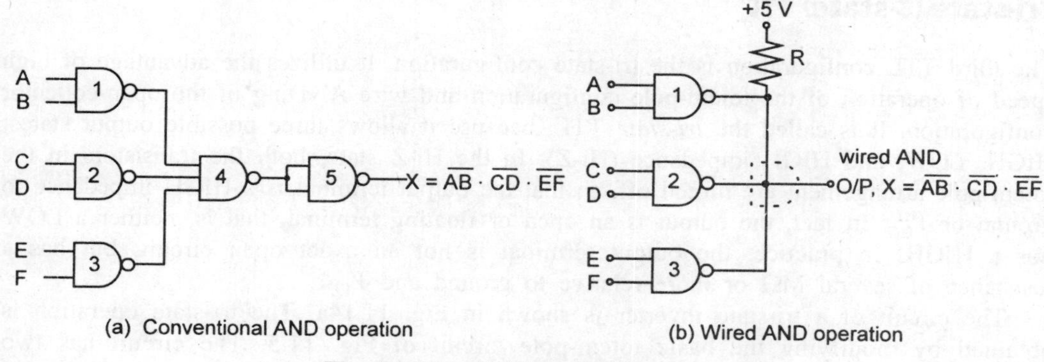

Fig. 11.12 ANDing of TTL gates.

The totem-pole outputs of gates cannot be wired ANDed, because when one output is HIGH and the other LOW and they are wired ANDed, a large current flows from supply to ground through Q_3 of the HIGH-state gate and Q_4 of the LOW-state gate. This is because Q_4 of the LOW-state gate acts as a very low resistance load on Q_3 of the HIGH-state gate (see Fig. 11.13). This large current can easily damage any of these transistors. The situation is even worse when more than two TTL outputs are tied together. The open-collector gates, on the other hand, may be wired ANDed without any problem.

Because of the absence of pull-up transistors, the wired-AND connections significantly reduce switching speeds. However, they are useful in reducing the chip count of a system when speed is not a consideration.

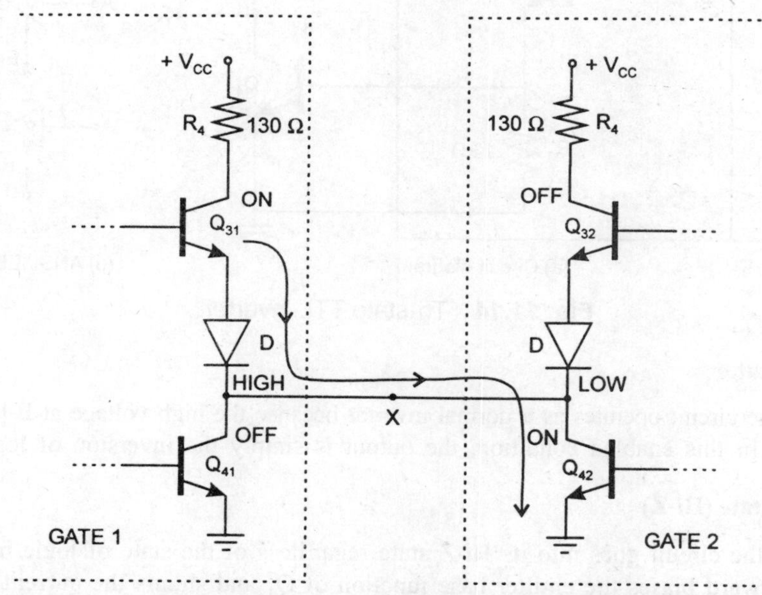

Fig. 11.13 Current flow in totem-pole gates when wired ANDed.

Tri-state (3-state) TTL

The third TTL configuration is the tri-state configuration. It utilizes the advantage of high speed of operation of the totem-pole configuration and wire ANDing of the open-collector configuration. It is called the *tri-state* TTL, because it allows three possible output states: HIGH, LOW, and HIGH impedance (Hi-Z). In the Hi-Z state, both the transistors in the totem-pole arrangement are turned off, so that the output terminal is a HIGH impedance to ground or V_{CC}. In fact, the output is an open or floating terminal, that is, neither a LOW nor a HIGH. In practice, the output terminal is not an exact open circuit, but has a resistance of several MΩ or more relative to ground and V_{CC}.

The circuit of a tri-state inverter is shown in Fig. 11.14a. The tri-state operation is obtained by modifying the basic totem-pole circuit of Fig. 11.5. The circuit has two inputs—A is the normal logic input and E is an enable input that can produce the Hi-Z.

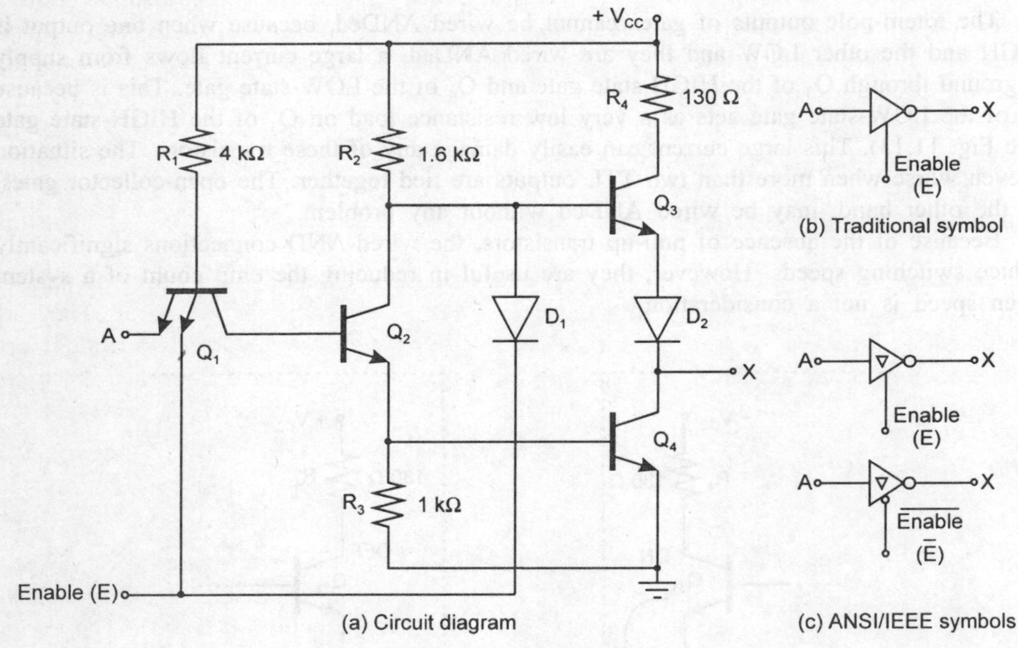

(a) Circuit diagram

Fig. 11.14 Tri-state TTL inverter.

The enabled state

With E = 1, the circuit operates as a normal inverter because the high voltage at E has no effect on Q_1 or Q_2. In this enabled condition, the output is simply the inversion of logic input A.

The disabled state (Hi-Z)

When E = 0, the circuit goes into its Hi-Z state regardless of the state of logic input A. The LOW at E forward biases the emitter base junction of Q_1 and shunts the current in R_1 away from Q_2, so that, Q_2 turns off, which in turn turns Q_4 off. The LOW at E also forward biases diode D_1 to shunt current away from the base of Q_3, and therefore, Q_3 also turns off. With both totem-pole transistors in the non-conducting state, the output terminal is essentially an open circuit.

There are many ICs that are designed with tri-state outputs. The advantage of the tri-state configuration is that the outputs of the tri-state ICs can be connected together without sacrificing the switching speed.

The traditional logic symbology has no special notation for tri-state outputs. Figure. 11.14c shows, the notation used in the ANSI/IEEE symbology to indicate a tri-state output. It is a 'triangle that points downwards'.

Buffer/Drivers

Any logic circuit that is called a *buffer*, a *driver*, or a *buffer/driver* is designed to have a greater output current and/or voltage capability than that of an ordinary logic circuit. Buffer/driver ICs are available with totem-pole outputs, open-collector outputs, or tri-state outputs. Some tri-state buffers also invert the signal as it goes through. They are called *inverting* tri-state buffers.

11.6 TTL SUBFAMILIES

Standard TTL, 74 Series

The standard TTL ICs, i.e. 74 series, offer a combination of speed and power dissipation suited for many applications. The 54 series is the counterpart of the 74 series. They are functionally equivalent to the 74 series, but can be operated over wider temperature and voltage ranges, as required by military specifications. Several other TTL series discussed below have also been developed. The standard TTL is now rarely used in new systems.

Low Power TTL, 74L Series

The low power TTL circuits, designated as the 74L series, have essentially the same basic circuit as the standard 74 series, except that all resistance values are increased (R_1 = 40 kΩ, R_2 = 20 kΩ, R_3 = 12 kΩ, and R_4 = 500 Ω). The larger resistors reduce the current and, therefore, the power requirement, but at the expense of reduction in speed. The power consumption of low power TTL is about 1/10 of that of standard TTL, but the standard TTL is more than three times faster than the low power TTL. The low power version is now not available in 7400 series. Low power Schottky TTL and CMOS versions of the 7400 series are now widely used instead.

High Speed TTL, 74H Series

The high speed TTL circuits, designated as the 74H series, have essentially the same basic circuit as the standard 74 series, except that smaller resistance values (R_1 = 2.8 kΩ, R_2 = 760 Ω, R_3 = 470 Ω, and R_4 = 58 Ω) are used and the emitter follower transistor Q_3 is replaced by a Darlington pair and emitter to base joining of Darlington pair ($Q_5 - Q_3$) is connected to ground through a resistance of 4 kΩ. The switching speed of the 74H series is approximately two times more than that of the standard TTL, as also the power consumption. Newer Schottky versions are superior in both speed and power consumption.

Schottky TTL, 74S Series

The standard TTL, low power TTL, and high speed TTL series operate using saturated switching. When a transistor is saturated, excess charge carriers will be stored in the base region and they must be removed before the transistor can be turned off. So, owing to storage time delay, the speed is reduced. The Schottky TTL 74S series reduces this storage time delay by not allowing the transistor to go into full saturation. This is accomplished by using a Schottky barrier diode (SBD) between the base and the collector of each transistor. Virtually, all modern TTL devices incorporate this so-called Schottky clamp. The SBD has a forward voltage of only 0.25 V. The circuits in the 74S series also use smaller resistance values to improve the speed of operation. The speed of the 74S series is twice that of the 74H series. Schottky TTL has more than three times the switching speed of standard TTL, at the expense of approximately doubling the power consumption. Figure 11.15 shows the circuit diagram of a two-input Schottky TTL NAND gate.

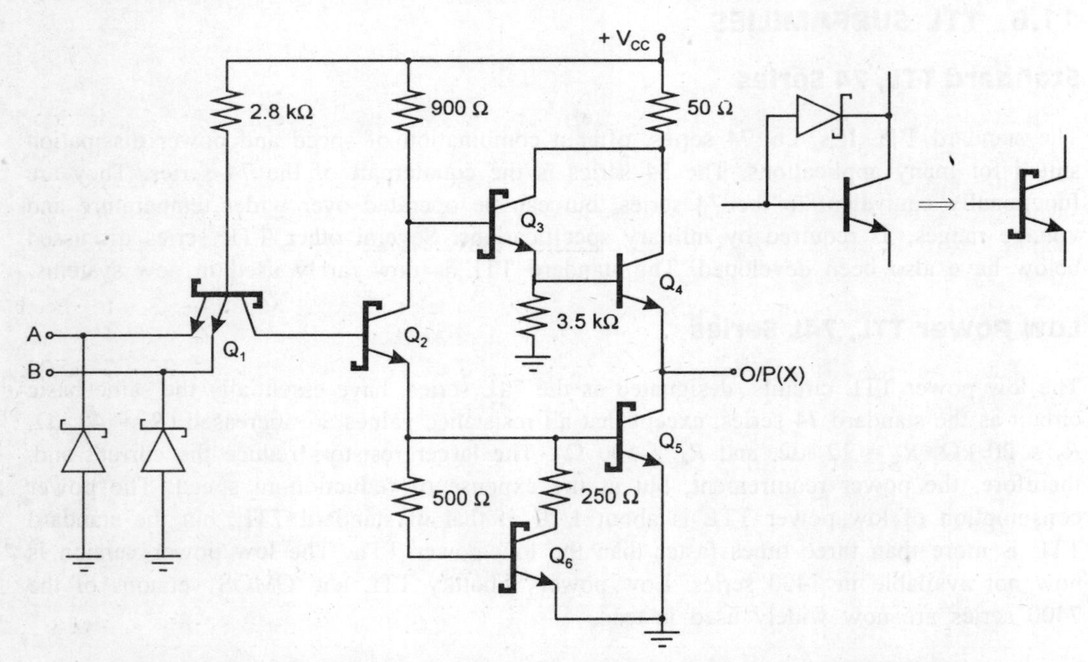

Fig. 11.15 Circuit diagram of a two-input Schottky TTL NAND gate.

Low Power Schottky TTL, 74LS Series

The 74LS series is a low-powered, slower-speed version of the 74S series. It uses the Schottky clamped transistor, but with larger resistance values than those in the 74S series.

The larger resistance values reduce the circuit power requirement but at the expense of reduction in speed. The switching speed of low power Schottky TTL is about the same as that of the standard TTL, but the power consumption is about 1/5 of the standard TTL. The 74LS NAND gate does not use the multiple emitter input transistor. Instead it uses diodes. This series is replacing the 74 series.

Advanced Schottky TTL, 74AS Series

This series has developed owing to recent innovations in IC design. It provides considerable improvement in speed over the 74S series and at a much lower power requirement. It is the fastest TTL series and its speed–power product is significantly lower than that of the 74S series. Its fan-out is larger than that of the 74S series because of its lower input current requirement. It is twice as fast and consumes less than half as much power as the 74S series.

Advanced Low Power Schottky TTL, 74ALS Series

This is a low power version of the advanced Schottky TTL. This series offers an improvement over the 74LS series in both speed and power dissipation. The 74ALS has the lowest speed–power product of all the TTL series, and it is very close to having the lowest gate power dissipation.

F(fast)TTL, 74F Series

This is the newest and fastest TTL series. Devices in this series have the letter 'F' inserted in their designations.

The ALS and AS technologies are the recent enhancements in Schottky TTL circuitry. Among other refinements, the advanced Schottky devices are fabricated with an improved doping technique and Schottky-clamped transistors provide improved isolation. These enhancements reduce capacitance and, thus, improve switching times.

Also, it is a more complex circuit design which uses additional active devices to speed up switching, reduce power consumption, and increase fan-out.

The inverter IC in different TTL sub-families is 7404, 74L04, 74H04, 74LS04, 74AS04, 74ALS04, and 74F04.

Typical TTL Series Characteristics

The typical characteristics of TTL subfamilies are shown in Table 10.2.

Table 11.2 The typical characteristics of TTL subfamilies

Performance rating	74	74L	74H	74S	74LS	74AS	74ALS
Propagation delay (ns)	9	33	6	3	9.5	1.7	4
Power dissipation (mW)	10	1	23	20	2	8	1.2
Speed-power product (pJ)	90	33	138	60	19	13.6	4.8
Max. clock rate (MHz)	35	3	50	125	45	200	70
Fan-out (same series)	10	20	10	20	20	40	20
Noise margin (V)	0.4	0.4	0.4	0.7	0.7	0.5	0.5

EXAMPLE 11.4 Determine the maximum average power dissipation and the maximum average propagation delay of a single gate of IC 7400.

Solution

From the data sheets of the 7400 NAND IC, the maximum values of I_{CCH} and I_{CCL} are 8 mA and 22 mA, respectively. The average I_{CC} is, therefore,

$$I_{CC}(\text{avg}) = \frac{I_{CCH} + I_{CCL}}{2} = \frac{(8 + 22)\,\text{mA}}{2} = 15 \text{ mA}$$

The average power is obtained by multiplying $I_{CC}(\text{avg})$ by V_{CC}. These I_{CC} values are obtained when V_{CC} has its maximum value of 5.25 V. Thus, we have:

The power drawn from the complete IC is, $P_D(\text{avg}) = 15 \text{ mA} \times 5.25 \text{ V} = 78.75 \text{ mW}$

The power drain of each NAND gate is, $\dfrac{P_D(\text{avg})}{4} = \dfrac{78.75 \text{ mW}}{4} = 19.7 \text{ mW}$

The maximum propagation delays for a 7400 NAND gate are

$$t_{PLH} = 22 \text{ ns} \quad \text{and} \quad t_{PHL} = 15 \text{ ns}$$

so that the average propagation delay is

$$t_{pd}(\text{avg}) = \frac{(22 + 15)\text{ ns}}{2} = 18.5 \text{ ns}$$

This is the worst case of maximum possible average propagation delay.

11.7 INTEGRATED INJECTION LOGIC (IIL OR I²L)

Integrated injection logic (IIL or I²L) or current injection logic (CIL) is the newest of the logic families, which is finding widespread use in LSI and VLSI circuits. It is not suitable for discrete gate ICs. The I²L logic gates are constructed using bipolar transistors only. The absence of resistors makes it possible to integrate a large number of gates on a single package. Complete microprocessors can be obtained on a single chip. The I²L circuits are easily fabricated and are economical. Their power consumption is also low. The speed–power product is constant and very small of the order of 4 pJ, comparable to advanced low power Schottky TTL. The I²L has, t_{pd} = 1 ns, P_D = 1 mw, NM = 0.35 V, fan-out = 8, and the relative cost is very low.

In I²L, since the currents are constant, no transients are produced as in TTL and MOS. It can easily be integrated on the same chip with bipolar analog circuits such as op-amps. By programming the injector currents, the propagation delay and power dissipation can be varied over a wide range. The disadvantage is that, it requires one more step in its manufacturing process than those used in MOS.

I²L Inverter

Since discrete gates are not available in I²L, the operation of an I²L inverter can be explained by considering the inverter of Fig. 11.16a that behaves in the same way as an I²L inverter.

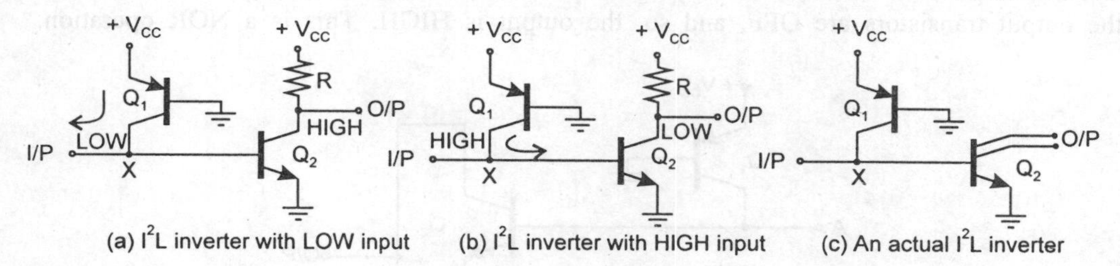

Fig. 11.16 I²L inverters.

The P–N–P transistor Q_1 serves as a constant current source that 'injects' current into node X. The direction in which the current flows after entering node X depends on the input level. A LOW input is a current sink. When the input is LOW, the injected current flows into the input, thus, diverting current from the base of Q_2. Transistor Q_2 is, therefore, OFF and the output is HIGH. If the input is HIGH, the injected current flows into the base of Q_2 turning it ON and making the output LOW as shown in Fig. 11.16b. Figure 11.16c shows an actual I²L inverter. The output transistor has two collectors (sometimes three), making it equivalent to two transistors with parallel bases and emitters. Thus, it produces two equal outputs. Instead of a collector resistor, the outputs are connected directly to the inputs of other I²L gates.

I²L NAND Gate

The I²L NAND gate shown in Fig. 11.17 is simply an inverter with inputs connected directly together at the inverter input. If, either input A, or input B, or both the inputs A and B are LOW (current sinks), the injected current flows into those inputs and Q_2 remains OFF (HIGH). If both the inputs are HIGH, the injected current turns on Q_2 making the output LOW. Thus, NAND operation is performed. The transistor Q_1 is called a current injector transistor, because when its emitter is connected to an external power source, it can supply current to the base of Q_2.

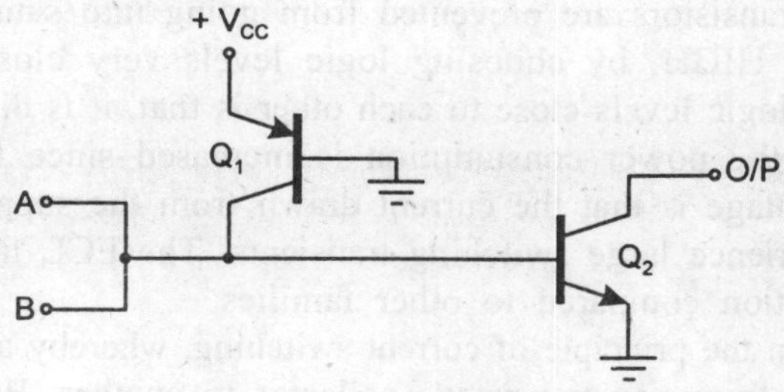

Fig. 11.17 Two-input I²L NAND gate.

I²L NOR Gate

The I²L NOR gate shown in Fig. 11.18 is simply two inverters with their outputs connected together. If either or both the inputs are HIGH, the corresponding output transistor is ON and the output is a current sink. So, the output is LOW. If both the inputs are LOW, both the output transistors are OFF, and so, the output is HIGH. This is a NOR operation.

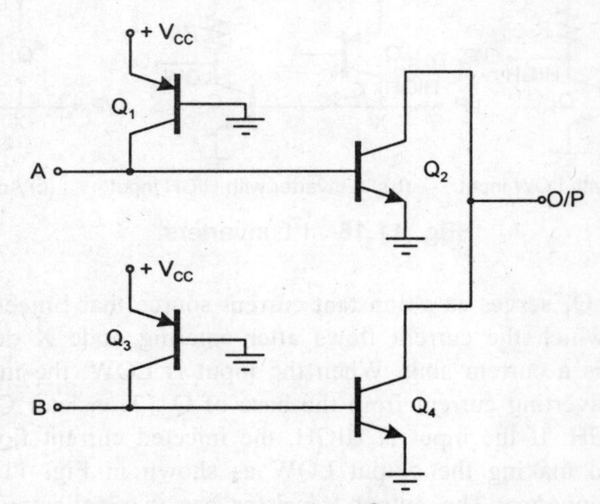

Fig. 11.18 Two-input I²L NOR gate.

11.8 EMITTER-COUPLED LOGIC (ECL)

Emitter-coupled logic (ECL), also called current-mode logic or current-steering logic, is the fastest of all logic families because of the following reasons:

1. It is a non-saturated logic, in the sense that the transistors are not allowed to go into saturation. So, storage time delays are eliminated and, therefore, the speed of operation is increased.
2. Currents are kept high, and the output impedance is so low that circuit and stray capacitances can be quickly charged and discharged.
3. The limited voltage swing.

The ECL is so named because of its use of BJTs that are coupled (joined) at their emitters. In ECL, the transistors are prevented from going into saturation when the input changes from LOW to HIGH, by choosing logic levels very close to each other. One disadvantage of having logic levels close to each other is that, it is difficult to achieve good noise immunity. Also, the power consumption is increased since the transistors are not saturated. But the advantage is that the current drawn from the supply is more steady and ECL gates do not experience large switching transients. The ECL family has considerably greater power consumption compared to other families.

The ECL operates on the principle of current switching, whereby a fixed bias current less than I_C(sat) is switched from one transistor's collector to another. Because of this current-mode operation, this logic form is also referred to as current-mode logic (CML). It is also called current-steering logic (CSL), because current is steered from one device to another. The ECL family is not as popular and widely used as the TTL and MOS, except in very high frequency applications where its speed is superior. It has the following drawbacks:

1. High cost
2. Low noise margin
3. High power dissipation
4. Its negative supply voltage and logic levels are not compatible with other logic families (making it difficult to use ECL in conjunction with TTL and MOS circuits).
5. Problem of cooling

Still, the ECL is used in superfast computers and high-speed special purpose applications. The ECL gates can be wired-ORed, no noise spikes are generated, and complementary outputs are also available. The important characteristics of ECL gates are:

1. Transistors never saturate. So, speed is high with t_{pd} = 1 ns.
2. Logic levels are negative, – 0.9 V for a logic 1 and – 1.7 V for a logic 0.
3. Noise margin is less, about 250 mV. This makes ECL unreliable for use in heavy industrial environment.
4. ECL circuits produce the output and its complement, and therefore, eliminate the need for inverters.
5. Fan-out is large because the output impedance is low. It is about 25.
6. Power dissipation per gate is large, P_D = 40 mW.
7. The total current flow in ECL is more or less constant. So, no noise spikes will be internally generated.

ECL OR/NOR Gate

Figure 11.19a shows a two-input ECL OR/NOR gate. Figure 11.19b shows its I/O characteristics. It has two outputs which are complements of each other. Transistors Q_2 and Q_{1A} form a differential amplifier. Transistors Q_{1A} and Q_{1B} are in parallel. Transistors Q_3 and Q_4 are emitter followers whose emitter voltages are the same as the base voltages (less than 0.8 V base to emitter drops). Inputs are applied to Q_{1A} and $Q_{1B,}$ and Q_2 is supplied with constant –1.3 V.

When the inputs A and B are both LOW, i.e. – 1.7 V, Q_2 is more forward biased than Q_{1A} and Q_{1B}, and so, Q_2 is ON and Q_{1A} and Q_{1B} are OFF. The value of R_2 is such that current flowing through Q_2 puts the collector at about – 0.9 V. Therefore, the emitter of Q_4 is at, – 0.9 – 0.8 = – 1.7 V, and so, the OR output is LOW. The base current of Q_3 passing through R_1 is very small. The value of R_1 is such that this current puts the collectors of Q_{1A} and Q_{1B} at about – 0.1 V. So, the emitter of Q_3 is at, – 0.1 – 0.8 = – 0.9 V, that is, the NOR output is HIGH.

When A is HIGH, or B is HIGH, or both A and B are HIGH, the corresponding transistors are ON, because they are more forward biased than Q_2, and Q_2 is OFF. So, the collectors of Q_{1A} and Q_{1B} are at – 0.9 V, which makes the NOR output = – 0.9 – 0.8 = – 1.7 V, i.e. a logic 0. Only the small base current of Q_4 flows through R_2. So, the collector of Q_2 is approximately at – 0.1 V, and therefore, the OR output is, – 0.1 – 0.8 = – 0.9 V, i.e. a logic 1. This shows that the above circuit works as a OR/NOR gate. Figure 11.19c shows the logic symbol a two-input ECL OR/NOR gate.

One advantage of the differential input circuitry in ECL gates is that, it provides common mode rejection—power supply noise common to both sides of the differential configuration is effectively cancelled out (differenced out). Also, since the ECL output is produced at an emitter follower, the output impedance is desirably low. As a consequence, the ECL gates not only have a large fan-out, but also are relatively unaffected by capacitive loads. Some ECL gates are available with multiple outputs, that are derived from multiple emitter transistors in the emitter-follower output. For example, one OR/NOR gate may have two OR outputs and two NOR outputs.

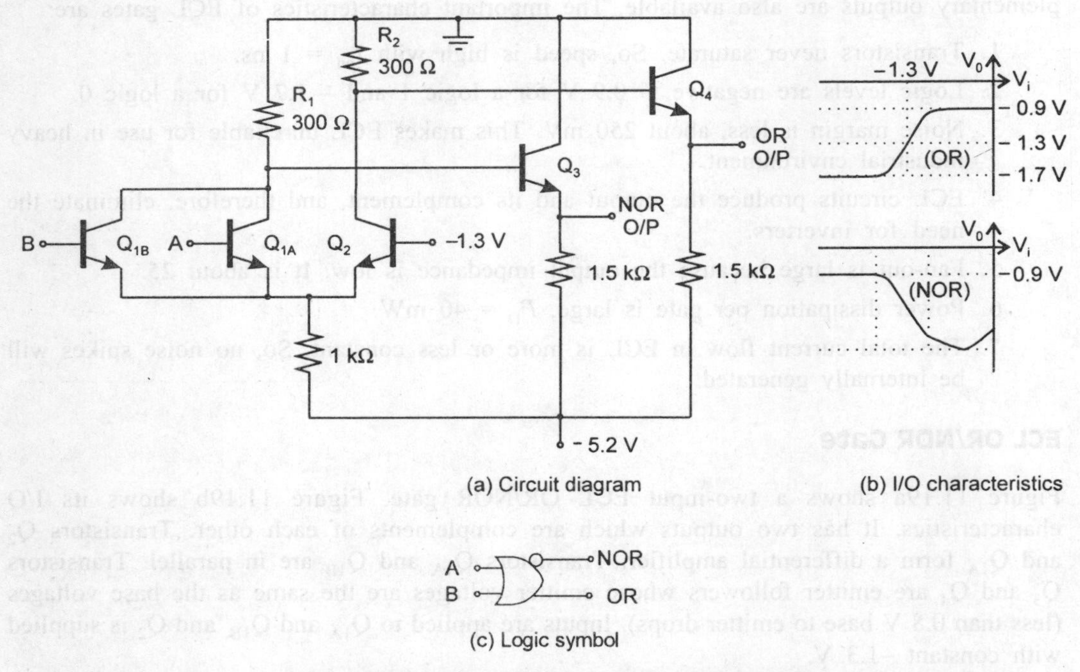

(a) Circuit diagram (b) I/O characteristics

(c) Logic symbol

Fig. 11.19 Two-input ECL OR/NOR gate

ECL Subfamilies

There are many ECL subfamilies. They differ in characteristics such as propagation delay, power dissipation per gate, and speed-power product. The ECL subfamilies do not include as wide a range of general purpose logic gates as do TTL and CMOS families. They do, however, include many complex special purpose circuits used in high speed digital data transmission, arithmetic units, and memories.

The first ECL series marketed by Motorola was the MECL-I series. It was followed by MECL-II series. Both these series are now obsolete. The MECL-III series carrying MC1600 numbers, the MECL10K series carrying MC10000 numbers, and the recent MECL10KH series with MC10H000 numbers are in use presently. The MECL10KH series has a t_{pd} of 1 ns and P_D of 25 mW, giving a speed–power product of 25 pJ (least of all ECL series).

Wired OR Connections

The ECL gates are available with open-emitter outputs, that is, with resistors in the output emitter followers omitted. The open-emitter outputs can be connected together directly, and the common emitter output terminal may be connected through an external resistor to a negative supply voltage (–5.2 V) as shown in Fig. 11.20a to perform a wired OR operation. The transistors labelled Q_3 are the output transistors of gates 1 and 2.

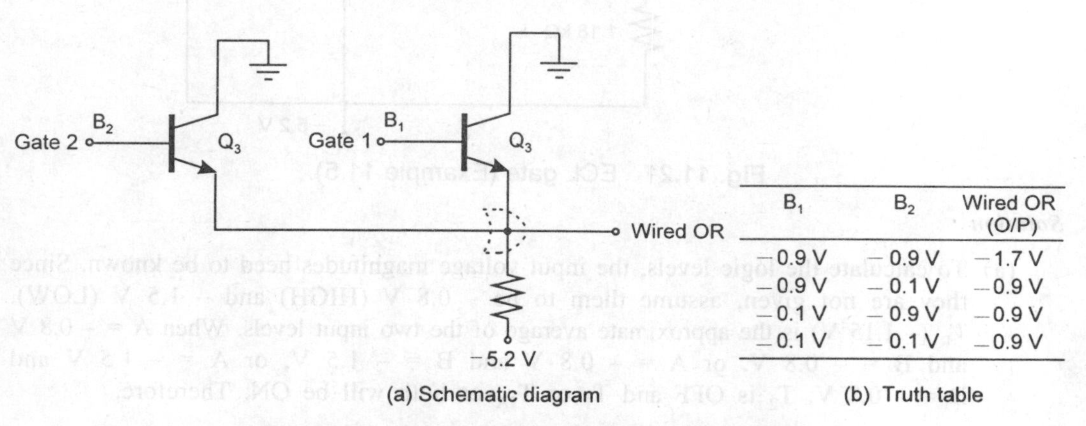

(a) Schematic diagram

B_1	B_2	Wired OR (O/P)
–0.9 V	–0.9 V	–1.7 V
–0.9 V	–0.1 V	–0.9 V
–0.1 V	–0.9 V	–0.9 V
–0.1 V	–0.1 V	–0.9 V

(b) Truth table

Fig. 11.20 Wired OR operation of ECL gates.

When the bases of both the transistors are at – 0.9 V, both the transistors conduct and make the common emitter voltage to be, – 0.9 V – 0.8 V = – 1.7 V. When both the bases are at – 0.1 V, again both the transistors conduct and make the output voltage to be, – 0.1 V – 0.8 V = – 0.9 V. When only one base is at – 0.1 V and the other at – 0.9 V, the output transistor with – 0.1 V base voltage conducts and makes the common emitter voltage – 0.9 V preventing the second transistor from conducting. Hence, the circuit provides OR operation.

Interfacing ECL Gates

Since ECL logic levels are so different from those of TTL and CMOS circuits, interfacing ECL gates with other logic families requires special level shifter circuits called *level translators*. Level translators are available in various ECL series to facilitate interfacing of ECL with other families.

EXAMPLE 11.5 What are the logic levels at the output of the ECL gate shown in Fig. 11.21?

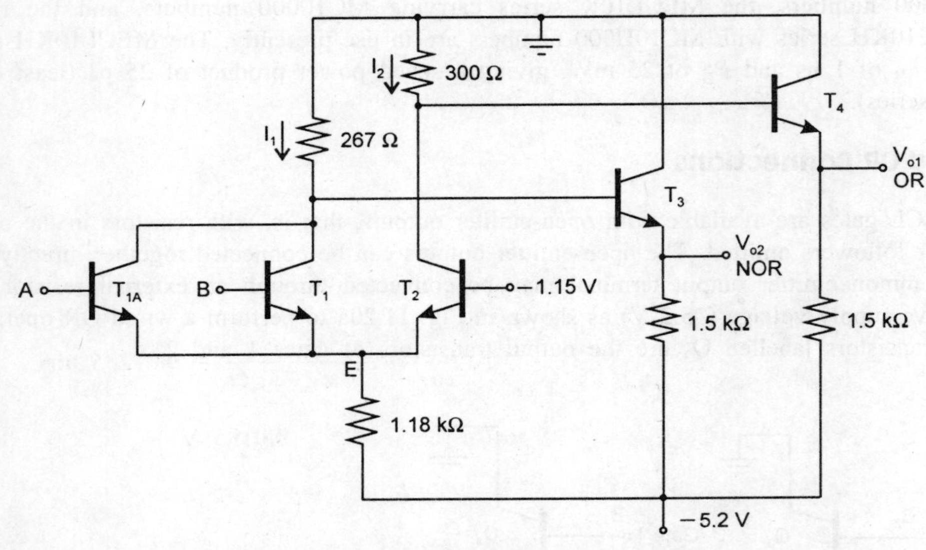

Fig. 11.21 ECL gate (Example 11.5).

Solution

(a) To calculate the logic levels, the input voltage magnitudes need to be known. Since they are not given, assume them to be − 0.8 V (HIGH) and − 1.5 V (LOW). V_R (− 1.15 V) is the approximate average of the two input levels. When A = − 0.8 V and B = − 0.8 V, or A = − 0.8 V and B = − 1.5 V, or A = − 1.5 V and B = − 0.8 V, T_2 is OFF and T_1 or T_{1A} or both will be ON. Therefore,

$$V_E = -0.8\text{ V} - 0.7\text{ V} = -1.5\text{ V}$$

and

$$I_E = \frac{[-1.5 - (-5.2)]\text{ V}}{1.18\text{ k}\Omega} \approx 3.1\text{ mA}$$

Therefore,

$$I_1 = I_E \frac{h_{FE}}{1 + h_{FE}} \approx 3.1\text{ mA}$$

$$V_{C1} = 0 - (267 \times 3.1 \times 10^{-3}) = -0.827\text{ V}$$

$$V_{O2} = (-0.827 - 0.7)\text{ V} = -1.527\text{ V}$$

V_{O2} is the OR output = − 1.527 V (logic 0)

When T_2 is OFF, I_2 is the small base current of T_4 given by

$$I_2 = \frac{[0 - 0.7 - (-5.2)]}{[300 + 1.5 \times 10^3 (1 + h_{FE})]\,\Omega} = \frac{4.5\text{ V}}{(300 + 1500 \times 41)\,\Omega} = 0.0728\text{ mA}$$

$$V_{C2} = 0 - (300 \times I_2) = (-300 \times 0.0728)\text{ V} = -0.0218\text{ V}$$

Therefore,

$$V_{O1} = (-0.0218 - 0.7)\text{ V} = -0.7218\text{ V (logic 1)}$$

When A = – 1.5 V and B = – 1.5 V, both T_1 and T_{1A} are OFF and T_2 is ON. When T_1 and T_{1A} are OFF, I_1 is the small base current of T_3 given by

$$I_1 = \frac{[0 - 0.7 - (-5.2)]\text{ V}}{[267 + 1.5 \times 10^3(1 + h_{FE})]\,\Omega} = \frac{4.5\text{ V}}{(267 + 1500 \times 41)\,\Omega} = 0.07285\text{ mA}$$

$$V_{C1} = 0 - (0.07285 \times 10^{-3} \times 267) = -0.0195\text{ V}$$

Therefore,

$$V_{O2} = -0.0195 - 0.7 = -0.7195\text{ V (logic 1)}$$

Since T_2 is ON, $V_E = -1.15 - 0.7 = -1.85$ V

Therefore,

$$I_E = \frac{[1.85 - (-5.2)]\text{ V}}{1.18\text{ k}\Omega} = 2.84\text{ mA}$$

and

$$I_2 = I_E \frac{h_{FE}}{1 + h_{FE}} \approx I_E = 2.84\text{ mA}$$

Therefore,

$$V_{C2} = 0 - (2.84 \times 10^{-3} \times 300) = -0.852\text{ V}$$

$$V_{O1} = -0.852 - 0.7 = -1.56\text{ V (logic 0)}$$

(b) To show that the transistors do not saturate, find the V_{CE} of the transistors. When T_2 is conducting, from the above calculations we see that its collector is at – 0.852 V and its emitter is at – 1.52 V. Therefore $V_{CE} = -0.852 - (-1.85) \approx 1$ V. Since $V_{CE} \approx 1$ V, the transistor is in the active region only. Similarly, when T_1 or T_{1A} or both are ON, $V_{C1} = -0.847$ V and $V_E = -1.5$ V. Therefore,

$$V_{CE} = -0.847 - (-1.5) = +0.653\text{ V}$$

So, the transistor is in the active region only.

(c) Noise margins:

For an ECL gate, the limits of transition region are – 1.1 V and – 1.25 V.

In the problem, we got logic HIGH as – 0.7218 V and logic LOW as – 1.52 V.

Therefore,

High level noise margin < 1 = – 0.73 V – (– 1.1 V) = + 0.37 V

Low level noise margin < 0 = – 1.25 V – (– 1.52 V) = + 0.27 V

These noise margins are typical and not worst case values.

11.9 METAL OXIDE SEMICONDUCTOR (MOS) LOGIC

The MOS logic is so named because it uses metal oxide semiconductor field effect transistors (MOSFETs). Compared to the bipolar logic families, the MOS families are simpler and inexpensive to fabricate, require much less power, have a better noise margin, a greater supply voltage range, a higher fan-out, and require much less chip area. But they are slower in operating speed and are susceptible to static charge damage. For MOS logic, t_{pd} = 50 ns, NM = 1.5 V (for + 5 V supply), P_D = 0.1 mW, and fan-out = 50 for frequencies greater than 100 Hz and it is virtually unlimited for dc or low frequencies. The propagation delay associated with MOS gates is large (50 ns) because of their high output resistance (100 kΩ) and capacitive loading presented by the driven gates.

The MOS logic is the simplest to fabricate and occupies very small space, because it requires only one basic element—an NMOS or a PMOS transistor. It does not require other elements like resistors and diodes, which occupy large space. Because of its ease of fabrication and lower power dissipation per gate P_D, it is ideally suited for LSI, VLSI, and ULSI for dedicated applications such as large memories, calculator chips, large microprocessors, etc. The operating speed of MOS is slower than that of TTL, so, they are hardly used in SSI and MSI applications. The greater packing density of MOS ICs results in higher reliability because of the reduction in the number of external connections.

Because of the very high impedance present at a MOSFET's input, the MOS logic families are more susceptible to static charge damage. The CMOS family is less susceptible to static charge damage.

There are presently two general types of MOSFETs—*depletion type* and *enhancement type*. The MOS digital ICs use enhancement MOSFETs exclusively. The MOSFETs can be of NMOS type or PMOS type. Most modern MOSFET circuitry is constructed using NMOS devices, because they operate at about three times the speed of their PMOS counterparts, and also have twice the packing density of PMOS.

Both NMOS and PMOS have greater packing density than that of CMOS, and are therefore, more economical than CMOS. The CMOS family has the greatest complexity and the lowest packing density of all the MOS families, but it possesses the important advantages of higher speed and much lower power dissipation. The CMOS can be operated at high voltage resulting in improved noise margin.

Symbols and Switching Action of NMOS and PMOS

Figure 11.22a shows the circuit symbol of NMOSFET. Figure 11.22b shows its equi-valent as a closed switch when it is ON and Fig. 11.22c shows its equivalent as an open switch when it is OFF.

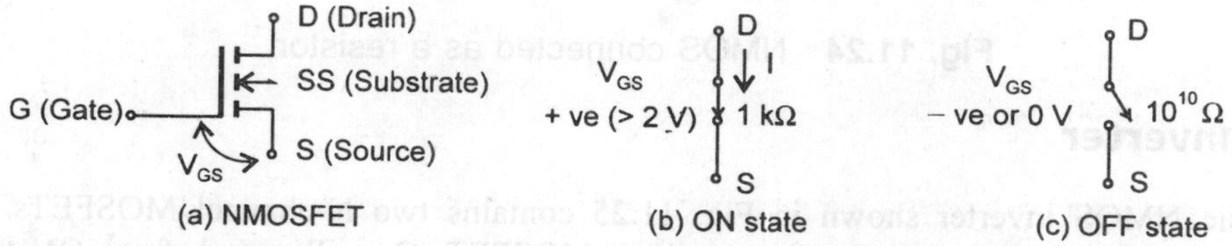

Fig. 11.22 Circuit symbol and ON and OFF equivalents of NMOSFET.

Figure 11.23a shows the circuit symbol of PMOSFET. Figure 11.23b shows its equivalent as a closed switch when it is ON and Fig. 11.23c shows its equivalent as an open switch when it is OFF.

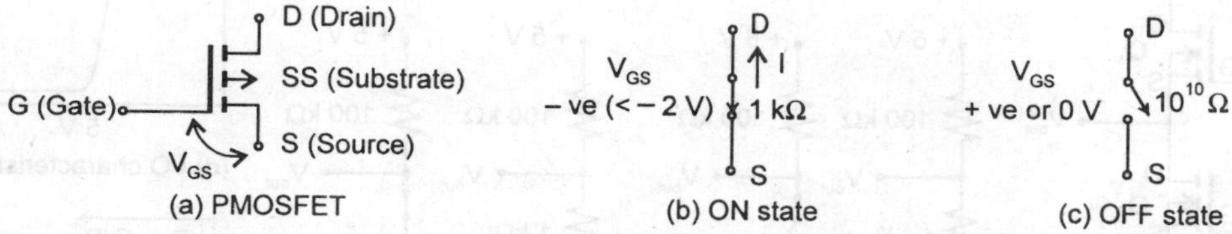

Fig. 11.23 Circuit symbol and ON and OFF equivalents of PMOSFET.

The arrow in the symbols of MOSFETs indicates either P or N channel. In the channel, the broken line between the source and the drain indicates that normally there is no conducting channel between these electrodes. The separation between the gate and the other terminals indicates the existence of very high resistance (10,000 MΩ) between the gate and the channel. The switch in a MOSFET is between the drain and source terminals. The gate-to-source voltage V_{GS} controls the switch. In an N-channel MOSFET, switch closes and current flows from drain to source when V_{GS} is positive, and switch opens when V_{GS} is negative or zero w.r.t. the source.

Resistor

A MOS transistor can be connected as a resistor as shown in Fig. 11.24. The value of the resistance presented by a resistor-connected NMOS device depends on the current through it. The gate is permanently connected to + 5 V, and so, it is always in the ON state and the transistor acts as a resistor of value R_{ON}. The load resistor is designed to have a narrower channel, so, its R_{ON} is much greater than the R_{ON} of the switching transistor. Typically, its R_{ON} = 100 kΩ.

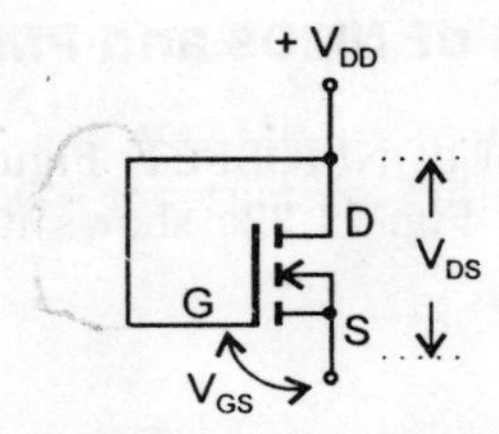

Fig. 11.24 NMOS connected as a resistor.

NMOS Inverter

The basic NMOS inverter shown in Fig. 11.25 contains two N-channel MOSFETs. Q_1 is called the load MOSFET and Q_2 the switching MOSFET. Q_2 will switch from ON to OFF in response to V_{in}. These two MOSFETs can be considered as resistors and the circuit as a potential divider.

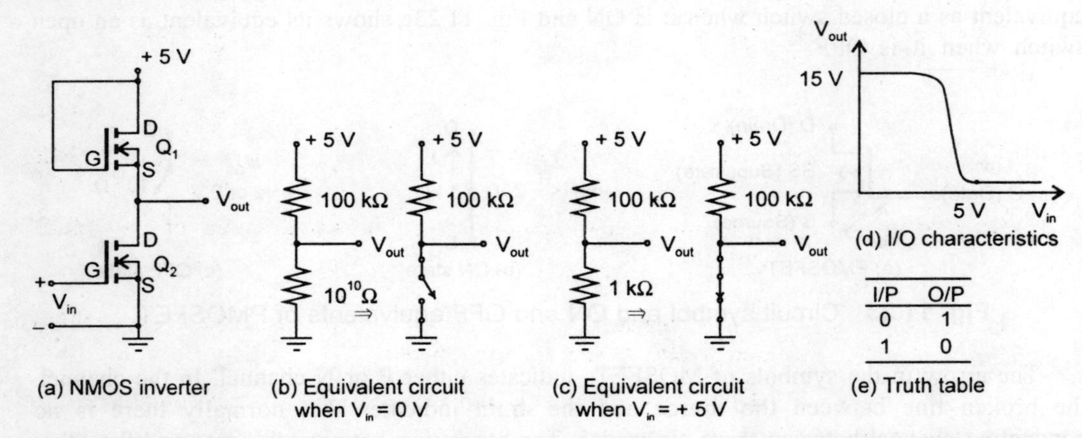

Fig. 11.25 Circuit diagram and equivalent circuits for various inputs of the NMOS inverter.

- When $V_{in} = 0$ V, Q_2 is OFF. So, its $R_{OFF} = 10^{10}\ \Omega$, and the equivalent circuit (b) results. Therefore,

$$V_{out} = \frac{V_{DD}\ R_{OFF}(Q_2)}{R_{ON}(Q_1) + R_{OFF}(Q_2)} \approx \frac{5 \times 10^{10}}{100 \times 10^3 + 10^{10}} \approx 5\text{ V}$$

- When $V_{in} = 5$ V, Q_2 is ON. So, its $R_{ON} = 1\ \text{k}\Omega$, and the equivalent circuit (c) results. Therefore,

$$V_{out} = \frac{V_{DD}\ R_{ON}(Q_2)}{R_{ON}(Q_1) + R_{ON}(Q_2)} = \frac{5 \times 1}{100 + 1} \approx 0\text{ V}$$

This shows that the above circuit acts as an inverter. The I/O characteristics and the truth table are shown in Figs. 11.25d and e, respectively.

NMOS NAND Gate

Figure 11.26 shows an NMOS two-input NAND gate and its equivalent circuits for different possible combinations of inputs in terms of resistance values of transistors in ON and OFF positions.

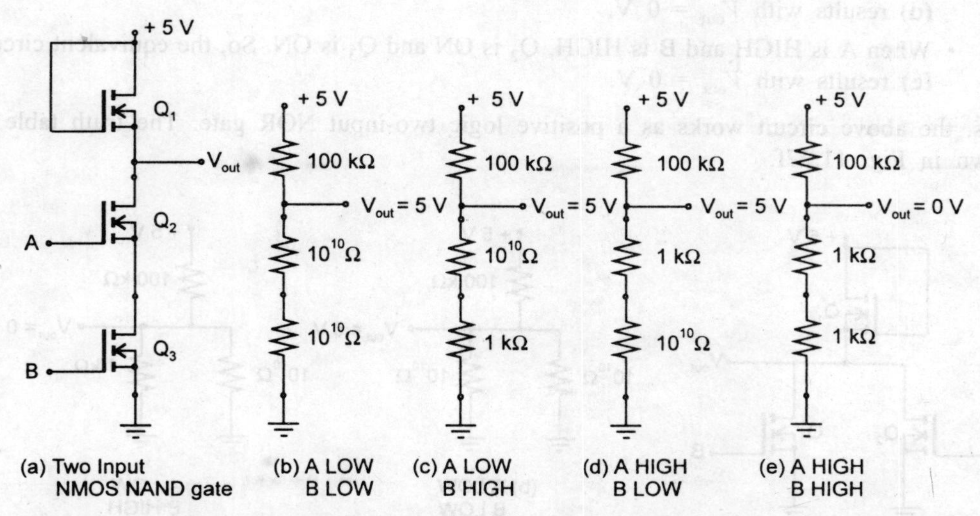

Fig. 11.26 Circuit diagram and equivalent circuits for various inputs of the NMOS NAND gate.

In the NMOS NAND gate shown, Q_1 is acting as a load resistor and Q_2 and Q_3 as switches controlled by input levels A and B, respectively.

- When both A and B are 0 V, both Q_2 and Q_3 are OFF. So, the equivalent circuit (b) results with V_{out} = + 5 V.
- When A = 0 V and B = + 5 V, Q_2 is OFF and Q_3 is ON. So, the equivalent circuit (c) results with V_{out} = + 5 V.
- When A = + 5 V and B = 0 V, Q_2 is ON and Q_3 is OFF. So, the equivalent circuit (d) results with V_{out} = + 5 V.
- When A = + 5 V and B = + 5 V, both Q_2 and Q_3 are ON. So, the equivalent circuit (e) results with V_{out} = 0 V.

Thus, the above circuit works as a positive logic two-input NAND gate.

NMOS NOR Gate

Figure 11.27 shows an NMOS two-input NOR gate and its equivalent circuits for different possible combinations of inputs in terms of resistance values of transistors in ON and OFF positions. Q_1 is the resistor-connected NMOS transistor that serves as a load and Q_2 and Q_3 are the switching transistors controlled by the inputs A and B, respectively.

- When A is LOW and B is LOW, Q_2 is OFF and Q_3 is OFF. So, the equivalent circuit (b) results with $V_{out} = +5$ V.
- When A is LOW and B is HIGH, Q_2 is OFF and Q_3 is ON. So, the equivalent circuit (c) results with $V_{out} = 0$ V.
- When A is HIGH and B is LOW, Q_2 is ON and Q_3 is OFF. So, the equivalent circuit (d) results with $V_{out} = 0$ V.
- When A is HIGH and B is HIGH, Q_2 is ON and Q_3 is ON. So, the equivalent circuit (e) results with $V_{out} = 0$ V.

Thus, the above circuit works as a positive logic two-input NOR gate. The truth table is shown in Fig. 11.27f.

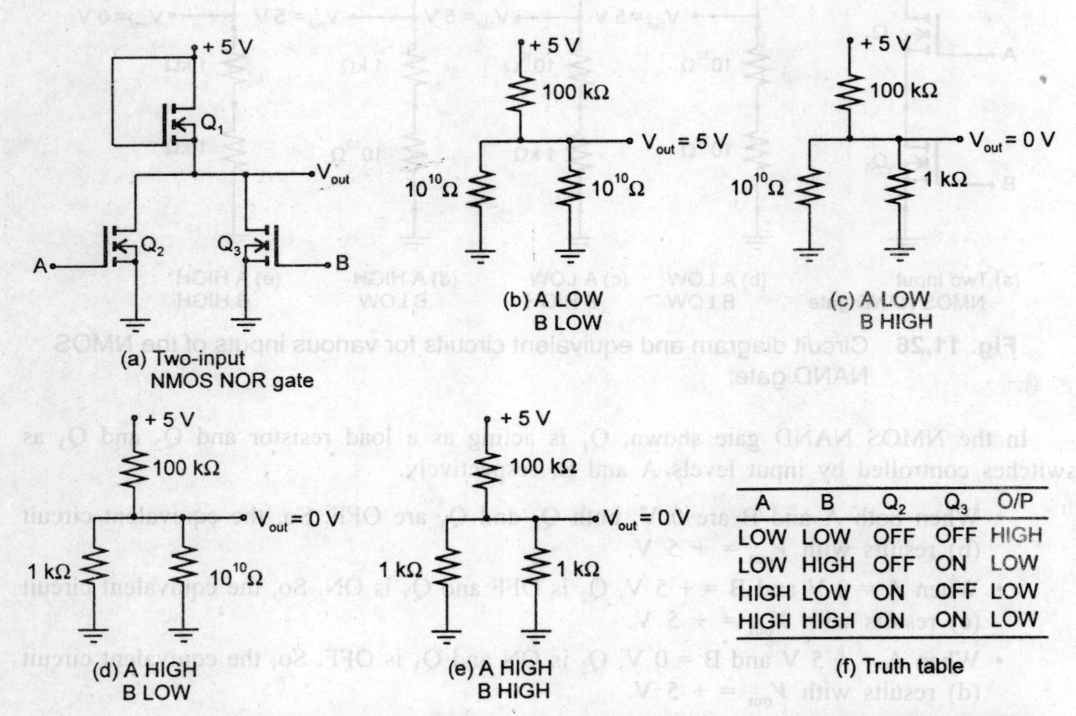

A	B	Q_2	Q_3	O/P
LOW	LOW	OFF	OFF	HIGH
LOW	HIGH	OFF	ON	LOW
HIGH	LOW	ON	OFF	LOW
HIGH	HIGH	ON	ON	LOW

(f) Truth table

Fig. 11.27 Circuit diagram and equivalent circuits for various inputs of the NMOS NOR gate.

11.10 COMPLEMENTARY METAL OXIDE SEMICONDUCTOR (CMOS) LOGIC

The CMOS logic family uses both P and N channel MOSFETs in the same circuit to realize several advantages over the PMOS and NMOS families. The CMOS family is faster and consumes less power than the other MOS families. These advantages are offset somewhat by the increased complexity of the IC fabrication process and a lower packing density. The

CMOS can be operated at higher voltages resulting in improved noise immunity. It is widely used for general purpose logic circuitry. The CMOS technology has been used to construct small, medium, and large scale ICs for a wide variety of applications ranging from general-purpose logic to microprocessors. Because of its extremely small power consumption, it is useful for applications in watches and calculators. The CMOS, however, cannot yet compete with MOS in applications requiring the utmost in LSI. The CMOS has very high input resistance. Thus, it draws almost zero current from the driving gate, and therefore, its fan-out is very high. Its output resistance is small (1 kΩ) compared to that of NMOS (100 kΩ). Hence, it is faster than NMOS. The speed of CMOS decreases with increase in load. In CMOS, there is always a very high resistance between the V_{DD} terminal and ground, because of the MOSFET in the current path. Hence, its power consumption is very low. The noise margin of CMOS is the same in both the LOW and HIGH states and it is 30% of V_{DD}, indicating that noise margin increases with an increase in power supply voltage. So in noisy environments, CMOS with large V_{DD} is preferred. However, an increase in V_{DD} results in the corresponding increase in P_D. The CMOS loses some of its advantages at high frequencies.

In MSI, the CMOS is also competitive with TTL. The CMOS fabrication process is simpler than that of the TTL and it has greater packing density, thereby permitting more circuitry in a given area and reducing the cost per function. The CMOS uses only a fraction of the power needed even for low power TTL and is, thus, ideally suited for applications requiring battery power or battery backup power. The CMOS is, however, generally slower than TTL.

CMOS Inverter

Figure 11.28 shows a CMOS inverter and its equivalent circuits for different inputs. It consists of an NMOS transistor Q_1 and a PMOS transistor Q_2. The input is connected to the gates of both the devices and the output is at the drain of both the devices. The positive supply voltage is connected to the source of the PMOS transistor Q_2, and the source of Q_1 is grounded.

- When V_{in} = 0 V (LOW), V_{GS2} = – 5 V, and V_{GS1} = 0 V. So, Q_2 is ON and Q_1 is OFF. Therefore, the switching circuit (b) results with V_{out} = 5 V.
- When V_{in} = + 5 V (HIGH), V_{GS2} = 0 V and V_{GS1} = + 5 V. So, Q_2 is OFF and Q_1 is ON. Therefore, the switching circuit (c) results with V_{out} = 0 V.

Thus, the above circuit acts as an inverter.

CMOS NAND Gate

Figure 11.29 shows a CMOS two-input NAND gate and its equivalent circuits for various input combinations. Here, Q_1 and Q_2 are parallel-connected PMOS transistors, and Q_3 and Q_4 are series-connected NMOS transistors.

- When A = 0 V and B = 0 V, V_{GS1} = V_{GS2} = – 5 V, V_{GS3} = V_{GS4} = 0 V. So, Q_1 is ON, Q_3 is OFF, Q_2 is ON and Q_4 is OFF. Thus, the switching circuit (b) results with V_{out} = + 5 V.

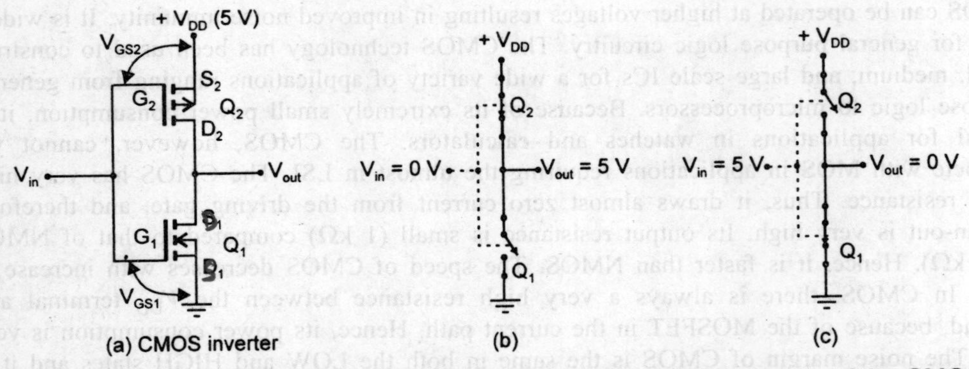

Fig. 11.28 Circuit diagram and equivalent circuits for various inputs of the CMOS inverter.

- When A = 0 V and B = + 5 V, V_{GS1} = – 5 V, V_{GS2} = 0 V, V_{GS3} = 0 V, V_{GS4} = 5 V. So, Q_1 is ON, Q_3 is OFF, Q_2 is OFF and Q_4 is ON. Thus, the switching circuit (c) results with V_{out} = + 5 V.
- When A = + 5 V and B = 0 V, V_{GS1} = 0 V, V_{GS2} = – 5 V, V_{GS3} = 5 V, V_{GS4} = 0 V. So, Q_1 is OFF, Q_3 is ON, Q_2 is ON and Q_4 is OFF. Thus, the switching circuit (d) results with V_{out} = + 5 V.
- When A = + 5 V and B = + 5 V, V_{GS1} = V_{GS2} = 0 V, V_{GS3} = V_{GS4} = 5 V. So, Q_1 is OFF, Q_3 is ON, Q_2 is OFF and Q_4 is ON. Thus, the switching circuit (e) results with V_{out} = 0 V.

Thus, the circuit works as a two-input NAND gate. The truth table is shown in Fig. 11.29f.

CMOS NOR Gate

Figure 11.30 shows a CMOS two-input NOR gate and its equivalent circuits for various input combinations. Here, the NMOS transistors Q_3 and Q_4 are connected in parallel and the PMOS transistors Q_1 and Q_2 in series.

The operation of the CMOS NOR gate can be explained as follows:

- When A = 0 V and B = 0 V, V_{GS1} = V_{GS2} = – 5 V, V_{GS3} = V_{GS4} = 0 V. So, Q_1 and Q_2 are ON, and Q_3 and Q_4 are OFF. Thus, the equivalent circuit (b) results with V_{out} = + 5 V.
- When A = 0 V and B = + 5 V, V_{GS1} = – 5 V, V_{GS2} = 0 V, V_{GS3} = 0 V, V_{GS4} = 5 V. So, Q_1 and Q_4 are ON, and Q_2 and Q_3 are OFF. Thus, the equivalent circuit (c) results with V_{out} = 0 V.
- When A = + 5 V and B = 0 V, V_{GS1} = 0 V, V_{GS2} = – 5 V, V_{GS3} = 5 V, V_{GS4} = 0 V. So, Q_1 and Q_4 are ON, and Q_2 and Q_3 are OFF. Thus, the equivalent circuit (d) results with V_{out} = 0 V.
- When A = + 5 V and B = + 5 V, V_{GS1} = V_{GS2} = 0 V, V_{GS3} = V_{GS4} = 5 V. So, Q_1 and Q_2 are OFF, and Q_3 and Q_4 are ON. Thus, the equivalent circuit (e) results with V_{out} = 0 V.

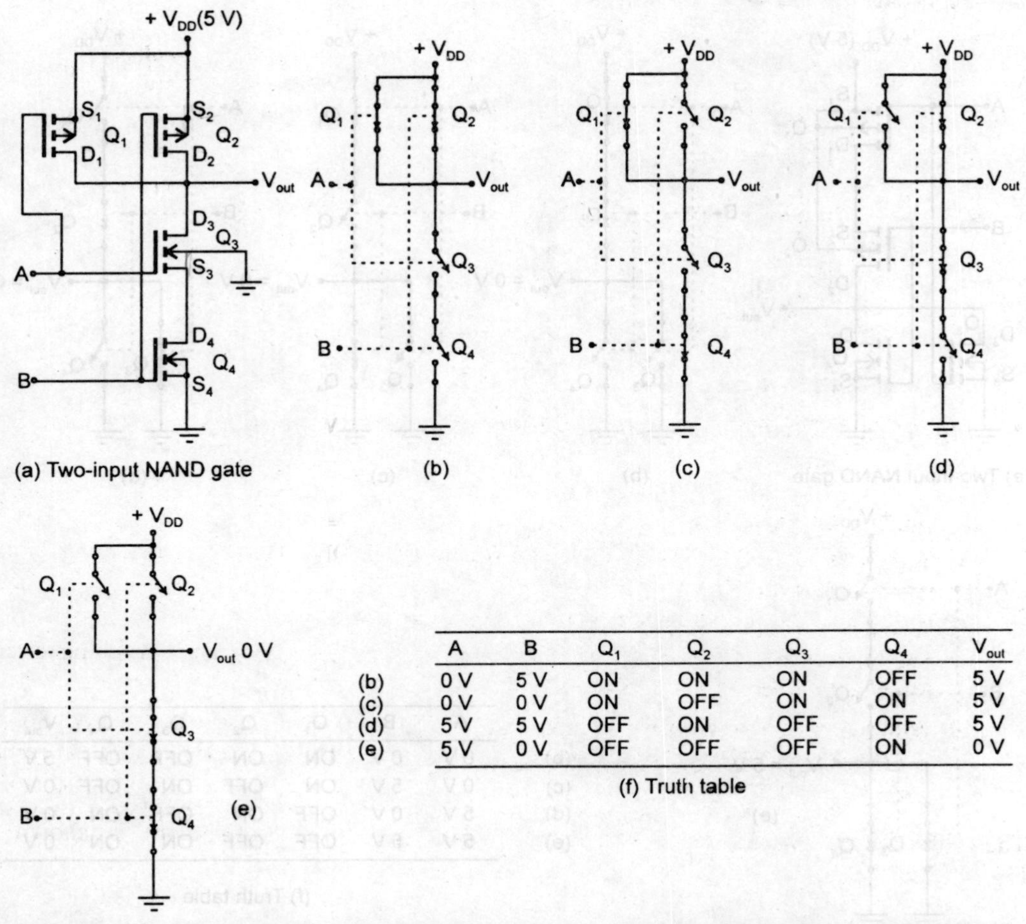

	A	B	Q_1	Q_2	Q_3	Q_4	V_{out}
(b)	0 V	5 V	ON	ON	ON	OFF	5 V
(c)	0 V	0 V	ON	OFF	ON	ON	5 V
(d)	5 V	5 V	OFF	ON	OFF	OFF	5 V
(e)	5 V	0 V	OFF	OFF	OFF	ON	0 V

(f) Truth table

Fig. 11.29 Circuit diagram and equivalent circuits for various inputs of the CMOS NAND gate.

The above analysis shows that the circuit works as a two-input NOR gate. The truth table is shown in Fig. 11.30f.

Buffered and Unbuffered Gates

Some metal gate CMOS circuits are available in buffered and unbuffered versions. The gates in buffered circuits have CMOS inverters in series with their outputs to suppress switching transients and to improve the sharpness of the voltage transition at the output. The gates discussed above are unbuffered gates.

Transmission Gate

A transmission gate is simply a digitally controlled CMOS switch. When the switch is open (OFF), the impedance between its terminals is very large. It is used to implement special

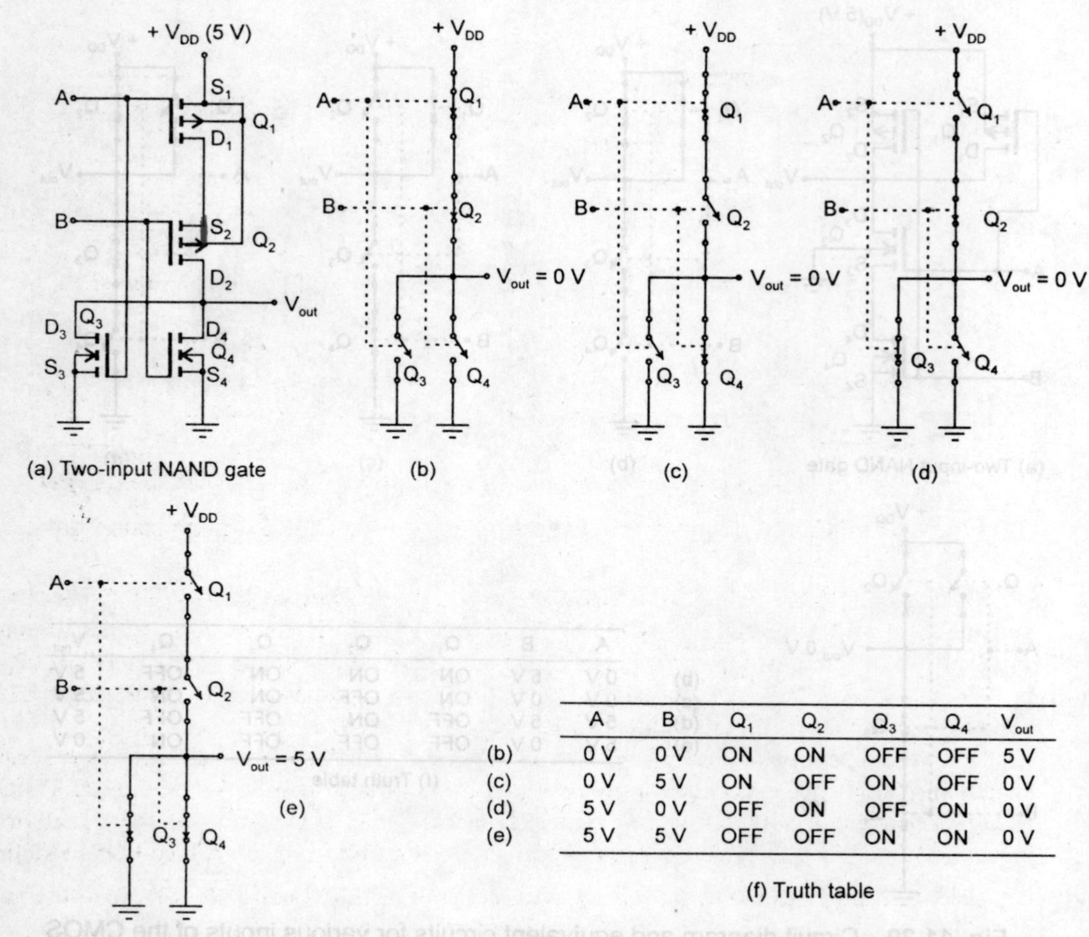

	A	B	Q_1	Q_2	Q_3	Q_4	V_{out}
(b)	0 V	0 V	ON	ON	OFF	OFF	5 V
(c)	0 V	5 V	ON	OFF	ON	OFF	0 V
(d)	5 V	0 V	OFF	ON	OFF	ON	0 V
(e)	5 V	5 V	OFF	OFF	ON	ON	0 V

(f) Truth table

Fig. 11.30 Circuit diagram and equivalent circuits for various inputs of the CMOS NOR gate.

logic functions. Since the CMOS gate can transmit signals in both directions, it is called a *bilateral transmission gate*. It is also called a bilateral switch. It is useful for digital and analog applications. The TTL and ECL gates are essentially unidirectional.

Figure 11.31 shows the schematic diagram and logic symbols of a CMOS transmission gate. The NMOS and PMOS transistors are connected in parallel. So, both polarities of input voltages can be switched. The CONTROL signal is connected to the NMOSFET and its inverse is connected to the PMOSFET. When the CONTROL is HIGH, the gate of PMOSFET Q_1 is LOW and the gate of NMOSFET Q_2 is HIGH. If the input (data) is LOW, V_{GS1} is 0 V and V_{GS2} is positive. So, Q_1 is OFF and Q_2 is ON. If the input is HIGH, V_{GS1} is negative and V_{GS2} is 0 V. So, Q_1 is ON and Q_2 is OFF. Thus, there is always one conducting path from input to output when the CONTROL is HIGH.

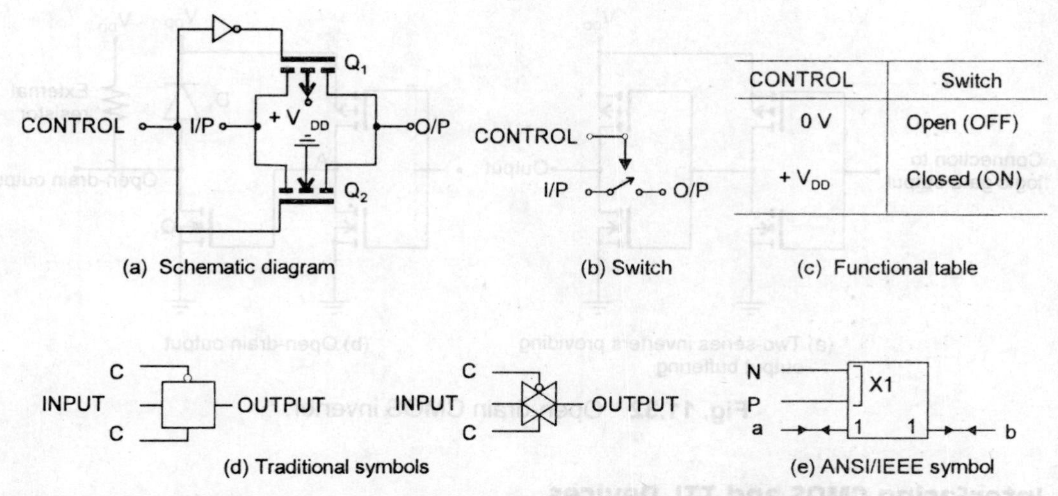

Fig. 11.31 Circuit diagram and logic symbols of the CMOS transmission gate.

On the other hand, when the CONTROL is LOW, the gate of PMOSFET Q_1 is HIGH and the gate of NMOSFET Q_2 is LOW. If the input (data) is LOW, V_{GS1} is positive and V_{GS2} is 0 V. Therefore, Q_1 is OFF and Q_2 is also OFF. If the input (data) is HIGH, V_{GS1} is 0 V and V_{GS2} is negative. So, again Q_1 is OFF and Q_2 is also OFF. Thus, there is no conducting path from input to output when the CONTROL is LOW.

So, we can conclude that when the CONTROL is HIGH, the circuit acts as a closed switch and allows the transmission of the signal from input to output. When the CONTROL is LOW, the circuit acts as an open switch and blocks the transmission of the signal from input to output. The CONTROL acts as an active-HIGH enabling signal. Active-LOW enabling is possible, if the CONTROL is connected to the gate of PMOS and $\overline{\text{CONTROL}}$ to the gate of NMOS.

Since the input and output terminals can be interchanged, the circuit can also transmit signals in the opposite direction. Hence, it acts as a bilateral switch.

Open Drain and High Impedance Outputs

The CMOS logic gates are available with open-drain outputs similar to their TTL counterparts with open-collector outputs. In these devices, the output stage consists only of an N-channel MOSFET whose drain is unconnected, since the upper P-channel MOSFET has been eliminated. An external pull-up resistor is needed to produce a HIGH state voltage level. Like open-collector outputs, the open-drain outputs can be wired ANDed. Figure 11.32a shows two inverters at the output of a CMOS gate that are used to provide the buffering. As shown in Fig. 11.32b, the open-drain output is obtained by omitting the PMOS transistor in the output inverter. Diode D_1 is connected internally to provide protection from electrostatic discharge.

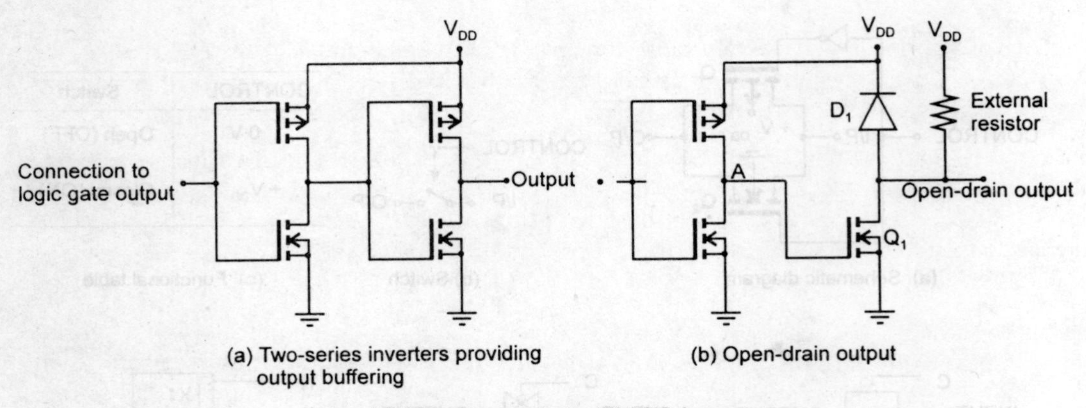

Fig. 11.32 Open-drain CMOS inverter.

Interfacing CMOS and TTL Devices

Specially designed ICs called level shifters are available to make devices from different logic families compatible with each other. The 74C series of CMOS is compatible pin for pin and function for function with TTL devices having the same number.

CMOS Series

4000/14000 series. It was the first CMOS series. The original 4000 series is now the 4000A series. The 4000B series is an improved version of the 4000A. The 4000B series has a higher output current capability than that of the 4000A series. The 4000A and 4000B series are still widely used despite the emergence of the new CMOS series.

74C series. This series is compatible pin for pin and function for function with TTL devices having the same number. Many, but not all, functions that are available in TTL are also available in this CMOS series. This makes it possible to replace some TTL circuits by an equivalent CMOS design. The performance characteristics of the 74C series are about the same as those of the 4000 series.

74HC series (High speed series). This is an improved version of the 74C series. High speed (10 times that of the 74C series) and higher output current capability are its main features. The speed of the devices in this series is compatible with that of the 74LS TTL series.

74 HCT series. This is also a high speed CMOS series and it is designed to be voltage compatible with TTL devices. In other words, it can be directly driven by a TTL output.

Operating and Performance Characteristics of CMOS

Supply voltage. The 4000 and 74C series can operate with V_{DD} values ranging from 3 to 15 volts. The 74HC and 74HCT series can operate with V_{DD} values ranging from 2 to 6 volts.

Voltage levels. When a CMOS output drives only a CMOS input and as a CMOS gate has an extremely high input resistance, the current drawn is almost zero and, therefore, the output voltage levels will be very close to zero for LOW state and V_{DD} for HIGH state, i.e. $V_{OL}(\max) = 0$ V, $V_{OH}(\min) = V_{DD}$. Usually, the input voltage levels are expressed as percentage of V_{DD} values, for example, $V_{IL}(\max) = 30\%$ of V_{DD}, $V_{IH}(\min) = 70\%$ of V_{DD}.

Power dissipation. When a CMOS circuit is in a static state, its power dissipation per gate is extremely small, but it increases with increase in operating frequency and supply voltage level. For dc, CMOS power dissipation is only 2.5 nW per gate when $V_{DD} = 5$ V, and it increases to 10 nW per gate when $V_{DD} = 10$ volts. With a V_{DD} of 10 V at a frequency of 100 KPPS, power dissipation is 0.1 mW/gate, and at 1 MHz, $P_D = 1$ mW.

Noise margins. Since for a CMOS gate, $V_{OL}(\max) = 0$ V, $V_{OH}(\min) = V_{DD}$ and $V_{IL}(\max)$ is 30% of V_{DD} and $V_{IH}(\min)$ is 70% of V_{DD}, the low level and high level noise margins will be the same (30% of V_{DD}) and increase with an increase in the value of V_{DD}. Of course, the higher values of V_{DD} result in higher power dissipations.

$$V_{NH} = V_{OH}(\min) - V_{IH}(\min) = V_{DD} - 70\% \text{ of } V_{DD} = 30\% \text{ of } V_{DD}$$

$$V_{NL} = V_{IL}(\max) - V_{OL}(\max) = 30\% \text{ of } V_{DD} - 0 \text{ V} = 30\% \text{ of } V_{DD}.$$

Fan-out. The CMOS fan-out depends on the permissible maximum propagation delay. For low frequencies (≤ 1 MHz), the fan-out is 50, and for high frequencies it will be less.

Switching speed. The speed of the CMOS gate increases with increase in V_{DD}. The 4000 series has $t_{pd} = 50$ ns at $V_{DD} = 5$ V and $t_{pd} = 25$ ns at $V_{DD} = 10$ V. The increase in V_{DD} results in increase in power dissipation too.

Unused inputs. The CMOS inputs should never be left disconnected. All CMOS inputs have to be tied either to a fixed voltage level (0 V or V_{DD}) or to another input.

Static charge susceptibility. The high input resistance of CMOS inputs makes CMOS gates prone to static charge build-up, that can produce voltages large enough to break down the dielectric insulation between the MOSFET gate and the channel. Most of the newer CMOS devices are protected against static charge damage by the inclusion of protective zener diodes on each input.

11.11 DYNAMIC MOS LOGIC

When power consumption and physical size are the prime design considerations as in digital watches and calculators, dynamic MOS logic is usually the family selected to meet these requirements. Each transistor used in a dynamic MOS circuit is identical to the other, and each can be fabricated in a very small amount of space on a chip. Consequently, large and very large scale integrations are possible.

In dynamic MOS logic, power consumption is minimized by relying on the inherent capacitance of the MOS transistors to store logic levels, i.e. to remain charged or discharged—and by using clock signals to turn on transistors for very brief intervals of time only. The clock signals turn transistors on to allow the capacitance to recharge or discharge at periodic intervals. Since a transistor is OFF during most of any given time interval, the average power consumption is quite small.

The NMOS transmission gate shown in Fig. 11.33a is a fundamental component of dynamic logic circuits. Because the NMOSFET is completely symmetrical, the drain and source terminals are indistinguishable, i.e. current can flow in either direction. In dynamic logic applications, there is a shunt capacitance at each of these terminals identified in the figure as C_1 and C_2.

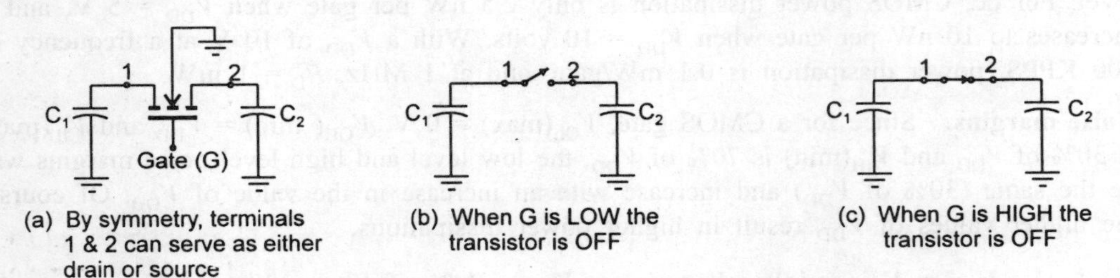

Fig. 11.33 NMOS transmission gate.

When the gate terminal G is LOW, the transistor will be OFF irrespective of the potentials at drain and source, i.e. irrespective of charges on C_1 and C_2, because the gate-to-source voltage may be either 0 or negative. Once the transistor is OFF, it acts as an open switch as shown in Fig. 11.33b and the charges on capacitors remain as they are. That is, no transfer of charge takes place, and therefore, no signal transmission takes place.

When G is HIGH, the transistor is ON and acts as a closed switch as shown in Fig 11.33c. If capacitors C_1 and C_2 are charged to the same level, no transfer of charge takes place. But if one capacitor is charged and the other discharged, transfer of charge takes place from one capacitor to the other, i.e. the input is transmitted to the output.

Dynamic MOS Inverter

Figure 11.34 shows a dynamic MOS inverter. The capacitance shown by dotted lines represents the inherent device (interelectrode) capacitance. The ϕ_1 and ϕ_2 are the control signals that are used to control the ON and OFF of Q_2 and Q_3. The two together are called

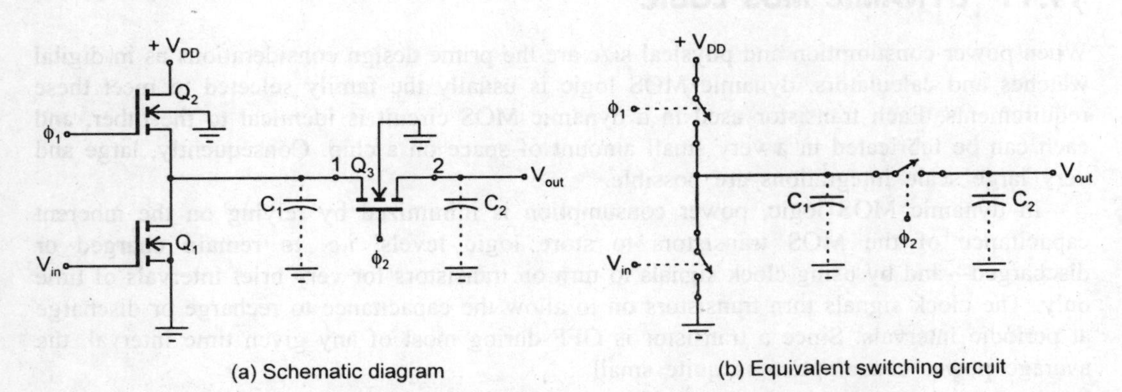

Fig. 11.34 Dynamic MOS inverter.

a two-phase non-overlapping clock, because ϕ_1 and ϕ_2 are never both HIGH at the same time. As in the case of a normal MOS inverter, Q_1 acts as a switching transistor and Q_2 as a load resistor. The only difference is that, in this case Q_2 acts as an active load only when clock ϕ_1 is HIGH. The rest of the time (i.e. when ϕ_1 is LOW), Q_2 is OFF and does not allow any current to pass through it. The transistor Q_3 acts as a transmission gate, i.e. it transfers charge only when clock ϕ_2 is HIGH. The rest of the time (i.e. when ϕ_2 is LOW), no transfer of charge takes place.

When V_{in} is LOW, Q_1 is OFF. When ϕ_1 goes HIGH, Q_2 conducts and C_1 is charged, but when ϕ_1 goes LOW, there is no path for C_1 to discharge, and so, C_1 remains charged. When ϕ_2 goes HIGH, this charge on C_1 is transferred to C_2, and so, V_{out} goes HIGH. Thus, a LOW at the input results in a HIGH at the output.

Suppose V_{in} is HIGH, when ϕ_1 goes HIGH, Q_2 conducts and Q_1 also turns on. So, C_1 cannot charge. When ϕ_2 goes HIGH, Q_3 acts as a closed switch and C_2 discharges into C_1. So, V_{out} goes LOW. V_{out} remains LOW when ϕ_2 is LOW. Thus, a HIGH at the input results in a LOW at the output. Therefore, the above circuit acts as an inverter.

The output of a dynamic logic gate is 'valid' only when ϕ_2 is HIGH. Thus, we can say that the gates are sampled at the frequency of ϕ_2. A sampled output becomes the input to other gates, whose responses become available only at the next sampling time. The disadvantage of dynamic logic is the complexity added by the clocking requirements. The capacitors need to be recharged periodically so that the charge on the capacitors does not decay very much. This process of recharging is called *refreshing*. The minimum clock frequency is, therefore, determined by the amount of time taken by the capacitance to decay significantly. A typical period is 1 ms, giving a minimum clock frequency of 1 kHz.

Dynamic NAND Gate

Figure 11.35 shows a dynamic two-input MOS NAND gate and its equivalent switching circuit. The only difference between this and the static NMOS NAND gate is that, the load MOSFET is clocked by ϕ_1 and a transmission gate is added at the output and the outputs are clocked through the transmission gate by ϕ_2.

When ϕ_1 goes HIGH, C_1 is charged according to the NAND logic of inputs A and B and when ϕ_2 goes HIGH, this charge is transferred from C_1 to C_2. So, the output V_{out} follows the NAND logic.

When either A is LOW, or B is LOW, or both A and B are LOW, the corresponding MOSFETs (Q_A and Q_B) will be OFF and no current passes through them. Thus, C_1 is charged when ϕ_1 goes HIGH, and this charge on C_1 is transferred to C_2, when ϕ_2 goes HIGH. Therefore, the output goes HIGH (C_1 remains discharged after ϕ_2 goes LOW).

Only when both A and B are HIGH, Q_A and Q_B will turn on when ϕ_1 goes HIGH and, therefore, no current flows through C_1 and it does not charge and remains in the discharged condition only. When ϕ_2 goes HIGH, C_2 discharges into C_1, and so the output goes LOW. Hence, this circuit works as a two-input NAND gate.

Dynamic NOR Gate

Figure 11.36 shows a dynamic two-input MOS NOR gate and its equivalent switching circuit. The only difference between this and the static NMOS NOR gate is that, the load

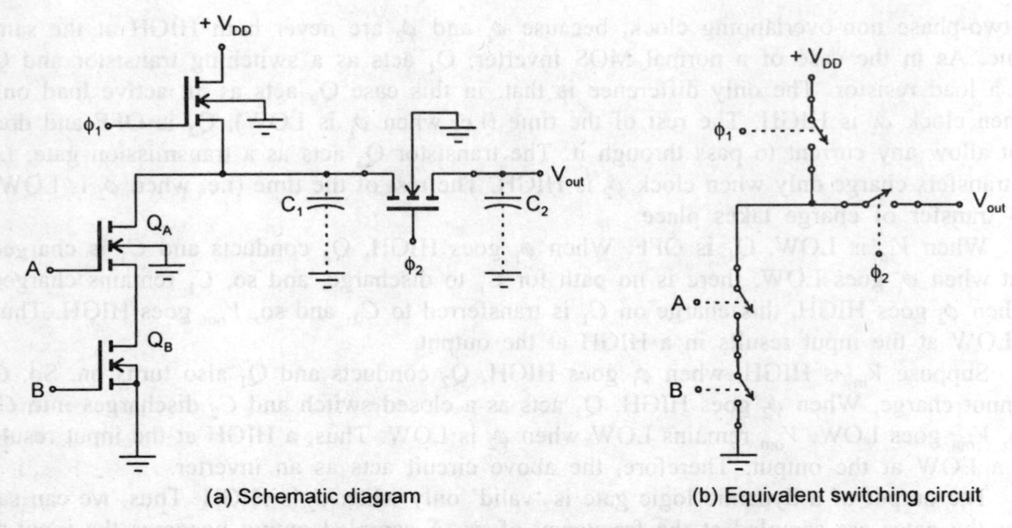

Fig. 11.35 Dynamic MOS NAND gate.

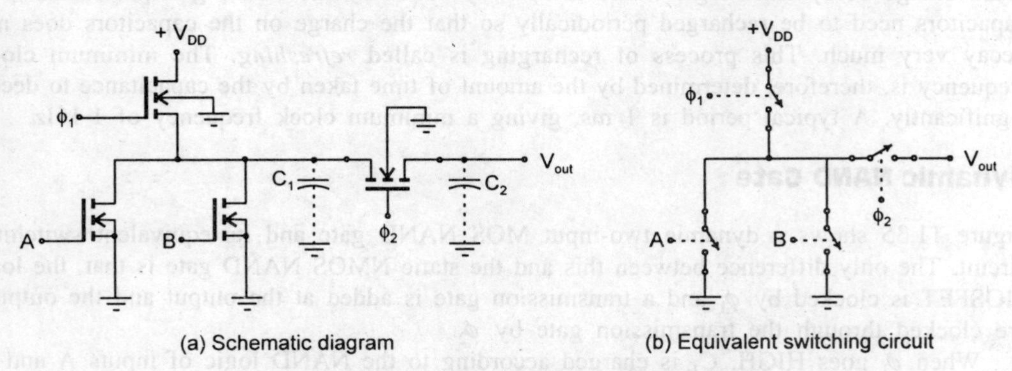

Fig. 11.36 Dynamic MOS NOR gate.

MOSFET is clocked by ϕ_1 and a transmission gate is added at the output and the outputs are clocked through the transmission gate by ϕ_2.

When ϕ_1 goes HIGH, C_1 is charged according to the NOR logic of inputs A and B, and when ϕ_2 goes HIGH, this charge is transferred from C_1 to C_2. So, the output V_{out} follows the NOR logic.

When both A and B are LOW, Q_A and Q_B will be OFF. So, C_1 charges when ϕ_1 goes HIGH and this charge on C_1 is transferred to C_2 when ϕ_2 goes HIGH, and so, the output goes HIGH. (C_1 remains discharged, after ϕ_2 goes LOW).

When either A is HIGH or B is HIGH or both A and B are HIGH, either Q_A or Q_B or both Q_A and Q_B will turn on when ϕ_1 goes HIGH, keeping C_1 in the discharged condition only. When ϕ_2 goes HIGH, the charge on C_2 is transferred to C_1, and so, the output goes LOW. Hence, this circuit works as a two-input NOR gate.

11.12 INTERFACING

Interfacing means connecting the output(s) of one circuit or system to the input(s) of another system with different electrical characteristics.

There are a number of logic families, each having its own strong points. In designing more complex digital systems, the designers utilize different logic families for different parts of the system in order to take advantage of the strong points of each family. When the designed parts are assembled, since the electrical characteristics of different logic families vary widely, interfacing circuits or logic level translators are used to connect the driver circuit belonging to one family to the load circuit belonging to another family.

TTL to ECL

The TTL is the most widely used logic family, but its speed of operation is not very high. The ECL is the fastest family. In some applications, the rate at which input data is to be handled may be much lower than the rate at which the output data is to be handled. Therefore, it becomes necessary to interconnect the two different logic systems, such as TTL and ECL. One such application is in the time division multiplexing of M digital signals to form a single digital signal. Although, the bit rate of each of the M signals may be handled using TTL, the bit rate of the composite signal is M times faster and may require ECL to process it.

ECL to TTL

Sometimes, the input data is at a faster rate, but the output data is at a slower rate like in demultiplexers. An ECL to TTL logic translator will be of use in such cases.

TTL to CMOS

The MOS and CMOS gates are slower than the TTL gates, but consume less space. Hence, there is an advantage in using TTL and MOS devices in combination.

The input current values of CMOS are extremely low compared with the output current capabilities of any TTL series. Thus, TTL has no problem in meeting the CMOS input current requirements. So, a level translator is used to raise the level of the output voltage of the TTL gate to an acceptable level for CMOS. The arrangement is shown in Fig. 11.37a, where the TTL output is connected to a + 5 V source with a pull-up resistor. The presence of the pull-up resistor will cause the TTL output to rise to approximately + 5 V in the HIGH state, thereby providing an adequate CMOS input.

If the TTL has to drive a high voltage CMOS, the pull-up resistor cannot be used to raise the level of the TTL output to the level of the CMOS input, since the TTL is sensitive to voltage levels. In such a case, an open collector buffer can be used to interface TTL to a high voltage CMOS as shown in Fig. 11.37b.

CMOS to TTL

The CMOS output can supply enough voltage and current to satisfy the TTL input requirements in the HIGH state. Hence, no special consideration is required for the HIGH

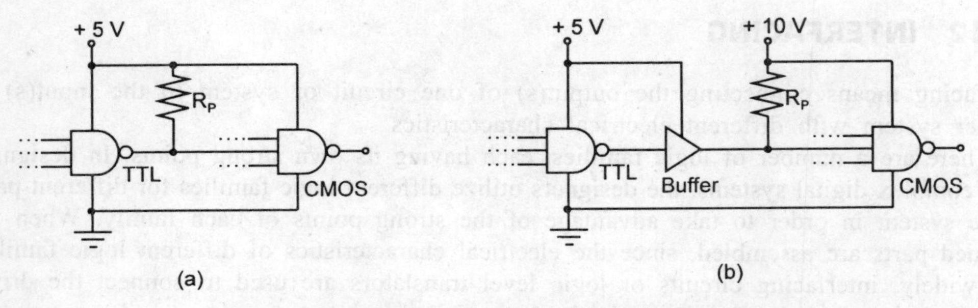

Fig. 11.37 TTL to CMOS interfacing.

state. But the TTL input current requirements at LOW state cannot be met directly. Therefore, an interface circuit with a LOW input current requirement and a sufficiently high output current rating is required. A CMOS buffer serves this purpose. The arrangement is shown in Fig. 11.38a.

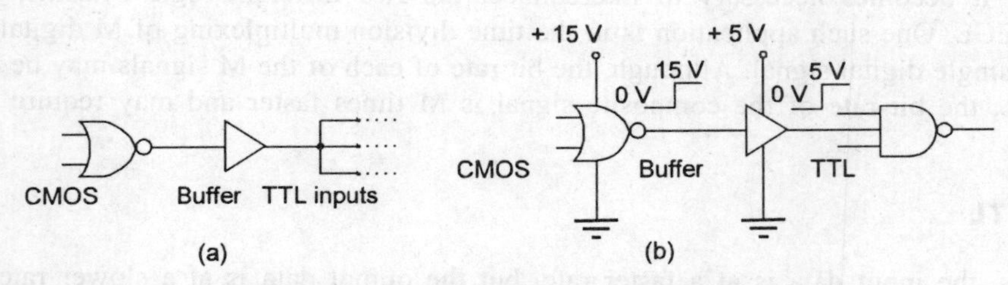

Fig. 11.38 CMOS to TTL interfacing.

When a high voltage CMOS has to drive a TTL gate, a voltage level translator that converts the high voltage input to a + 5 V output is used between CMOS and TTL as shown in Fig. 11.38b.

SUMMARY

- Propagation delay is the average transition delay time for a pulse to propagate from the input to the output of a switching circuit.
- The fan-in of a logic gate is defined as the number of inputs the gate can handle.
- The fan-out of a logic gate specifies the number of standard loads the output of a gate can drive without impairing its operation.
- A standard load is usually defined as the amount of current needed by an input of another gate of the same logic family.
- The noise margin represents the maximum noise signal that can be tolerated by a gate.
- The power dissipation of a logic gate is the supply power required by the gate to operate with 50% duty cycle at a specified frequency.
- The speed-power product is called the *figure of merit* of an IC family. It is the product of the propagation delay and the power dissipation per gate.

- The RTL, DTL, DCTL, and HTL logic families are now obsolete.
- The TTL is the most widely used logic family. There are eight TTL subfamilies.
- The TTL gates may be totem-pole type or open-collector type or tri-state type.
- The totem-pole outputs cannot be wired ANDed whereas open-collector outputs can be.
- A TTL circuit acts as a current sink in the LOW state and as a current source in the HIGH state.
- The TTL is the fastest of all saturated logic families. It is used in SSI and MSI ICs.
- Low power TTL uses large resistors to reduce the power requirements, which consequently reduce its speed of operation.
- High speed TTL uses smaller resistors and a Darlington pair in place of Q_3, to increase its speed.
- The speed of Schottky TTL is higher than that of standard TTL, because in Schottky TTL the transistors are not allowed to go into saturation by connecting a Schottky barrier diode between the base and the collector of each transistor.
- The totem-pole configuration has the advantages of high speed and low power dissipation but the disadvantages are those of generation of current spikes and the inability to be wired ANDed.
- In TTL gates, Q_3 is called the current-sourcing transistor or pull-up transistor because it pulls up the output voltage to its logic HIGH level and Q_4 is called the current-sinking transistor or pull-down transistor because it pulls down the output voltage to its logic LOW level.
- A tri-state gate has three states—HIGH, LOW, and HIGH impedance state. It utilizes the advantages of high speed of operation of the totem-pole configuration and wire ANDing of the open-collector configuration.
- The resistor R connected between V_{out} and + V_{CC} of an open-collector gate which pulls the V_{out} to V_{CC} level when Q_4 is OFF is called the pull-up resistor.
- The IIL has the highest packing density and is mainly suitable for VLSI and ULSI.
- The ECL is the fastest of all logic families and is mainly used in supercomputers. It is a non-saturated logic. Its logic levels are negative and complementary outputs are available as well.
- ECL gates, which are available in open-emitter configuration, can be wired ORed.
- The MOS logic is the simplest to fabricate.
- A MOS transistor can be connected as a resistor.
- Most modern MOSFET circuitry is constructed using NMOS devices because they are three times faster and two times denser than their PMOS counterparts.
- The MOS is ideally suited for LSI/VLSI/ULSI and for dedicated applications such as large memories, large microprocessor chips, etc.
- The MOS digital ICs use enhancement MOSFETs exclusively.
- The CMOS uses both N-channel and P-channel MOSFETs in the same circuit.
- The CMOS requires the least power, whereas the ECL requires the maximum power.

- The CMOS is ideally suited for battery-powered circuits.
- The CMOS has the greatest complexity and the lowest packing density of the MOS families, but it possesses the important advantages of higher speed and much less power dissipation.
- The CMOS can be operated at high voltage resulting in improved noise margin.
- The CMOS inputs should never be left disconnected. All CMOS inputs should be tied either to a fixed voltage level (0 V or V_{DD}) or to another input.
- The speed of the CMOS gate increases with increase in V_{DD}.
- The noise margin of CMOS gates is 30% of V_{DD}. Therefore, they are very much preferred in noisy environments.
- At higher frequencies, the CMOS loses some of its advantages over the other logic families.
- The fan-out of CMOS gates depends on the maximum permissible value of the propagation delay. The propagation delay of a CMOS gate increases with increase in load.
- Each CMOS load increases the propagation delay of the driving circuit by 3 ns, because each CMOS input typically presents a 5 pF load to ground.
- The CMOS has large fan-out because of its low output resistance.
- The CMOS is faster than NMOS because of its smaller output resistance than that of the MOS.
- The CMOS logic gates are available with open-drain outputs, similar to their TTL counterparts with open-collector outputs.
- The open-drain CMOS gates can be wired ANDed.
- A transmission gate is simply a digitally-controlled CMOS switch. It is a bilateral device.
- Both NMOS and PMOS are more economical than CMOS because of their greater packing density than that of CMOS.
- When power consumption and physical size are the prime design considerations, as in digital watches and calculators, the dynamic MOS logic is preferred.
- Interfacing means connecting the outputs of one circuit or system to the inputs of another system with different electrical characteristics.
- Level shifters are specially designed ICs which are used to make devices from different logic families compatible with each other.

QUESTIONS

1. Explain the following terms with reference to a gate: (a) propagation delay, (b) fan-in, (c) fan-out, (d) noise margin, and (e) speed-power product.
2. What do you mean by a standard load?
3. Name the different technologies that are used to manufacture ICs.
4. What are the merits and demerits of the TTL family?

5. Name the TTL outputs which can be wired ANDed and which cannot be.
6. Find the logic output of a TTL NAND gate that has all its inputs unconnected.
7. What are the two acceptable ways to handle unused inputs to an AND gate?
8. When does a TTL circuit act as a current source? As a current sink?
9. Name the three types of TTL gates.
10. Which TTL series is most suitable for a battery-powered circuit operating at 10 MHz?
11. What is wired ANDing?
12. Why shouldn't logic devices with totem-pole outputs be wired ANDed?
13. How do open-collector outputs differ from totem-pole outputs?
14. Why do open-collector outputs need a pull-up resistor?
15. Why are open-collector outputs generally slower than totem-pole outputs?
16. What factors are involved in determining the value for the pull-up resistor?
17. What are the three possible output states of a tri-state IC?
18. Why is low power TTL slower than the standard TTL?
19. What do you mean by Schottky TTL? Why is it faster than the standard TTL?
20. Which TTL series is the most suitable at high frequencies?
21. What are the characteristics of the ECL family?
22. Which is the non-saturated logic?
23. Which logic gives complementary outputs?
24. Which logic is preferred in superfast computers?
25. What determines the fan-out limitations of the MOS logic?
26. Why does the MOS family mostly use NMOS devices?
27. Why are MOS ICs especially sensitive to static charge?
28. What precautions need to be observed in the design and handling of MOS ICs?
29. What factors limit CMOS fan-out?
30. Which CMOS series are compatible pin for pin with TTL?
31. What should be done with unused CMOS inputs?
32. Describe what happens to each of the following CMOS characteristics as V_{DD} is increased? (a) noise margin (b) power dissipation (c) switching speed
33. In which application is CMOS ideally suited?
34. Which special CMOS circuit has no TTL or ECL counterpart?
35. What do you mean by a transmission gate?
36. Describe the operation of a CMOS bilateral switch. Is there any TTL bilateral switch?
37. Compare the following technologies:

    (a) Bipolar and CMOS (b) TTL and MOS (c) MOS and CMOS
38. Compare TTL, ECL, IIL, MOS and CMOS with respect to fan-in, fan-out, noise margin, t_{pd} and P_D.

39. What are the merits and demerits of various logic families?
40. What do you mean by dynamic MOS logic? When is it preferred?
41. What do you mean by interfacing?
42. Which is the fastest logic family? The slowest family?
43. Which family has the highest packing density? The lowest packing density?
44. Which logic is the simplest to fabricate? The most complex to fabricate?
45. Which logic family is suitable for SSI and MSI? For LSI and VLSI? For VLSI and ULSI?
46. Which logic family consumes the least power? The maximum power?
47. With the help of neat circuit diagrams, explain the working of:

    (a) A two-input TTL NAND gate
    (b) A two-input ECL OR/NOR gate
    (c) IIL NAND and NOR gates.

48. With the help of a neat circuit diagram, explain the working of:

    (a) A MOS inverter
    (b) A two-input MOS NAND gate
    (c) A two-input MOS NOR gate

49. With the help of a neat circuit diagram, explain the working of:

    (a) A CMOS inverter
    (b) A two-input CMOS NAND gate
    (c) A two-input CMOS NOR gate

50. When NAND gate inputs are tied together, they are always treated as single load on the signal source. True or False?
51. True or False?

    (a) The CMOS is faster but consumes more power than MOS families do.
    (b) The CMOS can pack more circuits on a chip than TTL can do.
    (c) The CMOS devices are ideal for battery-operated circuits.
    (d) In a CMOS inverter, both MOSFETs are ON at the same time.

52. True or False?

    (a) In ECL, the high speed of operation is obtained by preventing saturation of transistors.
    (b) ECL circuits usually have complementary outputs.
    (c) The noise margins for ECL circuits are larger than those for TTL circuits.
    (d) ECL circuits do not generate noise spikes during state transitions.
    (e) ECL devices require less power than that by standard TTL.
    (f) ECL can be easily used with TTL.

Chapter 12

SEQUENTIAL MACHINES

12.1 INTRODUCTION

Switching circuits are of two types—combinational and sequential. Combinational circuits are those in which the output at any instant is the function only of the present circuit inputs. Sequential circuits are those in which the output at any instant is the function not only of the present inputs to the circuit but also of the present state (i.e. the past history or stored information at that time) of the circuit. A combinational circuit gives the same output at any time for identical input conditions (because these circuits do not have memory), whereas for identical input conditions, a sequential circuit may give different outputs at different instants of time (because these circuits have memory). Logic gates are combinational circuits. The traffic light system and the lock on a safe which remember not only the combination numbers but also their sequence, are examples of sequential circuits. Sequential circuits may be of asynchronous or synchronous type. In synchronous sequential circuits, there is a master oscillator that provides regular timing pulses. The only time that events are permitted to occur, is during one of these timing pulses. Except for propagation delays, the circuit is inert at other times. Magnetic tape reader and synchronous counter are examples of synchronous systems. In asynchronous systems, events occur after the previous event is completed. There is no need to wait for a timing pulse. The ripple counter is an example of this system.

12.2 THE FINITE STATE MODEL

A finite state machine is an abstract model describing the synchronous sequential machine and its spatial counterpart, the iterative network. The behaviour of a finite state machine is described as a sequence of events that occur at discrete instants designated as $t = 1, 2, 3, \ldots$, etc. Suppose that the machine has been receiving input signals and also responding by producing output signals. If, now at time t, we were to apply an input signal $x(t)$ to the machine, the response $z(t)$ would depend on $x(t)$ as well as on the past inputs to the machine, and since a given machine might have an infinite varieties of possible histories, it would need an infinite capacity for storing them. Since in practice, it is impossible to implement machines which have infinite storage capabilities, we will concentrate on those machines whose past histories can affect their future behaviour only in a finite number of ways. These are called the *finite state machines*, that is, machines with a fixed number of states. These machines can distinguish among a finite number of classes of input histories. These classes of input histories are referred to as the internal states of the machine. Every finite state machine, therefore, contains a finite number of memory devices.

Figure 12.1 shows the schematic diagram of a synchronous sequential machine. The circuit has a finite number l of input terminals. The signals entering the circuit via these

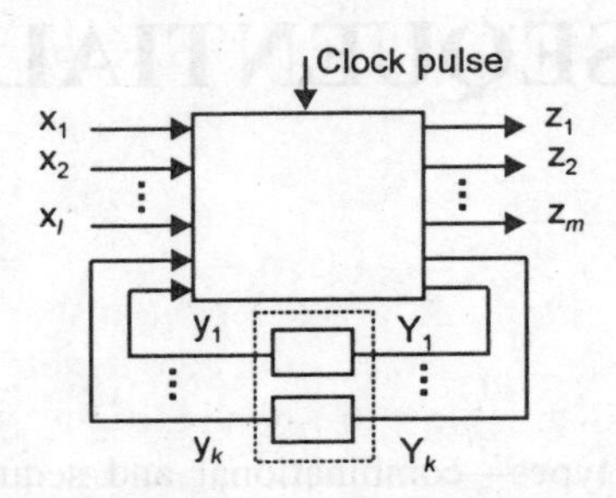

Fig. 12.1 Block diagram of a finite state model.

terminals constitute the set $[x_1, x_2, x_3,..., x_l]$ of input variables. Each input variable may take one of two possible values, a 0 or a 1. An ordered l-tuple of 0s and 1s is an input configuration. The set p of all possible combinations of l inputs, i.e. $p = 2^l$ is called an input alphabet I, and each one of these configurations is referred to as the symbol of the alphabet

$$I = (I_1, I_2,..., I_p)$$

Similarly, the circuit has a finite number m of output terminals which define the set $[z_1, z_2, z_3,..., z_m]$ of output variables. Each output variable is a binary variable. An ordered m-tuple of 0s and 1s is an output configuration. The set of $q = 2^m$ ordered m-tuples is called the output alphabet and is given by

$$O = [O_1, O_2, O_3,...,O_q]$$

Each output configuration is called a symbol of the output alphabet.

The signal value at the output of each memory element is referred to as the state variable and the set $[y_1, y_2,...,y_k]$ constitutes the set of state variables. The combination of values at the outputs of k memory elements $y_1, y_2,...,y_k$ defines the present internal state or the present state of the machine. The set S of $n = 2^k$, k-tuples constitutes the entire set of states of the machine.

$$S = [S_1, S_2, S_3,...,S_n]$$

The external inputs $x_1, x_2,...,x_l$ and the values of the state variables $y_1, y_2,..., y_k$ are supplied to the combinational circuit, which in turn produces outputs $z_1, z_2,..., z_m$ and the values $Y_1, Y_2,...,Y_k$. The values of the Ys which appear at the output of the combinational circuit at time t determine the state variables at time $t + 1$, and therefore, the next state of the machine.

Synchronization is achieved by means of clock pulses. The clock pulses may be applied to various AND gates to which input signal is applied. This allows the gates to transmit signals only at instants which coincide with the arrival of clock pulses.

State Diagram

The state diagram or state graph is a pictorial representation of the relationships among the present state, the input, the next state, and the output of the finite state sequential machine. The vertices of the graph represent the states of the machine. The directed arcs emanating from each vertex indicate the state transitions caused by various input symbols. The label on the directed arc indicates the input symbol that causes the transition, and the output symbol that is to be generated.

State Table

The state table is a tabular representation of the relationship between the present state, the input, the next state, and the output. Each column of the state table corresponds to one input symbol, and each row of the state table corresponds to one state. The entries corresponding to each combination of the input symbols and the present state specify the output that will be generated and the next state to which the machine will go. If there are p input symbols and n states in the state table of a machine, then such a machine can in practice be realized by a circuit with $l = [\log_2 p]$ input terminals and $k = [\log_2 n]$ memory elements where $[g]$ is the smallest integer larger than or equal to g.

Both the state diagram and the state table contain the same information and the choice between the two representations is a matter of convenience. Both have the advantage of being precise, unambiguous, and thus, more suitable for describing the operation of a sequential machine than that by any verbal description.

The succession of states through which a sequential machine passes and the output sequence which it produces in response to a known input sequence, are specified uniquely by the state diagram or by the state table and the initial state. The initial state refers to the state of the machine prior to the application of the input sequence and final state refers to the state of the machine after the application of the input sequence.

The state of a memory element is specified by the value of its output, which may assume either a 0 or a 1. The present state of the sequential machine indicates the present outputs of the memory elements used in the machine. The next state of the machine indicates the next outputs of the FFs that will be obtained when the present inputs are applied to the machine in the present state. The process of assigning the states of a physical device to the states of the sequential machine is known as *state assignment*. The output values of the physical device are referred to as *state variables*.

Transition and Output Table

The transition and output table can be obtained from the state table by modifying the entries of the state table to correspond to the states of the machine, in accordance with the selected state assignment. In this table, the next state and output entries are separated into two sections.

The entries of the next state part of this table define the necessary state transitions of the machine and, thus, specify the next values of the outputs of the FFs used. The next state part of the state table is called the *transition table*. The output part of the table indicates the output of the sequential machine for various input combinations applied to the machine, which is in the present state.

Excitation Table

The excitation table of a sequential machine gives information about the excitations or inputs required to be applied to the memory elements in the sequential circuit to bring the sequential machine from the present state to the next state. It also gives information about the outputs of the machine after application of the present inputs. The minimal expressions for the excitations of the FFs and outputs of the machine can be obtained by minimizing the expressions obtained from the excitation table using K-maps. The circuit can be realized using these minimal expressions.

12.3 MEMORY ELEMENTS

The various memory elements used in sequential machines are—D flip-flop, T flip-flop, S-R flip-flop, and J-K flip-flop. Each FF can store only one bit of information, i.e. a 0 or a 1. So, only one state variable is required to represent the output. Each FF has two states and both the states are stable, which implies that each FF can remain in any state indefinitely until it is directed by an input signal to do otherwise. The input signals to the FFs may be pulses or voltage levels, while the output signals of FFs are usually voltage levels.

D Flip-flop

The excitation requirements, the excitation table together with the logic symbol of the D flip-flop are shown in Fig. 12.2a to c.

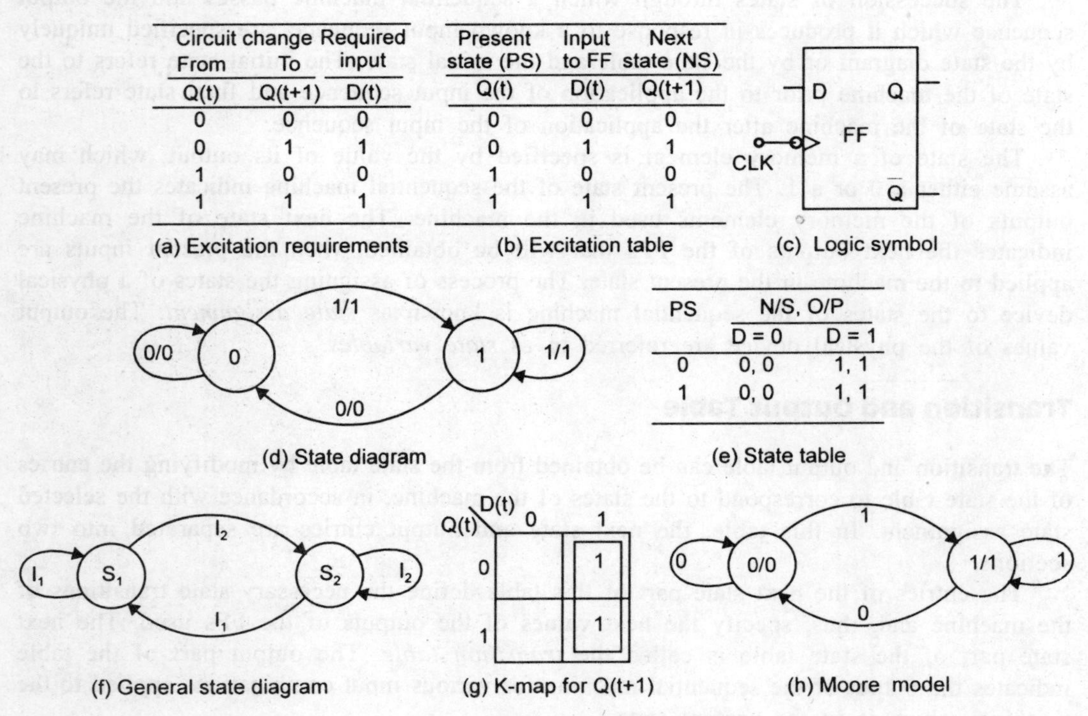

Circuit change From	To	Required input
Q(t)	Q(t+1)	D(t)
0	0	0
0	1	1
1	0	0
1	1	1

(a) Excitation requirements

Present state (PS)	Input to FF	Next state (NS)
Q(t)	D(t)	Q(t+1)
0	0	0
0	1	1
1	0	0
1	1	1

(b) Excitation table

PS	N/S, O/P D = 0	D = 1
0	0, 0	1, 1
1	0, 0	1, 1

(e) State table

Fig. 12.2 D flip-flop.

The state diagram and the state table are shown in Fig. 12.2(d) and (e), respectively. This state diagram is known as the Mealy model. Each node in the state diagram represents a particular state of the flip-flop (0 or 1). The labels on the arcs indicate the input/output, i.e. the input that is given when the FF is in a particular state and the corresponding output. The directions of arrows point to the next state the FF will go after the input is applied. If initially the FF is in a 0 state, a 0 input keeps the FF in the same state and a 1 input takes the FF to a 1 state. If initially the FF is in a 1 state, a 1 input keeps the FF in the same state, whereas a 0 input takes the FF to a 0 state.

The general state diagram, the K-map for the next state and the Moore model of the state diagram are shown in Fig. 12.2(f) to (h), respectively. In the general state diagram, S_1 and S_2 represent the states of the FF and I_1 and I_2 correspond to the input conditions. In the Moore model, the state code and the value of the output are written inside the circle. The directed line joining one node to the other, or looping back to the same node, has the value of the input written beside the line. In other words, each arc is labelled with the input condition that causes the transition and the circle contains the code for the internal state and the output state.

From the K-map we see that the next state of the D FF, $Q(t+1)$, can be expressed in terms of its present state $Q(t)$ and the present excitation $D(t)$ as

$$Q(t + 1) = D(t)$$

T Flip-flop

The T flip-flop toggles, i.e. changes state whenever the T input is a 1, and remains in the same state whenever the T input is a 0. The excitation requirements, the excitation table, and the logic symbol of the T FF are shown in Fig. 12.3a to c, respectively. The state diagram and the state table are shown in Fig. 12.3d and e, respectively.

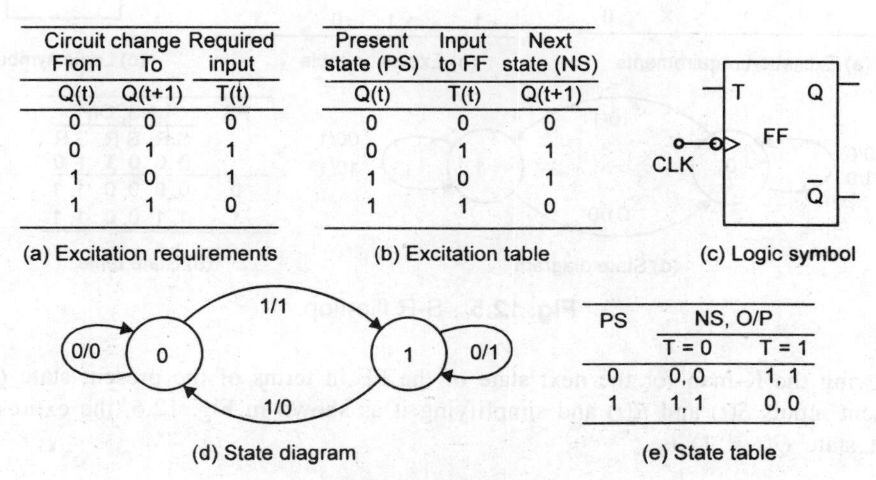

Circuit change From Q(t)	To Q(t+1)	Required input T(t)
0	0	0
0	1	1
1	0	1
1	1	0

(a) Excitation requirements

Present state (PS) Q(t)	Input to FF T(t)	Next state (NS) Q(t+1)
0	0	0
0	1	1
1	0	1
1	1	0

(b) Excitation table

(c) Logic symbol

(d) State diagram

PS	NS, O/P T = 0	NS, O/P T = 1
0	0, 0	1, 1
1	1, 1	0, 0

(e) State table

Fig. 12.3 T flip-flop.

Drawing the K-map for the next state of the T FF in terms of the present state $Q(t)$ and the present input $T(t)$ and simplifying it as shown in Fig. 12.4, the next state $Q(t+1)$ is

$$Q(t+1) = \overline{Q}(t)T(t) + Q(t)\overline{T}(t) = Q(t) \oplus T(t)$$

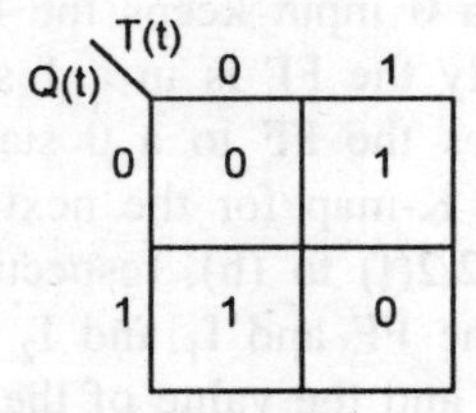

Fig. 12.4 K-map for $Q(t+1)$ of T flip-flop.

S-R Flip-flop

The excitation requirements, the excitation table and the logic symbol of the S-R FF are shown in Fig. 12.5a to c, respectively. The state diagram and the state table are shown in Fig. 12.5d and e, respectively.

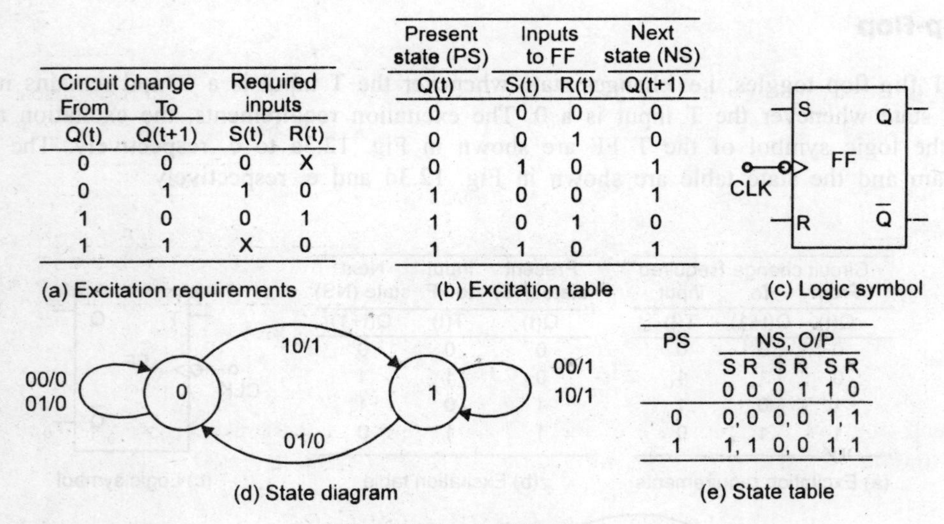

Circuit change From Q(t)	To Q(t+1)	Required inputs S(t)	R(t)
0	0	0	X
0	1	1	0
1	0	0	1
1	1	X	0

(a) Excitation requirements

Present state (PS) Q(t)	Inputs to FF S(t)	R(t)	Next state (NS) Q(t+1)
0	0	0	0
0	0	1	0
0	1	0	1
1	0	0	1
1	0	1	0
1	1	0	1

(b) Excitation table

(c) Logic symbol

(d) State diagram

PS	NS, O/P SR 00	SR 01	SR 10
0	0, 0	0, 0	1, 1
1	1, 1	0, 0	1, 1

(e) State table

Fig. 12.5 S-R flip-flop.

Drawing the K-map for the next state of the FF in terms of the present state $Q(t)$ and the present inputs $S(t)$ and $R(t)$ and simplifying it as shown in Fig. 12.6, the expression for the next state $Q(t+1)$ is

$$Q(t+1) = Q(t)\overline{R}(t) + S(t)$$

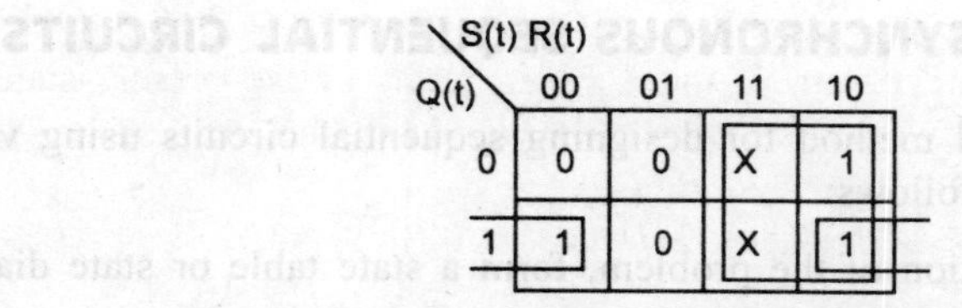

Q(t) \ S(t) R(t)	00	01	11	10
0	0	0	X	1
1	1	0	X	1

Fig. 12.6 K-map for $Q(t + 1)$ of S-R flip-flop.

J-K FLIP-fLOP

The excitation requirements, the excitation table and the logic symbol of the J-K FF are shown in Fig. 12.7a to c, respectively. The state diagram and the state table are shown in Fig. 12.7d and e, respectively.

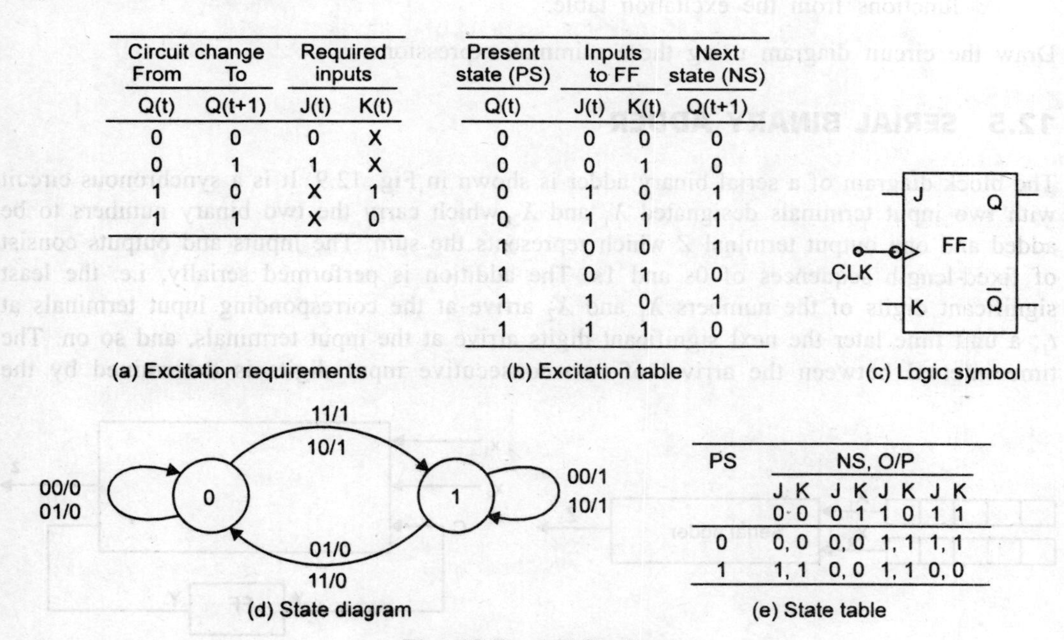

Circuit change From Q(t)	To Q(t+1)	Required inputs J(t)	K(t)
0	0	0	X
0	1	1	X
1	0	X	1
1	1	X	0

(a) Excitation requirements

Present state (PS) Q(t)	Inputs to FF J(t)	K(t)	Next state (NS) Q(t+1)
0	0	0	0
0	0	1	0
0	1	0	1
0	1	1	1
1	0	0	1
1	0	1	0
1	1	0	1
1	1	1	0

(b) Excitation table

(c) Logic symbol

(d) State diagram

PS	NS, O/P J K 0 0	J K 0 1	J K 1 0	J K 1 1
0	0, 0	0, 0	1, 1	1, 1
1	1, 1	0, 0	1, 1	0, 0

(e) State table

Fig. 12.7 J-K flip-flop.

The K-map for the next state of the J-K FF in terms of the present state $Q(t)$ and the present inputs $J(t)$ and $K(t)$ is shown in Fig. 12.8. The next state $Q(t + 1)$ is

$$Q(t + 1) = J(t)\overline{Q}(t) + Q(t)\overline{K}(t)$$

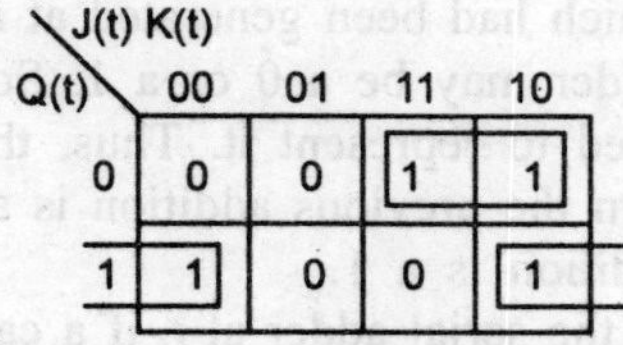

Q(t) \ J(t) K(t)	00	01	11	10
0	0	0	1	1
1	1	0	0	1

Fig. 12.8 K-map for $Q(t + 1)$ of J-K flip-flop.

12.4 SYNTHESIS OF SYNCHRONOUS SEQUENTIAL CIRCUITS

The main steps in the general method for designing sequential circuits using various types of memory elements are as follows:

1. From a word description of the problem, form a state table or state diagram which specifies the circuit performance.
2. Check this table to determine whether it contains any redundant states. Develop the state table so that it does not contain any redundant states.
3. Select a state assignment and derive the transition and output table.
4. Select the type of the memory element to be used and draw an excitation table.
5. Using K-maps, obtain the minimal expressions for the excitation and output functions from the excitation table.

Draw the circuit diagram using these minimal expressions.

12.5 SERIAL BINARY ADDER

The block diagram of a serial binary adder is shown in Fig. 12.9. It is a synchronous circuit with two input terminals designated X_1 and X_2 which carry the two binary numbers to be added and one output terminal Z which represents the sum. The inputs and outputs consist of fixed-length sequences of 0s and 1s. The addition is performed serially, i.e. the least significant digits of the numbers X_1 and X_2 arrive at the corresponding input terminals at t_1; a unit time later the next significant digits arrive at the input terminals, and so on. The time interval between the arrival of two consecutive input digits is determined by the

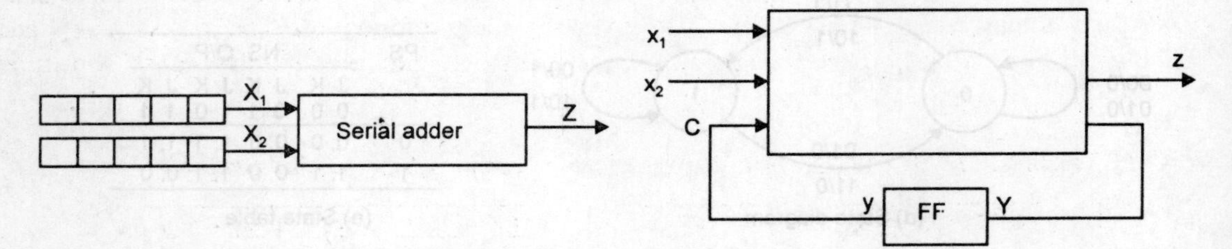

Fig. 12.9 Block diagram of the serial binary adder.

frequency of the circuit's clock. The delay within the combinational circuit is small with respect to the clock frequency and thus the sum digit arrives at the Z terminal immediately following the arrival of the corresponding input digits at the input terminals.

The output of the serial adder z_i at time t_i is a function of the inputs $x_1(t_i)$ and $x_2(t_i)$ at that time t_i and of a carry which had been generated at t_{i-1}. The carry which represents the past history of the serial adder may be a 0 or a 1. So, one FF is required to store it and one state variable is required to represent it. Thus, the circuit has two states. If one state indicates that the carry from the previous addition is a 0, the other state indicates that the carry from the previous addition is a 1.

Let A designate the state of the serial adder at t_i if a carry 0 was generated at t_{i-1}, and let B designate the state of the serial adder at t_i if a carry 1 was generated at t_{i-1}. The state of the adder at the time when the present inputs are applied is referred to as the *present state* (PS), and the state to which the adder goes as a result of the new carry value is referred to as the *next state* (NS).

The behaviour of a serial adder may be conveniently described by its state diagram and the state table as shown in Fig. 12.10. The state diagram shows that if the machine is in state A, i.e. carry from the previous addition is a 0, the inputs $x_1 = 0$, $x_2 = 0$ give sum 0 and carry 0. So, the machine remains in state A and outputs a 0. Inputs $x_1 = 0$ and $x_2 = 1$ give sum 1 and carry 0, so, the machine remains in state A and outputs a 1. Inputs $x_1 = 1$, $x_2 = 0$ give sum 1 and carry 0, so, the machine remains in state A and outputs a 1, but the inputs $x_1 = 1$, $x_2 = 1$ give sum 0 and carry 1, so, the machine goes to state B and outputs a 0.

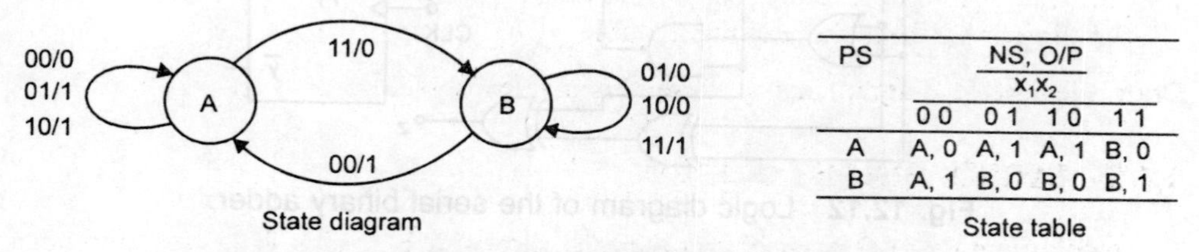

PS	NS, O/P			
	x_1x_2			
	0 0	0 1	1 0	1 1
A	A, 0	A, 1	A, 1	B, 0
B	A, 1	B, 0	B, 0	B, 1

State table

Fig. 12.10 State diagram and state table of the serial adder.

If the machine is in state B, i.e. carry from the previous addition is a 1, inputs $x_1 = 0$, $x_2 = 1$ give sum 0 and carry 1, so, the machine remains in state B and outputs a 0. Inputs $x_1 = 1$, $x_2 = 0$ give sum 0 and carry 1, so, the machine remains in state B and outputs a 0. Inputs $x_1 = 1$, $x_2 = 1$ give sum 1 and carry 1, so, the machine remains in state B and outputs a 1. Inputs $x_1 = 0$, $x_2 = 0$ give sum 1 and carry 0, so, the machine goes to state A and outputs a 0. The state table also gives the same information. The states, A = 0 and B = 1 have already been assigned, so, the transition and output table is as shown in Table 12.1.

Table 12.1 Transition and output table

PS	NS				O/P			
	x_1x_2				x_1x_2			
	00	01	10	11	00	01	10	11
0	0	0	0	1	0	1	1	0
1	0	1	1	1	1	0	0	1

Table 12.2 Excitation table

PS	I/P		NS	I/P to FF	O/P
y	x_1	x_2	y	D	z
0	0	0	0	0	0
0	0	1	0	0	1
0	1	0	0	0	1
0	1	1	1	1	0
1	0	0	0	0	1
1	0	1	1	1	0
1	1	0	1	1	0
1	1	1	1	1	1

To write the excitation table, select the memory element. Let us say, we want to use *D* flip-flop. The excitation table is shown in Table 12.2. Obtain the minimal expressions for *D* and *z* in terms of y, x_1, and x_2 by using K-maps as shown in Fig. 12.11. Implement the circuit using those expressions as shown in Fig. 12.12.

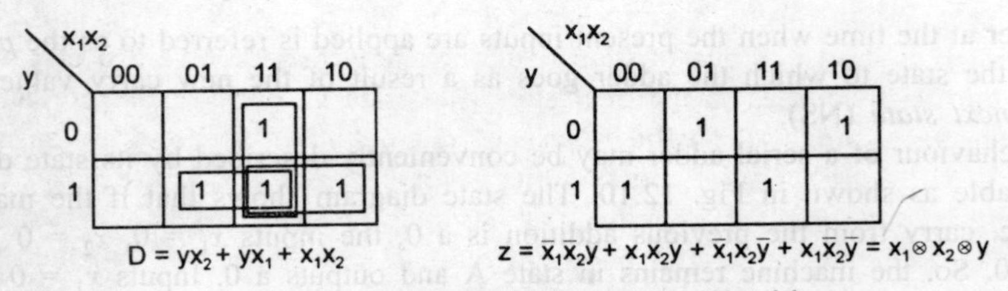

Fig. 12.11 K-maps for the serial adder.

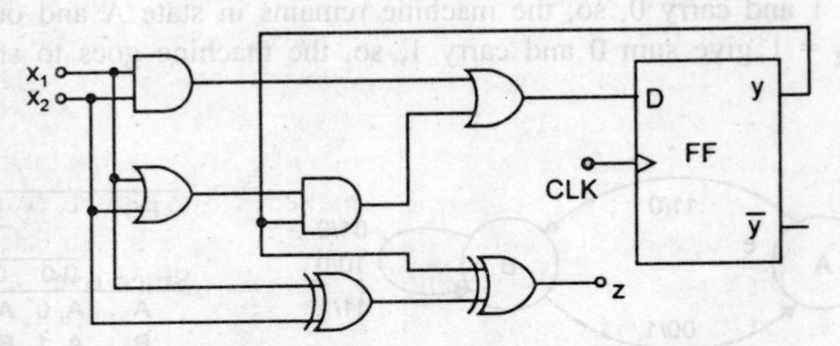

Fig. 12.12 Logic diagram of the serial binary adder.

12.6 THE SEQUENCE DETECTOR

A sequence detector is a sequential machine which produces an output 1 everytime the desired sequence is detected, and an output 0 at all other times.

Suppose we want to design a sequence detector to detect the sequence 1010 and say that overlapping is permitted, i.e. for example, if the input sequence is 01101010 the corresponding output sequence is 00000101. We can start the synthesis procedure by constructing the state diagram of the machine.

The state diagram and the state table of the sequence detector are shown in Fig. 12.13. At time t_1, the machine is assumed to be in the initial state designated arbitrarily as A. While in this state, the machine can receive input, either a 0 or a 1. If the input bit is a 0, the machine does not start the detection process because the first bit in the desired sequence is a 1. So, it remains in the same state and outputs a 0. Hence, an arc labelled 0/0 starting and terminating at A is drawn. If the input bit is a 1, the detection process starts, so, the machine goes to state B and outputs a 0. Hence, an arc labelled 1/0 starting at A and

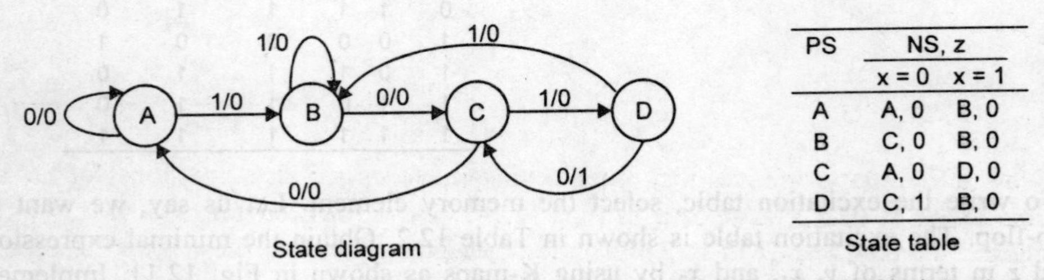

PS	NS, z	
	x = 0	x = 1
A	A, 0	B, 0
B	C, 0	B, 0
C	A, 0	D, 0
D	C, 1	B, 0

State table

Fig. 12.13 State diagram and state table of a sequence (1010) detector.

terminating at B is drawn. While in state B, the machine may receive a 0 or a 1 bit. If the bit is a 0, the machine goes to the next state, say state C, because the previous two bits are 10 which are a part of the valid sequence, and outputs a 0. So, draw an arc labelled 0/0 from B to C. If the bit is a 1, the two bits become 11 and this is not a part of the valid sequence. Since overlapping is permitted, the second 1 may be used to start the sequence, so, the machine remains in state B only and outputs a 0. So, an arc labelled 1/0 starting and terminating at B is drawn. While in state C, the machine may receive a 0 or a 1 bit. If it receives a 0, the last three bits received will be 100 and this is not a part of the valid sequence and also none of the last two bits can be used to start the new sequence. So, the machine goes to state A to restart the detection process and outputs a 0. Hence, an arc labelled 0/0 starting at C and terminating at A is drawn. If it receives a 1 bit, the last three bits will be 101 which are a part of the valid sequence. So, the machine goes to the next state, say state D, and outputs a 0. So, an arc labelled 1/0 is drawn from C to D. While in state D, the machine may receive a 0 or a 1. If it receives a 0, the last four bits become 1010 which is a valid sequence and the machine outputs a 1. Since overlapping is permitted, the machine can utilize the last two bits 10 to get another 1010 sequence. So, the machine goes to state C (if overlapping is not permitted or the machine has to restart after outputting a 1, the machine goes to state A). So, an arc labelled 0/1 is drawn from D to C. If it receives a 1 bit, the last four bits received will be 1011 which is not a valid sequence. So, the machine outputs a 0. Since the fourth bit 1 can become the starting bit for the valid sequence, the machine goes to state B. So, an arc labelled 1/0 is drawn from D to B.

There are four states, therefore, two state variables are required. Two state variables can have a maximum of four states. So, all states are utilized and thus there are no invalid states. Hence, there are no don't cares. Assign the states arbitrarily. Let A → 00, B → 01, C → 10, and D → 11 be the state assignment. With this assignment draw Table 12.3. Select the memory elements and draw the excitation Table 12.4.

Table 12.3 Transition and output table

PS	NS		O/P	
	x = 0	x = 1	x = 0	x = 1
A→00	00	01	0	0
B→01	10	01	0	0
C→10	00	11	0	0
D→11	10	01	1	0

Table 12.4 Excitation table

PS		I/P	NS		I/P to FFs		O/P
y_1	y_2	x	Y_1	Y_2	D_1	D_2	z
0	0	0	0	0	0	0	0
0	0	1	0	1	0	1	0
0	1	0	1	0	1	0	0
0	1	1	0	1	0	1	0
1	0	0	0	0	0	0	0
1	0	1	1	1	1	1	0
1	1	0	1	0	1	0	1
1	1	1	0	1	0	1	0

Draw the K-maps and simplify them to obtain the minimal expressions for D_1 and D_2 in terms of y_1, y_2, and x as shown in Fig. 12.14. The expression for z ($z = y_1, y_2, \bar{x}$) can be obtained directly from Table 12.4. The logic diagram based on these minimal expressions is shown in Fig. 12.15.

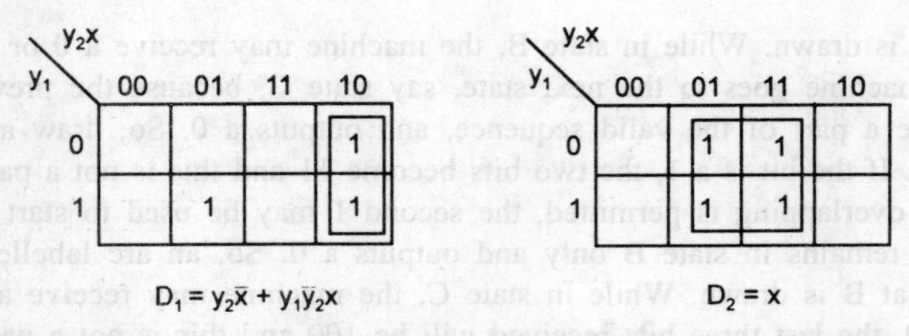

Fig. 12.14 K-maps for D_1 and D_2 of the sequence (1010) detector.

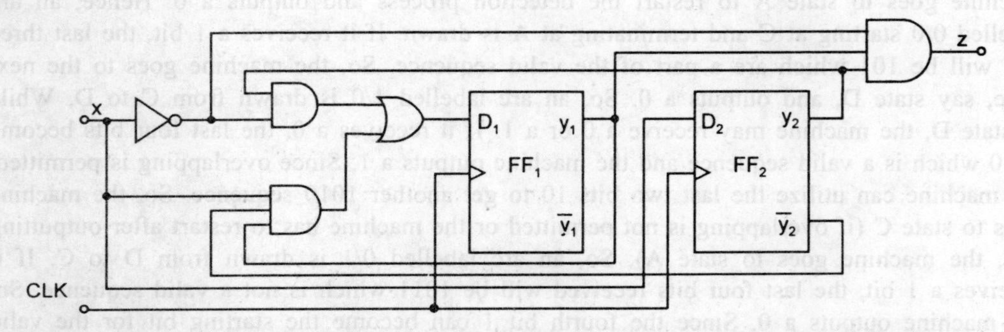

Fig. 12.15 Logic diagram of the sequence (1010) detector using *D* flip-flops.

12.7 PARITY-BIT GENERATOR

A serial parity-bit generator is a two-terminal circuit which receives coded messages and adds a parity bit to every *m* bits of the message, so that the resulting outcome is an error-detecting coded message. The inputs are assumed to arrive in strings of three symbols ($m = 3$) and the strings are spaced apart by single time units (i.e. the fourth place is a blank). The parity bits are inserted in the appropriate spaces so that the resulting outcome is a continuous string of symbols without spaces. For even parity, a parity bit 1 is inserted, if and only if the number of 1s in the preceding string of three symbols is odd. For odd parity, a parity bit 1 is inserted, if and only if the number of 1s in the preceding string of three symbols is even.

Odd Parity-bit Generator

The state diagram and the state table of an odd parity-bit generator are shown in Fig. 12.16. States B, D, and F correspond to even number of 1s out of 1, 2, and 3 incoming inputs, respectively. Similarly, states C, E, and G correspond to odd number of 1s out of 1, 2, and 3 incoming inputs, respectively. From either state F or state G, the machine goes to state A regardless of the input. In fact, the fourth input is a blank.

Since the state diagram contains seven states, three state variables are needed for an assignment. But since three state variables can have a total of eight states, one of the states will not be assigned and its entries in the corresponding state table may be considered as don't cares.

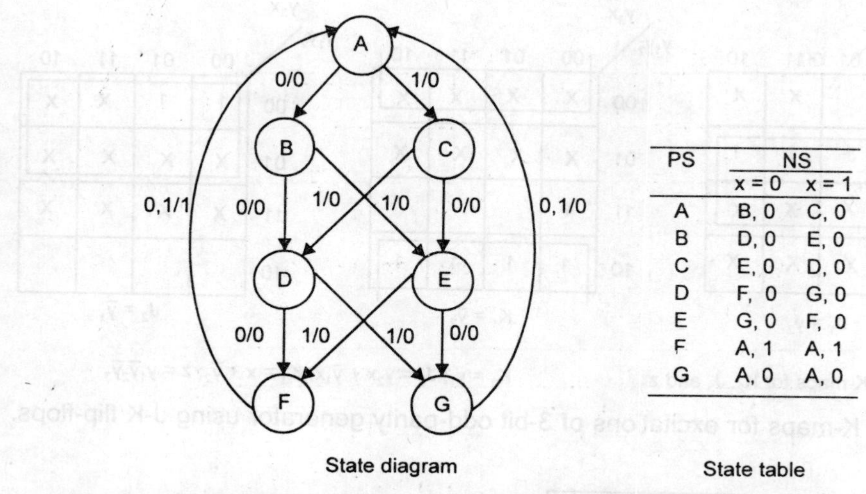

PS	NS	
	x = 0	x = 1
A	B, 0	C, 0
B	D, 0	E, 0
C	E, 0	D, 0
D	F, 0	G, 0
E	G, 0	F, 0
F	A, 1	A, 1
G	A, 0	A, 0

Fig. 12.16 State diagram and state table of a 3-bit odd-parity generator.

The state assignment is not unique. Many possible assignments are there. One possible assignment is A → 000, B → 010, C → 011, D → 110, E → 111, F → 100, and G → 101. With this assignment the transition and output table is given in Table 12.5.

Select the memory elements. Suppose J-K flip-flops are selected. For implementing the parity-bit generator using J-K flip-flops, draw the excitation table as shown in Table 12.6. The minimal expressions for excitations J_1, K_1, J_2, K_2, J_3, K_3 of flip-flops and the output of the odd-parity generator z in terms of the present state variables y_1, y_2, y_3, and the input x can be obtained using K-maps as shown in Fig. 12.17. The logic diagram based on those expressions is shown in Fig. 12.18.

Table 12.5 Transition and output table

PS	NS		O/P	
	x = 0	x = 1	x = 0	x = 1
000	010	011	0	0
010	110	111	0	0
011	111	110	0	0
110	100	101	0	0
111	101	100	0	0
100	000	000	1	1
101	000	000	0	0

Table 12.6 Excitation table

PS			I/P	NS			Present excitations required						O/P
y_1	y_2	y_3	x	Y_1	Y_2	Y_3	J_1	K_1	J_2	K_2	J_2	K_3	z
0	0	0	0	0	1	0	0	X	1	X	0	X	0
0	0	0	1	0	1	1	0	X	1	X	1	X	0
0	1	0	0	1	1	0	1	X	X	0	0	X	0
0	1	0	1	1	1	1	1	X	X	0	1	X	0
0	1	1	0	1	1	1	1	X	X	0	X	0	0
0	1	1	1	1	1	0	1	X	X	0	X	1	0
1	1	0	0	1	0	0	X	0	X	1	0	X	0
1	1	0	1	1	0	1	X	0	X	1	1	X	0
1	1	1	0	1	0	1	X	0	X	1	X	0	0
1	1	1	1	1	0	0	X	0	X	1	X	1	0
1	0	0	0	0	0	0	X	1	0	X	0	X	1
1	0	0	1	0	0	0	X	1	0	X	0	X	1
1	0	1	0	0	0	0	X	1	0	X	X	1	0
1	0	1	1	0	0	0	X	1	0	X	X	1	0

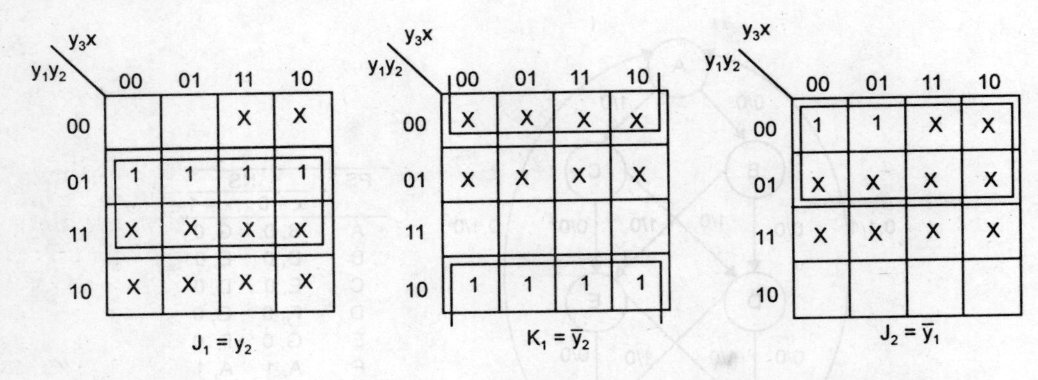

Fig. 12.17 K-maps for excitations of 3-bit odd-parity generator using J-K flip-flops.

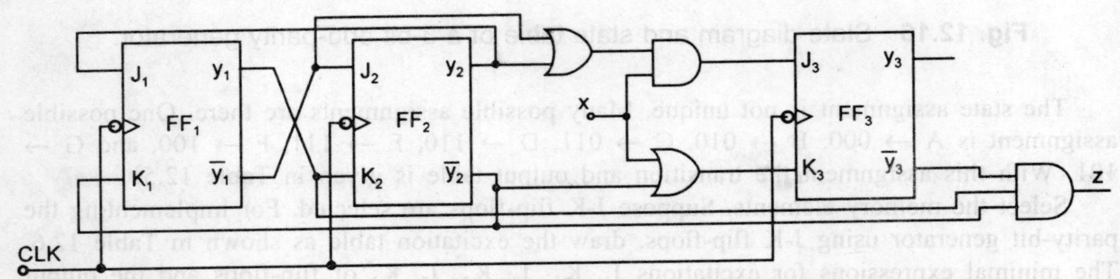

Fig. 12.18 Logic diagram of a 3-bit odd-parity generator using J-K flip-flops.

12.8 COUNTERS

Design of a 3-bit Gray Code Counter

The counter is to be designed with one input terminal (which receives pulse signals) and one output terminal. It should be capable of counting in the Gray system up to 7 and producing an output pulse for every 8 input pulses. After the count 7 is reached, the next pulse will reset the counter to its initial state, i.e. to a count of zero. The state assignment, the state diagram, and the state table of the 3-bit Gray code counter are shown in Fig. 12.19.

There are eight states for a 3-bit counter. So, three state variables are required which can give a maximum of eight possible states. So, no invalid states exist. The state assignment cannot be arbitrary, since the counter has to change the states in a definite manner. Hence, the state assignment is

$S_0 \to 000$, $S_1 \to 001$, $S_2 \to 011$, $S_3 \to 010$, $S_4 \to 110$, $S_5 \to 111$, $S_6 \to 101$, $S_7 \to 100$.

The transition and output table for the counter is shown in Table 12.7.

To complete the synthesis, select an appropriate set of memory elements. We select T flip-flops and draw the excitation table as shown in Table 12.8. The minimal expressions for excitation functions to T flip-flops, T_1, T_2 and T_3 in terms of the present state variables

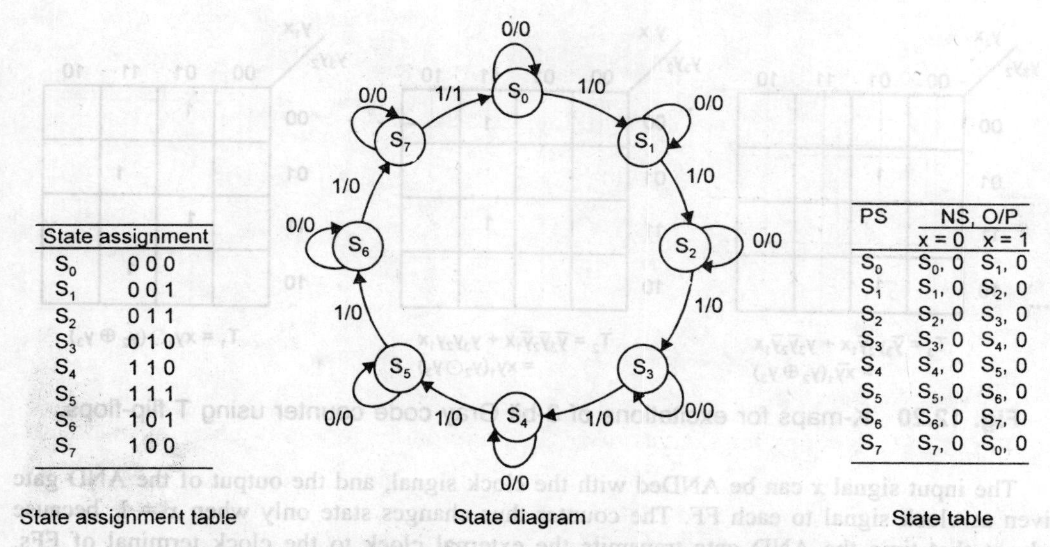

Fig. 12.19 State assignment, state diagram, and state table of a 3-bit Gray code counter.

y_1, y_2, y_3, and the input x can be obtained using K-maps as shown in Fig. 12.20. From the excitation table, the expression for output z is

$$z = y_3 \bar{y}_2 \bar{y}_1 x$$

Table 12.7 Transition and output table

PS	NS x = 0	NS x = 1	O/P x = 0	O/P x = 1
000	000	001	0	0
001	001	011	0	0
011	011	010	0	0
010	010	110	0	0
110	110	111	0	0
111	111	101	0	0
101	101	100	0	0
100	100	000	0	1

Table 12.8 Excitation table

PS y_3	y_2	y_1	I/P x	NS Y_3	Y_2	Y_1	Inputs to FFs T_3	T_2	T_1	O/P z
0	0	0	0	0	0	0	0	0	0	0
0	0	0	1	0	0	1	0	0	1	0
0	0	1	0	0	0	1	0	0	0	0
0	0	1	1	0	1	1	0	1	0	0
0	1	1	0	0	1	1	0	0	0	0
0	1	1	1	0	1	0	0	0	1	0
0	1	0	0	1	1	0	0	0	0	0
0	1	0	1	1	1	0	1	0	0	0
1	1	0	0	1	1	0	0	0	0	0
1	1	0	1	1	1	1	0	0	1	0
1	1	1	0	1	1	1	0	0	0	0
1	1	1	1	1	0	1	0	1	0	0
1	0	1	0	1	0	1	0	0	0	0
1	0	1	1	1	0	0	0	0	1	0
1	0	0	0	1	0	0	0	0	0	0
1	0	0	1	0	0	0	1	0	0	1

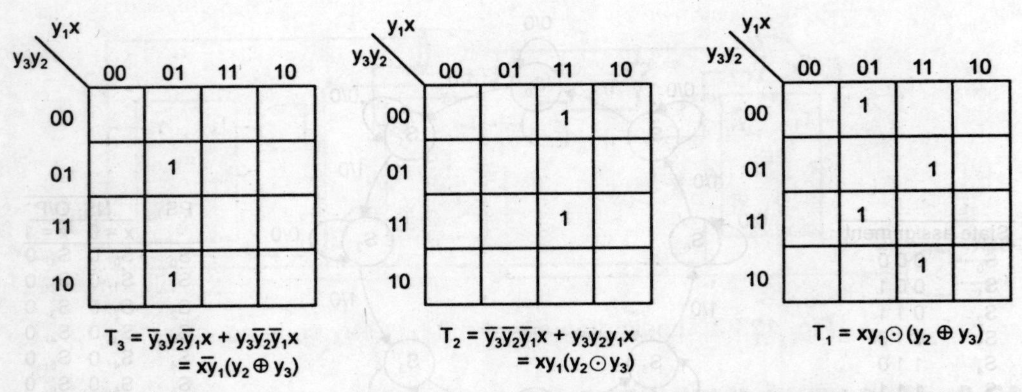

Fig. 12.20 K-maps for excitations of 3-bit Gray code counter using T flip-flops.

The input signal x can be ANDed with the clock signal, and the output of the AND gate given as clock signal to each FF. The counter thus changes state only when $x = 1$, because only at that time the AND gate transmits the external clock to the clock terminal of FFs. When $x = 0$, the AND gate is disabled, the FFs receive no clock and the counter remains in the previous state. The excitation table is shown in Table 12.9. The minimal expressions obtained from K-maps are shown in Fig. 12.21.

Table 12.9 Excitation table

PS			NS	O/P	Excitation		
y_3	y_2	y_1	x = 1, clock present		T_3	T_2	T_1
0	0	0	0 0 1	0	0	0	1
0	0	1	0 1 1	0	0	1	0
0	1	1	0 1 0	0	0	0	1
0	1	0	1 1 0	0	1	0	0
1	1	0	1 1 1	0	0	0	1
1	1	1	1 0 1	0	0	1	0
1	0	1	1 0 0	0	0	0	1
1	0	0	0 0 0	1	1	0	0

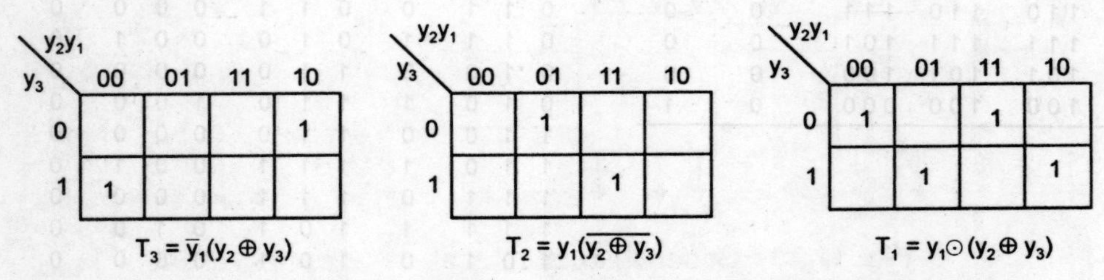

Fig. 12.21 K-maps for excitations when x is ANDed with clock.

The expressions in both the cases are the same, but the design is simpler in the second case. The logic diagram based on these expressions is shown in Fig. 12.22.

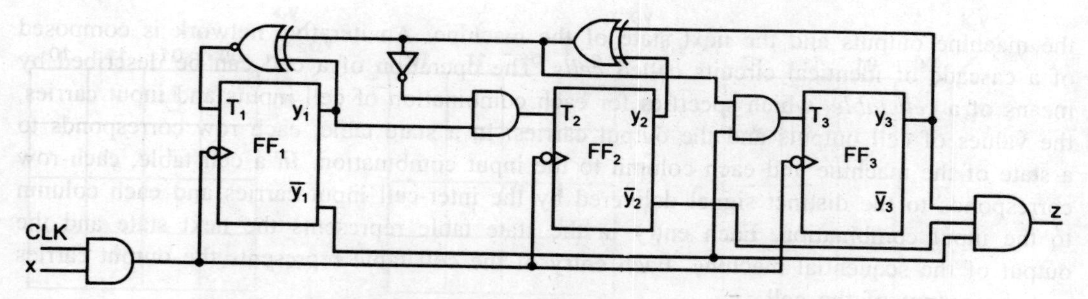

Fig. 12.22 Logic diagram of a 3-bit Gray code counter.

12.9 ITERATIVE NETWORKS

An iterative network is a digital structure composed of a cascade of identical circuits or cells. An iterative network may be sequential in nature, where each cell is a sequential circuit, e.g. the counter or a shift register which consists of a number of cascaded flip-flops, or it may be combinational in nature where each cell is a combinational circuit. The description and synthesis of combinational iterative networks are similar to those of synchronous sequential circuits. Every finite output sequence that can be produced sequentially by a sequential machine can also be produced spatially (or simultaneously) by a combinational iterative network. The number of cells in an iterative network must equal the length of the input patterns applied to it.

The general structure of an iterative network has the form shown in Fig. 12.23. The external inputs applied to the ith cell are designated as $x_{i1}, x_{i2},\ldots,x_{il}$, where the ith cell is counted from the left. The cell outputs are designated as $z_{i1}, z_{i2},\ldots,z_{im}$. In addition, each cell receives information from the preceding cell via the inter-cell carry leads $y_{i1}, y_{i2},\ldots,y_{ik}$ which are called the *input carries* and transmits information to the next cell via the inter-cell carry leads $Y_{i1}, Y_{i2},\ldots,Y_{ik}$ called the *output carries*. The inputs are applied to all the cells simultaneously and the outputs are generated instantaneously.

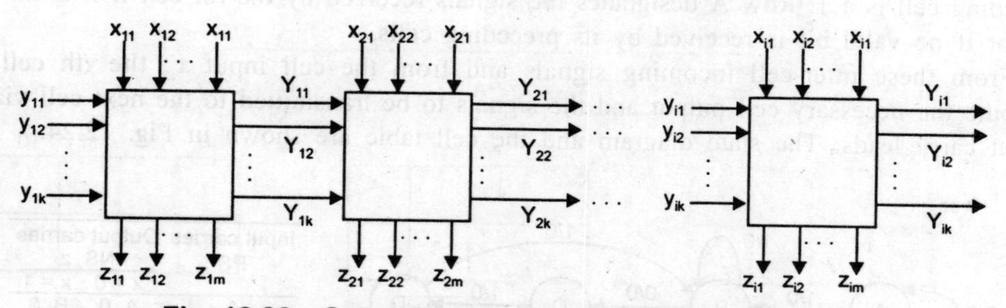

Fig. 12.23 General structure of an iterative network.

Analogy between Sequential Machines and Iterative Networks

The operation of a sequential machine can be described by its state table, which specifies for each combination of machine inputs and the present state of the machine, the values of the machine outputs and the next state of the machine. An iterative network is composed of a cascade of identical circuits called *cells*. The operation of a cell can be described by means of a *cell table*, which specifies for each combination of cell inputs and input carries, the values of cell outputs and the output carries. In a state table, each row corresponds to a state of the machine and each column to the input combination. In a cell table, each row corresponds to the distinct signal delivered by the inter-cell input carries and each column to the input combination. Each entry in the state table represents the next state and the output of the sequential machine. Each entry of the cell table represents the output carries and the output of the cell.

In general, if the same assignment is selected for the iterative network as for the sequential machine, then the logic circuit of the *i*th cell and the combinational logic of the sequential machine are identical. While in the sequential case, information is fed through delays, in the iterative network the entire computation is executed instantaneously by using many identical cells. The inputs are applied to all the cells simultaneously and the outputs are assumed to be generated instantaneously. The number of cells in an iterative network must equal the length of the input pattern applied to it.

Design of a Sequence Detector Using Iterative Networks

Let us design an iterative network to detect the pattern 1010, that is, we have to design an iterative network which consists of an arbitrarily large number of cells and whose typical cell contains a single input x_i and a single output z_i. The inputs are applied to all the cells simultaneously. The output z_i is 1 if and only if the input pattern of the four cells $i - 3$, $i - 2$, $i - 1$, and i is 1010, i.e. $x_{i-3} = 1$, $x_{i-2} = 0$, $x_{i-1} = 1$, and $x_1 = 0$.

The cell table must have four rows (or states) corresponding to the four possible distinct signals delivered by the inter-cell input carries. Row D designates the signals received by the *i*th cell, when the input pattern in the preceding three cells is 101. Row C designates the signals received by the *i*th cell, when the input pattern in the preceding two cells is 10. Row B designates the signals received by the *i*th cell, when the input to the preceding cell is a 1. Row A designates the signals received by the *i*th cell if it is the first cell or if no valid bit is received by its preceding cells.

From these inter-cell incoming signals and from the cell input x_i, the *i*th cell can compute the necessary cell output and the signals to be transmitted to the next cell via the output carry leads. The state diagram and the cell table are shown in Fig. 12.24.

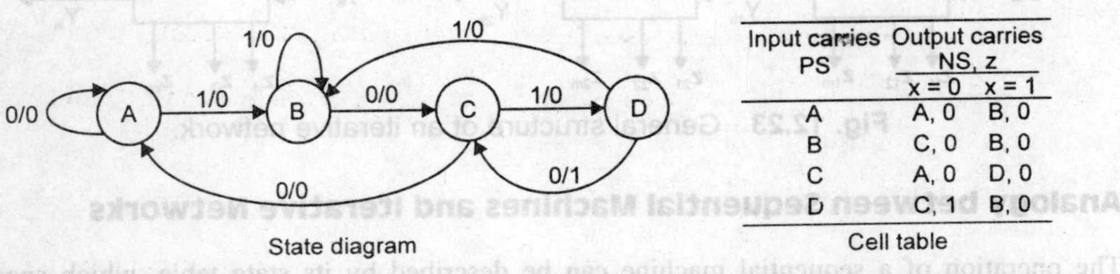

Input carries PS	Output carries NS, z x = 0	x = 1
A	A, 0	B, 0
B	C, 0	B, 0
C	A, 0	D, 0
D	C, 1	B, 0

Cell table

Fig. 12.24 State diagram and cell table of sequence (1010) detector.

If we assign the states A → 00, B → 01, C → 10, and D → 11, the transition and output table and the output carries and cell output table will be as shown in Table 12.10 and Table 12.1, respectively.

Table 12.10 Transition and output table

PS	NS		Output, z	
	$x_i = 0$	$x_i = 1$	$x_i = 0$	$x_i = 1$
00	00	01	0	0
01	10	01	0	0
10	00	11	0	0
11	10	01	1	0

Table 12.11 Output carries and cell output table

Input carries	I/P	Output carries	O/P
y_{i1} y_{i2}	x_i	Y_{i1} Y_{i2}	z_i
0 0	0	0 0	0
0 0	1	0 1	0
0 1	0	1 0	0
0 1	1	0 1	0
1 0	0	0 0	0
1 0	1	1 1	0
1 1	0	1 0	1
1 1	1	0 1	0

The minimal expressions for output carries Y_{i1} and Y_{i2} in terms of the input carries y_{i1} and y_{i2} and input x_i obtained by using K-maps and the combinational circuit for the ith cell based on those minimal expressions, are shown in Fig. 12.25. An iterative network to detect the pattern 101010 is shown in Fig. 12.26.

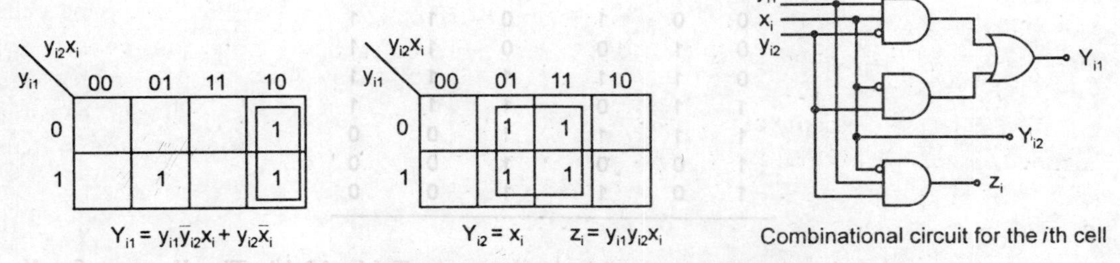

Fig. 12.25 K-maps and logic diagram for the *i*th cell.

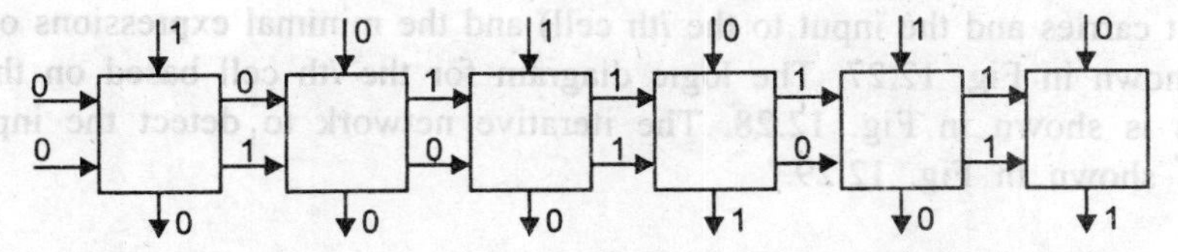

Fig. 12.26 Iterative network to detect the pattern 101010.

EXAMPLE 12.1 Design an n-cell network, where each cell has one cell input x_i and one cell output z_i such that $z_i = 1$, if and only if either one or two of the cell inputs $x_1\ x_2, \ldots, x_i$ equals 1.

Solution

The input carries (PS) to the ith cell may represent one of the four conditions: (1) none of its preceding cells received a 1, (2) only one of its preceding cells received a 1, (3) only two of its preceding cells received a 1, and (4) more than two of its preceding cells received a 1 input. Each one of these represents a state. So, the cell table of the *i*th cell must have at least four rows to distinguish the above four distinct states. Row A designates the state where none of the preceding cells received a 1 input. Similarly, rows B, C, and D designate the states where one, two, three, or more of the preceding cells received a 1 input. The cell table in terms of states A, B, C, and D and its equivalent for a state assignment of A → 00, B → 01, C → 11, and D → 10 are shown in Tables 12.12 and 12.13, respectively.

Table 12.12 Cell table (Example 12.1)

PS	NS, z_i	
	$x_i = 0$	$x_i = 1$
A	A,0	B,1
B	B,1	C,1
C	C,1	D,0
D	D,0	D,0

Table 12.13 Equivalent of the cell table (Example 12.1)

Input carries		Output carries		Output	
		$x_i = 0$	$x_i = 1$	$x_i = 0$	$x_i = 1$
A	00	00	01	0	1
B	01	01	11	1	1
C	11	11	10	1	0
D	10	10	10	0	0

Table 12.14 Output carries and cell output table (Example 12.1)

PS		I/P	NS		O/P
y_{i1}	y_{i2}	x_i	Y_{i1}	Y_{i2}	z_i
0	0	0	0	0	0
0	0	1	0	1	1
0	1	0	0	1	1
0	1	1	1	1	1
1	1	0	1	1	1
1	1	1	1	0	0
1	0	0	1	0	0
1	0	1	1	0	0

The output carries and cell output table is shown in Table 12.14. The K-maps for Y_{i1}, Y_{i2}, and z_i (i.e. for the output carries and the output of the *i*th cell) in terms of y_{i1}, y_{i2}, and x_i (the input carries and the input to the *i*th cell) and the minimal expressions obtained from them are shown in Fig. 12.27. The logic diagram for the *i*th cell based on those minimal expressions is shown in Fig. 12.28. The iterative network to detect the input sequence 0101010 is shown in Fig. 12.29.

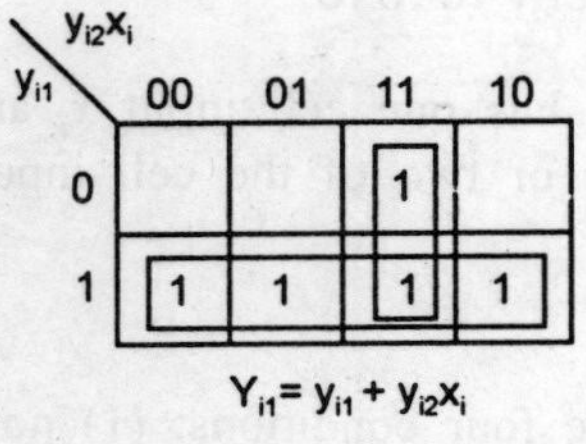

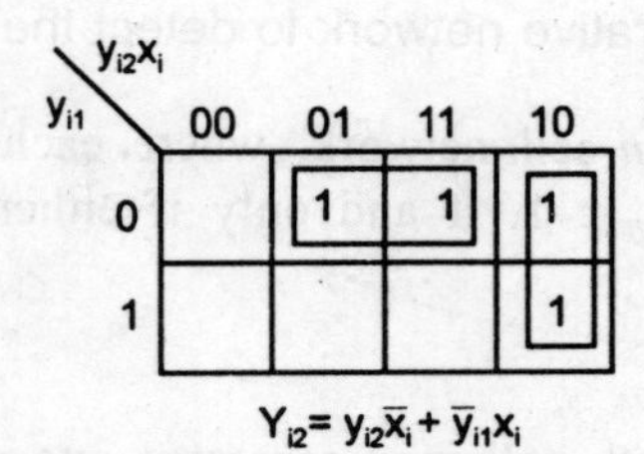

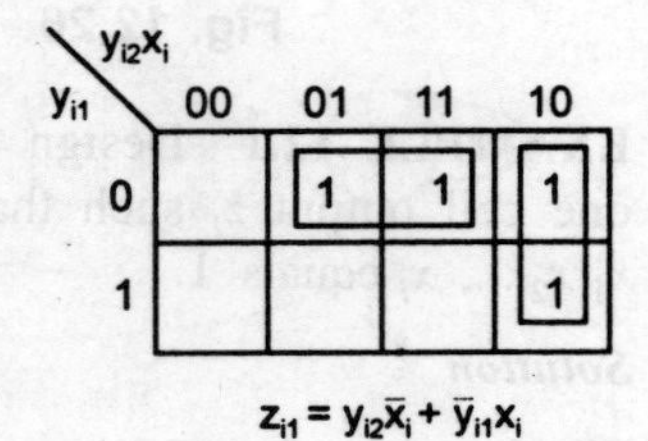

Fig. 12.27 K-maps (Example 12.1).

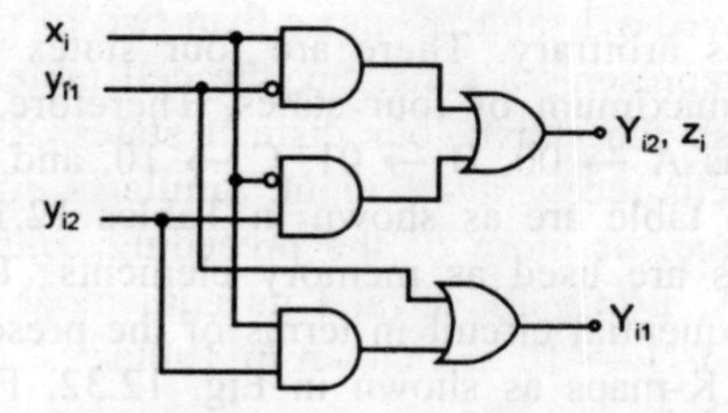

Fig. 12.28 Logic diagram of the *i*th cell (Example 12.1).

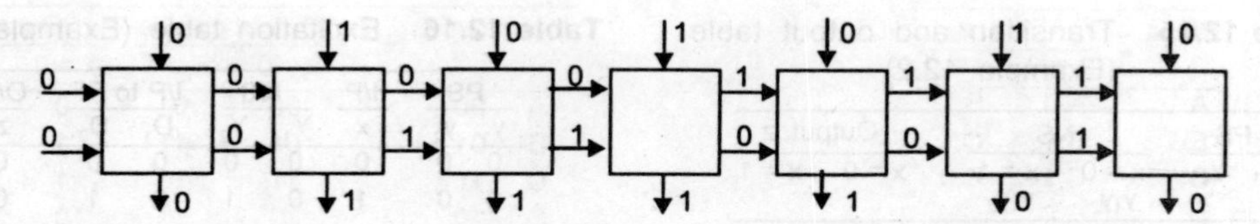

Fig. 12.29 Iterative network to detect the pattern 0101010 (Example 12.1).

EXAMPLE 12.2 A long sequence of pulses enters a 2-input 2-output synchronous sequential circuit which is required to produce an output $z = 1$, whenever the sequence 1111 occurs. Overlapping sequences are accepted. For example, if the input is 01011111, the required output is 00000011. Design the circuit.

Solution

The block diagram of the detector with one input terminal and one output terminal is shown in Fig. 12.30.

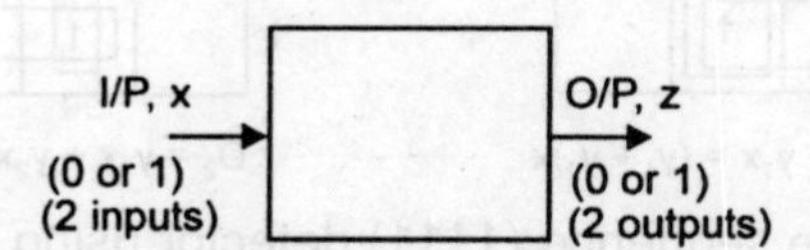

Fig. 12.30 Block diagram of a 2-input 2-output synchronous sequential machine.

Let the machine be initially in state A. The state diagram and the state table indicating the transition of states of the machine and, therefore, indicating its working are shown in Fig. 12.31.

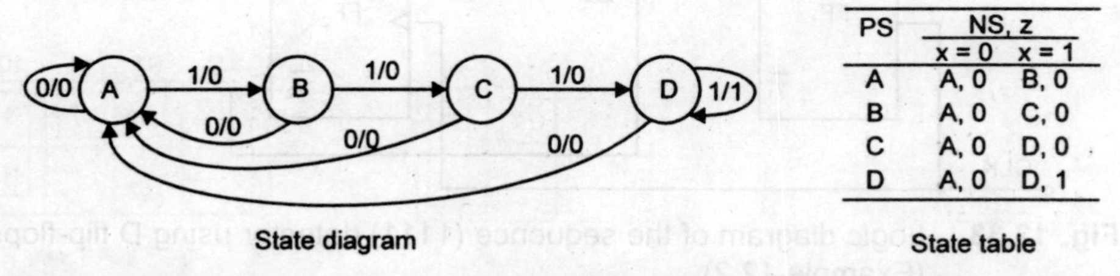

PS	NS, z	
	x = 0	x = 1
A	A, 0	B, 0
B	A, 0	C, 0
C	A, 0	D, 0
D	A, 0	D, 1

State table

Fig. 12.31 State diagram and state table of the sequence (1111) detector (Example 12.2).

The state assignment is arbitrary. There are four states. So, two state variables are needed, which can give a maximum of four states. Therefore, there are no invalid states. Let the states be assigned as A → 00, B → 01, C → 10, and D → 11. The transition and output table, and excitation table are as shown in Tables 12.15 and 12.16, respectively.

Let us say D flip-flops are used as memory elements. The minimal expressions for excitations of FFs of the sequential circuit in terms of the present state variables y_1, y_2, and input x are obtained using K-maps as shown in Fig. 12.32. From the excitation table the output z is given by $z = y_1 y_2 x$. The logic diagram of the detector based on those minimal expressions is shown in Fig. 12.33.

Table 12.15 Transition and output table (Example 12.2)

PS y_1 y_2	NS $x = 0$ y_1y_2	NS $x = 1$ y_1y_2	Output, z $x = 0$	Output, z $x = 1$
0 0	0 0	0 1	0	0
0 1	0 0	1 0	0	0
1 0	0 0	1 1	0	0
1 1	0 0	1 1	0	1

Table 12.16 Excitation table (Example 12.2)

PS y_1	PS y_2	I/P x	NS Y_1	NS Y_2	I/P to FF D_1	I/P to FF D_2	O/P z
0	0	0	0	0	0	0	0
0	0	1	0	1	0	1	0
0	1	0	0	0	0	0	0
0	1	1	1	0	1	0	0
1	0	0	0	0	0	0	0
1	0	1	1	1	1	1	0
1	1	0	0	0	0	0	0
1	1	1	1	1	1	1	1

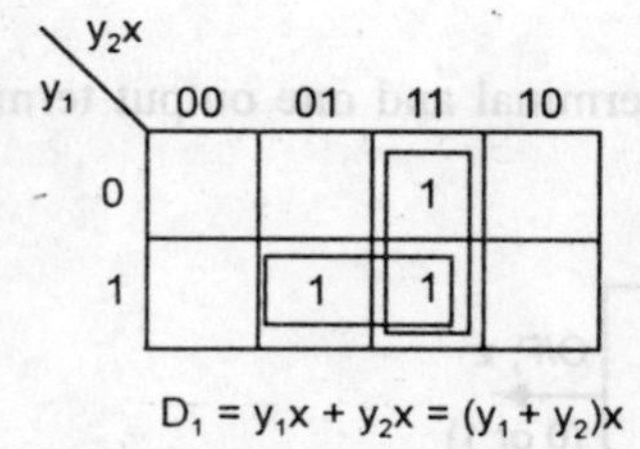

$D_1 = y_1x + y_2x = (y_1 + y_2)x$

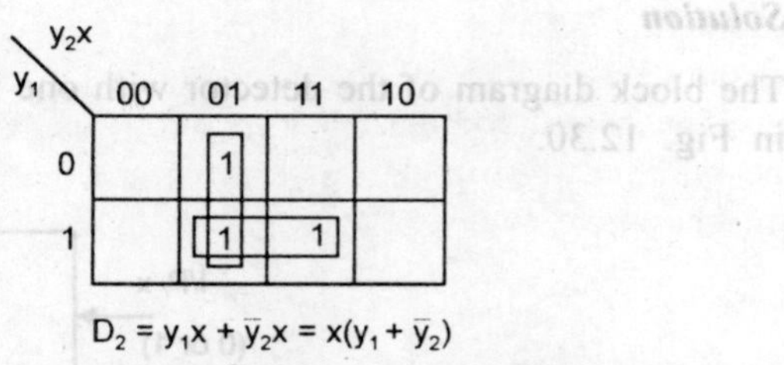

$D_2 = y_1x + \bar{y}_2x = x(y_1 + \bar{y}_2)$

Fig. 12.32 K-maps for the sequence (1111) detector using D flip-flops (Example 12.2).

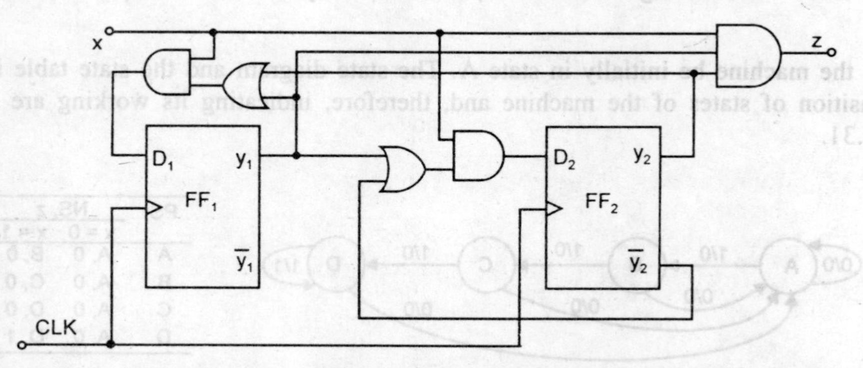

Fig. 12.33 Logic diagram of the sequence (1111) detector using D flip-flops (Example 12.2).

EXAMPLE 12.3 A synchronous sequential machine has a single control input x and the clock, and two outputs A and B. On consecutive rising edges of the clock, the code on A and B changes from 00 to 01 to 10 to 11 and repeats itself if $x = 1$; if at any time, $x = 0$, it holds to the present state. Draw the state diagram and implement the circuit using T flip-flops.

Solution

The block diagram of the sequential machine is shown in Fig. 12.34.

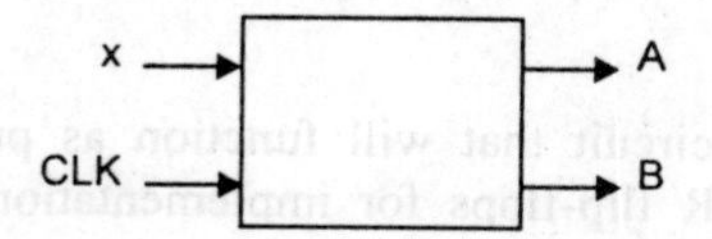

Fig. 12.34 Sequential machine (Example 12.3).

The given machine is nothing but a sequential circuit with two flip-flops. Let A and B be the outputs of the flip-flops. So, two state variables are required which can have a maximum of four states. There are, no invalid states present. Let the initial state be P.

The state diagram and the state table of the sequential machine with state assignment of

$$P \rightarrow 00,\ Q \rightarrow 01,\ R \rightarrow 10, \text{ and } S \rightarrow 11$$

are shown in Fig. 12.35. The transition and output table with this assignment is shown in Table 12.17. Let us select T flip-flops as memory elements. With T FFs, the excitation table is shown in Table 12.18.

The outputs of the machine are the same as the outputs of the flip-flops. The K-maps, the minimal expressions for excitations, and the logic diagram based on those expressions are shown in Fig. 12.36.

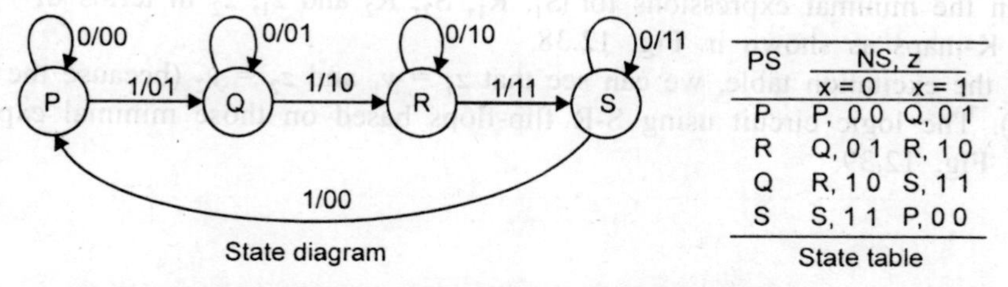

PS	NS, z	
	x = 0	x = 1
P	P, 0 0	Q, 0 1
R	Q, 0 1	R, 1 0
Q	R, 1 0	S, 1 1
S	S, 1 1	P, 0 0

State table

Fig. 12.35 State diagram and state table (Example 12.3).

Table 12.17 Transition and output table (Example 12.3)

PS		NS		O/P	
y_1	y_2	x = 0	x = 1	x = 0	x = 1
0	0	0 0	0 1	0 0	0 1
0	1	0 1	1 0	0 1	1 0
1	0	1 0	1 1	1 0	1 1
1	1	1 1	0 0	1 1	0 0

Table 12.18 Excitation table (Example 12.3)

PS		I/P	NS		Required excitation	
y_1	y_2	x	Y_1	Y_2	T_1	T_2
0	0	0	0	0	0	0
0	0	1	0	1	0	1
0	1	0	0	1	0	0
0	1	1	1	0	1	1
1	0	0	1	0	0	0
1	0	1	1	1	0	1
1	1	0	1	1	0	0
1	1	1	0	0	1	1

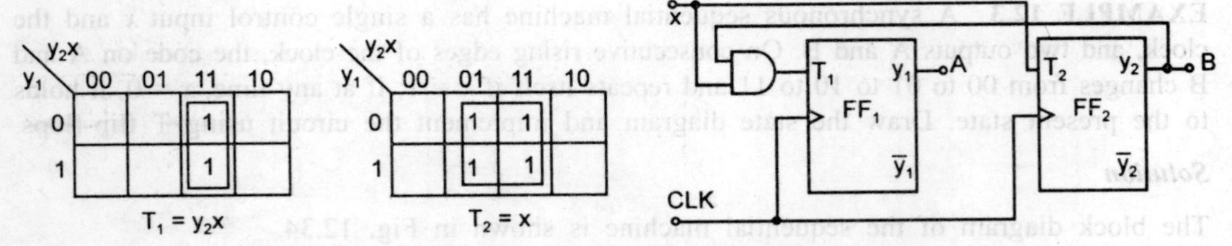

Fig. 12.36 K-maps and logic diagram (Example 12.3).

EXAMPLE 12.4 Design a circuit that will function as prescribed by the state diagram shown in Fig. 12.37. Use S-R flip-flops for implementation.

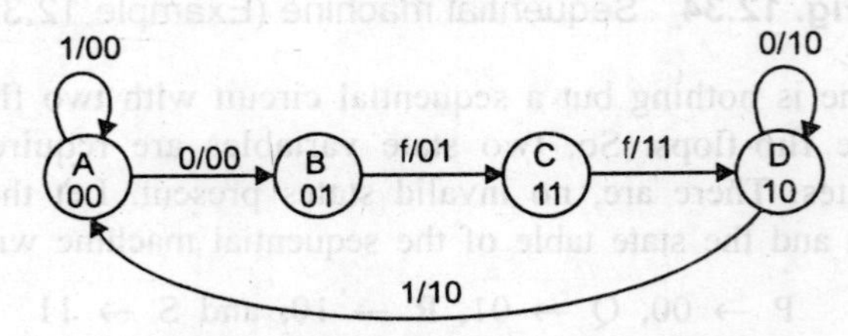

Fig. 12.37 State diagram (Example 12.4).

Solution

The states are already assigned. Write the transition and output table (Table 12.19) and the excitation table (Table 12.22) using S-R flip-flops. Two state variables are required. There are no invalid states.

Obtain the minimal expressions for S_1, R_1, S_2, R_2 and z_1, z_2 in terms of y_1, y_2 and x by using K-maps as shown in Fig. 12.38.

From the excitation table, we can see that $z_1 = y_1$ and $z_2 = y_2$ (because the entries are the same). The logic circuit using S-R flip-flops based on those minimal expressions is shown in Fig. 12.39.

Table 12.19 Transition and output table (Example 12.4)

PS		NS		O/P	
y_1	y_2	x = 0	x = 1	x = 0	x = 1
0	0	0 1	0 0	0 0	0 0
0	1	1 1	1 1	0 1	0 1
1	1	1 0	1 0	1 1	1 1
1	0	1 0	0 0	1 0	1 0

Table 12.20 Excitation table (Example 12.4)

PS		I/P	NS		Required excitation				O/P	
y_1	y_2	x	Y_1	Y_2	S_1	R_1	S_2	R_2	z_1	z_2
0	0	0	0	1	0	X	1	0	0	0
0	0	1	0	0	0	X	0	X	0	0
0	1	0	1	1	1	0	X	0	0	1
0	1	1	1	1	1	0	X	0	0	1
1	1	0	1	0	X	0	0	1	1	1
1	1	1	1	0	X	0	0	1	1	1
1	0	0	1	0	X	0	0	X	1	0
1	0	1	0	0	0	1	0	X	1	0

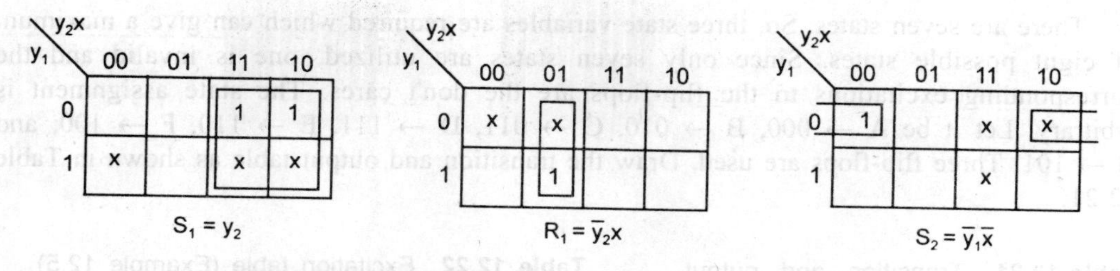

Also from K-maps for R_2, Z_1 and Z_2: $R_2 = y_1$ $Z_1 = y_1$ $Z_2 = y_2$

Fig. 12.38 K-maps (Example 12.4).

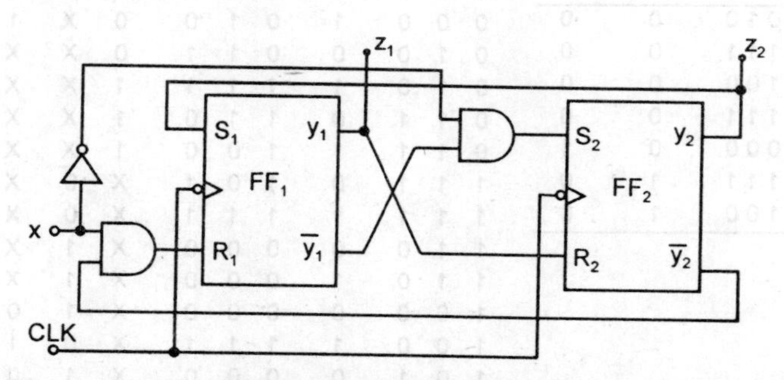

Fig. 12.39 Logic diagram of the sequential machine (Example 12.4).

EXAMPLE 12.5 Design a 2-input 2-output synchronous sequential circuit which produces an output $z = 1$, whenever any of the following input sequences 1100, 1010, or 1001 occurs. The circuit resets to its initial state after a 1 output has been generated.

Solution

Let the sequential machine be initially in state A. The state diagram and the state table are shown in Fig. 12.40.

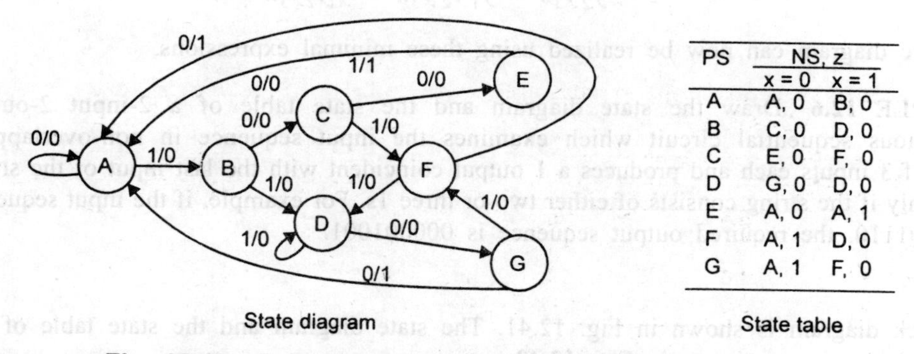

PS	NS, z	
	x = 0	x = 1
A	A, 0	B, 0
B	C, 0	D, 0
C	E, 0	F, 0
D	G, 0	D, 0
E	A, 0	A, 1
F	A, 1	D, 0
G	A, 1	F, 0

State table

Fig. 12.40 State diagram and state table (Example 12.5).

There are seven states. So, three state variables are required which can give a maximum of eight possible states. Since only seven states are utilized, one is invalid and the corresponding excitations to the flip-flops are the don't cares. The state assignment is arbitrary. Let it be A → 000, B → 010, C → 011, D → 111, E → 110, F → 100, and G → 101. Three flip-flops are used. Draw the transition and output table as shown in Table 12.21.

Table 12.21 Transition and output table (Example 12.5)

	PS	NS		O/P	
	$y_1\ y_2\ y_3$	x = 0	x = 1	x = 0	x = 1
A	0 0 0	000	010	0	0
B	0 1 0	011	111	0	0
C	0 1 1	110	100	0	0
D	1 1 1	101	111	0	0
E	1 1 0	000	000	0	1
F	1 0 0	000	111	1	0
G	1 0 1	000	100	1	0

Table 12.22 Excitation table (Example 12.5)

PS			I/P	NS			Present excitations required						O/P
y_1	y_2	y_3	x	Y_1	Y_2	Y_3	J_1	K_1	J_2	K_2	J_3	K_3	z
0	0	0	0	0	0	0	0	X	0	X	0	X	0
0	0	0	1	0	1	0	0	X	1	X	0	X	0
0	1	0	0	0	1	1	0	X	X	0	1	X	0
0	1	0	1	1	1	1	1	X	X	0	1	X	0
0	1	1	0	1	1	0	1	X	X	0	X	1	0
0	1	1	1	1	0	0	1	X	X	1	X	1	0
1	1	1	0	1	0	1	X	0	X	1	X	0	0
1	1	1	1	1	1	1	X	0	X	0	X	0	0
1	1	0	0	0	0	0	X	1	X	1	0	X	0
1	1	0	1	0	0	0	X	1	X	1	0	X	1
1	0	0	0	0	0	0	X	1	0	X	0	X	1
1	0	0	1	1	1	1	X	0	1	X	1	X	0
1	0	1	0	0	0	0	X	1	0	X	X	1	1
1	0	1	1	1	0	0	X	0	0	X	X	1	0

Select J-K FFs as memory elements. Draw the excitation table (Table 12.22). Draw the K-maps using the entries of the excitation table and obtain the minimal expressions for J_3, K_3, J_2, K_2, J_1, K_1, and z in terms of y_1, y_2, y_3, and x as follows:

$$J_1 = y_2x + y_3,\ K_1 = \bar{y}_2\bar{x} + y_2\bar{y}_3;\quad J_2 = \bar{y}_3x,\ K_2 = y_1\bar{y}_3 + y_1\bar{x} + \bar{y}_1y_3x$$

$$J_3 = \bar{y}_1y_2 + y_1\bar{y}_2x,\quad K_3 = \bar{y}_1 + \bar{y}_2;$$

$$z = \bar{y}_2y_3\bar{x} + y_1\bar{y}_2\bar{y}_3\bar{x} + y_1y_2\bar{y}_3x$$

The logic diagram can now be realized using these minimal expressions.

EXAMPLE 12.6 Draw the state diagram and the state table of a 2-input 2-output synchronous sequential circuit which examines the input sequence in non-overlapping strings of 3 inputs each and produces a 1 output coincident with the last input of the string if and only if the string consists of either two or three 1s. For example, if the input sequence is 010101110, the required output sequence is 000001001.

Solution

The block diagram is shown in Fig. 12.41. The state diagram and the state table of the sequential circuit are shown in Fig. 12.42.

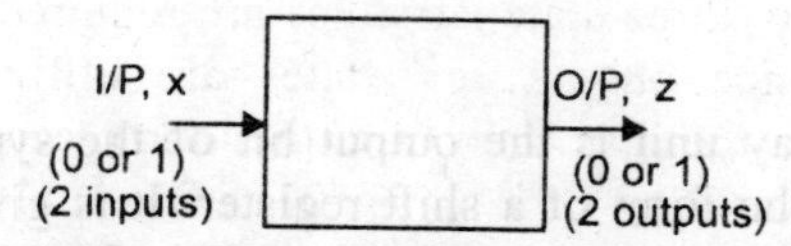

Fig. 12.41 Block diagram of the sequential machine (Example 12.6).

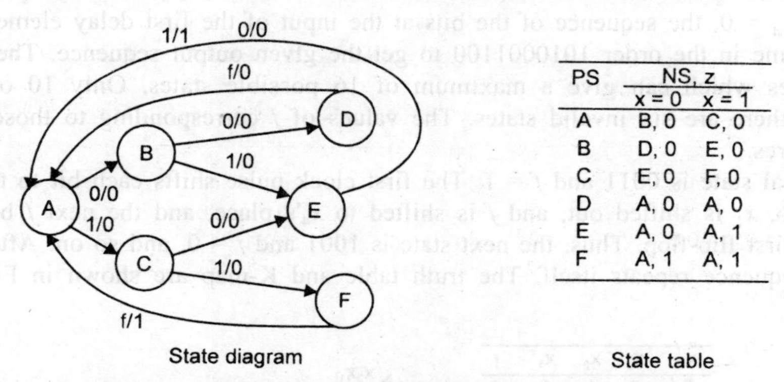

PS	NS, z x = 0	NS, z x = 1
A	B, 0	C, 0
B	D, 0	E, 0
C	E, 0	F, 0
D	A, 0	A, 0
E	A, 0	A, 1
F	A, 1	A, 1

Fig. 12.42 State diagram and state table (Example 12.6).

Initially the machine is in state A. It may receive the first bit either as a 0 or as a 1. Both are valid for the sequence. So, the machine may go to state B or C. While at B, the machine may receive the next bit as a 0 or a 1. If it is a 1, the last two bits are 01, which are part of the valid string. So, the machine goes to state E. If it is a 0, the last two bits are 00, which are not part of the valid string. But since the machine cannot start processing the next string, it goes to state D just to provide a time delay and then goes to A whether the third bit received is a 0 or a 1. While at E, if the next bit is a 0, the output is a 0; if it is a 1, the string is valid and, so, the machine outputs a 1. In both the cases, the machine goes to A. While at C, the machine may receive a 0 or a 1. If it is a 0, the machine goes to E because the first two bits are 10. If it is a 1, it goes to state F. The first two bits are now 11. While at F, whether the next bit is a 0 or a 1 the machine outputs a 1 and goes to state A, because both the sequences 110 and 111 are valid.

EXAMPLE 12.7 The synchronous circuit shown in Fig. 12.43 where D denotes a unit delay, produces a periodic binary output sequence. Assume that initially $x_1 = 1$, $x_2 = 1$, $x_3 = 0$, and $x_4 = 0$ and that the initial output sequence is 1100101000. Thereafter, the sequence repeats itself. Find a minimal expression for the combinational circuit $f(x_1, x_2, x_3, x_4)$. The clock need not be included in the expression although it is implicit.

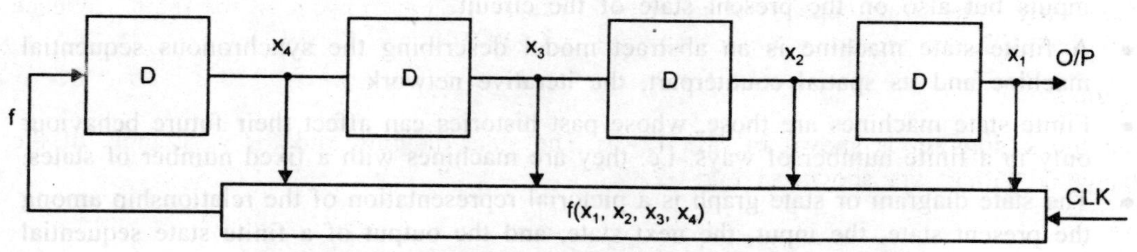

Fig. 12.43 Synchronous machine (Example 12.7).

Solution

The output x_1 of the last delay unit is the output bit of the synchronous circuit. There are four delay units arranged in the form of a shift register. It is given that the output sequence 1100101000 repeats itself. That means, the input to the first flip-flop which is the output of the combinational circuit must also be the same sequence. Since initially $x_1 = 1$, $x_2 = 1$, $x_3 = 0$, and $x_4 = 0$, the sequence of the bits at the input of the first delay element, i.e. bits of f must come in the order 1010001100 to get the given output sequence. There are four state variables which can give a maximum of 16 possible states. Only 10 of them are utilized, so, there are six invalid states. The values of f corresponding to those states are the don't cares.

The initial state is 0011 and $f = 1$. The first clock pulse shifts each bit to the right by one place. So, x_1 is shifted out, and f is shifted to x_4's place, and the next f becomes the input to the first flip-flop. Thus, the next state is 1001 and $f = 0$, and so on. After 10 clock pulses the sequence repeats itself. The truth table and K-map are shown in Fig. 12.44.

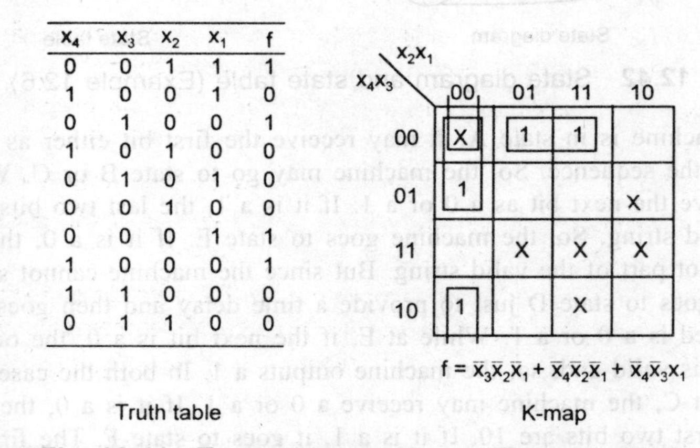

x_4	x_3	x_2	x_1	f
0	0	1	1	1
1	0	0	1	0
0	1	0	0	1
1	0	1	0	0
0	1	0	1	0
0	0	1	0	0
0	0	0	1	1
1	0	0	0	1
1	1	0	0	0
0	1	1	0	0

Truth table

Fig. 12.44 Truth table and K-map (Example 12.7).

SUMMARY

- In combinational circuits, the outputs at any instant of time depend only on the inputs applied at that instant of time.
- In sequential circuits, the outputs at any instant of time depend not only on the present inputs but also on the present state of the circuit.
- A finite state machine is an abstract model describing the synchronous sequential machine and its spatial counterpart, the iterative network.
- Finite state machines are those, whose past histories can affect their future behaviour only in a finite number of ways, i.e. they are machines with a fixed number of states.
- The state diagram or state graph is a pictorial representation of the relationship among the present state, the input, the next state, and the output of a finite state sequential machine.

- Each node in the state diagram represents a particular state of the machine. The labels on the arcs represent input/output and the direction of arrows point to the next state that the machine will go after the input is applied.
- The process of assigning the states of a physical device to the states of a sequential machine is known as state assignment.
- The state table is a tabular representation of the relationship among the present state, the input, the next state, and the output.
- The state table specifies the values of the machine outputs and the next state of the machine for each combination of the machine inputs and the present state of the machine.
- Both the state diagram and the state table contain the same information, and the choice between the two representations is a matter of convenience.
- The transition and output table can be obtained from the state table by modifying the entries of the state table to correspond to the states of the machine in accordance with the selected state assignment?
- The excitation table of a sequential machine gives information about the excitations or inputs required to be applied to the memory elements in the sequential circuit to bring the sequential machine from the present state to the next state.
- A sequence detector is a sequential machine which produces an output 1 every time the desired sequence is detected, and an output 0 at all other times.
- A serial binary adder is a synchronous circuit which adds the input numbers serially.
- A serial parity-bit generator adds a parity bit to every *m* bits of the message, so that the resulting outcome is an error-detecting coded message.
- An iterative network is a digital structure composed of a cascade of identical circuits or cells. Each cell may represent a sequential circuit or a combinational circuit.
- The inputs to all the cells of an iterative network are applied simultaneously and the outputs are assumed to be generated instantaneously.
- The number of cells in an iterative network must equal the length of the input pattern applied to it.
- The cell table specifies the values of the cell outputs and the output carries for each combination of the cell inputs and the input carries.
- If the same assignment is selected for an iterative network as for a sequential machine, then the logic circuit of the *i*th cell and the combinational logic of the sequential machine are identical.
- The output values of physical devices are referred to as state variables.

QUESTIONS

1. Distinguish between combinational and sequential circuits. List some applications of sequential circuits.
2. Distinguish between asynchronous and synchronous sequential circuits. List various memory elements used in sequential machines.

3. How is the state of a memory element specified?
4. What do you mean by the term 'state diagram'? What do the vertices, the directed arcs, and the labels on the arcs of a state diagram represent?
5. What do you mean by the Moore model and the Mealy model of the state diagram?
6. What do you mean by the term 'state table'? What does each row, column, and entry of the state table represent?
7. What do you mean by the term 'excitation table'? What information does it give?
8. What do you mean by the following terms?
   (a) Transition and output table (b) Finite state machine (c) Input alphabet
   (d) Output alphabet (e) Iterative network
9. Explain the following terms:
   (a) Finite state model (b) State assignment (c) State variables
10. What are the advantages of (a) the state diagram and (b) the state table? Compare them.
11. What are iterative networks? Discuss the analogy between sequential machines and iterative networks. How is a sequence detector designed using iterative networks?
12. Define cell of a network. How many cells are required in an iterative network?
13. Explain the terms: input carries and output carries.
14. What is a 'cell table'? What does each entry in the cell table represent?
15. What is a serial adder? Explain its working with the help of a state diagram.
16. What is a sequence detector? Explain its working with the help of a state diagram to detect any arbitrary sequence.
17. What do you mean by a parity generator? Explain its working with the help of a state diagram to generate odd parity.
18. Write the main steps in the synthesis of synchronous sequential circuits.

PROBLEMS

12.1 A long sequence of pulses enters a 2-input 2-output synchronous sequential circuit which is required to produce an output $z = 1$, whenever the sequence 1101 occurs. Overlapping sequences are accepted. Design the circuit.

12.2 Design a 3-bit counter which counts in the following sequence.

$$0 \rightarrow 2 \rightarrow 5 \rightarrow 3 \rightarrow 4 \rightarrow 0 \rightarrow 2 \rightarrow \text{..., etc.}$$

12.3 Design a circuit using S–R FFs that will function as per the state diagram shown below.

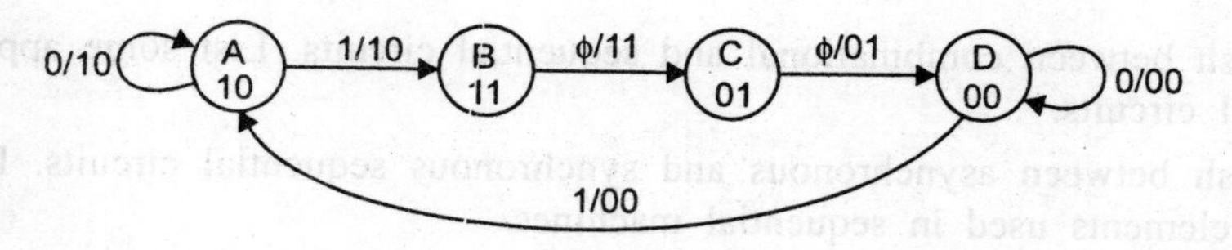

12.4 Design a 2-input 2-output synchronous sequential circuit which produces an output $z = 1$, whenever any of the following input sequences—1101, 1011, or 1001—occurs. The circuit resets to the initial state after a 1 output is generated.

12.5 A long sequence of pulses enters a 2-input 2-output synchronous sequential circuit which produces an output pulse $z = 1$, whenever the sequence 10010 occurs. The overlapping sequences are accepted. Draw the state diagram, select an assignment and show the excitation table.

12.6 Design a sequence detector which generates an output $z = 1$, whenever the string is 0110, and generates a 0 at all other times. The overlapping sequences are detected. Implement the circuit using D FFs.

12.7 Design a 3-bit up/down counter which counts up when the control signal M = 1 and counts down when M = 0.

12.8 Draw the state diagram and the state table for a 4-bit odd-parity generator.

12.9 Construct the state diagram and the state table for a 2-input machine, which produces an output $z = 1$, whenever the last string of five inputs contains exactly four zeros and the string starts with three zeros. Analysis of the next string does not start until the end of this string of five inputs, whether or not it produces a 1 output.

Chapter 13

ANALOG-TO-DIGITAL AND DIGITAL-TO-ANALOG CONVERTERS

13.1 INTRODUCTION

An analog quantity is one that can take on any value over a continuous range of values. It represents an exact value. Most physical variables are analog in nature. Temperature, pressure, light and sound intensity, position, rotation, speed, etc. are some examples of analog quantities.

A digital quantity takes on only discrete values. The value is expressed in a digital code such as a binary or BCD number.

When a physical process is monitored or controlled by a digital system such as a digital computer, the physical variables are first converted into electrical signals using transducers, and then these electrical analog signals are converted into digital signals using analog-to-digital converters (ADCs). These digital signals are processed by a digital computer and the output of the digital computer is converted into analog signals using digital-to-analog converters (DACs). The output of the DAC is modified by an actuator and the output of the actuator is applied as the control variable.

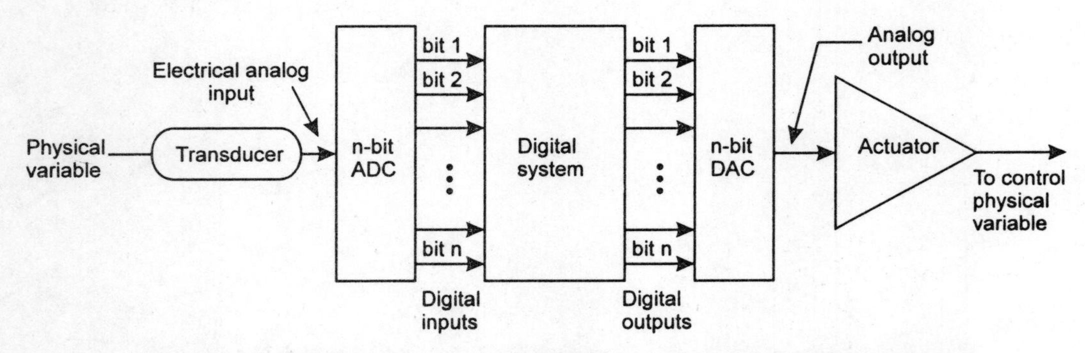

Fig. 13.1 Interfacing a digital computer to the analog world.

Figure 13.1 shows how ADCs and DACs function as interfaces between a completely digital system such as a digital computer and the analog world. This function has become increasingly more important as inexpensive microcomputers are being widely used for process control.

13.2 DIGITAL-TO-ANALOG (D/A) CONVERSION

Basically, D/A conversion is the process of converting a value represented in digital code, such as straight binary or BCD, into a voltage or current which is proportional to the digital value. Figure 13.2 shows the symbol for a typical 4-bit D/A converter. Each of the digital inputs A, B, C, and D can assume a value 0 or a 1, therefore, there are $2^4 = 16$ possible combinations of inputs. For each input number 0000, 0001,...,1111, the D/A converter outputs a unique value of voltage. The analog output voltage V_{out} is proportional to the input binary number, that is,

$$\text{Analog output} = K \times \text{digital input}$$

where K is the proportionality factor and is a constant value for a given DAC. The analog output can, of course, be current or voltage.

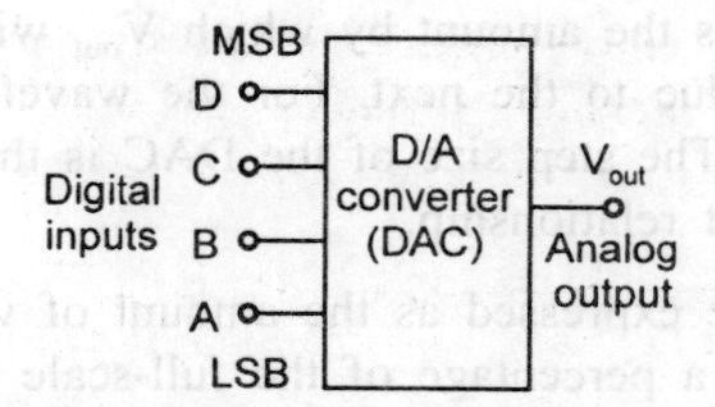

Fig. 13.2 Block diagram of a 4-bit DAC.

Strictly speaking, the output of a DAC is not a true analog quantity, because it can take on only specific values. In that sense, it is actually digital. Thus, the output of a DAC is a 'pseudo-analog' quantity. By increasing the number of input bits, the number of possible output values can be increased and also the step size (the difference between two successive output values) can be reduced, thereby producing an output that is more like an analog quantity. Figure 13.3 shows the output waveform of a DAC when it is fed by a 4-bit binary counter.

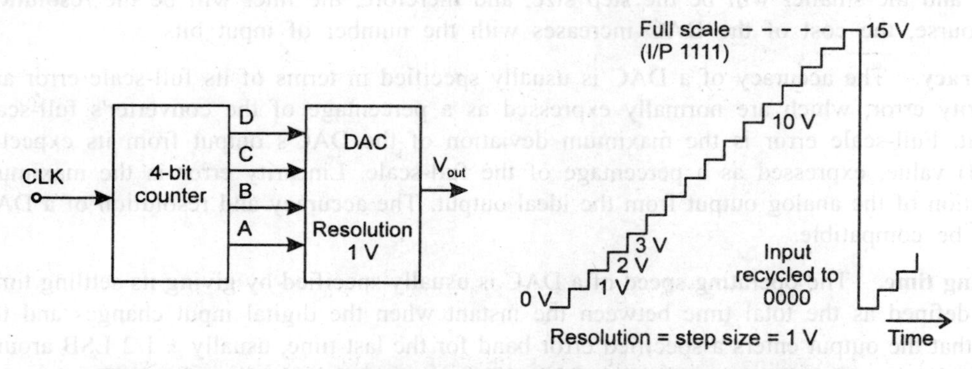

Fig. 13.3 Output waveform of a DAC fed by a binary counter.

When the binary counter is continually recycled through its 16 states by applying the clock signal, the DAC output will be a staircase waveform with a step size of 1 V. When the counter is at 0000, the output of the DAC is minimum (0 V). When the counter is at 1111, the output of the DAC is maximum (15 V). This is the full-scale output.

Digital-to-analog and analog-to-digital conversions form the very important aspects of digital data processing. Digital-to-analog conversion is a straightforward process and is considerably easier than the A/D conversion. In fact, a DAC is usually an integral part of any ADC.

Parameters of DAC

Resolution (step size). The resolution of a DAC is defined as the smallest change that can occur in an analog output as a result of a change in the digital input. The resolution of a DAC is also defined as the reciprocal of the number of discrete steps in the full-scale output of the DAC. The resolution is always equal to the weight of the LSB and is also referred to as the *step size*. The resolution or step size is the size of the jumps in the staircase waveform. Step size is the amount by which V_{out} will change as the digital input value is changed from one value to the next. For the waveform shown in Fig. 13.3, the resolution or step size = 1 V. The step size of the DAC is the same as the proportionality factor in the DAC input-output relationship.

Although resolution can be expressed as the amount of voltage or current per step, it is also useful to express it as a percentage of the full-scale output as

$$\% \text{ resolution} = \frac{\text{step size}}{\text{full scale}} \times 100\%$$

Since, full-scale = number of steps × step size, resolution can be expressed as

$$\% \text{ resolution} = \frac{1}{\text{total number of steps}} \times 100\%$$

In general, for an N-bit DAC, the number of different levels will be 2^N and the number of steps will be $2^N - 1$. The greater the number of bits, the greater will be the number of steps and the smaller will be the step size, and therefore, the finer will be the resolution. Of course, the cost of the DAC increases with the number of input bits.

Accuracy. The accuracy of a DAC is usually specified in terms of its full-scale error and linearity error, which are normally expressed as a percentage of the converter's full-scale output. Full-scale error is the maximum deviation of the DAC's output from its expected (ideal) value, expressed as a percentage of the full-scale. Linearity error is the maximum deviation of the analog output from the ideal output. The accuracy and resolution of a DAC must be compatible.

Settling time. The operating speed of a DAC is usually specified by giving its settling time. It is defined as the total time between the instant when the digital input changes and the time that the output enters a specified error band for the last time, usually ± 1/2 LSB around the final value after the change in digital input. It is measured as the time for the DAC output to settle within ± 1/2 step size of its final value. Generally, DACs with a current output will have shorter settling times than those with voltage outputs.

Offset voltage. Ideally, the output of a DAC should be zero when the binary input is zero. In practice, however, there is a very small output voltage under this situation called the *offset voltage*. This offset error, if not corrected, will be added to the expected DAC output for all input cases.

Monotonicity. A DAC is said to be monotonic if its output increases as the binary input is incremented from one value to the next. This means that the staircase output will have no downward steps as the binary input is incremented from 0 to full-scale value. The DAC is said to be non-monotonic, if its output decreases when the binary input is incremented.

EXAMPLE 13.1 Determine the resolution of (a) a 6-bit DAC and that of (b) a 12-bit DAC in terms of percentage.

Solution

(a) For the 6-bit DAC,

$$\% \text{ resolution} = \frac{1}{2^N - 1} \times 100 = \frac{1}{2^6 - 1} \times 100 = \frac{1}{63} \times 100 = 1.587\%$$

(b) For the 12-bit DAC,

$$\% \text{ resolution} = \frac{1}{2^N - 1} \times 100 = \frac{1}{2^{12} - 1} \times 100 = \frac{1}{4095} \times 100 = 0.0244\%$$

EXAMPLE 13.2 A 6-bit DAC has a step size of 50 mV. Determine the full-scale output voltage and the percentage resolution.

Solution

With 6 bits, there will be $2^6 - 1 = 63$ steps of size 50 mV each.
The full-scale output will, therefore, be

$$63 \times 50 \text{ mV} = 3.15 \text{ V}$$

$$\% \text{ resolution} = \frac{50 \text{ mV}}{3.15 \text{ V}} \times 100 = \frac{1}{63} \times 100 = 1.587\%$$

EXAMPLE 13.3 An 8-bit DAC produces $V_{out} = 0.05$ V for a digital input of 00000001. Find the full-scale output. What is the resolution? What is V_{out} for an input of 00101010?

Solution

$$\text{Full-scale output} = \text{Step size} \times \text{Number of steps}$$
$$= 0.05(2^8 - 1) = 0.05 \times 255 = 12.75 \text{ V}$$

$$\% \text{ resolution} = \frac{1}{2^N - 1} \times 100 = \frac{1}{255} \times 100 = 0.392\%$$

V_{out} for an input of $00101010 = 42 \times 0.05 = 2.10$ V

EXAMPLE 13.4 A certain 6-bit DAC has a full-scale output of 2 mA and a full-scale error of ± 0.5%. What is the range of possible outputs for an input of 100000?

Solution

The step size is 2 mA/63 = 31.7 μA. Since 100000 = 32_{10}, the ideal output should be 32 × 31.7 = 1014 μA. The error can be as much as ± 0.5% × 2 mA = 10 μA.

Thus the actual output can deviate by this amount from the ideal value of 1014 μA, and therefore, the actual output can be anywhere from 1004 μA to 1024 μA.

DAC Using BCD Input Code

The DAC we considered earlier used a binary input code. Many DACs use a BCD input code, where 4-bit code groups are used for each decimal digit. Figure 13.4 shows the diagram of an 8-bit (2-digit) converter of this type. Each 4-bit code group can range from 0000 to 1001, and so the BCD inputs represent decimal numbers from 00 to 99. Within each code group, the weights of the different bits are in the normal binary proportions (1, 2, 4, 8), but the relative weights of each code group are different by a factor of 10. Figure 13.4 shows the relative weights of the various bits. Note that the bits that make up the BCD code for the most significant digit (MSD) have a relative weight that is 10 times that of the corresponding bits of the LSD.

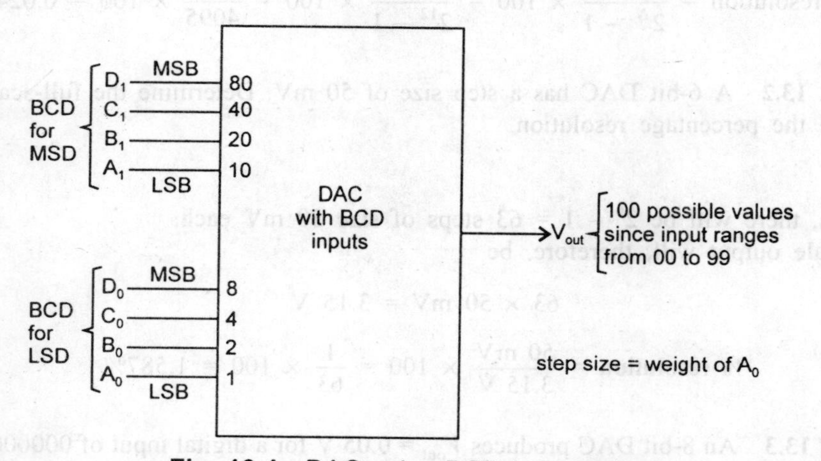

Fig. 13.4 DAC using BCD input code.

EXAMPLE 13.5 If the weight of A_0 is 0.2 V in Fig. 13.4 , find the following values: (a) step size, (b) full-scale output, (c) percentage resolution, and (d) V_{out} for $D_1C_1B_1A_1$ = 0110 and $D_0C_0B_0A_0$ = 0100.

Solution

(a) Step size is the weight of the LSB of the LSD, i.e. 0.2 V.

(b) There are 99 steps since there are two BCD digits. Thus, the full-scale output is 99 × 0.2 V = 19.8 V.

(c) The resolution is $\dfrac{\text{step size}}{\text{full-scale output}} \times 100\% = \dfrac{0.2}{19.8} \times 100\% \approx 1\%$

The exact weights of the various bits in the number in volts are:

$$\text{MSD: } D_1 = 16,\ C_1 = 8,\ B_1 = 4,\ A_1 = 2$$

$$\text{LSD: } D_0 = 1.6,\ C_0 = 0.8,\ B_0 = 0.4,\ A_0 = 0.2$$

(d) One way to find V_{out} for a given input is to add the weights of all the bits that are 1s. Thus, for an input of 01100100, we have

$$\begin{matrix} & C_1 & & B_1 & & C_0 \\ V_{out} = & 8\text{ V} & + & 4\text{ V} & + & 0.8\text{ V} = 12.8\text{ V} \end{matrix}$$

As the BCD input code represents 64_{10} and the step size is 0.2 V, V_{out} can also be determined as

$$V_{out} = (0.2\text{ V}) \times 64 = 12.8\text{ V}$$

EXAMPLE 13.6 A certain 12-bit BCD DAC has a full-scale output of 19.98 V. Determine (a) the percentage resolution and (b) the converter's step size.

Solution

(a) 12 BCD bits correspond to three decimal digits, i.e. decimal numbers from 000 to 999. Therefore, the output of the DAC has 999 possible steps from 0 V to 19.98 V. Thus, we have

$$\%\text{ resolution} = \frac{1}{\text{number of steps}} \times 100\% = \frac{1}{999} \times 100\% \approx 0.1\%$$

(b) $\text{Step size} = \dfrac{\text{full-scale output}}{\text{number of steps}} = \dfrac{19.98\text{ V}}{999} = 0.02\text{ V}$

Bipolar DACs

So far we have assumed that the binary input to a DAC is an unsigned number and the DAC output is a positive voltage or current. Such types of DACs are called unipolar DACs.

Bipolar DACs are designed to produce both positive and negative values. This is generally done by using the binary input as a signed number with the MSB as the sign bit (0 for + and 1 for –). Negative input values are often represented in 2's complement form, although the true magnitude form is also used by some DACs. For example, suppose we have a 6-bit bipolar DAC that uses the 2's complement system and has a resolution of 0.2 V. The binary input values range from 100000 (– 32) to 011111(+ 31) to produce analog outputs in the range from – 6.4 V to + 6.2 V. There are 63 steps (2^N – 1) of 0.2 V between these negative and positive limits.

13.3 THE *R*-2*R* LADDER TYPE DAC

This is the most popular DAC. It uses a ladder network containing series-parallel combinations of two resistors of values *R* and 2*R*. Hence the name. The operational amplifier configured as voltage follower is used to prevent loading. Figure 13.5 shows the circuit diagram of a *R*-2*R* ladder type DAC having a 4-bit digital input. When a digital signal $D_3D_2D_1D_0$ is applied at the input terminals of the DAC, an equivalent analog signal is produced at the output terminal. The operation of the DAC is as follows:

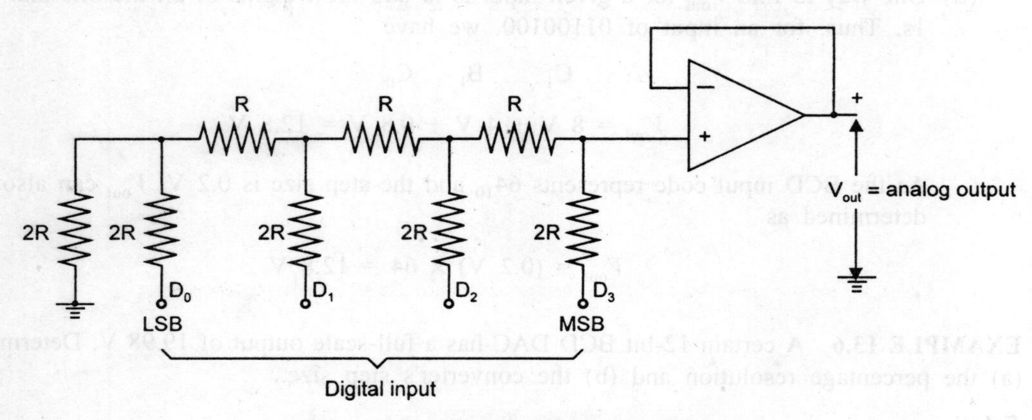

Fig. 13.5 *R*-2*R* ladder type DAC.

Case 1: When the input is 1000. Figure 13.6 illustrates the procedure to calculate V_{out} when the input is 1000. At the left end of the ladder, 2*R* is in parallel with 2*R*, so that the combination is equivalent to *R*. This *R* is in series with another *R* giving 2*R*. This 2*R* in parallel with another 2*R* is equivalent to *R*. Continuing in this manner, we ultimately find that $R'_{eq} = 2R$. The output voltage, $V_{out} = E/2$

Case 2: when the input is 0100. Figure 13.7 illustrates the procedure to calculate V_{out} when the input is 0100. Here, we find that to the left of terminal B, $R_{eq} = 2R$. The output voltage, $V_{out} = E/4$.

Case 3: when the input is 0010. Figure 13.8 illustrates the procedure to calculate V_{out} when the input is 0010. Here, R_{eq} to the left of C is 2*R*. The output voltage, $V_{out} = E/8$.

Case 4: when the input is 0001. Figure 13.9 illustrates the procedure to calculate V_{out} when the input is 0001. The output voltage, $V_{out} = E/16$.

In general, when the D_n input is a 1 and all other inputs are a 0, the output is $V_{out} = \frac{E}{2^{N-n}}$, where *N* is the total number of binary inputs. For example, if $E = 5$ V, then the output voltage when the input is 0100 is

$$\frac{5}{2^{4-2}} = \frac{5}{4} = 1.25 \text{ V}$$

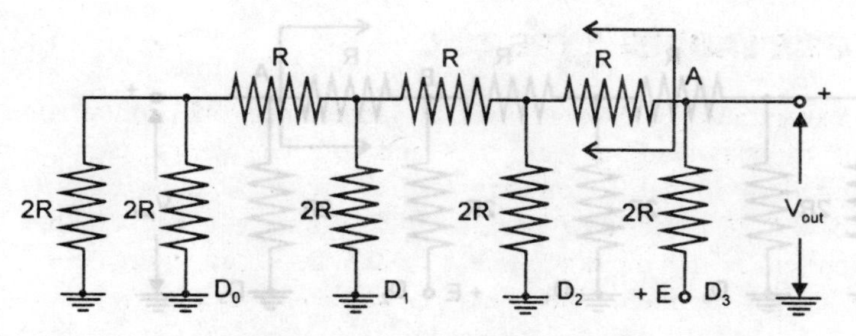

(a) When the input is 1000; D_0, D_1 and D_2 are grounded (0 V) and D_3 = + E.

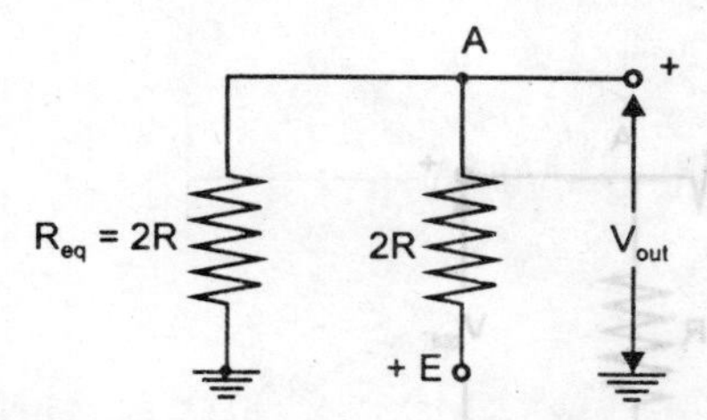

(b) The circuit equivalent to (a) when the circuit to the left of A is replaced by its equivalent resistance R_{eq}.

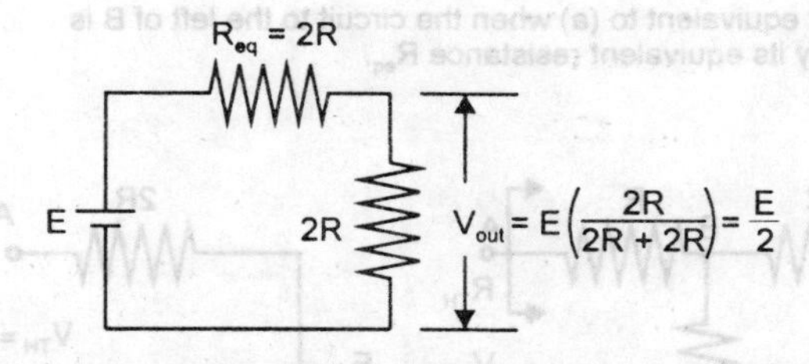

(c) Calculation of V_{out} using the voltage divider rule.

Fig. 13.6 Calculation of V_{out} when the input is 1000.

To find the output voltage corresponding to any input combination, apply the principle of superposition and simply add the voltages produced by the inputs where 1s are applied.

Typical values for R and $2R$ are 10 kΩ and 20 kΩ, respectively. For accurate conversion, the output of the R-$2R$ ladder network is connected to a high impedance circuit to prevent loading. An op-amp configured as a voltage follower can be used for this purpose as shown in Fig. 13.5.

The principal advantage of this converter is that resistors of only two values are required. Therefore, standard resistors can be used.

EXAMPLE 13.7 What are the output voltages caused by logic 1 in each bit position in an 8-bit ladder if the input level for 0 is 0 V and that for 1 is + 10 V?

Solution

The output voltage level caused by the D_n bit $= \dfrac{E}{2^{N-n}}$

(a) When the input is 0100, D_0, D_1 and D_3 are grounded (0 V) and D_2 = + E.

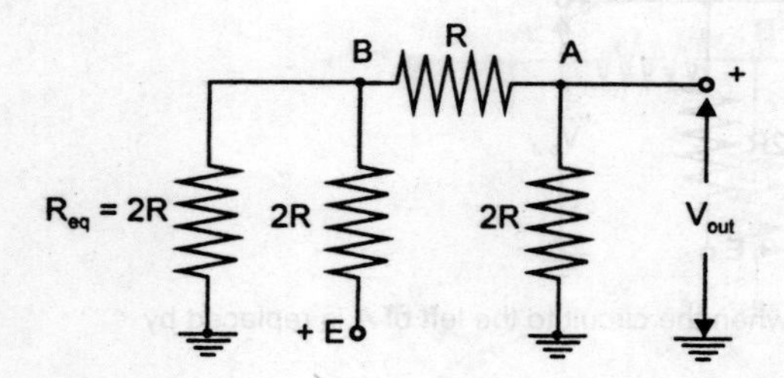

(b) The circuit equivalent to (a) when the circuit to the left of B is replaced by its equivalent resistance R_{eq}.

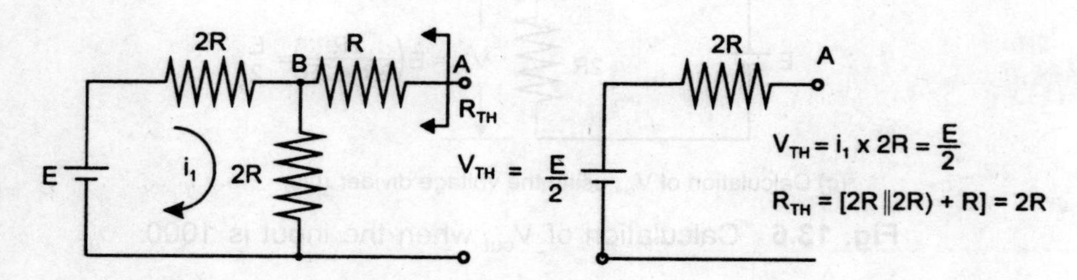

(c) Thevenin's equivalent to the left of A.

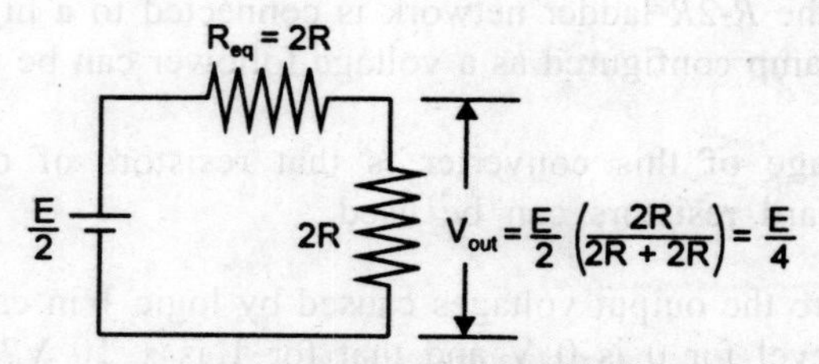

(d) Calculation of V_{out} using the voltage divider rule.

Fig.13.7 Calculation of V_{out} when the input is 0100.

R R R
C B A
2R 2R 2R 2R 2R
D_0 $+E$ D_1 D_2 D_3
V_{out}

(a) When the input is 0010, D_0, D_2 and D_3 are grounded (0 V) and D_1 = + E.

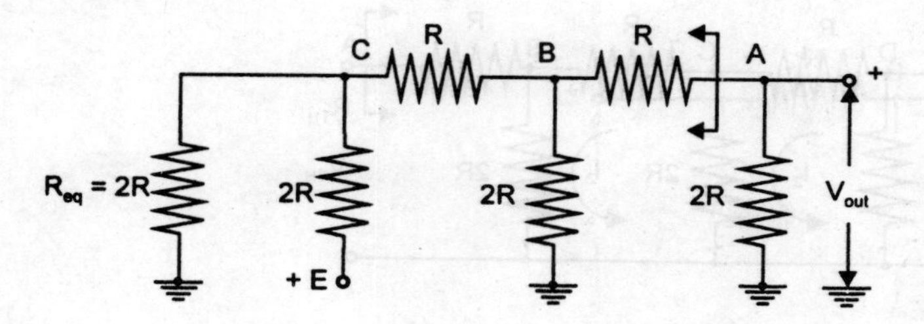

(b) The circuit equivalent to (a) when the circuit to the left of C is replaced by R_{eq} = 2R.

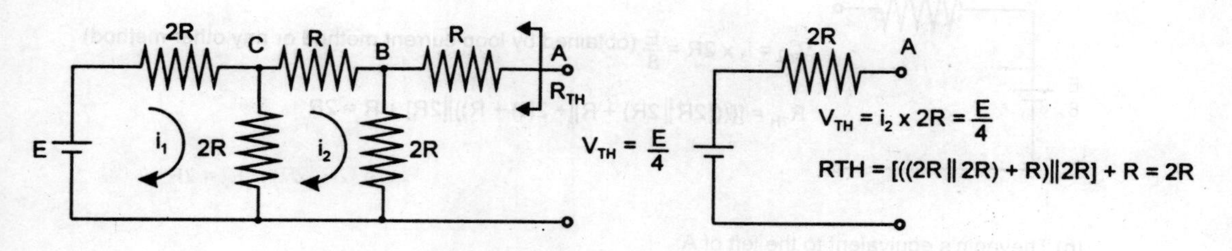

(c) Thevenin's equivalent to the left of A.

2R

$\frac{E}{4}$ 2R

$$V_{out} = \frac{E}{4}\left(\frac{2R}{2R + 2R}\right) = \frac{E}{8}$$

(d) Calculation of V_{out} using the voltage divider rule.

Fig. 13.8 Calculation of V_{out} when the input is 0010.

(a) When the input is 0001, D_1, D_2 and D_3 are grounded (0 V) and D_0 = + E.

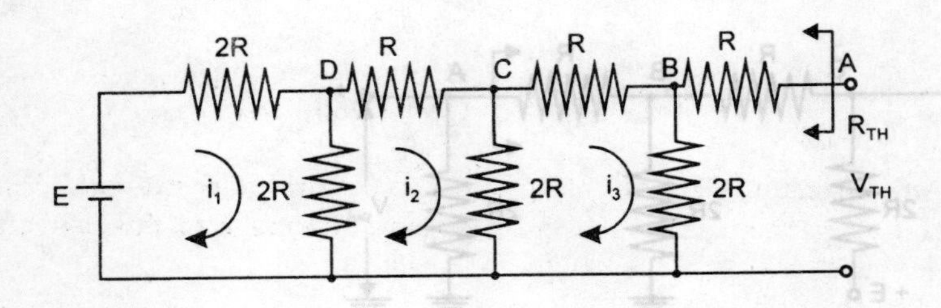

(b) The circuit equivalent to (a).

$V_{TH} = i_3 \times 2R = \frac{E}{8}$ (obtained by loop current method or any other method)

$R_{TH} = [\{(((2R \| 2R) + R \| + 2R) + R)\} \| 2R] + R = 2R$

(c) Thevenin's equivalent to the left of A.

$$V_{out} = \frac{E}{8}\left(\frac{2R}{2R + 2R}\right) = \frac{E}{16}$$

(d) Calculation of V_{out} using the voltage divider rule.

Fig. 13.9 Calculation of V_{out} when the input is 0001.

The output voltage level caused by the D_7 bit (MSB) $= \dfrac{E}{2^{8-7}} = \dfrac{E}{2^1} = \dfrac{10}{2} = 5$ V

The output voltage level caused by the D_6 bit $= \dfrac{E}{2^2} = \dfrac{10}{4} = 2.5$ V

The output voltage level caused by the D_5 bit $= \dfrac{E}{2^3} = \dfrac{10}{8} = 1.25$ V

The output voltage level caused by the D_4 bit $= \dfrac{E}{2^4} = \dfrac{10}{16} = 0.625$ V

$\vdots$

The output voltage level caused by the D_0 bit (LSB) $= \dfrac{E}{2^8} = \dfrac{10}{256} = 0.03906$ V

EXAMPLE 13.8 What is the resolution of a 9-bit DAC, which uses a ladder network? What is this resolution expressed as a percentage? If the full-scale output voltage of this converter is + 5 V, what is the resolution in volts?

Solution

The LSB in a 9-bit system has a weight of 1/512. Thus, this converter has a resolution of 1 part in 512. The resolution expressed as a percentage is, $(1/512) \times 100 \approx 0.2\%$. The voltage resolution is obtained by multiplying the weight of the LSB by the full-scale output voltage. Thus, the resolution in volts is, $(1/512) \times 5$ V = 9.8 mV.

13.4 THE WEIGHTED-RESISTOR TYPE DAC

The diagram of the weighted-resistor DAC is shown in Fig. 13.10. The operational amplifier is used to produce a weighted sum of the digital inputs, where the weights are proportional to the weights of the bit positions of inputs. Since the op-amp is connected as an inverting amplifier, each input is amplified by a factor equal to the ratio of the feedback resistance divided by the input resistance to which it is connected. The MSB D_3 is amplified by R_f/R, D_2 is amplified by $R_f/2R$, D_1 is amplified by $R_f/4R$, and D_0, the LSB is amplified by $R_f/8R$.

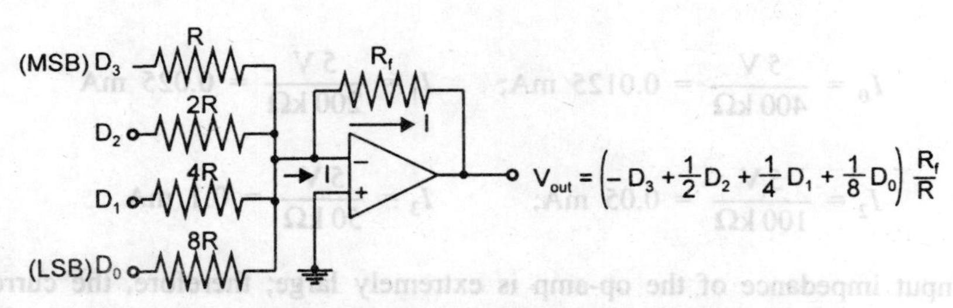

Fig. 13.10 Weighted-resistor type DAC.

The inverting terminal of the op-amp in Fig. 13.10 acts as a virtual ground. Since the op-amp adds and inverts,

$$V_{out} = -\left(D_3 + \frac{D_2}{2} + \frac{D_1}{4} + \frac{D_0}{8}\right) \times \left(\frac{R_f}{R}\right)$$

The main disadvantage of this type of DAC is, that a different-valued precision resistor must be used for each bit position of the digital input.

EXAMPLE 13.9 For the 4-bit weighted-resistor DAC shown in Fig. 13.10, determine (a) the weight of each input bit if the inputs are 0 V and 5 V, (b) the full-scale output, if $R_f = R = 1\ k\Omega$. Also, find the full-scale output if R_f is changed to 500 Ω.

Solution

(a) The MSB passes with a gain of 1, so, its weight = 5 V; the next bit passes with a gain of 1/2, so, its weight = 5/2 = 2.5 V; the following bit passes with a gain of 1/4, so, its weight = 5/4 = 1.25 V; the LSB passes with a gain of 1/8, so, its weight = 5/8 = 0.625 V.

(b) Therefore, the full scale output (when $R_f = R = 1\ k\Omega$) $= -\left(1 + \frac{1}{2} + \frac{1}{4} + \frac{1}{8}\right) \times 5$

$= 9.375$ V

The full-scale output, when R_f is changed to 500 Ω is

$$V_{out} = -\left(1 + \frac{1}{2} + \frac{1}{4} + \frac{1}{8}\right) \times \left(\frac{5}{2}\right) = -4.6875\ V$$

EXAMPLE 13.10 Determine the output of the DAC in Fig. 13.11a, if the sequence of the 4-bit numbers in Fig. 13.11b is applied to the inputs. D_0 is the LSB.

Solution

First, let us determine the output current I for each of the weighted inputs. Since the inverting input (–) of the op-amp is at 0 V (virtual ground) and a binary 1 corresponds to + 5 V, the current through any of the input resistors is 5 V divided by the resistance value. Thus,

$$I_0 = \frac{5\ V}{400\ k\Omega} = 0.0125\ mA; \qquad I_1 = \frac{5\ V}{200\ k\Omega} = 0.025\ mA$$

$$I_2 = \frac{5\ V}{100\ k\Omega} = 0.05\ mA; \qquad I_3 = \frac{5\ V}{50\ k\Omega} = 0.1\ mA$$

The input impedance of the op-amp is extremely large; therefore, the current into the op-amp is zero. Thus, the current, $I_0 + I_1 + I_2 + I_3$, has to go through the feedback resistance R_f. Since one end of R_f is at 0 V (virtual ground), the drop across R_f equals the output voltage.

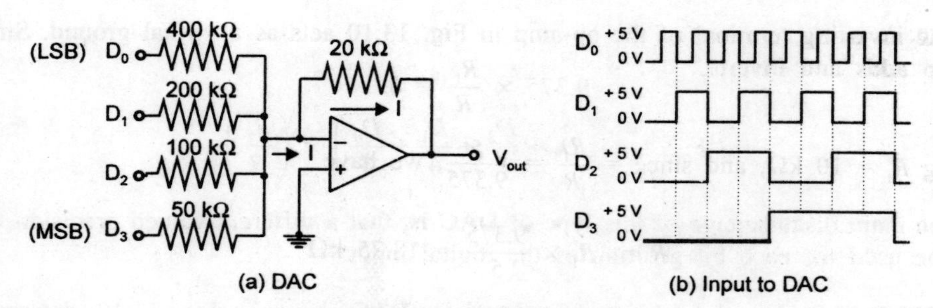

Fig. 13.11 Example 13.10.

The output voltage due to input D_0 is, 20 kΩ (– 0.0125 mA) = – 0.25 V
The output voltage due to input D_1 is, 20 kΩ (– 0.025 mA) = – 0.5 V
The output voltage due to input D_2 is, 20 kΩ (– 0.0 5 mA) = – 1 V
The output voltage due to input D_3 is, 20 kΩ (– 0.1 mA) = – 2 V

From Fig. 13.11b, the first input code is 0001. For this, the output voltage is – 0.25 V. The next input code is 0010, for which the output voltage is – 0.5 V. The next code is 0011 for which the output voltage is – 0.75, and so on. Each successive binary code increases the output voltage by – 0.25 V. So, for this particular straight binary sequence of the inputs, the output is a staircase waveform going from 0 V to – 3.75 V in – 0.25 V steps. The output waveform is shown in Fig. 13.12.

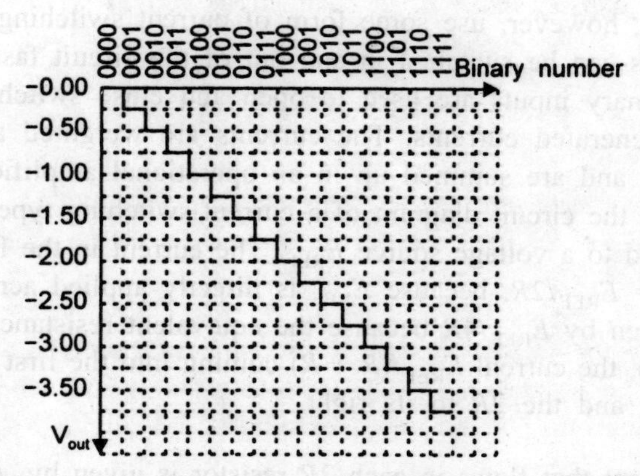

Fig. 13.12 Output waveform (Example 13.10).

EXAMPLE 13.11 Design a 4-bit weighted-resistor DAC whose full-scale output voltage is – 5 V. The logic levels are 1 = + 5 V and 0 = 0 V. What is the output voltage when the input is 1101?

Solution

The full-scale output voltage is the output voltage when the input voltage is maximum, i.e. 1111. Thus,

$$\text{The full-scale output} = -\left(5\text{ V} + \frac{5\text{ V}}{2} + \frac{5\text{ V}}{4} + \frac{5\text{ V}}{8}\right)\frac{R_f}{R} = -5\text{ V}$$

Therefore,

$$9.375 \times \frac{R_f}{R} = 5$$

Choosing $R_f = 10\ k\Omega$, and since $= \frac{R_f}{R} = \frac{5}{9.375}$, we have

$$R = R_f \times \frac{9.375}{5} = 18.75\ k\Omega$$

Thus,

$$2R = 37.5\ k\Omega; \quad 4R = 75\ k\Omega; \quad 8R = 150\ k\Omega$$

When the input is 1101, the output voltage is

$$V_{out} = -\left(5\ V + \frac{5\ V}{2} + 0 + \frac{5\ V}{8}\right)\frac{R_f}{R} = -\ 8.125 \times \frac{10}{18.75} = -\ 4.333\ V$$

13.5 THE SWITCHED CURRENT-SOURCE TYPE DAC

The *R*-2*R* ladder network DAC and the weighted-resistor DAC can be regarded as switched voltage-source DACs, because when an input position goes HIGH, the HIGH voltage is effectively switched into the circuit, where it is summed up with other input voltages. Most integrated circuit DACs, however, use some form of current switching rather than voltage switching, since currents can be switched in and out of the circuit faster than voltages. In the former type, the binary inputs are used to open and close switches that connect and disconnect internally generated currents. The currents are weighted according to the bit positions they represent and are summed up in an operational amplifier.

Figure 13.13 shows the circuit diagram of a current-switching type DAC. Note that an *R*-2*R* ladder is connected to a voltage source E_{REF}. The current in the first 2*R* resistor from supply is given by $I_3 = E_{REF}/2R$, because E_{REF} is directly applied across 2*R*. The current in the second 2*R* is given by $E_{REF}/4R$ because the equivalent resistance to the right of the second 2*R* is 2*R* and, so, the current $E_{REF}/(R + R)$ coming into the first *R* is equally divided between the second 2*R* and the 2*R* to its right.

In general, the current that flows in each 2*R* resistor is given by, $I_n = \left(\frac{E_{REF}}{R}\right) \times \frac{1}{2^{N-n}}$

where $n = 0, 1, \ldots, N - 1$ is the subscript for the current created by input D_n and *N* is the total number of inputs. Thus, each current is weighted by the bit position it represents.

The switches that connect the currents either to ground or to the input of the operational amplifier are controlled by the digital input. The op-amp sums up all those currents whose corresponding digital inputs are HIGH. The amplifier also serves as a voltage-to-current converter. It is connected in an inverting configuration and its output is $V_{out} = -\ I_T R$, where I_T is the sum of the currents that have been switched to its input.

If E_{REF} is an externally variable voltage, the output of the DAC is proportional to the product of the variable E_{REF} and the variable digital input. In that case, the circuit is called

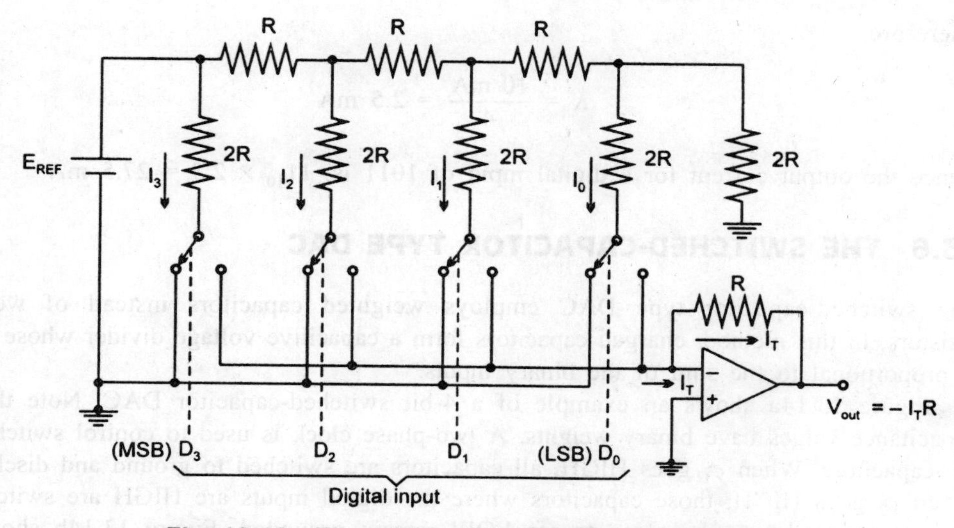

Fig. 13.13 The 4-bit switched current-source DAC.

a *multiplying* DAC and the output represents the product of an analog input E_{REF} and a digital input.

EXAMPLE 13.12 The switched current-source DAC in Fig. 13.13 has R = 5 kΩ and E_{REF} = 10 V. Find the total current delivered to the amplifier and the output voltage when the digital input is 1101.

Solution

Since the digital input is 1101, the total current into the amplifier, i.e. I_T is given by

$$I_T = I_3 + I_2 + I_0$$

$$= \left(\frac{E_{REF}}{2R}\right) + \left(\frac{E_{REF}}{4R}\right) + \left(\frac{E_{REF}}{16R}\right)$$

$$= \left(\frac{10}{10} + \frac{10}{20} + \frac{10}{80}\right) = \frac{130}{80} = \frac{13}{8} \text{ mA}$$

Therefore, the output voltage, $V_{out} = -I_T \times R = -\left(\frac{13}{8} \times 5\right) = -8.125$ V

EXAMPLE 13.13 In a 4-bit DAC, for a digital input of 0100 an output current of 10 mA is produced. What will be the output current for a digital input of 1011?

Solution

Output current = 10 mA for a digital input of 0100, i.e. 4_{10}.

Analog output = $K \times$ digital input

Therefore,

$$K = \frac{10 \text{ mA}}{4} = 2.5 \text{ mA}$$

Hence the output current for a digital input of 1011 is, $11_{10} \times 2.5 = 27.5$ mA

13.6 THE SWITCHED-CAPACITOR TYPE DAC

The switched-capacitor type DAC employs weighted capacitors instead of weighted resistors. In this method, charged capacitors form a capacitive voltage divider whose output is proportional to the sum of the binary inputs.

Figure 13.14a shows an example of a 4-bit switched-capacitor DAC. Note that the capacitance values have binary weights. A two-phase clock is used to control switching of the capacitors. When ϕ_1 goes HIGH, all capacitors are switched to ground and discharged. When ϕ_2 goes HIGH, those capacitors where the digital inputs are HIGH are switched to E_{REF}, whereas those whose inputs are LOW remain grounded. Figure 13.14b shows the equivalent circuit when ϕ_2 is HIGH and the digital input is 1101.

We see that the capacitors whose digital inputs are a 1, are in parallel and the capacitors whose digital inputs are a 0, are in parallel with $C/8$. The circuit is redrawn in Fig. 13.14c, where each set of the parallel capacitors is replaced by its equivalent capacitance. The output of the capacitive voltage divider is

$$V_{out} = E_{REF}\left(\frac{\frac{13}{8}C}{2C}\right) = \frac{13}{16}E_{REF}$$

In general, for any binary input,

$$V_{out} = \left(\frac{C_{EQ}}{2C}\right) \times E$$

where $2C$ is the sum of all the capacitance values in the circuit and C_{EQ} is the sum of all the capacitors whose digital inputs are HIGH.

The analog output is proportional to the digital input. When the input is 0000, the positive terminal of E_{REF} is effectively open-circuited as shown in Fig. 13.14d; so, the output is 0 V.

Switched-capacitor technology has been developed for implementing analog functions in integrated circuits, particularly MOS circuits. It is used to construct filters, amplifiers, and many other special devices. The principal advantage of this technology is that, small capacitors of the order of a few picofarads can be constructed in the integrated circuits to perform the function of the much larger capacitors that are normally needed in low-frequency analog circuits.

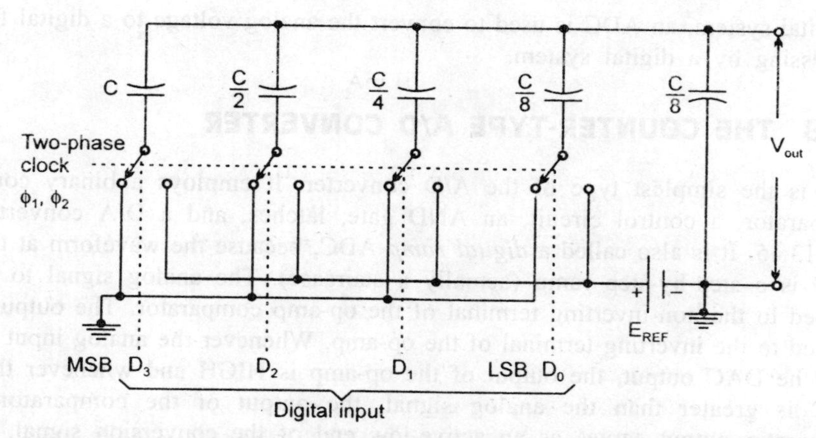

(a) All capacitors are switched to ground by ϕ_1. Those capacitors whose digital inputs are HIGH are switched to E_{REF} by ϕ_2.

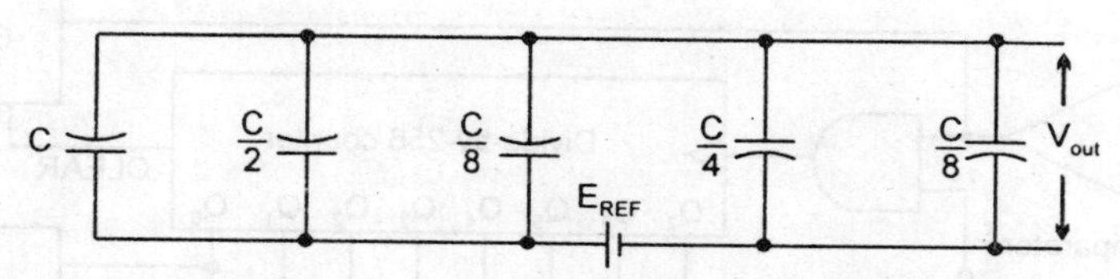

(b) Equivalent circuit when the input is 1101. The capacitors switched to E_{REF} are in parallel as are the ones connected to ground.

$$\frac{13C}{8} \qquad \frac{3C}{8} \qquad V_{out} = E_{REF}\left(\frac{\frac{13}{8}C}{2C}\right) = \frac{13}{16}E_{REF}$$

E_{REF}

(c) Circuit equivalent to (b). The output is determined by a capacitor voltage divider.

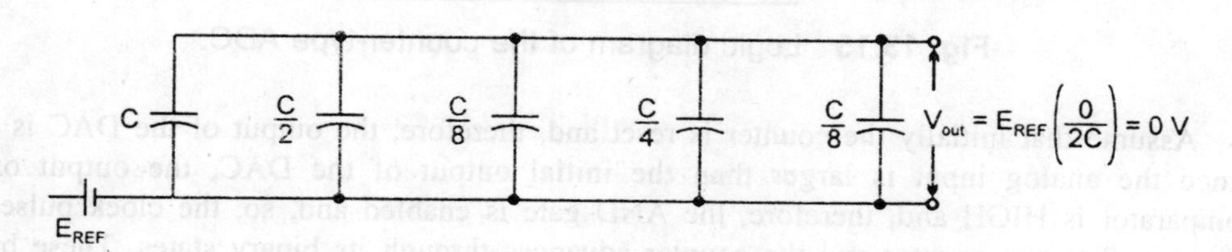

(d) Equivalent circuit when the input is 0000.

Fig. 13.14 The switched-capacitor type DAC.

13.7 ANALOG-TO-DIGITAL CONVERSION

An analog-to-digital converter (A/D converter or ADC) produces a digital output, that is proportional to the value of the input analog signal. When an analog signal is processed by a digital system, an ADC is used to convert the analog voltage to a digital form suitable for processing by a digital system.

13.8 THE COUNTER-TYPE A/D CONVERTER

This is the simplest type of the A/D converter. It employs a binary counter, a voltage comparator, a control circuit, an AND gate, latches, and a D/A converter as shown in Fig. 13.15. It is also called a *digital ramp* ADC, because the waveform at the output of the DAC is a step-by-step ramp (actually a staircase). The analog signal to be converted is applied to the non-inverting terminal of the op-amp comparator. The output of the DAC is applied to the inverting terminal of the op-amp. Whenever the analog input signal is greater than the DAC output, the output of the op-amp is HIGH and whenever the output of the DAC is greater than the analog signal, the output of the comparator is LOW. The comparator output serves as an active-low end of the conversion signal.

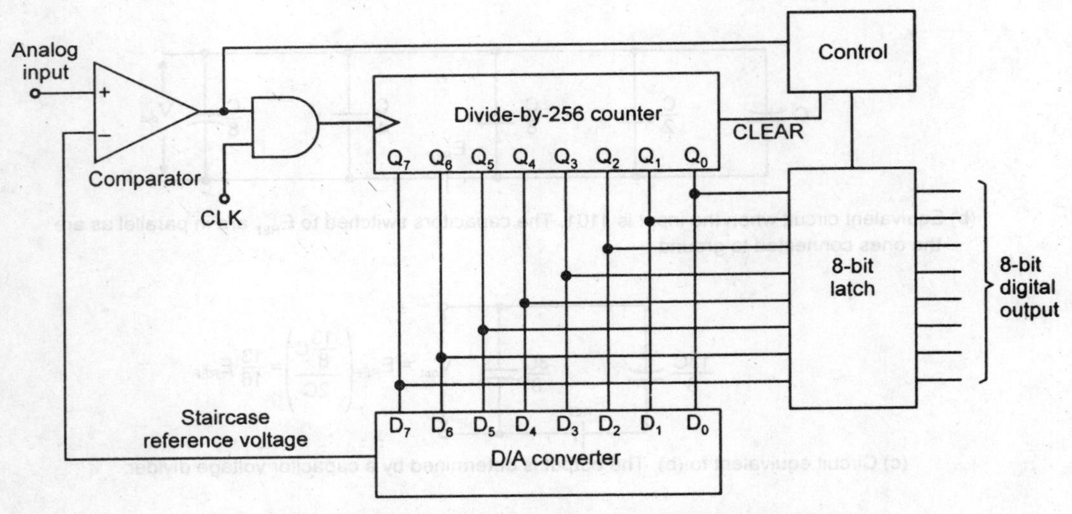

Fig. 13.15 Logic diagram of the counter-type ADC.

Assume that initially the counter is reset and, therefore, the output of the DAC is zero. Since the analog input is larger than the initial output of the DAC, the output of the comparator is HIGH and, therefore, the AND gate is enabled and, so, the clock pulses are transmitted to the counter and the counter advances through its binary states. These binary states are converted into reference analog voltage (which is in the form of a step) by the DAC. The counter continues to advance from one state to the next, producing successively larger steps in the reference voltage. When the staircase output voltage reaches the value of the analog signal, the comparator outputs a LOW, and the AND gate is disabled; so, the clock pulses do not reach the counter and the counter stops. The count it reached is the digital output proportional to the analog input. The control logic loads the binary count into the latches and resets the counter, thus, beginning another count sequence to sample the input value. The cycle thus repeats itself.

The resolution of this ADC is equal to the resolution of the DAC it contains. The resolution can also be thought of as the built-in error and is often referred to as the *quantization error*. Thus,

$$\text{Resolution} = \frac{\text{FSR}}{2^N}$$

where FSR is the full-scale reading and N is the number of bits in the counter.

As in the DAC, the accuracy is not related to the resolution, but is dependent on the accuracy of the circuit components such as comparator, the DAC's precision resistors, etc.

Figure 13.16 illustrates the output of a 4-bit DAC in an ADC over several cycles when the analog input is a slowly varying voltage.

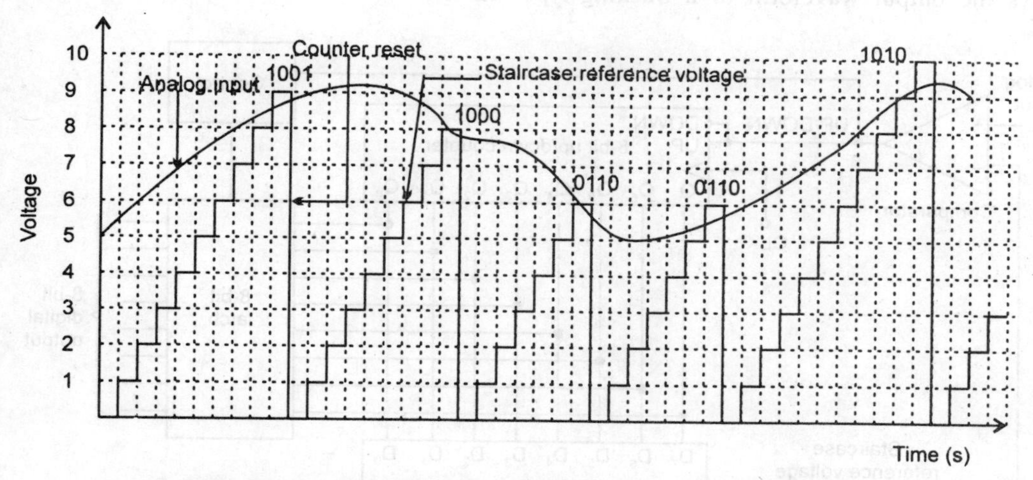

Fig. 13.16 Output waveform of the counter-type ADC.

The principal disadvantage of this type of converter is that, the conversion time depends on the magnitude of the analog input. The larger the input, the more will be the number of clock pulses that must pass to reach the proper count, and, therefore, the larger will be the conversion time. For each conversion, the counter has to start from reset only and count up to the point at which the staircase reference voltage reaches the analog input voltage. This type of converter is considered quite slow in comparison with the other types.

13.9 THE TRACKING-TYPE A/D CONVERTER

The counter-type ADC is slow, because the counter resets itself after each conversion. The tracking-type ADC uses an up/down counter and is faster than the counter-type ADC, because the counter is not reset after each sample, but tends to track the analog input, i.e. counts up or down from its last count to its new count. Thus, the total number of clock pulses required to perform a conversion is proportional to the change in the analog input between counts, rather than to its magnitude. Since the count more or less keeps up with the changing analog input, this type of converter is called a *tracking converter*.

Figure 13.17 shows the logic diagram of a tracking type-ADC. As long as the D/A output reference voltage is less than the analog input, the comparator output is HIGH putting the counter in the up mode and causing it to produce an up-sequence of binary counts. This causes an increasing value of the reference voltage out of the D/A converter, which continues until the staircase reaches the value of the input analog voltage. When the reference voltage equals the analog input, the comparator output switches LOW and puts the counter in the down mode, causing it to back up one count. If the analog input is now decreasing, the counter will continue to back down in its sequence and effectively track the input. If the analog input remains constant, the counter keeps on changing from up-to-down-to-up continuously and, therefore, the output of the ADC keeps on oscillating about the constant analog input. This is a disadvantage of this type of converter. Figure 13.18 shows the output waveform of a tracking-type ADC.

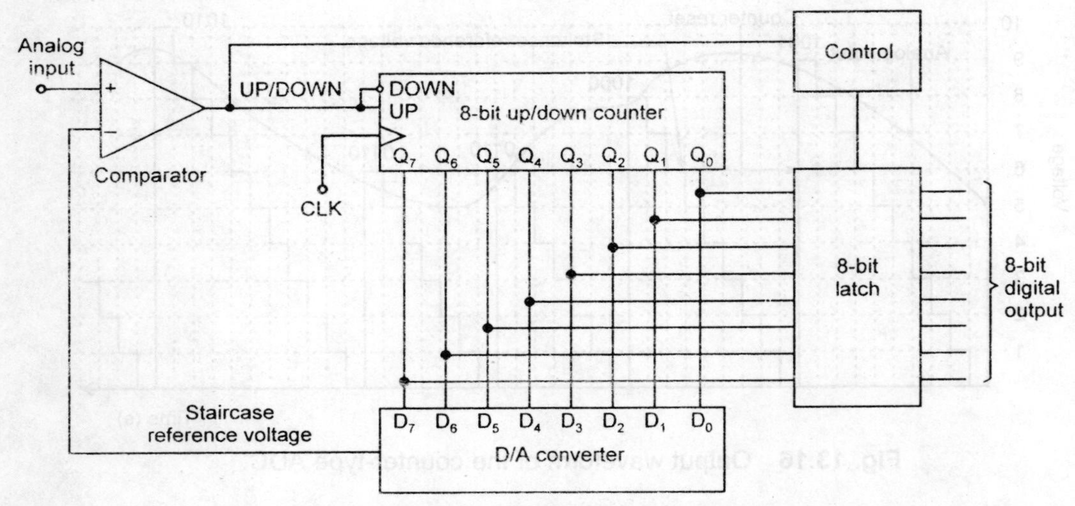

Fig. 13.17 Logic diagram of the tracking-type ADC.

The conversion time of this ADC is the time interval between the starting of the conversion and the time the comparator outputs a LOW (stopping the count). That is, the conversion time is the time required to convert a single analog input to a digital output.

$$t_c(\max) = (2^N - 1) \text{ clock cycles}$$

$$= (2^N - 1) \times \text{time for 1 cycle}$$

$$\text{The average conversion time} = \frac{t_c(\max)}{2}$$

The ADC must perform at a rate equal to at least twice the frequency of the highest component of the input.

EXAMPLE 13.14 Determine the maximum conversion time that an ADC can have, if it is used to convert signals in the range of 1 kHz to 50 kHz.

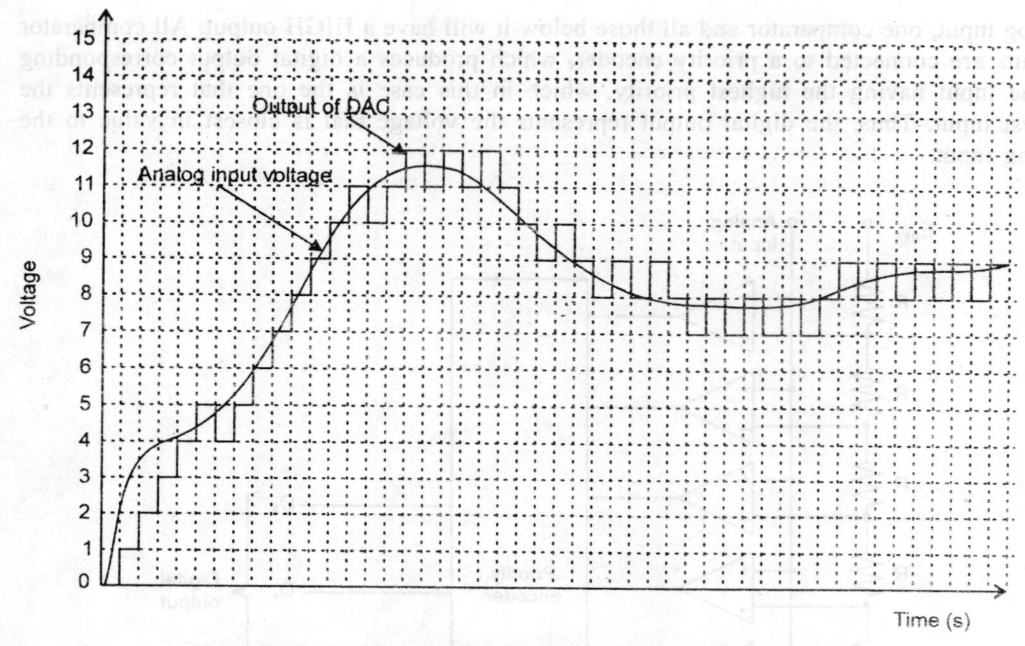

Fig. 13.18 Output waveform of the tracking-type ADC.

Solution

Since the highest input frequency is 50 kHz, conversions should be performed at the rate of $2 \times 50 \times 10^3 = 100 \times 10^3$ conversions/s. The maximum allowable conversion time is, therefore, equal to $1/(100 \times 10^3) = 10\ \mu s$.

EXAMPLE 13.15 An ADC has a total conversion time of 200 μs. What is the highest frequency that its analog input should be allowed to contain?

Solution

The highest frequency that the analog signal can contain is

$$\frac{1}{2 \times \text{conversion time}} = \frac{1}{2 \times 200\ \mu s} = 2.5 \text{ kHz}$$

13.10 THE FLASH-TYPE A/D CONVERTER

The flash (or simultaneous or parallel) type A/D converter is the fastest type of A/D converter. This type of converter utilizes the parallel differential comparators, that compare reference voltages with the analog input voltage. The main advantage of this type of converter is that the conversion time is less, but the disadvantage is that, an *n*-bit converter of this type requires $2^n - 1$ comparators, 2^n resistors, and a priority encoder. Figure 13.19 shows a 3-bit flash type A/D converter which requires $7(= 2^3 - 1)$ comparators. A reference voltage E_{REF} is connected to a voltage divider that divides it into seven equal increment levels. Each level is compared to the analog input by a voltage comparator. For any given analog input, one comparator and all those below it will have a HIGH output. All comparator outputs are connected to a priority encoder, which produces a digital output corresponding to the input having the highest priority, which in this case is the one that represents the largest input. Thus, the digital output represents the voltage that is closest in value to the analog input.

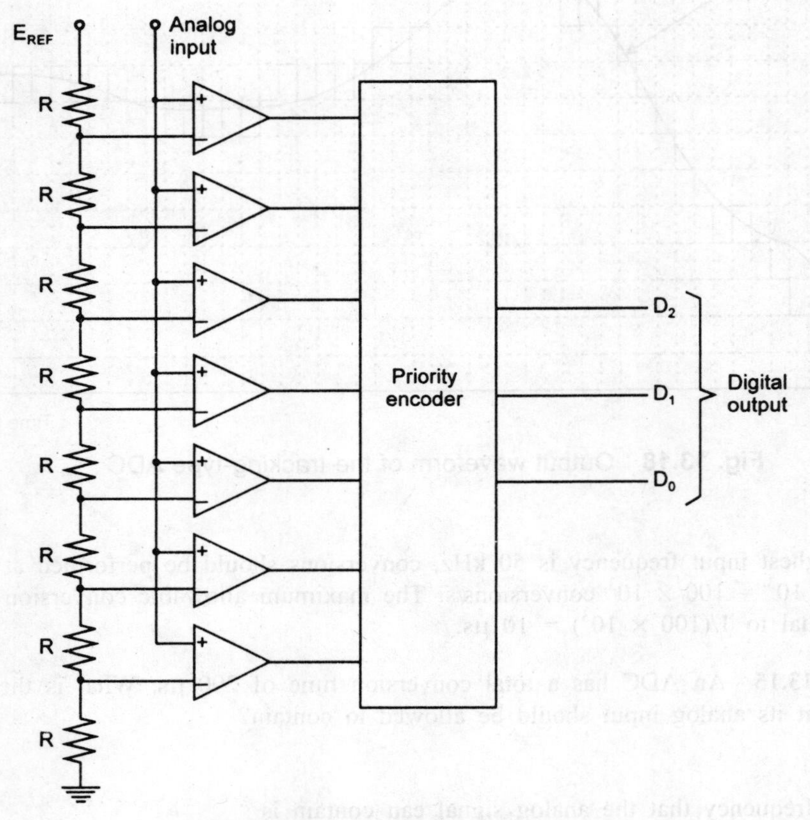

Fig. 13.19 The flash-type ADC.

The voltage applied to the inverting terminal of the uppermost comparator in Fig. 13.19 is (by voltage divider action),

$$\left(\frac{7R}{7R+R}\right) \times E_{REF} = \frac{7}{8} \times E_{REF}$$

Similarly, the voltage applied to the inverting terminal of the second comparator is

$$\left(\frac{6R}{7R+R}\right) \times E_{REF} = \frac{6}{8} \times E_{REF}$$

and so forth. The increment between voltages is, $\frac{1}{8} \times E_{REF}$.

The flash converter uses no clock signal, because there is no timing or sequencing period. The conversion takes place continuously. The only delays in the conversion are in the comparators and the priority encoders.

Figure 13.20 shows the block diagram of a modified flash A/D converter. To perform an 8-bit conversion, it requires two 4-bit flash converters. So, an 8-bit conversion can be done by using 30[$= 2 \times (2^4 - 1)$] comparators instead of 255($= 2^8 - 1$) comparators. One 4-bit flash converter is used to produce the four most significant bits (MSBs). Those four bits are converted back to an analog voltage by a D/A converter and this voltage is subtracted from the analog input. The difference between the analog input and the analog voltage corresponding to the four most significant bits, is an analog voltage corresponding to the four least significant bits (LSBs). Therefore, that voltage is converted to the four least significant bits by another 4-bit flash converter.

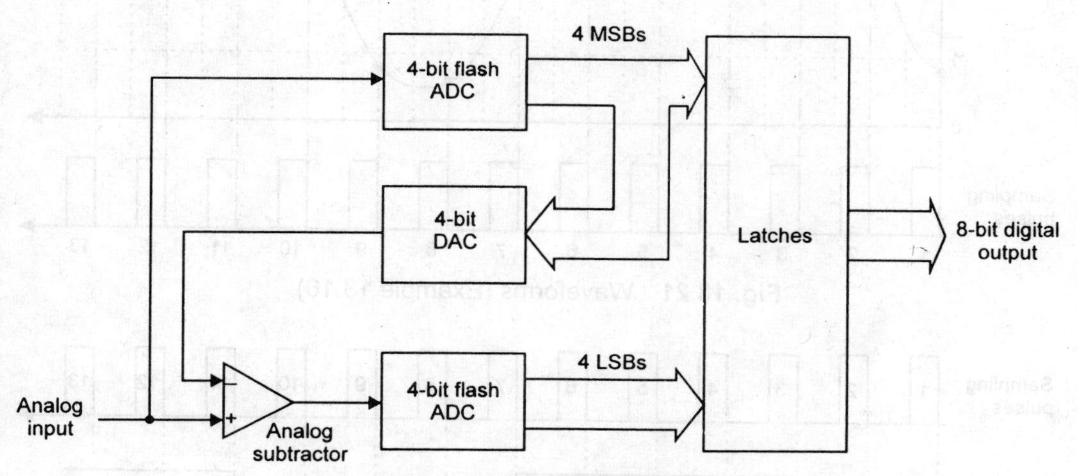

Fig. 13.20 Modified flash ADC.

EXAMPLE 13.16 Determine the digital output of a 3-bit simultaneous A/D converter for the analog input signal and the sampling pulses (encoder enable) shown in Fig. 13.21. V_{REF} = + 8 V.

Solution

The resulting A/D output sequence is listed as follows and shown in Fig. 13.22 in relation to the sampling pulses.

000, 010, 101, 110, 110, 100, 010, 000, 000, 011, 101, 110.

13.11 THE DUAL-SLOPE TYPE A/D CONVERTER

The dual-slope converter is one of the slowest converters, but is relatively inexpensive because it does not require precision components such as a DAC or VCO. Another advantage of the dual-slope ADC is its low sensitivity to noise, and to variations in its component values caused by temperature changes. Because of its large conversion time, the

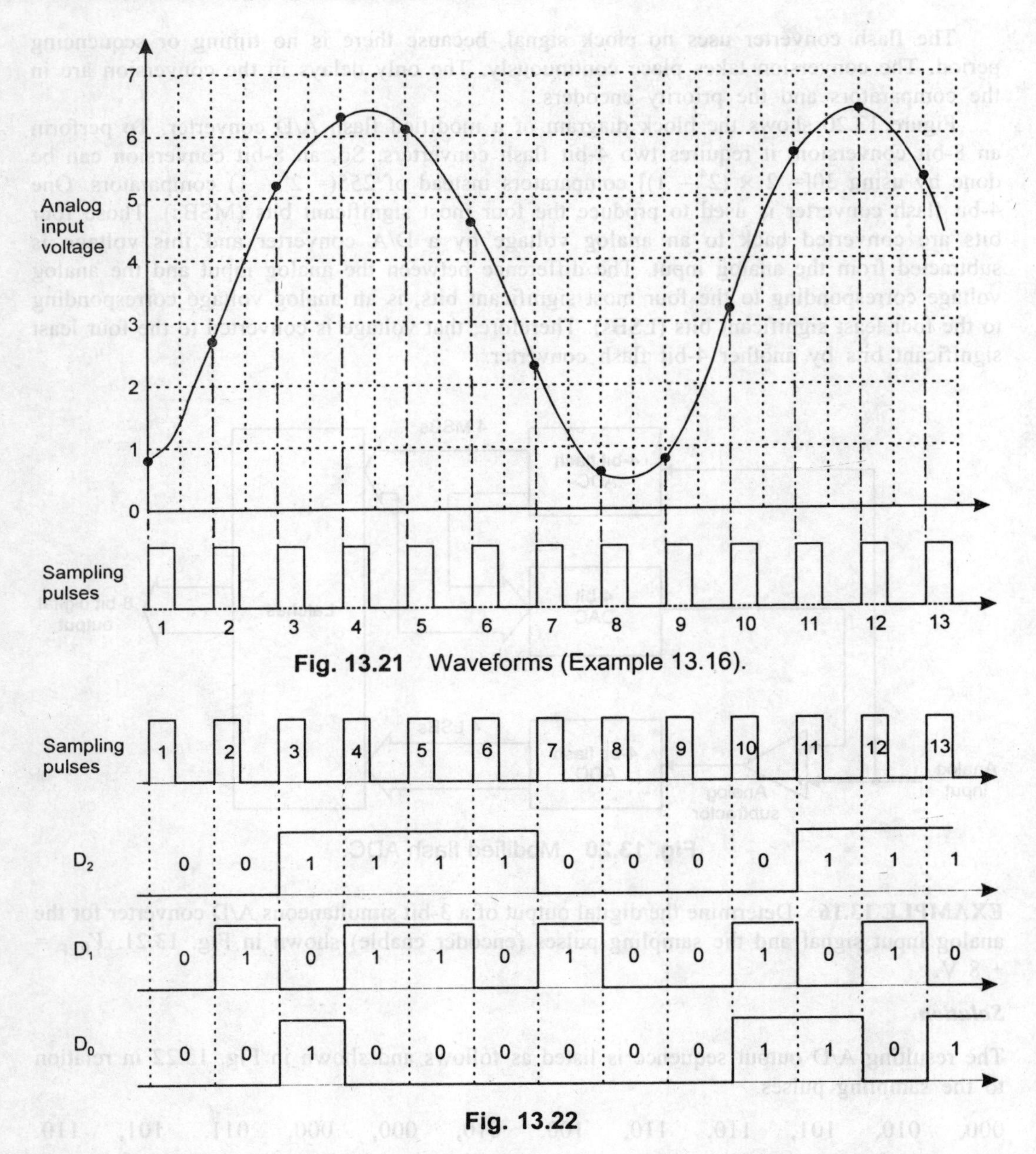

Fig. 13.21 Waveforms (Example 13.16).

Fig. 13.22

dual-slope ADC is not used in any data acquisition applications. The major applications of this type of converter are in digital voltmeters, multimeters, etc. where slow conversions are not a problem. Since it is not fast enough, its use is restricted to signals having low to medium frequencies.

A dual-slope ADC uses an operational amplifier to integrate the analog input. The output of the integrator is a ramp, whose slope is proportional to the input signal E_{in}, since the components R and C are fixed. If the ramp is allowed to continue for a fixed time, the voltage it reaches in that time, depends on the slope of the ramp and, therefore, on the value of E_{in}. The basic principle of the integrating ADC is that, the voltage reached by the ramp controls the length of time that the binary counter is allowed to count. Thus, a binary number proportional to the value of E_{in} is obtained. In the dual-slope ADC, two integrations are performed.

Figure 13.23 shows the functional block diagram of a dual-slope ADC. Assume that the counter is reset and the output of the integrator is zero. A conversion begins with the switch

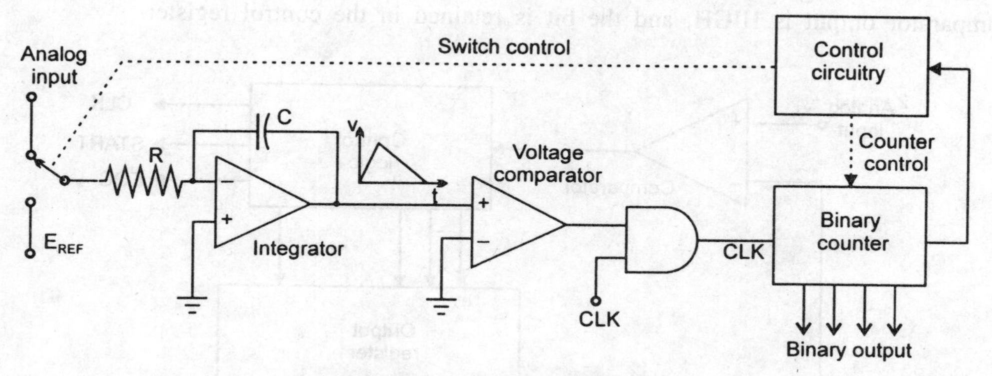

Fig. 13.23 The dual-slope ADC.

connected to the analog input. Assume that the input is a negative voltage and is constant for a period of time; so, the output of the integrator is a positive ramp. The ramp is allowed to continue for a fixed time and the voltage it reaches in that time is directly dependent on the analog input. The fixed time is controlled by sensing the time when the counter reaches a particular count. At that time, the counter is reset and the control circuitry causes the switch to be connected to a reference voltage E_{REF}, having a polarity opposite to that of the analog input; in this case a positive reference voltage. Therefore, the output of the integrator is a negative going ramp, beginning from the positive value it reached during the first integration. The AND gate is enabled and the counter starts counting. When the ramp reaches 0 V, the voltage comparator switches to LOW, inhibiting the clock pulses and the counter stops counting. The binary count is latched, thus, completing one conversion. The count it contains at that time is proportional to the time required for the negative ramp to reach zero, which is proportional to the positive voltage reached during the first integration, which in turn is proportional to the analog input.

The accuracy of the converter does not depend on the values of the integrator components or upon any changes in them. The accuracy does depend on E_{REF}; so, the reference voltage should be very precise.

13.12 THE SUCCESSIVE-APPROXIMATION TYPE ADC

The successive-approximation (SA) converter is one of the most widely used type of ADCs. It has a much shorter conversion time than the other types, with the exception of the flash type. It also has a fixed conversion time which is not dependent on the value of the analog input.

Figure 13.24 shows a basic block diagram of a 4-bit successive-approximation type ADC. It consists of a DAC, an output register, a comparator, and control circuitry or logic. The basic operation is as follows: The bits of the DAC are enabled one at a time, starting with the MSB. As each bit is enabled, the comparator produces an output that indicates whether the analog input voltage is greater or less than the output of the DAC V_{AX}. If the D/A output is greater than the analog input, the comparator output is LOW, causing the bit in the control register to reset. If the D/A output is greater than the analog input, the comparator output is HIGH, and the bit is retained in the control register.

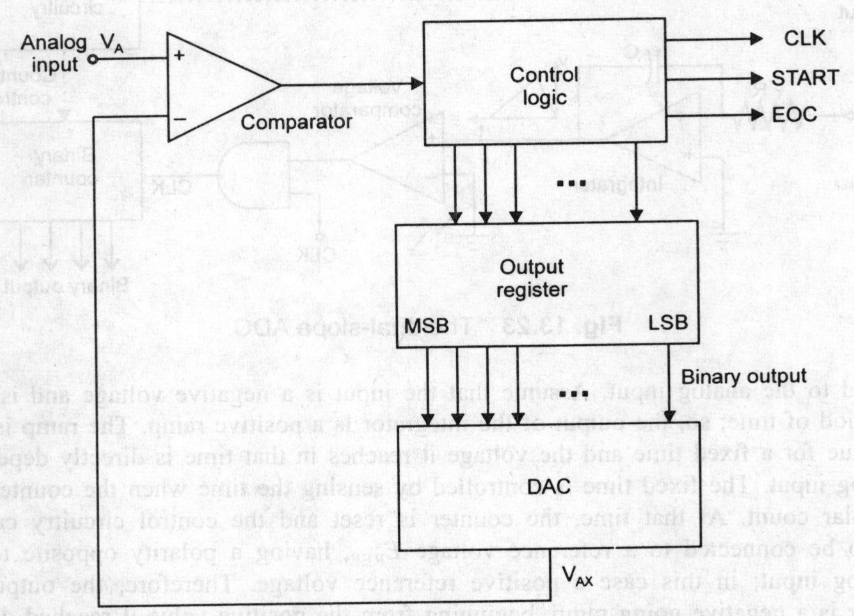

Fig. 13.24 The successive-approximation type ADC.

The system enables the MSB first, then the next significant bit, and so on. After all the bits of the DAC have been tried, the conversion cycle is complete. The processing of each bit takes one clock cycle; so, the total conversion time for an *N*-bit SA-type ADC will be *N* clock cycles. That is,

$$t_c \text{ for SAC} = (N \times 1) \text{ clock cycles}$$

The conversion time will be the same regardless of the value of V_A. This is because the control logic has to process each bit to see whether a 1 is needed or not.

The method is best explained by an example. Let us assume that the output of the DAC ranges from 0 V to 15 V as its binary input ranges from 0000 to 1111, with 0000 producing 0 V, and 0001 producing 1 V, and so on. Suppose that the unknown analog input voltage

V_A is 10.3 V. On the first clock pulse, the output register is loaded with 1000, which is converted by the DAC to 8 V. The voltage comparator determines that 8 V is less than the analog input (10.3 V); so, the control logic retains that bit. On the next clock pulse, the control circuitry causes the output register to be loaded with 1100. The output of the DAC is now 12 V, which the comparator determines as greater than the analog input. Therefore, the comparator output goes LOW. The control logic clears that bit; so, the output goes back to 1000. On the next clock pulse, the control circuitry causes the output register to be loaded with 1010. The output of the DAC is now 10 V, which the comparator determines as less than the analog input. Thus, on the next clock pulse, the control logic causes the output register to be loaded with 1011. The output of the DAC is now 11 V, which the comparator determines as greater than the analog input; so, the control logic clears that bit. Now the output of the ADC is 1010 which is the nearest integer value to the input (10.3 V).

At this point, all of the register bits have been processed, the conversion is complete and the control logic activates its $\overline{\text{EOC}}$ output to signal that the digital equivalent of V_A is now in the output register.

EXAMPLE 13.17 Compare the maximum conversion periods of an 8-bit digital ramp ADC and an 8-bit successive approximation ADC if both utilize a 1 MHz clock frequency.

Solution

For the digital-ramp converter, the maximum conversion time is

$$(2^N - 1) \times (1 \text{ clock cycle}) = 255 \times 1\ \mu\text{s} = 255\ \mu\text{s}$$

For an 8-bit successive-approximation converter, the conversion time is always 8 clock periods, i.e. $8 \times 1\ \mu\text{s} = 8\ \mu\text{s}$.

Thus, the successive-approximation conversion is about 30 times faster than the digital-ramp conversion.

EXAMPLE 13.18 An 8-bit SAC has a resolution of 30 mV. What will its digital output be for an analog input of 2.86 V?

Solution

Since, $\dfrac{2.86 \text{ V}}{30 \text{ mV}} = 95.3$, the step 95 would produce 2.85 V and step 96 would produce 2.88 V. The SAC always produces a final output, that is, at the step below the analog input. Therefore, for the case of $V_A = 2.86$ V, the digital result would be $95_{10} = 01011111_2$.

A Specific A/D Converter

The ADC 0801 is an example of a successive-approximation type analog-to-digital converter. The pin diagram is shown in Fig. 13.25. This device operates from a + 5 V supply, and has a resolution of 8 bits with a conversion time of 100 μs. Also, it has guaranteed monotonicity and an on-chip clock generator. The data outputs are tri-stated so that it can be interfaced with a microprocessor bus system. The two analog inputs are V_{IN+} and V_{IN-}.

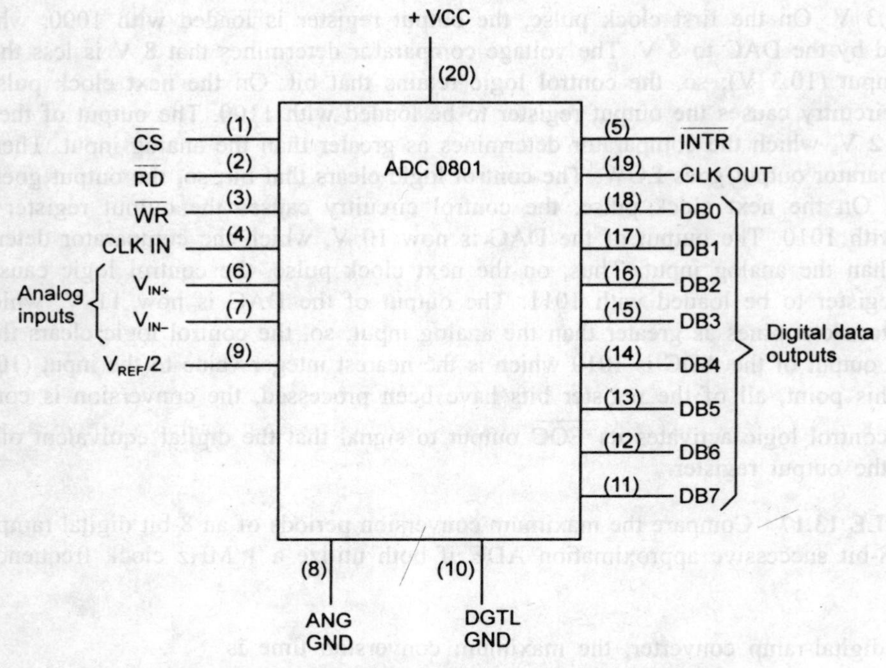

Fig. 13.25 Pin configuration of the ADC 0801.

$\overline{\text{CS}}$ (Chip Select): This input has to be in active-LOW state, for $\overline{\text{RD}}$ and $\overline{\text{WR}}$ inputs to have any effect. With $\overline{\text{CS}}$ HIGH, the digital outputs are in the Hi-Z state and no conversion takes place.

$\overline{\text{RD}}$ (Output Enable): This input is used to enable the digital output buffers. With $\overline{\text{CS}}$ = $\overline{\text{RD}}$ = LOW, the digital output pins have logic levels representing the results of the last A/D conversion.

$\overline{\text{WR}}$ (Start Conversion): A LOW pulse is applied to this input to signal the start of a new conversion.

$\overline{\text{INTR}}$ (End of Conversion): This output signal will go HIGH at the start of a conversion and return LOW to indicate the end of the conversion.

$V_{ref}/2$: This is an optional input that can be used to reduce the internal reference voltage and thereby change the analog input range that the converter can handle.

CLK OUT: A resistor is connected to this pin to use the internal clock. The clock signal appears on this pin.

CLK IN: It is used for the external clock input, or for the capacitor connection when the internal clock is used.

Voltage-to-Frequency ADC

The voltage-to-frequency ADC is simpler than other ADCs, because it does not use a DAC. Instead, it uses a linear voltage controlled oscillator (VCO) that produces an output frequency proportional to its input voltage. The analog voltage is applied to the VCO to generate an output frequency. This frequency is fed to a counter, to be counted for a fixed time interval. The final count is proportional to the value of the analog input.

To illustrate this, suppose the VCO generates a frequency of 5 kHz for each volt of input (i.e. 1 V produces 5 kHz, 1.5 V produces 7.5 kHz, 2.6 V produces 13 kHz). If the analog input is 4.65 V, then the VCO output will be a 23.25 kHz signal that clocks a counter for, say, 10 ms. After the 10 ms counting interval, the counter will hold the count of 232.

Although this is a simple method of conversion, it is difficult to achieve a high degree of accuracy because of the difficulty in designing VCOs with accuracies better than 0.1 per cent.

One of the main applications of this type of converter is in noisy industrial environments, where small analog signals have to be transmitted from transducer circuits to a control computer. The small analog signals can get drastically affected by noise, if they are transmitted directly to the control computer. A better approach is to feed the analog signal to a VCO, which generates a digital signal whose output frequency changes according to the analog input. The digital signal is transmitted to the computer and will be much less affected by noise. Circuitry in the control computer will count the digital pulses to produce a digital value, equivalent to the original analog input.

SUMMARY

- An analog quantity can take on any value over a continuous range of values.
- A digital quantity takes on only discrete values, expressed as binary numbers or codes.
- Digital processing of analog signals requires A/D and D/A converters.
- ADCs and DACs function as interfaces between a completely digital system, such as a digital computer, and the analog world.
- D/A conversion is the process of converting a value represented in a digital code to a voltage or current proportional to the digital value.
- The output of DAC is not a true analog quantity; it is a pseudo-analog quantity.
- The resolution of a DAC is defined as the smallest change that can occur in the analog output as a result of the change in the digital input.
- The resolution of a DAC is equal to the step size, being equal to the weight of the LSB.
- The offset voltage of a DAC is the small voltage that appears at its output, when its binary input has all 0s.
- The output of a monotonic DAC increases when its binary input is incremented.
- The settling time of a DAC is the time taken by its output to settle down to within ± 1/2 step size of its final value after the application of the digital input.
- The step size of DAC is the same as the proportionality factor in the DAC I/O relationship.
- The full-scale error is the maximum deviation of the DAC's output from its expected value, expressed as a percentage of full-scale.

- The *R*-2*R* ladder type DAC is the most popular DAC.
- The disadvantage of the weighted-resistor DAC is, that a different-valued precision resistance is needed for each bit position of the digital input.
- Switched-current DACs are faster than the switched-voltage DACs.
- A/D conversion is the process of converting an analog input voltage to a number of equivalent digital output levels.
- An ADC produces a digital output proportional to the value of the input analog signal.
- The process of approximation used in digitizing samples is called quantization.
- Linearity error is the maximum deviation in step size from the ideal step size.
- The counter-type ADC is the simplest and the flash-type the fastest. The counter-type ADC is somewhat slower, but can digitize high resolution signals.
- The tracking-type ADC is also called the continuous-conversion type ADC.
- The flash converter is the fastest and the most expensive ADC.
- The dual-slope ADC is the slowest type, but it is cheap and offers excellent accuracy in a relatively inexpensive circuit. This ADC is widely used in digital voltmeters.
- The successive-approximation type ADC is the most widely used type of ADC.

QUESTIONS

1. Define the following parameters of DACs: (a) resolution, (b) accuracy, (c) monotonicity, (d) settling time, (e) offset voltage, (f) conversion time, (g) percentage resolution, (h) full-scale error, (i) linearity error, and (j) step-size.
2. How many different output voltages can an 8-bit DAC produce?
3. What is the advantage of a smaller (finer) resolution?
4. What is the advantage of the *R*-2*R* ladder DAC over the weighted-resistor type DAC?
5. A certain 6-bit DAC uses binary weighted resistors. If the MSB resistor is 40 kΩ, what is the LSB resistor?
6. Why are voltage DACs generally slower than current DACs?
7. List the various types of DACs and ADCs. Name the most widely used DAC.
8. Why does the conversion time increase with the value of the analog input voltage in a counter-type ADC?
9. What is the major advantage and disadvantage of a flash-type ADC? What is the other name of the flash-type ADC?
10. Does a flash-type ADC contain a DAC?
11. What is the main advantage and disadvantage of a SAC over a digital-ramp ADC?
12. How does the up/down digital-ramp ADC improve upon the digital-ramp ADC?
13. Give one advantage and one disadvantage of the digital-ramp ADC.
14. Give two advantages and one disadvantage of the dual-slope ADC.

15. Name three types of ADC that do not use a DAC.

16. Which is the fastest ADC and why?

17. What is the other name of the counter-type ADC?

18. With the help of neat diagrams explain the working of the following DACs and ADCs.

(a) *R*-2*R* ladder network type DAC
(b) Weighted-resistor type DAC
(c) Current-switching type DAC
(d) Switching-capacitor type DAC
(e) Current-type ADC
(f) Tracking-type ADC
(g) Flash-type ADC
(h) Dual-slope ADC
(i) Successive-approximation type ADC.

PROBLEMS

13.1 A 5-bit DAC produces an output of 0.1 V for a digital input of 00001. What is the full-scale output? Find the output for an input of 10101.

13.2 The logic levels used in a 6-bit *R*-2*R* ladder DAC are: 1 = 5 V and 0 = 0 V. Find the output voltage for inputs (a) 010110, and (b) 101011.

13.3 The logic levels used in an 8-bit *R*-2*R* ladder DAC are: 0 = 0 V and 1 = + 5 V. What is the binary input when the analog output is 4 V?

13.4 Design a 6-bit weighted-resistor DAC whose full-scale output voltage is – 12 V. Logic levels are 1 = + 5 V and 0 = 0 V. What is the output voltage when the input is 1011?

13.5 How many bits are required for a DAC, so that its full-scale output is 12.6 V and resolution 20 mV?

13.6 A 6-bit DAC has an output current of 20 mA for a digital input of 101100. What will be the output current for an input of 010110?

13.7 In the switched current-source DAC shown in Fig. 13.13, R =10 kΩ and E_{REF} = 15 V. Find the current in each 2*R* resistor when it is connected to E_{REF}.

13.8 An 8-bit switched current-source DAC of the design shown in Fig. 13.13, has R = 5 kΩ and E_{REF} = 20 V. Find the total current I_T delivered to the amplifier and the output voltage when the input is 01110100.

13.9 The maximum conversion time of a tracking-type ADC is 100 ns. At what frequency is it clocked?

13.10 A flash-type 5-bit ADC has a reference voltage of 20 V. How many voltage comparators does it have? How many resistors does it have? What is the increment between the voltages applied to the comparators?

13.11 The resolution of a 12-bit ADC is 10 mV. What is its full-scale range?

13.12 The frequency components of the analog input to an ADC range from 50 Hz to 50 kHz. What is the maximum total conversion time that the converter can have?

13.13 An 8-bit SAC has a resolution of 15 mV. What will its digital output be for an analog input of 2.65 V?

Chapter 14

MEMORIES

14.1 THE ROLE OF MEMORY IN A COMPUTER SYSTEM

Program and Data Memory

Basically, memory is a means for storing data or information in the form of binary words. It is made up of storage locations in which numeric or alphanumeric information or programs (sets of instructions that a computer executes to achieve a desired result) may be stored. Memory used to store data is called *data memory*, and memory used to store programs is called *program memory*. Small special-purpose computer systems may have little or no data memory, whereas a large portion of memory of a general-purpose computer system is usually reserved for data storage.

Computers which store programs in their memory are called *stored-program type computers*. Virtually, all modern computers are of the stored-program type. In these computers, programs are stored as a set of *machine language instructions,* in binary codes. Each memory location is identified by an *address*. The number of storage locations can vary from a few in some memories to millions in others.

Each storage location can accommodate one or more bits. Generally, the total number of bits a memory can store is its *capacity*. Sometimes the capacity is specified in terms of bytes. Memories are made up of storage elements (FFs or capacitors in semiconductor memories and magnetic domains in magnetic memories), each of which stores one bit of data. A storage element is called a *cell*.

Magnetic tapes and magnetic disks are popular mass storage devices that cost less per bit than the internal memory devices. A newer entry into the mass memory category is the magnetic bubble memory (MBM), a semiconductor memory that uses magnetic principles to store millions of bits on one chip. The MBM is relatively slow and cannot be used as an internal memory.

Main and Peripheral Memory

Computer memories may also be classified as *main* or *peripheral*. The main memory is an internal part of the computer and is very fast. The peripheral memory also called the *auxiliary* memory is a typically add-on memory with very large storage capacity, but it is much slower than the main memory. It often serves as the data memory for storing very large quantities of data. The main memory that serves as the program memory is in constant communication with the CPU during program execution. The program to be currently executed and any data used by the program are stored in the main memory. Semiconductor memories are well-suited as the main memory because of their high speed of operations.

A semiconductor memory is typically constructed from semiconductor IC devices, whereas the peripheral memory consists of magnetic tapes or disks. In older computers, the main memory was constructed from tiny electromagnets called magnetic cores and the term *core memory* or simply *core* was used synonymously with the main memory. Mass storage refers to memory that is external to the main computer (magnetic tapes and disks) and has the capacity to store millions of bits of data without the need for electrical power. Mass memory is normally much slower than the main memory and is used to store information (programs, data, etc.) not currently being used by the computer. The required information is transferred to the main memory from the mass memory when the computer actually needs it.

14.2 MEMORY TYPES AND TERMINOLOGY

Memory Organization and Operation

All memory, regardless of its type or use, consists of locations for storing binary information or bits. Each location is identified by an *address*. A *word* is the fundamental group of bits used to represent one entity of information such as one numerical value. The word size—the number of bits in a word—varies among computer systems and may range from 4 to 64 or more bits. The word size is usually expressed as a certain number of bytes. For example, a 16-bit word is 2 bytes.

A memory location is thus a set of devices capable of storing one word. For example, each memory location in an 8-bit microcomputer (one that uses 8-bit words) might consist of eight latches. Each *latch*, stores one bit of a word, and is referred to as a *cell*. The capacity or size of a memory is the total number of bits or bytes or words that it can store. For convenience, the size of a memory is expressed as a multiple of 2^{10} = 1024, which is abbreviated K. For example, a memory of size 2^{11} = 2048 is said to be 2K. A memory of size 2^{14} (16,384) is 16K, and a memory of size 2^{16} (65,536) is 64K.

Every memory system requires several different types of input and output lines to perform the following functions.

1. Select the address in memory that is to be accessed for a read or write operation.
2. Select either a read or a write operation to be performed.
3. Supply the input data to be stored in memory during a write operation.
4. Hold the output data coming from memory during a read operation.
5. Enable (or disable) the memory, so that it will (or will not) respond to the address inputs and read/write command.

Figure 14.1a illustrates these basic functions in a simplified diagram of a 32 × 4 memory that stores 32 4-bit words. Since the word size is 4-bits, there are four data input lines I_0 to I_3 and four data output lines O_0 to O_3. During a write operation, the data to be stored in memory have to be applied to the data input lines. During a read operation, the word being read from memory appears at the data output lines.

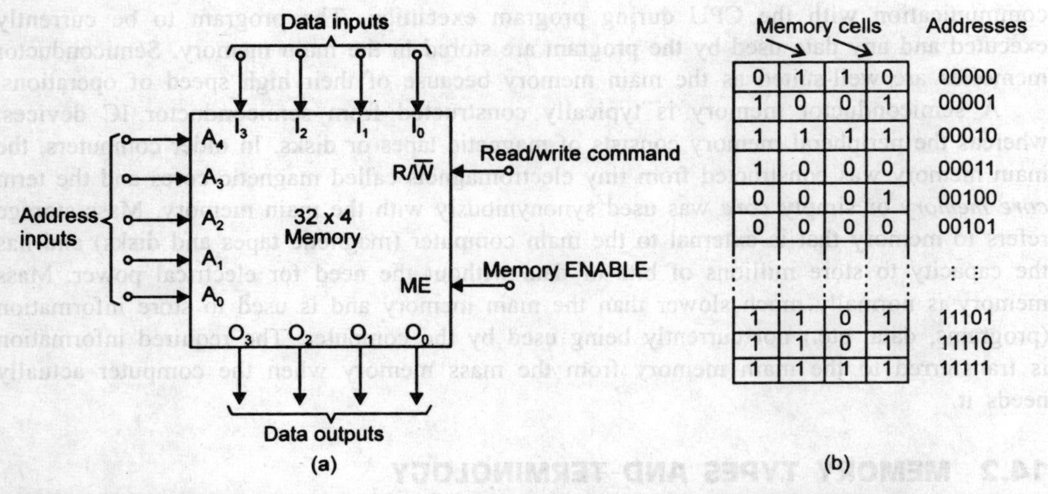

Fig. 14.1 (a) Diagram of a 32 × 4 memory and (b) arrangement of memory cells.

Address inputs

Since this memory stores 32 words, it has 32 different storage locations and, therefore, 32 different binary addresses ranging from 00000 to 11111 (0 to 31 decimal). Thus, there are five address inputs A_0 to A_4. To access one of the memory locations for a read or write operation, the 5-bit address code for that particular location is applied to the address inputs. In general, N address inputs are required for a memory that has a capacity of 2^N words.

We can visualize the memory of Fig. 14.1a as an arrangement of 32 registers with each register holding a 4-bit word as illustrated in Fig. 14.1b. Each address location is shown containing four memory cells that hold 1s and 0s to make up the data word stored at that location.

The R/$\overline{\text{W}}$ input

The read/write (R/$\overline{\text{W}}$) input line determines the memory operation that would take place. Some memory systems use two separate inputs, one for read and one for write. When a single R/$\overline{\text{W}}$ input is used, the read operation takes place for R/$\overline{\text{W}}$ = 1 and the write operation takes place for R/$\overline{\text{W}}$ = 0.

Memory ENABLE

Many memory systems have some means of completely disabling all or part of the memory so that it does not respond to the other inputs. This is represented in Fig. 14.1a as the memory ENABLE input, although it can have different names in the various memory systems. Here it is shown as an active-HIGH input that enables the memory to operate normally when it is kept HIGH. A LOW on this input disables the memory, preventing it to respond to address and R/$\overline{\text{W}}$ inputs. This type of input is useful when several memory modules are combined to form a larger memory.

EXAMPLE 14.1 A certain memory has a capacity of 8K × 16.

(a) How many data input and data output lines does it have?

(b) How many address lines does it have?

(c) What is its capacity in bytes?

Solution

(a) Since the word size is 16 bits, data input and data output will be 16 lines each.

(b) The memory stores 8K = 8 × 1024 = 8192 words. Thus, there are 8192 memory addresses. Since 8192 = 2^{13}, it requires 13 address lines.

(c) A byte is 8-bits. This memory therefore, has a capacity of 16K bytes.

Reading and Writing

The process of storing data in memory is called *writing* in memory. Retrieving data from memory is called *reading* memory. Figure 14.2 illustrates how reading from and writing into memory is accomplished in an 8-bit microprocessor system. The microprocessor serves as the central processing unit (CPU) for the computer. It contains an arithmetic/logic unit (ALU), numerous registers, and logic circuitry, that it uses to perform read and write operations as well as to execute programs. Note the control signals labelled *read* and *write*. The CPU activates these when a read or a write operation is to be performed. The wide two-headed arrow represents an 8-bit data bus consisting of eight lines on which data bits

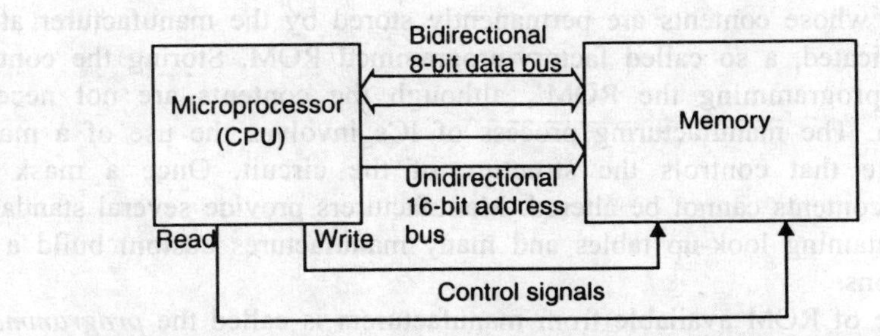

Fig. 14.2 Reading and writing of data in a microcomputer system.

D_0 through D_7 are transmitted. It is called a ***bidirectional data bus***, because words can be transmitted from the CPU to memory (when writing) or from memory to the CPU (when reading). The unidirectional address bus is the set of lines over which the CPU transmits the address bits corresponding to the memory address to be read or written into. In the example shown, the address bus is a 16-bit bus (A_0 through A_{15}) meaning that the CPU can access (read or write into) up to 2^{16} = 65,536 different memory addresses. The following is a typical sequence of events, during which a byte is read from one memory location and written into another.

1. The CPU activates the *read* control and transmits the 16-bit address, say, $000A_{16}$ to memory via the address bus.

2. As a result of step 1, the 8-bit word stored in address $000A_{16}$, say, 45_{16} is placed on the data bus and transmitted to the CPU.
3. The CPU activates the *write* control and transmits the 16-bit address, say, $000B_{16}$ to memory. It also transmits the 8-bit data word 45_{16} to memory via the data lines.
4. As a result of step 3, the data word 45_{16} is stored at address $000B_{16}$ (the original contents of that address are lost).

RAMs, ROMs, and PROMs

The type of memory we have discussed is called the read/write memory (RWM) because it can be accessed for both these kinds of operations. On the other hand, read only memory (ROM) cannot be written into. It, is therefore, used for permanent storage of programs or data in dedicated applications.

The data stored in RWMs constructed from semiconductor devices will be lost if power is removed. Such memory is said to be *volatile*. But ROM is non-volatile. Random access memory (RAM) is the memory whose memory locations can be accessed directly and immediately. By contrast, to access a memory location on a magnetic tape, it is necessary to wind or unwind the tape and go through a series of addresses before reaching the address desired. Therefore, a tape is called the *sequential access memory*. It is conventional to use the term RAM to mean read/write memory, in contrast to ROM. However, most ROMs are random access (to read only) in the sense we have described.

There are three types of ROMs available in ICs. The term ROM by itself generally refers to an IC memory whose contents are permanently stored by the manufacturer at the time the circuit is fabricated, a so called factory-programmed ROM. Storing the contents of a ROM is called 'programming the ROM', although the contents are not necessarily a computer program. The manufacturing process of ICs involves the use of a mask which is like a template that controls the structure of the circuit. Once a mask ROM is manufactured, its contents cannot be altered. Manufacturers provide several standard ROMs such as those containing look-up tables and many manufactures custom build a ROM to user's specifications.

A second type of ROM available from manufacturers is called the *programmable read only memory* (PROM). A PROM is basically a blank ROM that can be programmed by a user by using a special PROM programming apparatus. It is thus said to be field programmable. Like a ROM, once a PROM is programmed, its contents cannot be altered.

A third type of ROM available from manufacturers is called the *erasable programmable read only memory* (EPROM). An EPROM can be field programmed and also subsequently reprogrammed. That is, the contents can be *erased* and new contents stored in their place. The contents of an EPROM are still referred to as permanent, in the sense that they are non-volatile and can be erased and reprogrammed only through the use of special equipment. Actually, there are two types of EPROMs—those that can be erased by exposing them to UV light and those that can be erased by subjecting them to certain electrical voltages. The latter type is called an *electrically erasable* PROM or EEPROM. It is also called an *electrically alterable* PROM (EAPROM), or an *electrically erasable* ROM (EEROM), or an *electrically alterable* ROM (EAROM).

Constituents of Memories

We know that there are two principal types of transistors used in the manufacture of digital ICs. They are BJTs and MOSFETs. The IC memories are available in both of these technologies. However, not every type of memory we have discussed is available in bipolar technology.

Because MOSFETs are more easily manufactured than BJTs and occupy much less space in an IC chip, that technology has become more dominant in the production of large capacity integrated circuit memories. However, the BJT memory is considerably faster than the MOS memory and the BJT ICs are used in applications where high speed is more important than the large storage capacity. Memory circuits constructed with the I^2L technology are also available. These feature higher speeds than those of the MOS memories and greater capacity per circuit than that of the BJT memories.

In Chapter 11, we discussed dynamic logic in which logic levels are preserved as charge (or absence of charge) on capacitances. Dynamic memory is constructed with this technology. Its principal advantages comprise very large number of memory cells in a single circuit and very low power consumption. However, such circuitry must be periodically refreshed to replenish the stored charges. This requirement adds somewhat to the complexity of the systems that incorporate dynamic logic. Dynamic memory is available only in MOSFET circuits. To distinguish between the dynamic memory and the memory that utilizes conventional storage devices such as latches, the latter type is termed *static memory*. Static memory is available in both BJT and MOSFET technologies. To distinguish between static RAM and dynamic RAM, the terms SRAM and DRAM are sometimes used. Figure 14.3 summarizes the technologies used in the manufacture of IC ROMs and RAMs (read/write memories).

14.3 READ ONLY MEMORY (ROM)

ROM Organization

A typical block diagram for a ROM is shown in Fig. 14.4. It has three sets of signals: address inputs, control input, and data outputs. We can see that this ROM can store 16 words since it has $2^4 = 16$ possible addresses, and each word contains eight bits since there are eight data outputs. Thus, this is a 16×8 ROM. Another way to describe the ROM capacity is to say that it stores 16 bytes of data. The data outputs of most ROM ICs are tristate outputs to permit the connection of many ROM chips to the same data bus for memory expansion.

The control input $\overline{CS}$ stands for chip select. This is essentially an enable input that enables or disables the ROM outputs. Some manufacturers use different labels for the control input, such as $\overline{CE}$ (chip enable) or $\overline{OE}$ (output enable). Many ROMs have two or more control inputs that have to be active in order to enable the data outputs. In some ROM ICs, one of the control inputs (usually the $\overline{CE}$) is used to place the ROM in a low-power standby mode when it is not being used. This reduces the current drive from the system power supply. The $\overline{CS}$ input shown in Fig. 14.4 is active-LOW, therefore, it must be in the

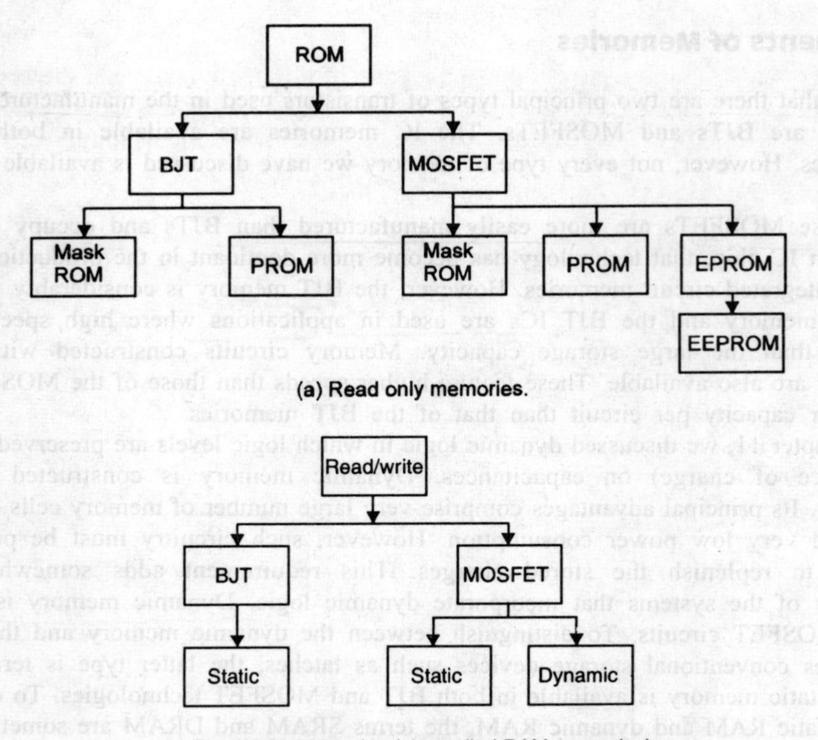

Fig. 14.3 Technologies used in the fabrication of IC memories.

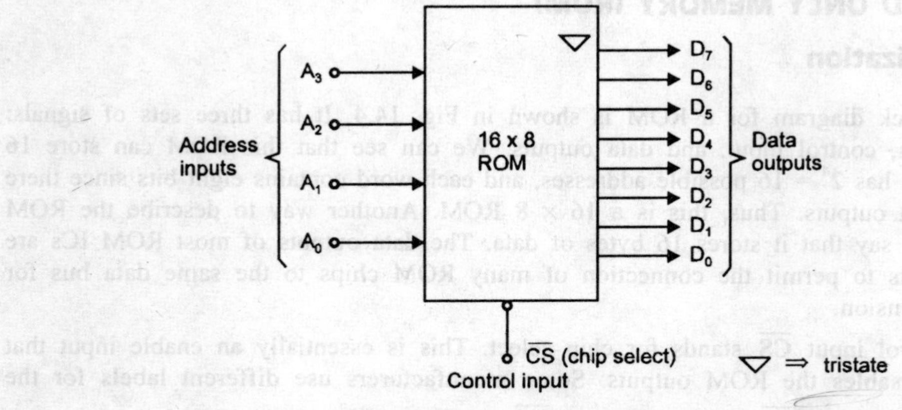

Fig. 14.4 ROM block diagram.

low state to enable the ROM data to appear at the data outputs. Notice that there is no $R/\overline{W}$ (read/write) input, because data cannot be written into the ROM during its normal operation. If $\overline{CS}$ is kept HIGH, the ROM outputs will be disabled and remain in the Hi-Z state.

ROM Timing

There is a propagation delay between the application of a ROM's inputs and the appearance of the data outputs during a read operation. This time delay, called the access time t_{ACC}, is a measure of the ROM's operating speed. Access time is described graphically by the waveform in Fig. 14.5.

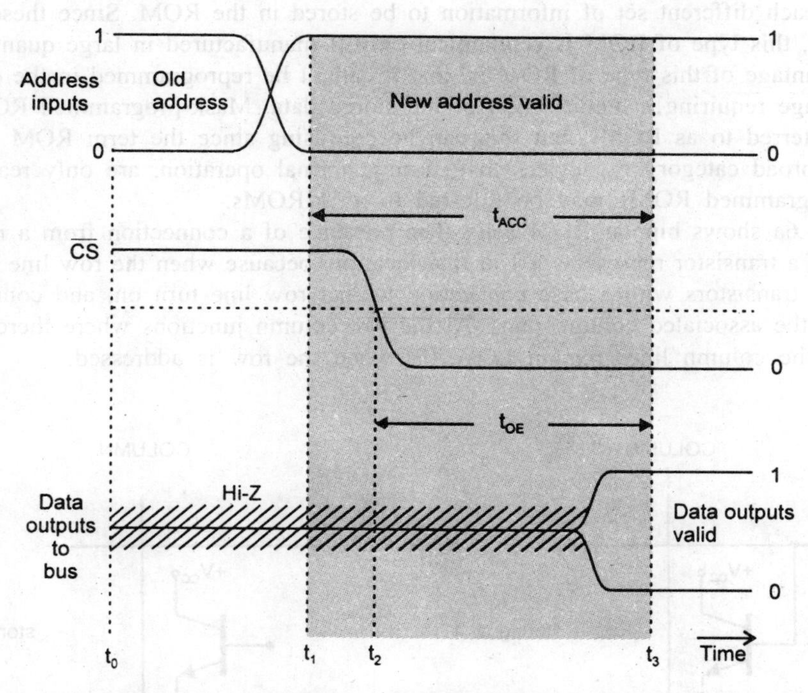

Fig. 14.5 Typical timing diagram for a ROM read operation.

The top waveform represents the address inputs, the middle waveform is an active-LOW chip select, $\overline{CS}$, and the bottom waveform represents the data outputs. At time t_0, the address inputs are all at some specified level, some HIGH and some LOW. The $\overline{CS}$ is HIGH, so that the ROM data outputs are in their Hi-Z state (represented by the hatched line).

Just prior to t_1, the address inputs are changing to a new address for a new read operation. At t_1, the new address is valid, that is, each address input is a valid logic level. At this point, the internal ROM circuitry begins to decode the new address inputs to select the register which is to send its data to the output buffers. At t_2, the $\overline{CS}$ input is activated to enable the output buffers. Finally at t_3, the outputs change from the Hi-Z state to the valid data that represent the data stored at the specified address. The output enable time, t_{OE} is the delay between the $\overline{CS}$ input and the valid data output.

Types of ROMs

The mask-programmed ROM (MROM)

The mask-programmed ROM has its storage locations written into (programmed) by the manufacturer according to the customer's specifications. A photographic negative called a *mask* is used to control the electrical interconnections on the chip. A special mask is required for each different set of information to be stored in the ROM. Since these masks are expensive, this type of ROM is economical only if manufactured in large quantities. A major disadvantage of this type of ROM is, that it cannot be reprogrammed in the event of a design change requiring a modification of the stored data. Mask-programmed ROMs are commonly referred to as ROMs, but this can be confusing since the term ROM actually represents a broad category of devices that, during normal operation, are only read from. So, mask-programmed ROMs may be referred to as MROMs.

Figure 14.6a shows bipolar ROM cells. The presence of a connection from a row line to the base of a transistor represents a 1 at that location, because when the row line is taken HIGH all the transistors with a base connection to that row line turn on, and connect the HIGH (1) to the associated column lines. At the row/column junctions where there are no connections, the column lines remain LOW (0) when the row is addressed.

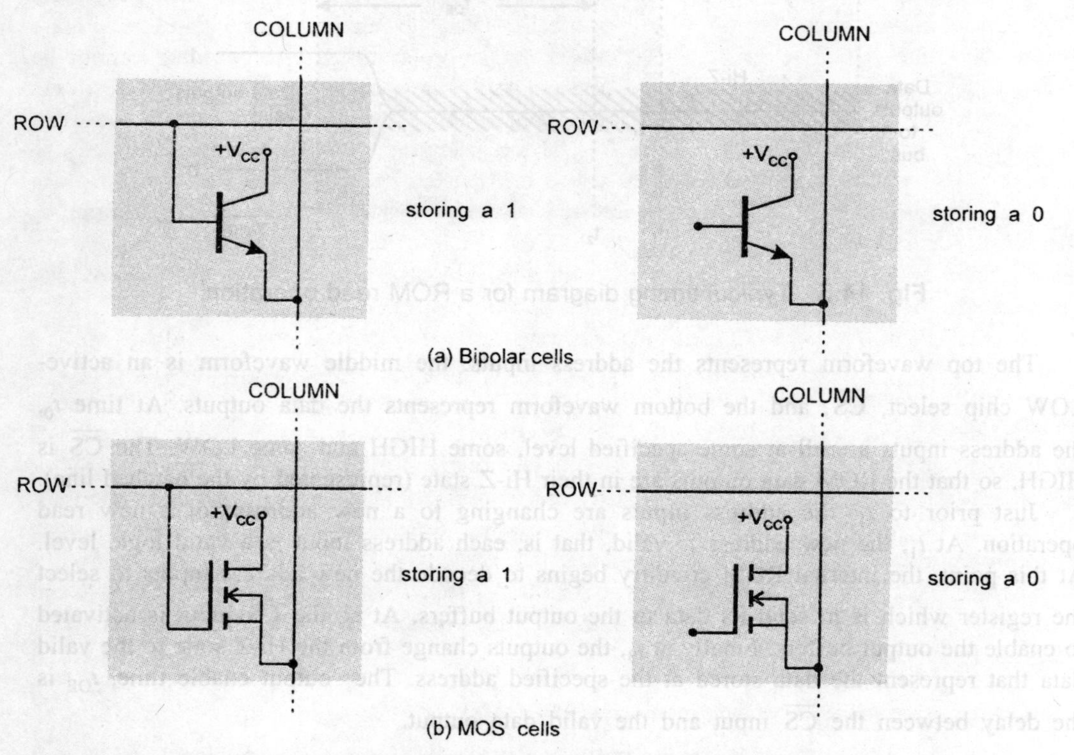

Fig. 14.6 Bipolar and MOS ROM cells.

Figure 14.6b illustrates MOS ROM cells. They are basically the same as the bipolar ROM cells, except that they are made with MOSFETs. The presence or absence of a gate connection at a junction permanently stores a 1 or a 0 as shown.

Programmable read only memories (PROMs)

Mask-programmed ROMs are not economically feasible unless produced in large quantities. On the other hand, the field programmable ROM or PROM is ideally suited for development work and small production quantities. The PROMs are not programmed during the manufacturing process, but are custom-programmed by the user. Once programmed, however, a PROM is like an MROM, which cannot be erased and reprogrammed. Thus, if the program in the PROM is faulty or has to be changed, it has to be thrown away. For this reason, these devices are often referred to as 'one-time programmable' ROMs. PROMs are often used during testing and refining new system designs because they can be replaced with relatively little effort and cost.

The PROM structure is very similar to the MROM, but the PROMs are manufactured with fusible links that can be selectively 'blown' by the user to create open circuits in a cell array. Figure 14.7a shows an array of bipolar cells having fusible links in series with the emitters of all transistors. Since every collector is connected to V_{CC}, 1s are stored at every memory location. Using a special apparatus called a PROM programmer, a user causes a large current to flow through the transistors at cells, where a 0 is to be stored. The current blows open the fuses in those cells, thus, disconnecting the emitters. Figure 14.7b shows an example. Once a PROM has been programmed, its contents cannot be changed (except to store additional 0s) because the burned-open links cannot be restored.

Fusible links are constructed by using a metal such as nichrome or titanium–tungsten, or by using polycrystalline silicon. A PROM programmer subjects the selected cells to current pulses whose amplitudes and durations are closely controlled, as necessary to burn the type of links used. A typical commercially available programmer is fully automatic, in the sense that the data to be stored in the PROM can first be loaded into the programmer, which then automatically sequences through the memory addresses of the PROM and supplies the necessary current pulses at the correct cell locations.

Bipolar PROMs are available in a wide variety of organizations, including 32 × 8 (74S188/288), 256 × 4 (74S287/387), 512 × 4 (74S570/71), 256 × 8 (74LS471), 512 × 8 (74S472/73), 1024 × 4 (74S72/73), 1024 × 8 (77S180 series), 2048 × 4 (77S184 series), 2048 × 8 (77S190 series), 4096 × 4 (77S195 series), and 4096 × 8 (77S321 series). All have enabling/chip select inputs, and different versions are available with open collector or 3-state outputs to facilitate expansion. Many are manufactured with 0s stored in all the cells instead of 1s. Most MOSFET PROMs are erasable (EPROMs); but some large capacity PROMs are also available. Examples include, the 128K × 8 TMX 27PC010 and the 64K × 16 TMX 27PC210. The TMS 27PC256 is a CMOS PROM with a capacity of 32K × 8 and a standby power dissipation of only 1.4 mW. It is available with maximum access times ranging from 120 to 250 ns.

Erasable programmable read only memories (EPROMs)

Like the fusible-link PROM, the erasable PROM (EPROM) is also useful for developmental and experimental work, because its contents can be programmed with relative ease and at

(a) A bipolar PROM array before programming. All fusible links are intact, so a 1 is stored in every cell.

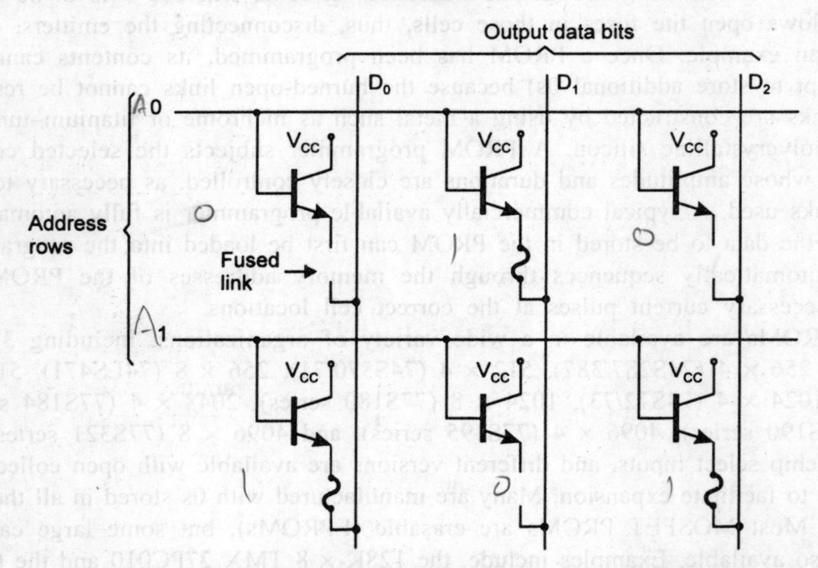

(b) Example of the PROM array after programming. The fusible links have been permanently burned to store 0s at the selected locations. The word 010 is stored at address 0 and 101 at address 1.

Fig. 14.7 A bipolar PROM array before and after programming.

small cost. The principal feature of the EPROM is, that its contents can be erased and reprogrammed, enabling the same device to be used repeatedly. Once programmed, the EPROM is non-volatile meaning that it would hold its stored data indefinitely. The term ultraviolet (UV) erasable PROM is often used with EPROM, since that is the mechanism used to erase it.

The process of programming an EPROM involves the application of special voltage levels (typically in the 10 to 25 V range) to the appropriate chip inputs for a specified amount of time (typically 50 ms per address location). The programming is usually performed by a special circuit, which is outside the circuit in which the EPROM would eventually be working. The complete programming process can take up several minutes for one EPROM chip. The EPROM cells are constructed from MOSFETs whose gates have a special design.

To erase an EPROM, the cells must be exposed to relatively intense UV radiation. The EPROMs have a transparent quartz window to permit exposure of the cells to UV radiation. This erasing process typically requires 15 to 30 minutes of exposure to UV rays. Unfortunately, there is no way to erase only the selected cells; the UV light erases all cells at the same time; so an erased EPROM stores all 1s. Once erased, the EPROM can be reprogrammed. Once an EPROM has been programmed, an opaque label is placed over the window to prevent stray radiation from causing gradual erasure.

The EPROMs are available in a wide range of capacities and access times; devices with a capacity of 128K × 8 and an access time of 45 ns are commonplace. A popular line of commercially available EPROMs is the 27XX and the 27CXX series, in both NMOS and CMOS (27CXX) versions. The XX designates the capacity in K bits. All devices in the series store 8-bit words. For example, the 2732 is a 4K × 8 NMOS EPROM.

Electrically erasable PROMs (EEPROMs)

As noted earlier, the EPROMs have two major disadvantages. First, they have to be removed from their sockets in order to be erased and reprogrammed. Second, the eraser removes the complete memory contents; this necessitates complete reprogramming even when only one memory word has to be changed. These disadvantages are overcome in EEPROMs.

The electrically erasable PROM (EEPROM) represents the most recent development in the evolution of ROM technology. Unlike the UV erasable PROM, the EEPROM does not necessarily have to be removed from a circuit and exposed to a different environment to be erased. Many EEPROMs can be both erased and programmed with modest power requirements, so it is possible to integrate erasing and reprogramming circuitry into the system utilizing the memory. Although these capabilities bring the EEPROM a step closer to becoming a non-volatile R/$\overline{\text{W}}$ memory, the time required to 'write' (erase and reprogram) is still in the millisecond range, far less than that required by the conventional R/$\overline{\text{W}}$ memory. Another advantage of the EEPROM over the EPROM is that, it is possible to erase and restore a single byte in an array. Also, the EEPROM can be erased in a very short time (10 ms compared to 30 minutes required for EPROM in external UV light) and programmed rapidly (it requires only a 10 ms programming pulse for each data word compared with a 50 ms pulse required for an EPROM).

The Intel 2816 was the original EEPROM. It was introduced by the Intel Corporation in 1981 with 2K × 8 capacity and an access time of 250 ns. The 2864 is an 8K × 8 EEPROM which contains on-chip circuitry that generates the high voltages for the eraser and programming operations so that the chip requires only a + V_{CC} power supply. This makes the 2864 almost as easy to use as the static RWM devices. Of course, unlike the static RWM, the EEPROM is non-volatile and will hold all written data when power is turned off. On the other hand, the static RWM has less complex internal circuitry and much faster access time.

Applications of ROMs

ROMs can be used in any application requiring non-volatile data storage, where the data rarely or never changes. We briefly describe below some of the most common application areas.

Microcomputer program storage (firmware)

At present, microcomputer firmware is the most widespread application of ROMs. Some personal and business microcomputers use ROMs to store their operating system programs and their language interpreters (e.g. BASIC), so that the computer can be used immediately after power is turned on. Products that include a microcomputer to control their operation use ROMs to store the control programs. Examples are, electronic games, electronic cash registers, electronic scales, and microcomputer-controlled automobile fuel injection systems. The microcomputer programs that are stored in ROMs are referred to as *firmware*, because they are not subject to change. Programs that are stored in RWMs are referred to as *software*, because they can be easily altered.

Bootstrap memory

Many microcomputers and most large computers do not have their operating system programs stored in ROMs. Instead, these programs are stored in external mass memory, usually the magnetic disk. How, then, do these computers know what to do when they are powered on? A relatively small program, called the *bootstrap program*, is stored in a ROM (the term *bootstrap* comes from the idea of pulling oneself up by one's own boot straps). When the computer is powered on, it will execute the instructions that are in its bootstrap program. These instructions typically cause the CPU to initialize the system hardware. The bootstrap program then loads the operating system programs from mass storage (disk into its main internal memory). At that point, the computer begins executing the operating system program and is ready to respond to the user commands. This start-up process is often called 'booting up' the system.

Data tables

ROMs are often used to store tables of data that do not change. Some examples are trigonometric tables (i.e. sine, cosine, etc.) and code conversion tables.

Several standard ROM look-up tables are available with trigonometric functions. The National Semiconductor MM 4220BM stores the sine function for angles between 0 and 90 degrees. This ROM is organized as a 128×8, with seven address inputs and eight data outputs. The address inputs represent the angle in increments of approximately 0.7 degree. For example, the address 0000000 is for 0°, the address 0000001 is for 0.7°, the address 0000010 is for 1.41°, and so on up to address 111111 which is 89.3°. When an address is applied to the ROM, the data outputs will represent the approximate sine of the angle.

For example, with input address 1000000 (representing approximately 45°), the data output will be 10110101. Since, the sine is 45°, these data are interpreted as a fraction, 0.10110101, which when converted to decimal equals to 0.707 (the sine of 45°).

Data converters

The data converter circuit takes data expressed in one type of code and produces an output expressed in another type. Code conversion is needed, for example, when a computer is outputting data in straight binary code and it is required to convert it to BCD in order to display it on seven-segment LED readouts.

One of the easiest methods of code conversion uses a ROM programmed such that the application of a particular address (the old code) produces a data output that represents the equivalent one in the new code. The 74185 is a TTL ROM that stores the binary-to-BCD code conversion for a 6-bit binary input. To illustrate, a binary address input of 100110 (decimal 38) will produce a data output of 00111000, which is the BCD code for decimal 38.

Character generators

If you have ever looked closely at the alphanumeric characters (letters, numbers, etc.) printed on a computer's video display screen, you might have noticed that each is generally made up of a group of dots. Depending on the character being displayed, some dot positions are made bright while others are dark. Each character is made to fit into a pattern of dot positions, usually arranged as a 5 × 7 or 7 × 9 matrix. The pattern of dots for each character can be represented as a binary code (i.e. bright dot = 1, dark dot = 0).

A character generator ROM stores the dot pattern codes for each character at an address corresponding to the ASCII code for that character. For example, the dot pattern for the letter A would be stored at address 1000001, where 1000001 is the ASCII code for uppercase A. Character generator ROMs are used extensively in any application that displays or prints out alphanumeric characters.

Function generator

The function generator is a circuit that produces waveforms such as sine waves, saw tooth waves, triangular waves, and square waves. Figure 14.8 shows how a ROM look-up table and DAC are used to generate a sine wave output signal. The ROM stores 256 different 8-bit values, each one corresponding to a different waveform value, i.e. a different voltage point on the sine wave. The 8-bit counter is continuously pulsed by a clock signal to provide sequential address inputs to the ROM. As the counter cycles through the 256 different addresses, the ROM outputs the 256 data points to the DAC. The DAC output will be a waveform that steps through the 256 different analog voltage values corresponding to the data points. A low-pass filter smoothes out the steps in the DAC output to produce a smooth waveform. Circuits such as these are used in some commercial function generators. The same idea is used in some speech synthesizers where the digitized speech waveform values are stored in a ROM.

Figure 14.9 shows the pin diagrams of ROM ICs 74184 and 74185, a 6-bit BCD-to-binary converter and a 6-bit binary-to-BCD converter, respectively.

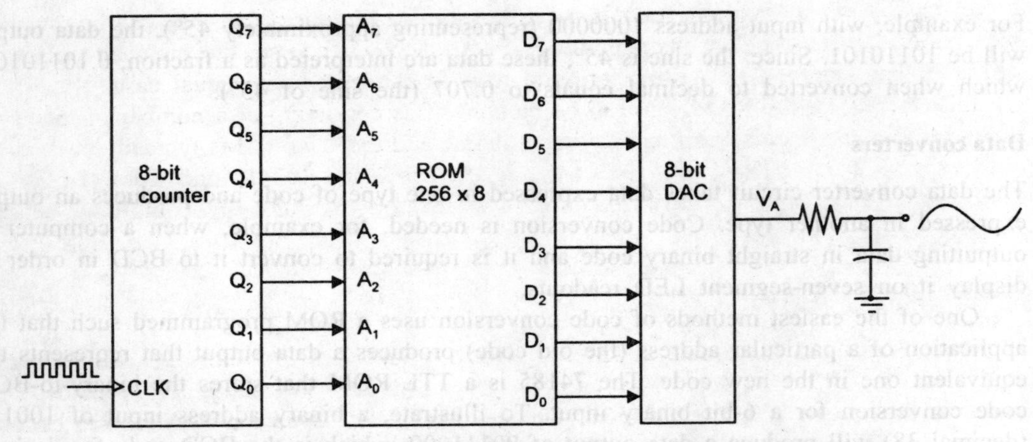

Fig. 14.8 Function generator using a ROM and DAC.

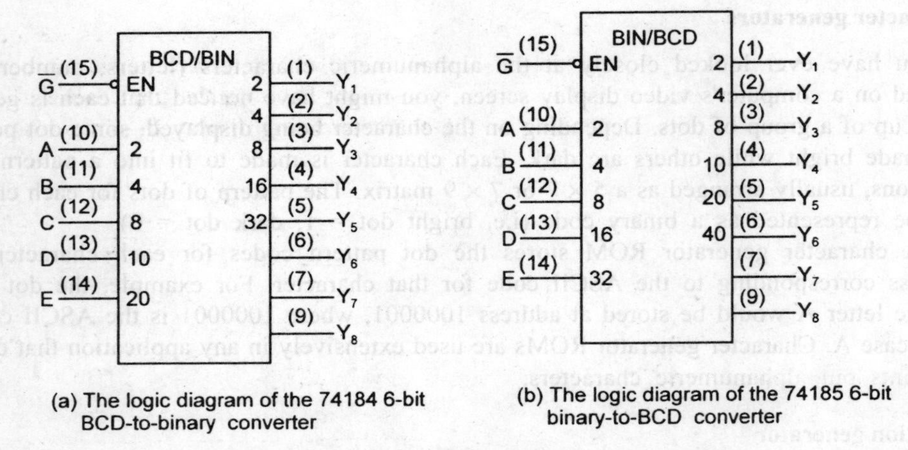

(a) The logic diagram of the 74184 6-bit BCD-to-binary converter

(b) The logic diagram of the 74185 6-bit binary-to-BCD converter

Fig. 14.9 Pin diagrams of 74184 and 74185.

The use of ROMs as binary-to-BCD converters and BCD-to-binary converters, and also how to use them to obtain the 9's complement and 10's complement of BCD numbers, is illustrated in Figs. 7.28, 7.29, 7.30, and 7.31 in Chapter 7 (Combinational Circuits).

14.4 SEMICONDUCTOR RAMs

When the term RAM is used with semiconductor memories, it is usually taken to mean read/write memory (RWM) as opposed to ROM. The RAMs are used in computers for the temporary storage of programs and data. The contents of many RAM address locations are read from and written to as the computer executes a program. This requires fast read and write cycle times for the RAM so as not to slow down the computer operation. A major disadvantage of RAMs is that they are volatile and lose all stored information if power is interrupted or turned off. Some CMOS RAMs, however, use such small amounts of power in the standby mode (when no read or write operations take place) that they can be powered from batteries whenever the main power is interrupted. Of course, the main advantage of RAMs is that they can be written into and read from rapidly with equal ease.

Like the ROM, the RAM can also be thought of as consisting of a number of registers, each storing a single data word and having a unique address. The RAMs typically come with word capacities of 1K, 4K, 8K, 16K, 64K, 128K, 256K and 1024K, and word sizes of 1, 4, or 8-bits. The word capacity and word size can be expanded by combining several memory chips.

Static RAMs (SRAMs)

The static RAM can store data as long as power is applied to the chip. Its memory cells are essentially flip-flops that will stay in a given state (store a bit) indefinitely, provided that power to the circuit is not interrupted. On the other hand, dynamic RAMs store data as charges on capacitors. With dynamic RAMs, the stored data will gradually disappear because of capacitor discharge, therefore, it is necessary to periodically refresh the data (i.e. recharge the capacitors).

Static RAMs (SRAMs) are available both in bipolar and MOS technologies, although the vast majority of applications use NMOS or CMOS RAMs. As stated earlier, the bipolars have the advantage of speed (though NMOS is gradually closing the gap), while the MOS devices have much greater capacities and lower power consumption. Figure 14.10 shows,

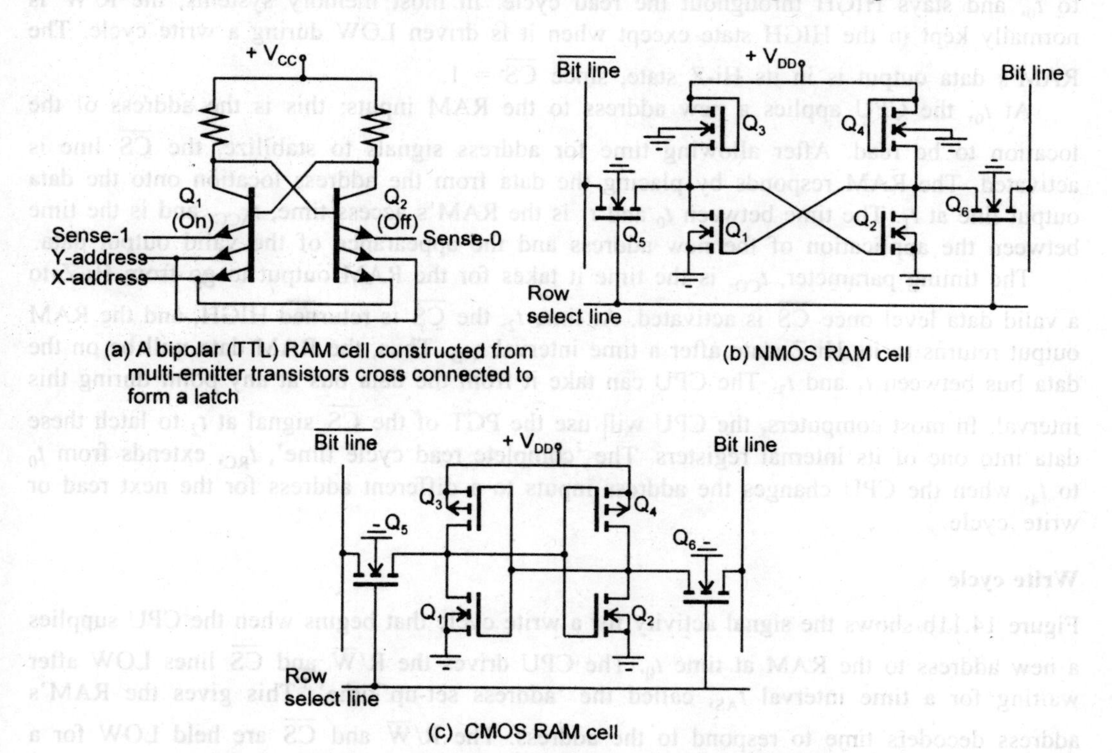

(a) A bipolar (TTL) RAM cell constructed from multi-emitter transistors cross connected to form a latch

(b) NMOS RAM cell

(c) CMOS RAM cell

Fig. 14.10 The bipolar, NMOS, and CMOS RAM cells.

for comparison, a typical bipolar static memory cell, a typical NMOS static memory cell, and a typical CMOS static memory cell. The bipolar cell contains two multi-emitter transistors and two resistors. The NMOS cell contains four *N*-channel MOSFETs. The CMOS cell requires two CMOS FETs.

The RAM ICs are most often used as the internal memory of a computer. The CPU continually performs read and write operations on this memory at a very fast rate constrained only by the limitations of the CPU. The memory chips that are interfaced to the CPU have to be, therefore, fast enough to respond to the CPU read and write commands. Figure 14.11 shows the timing diagrams of a typical RAM chip for a complete read cycle and a complete write cycle.

Read cycle

The waveforms in Fig. 14.11a show how the address, $R/\overline{W}$ and $\overline{CS}$ inputs behave during a memory read cycle. The CPU supplies these input signals to the RAM when it wants to read data from a specific location. Although a RAM may have many address inputs coming from the CPU's address bus, for clarity, the figure shows only two.

The read cycle begins at time t_0. Prior to that, the address inputs will be whatever address is on the address bus from the preceding operation. Since the RAM's chip select is not active, it will not respond to this *old* address. Note that, the $R/\overline{W}$ line is HIGH prior to t_0, and stays HIGH throughout the read cycle. In most memory systems, the $R/\overline{W}$ is normally kept in the HIGH state except when it is driven LOW during a write cycle. The RAM's data output is in its Hi-Z state, since $\overline{CS} = 1$.

At t_0, the CPU applies a new address to the RAM inputs; this is the address of the location to be read. After allowing time for address signals to stabilize, the $\overline{CS}$ line is activated. The RAM responds by placing the data from the address location onto the data output line at t_1. The time between t_0 and t_1 is the RAM's access time, t_{ACC}, and is the time between the application of the new address and the appearance of the valid output data.

The timing parameter, t_{CO}, is the time it takes for the RAM output to go from Hi-Z to a valid data level once $\overline{CS}$ is activated. At time t_2, the $\overline{CS}$ is returned HIGH, and the RAM output returns to its Hi-Z state after a time interval t_{OD}. Thus, the RAM data will be on the data bus between t_1 and t_3. The CPU can take it from the data bus at any point during this interval. In most computers, the CPU will use the PGT of the $\overline{CS}$ signal at t_2 to latch these data into one of its internal registers. The 'complete read cycle time', t_{RC}, extends from t_0 to t_4, when the CPU changes the address inputs to a different address for the next read or write cycle.

Write cycle

Figure 14.11b shows the signal activity for a write cycle that begins when the CPU supplies a new address to the RAM at time t_0. The CPU drives the $R/\overline{W}$ and $\overline{CS}$ lines LOW after waiting for a time interval t_{AS}, called the 'address set-up time'. This gives the RAM's address decoders time to respond to the address. The $R/\overline{W}$ and $\overline{CS}$ are held LOW for a time interval t_W, called the 'write time interval'.

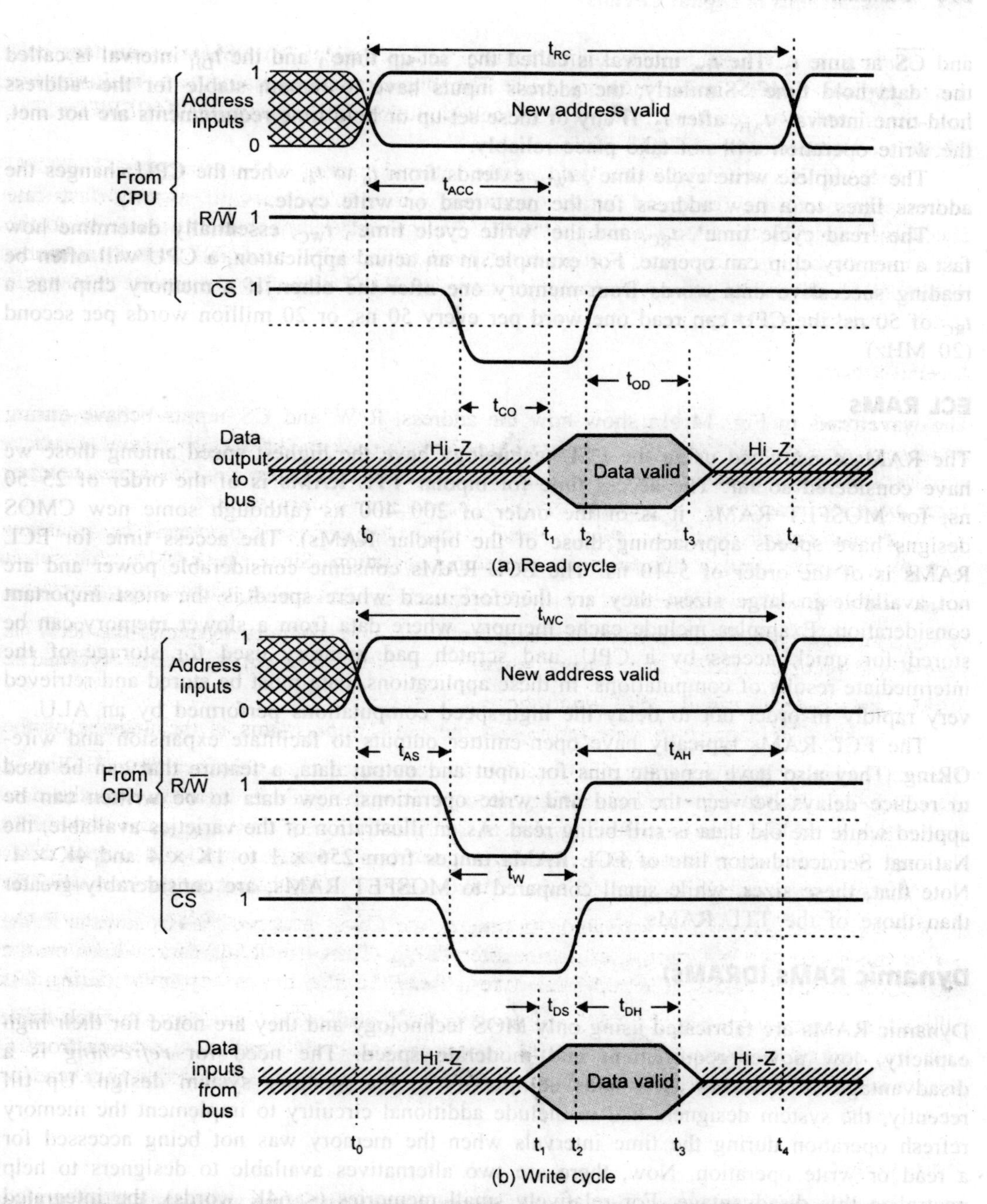

Fig. 14.11 Typical timings for SRAM: (a) read cycle and (b) write cycle.

During this write time interval, at time t_1, the CPU applies valid data to the data bus to be written into the RAM. These data have to be held at the RAM input for at least a time interval t_{DS} prior to, and for at least a time interval t_{DH} after, the deactivation of the R/$\overline{W}$

and $\overline{CS}$ at time t_2. The t_{DS} interval is called the 'set-up time', and the t_{DH} interval is called the 'data hold time'. Similarly, the address inputs have to remain stable for the 'address hold-time interval' t_{AH}, after t_2. If any of these set-up or hold time requirements are not met, the write operation will not take place reliably.

The 'complete write cycle time', t_{WC}, extends from t_0 to t_4, when the CPU changes the address lines to a new address for the next read or write cycle.

The 'read cycle time', t_{RC}, and the 'write cycle time', t_{WC}, essentially determine how fast a memory chip can operate. For example, in an actual application, a CPU will often be reading successive data words from memory one after the other. If a memory chip has a t_{RC} of 50 ns, the CPU can read one word per every 50 ns, or 20 million words per second (20 MHz).

ECL RAMs

The RAMs constructed using the ECL technology have the highest speed among those we have considered so far. The access time for bipolar TTL RAMs is of the order of 25–50 ns; for MOSFET RAMs, it is of the order of 200–400 ns (although some new CMOS designs have speeds approaching those of the bipolar RAMs). The access time for ECL RAMs is of the order of 5–10 ns. The ECL RAMs consume considerable power and are not available in large sizes; they are therefore used where speed is the most important consideration. Examples include cache memory, where data from a slower memory can be stored for quick access by a CPU, and scratch pad memory used for storage of the intermediate results of computations. In these applications, data must be stored and retrieved very rapidly in order not to delay the high-speed computations performed by an ALU.

The ECL RAMs typically have open-emitter outputs to facilitate expansion and wire-ORing. They also have separate pins for input and output data, a feature that can be used to reduce delays between the read and write operations; new data to be written can be applied while the old data is still being read. As an illustration of the varieties available, the National Semiconductor line of ECL RAMs ranges from 256 × 1 to 1K × 4 and 4K × 1. Note that, these sizes, while small compared to MOSFET RAMs, are considerably greater than those of the TTL RAMs.

Dynamic RAMs (DRAMs)

Dynamic RAMs are fabricated using only MOS technology and they are noted for their high capacity, low power requirement and moderate speed. The need for *refreshing* is a disadvantage of DRAMs as it adds complexity to the memory system design. Up till recently, the system designers had to include additional circuitry to implement the memory refresh operation during the time intervals when the memory was not being accessed for a read or write operation. Now, there are two alternatives available to designers to help neutralize this disadvantage. For relatively small memories (< 64K words), the integrated RAM (iRAM) provides a solution. An iRAM is an IC that includes the refresh circuitry on the same chip as the memory cell array. The result is a chip that externally operates just like a static RAM chip—you apply the address and collect the data—but internally uses a high density DRAM structure. The designer is not concerned with the memory refresh operation; it is done internally and automatically.

For larger memory systems (> 64K), a more cost-effective approach uses LSI chips called *dynamic memory controllers* which contain all of the necessary logic for refreshing the DRAM chips that make up the system. This greatly reduces the amount of extra circuitry in a dynamic RAM system.

For applications, where speed and reduced complexity are more critical than space and power considerations, static RAMs are still the best. They are generally faster than DRAMs and require no refresh operation. They are simpler to design, but cannot compete with higher capacity and lower power requirements of DRAMs.

Because of their simple cell structure, the DRAMs typically have four times the density of SRAMs. This allows four times as much memory capacity to be placed on a single board, or alternatively, requires one-fourth as much board space for the same amount of memory. The cost per bit of DRAM storage is typically one-fifth to one-fourth that of SRAMs. A further cost saving is realized because the lower power requirement of a DRAM, typically one-sixth to one-half that of a SRAM, allows the use of smaller and less expensive power supplies.

The main applications of SRAMs are in areas where only small sizes of memory are needed (up to 64K) or where high speed is required. Many microprocessor controlled instruments and appliances have very small memory capacity requirements. Some instruments such as digital storage oscilloscopes and logic analyzers require very high speed memory; for such applications, SRAMs are normally used.

The main internal memory of most personal computers is DRAM, because of its high capacity and low power consumption. These computers, however, sometimes use small amounts of SRAM as well, for functions requiring maximum speed such as video graphics and look-up tables.

Early DRAMs had four NMOS transistors per cell and relied on inherent gate capacitance of the transistors for storage of charge. The 4-transistor cell was complex, because transistors were needed to buffer and sense the tiny charges stored on the gate capacitance. A later improvement was a 3-transistor cell and the most current design is a single-transistor cell shown in Fig. 14.12. The single-transistor cell is the ultimate in simplicity. The transistor serves as a transmission gate controlled by the address line. To read, the address line is made HIGH, turning on the transmission gate, and the capacitor voltage appears on the bit line. To write, the address line is again made HIGH, and the voltage on the bit lines charges or discharges the capacitor through the transmission gate. Read out is destructive, so every read operation must be followed by a write operation.

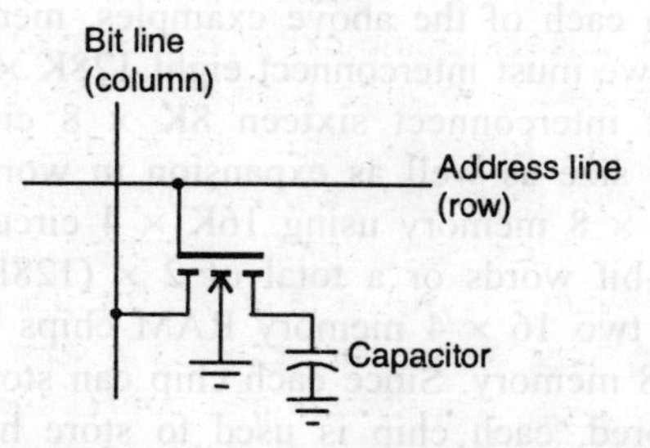

Fig. 14.12 The single-transistor dynamic memory cell.

Multiplexing

In order to reduce the number of pins on high capacity DRAM chips, manufacturers utilize address multiplexing whereby each address input pin can accommodate two different address bits.

DRAM Refreshing

DRAM cells have to be refreshed periodically as otherwise the stored data will be lost. In dynamic memories, refreshing can be accomplished by reading, since data must be automatically written back into the cells that are read. The maximum time between refresh cycles is typically 2, 4, or 8 ms in modern RAMs. Although it would be possible to read every cell in succession and thus refresh the entire memory during each refresh cycle, most dynamic memories are refreshed one entire row at a time in order to reduce the number of read operations that must take place to refresh the complete memory. In burst mode refreshing, each row of cells is refreshed in succession, with all normal memory operations suspended until all rows have been refreshed. Alternatively, row refreshing can be interspersed with other memory operations. In either case, refresh control circuitry is necessary to synchronize refresh cycles and to ensure that every row is refreshed within the specified time. Most manufacturers of dynamic memory ICs have developed special ICs to handle the refresh operation as well as the address multiplexing needed by the DRAM systems. These ICs are called *dynamic RAM controllers*. Some DRAMs have built-in refresh control circuitry and are said to be pseudo-dynamic (or semi-dynamic), because a user is not required to provide external hardware to accomplish the refresh task.

14.5 MEMORY EXPANSION

In most IC memory applications, the required memory capacity or word size cannot be satisfied by one memory chip. Instead, several memory chips are combined to provide the desired capacity and the word size. The process of increasing the word size or capacity by combining a number of IC chips is called *memory expansion*. In a *bit-organized* memory, each IC stores 1-bit of each word. For example, in a bit-organized 128K × 8 memory, the LSB of each word is stored in one 128K × 1 circuit, the next LSB in another 128K × 1 circuit, and so forth. A memory in which every bit of a word is stored in each circuit, is said to be *word organized*. For example, a word-organized 128K × 8 memory might consist of sixteen 8K × 8 circuits. In each of the above examples, memory expansion is required. In the bit-organized memory, we must interconnect eight 128K × 1 circuits and in the word-organized memory, we must interconnect sixteen 8K × 8 circuits. Some organizations require an expansion in word size as well as expansion in word capacity. For example, if we wish to construct a 128K × 8 memory using 16K × 4 circuits, we would need two of the latter for each 16K of 8-bit words or a total of 2 × (128K/16K) = 16 circuits.

Figure 14.13 shows how two 16 × 4 memory RAM chips with common I/O lines are combined to produce a 16 × 8 memory. Since each chip can store 16 4-bit words and since 16 8-bit words are to be stored, each chip is used to store half of each word. In other words, RAM 0 stores the four higher order bits of each of the 16 words, and RAM 1 stores

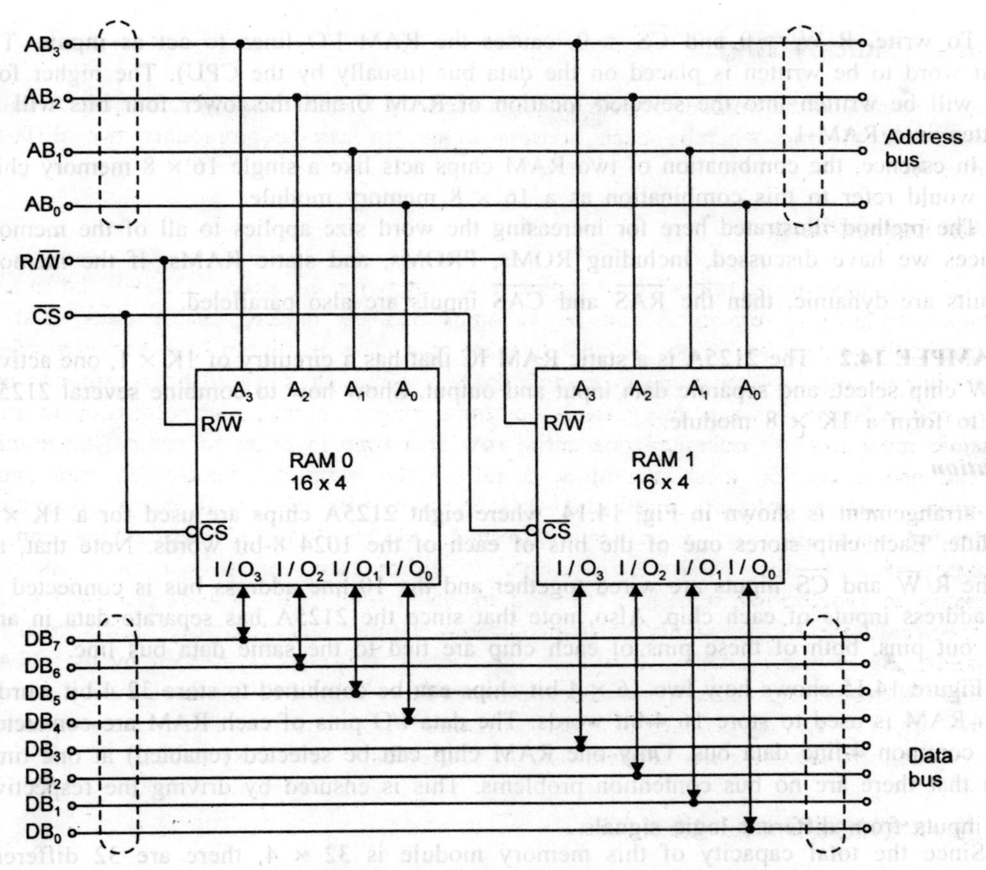

Fig. 14.13 Combining two 16 × 4 RAMs for a 16 × 8 module.

the four lower order bits of each of the 16 words. A full 8-bit word is available at the RAM outputs connected to the data bus.

Any one of the 16 words is selected by applying the appropriate address code to the four line address bus (AB_3, AB_2, AB_1 and AB_0). The address lines typically originate at the CPU. Note that, each address bus line is connected to the corresponding address input of each chip. This means that once an address code is placed on the address bus, the same address code is applied to both chips such that the same location on each chip is accessed at the same time.

Once the address is selected, we can read or write at this address under the control of the common $R/\overline{W}$ and $\overline{CS}$ line. To read, $R/\overline{W}$ must be HIGH and $\overline{CS}$ must be LOW. This causes the RAM I/O lines to act as outputs. RAM 0 places its selected 4-bit word on the upper four data bus lines and RAM 1 places its selected 4-bit word on the lower four data bus lines. The data bus then contains the full selected 8-bit word which can now be transmitted to some other device (usually a register in the CPU).

To write, $R/\overline{W} = 0$ and $\overline{CS} = 0$, causes the RAM I/O lines to act as inputs. The 8-bit word to be written is placed on the data bus (usually by the CPU). The higher four bits will be written into the selected location of RAM 0 and the lower four bits will be written into RAM-1.

In essence, the combination of two RAM chips acts like a single 16 × 8 memory chip. We would refer to this combination as a 16 × 8 memory module.

The method illustrated here for increasing the word size applies to all of the memory devices we have discussed, including ROMs, PROMs, and static RAMs. If the memory circuits are dynamic, then the $\overline{RAS}$ and $\overline{CAS}$ inputs are also paralleled.

EXAMPLE 14.2 The 2125A is a static RAM IC that has a circuitry of 1K × 1, one active-LOW chip select, and separate data input and output. Show how to combine several 2125A ICs to form a 1K × 8 module.

Solution

The arrangement is shown in Fig. 14.14, where eight 2125A chips are used for a 1K × 8 module. Each chip stores one of the bits of each of the 1024 8-bit words. Note that, all of the $R/\overline{W}$ and $\overline{CS}$ inputs are wired together and the 10-line address bus is connected to the address inputs of each chip. Also, note that since the 2125A has separate data in and data out pins, both of these pins of each chip are tied to the same data bus line.

Figure 14.15 shows how two 16 × 4-bit chips can be combined to store 32 4-bit words. Each RAM is used to store 16 4-bit words. The data I/O pins of each RAM are connected to a common 4-line data bus. Only one RAM chip can be selected (enabled) at one time such that there are no bus contention problems. This is ensured by driving the respective $\overline{CS}$ inputs from different logic signals.

Since the total capacity of this memory module is 32 × 4, there are 32 different addresses. This requires five address bus lines. The upper address line AB_4 is used to select one RAM or the other (via the $\overline{CS}$ inputs) which is to be read from or written into. The other four address lines AB_0 to AB_3 are used to select one memory location out of 16 from the selected RAM chip.

To illustrate, when $AB_4 = 0$, the $\overline{CS}$ of RAM-0 enables this chip for read or write. Then, any address location in RAM-0 can be accessed by AB_3 through AB_0. The latter four address lines can range from 0000 to 1111 to select the desired location. Thus, the range of addresses representing locations in RAM-0 is $AB_4\ AB_3\ AB_2\ AB_1\ AB_0$ = 00000 to 01111.

Note that, when $AB_4 = 0$, the $\overline{CS}$ of RAM 1 is HIGH, therefore, its I/O lines are disabled (high impedance) and cannot communicate (give or take data) with the data bus.

When $AB_4 = 1$, the roles of RAM-0 and RAM-1 are reversed. The RAM-1 is now enabled and the lines AB_3 to AB_0 select one of its locations.

Thus, the range of addresses selected in RAM-1 is $AB_4\ AB_3\ AB_2\ AB_1\ AB_0$ = 10000 to 11111.

EXAMPLE 14.3 It is desired to combine several 1K × 8 PROMs to produce a total capacity of 4K × 8. How many PROM chips are required? Show the arrangement.

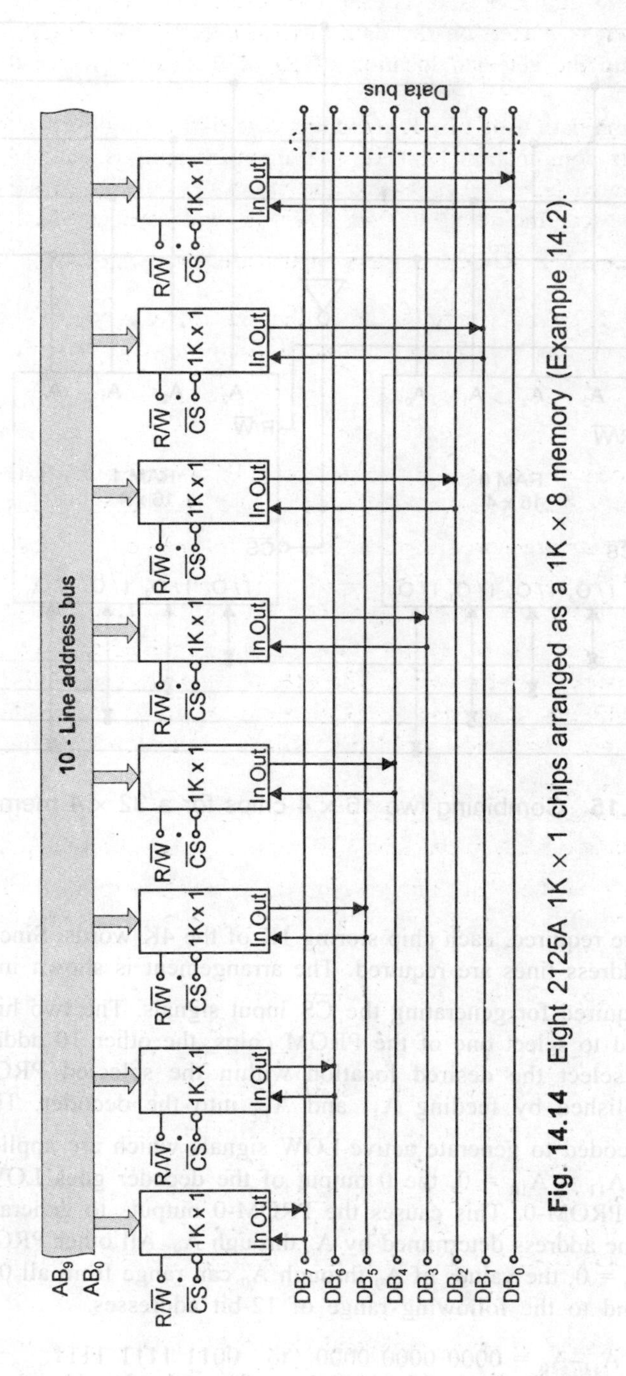

Fig. 14.14 Eight 2125A 1K × 1 chips arranged as a 1K × 8 memory (Example 14.2).

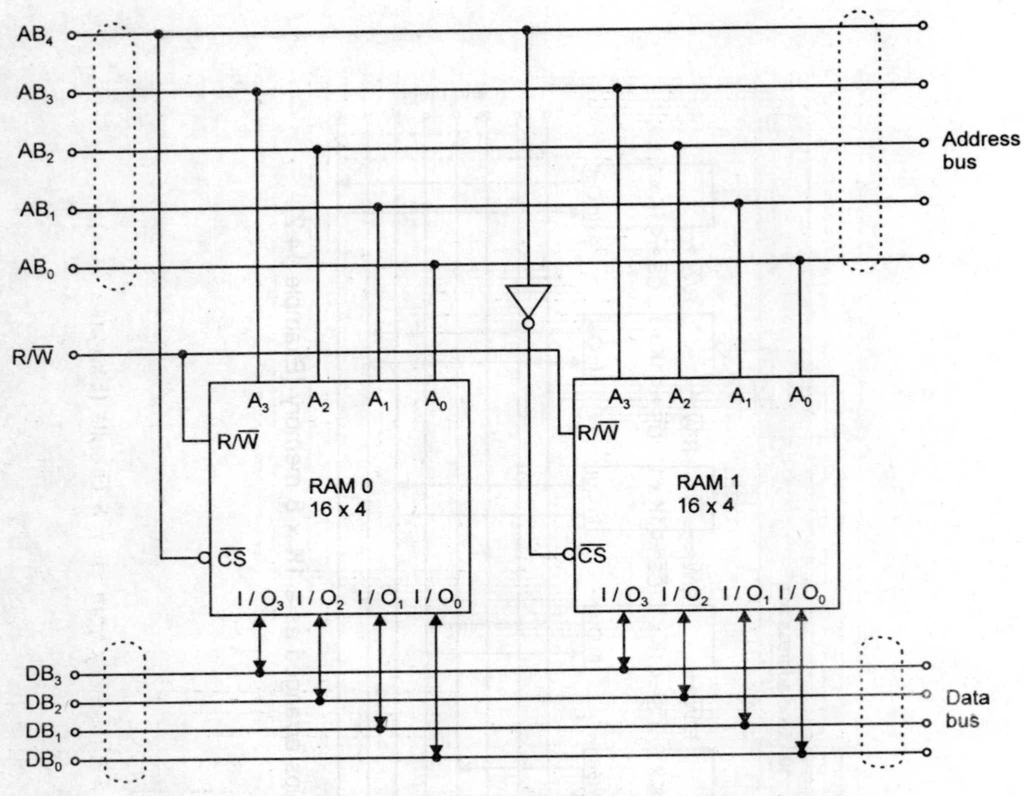

Fig. 14.15 Combining two 16 × 4 chips for a 32 × 4 memory.

Solution

Four PROM chips are required, each chip storing 1K of the 4K words. Since 4K = 4 × 1024 = 4096 = 2^{12}, 12 address lines are required. The arrangement is shown in Fig. 14.16. The decoder circuit is required for generating the $\overline{CS}$ input signals. The two highest order lines A_{11} and A_{10} are used to select one of the PROM chips, the other 10 address bus lines go to each PROM to select the desired location within the selected PROM. The PROM selection is accomplished by feeding A_{11} and A_{10} into the decoder. The four possible combinations are decoded to generate active-LOW signals which are applied to $\overline{CS}$ inputs. For example, when $A_{11} = A_{10} = 0$, the 0 output of the decoder goes LOW (all others are HIGH) and enables PROM-0. This causes the PROM-0 outputs to generate the data word internally stored at the address determined by A_0 through A_9. All other PROMs are disabled.

While $A_{11} = A_{10} = 0$, the values of A_9 through A_0 can range from all 0s to all 1s. Thus, PROM-0 will respond to the following range of 12-bit addresses.

$$A_{11}–A_0 = 0000\ 0000\ 0000 \quad \text{to} \quad 0011\ 1111\ 1111.$$

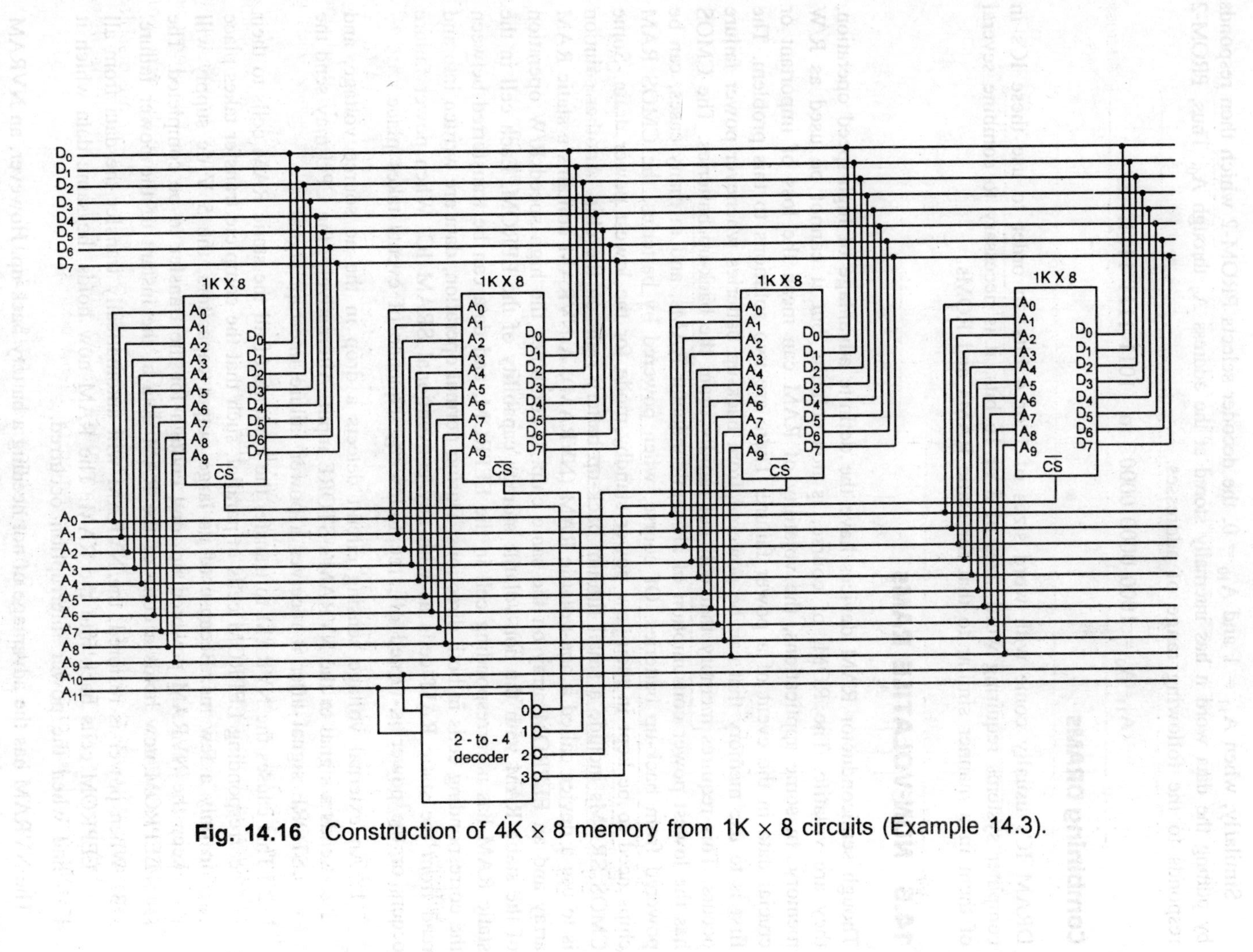

Fig. 14.16 Construction of 4K × 8 memory from 1K × 8 circuits (Example 14.3).

Similarly, when $A_{11} = 1$ and $A_{10} = 0$, the decoder selects PROM-2 which then responds by putting the data word it has internally stored at the address A_9 through A_0. Thus, PROM-2 responds to the following range of addresses.

$$A_{11}–A_0 = 1000\ 0000\ 0000 \quad \text{to} \quad 1011\ 1111\ 1111.$$

Combining DRAMs

DRAM ICs usually come with word sizes of 1 or 4 bits. In order to use these ICs in computer systems requiring word sizes of 8 or 16 bits, it is necessary to combine several of them in a manner similar to that for static RAMs and ROMs.

14.6 NON-VOLATILE RAMs

Though semiconductor RAM devices have the definite advantage of high-speed operation, they are volatile. The ROM, of course, is non-volatile, but it cannot be used as R/W memory. In some applications, the volatility of RAM can mean the loss of important or crucial data in the event of a power failure. There are two solutions to this problem. The first is to use memory that can be powered from back-up batteries whenever power failure occurs. This requires memory that will not rapidly drain the back-up batteries. The CMOS has the lowest power consumption of all semiconductor RAMs, and in many cases, can be powered from back-up batteries. Of course, when powered by batteries, the CMOS RAM chips need to be kept in their low power standby mode for the lowest power drain. Some CMOS SRAMs include a small lithium back-up battery right on the chip. Another solution is to use a device called a non-volatile RAM (NVRAM). A NVRAM contains a static RAM array and an EEPROM array on the same chip. It combines the high speed R/W operation of the static RAM with the non-volatile storage capability of the EPROM. Each cell in the static RAM has a corresponding cell in the EEPROM, and data can be transferred between the corresponding cells in both directions. During normal operation, data are written into and read from the static RAM cells as with any conventional SRAM IC. When power failure occurs or the power is turned off, the following sequence of events takes place.

1. An external voltage sensing circuit detects a drop in the ac source voltage, and sends a signal to the NVRAM's STORE input. Alternatively, the CPU may send the STORE signal after it receives a power failure interrupt signal.
2. This causes the NVRAM to transfer the contents of all the static RAM cells to their corresponding EEPROM cells in parallel, such that the complete transfer takes place in only a few ms. Because of its large output capacitors, the 5 V dc supply will keep the NVRAM powered up long enough for the transfer to be completed. The EEPROM now holds a copy of the RAM data at the instant of the power failure.
3. When power is restored, the NVRAM will automatically transfer the data from all EEPROM cells back into the RAM. The RAM now holds the same data which it had when the power interruption occurred.

The NVRAM has the advantage of not needing a battery back-up. However, an NVRAM is more complex than a normal memory chip because it contains both RAM and EEPROM cells as well as the circuitry needed to transfer data between the two. For this reason, the NVRAMs are not available in very high capacities. When a high capacity non-volatile memory is required, it is best to use CMOS RAM with battery back-up.

14.7 SEQUENTIAL MEMORIES

The semiconductor memories that we have discussed so far are random access memories. The high speed operation of random access devices makes them suitable for use as the internal memory of the computer. Sequential access semiconductor memories utilize shift registers to store data that can be accessed in a sequential manner. Although they are not useful as internal computer memory because of their relatively slow speed, shift register memories find application in areas where sequential repetitive data are required. A primary example is the storage and sequential transmission of the ASCII-coded data for the characters on a video display. This data have to be supplied to the video display circuits periodically in order to refresh the displayed image on the screen. By using shift registers, the stored data can be recirculated to refresh the screen image periodically. Shift register memories are also used in digital storage oscilloscopes and logic analyzers.

Recirculating Shift Registers

Figure 14.17a shows the block diagram of a recirculating shift register. There can be any number of FFs in a shift register, though four are shown in this illustration. Data enter the shift register from the serial input D_S, which shifts into Q_3, Q_3 shifts into Q_2, Q_2 into Q_1, and Q_1 into Q_0. The Q_0 output is recirculated back to the serial input through some controller. This logic provides two modes of operation that are controlled by the regulation input REC. The level at REC determines the source of the data that will reach the serial input.

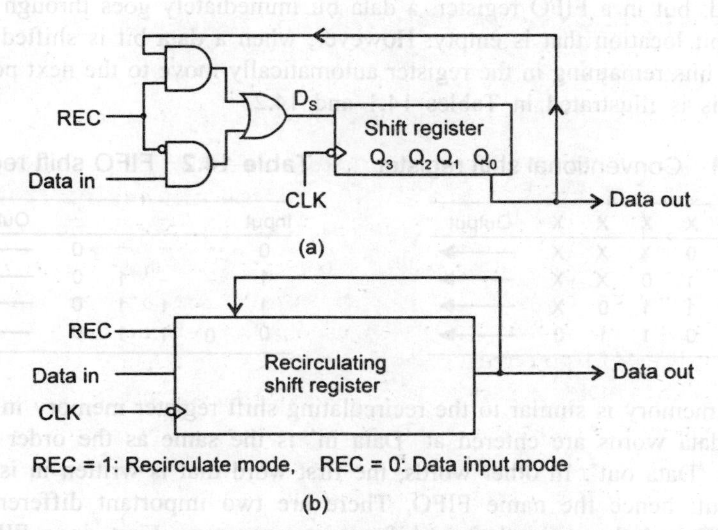

Fig. 14.17 Block diagram of a recirculating shift register.

Recirculate mode (REC = 1)

In this mode, the upper AND gate is enabled and Q_0 output (Data out) is applied to D_S. As clock pulses are applied, the data in the shift register will recirculate as shown below.

$Q_3 \rightarrow Q_2 \rightarrow Q_1 \rightarrow Q_0 \rightarrow$ Data out (with Q_0 fed back to Q_3)

While recirculating in the register, the data also appears at 'Data out', one bit at a time. In this mode, 'Data in' is inhibited and has no effect on the register data.

Data input mode (REC = 0)

In this mode, the lower AND gate is enabled and the 'Data in' signal is applied to D_S. As clock pulses are applied, the data will shift as shown below.

Data In $Q_3 \rightarrow Q_2 \rightarrow Q_1 \rightarrow Q_0 \rightarrow$ Data out

There is no recirculation of data, since the upper AND gate is inhibited by REC = 0. This mode is used to enter new data at 'Data in' for storage in the register.

Figure 14.17b is a simplified symbol that we will use for the circulating shift register. The control logic is understood to be built into the symbol.

First In First Out (FIFO) Memories

The FIFO is also a sequential access memory formed by an arrangement of shift registers. There is an important difference between a conventional register and a FIFO memory register. In a conventional register, a data bit moves through the register only as new data bits are entered; but in a FIFO register, a data bit immediately goes through the register to the rightmost bit location that is empty. However, when a data bit is shifted out by a shift pulse, the data bits remaining in the register automatically move to the next position towards the output. This is illustrated in Tables 14.1 and 14.2.

Table 14.1 Conventional shift register

Input	X	X	X	X	Output
0	0	X	X	X	→
1	1	0	X	X	→
1	1	1	0	X	→
0	0	1	1	0	→

Table 14.2 FIFO shift register

Input	–	–	–	–	Output
0	–	–	–	0	→
1	–	–	1	0	→
1	–	1	1	0	→
0	0	1	1	0	→

The FIFO memory is similar to the recirculating shift register memory in that, the order in which the data words are entered at 'Data in' is the same as the order in which they are read out at 'Data out'. In other words, the first word that is written in is the first word that is read out; hence the name FIFO. There are two important differences, however, between a FIFO and the recirculating shift register memory. First, in a FIFO, the output data are not recirculated; once the output data are shifted out they are lost. Second, in a

FIFO, the operation of shifting data into the memory is completely independent from the operation of shifting data out of the memory. In fact, the rate at which data are shifted in, is usually much different from the rate at which they are shifted out.

Applications

One important application area of the FIFO is the case in which two systems of different data rates must communicate. Data can be entered into a FIFO at one rate and be put out at another rate. An example of this is the data transfer from a computer to a printer. The computer can send data to the printer at a much more rapid rate than the printer can accept it and print it out. A FIFO can act as a data-rate buffer between the computer and the printer by accepting data form the computer at a faster rate and storing it; the data are then shifted out to the printer at a slow rate. The FIFO can also be used as a data-rate buffer for the transmission of data from a relatively slow device like a keyboard to a much faster device like a computer. In this case, the FIFO accepts data from the keyboard at a slow rate and stores them. The data are then shifted out to the computer at a faster rate. In this way, the computer can be performing other tasks while the FIFO is slowly being filled with data. Other applications are: (a) data input at an unsteady rate can be stored and retransmitted at a constant rate by using a FIFO; (b) data output at a steady rate can be stored and then outputted in even bursts; (c) data received in bursts can be stored and retransmitted into a steady-rate output. Figure 14.18 shows the use of the FIFO as data-rate buffers.

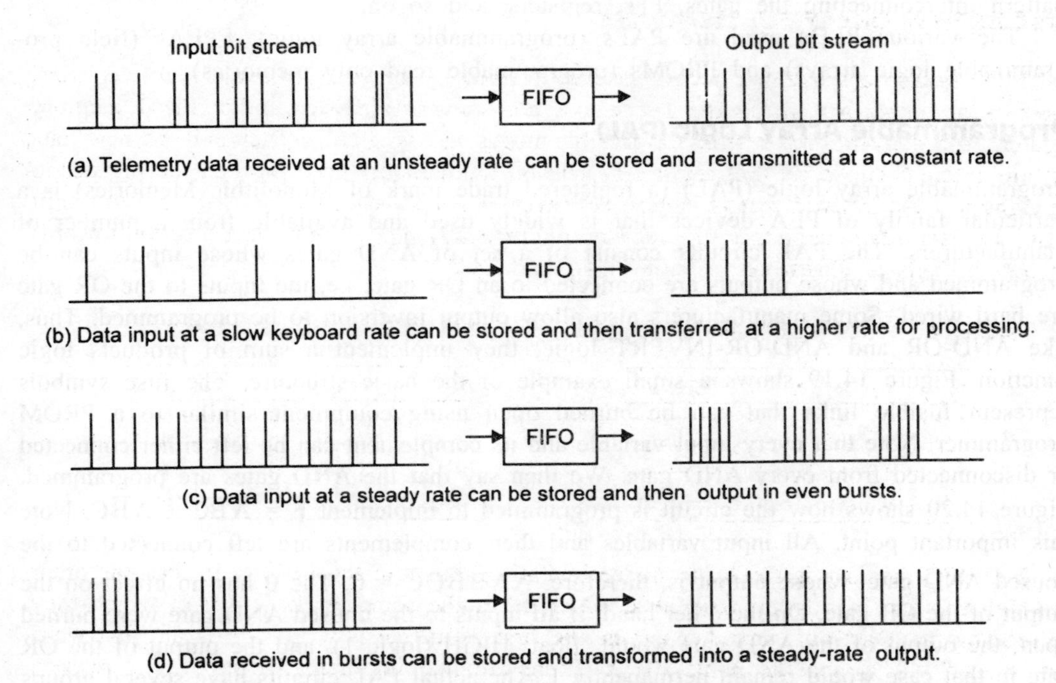

(a) Telemetry data received at an unsteady rate can be stored and retransmitted at a constant rate.

(b) Data input at a slow keyboard rate can be stored and then transferred at a higher rate for processing.

(c) Data input at a steady rate can be stored and then output in even bursts.

(d) Data received in bursts can be stored and transformed into a steady-rate output.

Fig. 14.18 The FIFO used in data-rate buffering applications.

14.8 PROGRAMMABLE LOGIC DEVICES

Logic designers have a wide range of standard ICs available to them with numerous logic functions and logic circuit arrangements on a chip. In addition, these ICs are available from many manufacturers and at a reasonably low cost. For these reasons, designers have been interconnecting standard ICs to form an almost endless variety of different circuits and systems.

However, there are some problems with circuit and system designs that use only standard ICs. Some system designs might require hundreds or thousands of these ICs. Such large numbers of ICs require not only a large circuit board space but also a great deal of time and cost in inserting, soldering, and testing the ICs. The programmable logic devices (PLDs) present the designers with a way to replace a number of standard ICs with a single IC, thereby reducing chip count and cost. They are used to implement specific logic functions. The principal advantage of PLDs is that in many applications, they can replace a number of other circuits. A PLD is an IC that contains a large number of gates, FFs, and registers that are interconnected on the chip. Many of the connections, however, are fusible links that can be broken. The IC is said to be programmable because the specific function of the IC for a given application is determined by the selective breaking of some of the interconnections while leaving others intact. The 'fuse blowing' process can be done either by the manufacturer in accordance with the customer's instructions, or by the customer himself. This process is called 'programming' because it produces the desired circuit pattern interconnecting the gates, FFs, registers, and so on.

The various PLDs used are PALs (programmable array logic), FPLAs (field programmable logic arrays) and PROMs (programmable read only memories).

Programmable Array Logic (PAL)

Programmable array logic (PAL) (a registered trade mark of Monolithic Memories) is a particular family of PLA devices that is widely used and available from a number of manufacturers. The PAL circuits consist of a set of AND gates whose inputs can be programmed and whose outputs are connected to an OR gate, i.e. the inputs to the OR gate are hard wired. Some manufacturers also allow output inversion to be programmed. Thus, like AND-OR and AND-OR-INVERT logic, they implement a sum of products logic function. Figure 14.19 shows a small example of the basic structure. The fuse symbols represent fusible links that can be burned open using equipment similar to a PROM programmer. Note that every input variable and its complement can be left either connected or disconnected from every AND gate. We then say that the AND gates are programmed. Figure 14.20 shows how the circuit is programmed to implement $F = \bar{A}BC + A\bar{B}C$. Note this important point. All input variables and their complements are left connected to the unused AND gate, whose output is, therefore, $A\bar{A}B\bar{B}C\bar{C} = 0$. The 0 has no effect on the output of the OR gate. On the other hand, if all inputs to the unused AND gate were burned open, the output of the AND gate would 'float' HIGH (logic 1), and the output of the OR gate in that case would remain permanently 1. The actual PAL circuits have several groups of AND gates, each group providing inputs to separate OR gates.

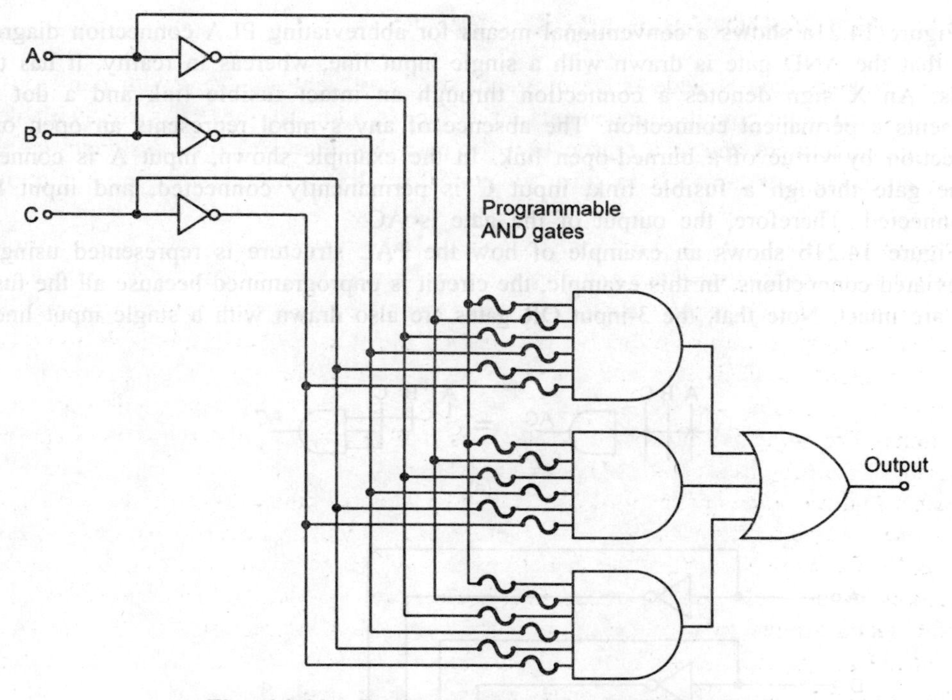

Fig. 14.19 Basic structure of a PAL circuit.

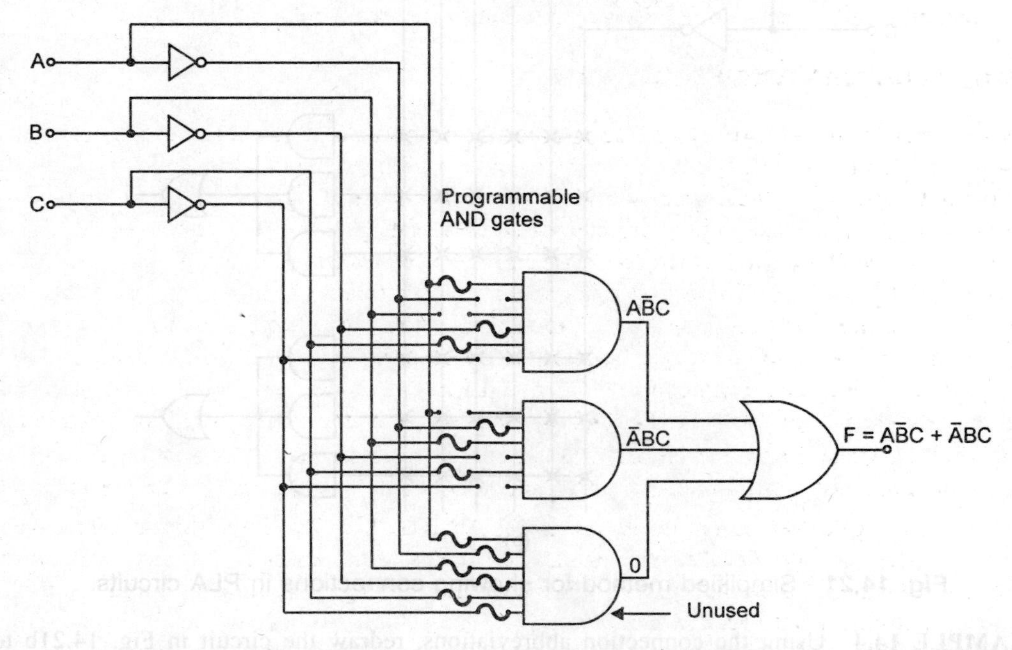

Fig. 14.20 The circuit of Fig. 14.19 programmed to implement $F = A\bar{B}C + \bar{A}BC$.

Figure 14.21a shows a conventional means for abbreviating PLA connection diagrams. Note that the AND gate is drawn with a single input line, whereas in reality, it has three inputs. An X sign denotes a connection through an intact fusible link and a dot sign represents a permanent connection. The absence of any symbol represents an open or no connection by virtue of a burned-open link. In the example shown, input A is connected to the gate through a fusible link, input C is permanently connected, and input B is disconnected. Therefore, the output of the gate is AC.

Figure 14.21b shows an example of how the PAL structure is represented using the abbreviated connections. In this example, the circuit is unprogrammed because all the fusible links are intact. Note that, the 3-input OR gates are also drawn with a single input line.

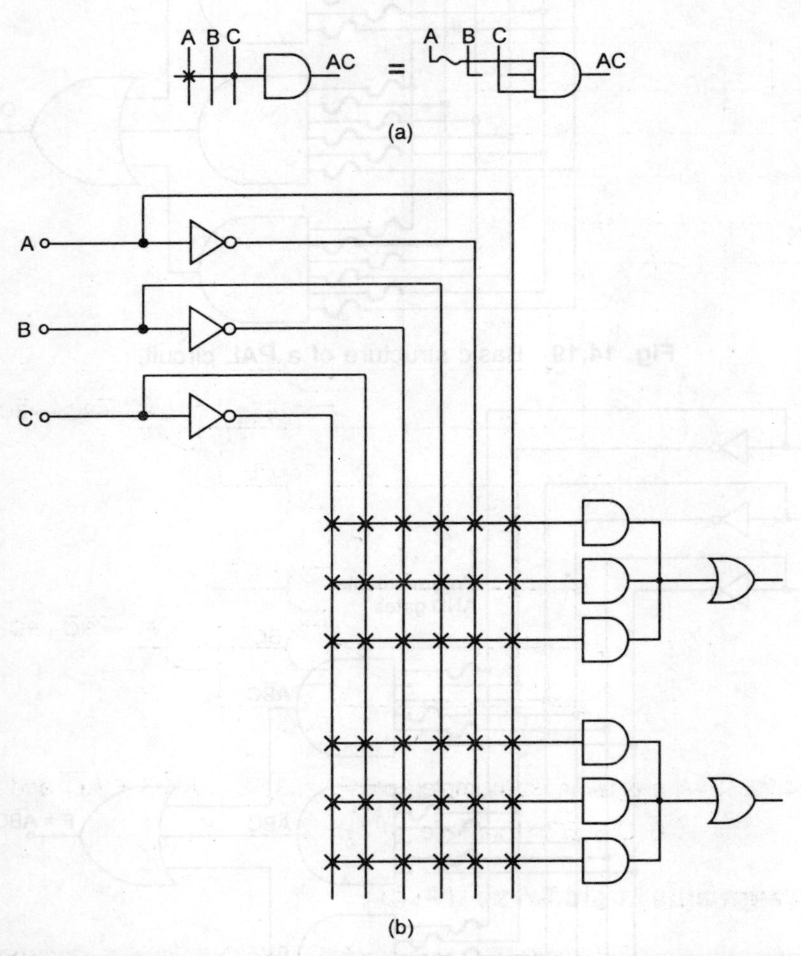

Fig. 14.21 Simplified method for showing connections in PLA circuits.

EXAMPLE 14.4 Using the connection abbreviations, redraw the circuit in Fig. 14.21b to show how it can be programmed to implement $F_1 = \overline{A}BC + A\overline{C} + A\overline{B}C$ and $F_2 = \overline{A}\,\overline{B}\,\overline{C} + BC$.

Solution

The redrawn circuit to implement the given functions is shown in Fig. 14.22. Note that one unused AND gate has all its links intact. Such a diagram is sometimes called the *fuse map*.

An example of an actual PAL IC is the PAL 18L8A from Texas Instruments. It is manufactured using low power Schottky technology and has ten logic inputs and eight output functions. Each output OR gate is hard wired to seven AND gate outputs and, therefore, it can generate functions that include up to seven terms. An added feature of this particular PAL is that six of the eight outputs are fed back into, AND array, where they can be connected as inputs to any AND gate. This makes the device very useful in generating all sorts of combinational logic.

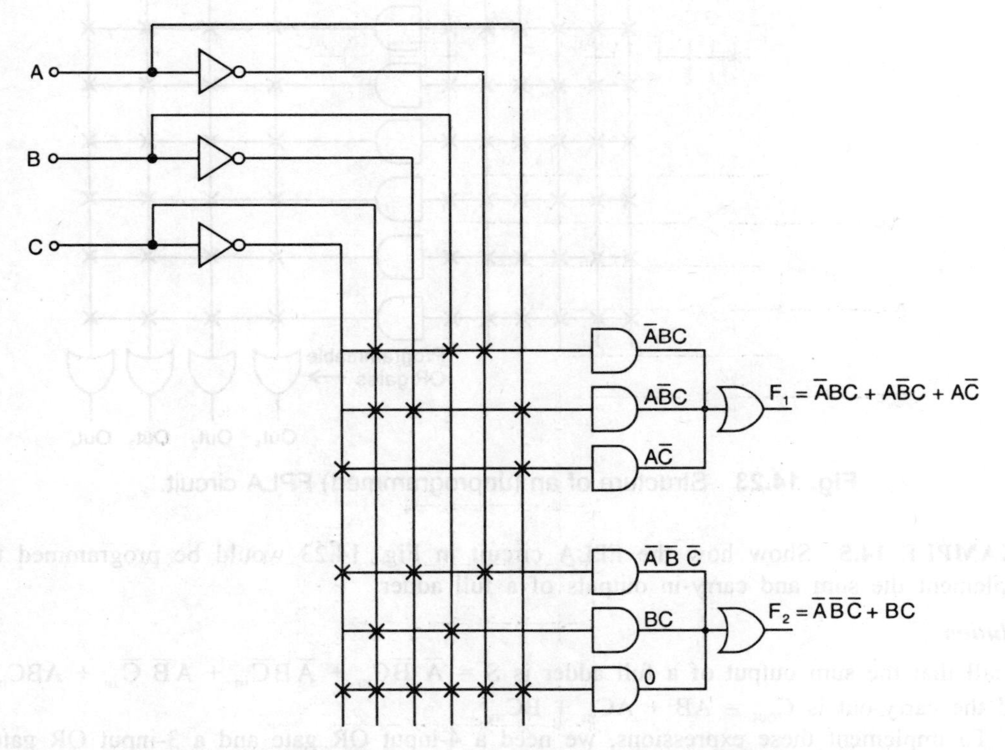

Fig. 14.22 PLA programmed to implement $F_1 = \bar{A}BC + A\bar{B}C + A\bar{C}$ and $F_2 = \bar{A}\ \bar{B}\ \bar{C} + BC$ (Example 14.4).

Field Programmable Logic Array (FPLA)

The FPLA represents another type of programmable logic but with a slightly different architecture. The FPLA combines the characteristics of the PROM and the PAL by providing both a programmable OR array and a programmable AND array. This feature makes it the most versatile of the three PLDs. However, it has some disadvantages. Because it has two sets of fuses, it is more difficult to manufacture, program and test it than a PROM or PAL. Figure 14.23 demonstrates the FPLA structure, with every fusible link intact.

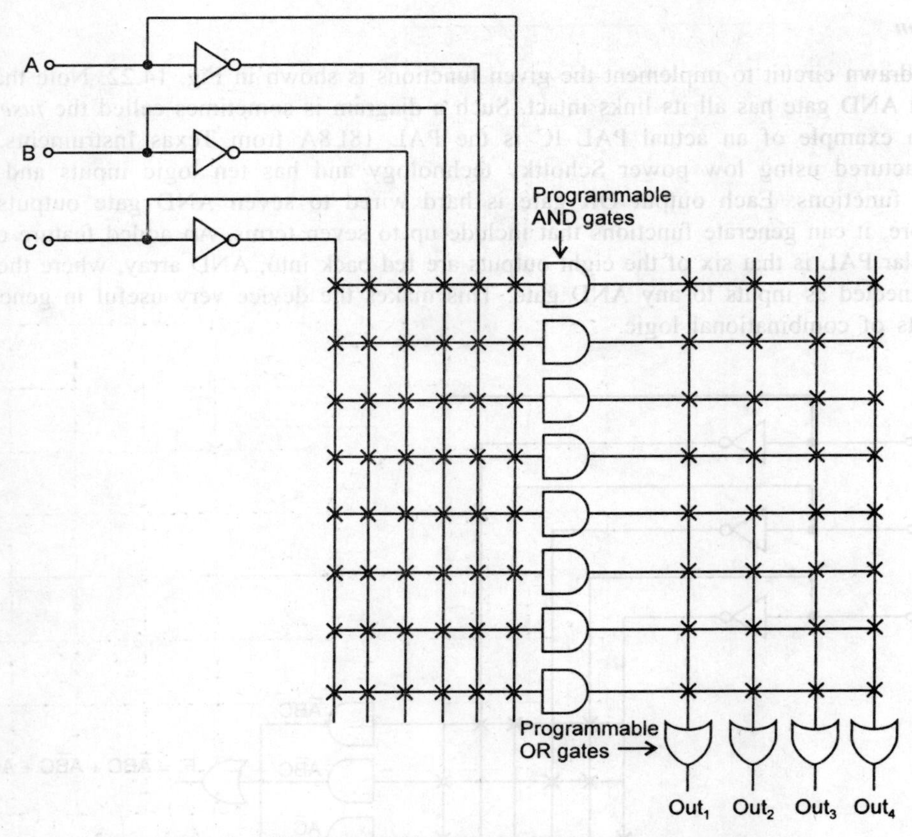

Fig. 14.23 Structure of an (unprogrammed) FPLA circuit.

EXAMPLE 14.5 Show how the FPLA circuit in Fig. 14.23 would be programmed to implement the sum and carry-in outputs of a full adder.

Solution

Recall that the sum output of a full adder is $S = \overline{A}\,\overline{B}C_{in} + \overline{A}B\overline{C}_{in} + A\overline{B}\,\overline{C}_{in} + ABC_{in}$ and the carry-out is $C_{out} = AB + AC_{in} + BC_{in}$

To implement these expressions, we need a 4-input OR gate and a 3-input OR gate. Since the inputs to the OR gates of the FPLA can be programmed, we can implement the given expressions as shown in Fig. 14.24.

An example of the actual FPLA is the Texas Instruments TIFPLA 840, which is specified to be a 14 × 32 × 6 FPLA. It has 14 input variables, 32 AND gates to generate products of these variables and 6 OR gates that can output any combination of the 32 AND gates. This IC also includes a programmable output polarity feature that permits the designer the option of inverting any of the outputs. This is illustrated in Fig. 14.25 where an X-OR gate and a polarity fuse have been added to the OR gate output. The X-OR gate will pass the OR gate output to Q_3 with no inversion if the fuse is intact (logic 0). If the fuse is blown (logic 1), the X-OR gate will invert the OR gate output. This programmable output polarity is available on many PAL and FPLA devices.

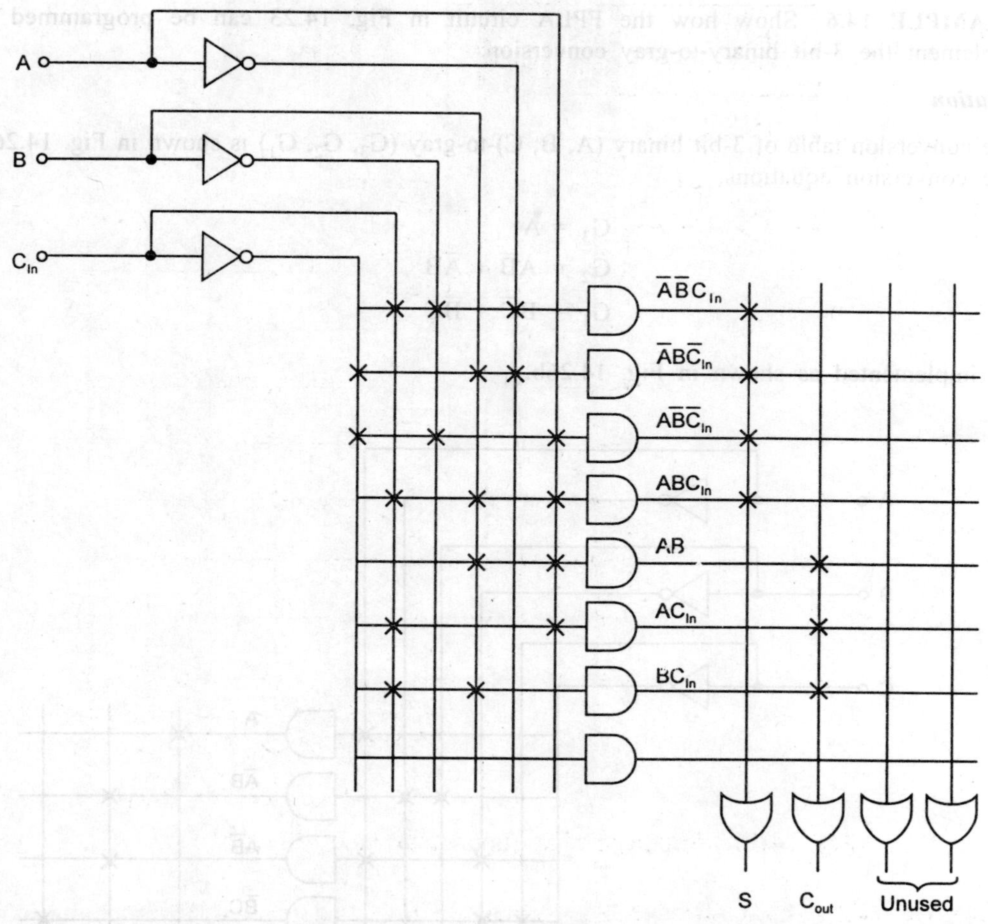

Fig. 14.24 FPLA circuit programmed to implement the sum and carry terms of a full adder (Example 14.5).

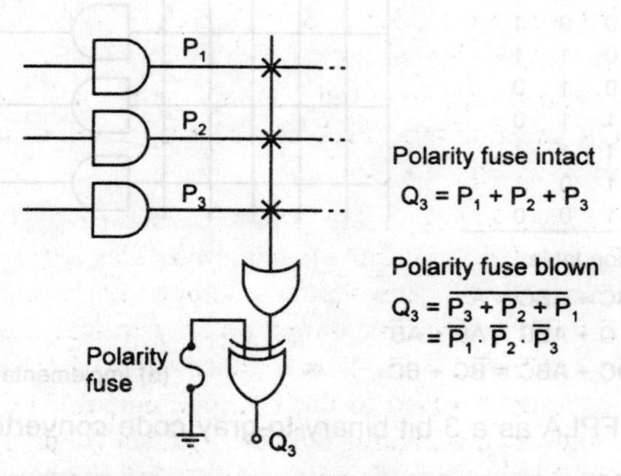

Fig. 14.25 PLD with a programmable polarity feature.

EXAMPLE 14.6 Show how the FPLA circuit in Fig. 14.23 can be programmed to implement the 3-bit binary-to-gray conversion.

Solution

The conversion table of 3-bit binary (A, B, C)-to-gray (G_3, G_2, G_1) is shown in Fig. 14.26a. The conversion equations,

$$G_3 = A$$
$$G_2 = A\overline{B} + \overline{A}B$$
$$G_1 = B\overline{C} + \overline{B}C$$

are implemented as shown in Fig. 14.26b.

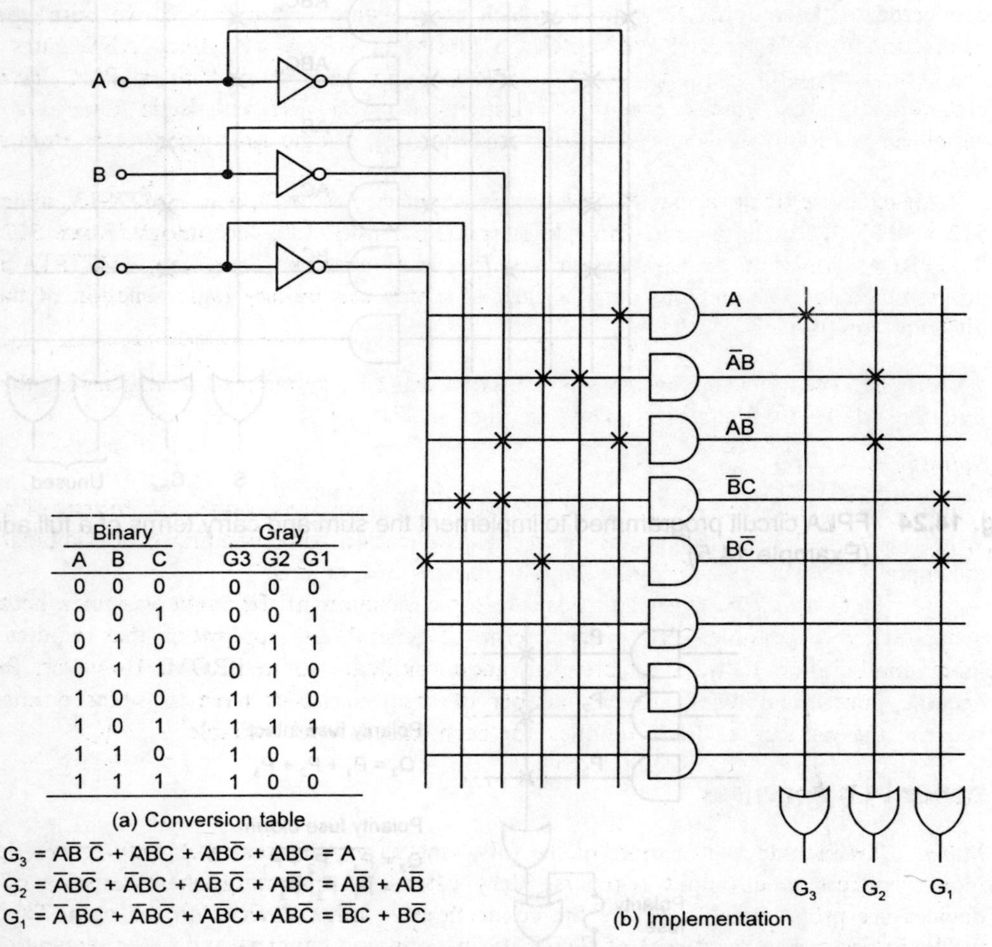

Binary			Gray		
A	B	C	G3	G2	G1
0	0	0	0	0	0
0	0	1	0	0	1
0	1	0	0	1	1
0	1	1	0	1	0
1	0	0	1	1	0
1	0	1	1	1	1
1	1	0	1	0	1
1	1	1	1	0	0

(a) Conversion table

$$G_3 = A\overline{B}\,\overline{C} + A\overline{B}C + AB\overline{C} + ABC = A$$
$$G_2 = \overline{A}B\overline{C} + \overline{A}BC + A\overline{B}\,\overline{C} + A\overline{B}C = A\overline{B} + A\overline{B}$$
$$G_1 = \overline{A}\,\overline{B}C + \overline{A}B\overline{C} + A\overline{B}C + AB\overline{C} = \overline{B}C + B\overline{C}$$

Fig. 14.26 Use of FPLA as a 3-bit binary-to-gray code converter (Example 14.6).

Programmable ROM (PROM)

A programmable ROM can be viewed as a type of programmable logic array and thus used for that purpose. The address inputs to the PROM serve as logic variable inputs and the data output as the node where the output of a logic function is realized. For example, stating that a 1 is stored at address 1001 is the same as stating that the logic function being implemented equals 1 when the input combination is $A\overline{B}\overline{C}D$. In both the cases, the output will be a 1 when the input is 1001. When we regard a PROM as a PLA, we realize that the AND gates are not programmed. In effect, an AND gate is already in place for every possible combination of the inputs corresponding to every possible address of the PROM. Therefore, to program a PROM as a PLA, we must have a truth table that specifies the value of the function being implemented for every possible combination of the inputs. For each combination where F = 1, we leave the output of the corresponding AND gate connected to the output OR gate. For each combination where F = 0, we burn open the connection to the OR gate. We see that a PROM is a PLA with fixed AND gates and a programmable OR gate. An M × N PROM can be regarded as a PLA having N programmable OR gates, capable of implementing N different logic functions of M variables. A PROM is ideally suited for implementing a logic function directly from a truth table.

An example of an actual PROM that is often used as a PLD is AM27S13, which is a 512 × 4 PROM manufactured using high speed Schottky TTL technology. Since $512 = 2^9$, this PROM has nine address inputs and four data outputs. Thus, the AM27S13 can be programmed to generate four outputs each of which can be any logic function of the nine different inputs.

EXAMPLE 14.7 Show how an 8 × 1 PROM can be programmed to implement the logic function whose truth table is shown in Fig. 14.27.

Solution

Figure 14.27 shows the programmed PROM in the simplified connection format of a PLA. A logic 1 or a 0 is stored at every address combination corresponding to a combination of the input variables for which the function equals a 1 or a 0.

The PROM can generate any possible logic function of the input variables because it generates every possible AND product term. In general, any application that requires every input combination to be available is a good candidate for a PROM. However, PROMs become impractical when a large number of input variables have to be accommodated, because the number of fuses doubles for each added input variable.

Other PLD Features

Many PLDs include one or more of the following as part of their architecture: FFs, latches, input registers, and output registers. Very often, the operating characteristics of these devices are programmable, as are the connections to other devices on the chip. This gives the logic designer a great deal of flexibility in designing counters and other sequential logic circuits. This type of PLD is sometimes called a *programmable logic sequencer*.

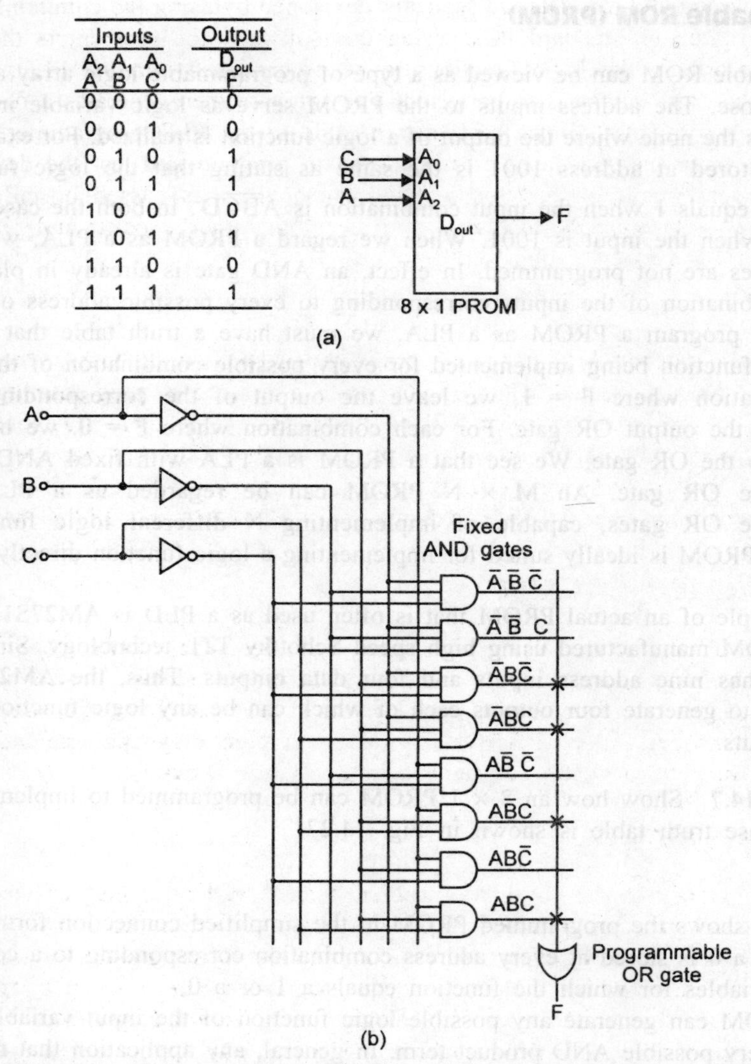

Inputs			Output
A_2	A_1	A_0	D_{out}
A	B	C	F
0	0	0	0
0	0	1	0
0	1	0	1
0	1	1	1
1	0	0	0
1	0	1	1
1	1	0	0
1	1	1	1

Fig. 14.27 Programming a PROM to implement a truth table (Example 14.7).

Programming

When PLDs were first introduced, the logic designer would develop a *fuse map* that showed which fuses to blow. The manufacturer would then program the device according to the fuse map, test it, and return it the designer. In recent years, the availability of relatively inexpensive programming equipment has made it convenient for users to program their own PLDs. There are universal programmers in the market that can program the most common PROMs, PALs, and FPLAs. The device to be programmed is plugged into a socket on the programmer and the programmer programs and tests the device according to data that have been supplied by the user.

The programming and test data are typically developed by using the commonly available software that will run on standard PCs. Using this software, the user enters the data into the computer describing the logic functions to be programmed into the PLD, as well as information on how the device is to be tested. The software then generates a fuse map and the test data in a form that can be sent over a cable to the PLD programmer's memory. Once the programmer has the data, it can proceed to program and test the device. When finished, the programmer will indicate whether the device has passed or failed the test procedure. If it passes, it can be removed from the programmer's socket and placed in the prototype circuit for further testing.

Erasable PLDs

The PLDs we have been talking about are programmed by blowing fuses. Once a fuse is blown, it cannot be reconnected. Thus, if you make a mistake in programming or if you want to change the design, the device will no longer be useful. This drawback has been addressed by several manufacturers who have developed PLDs that can be erased and programmed over and over. These are called *erasable programmable logic devices* (EPLDs). These devices are programmed and erased much like EPROMs and EEPROMs.

14.9 MAGNETIC MEMORIES

So far we have discussed semiconductor memory devices whose cells store data in the form of electrical charge or voltage. We will discuss now some storage devices whose basic storage mechanism is magnetic rather than electronic. The characteristic common to all magnetic storage devices is their non-volatility. Magnetic core, magnetic tape and disk, hard drives, floppy disk, etc. are some of the magnetic memory devices.

Magnetic Core Memory

Magnetic core memory is a non-volatile random access read/write memory that was the predecessor to semiconductor memory. Almost all early computers used magnetic core as their primary internal memory before the wide availability of semiconductor memory. The basic magnetic core memory cell is a small dough-nut shaped core made of a ferromagnetic material. These cores are called ferrite cores and typically have a diameter of 0.05 inch. A small wire is threaded through the centre of the core. When a current pulse is passed through this wire, a magnetic flux is set up in the core in a direction that depends on the direction of the current. Because of the core's magnetic retentivity, it stays magnetized even after the current pulse is terminated. In other words, it is non-volatile. The two directions of magnetization are used to represent 1 and 0, respectively.

Stored data are read from a core by magnetizing it in the 0 direction (i.e. writing a 0) and using a second threaded wire as a sense wire. The size of the voltage induced in the sense wire will be greater if the core is initially in 1 state rather than in 0 state. The reading of a core is said to be destructive, since it always leaves it in 0 state. Thus, the data must be rewritten after a read operation.

Magnetic core memory systems have access times from 100 ns to 500 ns and can still be found in some old minicomputers and mainframes. But for all practical purposes, their large physical size and complex interface circuitry have made them virtually obsolete.

Magnetic Disk Memory

A magnetic disk is a flat, circular plate that is coated with a magnetic material. Binary data is stored on the disk by magnetizing tiny regiments of the surface, and data is read from the disk by sensing that magnetization. Magnetic disk memories are used primarily for peripheral or auxiliary memory and for mass storage, because their access times are slower than those of the semiconductor memories. They are less costly per bit of storage capacity than the semiconductor memories and they have become the dominant type of peripheral memory in computer systems of all sizes. The disks themselves range in size from a 2-inch diameter 'floppy' disk used with microcomputer systems to a 14-inch diameter hard disk used in larger systems. Floppy disks made of plastic are inexpensive and readily portable. Hard disks made of aluminum have much greater storage capacity than that of floppies and are used in complex and more expensive systems where higher speeds and larger capacities are required.

Read and write mechanisms

The basic principle of writing and reading a disk is the same whether the disk is small or large, hard or floppy. These devices use a magnetic surface moving past a read/write head to store and retrieve data.

A simplified diagram of the magnetic surface read/write operation is shown in Fig. 14.28. A data bit (1 or 0) is written on the magnetic surface by magnetizing a small

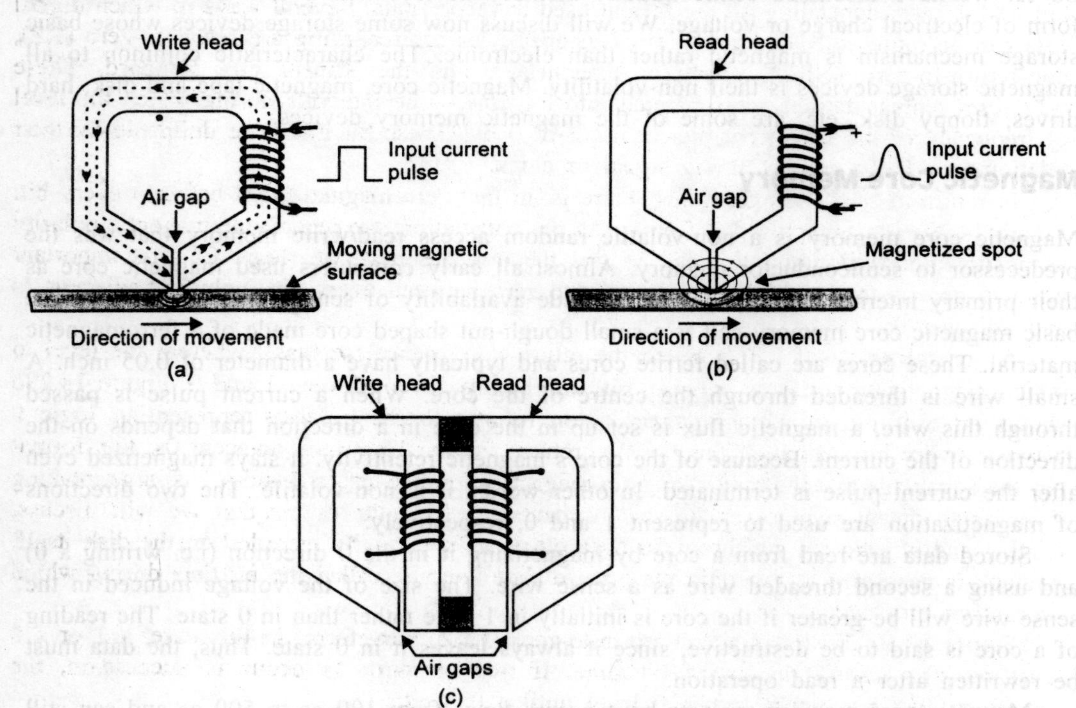

Fig. 14.28 Read/write function on a magnetic surface.

segment of the surface as it moves by the write head. The direction of the magnetic flux lines is controlled by the direction of the current pulse in the winding as shown in Fig. 14.28a. At the air gap in the write head, the magnetic flux takes a path through the surface of the storage device. This magnetizes a tiny spot on the surface in the direction of the field. A magnetized spot of one polarity represents a binary 1, and one of the opposite polarity represents a binary 0. Once a spot on the surface is magnetized, it remains there until written over with an opposite magnetic field. Thus, magnetic storage is non-volatile. To write a sequence of 1s and 0s, the disk is rotated and the winding is driven by a sequence of current pulses, each pulse having one direction or the other. Thus, 0s and 1s are stored in a sequence around a circular path on the disk. To read data on a disk, a read head similar in construction to the write head is used. When the magnetic surface passes a read head, the magnetized spots produce magnetic fields in the read head which induce voltage pulses in the winding. The polarity of these pulses depends on the direction of the magnetization spots and indicates whether the stored bit is a 1 or a 0. This is illustrated in Fig. 14.28b. Very often, the read and write heads are combined into a single unit as shown in Fig. 14.28c.

Magnetic Recording Formats

Our description of bit storage as a sequence of discrete dots having one magnetic polarity or the other implied that the regions between the dots were unmagnetized. That may or may not be the case, depending on the format used to record data. Several ways in which digital data can be represented for purposes of magnetic surface recording are return-to-zero (RZ), non-return-to-zero (NRZ), bi-phase, Manchester, and the Kansas city standard. These waveform representations are separated into bit times—the intervals during which the level or frequency of the waveform indicates a 1 or 0 bit. These bit times are definable by their relation to a basic system timing signal or clock.

In return-to-zero (RZ) recording, there is, in fact zero magnetization between every bit. Figure 14.29a illustrates the RZ format. As per convention, we regard one magnetic polarity as positive (corresponding to logic 1) and the other as negative (logic 0). The important point to note is that magnetization returns to zero between every bit, including adjacent 1s and adjacent 0s.

A generalization of RZ is called the return-to-reference (or bias), wherein the level to which magnetization returns between bits can be any level between 1 and 0. Figure 14.29b shows an example in which magnetization returns to the polarity representing logic 0 between every bit. Although there is no 'return' involved between adjacent 0s, this format has the advantage that the total change in magnetization between a 0 and a 1 is large, making it easier to detect such cases. In this format as well as the others that we will discuss, reading and writing are synchronized by a clock signal, usually recorded on the disk itself. A clock is necessary to ascertain the precise time interval, called the *bit time* during which successive bits occur.

Figure 14.30 illustrates a non-return-to-zero (NRZ) waveform. In this case, a 1 or a 0 level remains during the entire bit time. If two or more 1s occur in succession, the waveform does not return to the 0 level until a 0 occurs.

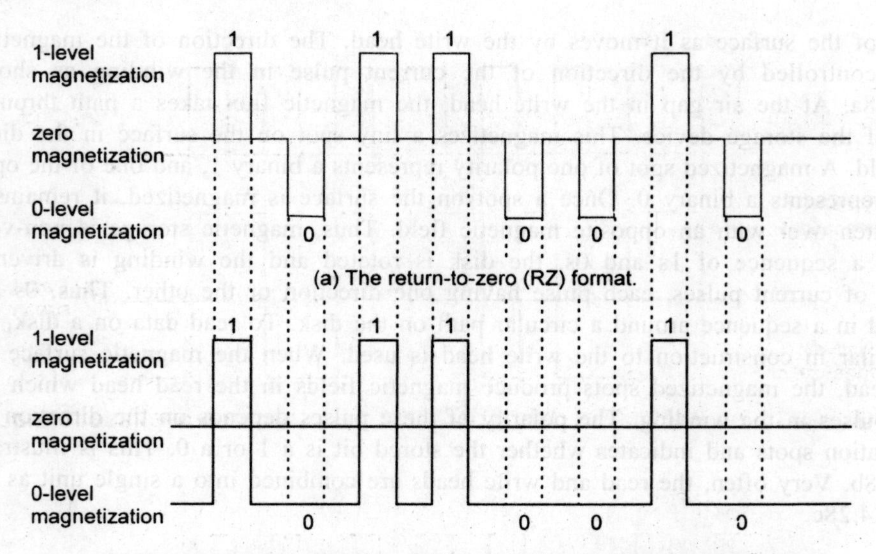

Fig. 14.29 Return-to-zero and return-to-bias formats for recording data.

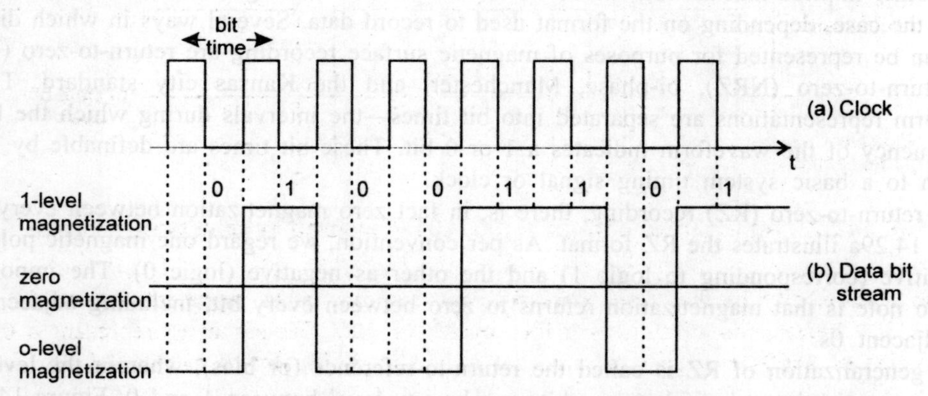

Fig. 14.30 An example of data (01001101) recorded in an NRZ format.

Figure 14.31 is an illustration of a bi-phase waveform. In this type, a 1 is a HIGH level for the first half of a bit time and a LOW level for the second half, so a HIGH-to-LOW transition occurring in the middle of a bit time is interpreted as a 1. A 0 is represented by a LOW level during the first half of a bit time followed by a HIGH level during the second half, so a LOW-to-HIGH transition in the middle of a bit time is interpreted as a 0.

Manchester is another type of phase encoding in which a HIGH-to-LOW transition at the start of a bit time represents a 0 and no transition represents a 1. Figure 14.32 illustrates a Manchester waveform.

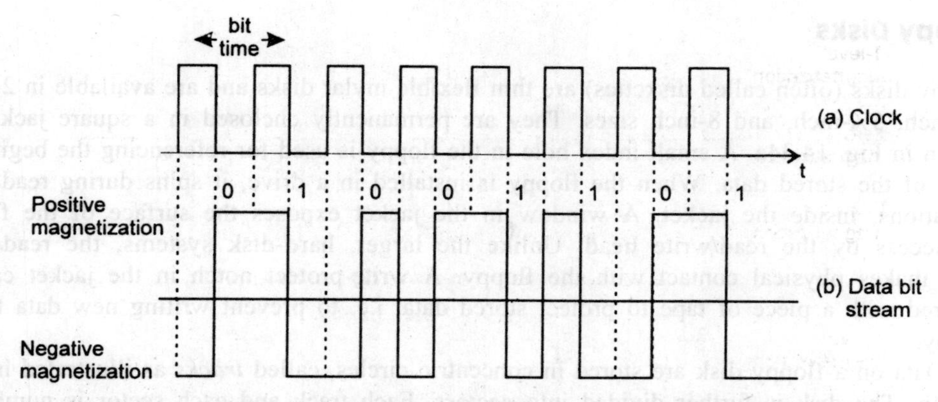

Fig. 14.31 An example of data (01001101) recorded in a bi-phase format.

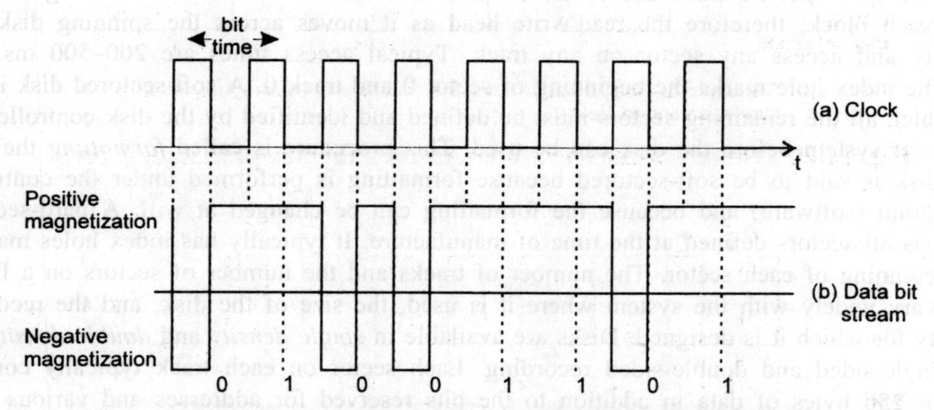

Fig. 14.32 An example of data (01001101) recorded in Manchester format.

The Kansas city method uses two different frequencies to represent 1s and 0s. The standard 300 bits/second version uses four cycles of 1200 Hz signal to represent a 0 and eight cycles of 2400 Hz signal to represent a 1. This is illustrated in Fig. 14.33.

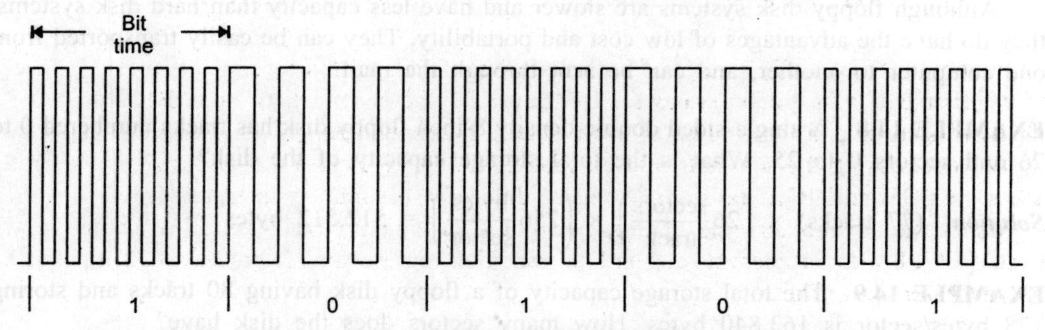

Fig. 14.33 An example of data (10101) recorded in Kansas city format.

Floppy Disks

Floppy disks (often called diskettes) are thin flexible mylar disks and are available in 2-inch, 3½-inch, 5¼-inch, and 8-inch sizes. They are permanently enclosed in a square jacket, as shown in Fig. 14.34a. A small index hole in the floppy is used for referencing the beginning point of the stored data. When the floppy is installed in a drive, it spins during read/write operations, inside the jacket. A window in the jacket exposes the surface of the floppy for access by the read/write head. Unlike the larger, hard-disk systems, the read/write head makes physical contact with the floppy. A write-protect notch in the jacket can be covered with a piece of tape to protect stored data, i.e. to prevent writing new data to the floppy.

Data on a floppy disk are stored in concentric circles, called *tracks* as illustrated in Fig. 14.34b. The disk is further divided into *sectors*. Each track and each sector is numbered, therefore, the portion of a particular track within a particular sector defines one *block* of data having a specific track and sector address. The address is stored at the beginning of each such block; therefore the read/write head as it moves across the spinning disk, can identify and access any sector on any track. Typical access times are 200–500 ms.

The index hole marks the beginning of sector 0 and track 0. A soft-sectored disk is one for which all the remaining sectors must be defined and identified by the disk controller and computer system before the disk can be used. This procedure is called *formatting* the disk. The disk is said to be soft-sectored because formatting is performed under the control of a program (software) and because the formatting can be changed at will. A hard-sectored disk has all sectors defined at the time of manufacture. It typically has index holes marking the beginning of each sector. The number of tracks and the number of sectors on a floppy disk vary widely with the system where it is used, the size of the disk, and the recording density for which it is designed. Disks are available in *single density* and *double density* and for single-sided and double-sided recording. Each sector on each track typically contains 128 or 256 bytes of data in addition to the bits reserved for addresses and various other identification and error-checking functions.

Floppy disk capacities typically range from 100 K to a few MB. Floppies have access times about 10 times larger and data rates about 10 times slower than those of hard disks. The average access time of floppy drives is 100 to 500 ms, and the average data transfer rates range from 250K/s to 1Mbits/s. When inserted into the disk drive unit, a floppy disk is rotated at a fixed speed of 300 to 360 rpm, which is much slower than the speed of the hard disks.

Although floppy disk systems are slower and have less capacity than hard disk systems, they do have the advantages of low cost and portability. They can be easily transported from one computer to another, and can be sent through the mail.

EXAMPLE 14.8 A single-sided double-density 8-inch floppy disk has tracks numbered 0 to 76 and sectors 0 to 25. What is the total storage capacity of the disk?

Solution $(77 \text{ tracks}) \times \left(26 \dfrac{\text{sectors}}{\text{track}}\right) \times \left(256 \dfrac{\text{bytes}}{\text{sector}}\right) = 512{,}512 \text{ bytes}$

EXAMPLE 14.9 The total storage capacity of a floppy disk having 80 tracks and storing 128 bytes/sector is 163,840 bytes. How many sectors does the disk have?

Solution

163,840 = (80) × (number of sectors) × (128). Therefore, the number of sectors = 16.

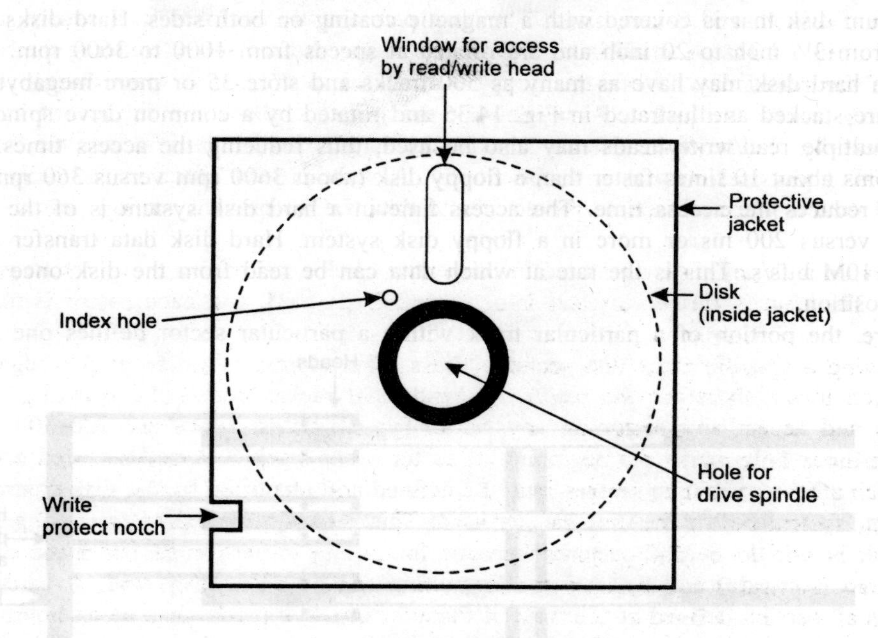

(a) The floppy disk

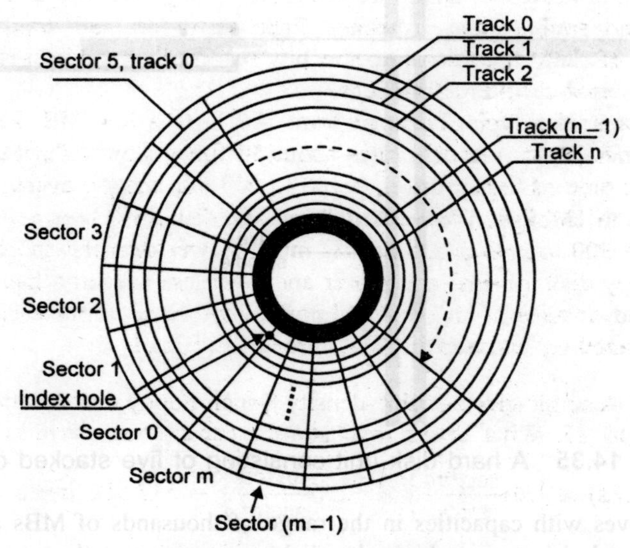

(b) Tracks and sectors on a floppy disk

Fig. 14.34 The floppy disk.

Hard Disk Systems

In a hard disk system, the data are recorded in concentric tracks and sectors on an aluminum disk that is covered with a magnetic coating on both sides. Hard disks come in sizes from 3½ inch to 20 inch and are rotated at speeds from 1000 to 3600 rpm. A single 14-inch hard disk may have as many as 500 tracks and store 35 or more megabytes. Hard disks are stacked as illustrated in Fig. 14.35 and rotated by a common drive spindle. Note that, multiple read/write heads may also be used, thus reducing the access times. A hard disk spins about 10 times faster than a floppy disk (about 3600 rpm versus 360 rpm) which further reduces the access time. The access time in a hard disk system is of the order of 20 ms versus 200 ms or more in a floppy disk system. Hard disk data transfer rates are 1M to 10M bits/s. This is the rate at which data can be read from the disk once the head is in position.

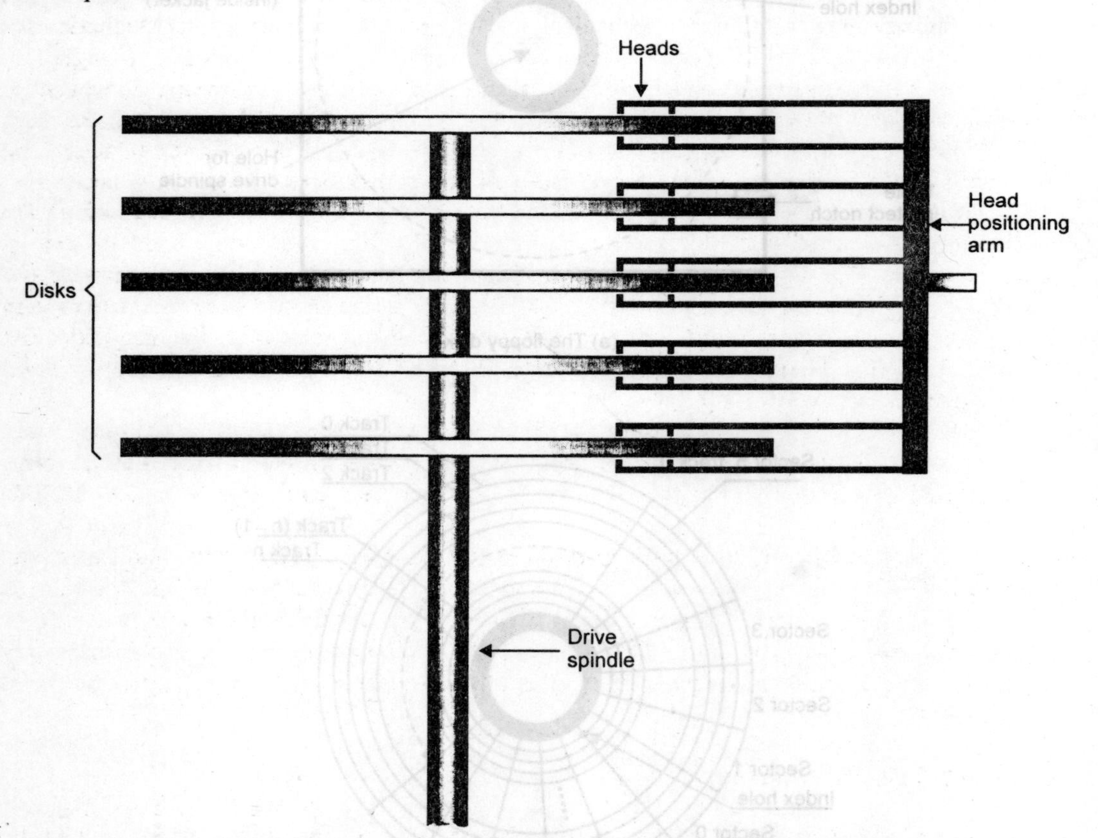

Fig. 14.35 A hard disk unit consisting of five stacked disks.

Multi-disk drives with capacities in the range of thousands of MBs are available. Most hard disks are fixed drives in which the disks are permanently mounted on the drive mechanism and cannot be removed in normal usage. Some hard disk systems use removable disks called *disk cartridges* or *disk packs*.

The high speed of rotation of hard disks produces a cushion of air a few micro inches above the disk surface. The read/write head floats on this cushion, such that it never touches the disk surface. It is thus called the *flying head*. This prevents wearing down of the disk surface and the read/write head. Since the clearance is very small, smoke or dust particles pose a serious threat of contamination. For this reason, most hard disk drives blow filtered air across the disks to keep them relatively clean. The disks are sealed in protective cases and cannot ordinarily be handled by a user.

Magnetic Tape Memory

The primary role of the magnetic tape is mass storage and back-up—the duplication and preservation of data stored in other media like hard disks. Back-up is important in many systems because, critical records, computer programs and/or scientific data are susceptible to loss through power failures, mechanical malfunctions or human error. Magnetic tape storage is non-volatile and has immense capacity at relatively low cost per bit. A single tape may be of very long length and can store thousands of megabyte. The tape is wound on reels that must be unwound to read or write at a specific location. The principal drawback of the tape memory is the long access time required to find a specific block of data. This drawback is not a disadvantage when the tape is used as disk back-up. The data is not read from the tape very often, and when read, the tape will be read from the beginning. The technology of magnetic storage and read/write mechanism is quite similar to that of magnetic disks. The adjacent regions on the surface of the tape are magnetized with one polarity or the other to represent 1s and 0s, and read/write heads are used to sense or alter the magnetization. The recording formats discussed in connection with the magnetic disk memory can also be used for magnetic tape storage, the NRZ format being the most common.

Tapes used in large systems, typically, have nine parallel tracks and a read/write head for each track. Seven of the tracks are used for storing data in ASCII code and the other two for polarity and timing. Storage density on a single track can range from 200 to 1600 bits per inch.

Audio cassette tapes are sometimes used for auxiliary storage in small personal computer systems, although floppy disks have already displaced the tape in that role. A frequency modulation (FM) format is used for recording on the cassette tape, whereby 0s and 1s are represented by pulse trains having different frequencies. In the Kansas city standard FM, a 0 is represented by four cycles of a 1200 Hz signal and a 1 by eight cycles of a 2400 Hz signal.

Magnetic Bubble Memory

Magnetic bubble memory (MBM) is a semiconductor memory that stores binary data in the form of tiny magnetic domains (bubbles) on a thin film of magnetic material. It is constructed from a magnetic material (yitrium-iron garnet) that has the property whereby small cylindrically shaped domains called *bubbles* can be created by subjecting the material to a strong externally applied magnetic field. The presence of a magnetic bubble in a specific position represents a stored 1, and its absence represents a stored 0. Continuously changing magnetic fields are used to move the bubbles around in loops inside the magnetic material much like recirculating shift registers. The data circulates past a pick-up point where it is available to the outside world at the rate of typically 50,000 bits per second. Clearly, the MBMs are sequential access devices.

The MBMs are better suited for the external mass storage function that has been previously dominated by magnetic tape and disk storage. At present, the MBMs are about 100 times faster than the floppy disks but somewhat more costly because of the required support circuitry. As and when their cost comes down, we might see more and more MBM systems being used in place of the slower, less reliable floppy disk systems.

The MBMs are compact and dissipate very little power (typically 1 μW/bit). They are capable of storing large amounts of data. They are also non-volatile. When power goes off, the bubbles simply remain in their fixed positions. When power is restored, the bubbles start circulating again around their loops. Unlike other non-volatile semiconductor memories such as ROMs, PROMs, EPROMs and EEPROMs, the MBMs can be written into and read from with equal ease. Because of these features, the MBMs are competitive with semiconductor memories in many applications. The main disadvantage is that it takes much longer to get data into and out of an MBM compared to semiconductor memories, because MBMs are serially accessed rather than randomly accessed. The typical access times are in the range of 1 to 20 ms. Therefore, an MBM is not suitable for use as the main internal memory. Compared with non-volatile tape and disk memories, an MBM system has no moving parts and is, therefore, quieter and more reliable.

Major/Minor loop architecture

Loops in an MBM are in the form of continuous, elongated paths of Chevron patterns. The major/minor loop arrangement consists basically of one major loop and many minor loops as shown in Fig. 14.36.

The minor loops are essentially the memory cell arrays that store the data bits. The major loop is primarily a path to get data from the minor loops to the output during read and to get data from the input to the minor loops during write.

Five control functions are used in the read/write cycles of an MBM. These are generation, transfer, replication, annihilation, and detection.

The read cycle. Data are read from an MBM in blocks called *pages*. Basically a 'page' consists of a number of bits equal to the number of minor loops. For example, a typical MBM may have 256 minor loops, each of which is capable of storing 600 bits. In this case, the page size would be 256 bits.

Each bit in a given page occupies the same relative location in each minor loop. During read, all of the bits in a page of data are shifted to the transfer gates and on to the major loop at the same time. Then, they are serially shifted around the major loop to the replicator/annihilator where each bubble is 'stretched' by the replicator until it 'splits' into two bubbles. One of the replicated bubbles (or no bubble as the case may be) is transferred to the detector where the presence or absence of the bubble is sensed and translated to the appropriate logic level to represent a 1 or a 0. The other replicated bubble continues along the major loop and is transferred back on to the appropriate minor loop for storage. The process continues until each bit in the page is read. The replication/detection process results in a non-destructive read-out.

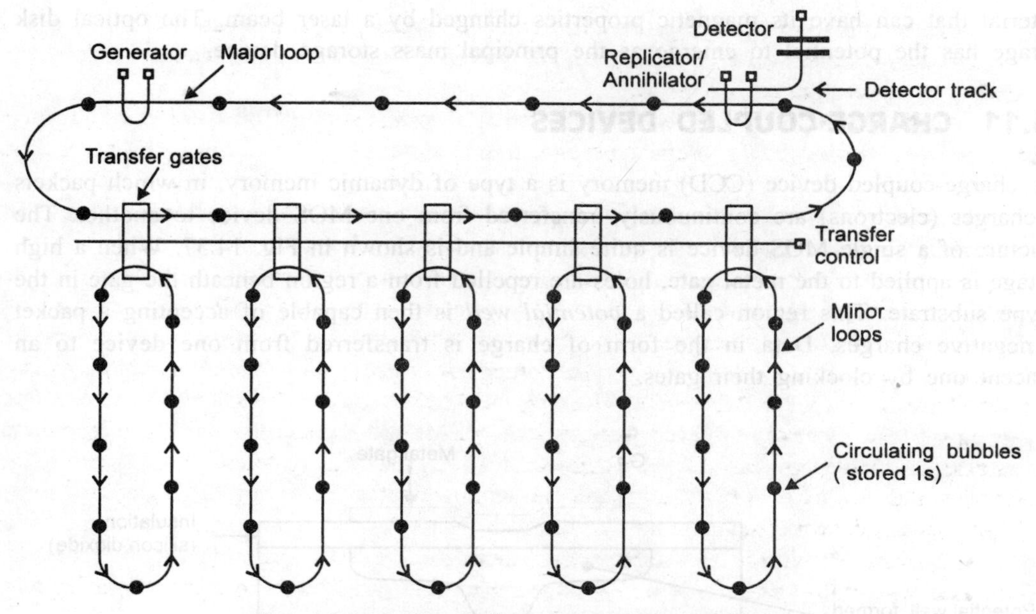

Fig. 14.36 Circulation of magnetic bubbles.

The write cycle. Before a new page of data can be written into an address, the data currently stored at that address must be annihilated. This is done by a destructive read-out in which the bubbles are not replicated. Now, the new page of data is produced by the generator one bit at a time. Each bit is injected on to the major loop until the entire page of data has been entered. It is then serially moved into position and transferred on to the minor loops for storage.

14.10 OPTICAL DISK MEMORY

The optical disk memory is the latest mass memory technology that has a promising future. Its operation is based on the reflection or scattering of an extremely narrow laser beam off a disk that has microscopic pits or bubbles representing logic 1s 'burned' on its surface. One strong point in favour of the optical disk memory is its high capacity—4½ inch disk with 650 MB capacity is readily available. Other advantages are its relatively low cost and its immunity to dust. Its access times and data transfer rates are comparable to the newest hard disk systems.

The optical disk systems are available in three basic types based on their writability. Disks that can only be read from are called optical ROM (OROM) or compact disc ROM (CD-ROM). They are used to store large fixed databases such as dictionaries or encyclopaedias. An optical disk that can be written to once is called a write-once read-many (WORM) disk. The CD-ROMs and WORMs cannot be erased. The read/write optical disk can be written and rewritten as often as desired, and therefore, operates like a magnetic hard disk. It uses a different disk surface from other types. Its surface is coated with a magnetic material that can have its magnetic properties changed by a laser beam. The optical disk storage has the potential to emerge as the principal mass storage device.

14.11 CHARGE-COUPLED DEVICES

The charge-coupled device (CCD) memory is a type of dynamic memory, in which packets of charges (electrons) are continuously transferred from one MOS device to another. The structure of a single MOS device is quite simple and is shown in Fig. 14.37. When a high voltage is applied to the metal gate, holes are repelled from a region beneath the gate in the P-type substrate. This region called a *potential well* is then capable of accepting a packet of negative charges. Data in the form of charge is transferred from one device to an adjacent one by clocking their gates.

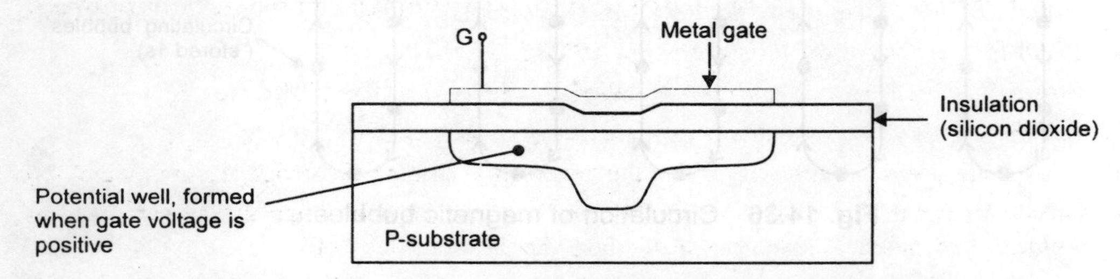

Fig. 14.37 Structure of a single MOS device in a CCD memory.

The CCD memory is inherently serial. Practical memories are constructed in the form of shift registers, each shift register being a line of CCDs. By controlling the timing of the clock signals applied to the shift registers, data can be accessed one bit at a time from a single register or several bits at a time from multiple registers. The principal advantage of the CCD memory is that, its single cell structure makes it possible to construct large capacity memories at low cost. On the other hand, like other dynamic memories, it must be periodically refreshed and driven by rather complex, multi-phase clock signals. Since data are stored serially, the average access time is long compared with the semiconductor RAM memory.

SUMMARY

- Memory is a means for storing binary words. Computer memories may be classified as main or peripheral.
- The main memory is an integral part of the computer. Being semiconductor memory, it is very fast.
- Peripheral memory, also called the *auxiliary* memory, is typically add-on memory having very large capacity. It is much slower than the main memory. It is also referred to as *mass memory*.

- Stored-program computers are those in which programs are stored as a set of machine language instructions—binary codes—in memory.
- The size of memory is normally specified by the number of words (m) and the number of bits per word (n), such as $m \times n$.
- Data are stored in memory by a process called *writing*.
- Data are retrieved from memory by a process called *reading*.
- The *capacity* of a memory is the total number of bits it can store.
- A *word* is a complete unit of binary data which can be of any number of bits.
- The *word size* is the number of bits in a word.
- Semiconductor memories can be either bipolar or MOS.
- ROMs are read only memories in which data are permanently or semi-permanently stored. ROMs may be masked ROMs (MROMs), PROMs, EPROMs or EEPROMs.
- PROMs are 'one-time programmable ROMs'.
- Storing the contents of a ROM is called *programming* the ROM.
- RAMs (random access memories) are read/write memories.
- RAMs are used in computers for the temporary storage of programs and data.
- *Non-volatility* means that the memory does not lose its data when the power goes off. *Volatility* means that data are lost when the power goes off.
- A number that identifies the location of a word in a memory is called the *address*.
- ROM normally means the mask programmed ROM, which is programmed by the manufacturer. It cannot be programmed by the user.
- Microcomputer programs stored in ROMs are referred to as *firmware*, because they are not subject to change. Programs stored in RWMs are referred to as *software*, because they can be easily altered.
- 'Memory access time' is the time required for valid data to appear on outputs after the activation of the appropriate inputs.
- PROMs are user programmable ROMs. EPROMs are erasable PROMs.
- RAMs can be either static or dynamic. A static RAM has latch storage cells. A dynamic RAM has capacitive storage cells.
- The ECL RAMs are used as cache memory and scratch-pad memory.
- Static memory is available in both BJT and MOS technologies, whereas dynamic memory is constructed using only the MOS technology.
- DRAMs have high capacity, need low power and operate at moderate speeds.
- Periodic recharging of the memory cells of a DRAM is called *refreshing* the DRAM.
- DRAM controllers are used to refresh large DRAM chips, whereas for small memories the iRAM provides the solution.
- A NVRAM contains a static RAM array and an EEPROM array on the same chip.
- Memory may be bit-organized or word-organized. In bit-organized memory, each IC stores one bit of each word; in word-organized memory all the bits of a word are stored in a single IC.

- PLAs are programmable logic arrays in which AND–OR gate arrays can be mask programmed to produce specified logic functions.
- PLDs are of three types: PALs, FPLAs and PROMs. PALs have programmable AND gates and fixed OR gates. FPLAs have both programmable AND gates and programmable OR gates. PROMs have fixed AND gates and programmable OR gates.
- Memory may be sequential access type or random access type. In a sequential access memory, the access time is not constant, but varies depending on the address of the location. In a random access memory, the access time is the same for every memory location.
- FIFOs are first-in first-out memories in which the first bit written in is the first bit read out.
- Semiconductor memories are faster than magnetic memories, but costlier in terms of per bit of storage.
- Floppy disks are made of plastic. They are inexpensive and readily portable.
- In a magnetic bubble memory (MBM), data are stored as tiny magnetic domains called *bubbles*. Bubble memories are serially accessed.
- Though an MBM is costlier than a floppy disk, it is faster.
- The optical disk memory is the newest mass memory technology. The optical disk systems are available in three basic types based on their writability—OROM or CD-ROM, WORM and Read/Write.
- CCD memories use channels of semiconductor capacitors to store charges representing data bits. The CCD memory is a sequential access memory. It is a dynamic memory.

QUESTIONS

1. Explain the meaning of the following types of memories:

   (a) Memory (b) RAM (c) ROM
   (d) PROM (e) EPROM (f) EEPROM
   (g) Mask ROM (h) Volatile memory (i) Non-volatile memory
   (j) Static memory (k) Dynamic memory (l) Main memory
   (m) Peripheral memory.

2. How are ROMs and RAMs classified?
3. Define the term mask-programmable.
4. What are the principal differences between the BJT memory and the MOSFET memory, in terms of their speed and size?
5. Which memory technology needs the least power?
6. Define the following terms:

   (a) Memory cell (b) Address (c) Byte
   (d) Access time (e) Memory word.

7. A certain memory has a capacity of 32K × 16. How many bits are there in each word? How many words are being stored? And how many memory cells does this memory contain?

8. A certain memory stores 8K 16-bit words. How many data input lines, data output lines, and address lines does it have? What is its capacity in bytes?

9. How many address inputs, data inputs, and data outputs are required for a 32K × 12 memory?

10. How does a PROM differ from an MROM? Can it be erased and reprogrammed? How is an EPROM erased?

11. Describe the procedure for reading from a ROM.

12. Describe the advantages of an EEPROM over an EPROM.

13. Define the following terms: (a) firmware, (b) software, and (c) hardware.

14. Describe how a computer uses the bootstrap program.

15. How does a static RAM cell differ from a dynamic RAM cell?

16. List the drawbacks and advantages of dynamic RAM vis-a-vis the static RAM.

17. Name the RAM timing parameters which determine its operating speed.

18. What are the ways of achieving non-volatile RAM storage?

19. How many memory cells does a 512 × 8 NVRAM contain?

20. What is the benefit of address multiplexing?

21. A 64K × 1 dynamic memory chip uses multiplexed address inputs. How many address inputs does this chip have?

22. How many 32 × 4 RAMs are needed to form a 32 × 8 memory? Is this an example of word capacity expansion or word length expansion?

23. Determine how many 16K × 4 memory circuits would be required to construct each of the following memories.

(a) 256K × 8 (b) 128K × 16 (c) 1M × 4.

24. A 1M × 8 memory is to be constructed by using 64K × 8 circuits, each of which has an active-LOW chip select input.

(a) How many 64K × 8 circuits are required?
(b) What type of decoder is necessary to select individual 64K × 8 circuits?
(c) What are the address inputs to the decoder?

25. Draw a logic diagram showing how to interconnect 2K × 8 memory circuits to obtain a 4K × 8 memory? Each circuit has an active-LOW chip select input and common data-in/data-out pins.

26. Draw a logic diagram showing how to interconnect 128 × 4 static RAMs to construct a 256 × 8 memory. Each circuit has an active-LOW chip select input and common data-in/data-out pins.

27. Explain the difference between sequential access memory and random access memory.

28. How does a FIFO differ from a recirculating shift register memory?

29. What is a data rate buffer?

30. How does a PLA differ from a ROM?

31. What is a PLD? What does an 'x' represent on a PLD diagram? What does a 'dot' represent on a PLD diagram?

32. How does the architecture of a PAL differ from that of a PROM?

33. How does the architecture of a FPLA differ from those of PROM and PAL?

34. Show how the PLA in Fig. 14.21 would be programmed to implement the functions

$$F_1 = A\overline{BC} + ABC \quad \text{and} \quad F_2 = ABC + \overline{AB} + A\overline{C}.$$

35. Show how the FPLA circuit in Fig. 14.23 would be programmed to implement

$$F_1 = A\overline{B}C + AB\overline{C}, \qquad F_2 = A\overline{BC}BC + \overline{A}BC + A\overline{B}C \quad \text{and} \quad F_3 = \overline{AB}$$

36. Show how the FPLA circuit in Fig. 14.23 would be programmed to implement

$$F_1 = \overline{ABC} + \overline{B}C + ABC \quad \text{and} \quad F_2 = F_1$$

37. Using the simplified connection format of a PLA, show how an 8 × 1 PROM should be programmed to implement the logic function

$$F = \Sigma(1, 4, 5, 7)$$

38. Show how an 8 × 1 PROM should be programmed to serve as a look-up table for the odd parity-bit of a 3-bit number? Use the simplified connection format of a PLA.

39. What property distinguishes a stored 1 from a stored 0 on a magnetic disk?

40. What are the main advantages and disadvantages of floppy disk storage compared to hard disk storage?

41. What is the major application of tape memory?

42. What is a fixed hard disk drive?

43. How are 1s and 0s represented in an MBM? What is the disadvantage of MBMs vis-a-vis semiconductor memories? What is a magnetic bubble? How is it created?

44. What are the chief advantages of optical disk memory? List the three types of optical disk storage.

45. How are data stored in a CCD memory? What are the principal advantages and disadvantages of the CCD memory?

Appendix

COMMONLY USED TTL ICs

Number	Description
7400	Quad 2-input NAND gates
7401	Quad 2-input NAND gates (open collector)
7402	Quad 2-input NOR gates
7403	Quad 2-input NOR gates (open collector)
7404	Hex inverters
7405	Hex inverters (open collector)
7406	Hex inverter buffers/drivers
7407	Hex buffer-drivers (open collector high voltage outputs)
7408	Quad 2-input AND gates
7409	Quad 2-input AND gates (open collector)
7410	Triple 3-input NAND gates
7411	Triple 3-input AND gates
7412	Triple 3-input NAND gates (open collector)
7413	Dual 4-input NAND Schmitt trigger
7414	Hex Schmitt-trigger inverters
74H15	Triple 3-input AND gate (open collector)
7416	Hex inverter buffer drivers
7417	Hex buffer drivers
7420	Dual 4-input NAND gates
7421	Dual 4-input AND gates
7422	Dual 4-input NAND gates (open collector)
7423	Dual 4-input NOR gates with strobe
7425	Dual 4-input NOR gates
7426	Quad 2-input TTL-MOS interface NAND gates
7427	Triple 3-input NOR gates
7428	Quad 2-input NOR buffers
7430	8-input NAND gate
74LS31	Delay elements
7432	Quad 2-input OR gates
7433	Quad 2-input NOR buffers (open collector)
7437	Quad 2-input NAND buffers
7438, 39	Quad 2-input NAND buffers (open collector)
7440	Dual 4-input NAND buffers
7441, 42	BCD-to-Decimal decoder

Number	Description
7443, 43A	Excess-3-to-Decimal decoder
7444, 44A	Gray-to-Decimal decoder
7445	BCD-to-Decimal decoder/driver
7446	BCD-to-Seven segment decoder/drivers (30 V output)
7447	BCD-to-Seven segment decoder/drivers (15 V output)
7448	BCD-to-Seven segment decoder/driver
7450	Expandable dual 2-input 2-wide AOI gates
7451	Dual 2-input 2-wide AOI gates
7452	Expandable 2-input 4-wide AND/OR gates
7453	Expandable 2-input 4-wide AOI gates
7454	2-input 4-wide AOI gates
7455	Expandable 4-input 2-wide AOI gates
7459	Dual 2/3-inputs 2-wide AOI gates
7460	Dual 4-input expanders
7461	Triple 3-input expanders
7462	2-2-3-3 input 4-wide A/O expanders
7464	2-2-3-4 input 4-wide AOI gates
7465	4-wide AOI gates (open collector)
7470	AND gated positive edge-triggered J-K FF with PRESET and CLEAR
7471	JK master-slave FF with preset
7472,74H72	AND gated J-K master-slave FF with PRESET and CLEAR
7473	Dual J-K master-slave FF with active-LOW CLEAR
7474	Dual positive edge-triggered D FF
7475	Quad D-latch
7476	Dual positive edge-triggered J-K flip-flops
74LS76A	Dual negative edge-triggered J-K flip-flops
7480	Gated full-adder
7482	2-bit binary full-adder
7483	4-bit binary full-adder with fast carry
7485	4-bit magnitude comparator
7486	Quad X-OR gate
7489	64-bit random access read/write memory
7490	Decade counter
7491	8-bit shift register
7492	Divide-by-12 counter
7493	4-bit binary counter
7494	4-bit shift register
7495	4-bit bidirectional shift register
7496	5-bit parallel-in parallel-out shift register
74100	4-bit bistable latch
74H103	Dual J-K negative edge-triggered FF with CLEAR
74104	J-K master-slave flip-flop
74105	J-K master-slave flip-flop
74H106	Dual J-K negative edge-triggered FF with PRESET and CLEAR
74107	Dual J-K master-slave FF with CLEAR

Number	Description
74LS107A	Dual J-K negative edge-triggered FF
74109	Dual J-K positive edge-triggered FF
74110	AND gated J-K master-slave FF with data lock-out
74111	Dual J-K master-slave FF with data lock-out
74116	Dual 4-bit latches with CLEAR
74121	Non-retriggerable one-shot
74122, 23	Retriggerable one-shot with CLEAR
74125, 26	3-state quad bus-buffer
74132	Quad 2-input NAND Schmitt trigger
74136	Quad 2-input X-OR gate
74LS138	3-line to 8-line decoder/demultiplexer
74141	BCD-to-Decimal decoder/driver
74142	BCD counter/latch/decoder/ driver
74145	BCD-to-Decimal decoder/driver
74147	Decimal-to-BCD priority encoder
74148	Octal-to-Binary priority encoder
74150	16-line to 1-line multiplexer
74151	8-channel digital multiplexer
74152	8-channel data selector/MUX
74153	Dual 4-line to 1-line multiplexer
74154	4-line to 16-line decoder/D MUX
74155,156	Dual 2-line to 4-line demultiplexer
74157	Quad 2-line to 1-line data selector
74160	Decade counter with asynchronous CLEAR
74161, 62, 63	Synchronous 4-bit counter
74164	8-bit parallel-out serial shift register
74165,66	Parallel load 8-bit serial shift register
74S168	Synchronous up/down decade counter
74S169	Synchronous up/down binary counter
74173	4-bit D type 3-state register
74174	Hex D flip-flops with CLEAR
74175	Quad D flip-flops with CLEAR
74176	35 MHz presettable decade counter
74177	35 MHz presettable binary counter
74179	4-bit parallel access shift register
74180	8-bit odd/even parity generator/checker
74181	Arithmetic logic unit
74182	Look-ahead carry generator
74184	BCD-to-Binary converter
74185	Binary-to-BCD converter
74189	3-state 64-bit RAM
74190	Synchronous up/down decade counter
74191,93	Synchronous binary up/down counter
74192	Synchronous BCD up/down counter
74194	4-bit bidirectional universal shift register

Number	Description
74195	4-bit parallel access shift register
74196	Presettable decade counter
74197	Presettable binary counter
74198	8-bit shift register
74221	Dual one-shot with Schmitt trigger inputs
74246, 47, 48	BCD-to-Seven segment decoder/driver
74251	3-state 8-channel multiplexer
74LS253	Dual 4-to-1 data MUX with 3-state output
74257	Quad 2-1 multiplexer
74259	8-bit addressable latch
74LS266	Quad 2-input X-NOR gates
74279	Quad latches
74LS280	9-bit odd/even parity generator/checker
74283	4-bit binary full-adder with fast carry
74284, 285	3-state 4-bit by 4-bit parallel binary multipliers
74290	Decade counter
74293	4-bit binary counter
74295	4-bit bidirectional shift register with 3-state outputs
74LS320	Crystal controlled oscillator
74365, 66, 67	3-state hex buffers
74390	Individual clocks with flip-flops
74393	Dual 4-bit binary counter

GLOSSARY

Active-HIGH (LOW) input Input which is normally LOW (HIGH) and goes to HIGH (LOW) when circuit operation is required.

Active logic level Logic voltage level at which a circuit is considered active.

A/D converter The circuit which converts an analog signal into its digital form.

Address The number that uniquely identifies the location of a word in memory.

Alphanumeric codes The codes that represent numbers, alphabetic characters, and symbols.

Analog Being continuous or having a continuous range of values as opposed to a discrete set of values.

Analog system Interconnection of devices designed to manipulate physical quantities that are represented in analog form.

AND gate A digital logic circuit used to implement the AND operation. The output of this circuit is a 1 only when each one of its inputs is a 1.

ANSI American National Standards Institute.

Anti-coincidence gate An X-OR gate which outputs a HIGH only when its two inputs differ.

Arbitration Selection of the input with the highest priority out of a number of inputs which are simultaneously high in an encoder.

Arithmetic Logic Unit (ALU) A digital circuit used in computers to perform arithmetic and logic operations.

ASCII code American Standard Code for Information Interchange. A seven-bit alphanumeric code used by most computer manufacturers.

Asserted level A term synonymous with active level, the level of the signal required to initiate the process.

Astable multivibrator A digital circuit with no stable state; it oscillates between two quasi-stable states.

Asynchronous Having no fixed time relationship.

Asynchronous counter A type of counter in which the external clock signal is applied only to the first FF and the output of each FF serves as the clock input to the next FF in the chain.

Asynchronous inputs Also called the overriding inputs of FFs. They can affect the operation of the FF independent of the lock and synchronous inputs.

Asynchronous transfer Transfer of data performed without the aid of clock.

Backplane Electrical connection common to all segments of the LCD.

Base The number of symbols in a number system. Also, one of the three regions in a BJT.

BCD Binary coded decimal, a digital code.

BCD adder An adder containing two 4-bit parallel adders and a correction detector circuit.

BCD counter A mod-10 counter that counts from 0000_2 to 1001_2.

Bidirectional shift register Shift-left, shift-right, shift register, in which data can be shifted in either direction.

Bilateral switch A CMOS circuit which acts like a single-pole, single-throw, switch controlled by an input logic level.

Binary Having two values or states.

Binary counter A counter in which the states of FFs represent the binary number equivalent to the number of pulses that have occurred at the input of the counter.

Binary multiplier A digital circuit capable of performing the arithmetic operation of multiplication on two binary numbers.

Binary number system A number system with only two values, 0 and 1.

Binary point A mark which separates the integer and fractional parts of a binary number.

Bipolar DAC A DAC whose output can assume a positive or negative value depending on the signed binary input

Bipolar ICs The ICs which use BJTs as the main circuit elements.

Bistable multivibrator A multivibrator which can remain indefinitely in any one of its two states. It is commonly known as flip-flop.

Bit *Bi*nary Digi*t*, a 1 or a 0.

Block parity A method of providing parity rowwise and columnwise for a block of information words.

Boolean algebra It is the study of mathematical logic.

Bubbled AND gate The AND gate with inverted inputs. It performs the NOR operation.

Bubbled OR gate The OR gate with inverted inputs. It performs the NAND operation.

Buffer driver A circuit with greater output current and/or voltage capability than an ordinary logic circuit.

Buffer register The register that holds digital data temporarily.

Byte A group of 8 bits.

Capacity The amount of storage space in a memory expressed as number of bits or number of words.

Carry A digit that is carried to the next column when two digits are added.

CCD Charge-coupled device, a type of semiconductor technology.

Cell A single storage element in a memory.

Cell table A table specifying the values of cell outputs and output carries for each combination of cell inputs and input carries.

Check sum A special data word that is derived from the addition of all other data words. It is used for error checking purposes.

Circulating shift register A shift register in which the output of the last FF is connected as the input to the first FF.

CLEAR An asynchronous FF input that makes $Q = 0$ instantaneously.

Clock The basic timing signal in a digital system.

Clock skew Arrival of a clock signal at different times at the clock inputs of different FFs as a result of propagation delays.

Clock transition times Minimum rise and fall times for the clock signal transitions.

CMOS Complementary Metal Oxide Semiconductor. The IC technology which uses both NMOS and PMOS FETs as the principal circuit elements.

Code A combination of binary digits that represents information such as numbers, alphabets, and other symbols.

Code converter An electronic digital circuit that converts one type of coded information into another coded form.

Coincidence gate The X-NOR gate which outputs a HIGH only when its two inputs are the same.

Combinational logic circuit A combination of logic devices having no storage capability and used to generate a specified function.

Complement An invert function in Boolean algebra

Computer word The group of bits used as the primary unit of information in a computer.

Control inputs Input signals synchronized with the active clock transition that determine the output state of a FF.

Cyclic code A code in which the successive code words differ in only one bit.

D/A converter The circuit which converts a digital input into an analog output.

Data Information in numeric, alphanumeric, or other form.

Data lock-out FFs A master-slave FF that has a dynamic clock input.

DC CLEAR Asynchronous FF input used to clear the FF.

DC SET Asynchronous FF input used to set the FF.

De Morgan's theorems (1) The complement of an OR operation on variables is equal to an AND operation on the complemented variables. (2) The complement of an AND operation on variables is equal to the OR operation on complemented variables.

Decade counter A digital counter having 10 different states.

Decimal number system The number system with 10 different symbols.

Decoder A digital circuit that converts coded information into familiar form.

Demultiplexer (DMUX) The logic circuit that channels its data input to one of several data outputs.

Dependency notation A notational system for logic symbols that specifies the input and output relationships.

Digit A symbol representing a given quantity in a number system.

Digital ICs Self-contained digital circuits, made by using one of several IC technologies.

Digital system A combination of devices designed to manipulate physical quantities that are represented in digital form.

DIP Dual-in-line-package; the most common type of IC package.

Dominating column In a prime implicants chart, a column which has a 'X' in every row in which another column has a 'X'.

Dominating row In a prime implicants chart, a row which has a 'X' in every column in which another row has a 'X'.

Don't care A condition in a logic circuit in which the output is independent of the state of a given input.

DRAM Dynamic RAM: a type of semiconductor memory that stores data as capacitor charges that need to be refreshed periodically.

Dynamic shift register A register in which data continually circulates through the register under the control of a clock.

EAPROM Electrically alterable programmable read only memory.

EBCDIC code Extended Binary Coded Decimal Interchange Code: an alphanumeric code.

ECL Emitter Coupled Logic. Also referred to as current mode logic (CML).

Edge detector A circuit that produces a narrow positive spike, coincident with the active transition of a clock pulse.

Edge-triggered FF A type of FF which is activated by the clock signal transition.

EEPROM Electrically erasable programmable read only memory.

Encoder A digital circuit that converts information into coded form.

EPROM Erasable programmable read only memory.

Excess-3 code A digital code in which each of the decimal digits is represented by a 4-bit code derived by adding 3 to each of the digits.

Excess-3 gray code A digital code in which each decimal digit is encoded with a gray code pattern of the decimal digit that is greater by 3.

Excitation table A table showing the required input conditions for each possible state transition of a device/machine.

Exclusive-NOR gate A logic circuit which produces a 1 output only when its inputs coincide.

Exclusive-OR gate A logic circuit which produces a 1 output only when its inputs are different.

Fan-out The maximum number of equivalent gate inputs that the output of a gate can drive without impairing its own function.

Finite state machine An abstract model describing the synchronous sequential machine and its spatial counterpart, the iterative network.

Flat package A type of IC package.

Flip-flop A memory device capable of storing a logic level.

Floppy disk Flexible magnetic disk used for mass storage.

Frequency counter A circuit that can measure and display the frequency of a signal.

Full-adder A digital circuit that adds two binary digits and an input carry, and produces a sum digit and an output carry.

Full-subtractor An digital circuit that subtracts one bit from another bit considering previous borrow.

Function generator A circuit that produces a variety of waveforms.

Gate A circuit that performs a specified logic operation.

Gated latch A latch that responds to the input only when its ENABLE is HIGH.

Glitch A voltage or current spike of short duration which is usually unwanted.

Gray code A unit distance code in which the successive code words differ in only one bit.

Half-adder A digital circuit which can add only two bits, and produces a sum bit and a carry bit.

Half-subtractor A circuit which can subtract one bit from another.

Hard disk A rigid metal magnetic disk used for mass storage.

Hexadecimal number system A number system consisting of 16 symbols, 0–9 and A–F.

Hold time The time interval for which the control signal has to be maintained at the input terminal of the FF after the clock termination in order to obtain reliable operation.

Hybrid circuits Circuits containing both integrated and discrete components.

Hybrid counter A synchronous counter whose output drives the clock input of another counter.

IC Integrated circuit: a type of circuit in which all the components are integrated on a single silicon chip of very small size.

Inhibit circuits Logic circuits that control the passage of an input signal through to the output.

Interfacing Joining of dissimilar devices in such a way that they are able to function in a compatible and coordinated manner; connection of the output of a system to the input of a different system with different electrical characteristics.

Invert It causes a logic level to go to the opposite state.

Inverter A logic circuit that implements the NOT operation.

Iterative network A digital structure composed of a cascade of identical circuits or cells.

J-K FF A type of FF that can operate in the 'no change', 'set', 'reset' and 'toggle' modes.

Johnson counter A type of shift register counter in which the inverted output of the last FF is connected as data input to the first FF.

Karnaugh map An arrangement of cells representing the combinations of variables in a Boolean expression and used for a systematic simplification of the expression.

Latch A non-clocked FF.

Linearity error The maximum deviation in step size from the ideal step size in a DAC.

Lock-out The state of a counter when successive clock pulses take the counter only from one invalid state to another invalid state.

Logic circuit A circuit that behaves according to a set of logic rules.

Logic level State of a voltage variable. States HIGH and LOW correspond to the two usable voltage levels of a digital device.

Look-ahead carry A method of binary addition whereby carries from the preceding stages are anticipated, thus avoiding the carry propagation delays.

Looping Combining of adjacent squares in a K-map containing 1s (0s) for the purpose of simplification of a SOP (POS) expression.

LSB The least significant bit, i.e. the rightmost bit in the binary number.

LSD The least significant digit, i.e. the digit that carries the least weight in a particular number.

LSI Large scale integration. A level of integration in which 100 to 9999 gates are integrated on a single chip.

Magnetic bubble A tiny magnetic region in magnetic material created by an external magnetic field.

Magnetic bubble memory (MBM) Solid state, non-volatile, sequential access, mass storage memory device consisting of tiny magnetic domains (bubbles).

Magnetic core memory Non-volatile random access memory made up of small ferrite cores.

Magnetic disk memory Mass storage memory that stores data as magnetized spots on a rotating flat disk surface.

Magnetic tape memory Mass storage memory that stores data as magnetized spots on an iron-coated plastic tape.

Magnitude comparator A digital circuit used to compare the magnitudes of two binary numbers and indicate whether they are equal or not, and if not which one has larger magnitude.

Mass storage Storage of large amounts of data, not part of the computer's internal memory.

Master slave FF A type of FF in which the input data affects the first of its two FFs at the leading edge of the clock pulse and then the contents of the first FF appear at the output of the second FF at the trailing edge of clock pulse.

Maxterm The product term in the standard POS form.

Memory array An arrangement of memory cells.

Memory cell An individual storage element in a memory.

Memory word Groups of bits in memory that represent instructions or data of some type.

Microprocessor A large-scale integrated circuit that can be programmed to perform arithmetic and logic functions and to manipulate data.

Minterm The sum term in the standard SOP form.

Modified modulus counter A counter that does not sequence through all of its natural states.

Modulus The maximum number of states in a counter sequence.

Monostable multivibrator A multivibrator having only one stable state. The other one is a quasi-stable state. It is also called one-shot.

Monotonicity A property whereby the output of a DAC either increases or stays at the same value, but never decreases as the input is increased.

MSB The most significant bit. The leftmost bit in a binary number.

MSD The most significant digit. The digit that carries the most weight in a particular number, i.e. the extreme left digit in the number.

MSI Medium scale integration. A level of integration in which 12–99 gate circuits are fabricated on a single chip.

Multiplex To put information from several sources on to a single line or transmission path.

Multiplexer (MUX) A digital circuit, depending on the status of its control inputs, that channels one of several data inputs to its output.

NAND gate The logic gate that outputs a 0 only when all its inputs are 1s. It gives the complement of the AND output.

NAND-gate latch Basic flip-flop constructed using two cross-coupled NAND gates.

Negative logic The system of logic in which a LOW represents a 1 and a HIGH represents a 0.

NMOS N-channel metal oxide semiconductor.

Noise Unwanted (spurious) signals.

Noise immunity The ability of a circuit to tolerate noise voltages on its inputs.

Noise margin Quantitative measure of the noise immunity. It is the maximum noise voltage that can be added at the input of a gate without affecting its operation.

Non-retriggerable one-shot A type of one-shot that does not respond to a trigger input signal while in its quasi-stable state.

Non-volatile memory Memory that is able to keep its information stored in the event of failure of electrical power.

NOR gate A logic circuit that outputs a 1 only when each one of its inputs is a 0. It is equivalent to an OR gate followed by an inverter.

NOR-gate latch A flip-flop constructed using two cross-coupled NOR gates.

NOT circuit A logic circuit that inverts its only input.

Octal number system A number system with 8 digits (0, 1, 2, 3, 4, 5, 6, 7).

Octet A group of 8 1s or 0s that are adjacent to each other in a Karnaugh map.

Offset error The voltage present at the output of a DAC when the input is all 0s.

One's complement form A form of representation obtained by complementing each bit of a binary number.

One-shot A multivibrator with only one stable state—the other name of monostable multivibrator.

Open-collector gates The TTL gates which use only one transistor with a floating collector in the output structure.

OR gate A logic circuit that outputs a HIGH whenever one of its inputs is a HIGH.

Override inputs Asynchronous inputs in a FF which override the effects of all other input signals.

Oscillator An electronic circuit that switches back and forth between two states.

Parallel adder A digital circuit with full-adders that adds all the bits from two numbers simultaneously.

Parallel counter A counter in which all the FFs are triggered simultaneously—the other name of synchronous counter.

Parallel data transfer The operation by which the entire contents of a register are transferred to another register.

Parallel transmission Simultaneous transfer of all bits of a binary number from one place to another.

Parallel-in, parallel-out, shift register A type of shift register that can be loaded with parallel data and has also parallel outputs available.

Parallel-in, serial-out, shift register A type of shift register that can be loaded with parallel data but has only one serial output terminal.

Parity bit An additional bit that is attached to each code group so that the total number of 1s being transmitted is either odd or even.

Parity checker A circuit that is used to check parity among the group of bits received.

Parity generator A digital circuit that takes a set of data bits and produces the correct parity bit for the data.

Percentage resolution The ratio of the step size to the full-scale value of a DAC. It can also be defined as the reciprocal of the maximum number of steps of a DAC.

PMOS P-channel metal oxide semiconductor.

Positional-value system The system in which the value of a digit is dependent on its relative position.

Positive logic system The system of logic in which a HIGH represents a 1 and a LOW represents a 0.

PRESET Asynchronous input used to instantaneously set Q = 1.

Presettable counter A counter that can be PRESET to any initial count either synchronously or asynchronously.

Prime implicant A term which cannot be combined further in the tabular method.

Priority encoder An encoder that produces a coded output corresponding to the highest-valued input when two or more inputs are applied simultaneously.

PROM A ROM that can be programmed by the user. It cannot be erased and reprogrammed.

Propagation delay The time interval between the occurrence of an input transition and the corresponding output transition.

Pulse A sudden change from one level to another followed by a sudden change back to the original level.

Quantization error The error caused by non-zero resolution of an ADC; it is an inherent error of the device.

Quasi-stable state A temporary stable state. A monostable circuit is triggered to this state before returning to the normal stable state. Both the states of an astable multivibrator are quasi-stable.

Quine McClusky method A tabular method used to minimize Boolean expressions.

Race A condition in a logic network in which the differences in propagation times through two or more signal paths in the network can produce an erroneous output.

Radix The base of a number system. The number of symbols in a given number system.

RAM Random access memory—a memory in which the access time is the same for all locations.

Read The process of retrieving information from a memory.

Read/write memory A memory that can be both read from and written into.

Reflected code A code with the property that an N-bit code can be obtained by reflecting an N – 1 bit code about an axis at the end of the code and putting 0s above the axis and 1s below the axis.

Refresh The process of renewing the contents of a dynamic memory.

Register A group of flip-flops capable of storing data.

RESET The state of a FF, register or counter when 0s are stored. This term is synonymous with CLEAR.

Resolution In a DAC, the smallest change that can occur in the output for a change in the digital input. Also, called the step size. In an ADC, the smallest amount by which the analog input must change to produce a change in the digital output.

Retriggerable one-shot A type of one-shot that will respond to a trigger input signal while in its quasi-stable state.

Ring counter Serial-in, serial-out, shift register in which the output of the last FF is connected to the input of the first FF.

Ripple counter A counter in which the external clock signal is applied to the first FF and then the clock input to every other FF is the output of the preceding FF.

ROM Read only memory.

Schmitt trigger A digital circuit that converts a slow-changing signal into a fast-changing signal.

Schottky TTL A TTL subfamily that uses the basic standard TTL circuit except that it uses a Schottky barrier diode between the base and collector of each transistor.

Self-complementing code A code in which the code word of the 9's complement of *N*, i.e. of 9 – *N* can be obtained from the code word of *N* by interchanging all 0s and 1s.

Sequential code A code in which each succeeding code word is one binary number greater than its preceding code word.

Sequence detector A digital circuit used to detect a sequence in the input.

Sequential logic A system of logic in which the logic output states and the sequence of operations depend on both the present and the past input conditions.

Serial adder A type of adder which adds two numbers by taking the bits from them serially with a carry.

Serial data transmission Transfer of data from one place to another one bit at a time on a single line.

Serial-in, parallel-out, shift register A type of shift register that can be loaded with data serially, but has parallel outputs available.

Serial-in, serial-out, shift register A type of shift register that can be loaded with data serially, and also has a serial output terminal available.

SET The state of a flip-flop when it is in the binary 1 state.

Set-up time The time interval for which the control signals must be held constant at the input terminals of a FF, prior to the arrival of the triggering edge of the clock pulse.

Settling time The time taken by the output of a DAC to rise and settle within one-half step size of its full scale value as the input is changed from all 0s to all 1s.

Shift register A digital circuit capable of storing and shifting binary data.

Sign bit A binary digit that is inserted at the leftmost position of a binary number to indicate whether that number represents a positive or negative quantity.

Speed power product Numerical value (in joules) often used to compare different logic families. It is obtained by multiplying the propagation delay (ns) by the power dissipation (mW) of a logic gate.

SSI Small scale integration (less than 12 gates per chip).

Stage One storage element in a register or counter.

Standard SOP form A form of Boolean expression in which each product term contains all the variables of the function and all these product terms are summed together.

Standard POS form A form of Boolean expression in which each sum term contains all the variables of the function and all these sum terms are multiplied together.

State assignment The process of assigning the states of a physical device to the states of a sequential machine.

State diagram A picture showing the relationship between the present state, the input, the next state and the output of a sequential machine.

State machine Any sequential circuit exhibiting a specified sequence of states.

State table A table showing the relationship between the present state, the input, the next state and the output of a sequential machine.

State variables The output values of the physical device in a sequential machine.

SRAM (Static RAM) A RAM that stores information in FF cells which do not have to be refreshed unlike those of the DRAM.

Storage The memory capability of a digital device. Also, the process of storing digital data for later use.

Straight binary coding Representation of a decimal number by its equivalent binary number.

Strobing A technique often used to eliminate decoding spikes.

Substrate A piece of semiconductor material over which the components are fabricated in an IC.

Sum of products form A form of Boolean expression, that is, the ORing of ANDed terms.

Synchronous Having a fixed time relationship.

Synchronous counter A counter in which the circuit outputs can change states only on the transitions of a clock.

Synchronous inputs Input signals synchronized with the active clock transition that determine the output state of a FF.

Synchronous systems Systems in which the circuit outputs can change states only on the transition of a clock.

Synchronous transfer Data transfer performed by using synchronous and clock inputs of a FF.

Terminal count The final state of a counter sequence.

Timing diagram Depiction of logic levels as related to time.

Toggle mode The mode in which a FF changes states for each clock pulse.

Totem-pole A term used to describe the way in which two bipolar transistors are connected one above the other at the outputs of most TTL gates.

Transition A change from one level to another.

Transition and output table A table showing the state transitions and the output of the sequential machine in terms of the assigned states.

Transmission gate A digitally controlled bi-lateral CMOS switch.

Trigger A pulse used to initiate a change in the state of a logic circuit.

Tri-state A type of output structure which allows three types of output states—HIGH, LOW and high-impedance.

Truth table A tabular representation of the outputs of a logic circuit for all possible combinations of inputs.

TTL (Transistor Transisitor Logic) An IC technology that uses the bipolar transistor as the basic circuit element.

ULSI Ultra large scale integration (more than 100,000 gates per chip).

Unipolar ICs The ICs in which the MOSFETs are the main circuit elements.

Unit distance code A code having the property such that the bit patterns for two consecutive numbers differ in one bit position only.

Unit load The current drawn by the input of a logic gate when connected to the output of an identical gate.

Universal gates The gates using which the basic logic functions (AND, OR and NOT) can be realized.

Universal shift register Shift-right, shift-left, shift register which can input and output data either serially or parallely.

Up-counter A counter that counts upwards from zero to a maximum count.

UVEPROM Ultra violet erasable programmable read only memory.

Variable modulus counter A counter in which the maximum number of states can be changed.

VLSI Very large scale integration (10,000 to 99,999 gates per chip).

Volatile memory Memory requiring electrical power to keep information stored.

Voltage level translator A circuit that takes one set of input voltage levels and translates it to a different set of output voltage levels.

Weight The positional value of a digit in a number.

Weighted code A digital code that utilizes weighted numbers as the individual code words.

Wired AND A term used to describe the logic function created when open-collector outputs are tied together.

Word A group of bits representing a complete piece of digital information.

Wrapping around Folding at the centre of a Karnaugh map such that the edges coincide.

Writing The process of storing information in a memory.

ANSWERS TO PROBLEMS

CHAPTER 2

2.1 (a) 188 (b) – 522 (c) 4427.70 (d) –389.3

2.2 (a) 11 (b) 109 (c) 13.75 (d) 110.375

2.3 (a) 100101 (b) 11100 (c) 11000101.1011
(d) 11001101.000011

2.4 (a) 101000 (b) 1001110.111 (c) 110000 (d) 110001.11

2.5 (a) 0110 (b) 1011 (c) 101.01 (d) 1.10

2.6 (a) 1000001 (b) 110010 (c) 1001011.101 (d) 110111

2.7 (a) 11.0101 (b) 110 (c) 110 (d) 1011.00111

2.10 (a) 1111 1101 1011 (b) 1111 0101 0011 (c) 1011 1110 . 1000
(d) 0011 1010.1000

2.11 (a) 1111 1001 1110 (b) 1111 0001 1111 (c) 1111 0011 0010 . 0011
(d) 1111 1110 0010 . 1001

2.12 (a) 17 (b) – 48 (c) 78.6 (d) –52.75

2.13 (a) 35 (b) –38 (c) 46.25 (d) –39.25

2.14

(a)	0000 1110	0000 1110	0000 1110
(b)	0001 1011	0001 1011	0001 1011
(c)	0001 01101	0001 01101	0001 01101
(d)	1001 0001	1110 1110	1110 1111
(e)	1010 0101	1101 1010	1101 1011
(f)	1100 1100	1011 0011	1011 0100

2.15 (a) AE (b) 41D (c) 3F2.98 (d) C2C.598

2.16 (a) 1253 (b) 41375 (c) 2367.52 (d) 136160.034

2.17 (a) 307 (b) 1070 (c) 1071.81 (d) 3420.06

2.18 (a) 437 (b) 7564 (c) 644.463 (d) 20434.360

2.19 (a) 460 (b) 3322 (c) 44.55 (d) 720.65

2.20 (a) 25 (b) 265 (c) 126.67 (d) 125.7

2.21 (a) 230 (b) 1623 (c) 73.31 (d) 4427.70

2.22 (a) 66.346 (b) 351.122 (c) 17.311 (d) 366.05

2.23 (a) 51 (b) –365 (c) 104.7 (d) 442.04

2.24 (a) 1100 0010 0000 (b) 1111 0010 1001 0111
(c) 1010 1111 1001.1011 0000 1101
(d) 1110 0111 1001 1010.0110 1010 0100

2.25 (a) 16 (b) 2DB (c) 1B7.78 (d) 1B6D.B4

2.26	(a) 2742	(b) 11959	(c) 41103.879	(d) 36423.629
2.27	(a) 1C4	(b) 12BC	(c) 4E0.8E5	(d) 22FD.C
2.28	(a) 161F	(b) 1585B	(c) 51D	(d) 1AD84.6E
2.29	(a) 19A	(b) 389	(c) C371.D2	(d) 1DE3.C5
2.30	(a) 1DFE	(b) 27A34	(c) 1E12E.94	(d) 664.DF
2.31	(a) E0.C	(b) 726	(c) 418.0E	(d) 5746.451
2.32	(a) 60	(b) 371	(c) 9976.FB	(d) 3AF6.2

CHAPTER 3

3.1	(a) 0010 1001 0110	(b) 0001 0101 0111.0101	(c) 0100 0010 0010 1000.0101	
	(d) 1000 0000 1001	(e) 0011 0111.0101 0010	(f) 0010 0000 0100 0000.0000 1000	
3.2	(a) 0010 0000 1011	(b) 0010 1100 0011 0000	(c) 1100 1111 1101	(d) 1110 0000
	(a) 0011 0000 1000	(b) 0011 1010 0101 0000	(c) 1010 1111 1100	(d) 1110 0000
	(a) 0110 0000 1011	(b) 0110 1010 0101 0000	(c) 1010 1111 1001	(d) 1000 0000
3.3	(a) 839	(b) 578.9	(c) 697.4	(d) 8603.8
3.4	(a) 1100 1010	(b) 1001 1000 0110	(c) 1011 0100 0101.1100	
	(d) 0100 1100	(e) 0101 1000 0100 1000	(f) 0101 0011 1000 0110.0011 1001	
3.5	(a) 859	(b) 374.5	(c) 695.4	(d) 4526.5
3.6	(a) 771	(b) 897	(c) 363.3	(d) 399.35
3.7	(a) 655	(b) 62	(c) 208.7	(d) 215.9
3.8	(a) 0101	(b) 1010	(c) 11110	(d) 1010000
	(e) 1100 10001			
3.9	(a) 1111	(b) 1011111	(c) 101101101	(d) 11001
	(e) 10101010			
3.10	(a) 1010	(b) 110110	(c) 111101001	(d) 1110
	(e) 1011010			
3.11	(a) Error	(b) No error	(c) No error	(d) Error
	(e) Error			
3.12	(a) 00110 11000	(b) 00101 01100 01001	(c) 10001 11000 10010	
	(a) 01 00100 10 10000	(b) 01 00010 10 00001 01 01000	(c) 10 00010 10 10000 10 00100	
	(a) 00011 10000	(b) 00001 11111 00111	(c) 11110 10000 10010	
	(a) 0000000010 1000000000	(b) 000000001 000010000 0000001000	(c) 0001000000 1000000000 0010000000	
3.14	(a) 0101010	(b) 0101101	(c) 0101101	(d) 1001011
	(e) 1111111			

3.15 1001011001100111000011010101

3.16 (a) 100 0010 100 1001 101 0010 101 0100 100 1000
(a) 1100 0010 1100 1001 1101 1001 1110 0001 1100 1000
(b) 100 0001 100 1011 011 0100 011 0111
(b) 1100 0001 1101 0010 1111 0100 1111 0111

CHAPTER 5

5.2	(a) ABC	(b) ABC	(c) 0	(d) 0
	(e) 0	(f) 0		

5.3 (a) $P + Q$ (b) 1 (c) 1 (d) 1
(e) 1 (f) P

5.4 (a) Y (b) 0 (c) 0 (d) XY
(e) 1 (f) $X\bar{Y} + XZ + \bar{Y}Z$

5.5 (a) XYZ (b) XY (c) 0 (d) ABC

5.6 (a) A (b) 0 (c) $AB + EF$ (d) ABC
(e) $\bar{A}B$ (f) $A + B + C$ (g) $\bar{B}\bar{C}$ (h) $\bar{B} + \bar{C}$
(i) $W(X + \bar{Y})$ (j) 1 (k) 1 (l) $\bar{C}$

5.8 (a) $\bar{P} + \bar{Q}\bar{R}$ (b) $\bar{P}Q + R\bar{S}$ (c) $\overline{(A + B)}\,\overline{(C + D)} + \overline{(E + F)}\,\overline{(G + H)}$
(d) $\bar{A}BC\bar{D}$

5.9 (a) $AC + BC + AD + BD$ (b) $ABC + ACD$ (c) $AD + \bar{B}CD$
(d) $AB + CD$

CHAPTER 6

6.1 (a) $\bar{A}B\bar{C} + A\bar{B}\bar{C} + A\bar{B}C + AB\bar{C} + ABC$

(b) $\bar{A}\bar{B}\bar{C} + \bar{A}\bar{B}C + \bar{A}B\bar{C} + A\bar{B}C + AB\bar{C} + ABC$

(c) $\bar{A}\bar{B}\bar{C}\bar{D} + \bar{A}\bar{B}C\bar{D} + \bar{A}\bar{B}CD + \bar{A}B\bar{C}\bar{D} + \bar{A}BC\bar{D} + \bar{A}BCD + A\bar{B}\bar{C}\bar{D} +$
$A\bar{B}C\bar{D} + A\bar{B}CD + AB\bar{C}\bar{D} + ABC\bar{D} + ABCD$

(d) $A\bar{B}CD\bar{E} + A\bar{B}CDE + AB\bar{C}\bar{D}\bar{E} + AB\bar{C}D\bar{E} + ABC\bar{D}\bar{E} + ABCD\bar{E} + ABCDE$

6.2 (a) $(A + B + C)(A + B + \bar{C})(A + \bar{B} + C)(A + \bar{B} + \bar{C})(\bar{A} + B + \bar{C})$

(b) $(A + \bar{B} + C + D)(A + \bar{B} + C + \bar{D})(A + \bar{B} + \bar{C} + D)(\bar{A} + B + C + D)(\bar{A} + \bar{B} + C + D)$
$(\bar{A} + B + \bar{C} + D)(\bar{A} + \bar{B} + \bar{C} + D)$

(c) $(A + B + C + \bar{D})(A + B + \bar{C} + \bar{D})(\bar{A} + \bar{B} + C + D)$

(d) $(\bar{A} + B + C + D)(\bar{A} + B + C + \bar{D})(\bar{A} + B + \bar{C} + D)(\bar{A} + B + \bar{C} + \bar{D})$
$(A + B + C + D)(A + B + \bar{C} + D)(A + B + \bar{C} + \bar{D})(A + \bar{B} + C + D)$
$(A + \bar{B} + \bar{C} + D)(\bar{A} + \bar{B} + \bar{C} + D)(\bar{A} + \bar{B} + \bar{C} + \bar{D})$

6.3 (a) 17 (b) 16 (c) 8 (d) 16
(e) 13 (f) 11

6.4 (a) $AC + B\bar{C}$ (b) $B + C + D$ (c) $A + C$ (d) $B + C$

6.5 (a) A (b) $A(B + \bar{C})(B + \bar{D})$ (c) B

6.6 $\bar{A}\bar{B}C + BD + AB + AD$

6.7 $(A + D)(A + C)(\bar{A} + \bar{B})(\bar{A} + \bar{D})$

6.8 $\bar{A}\bar{C}\bar{D} + AC$

6.9 (a) $(C + D)(A + B)$ (b) $\bar{B} + \bar{A}\bar{D}$ (c) $\bar{B}\bar{D} + \bar{A}C + A\bar{C}D$

(d) $(B + \bar{C} + \bar{D})(\bar{A} + B + D)(\bar{B} + C + \bar{D})(\bar{B} + \bar{C} + D)$

(e) $\bar{A}\bar{C} + BC + A\bar{B}D$ (f) $A(B + C + D)$

6.10 (a) $(B + D)(\bar{B} + \bar{E})(\bar{A} + B + \bar{C})(A + \bar{C} + \bar{D})(A + B + C)$

(b) $\bar{C}D + \bar{B}\bar{C}\bar{E} + A\bar{B}\bar{D} + \bar{A}BC\bar{D}$

(c) $AC\bar{D} + \bar{C}D\bar{E}\bar{F} + \bar{A}\bar{B}\bar{D}\bar{E} + \bar{B}\bar{D}EF + BD\bar{E}\bar{F} + \bar{A}\bar{B}\bar{C}DF + \bar{A}\bar{B}CDE + \bar{A}B\bar{C}\bar{D}F$

(d) $C\bar{D}\bar{E} + C\bar{D}\bar{F} + \bar{A}\bar{B}\bar{E} + \bar{B}\bar{C}D\bar{F} + AB\bar{D}\bar{E} + AB\bar{D}\bar{F} + \bar{A}C\bar{E}F + B\bar{D}E\bar{F} + A\bar{B}\bar{C}\bar{D}F$

6.11 (a) $(\bar{A} + C + D)(\bar{A} + B + \bar{D})(A + \bar{C} + D)$ (b) $(A + \bar{B} + C)(C + \bar{D})(\bar{A} + \bar{B} + \bar{C} + D)$

(c) $(\bar{A} + D)(\bar{A} + B)(B + \bar{C} + D)$ (d) $\bar{A}(\bar{B} + \bar{C})$

(e) $(\bar{A} + \bar{B} + C)(A + C + \bar{D})(A + \bar{B} + \bar{C})(\bar{A} + \bar{C} + \bar{D})$

(f) $B\bar{C} + B\bar{D} + \bar{A}\bar{C}D$

6.12 (a) $f_{1m} = \bar{A}C\bar{D} + A\bar{B}\bar{D} + \bar{A}\bar{C}D + A\bar{B}\bar{C}$

$f_{2m} = \bar{A}C\bar{D} + A\bar{B}\bar{D} + B\bar{C}\bar{D} + ABCD$

$f_c = \bar{A}C\bar{D} + A\bar{B}\bar{D}$

(b) $f_{1m} = \bar{A}\bar{D} + \bar{B}\bar{C} + A\bar{B}D$

$f_{2m} = \bar{A}\bar{D} + \bar{B}\bar{C}D + \bar{A}B + B\bar{C}\bar{D}$

$f_c = \bar{A}\bar{D}$

(c) $f_{1m} = \bar{A}\bar{B}\bar{C} + \bar{A}\bar{D} + BC + AC\bar{D}$

$f_{2m} = \bar{A}B\bar{C} + A\bar{B}D + \bar{B}CD + AC\bar{D}$

$f_c = AC\bar{D}$

(d) $f_{1m} = \bar{B}C + ABC + \bar{A}D$

$f_{2m} = \bar{A}\bar{C} + BD + ABC$

$fc = ABC$

6.13 (a) $AC + BC\bar{D} + ABD$ (b) $AC\bar{D} + AB\bar{E} + \bar{A}BC$

(c) $\bar{B}C + \bar{A}CD + B\bar{C}\bar{D} + AB\bar{C}$ (d) $BD + C\bar{D} + \bar{A}C$

(e) $A\bar{B}D + BC\bar{D} + \bar{A}E + \bar{B}\bar{C} + \bar{A}BC$ (f) $\bar{B}CD + AB\bar{D} + B\bar{C}D$

6.14 (a) $A\bar{D}E + BCF + AB\bar{D} + \bar{A}\bar{B}\bar{D} + \bar{A}BCD$

(b) $A\bar{C}\bar{D}\bar{F} + B\bar{C}\bar{D}F + ABCE + ACDE + BCDG + \bar{A}\bar{B}\bar{D} + \bar{B}C\bar{D}$

(c) $C\bar{D}F + A\bar{B}C\bar{D}G + BC\bar{F} + \bar{B}C\bar{D}H + \bar{A}B\bar{C}G + \bar{A}\bar{B}CD$

(d) $BC\bar{D}\bar{E} + \bar{A}BCG + \bar{B}C\bar{D}F + \bar{B}\bar{C}\bar{D}E + A\bar{B}\bar{D} + \bar{A}\bar{B}\bar{C}DH$

(e) $\bar{A}CDE + \bar{A}BD\bar{E} + AB\bar{D} + AC\bar{D}F$

(f) $\bar{A}\bar{B}\bar{D}E + \bar{A}\bar{B}\bar{C} + \bar{B}C\bar{D}G + ABDF + ABC$

CHAPTER 7

7.1 If A, B, C and D are the BCD inputs and f_3, f_2, f_1 and f_0 are the 2421 outputs, then

$f_3 = A + BC + BD$, $f_2 = A + BC + B\bar{D}$, $f_1 = A + \bar{B}C + B\bar{C}D$, $f_0 = D$

7.2 If A, B, C and D are the 2421 inputs and f_4, f_3, f_2, f_1 and f_0 are the 51111 outputs, then

$f_4 = A$, $f_3 = B$, $f_2 = AB + BC + BD + \bar{A}CD$, $f_1 = \bar{A}B + A\bar{C} + BC$, $f_0 = \bar{A}B + \bar{A}C + \bar{A}D + BCD$

7.3 If A, B, C, and D are the excess-3 code inputs, the decimal digits are given by

$D_0 = \bar{A}\bar{B}CD$ $\quad D_1 = \bar{A}B\bar{C}\bar{D}$ $\quad D_2 = \bar{A}B\bar{C}D$ $\quad D_3 = \bar{A}BC\bar{D}$

$D_4 = \bar{A}BCD$ $\quad D_5 = A\bar{B}\bar{C}\bar{D}$ $\quad D_6 = A\bar{B}\bar{C}D$ $\quad D_7 = A\bar{B}C\bar{D}$

$D_8 = A\bar{B}CD$ $\quad D_9 = AB\bar{C}\bar{D}$

7.4 If A, B, C and D are the 2421 code bits, the decimal digits are the given by

$D_0 = \bar{A}\bar{B}\bar{C}\bar{D}$ $\quad D_1 = \bar{A}\bar{B}\bar{C}D$ $\quad D_2 = \bar{A}\bar{B}C\bar{D}$ $\quad D_3 = \bar{A}\bar{B}CD$

$D_4 = \bar{A}B\bar{C}\bar{D}$ $\quad D_5 = A\bar{B}CD$ $\quad D_6 = AB\bar{C}\bar{D}$ $\quad D_7 = AB\bar{C}D$

$D_8 = ABC\bar{D}$ $\quad D_9 = ABCD$

7.5 If A, B, C and D are the 2421 code bits, $f_e = (A \oplus B) + (C \oplus D)$

7.6 If A, B, C and D are the 3321 BCD code bits, $f_0 = (A \oplus B) + (C \odot D)$

7.7 (a) If A, B, C and D are the 5211 inputs, and f_3, f_2, f_1 and f_0 are the 2421 outputs,

$f_3 = A$, $\quad f_2 = AB + C\bar{D} + BC$, $\quad f_1 = \bar{A}\bar{B}C + AB + B\bar{C}D + A\bar{B}\bar{C}$, $\quad f_0 = \bar{C}D + AD + A\bar{C}$

(b) If A, B, C, and D are the binary inputs and f_6, f_5, f_4, f_3, f_2, f_1 and f_0 are the excess-3 outputs, then

$f_6 = AB + AC$, $\quad f_5 = \bar{A} + \bar{B}\bar{C}$, $\quad f_3 = A\bar{B}\bar{C} + \bar{A}BD + \bar{A}BC + BCD$,

$f_2 = \bar{B}D + B\bar{C}\bar{D} + AB\bar{D} + AB\bar{C} + \bar{A}\bar{B}C$,

$f_1 = \bar{A}\bar{C}\bar{D} + \bar{A}CD + AC\bar{D} + A\bar{B}\bar{D} + AB\bar{C}D$

(c) If A, B, C, and D are the BCD inputs and G_3, G_2, G_1 and G_0 are the gray code outputs, then

$G_3 = A$, $\quad G_2 = A + B$, $\quad G_1 = B\bar{C} + \bar{B}C$, $\quad G_0 = A\bar{D} + C\bar{D} + \bar{A}\bar{C}D$

7.8 (a) $f = A \oplus B \oplus C$

(b) $f = \overline{A \oplus B \oplus C}$

CHAPTER 8

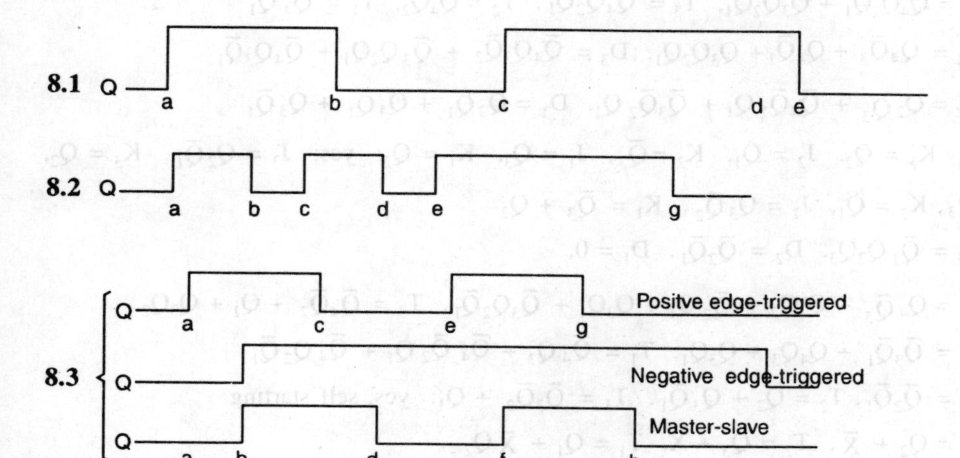

8.4

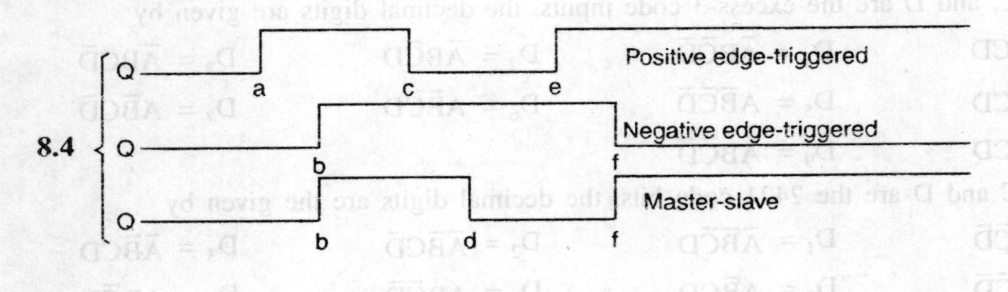

8.5

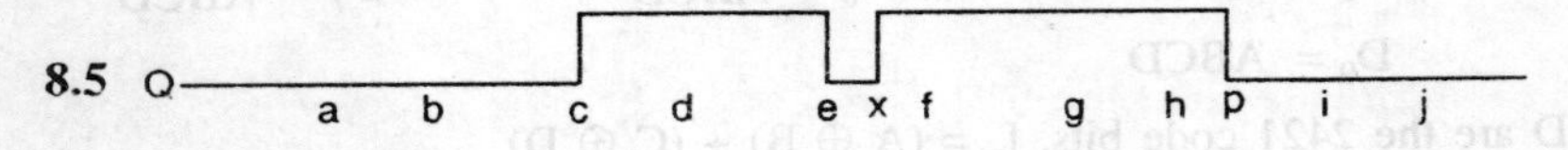

8.6

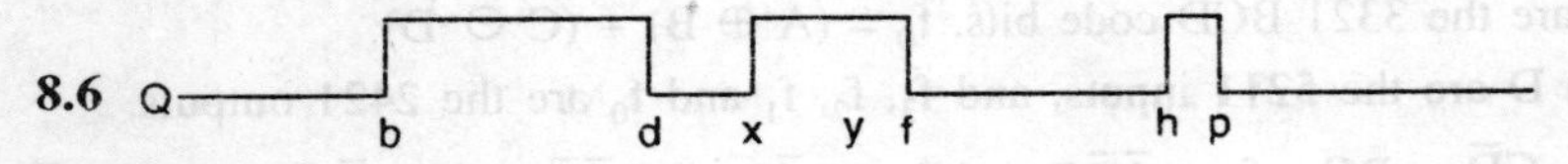

8.7 0 1 0 0 0 1 1 1

8.8 Let $C = 0.1\ \mu F$, then $R_A = 1.47\ k\Omega$

8.10 Let $R_{ext} = 10\ k\Omega$, then $C_{ext} = 714$ pF

8.11 Let $C = 1\ \mu F$, then $R = 900\ \Omega$

CHAPTER 10

10.1 (a) Connect all Js and Ks to 1. Provide a RESET pulse $\bar{R} = \overline{Q_3Q_2Q_1}$.

(b) Connect all Js and Ks to 1. Provide a RESET pulse $\bar{R} = \overline{Q_4Q_3}$.

(c) Design equations are $J_4 = Q_3Q_2Q_1$, $K_4 = 1$, $J_3 = Q_2Q_1$, $K_3 = Q_4$, $J_2 = Q_1$, $K_2 = Q_1$, $J_1 = \bar{Q}_4$, $K_1 = 1$

(d) $S_3 = Q_2Q_1$, $R_3 = Q_2\bar{Q}_1$, $S_2 = \bar{Q}_2Q_1$, $R_2 = Q_3Q_2 + Q_2Q_1$, $S_1 = \bar{Q}_2\bar{Q}_1 + \bar{Q}_3\bar{Q}_1$, $R_1 = Q_1$

(e) $T_4 = Q_4Q_2Q_1 + Q_3Q_2Q_1$, $T_3 = \bar{Q}_4Q_2Q_1$, $T_2 = Q_4Q_2$, $T_1 = \bar{Q}_4Q_1$

(f) $D_4 = Q_4\bar{Q}_3 + Q_4\bar{Q}_1 + Q_3Q_2Q_1$, $D_3 = \bar{Q}_4Q_3\bar{Q}_2 + \bar{Q}_3Q_2Q_1 + \bar{Q}_4Q_3\bar{Q}_1$

$D_2 = Q_2\bar{Q}_3 + \bar{Q}_4\bar{Q}_2Q_1 + \bar{Q}_3\bar{Q}_2Q_1$, $D_1 = Q_2\bar{Q}_1 + Q_4\bar{Q}_1 + Q_3\bar{Q}_1$

10.2 $J_3 = 1$, $K_3 = Q_2$, $J_2 = Q_1$, $K_2 = \bar{Q}_3$, $J_1 = Q_3$, $K_1 = Q_2$; yes; $J_3 = Q_2\bar{Q}_1$, $K_3 = Q_2$, $J_2 = \bar{Q}_3$, $K_2 = Q_1$, $J_1 = Q_3\bar{Q}_2$, $K_1 = \bar{Q}_3 + Q_2$

10.3 (a) $D_3 = \bar{Q}_1Q_2Q_3$, $D_2 = \bar{Q}_2\bar{Q}_1$, $D_1 = 0$,

(b) $T_4 = Q_4\bar{Q}_3 + Q_4Q_2 + \bar{Q}_2Q_1 + Q_3Q_1 + \bar{Q}_3Q_2\bar{Q}_1$, $T_3 = \bar{Q}_4\bar{Q}_2 + Q_1 + Q_3Q_2$

$T_2 = \bar{Q}_3\bar{Q}_1 + Q_4Q_3 + Q_3Q_1$, $T_1 = \bar{Q}_4Q_3 + \bar{Q}_4\bar{Q}_2Q_1 + \bar{Q}_4Q_2\bar{Q}_1$

(c) $T_3 = \bar{Q}_2\bar{Q}_1$, $T_2 = Q_2 + Q_3\bar{Q}_1$, $T_1 = \bar{Q}_3\bar{Q}_2 + Q_1$; yes, self starting

(d) $T_3 = Q_2 + \bar{X}$, $T_2 = Q_3 + X$, $T_1 = Q_1 + \bar{X}Q_2$

10.4 (a) $J_3 = Q_2,\ K_3 = Q_3,\ J_2 = \bar{Q}_3\bar{Q}_1 + Q_3Q_1,\ K_2 = Q_1,\ J_1 = 1,\ K_1 = Q_1 + Q_3$

(b) $T_3 = Q_2\bar{Q}_1,\ T_2 = \bar{Q}_3 + Q_2,\ T_1 = Q_3 + Q_2Q_1$

(c) $f_1 = \bar{Q}_3\bar{Q}_2\bar{Q}_1 + Q_3Q_1 + Q_3Q_2,\ f_2 = Q_3\bar{Q}_1 + \bar{Q}_2\bar{Q}_1 + \bar{Q}_3Q_2Q_1$ with a 3-bit ripple counter.

(d) $f = \overline{(Q_4Q_2)}$, with a 4-bit shift register.

(e) $f = \bar{Q}_4 + \bar{Q}_2\bar{Q}_1 + \bar{Q}_3Q_2$ with a 4-bit shift register.

10.5 $D_3 = Q_3\bar{Q}_2 + \bar{Q}_3\bar{Q}_1,\ D_2 = \bar{Q}_2,\ D_1 = Q_3 + Q_2Q_1$

10.6 $f = \overline{(Q_2Q_1)}$ with a 3-bit ripple counter.

10.7 $S_4 = Q_3Q_2Q_1M + \bar{Q}_4\bar{Q}_3\bar{Q}_2\bar{Q}_1$ $\quad S_3 = \bar{Q}_3Q_2Q_1M + Q_4\bar{Q}_1\bar{M}$

$S_2 = \bar{Q}_4\bar{Q}_2Q_1M + (Q_4\bar{Q}_1 + Q_3\bar{Q}_2Q_1)\bar{M}$ $\quad S_1 = \bar{Q}_1M + \bar{Q}_1\bar{M}$

$R_4 = Q_1M + Q_4\bar{Q}_1\bar{M}$ $\quad R_3 = Q_3Q_2Q_1M + Q_3\bar{Q}_2\bar{Q}_1\bar{M}$

$R_2 = Q_2Q_1M + Q_2\bar{Q}_1\bar{M}$ $\quad R_1 = Q_1M + Q_1\bar{M}$

10.8 $D_1 = \bar{Q}_3\bar{Q}_2M + \bar{Q}_3\bar{Q}_2\bar{Q}_1\bar{M}$ $\quad D_2 = \bar{Q}_3M + Q_3\bar{M} + Q_2\bar{Q}_1\bar{M}$

$D_3 = Q_2\bar{Q}_1M$

CHAPTER 12

Design equations for the problems are as given below.

12.1 $J_1 = y_2x,\ K_1 = y_2,\ J_2 = y_1\bar{x} + \bar{y}_1x,\ K_2 = \bar{y}_1 + \bar{x},\quad z = y_1y_2x$

12.2 $D_3 = y_2x + y_3\bar{x},\ D_2 = \bar{y}_3y_2\bar{x} + y_3y_2x + \bar{y}_3\bar{y}_2x,\ D_1 = y_3y_1 + y_1\bar{x} + y_2\bar{y}_1x$

$z = y_3\bar{y}_1x$

12.3 $S_1 = \bar{y}_2x,\ R_1 = y_2,\ S_2 = y_1x,\ R_2 = \bar{y}_1,\ z_1 = y_1,\ z_2 = y_2$

12.4 $T_1 = y_2\bar{x} + y_2y_3 + y_1x + y_1y_3,\ T_2 = y_1y_2 + \bar{y}_1y_3x + \bar{y}_2y_3\bar{x} + y_1\bar{x} + \bar{y}_3y_2\bar{x}$

$T_3 = \bar{y}_1\bar{y}_2x + y_2y_3\bar{x} + y_1y_3x,\ z = y_1x$

12.5 $S_3 = y_2y_3x,\ R_3 = y_1,\ S_2 = y_1\bar{x} + \bar{y}_2y_3\bar{x},\ R_2 = y_2y_3 + x,\ S_1 = y_2\bar{y}_3 + \bar{y}_3x,$
$R_1 = y_2y_3 + y_3\bar{x},$

12.6 $J_1 = y_2x,\ K_1 = y_2 + \bar{x},\ J_2 = \bar{y}_1\bar{x} + y_1x,\ K_2 = 1,\ z = y_1y_2\bar{x}$

12.7 $J_1 = K_1 = 1,\ J_2 = K_2 = y_1M + \bar{y}_1\bar{M},\ J_3 = K_3 = y_1y_2M + \bar{y}_1\bar{y}_2\bar{M}$
M = clock ANDed with control x and applied to J-K FFs.

12.8 $D_1 = \bar{y}_1\bar{y}_2\bar{y}_3\bar{x} + \bar{y}_3y_4\bar{x} + \bar{y}_2y_4\bar{x} + \bar{y}_4y_3x + \bar{y}_4y_2x$

$D_2 = \bar{y}_2y_3x + y_2\bar{y}_3\bar{x} + \bar{y}_3y_4\bar{x} + \bar{y}_4y_3x + \bar{y}_2\bar{y}_4y_1x$

$D_3 = y_2\bar{y}_3\bar{x} + \bar{y}_4y_2x + \bar{y}_2y_3\bar{x} + \bar{y}_2y_4x;\ D_4 = y_2y_3\bar{y}_4x + y_2\bar{y}_3y_4x$

12.9 $J_1 = y_2y_3,\ K_1 = 1,\ J_2 = \bar{y}_1y_3\bar{x},\ K_2 = x + y_3,\ J_3 = \bar{y}_1\bar{x},\ K_3 = \bar{x} + \bar{y}_2,\ z = y_1y_3\bar{x}$

CHAPTER 13

13.1 3.1 V; 2.1 V

13.2 (a) 55/32 V (b) 215/64 V

13.3 11001100

13.4 R_f = 10 kΩ, R = 7.8125 kΩ, $2R$ = 15.625 kΩ, $4R$ = 31.25 kΩ, $8R$ = 62.4 kΩ

13.5 6 bits

13.6 10 mA

13.7 I_3 = 0.75 mA, I_2 = 0.375 mA, I_1 = 0.1875 mA, I_0 = 0.09375 mA

13.8 1.8125 mA, 9.0625 V

13.9 f_{max} = 5 MHz

13.10 31 comparators and 32 resistors; 0.5 V assuming R = 1 kΩ and the last comparator is connected to V_{CC} through 9 kΩ.

13.11 40.95 V

13.12 10 μs

13.13 10110000

INDEX

Active-HIGH gates, 146
Active-LOW notation, 146
Active pull-up, 474
Adder, 5
  BCD, 249
  full, 234
  half, 233
  IC parallel, 244
  look-ahead carry, 242
  parallel binary, 240
  ripple carry, 241
  serial, 248
  serial binary, 522
Adder/subtractor, 246
Addition, 5
Adjacency, 163
All or nothing gate, 90
Analog
  circuit, 1
  device, 1
  ICs, 10
Analog-to-digital converters (ADCs), 546
  counter type, 564
  digital ramp type, 564
  dual-slope type, 569
  flash type, 567
  parallel type, 567
  simultaneous type, 567
  successive-approximation type, 571
  tracking type, 565
  voltage-to-frequency type, 574
ANSI/IEEE symbols, 104, 106
Anti-coincidence gate, 98
Any or all gate, 91
AOI logic, 94
Applications of flip-flops, 343
  counting, 344
  frequency division, 344
  parallel data storage, 343
  parallel-to-serial conversion, 344
  serial data storage, 343
  serial-to-parallel conversion, 344
  transfer of data, 343
Arithmetic circuits, 233
Asserted levels, 146
Astable multivibrator, 348, 355
  using 555 timer, 355
  using inverters, 355
  using Op-amps, 357
  using Schmitt trigger, 354
Asynchronous
  inputs, 326
  latches, 332
  systems, 334
Axioms, 121

Backplane, 280
Base, 17
Basic building blocks, 149
Basic gates, 89
Basic operations, 121
BCD arithmetic, 64
Binary codes, 62
  8421 BCD, 64
  alphanumeric, 82
  ASCII, 82
  BCD, 62
  biquinary, 78
  cyclic, 63
  EBCDIC, 84
  error correcting, 79
  error detecting, 75
  gray, 62, 63, 71
  Hamming, 80
  Johnson, 78
  natural, 64
  negatively-weighted, 62

non-weighted, 62
numeric, 62
positively-weighted, 62
reflective, 71
ring counter, 79
self-complementing, 63
sequential, 63
unit distance, 63
weighted, 62
XS-3, 67
XS-3 Gray, 74
Binary multipliers, 251, 254
Binary number system, 20
addition, 26
division, 29
multiplication, 28
subtraction, 26
Bipolar ICs, 11
Bistable multivibrator, 309
Bit, 20
Boolean algebra, 120
Boolean algebraic laws, 121
absorption, 128
AND, 122
associative, 123
commutative, 122
complementation, 122, 127
consensus, 129
De Morgan's, 130
distributive, 124
double negation, 127
idempotence, 126
identity, 127
null, 128
OR, 122
transposition, 130
Branching method, 217
Bubbled AND gate, 97
Bubbled OR gate, 95
Buffer/driver, 483
Buffer register, 369

Canonical POS form, 156, 159
Canonical SOP form, 156, 157
Carry flag, 30
Carry generate function, 244
Carry-in, 243
Carry propagate function, 244
Carry-out, 5, 234, 243
Cascading of
BCD adders, 251
parallel adders, 235, 245
ripple counters, 400
synchronous counters, 453
Cathode ray tube, 275
Cell, 162, 532, 578
Cell table, 532
Check sums, 76
Chip, 10
Chip select, 583, 594, 595
Circuit design, 2
Clipping circuits, 376
Clock skew, 332
Clock transition times, 331
CMOS, 498
bilateral switch, 503
buffered and unbuffered gates, 501
inverter, 499
NAND gate, 499
NOR gate, 500
open-drain outputs, 503
transmission gate, 501
Code converters, 253
binary-to-BCD, 259
binary-to-gray, 256
gray-to-binary, 257
Code converters using ICs, 259
Combinational circuits, 309
Common-anode type LED display, 276
Common-cathode type LED display, 276
Comparators, 265
Complement arithmetic
1's complement, 37
2's complement, 35
7's complement, 44
8's complement, 44
9's complement, 18
10's complement, 18
15's complement, 51
16's complement, 51
Computation of total gate inputs, 161
Computer method of division, 30
Computer method of multiplication, 28
Control unit, 12
Controlled buffer register, 369
Controlled inverter, 96, 98
Conversion from
binary to decimal, 21
binary to gray, 73

binary to hexadecimal, 46
binary to octal, 41
decimal to binary, 23
decimal to hexadecimal, 48
decimal to octal, 41
gray to binary, 74
hexadecimal to binary, 47
hexadecimal to decimal, 47
hexadecimal to octal, 50
octal to binary, 40
octal to decimal, 41
octal to hexadecimal, 49
Conversion of flip-flops, 339
D to J-K, 342
D to S-R, 340
J-K to D, 342
J-K to S-R, 340
J-K to T, 342
S-R to D, 341
S-R to J-K, 339
Conversion time of ADC, 565
Counters, 8, 389
applications, 455
asynchronous, 389, 391
design of, 394
basic ring, 438
BCD, 420
decade, 407
digital, 389
divide-by-N, 390
full modulus, 390
hybrid, 412
Johnson, 439
modulus, 390
presettable, 432
programmable, 431
ring, 385, 438
ripple, 389, 397, 404
self-starting, 414, 422
serial, 389
shift register, 437
shortened modulus, 390
synchronous, 389, 408
design of, 413
IC, 428
twisted ring, 439
Crystal-controlled clock generator, 358
Current mode logic, 488
Current steering logic, 488

Data lock-out FF, 338
Data selector, 290
Decimal number system, 17
Decoders, 6, 269
1-of-8, 271
3-line to 8-line, 270
applications, 274
BCD-to-decimal, 273
BCD-to-seven segment, 275, 277
Decoding, 6
of ripple counters, 400
Demorganization, 132
Demultiplexers, 7, 299
1-line to 4-line, 300
1-line to 8-line, 300
Demultiplexing, 7
Derived operations, 121
Digit, 17
Digital
circuits, 1
clock, 455
computer, 12
ICs, 10
quantity, 546
systems, 1
Digital-to-analog (D/A) conversion, 547
Digital-to-analog converters (DACs), 546
bipolar, 551
parameters of, 548
accuracy, 548
monotonicity, 549
offset voltage, 549
resolution, 548
settling time, 548
R-2*R* ladder type, 552
switched-capacitor type, 562
switched current-source type, 560
using BCD input code, 550
weighted-resistor type, 557
Diode matrix, 284
Direct logic, 442
Discrete
AND gate, 90
NAND gate, 96
NOR gate, 97
NOT gate, 94
OR gate, 92
Display devices, 275
Divider, 6
Dominating columns, 214
Dominating rows, 214
Don't care
combinations, 179, 213
conditions, 186
terms, 211

Double dabble method, 23
Double precision, 39
Dual-in-line package, 10
Duality, 133
Duals, 134
Duty cycle, 4
Dynamic
  inverter, 383
  MOS inverter, 506
  MOS logic, 505
  MOS registers, 383
  NAND gate, 507
  NOR gate, 507
  RAMs, 596
    address multiplexing, 598
    combining, 604
    refreshing, 598
  shift register, 383

ECL RAMs, 596
Edge-triggered flip-flops, 317
  D, 321
  J-K, 323
  S-R, 319
  T, 326
EEPROM, 589
Encoders, 6, 281
  decimal-to-BCD, 283, 285
  keyboard, 284
  octal-to-binary, 282
  priority, 285
    decimal-to-BCD, 285
    octal-to-binary, 287
Encoding, 6
End around carry, 18, 37, 44, 52
EPLD, 617
EPROM, 587
Equality comparator, 267
Even parity checker, 262
Excess-n notation, 40
Excess-3 arithmetic, 67
Excitation table, 414, 518
Exponent, 39

Fall time, 4
Falling edge, 4
Fan-in, 467
Fan-out, 468
Fast (F) TTL, 471
Finite state model, 515
Firmware, 590
Flat package, 10
Flip-flops, 309
  maximum clock frequency, 331
  non-clocked, 310
  operating characteristics, 329
    hold time, 330
    power dissipation, 332
    propagation delay time, 329
    set-up time, 330
Floating point numbers, 39
Floppy disk, 622
FPLA, 611
Frequency counter, 458
Frequency stability, 358
Full-subtractor, 238

Gated latches, 315
  D, 316
  S-R, 315
Gating, 353
Generation of narrow spikes, 318
Giant scale integration, 466

Half-subtractor, 236
Hard disk, 624
Hex dabble method, 48
Hexadecimal number system, 45
  addition, 50
  division, 54
  multiplication, 54
  subtraction, 51
Hybrid logic, 141, 181
Hybrid systems, 2
Hysterisis loop, 347

Illegal code, 64
Inclusive OR gate, 91
Incompletely specified functions, 193
Index of the term, 207
Indirect logic, 445
Inequality detector, 98
Inhibit
  carries, 531
  circuits, 100
  unit, 12
Instruction code, 40

Interfacing, 536
  CMOS to TTL, 504
  ECL to TTL, 509
  TTL to CMOS, 509
  TTL to ECL, 509
Inverter, 93
Iterative network, 531

J-K flip-flop with active-LOW PRESET and CLEAR, 345

Karnaugh map, 156
  five-variable, 174
  four-variable, 171
  six-variable, 176
  three-variable, 167
  two-variable, 161

Latch, 310
  active-HIGH input, 310
  active-LOW input, 310
LCD displays
  dynamic scattering type, 279
  field effect type, 279
  operation of, 279
Leading edge, 3
Least significant digit, 17
LEDs, 275, 278
Level
logic, 89
  shifters, 504
  translator, 492
Levels of integration, 10
  LSI, 11
  MSI, 11
  SSI, 10
  ULSI, 11
  VLSI, 11
Light emitting segments, 275
Linear sequence generator, 449
Liquid crystal cells, 279
  reflective type, 279
  transmittive type, 279
Literals, 208
Loading factor, 468
Lock-out, 390
  elimination of, 423

Logic
  circuits, 1
  design, 2, 89
  diagram, 137
  levels, 3
  operations, 121
  race, 181
Logic families, 471
  CMOS (*see* CMOS)
  ECL, 488
    interfacing, 492
    subfamilies, 491
    wired OR operation, 491
  I^{2}L, 486
  MOS, 494
  TTL, 471
    characteristics, 485
    current sinking, 474
    current sourcing, 474
    HIGH-state fan-out, 475
    loading, 474
    LOW-state fan-out, 475
    open collector output, 479
    passive pull-up, 480
    subfamilies, 483
    totem-pole output, 473
    tri-state, 482
    unit load, 476
    wired AND operation, 480
Logic gates, 89
  AND gate, 90
  NAND gate, 94
  NOR gate, 96
  NOT gate, 93
  OR gate, 91
  X-NOR gate, 99
  X-OR gate, 98
Look-ahead carry, 411
Lower trigger level, 347

Magnetic recording formats, 619
Mainframe, 13
Mantissa, 39
Master reset, 432
Master-slave flip-flops, 333
  D, 336
  J-K, 337
  S-R, 334
  with data lock-out, 339
Maxterm, 156

Mealy model, 519
Memory, 578
  add on, 578
  auxiliary, 578
  bit organized, 578
  charge coupled, 628
  data, 578
  expansion of, 598
  FIFO, 606
    data-rate buffer, 607
  internal, 578
  loop arrangement, 626
  magnetic, 617
  magnetic bubble, 578, 625
    major/minor loop arrangement, 626
  magnetic core, 617
  magnetic disk, 618
  magnetic tape, 625
  main, 578
  mass, 578
  optical disk, 627
  peripheral, 578
  program, 578
  random access, 582
  read only, 582
  read/write, 582
  sequential access, 582, 605
  volatile, 582
  word organized, 598
Memory unit, 12
Microcomputer, 13
Minicomputer, 13
Minterm, 156
Modulus arithmetic, 35
Monolithic IC, 10
Monostable multivibrator, 348
  applications, 353
Moore model, 519
MOS resistor, 495
Most significant digit, 17
Multiple output minimization, 184, 220
Multiplexers, 290
  2-input, 290, 291
  4-input, 292
  8-input, 293
  16-input, 293
  applications, 294
Multiplexing, 290
  seven segment displays, 297
Multiplier, 6
Multiplier quotient register, 28

NAND logic, 142
Negative AND gate, 97
Negative edge-triggered FF, 318
Negative logic system, 3, 89
Negative OR gate, 95
Nibble, 46
NMOS
  inverter, 496
  NAND gate, 497
  NOR gate, 497
Noise, 2
Noise margin, 469
Non-volatile RAMs, 604
NOR logic, 142

Octal number system, 40
  addition, 43
  division, 46
  multiplication, 45
  subtraction, 43
One-shot, 348
  non-retriggerable, 350
  triggerable, 350
Optical ROM, 628
Output carries, 531
Output unit, 13

Pair, 163
Parallel data transfer, 368
Parity, 75
  block, 76
  even, 75
  odd, 75
  two-dimensional, 76
Parity bit generator/checker, 259, 263, 526
Parity word, 77
Perfect induction, 120
Period, 4
Positional-weighted system, 17
Positive edge-triggered FF, 318
Positive logic system, 3, 89
Postulates of Boolean algebra, 121
Power dissipation in logic gates, 467
Prime implicant chart, 212, 222
Prime implicants, 208, 220
  essential, 213
Priority encoders
  decimal-to-BCD, 285
  octal-to-binary, 287

Program, 2
Programmable array logic, 608
Programmable logic devices (PLDs), 608
  erasable, 617
PROM, 587, 615
Propagation delay in gates, 466
Propositional, 5
Pseudo analog quantity, 548
Pull-down transistor, 474
Pull-up transistor, 474
Pulse, 3
Pulse train generators, 442
  using shift registers, 447
Pulse triggered FFs, 334
Pulsed operation, 109

Quad, 164
Quartz crystal oscillator, 358
Quasi-stable state, 348
Quine-McClusky method, 206

Radix, 17
Read/Write input, 580
Recirculating shift register, 605
Register, 368
Retriggerable IC one-shot, 350
Ripple carry
  adder, 241
  output, 429
Rise time, 4
Rising edge, 4
ROM, 583
  applications of, 590
  block diagram, 584
  mask programmed, 586
  organization of, 583
  programming the, 582
  timing, 585
  types of, 586

Schmitt trigger, 347
Secondary essential prime implicants, 214
Sequence detector, 524
  design using iterative networks, 532
Sequential circuits, 309, 515
  synchronous, 522
Sequential machines, 515
  counters, 528
  memory elements, 518
Serial binary adder, 522
Serial data transfer, 368
Serial/parallel data conversion, 385
Serial parity-bit generator, 526
Seven-bit Hamming code, 80
Seven segment display, 276
Shift registers
  applications, 384
  bidirectional, 381
  data transmission in, 371
  dynamic, 383
  parallel-in, parallel-out, 379, 380
  parallel-in, serial-out, 377, 378
  serial-in, parallel-out, 376, 377
  serial-in, serial-out, 372
  static, 383
  universal, 381, 382
Sign bit, 31
Sign magnitude form, 31
Single-shot, 348
Speed power product, 470
S-R latch, 310
  NAND gate, 313
  NOR gate, 311
Stable state, 309, 348
Standard POS form, 156
Standard SOP form, 156
State diagram, 517
State table, 517
Static RAMs, 593
  read cycle, 594, 595
  write cycle, 594, 595
Strobing, 402
Substrate, 10
Subtraction, 5
Subtractor, 5
Sum-of-weights method, 23
Switching circuits, 1, 309
System design, 2

Tabular method, 206
Terminal count, 390
Threshold voltage, 466
Time delays, 384
Time race, 332
Timing diagrams, 109
Trailing edge, 4

Transition and output table, 517
Transition time, 1
Triple precision, 39
Truth table, 89
Two-level logic, 181
Two-state devices, 20

UART, 385
Unasserted, 147
Unipolar ICs, 11
Universal
  building blocks, 94
  gates, 94
  logic, 180

Upper trigger level, 347

Variable mapping, 189

Wired AND operation (*see* TTL)
Wired OR operation (*see* FCL)
Word, 579
Wordsize, 579
Wrapping around, 168

XS-3 arithmetic, 67

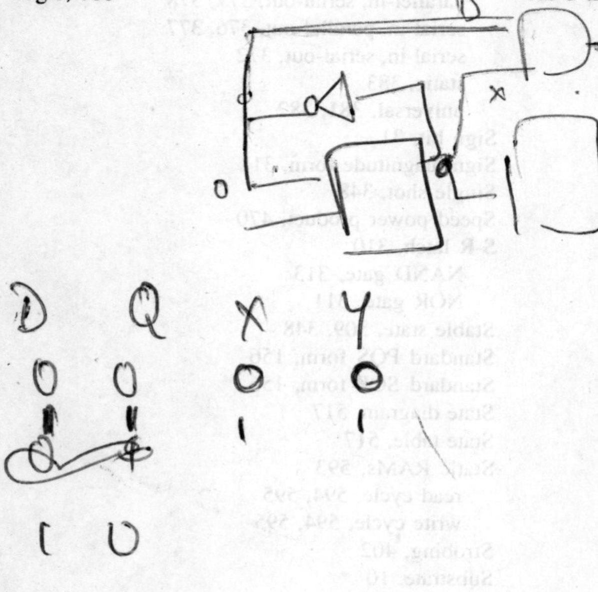